CW00516052

Waste

WASTE: A Handbook for Management

Edited by

TREVOR M LETCHER

DANIEL A VALLERO

AMSTERDAM • BOSTON • HEIDELBERG • LONDON
NEW YORK • OXFORD • PARIS • SAN DIEGO
SAN FRANCISCO • SINGAPORE • SYDNEY • TOKYO

Academic Press is an imprint of Elsevier

Academic Press is an imprint of Elsevier
30 Corporate Drive, Suite 400, Burlington, MA 01803, USA
525 B Street, Suite 1900, San Diego, CA 92101-4495, USA
The Boulevard, Kidlington, Oxford, OX51GB, UK
Radarweg 29, PO Box 211, 1000 AE Amsterdam, The Netherlands

First edition 2011

Library of Congress Cataloging-in-Publication Data
Waste: a handbook for management/edited by Trevor Letcher, Daniel Vallero.
p. cm.
Summary: "Preface Waste: A Handbook of Waste Management is designed to be a resource for the designer, practitioner, researcher, teacher and student. Waste is one of those entities which is defined by everyone, but not truly and completely understood by anyone. The scientist, engineer and consumer each define waste correctly, yet differently. The overriding challenge for the authors was to provide some uniformity, yet allow for the diversity of the various aspects of waste in this handbook. After all, a handbook is often the beginning of the search for information, certainly not the end. We have borrowed from all scientific disciplines, and a few humanities, to arrive at a balanced resource. And, indeed, balance is the key to environmental science and engineering" — Provided by publisher.

Includes bibliographical references and index.
ISBN 978-0-12-381475-3 (hardback)
1. Factory and trade waste–Handbooks, manuals, etc. 2. Refuse and refuse disposal–Handbooks, manuals, etc. I. Letcher, T. M. (Trevor M.) II. Vallero, Daniel A.
TD897.W3178 2011
363.72′8–dc22

2010050455

British Library Cataloguing in Publication Data
A catalogue record for this book is available from the British Library

For information on all Academic publications
visit our web site at books.elsevier.com

ISBN: 978-0-12-381475-3

Contents

Preface

WASTE: A Handbook for Management is designed to be a resource for the designer, practitioner, researcher, teacher, and student. Waste is one of those entities that is defined by everyone but not truly and completely understood by anyone. The scientist, engineer, and consumer each define waste correctly, yet differently. The overriding challenge for the authors was to provide some uniformity, yet allow the diversity of the various aspects of waste in this handbook. After all, a handbook is often the beginning of the search for information, certainly not the end. We have borrowed from all scientific disciplines, and a few humanities, to arrive at a balanced resource. And, indeed, balance is the key to environmental science and engineering.

Obviously, no single resource is sufficiently comprehensive on its own in dealing with waste, so each of the chapters is richly annotated and sourced to give the reader ample avenues for further investigation.

The original plan for this book (T M Letcher) was that it should be a collection of chapters highlighting the many waste streams in our society together with details of the recycling of waste. When Dan Vallero came on board, as coeditor, the book developed into a handbook, taking into account the added perspective of management of waste, thus opening it up to a very much wider audience.

As the editors of this book, we share many things in common that have added to this project's gratification. We both love teaching and research. We respect and adhere to the scientific method. We are blessed with a diverse network of colleagues who readily share their expertise, as evidenced by the chapter authors. We consider ourselves pragmatic and strive to apply the sciences to better humankind. However, we are also quite different in many ways. One is a thermodynamicist and one an engineer. One has resided in Africa and Europe and the other in North America. One leads fellow chemists into previously uncharted areas of global concerns and the other leads (and follows) engineers into daunting areas of professional ethics and sustainability.

Like other handbooks, the text seeks to include best practices and proper management of wastes before, during, and after they are generated. This has been particularly challenging, since there is never unanimity and rarely consensus on how best to manage any waste. Thus, key to the usefulness of this handbook is its attention to the scientific challenge of how to achieve effective and sustainable solutions to one of the most important and dynamic of society's problems; that is, how best to manage and eliminate waste in its various forms, sectors, and streams.

Our main objectives in having all the issues related to wastes in one volume, are to:

- create a source book for easy consultation and direction in further studies;
- create a book that will lead to helpful comparisons among different waste streams, leading to SYNERGISTIC solutions; and,

- above all, help in developing a better and more informed society.

This handbook will be of particular interest to

- students and lecturers (as a textbook) in science, engineering and environmental studies;
- researchers (the references will guide them to the very latest works);
- decision making officials and parliamentarians and leaders in society who need to be aware of the serious problems created by waste;
- editors and journalists who also need to know the latest issues on waste;
- captains of industry, technicians and maintenance personnel who need to be aware of the problems in their areas and related area; and,
- all interested parties, even the more casual readers, enhancing their awareness of the enormity of the problems surrounding waste.

That said, perhaps the central value of this book, and where it differs from other books of similar theme, is its pragmatic perspective. Each chapter is written by an expert scientist or engineer working in the specific field tapped to address a particular waste challenge. The book highlights the severity of each of the problems and offers the best solutions and recycling processes. In fact, one of the key values of this book is in what it cannot say. That is, the chapter authors are upfront about uncertainties and areas of need of advancement and future research.

The challenge of any compendium is to achieve a balance of uniformity of style and format with sufficient latitude for chapter authors to cover their subject matter sufficiently. We believe that we have struck this balance. The units and notation are given as more than one type (e.g., tons and tonnes) when these are not used uniformly by professionals or by scientific disciplines. However, the English language varies for authors, depending on the dialect of the nation where they practice. For example, the reader will note that some authors use 'authorize' whereas others use 'authorise'. Indeed, the terms are only different in spelling and have identical meanings. We believe this cultural richness and authenticity is more important than internal consistency. After all, as Emerson warned, "a foolish consistency is the hobgoblin of little minds, adored by little statesmen and philosophers and divines." [2]

Waste is one of the most important issues facing the world today. Our disregard for the problems generated by waste will compromise our children's lives, resulting in a legacy that our generation we will not be proud of. There is now a desperate need for an honest appraisal of the stresses that increase with geopolitical change, such as a shifting and growing population. We hope and expect that increasing awareness will enhance the likelihood of innovation and successful solutions.

STRUCTURE

The book is divided into three parts:

- **Part 1: Introduction to Waste Management** includes a history of waste management, an introduction to green engineering and sustainable design as related to waste management, waste regulations and waste collection;
- **Part 2: Waste Streams** is a collection of 25 chapters on the most important waste streams in our society and
- **Part 3: Best Practice and Management** contains chapters on landfills, yesterday, today and tomorrow, pollution management and responsible care, and risk management and accountability of waste.

THE AUTHORS

This book is blessed with an unprecedented cadre of chapter authors. They are listed in order of their contributions:

Prof. Trevor M. Letcher is Emeritus Professor of Chemistry at the University of KwaZulu-Natal, Durban and a Fellow of the Royal Society of Chemistry. He is a Director of the International Association of Chemical Thermodynamics and his research involves the thermodynamics of liquid mixtures and energy from landfill. He has more than 250 publications in peer review journals and edited and co-edited eight books in his research fields. His latest books are *Heat Capacities* (2010) and *Climate Change* (2009).

Prof. Daniel A. Vallero has conducted research, prepared environmental assessments, and advised US legislative on environmental issues, including climate change and acid rain and risks posed by chemicals. At Duke University, he teaches courses in ethics, sustainable design and green engineering. He is an advisor to the National Academy of Engineering's Online Ethics Committee and a member of the National Institute of Engineering Ethics. As an authority on the environmental measurement and modeling, Dr. Vallero has conducted research related to the emergency response and homeland security, notably leading exposure studies in the urban dispersion program in New York City. He is the author of nine engineering textbooks – the most recent is, *Environmental Biotechnology: A Biosystems Approach* (2008).

Dr Valerie L. Shulman is the Secretary General of the European Tyre Recycling Association (ETRA). The research she initiated in 1989 formed the nucleus of ETRA which was founded five years later. ETRA has over 250 member companies and organisations in 46 countries, worldwide. She was a contributor to the Basel Convention Guidelines on Use Tyres (1999) and has represented the recycling industries at Waste Policy hearings in the European Parliament. She has participated in numerous research projects, prepared more than 100 articles in scientific and industry journals and authored or co-authored several books on tyre recycling technologies, materials and applications.

Mr Fraser McLeod is a Research Fellow at the University of Southampton and a member of its Transportation Research Group within the School of Civil Engineering and the Environment. He has over 20 years of research experience in various areas of transport, including intelligent transport systems, bus priority, freight logistics and waste collection.

Dr Tom Cherrett is a Senior Lecturer of Freight and Logistics at the University of Southampton. His main research areas include developing sustainable strategies for the collection and disposal of wastes, goods distribution in urban areas, incident detection and journey time estimation.

Prof. Geoffrey Blight is Professor Emeritus at the University of the Witwatersrand and for the past 50 years has specialised in the engineering of waste management, with an emphasis on the problems of developing countries having semi-arid to arid climates. There are many similarities between the management of mine and municipal solid waste, although practitioners in the two fields largely work in isolation of each other. Blight's work has pointed out the commonalities in the two fields and provided rational solutions to many of the technical problems.

Prof. Thomas Pretz studied mining and specialised in processing, and in 1997 he was appointed Professor in the Department of Processing and Recycling, Rheinisch Westfälische Technische Hochschule (RWTH Aachen University), Germany. He has worked in the industry.

Dr Jörg Julius trained as an Electrical Engineer and is presently Senior Engineer at the Department of Processing and Recycling, RWTH Aachen University, Germany. He has worked in industry as sales director and also as technical director.

Dr John E. Marra is Associate Laboratory Director of Strategic Initiative Development at the Savannah River National Laboratory, Aiken, South Carolina, USA. He studied ceramic science engineering at New York State College of Ceramics and Ohio State University. He has held various technical staff and management positions at the US Department of Energy's Savannah River Site.

Dr Ronald A. Palmer is a Senior Research Associate at the Institute of Clean Energy Technology at Mississippi State University, Starkville, Mississippi. He studied ceramic engineering at the New York State College of Ceramics and has worked in high-level waste vitrification process development for West Valley Nuclear Services, West Valley, New York.

Prof. Agamuthu Pariathamby is a Professor in the Institute of Biological Sciences in University of Malaya. He is also the chairman of the Malaysian Society of Solid Waste Management and Environment, Editor-in-Chief of *Malaysian Journal of Science* (MJS) and Associate Editor of *Waste Management and Research* (WM&R). His main area of expertise is solid and hazardous waste management.

Prof. Ralf Otterpohl is Professor and Director of the Institute of Wastewater Management and Water Protection at Hamburg University of Technology, Germany. He is also co-owner of the consultancy Otterwasser GmbH, which specialises in computer simulation of large wastewater treatment plants and innovative water projects in Europe, the Middle East and Africa. He is co-chair of the specialist group 'Resources Oriented Sanitation' of the International Water Association.

Mr Christopher Buzie is an Environmental Engineer specialising in the development and planning of sustainable sanitation systems with a focus on developing countries. He has played a key role in European Community projects devoted to developing countries. Presently, he is the team leader of the research group 'Integrated Research on Agriculture and Sustainable Sanitation' (IRASS) at the Institute of Waste Water Management and Water Protection at the Hamburg University of Technology (TUHH).

Prof. Gary M. Scott is Professor and Chair of the Department of Paper and Bioprocess Engineering at the State University of New York – College of Environmental Science and Forestry, Syracuse. His research involves the use of biotechnological processing in the pulp and paper industry. He has investigated the use of a fungal pre-treatment for the production of mechanical pulp and worked on the scale-up of the process to the semi-commercial scale.

Dr John Butler is an independent consultant helping small business to assess the environmental impacts of their operations and plan for more environmentally benign outcomes. His research is primarily in the field of waste management and using life cycle assessment (LCA) to identify more environmentally benign options in the implementation of waste management policies.

Dr Paul D. Hooper is a Senior Enterprise Fellow in the Department of Environmental and Geographical Sciences at Manchester Metropolitan University and a member of the Centre for Air Transport and the Environment. His research focuses on the business response to the sustainable development agenda. With respect to waste issues he has coordinated research into the small and medium enterprises (SME) response to waste minimisation and the use of life cycle analysis (LCA) to optimise recycling systems.

Prof. Andreas Bartl is an Assistant Professor at Vienna University of Technology, Institute of Chemical Engineering. His research activities include grinding and characterisation of (short) fibres, fibre recycling and silicate filaments.

Dr Rebecca Slack is the Project Coordinator for water@leeds, an interdisciplinary water research centre at the University of Leeds. Her main research interests concern the fate of anthropogenic chemicals in the environment and environmental exposure assessment principally but not exclusively via the aquatic environment. The areas of cross-cutting research that she is active in, include water quality, environmental impact assessment and solid waste management, particularly modelling behaviour of landfill leachates in groundwater.

Prof. Marian Chertow leads the industrial environmental management program as well as the program on solid waste policy at the Yale School of Forestry and Environmental Studies. She has worked in waste management and recycling in the public, private and not-for-profit sectors and was an early adopter of 'industrial ecology', which offers a systems approach to questions concerning materials, energy and waste.

Ms Jooyoung Park is a Ph.D. student at the Yale School of Forestry and Environmental Studies and the Center for Industrial Ecology. She has studied environmental engineering at Seoul National University, South Korea. Her interests are in bridging resource and waste management and addressing inter-firm behaviours for waste reuse.

Mr Mohamed Osmani is a Senior Lecturer in Architecture and Sustainable Construction at the Department of Civil and Building Engineering at Loughborough University. He is an architect by training and has developed a significant portfolio of research projects on resource efficiency, winning more than £2 million funding from a range of sponsors in the last six years. He is currently supervising a range of undergraduate, post-graduate, doctoral research studied, and funded projects covering resource and energy efficiency in buildings, and construction waste minimisation, recovery and optimization. He has been a member of a number of committees and task groups including the House of Lords Waste Enquiry, the UK Green Building Council 'Vision for Sustainable Built Environment', The Office of Government Commerce 'Construction and Refurbishment: Building a Future' and British Standards Institution. He is the chairman of the Construction Industry Research and Information Association (CIRIA) Sustainability Advisory Panel, which provides focus on cross-cutting sustainability issues and challenges facing the construction industry. He is regularly invited to present keynote papers at both industry- and academia-focused national and international events and conferences.

Dr Paola Lettieri is a Reader and Associate Professor in chemical engineering at University College London. She is a Fellow of IChemE, serves on the committee of the IChemE Particle Technology and the Sustainability subject groups and is a committee member of the EFCE section on Process Engineering for Sustainable Energy. She heads the Fluidization Research Group, at UCL. The research activities focus on industrial applications and sustainable development, including experimentation, computational fluid dynamics (CFD) simulations, energy from biomass and waste and recycling of plastics. She has published over 80 refereed papers and co-authored five book chapters.

Mr Sultan M. Al-Salem is currently a Ph.D. student at University College London. His research investigates the thermo-chemical treatment of plastic solid waste focusing in particular on the pyrolysis and the thermal cracking kinetic modelling of different plastic materials.

Dr José Vinicio Macías-Zamora is a Lecturer and Researcher at the Instituto de Investigaciones Ocean, University of Baja California in Mexico. His research interest is in marine pollution and environmental problems. He regularly publishes in prestigious peer review journals related to environmental pollution and toxicology and marine chemistry.

Mr Jirang Cui is a doctorate candidate at the Department of Material Science and Engineering, Norwegian University of Science and Technology, Norway. He is working in the area of recycling electronic and automotive waste and recycling aluminium scrap.

Prof. Hans J. Roven is Professor of Physical Metallurgy at the Department of Material Science and Engineering, Norwegian University of Science and Technology, Norway. He is member of Royal Norwegian Society of Sciences and Letters (DKVNS) and Norwegian Academy of Technological Sciences (NTVA). His research involves the recycling of light metal scrap and severe plastic deformation (SPD) of light metals.

Dr Ash M. Genaidy has had more than two and half decades of research experience and this has established him as a leading authority in the fields of sustainability, environmental health and safety, risk assessment, biomechanics and ergonomics. He is a well-known presenter and has published extensively on these topics in leading journals. Currently he works as Chief Technical Officer at WorldTek Inc., an innovation-based consulting and research firm at Cincinnati, Ohio.

Dr Reynold Sequeira is a Researcher working with WorldTek Inc. of Cincinnati, Ohio. He has published a number of journal articles in the field of sustainability, environmental health and safety, risk assessment and ergonomics. One of his interests is battery waste.

Mr Andrew L. Shannon is an Environmental Quality Analyst in the Medical Waste Regulatory Program for the State of Michigan, Department of Natural Resources and Environment. He has studied at Grand Valley State University in Allendale, Michigan and is a Registered Environmental Health Specialist through the National Environmental Health Association (United States).

Dr Anne Woolridge is the Principal Waste Management Consultant at the Independent Safety Services Limited, Dabell Avenue, Blenheim Industrial Estate, Bulwell, Nottinghamshire. As a Chartered Waste Manager and Dangerous Goods Safety Adviser, she has an international presence in the field of healthcare waste management which she approaches from a multi-disciplinary viewpoint. Her research field is the development of sustainable systems for the management of these wastes. She has an extensive practical and theoretical understanding of the healthcare waste management process from the generation of waste through to each of the final disposal options.

Prof. Ramachandra Murthy Nagendran is Professor of Environmental Science at the Centre for Environmental Studies at Anna University, Chennai. He is an international environmental management consultant, trainer and advisor in the area of environmental ecology. His academic interests include industrial ecology, ecological engineering and climate change.

Dr Victor F. Medina is a Research Environmental Engineer at the US Army Engineer Research and Development Center in Vicksburg, MS. He has over 15 years of experience in addressing environmental issues and concerns related to the military and over 20 years of experience in environmental research and management in general.

Mr Scott A. Waisner is a Research Environmental Engineer at the US Army Engineer Research and Development Center. He has spent the past 17 years addressing and developing solutions to environmental contamination and military waste disposal issues for the US Army.

Mr Gene Stansbery is the Program Manager of NASA's Orbital Debris Program Office at the Johnson Space Center. He has been involved in orbital debris research since 1986. He was NASA's technical lead for the very successful Haystack radar debris measurements. These measurements first characterised the 1-cm orbital debris environment. He is also a private pilot and owns and flies his own Chinese military trainer aircraft.

Mr Daniel J. Vallero P. E. is a Civil Engineer in the City of Durham, North Carolina's Engineering Section of the Public Works Department. He reviews and ensures the adequacy of projects with respect to zoning, site plans, preliminary plats, construction drawings, building permits and final plats, ensuring that a project adheres to road standards, sidewalk, water system, fire protection systems, sanitary sewer system and storm-water drainage and conveyance systems. He has practiced in both the private and public sectors, focusing on various aspects of land development, including measures to prevent water pollution, control erosion and ensure effective use of land.

Dr Nicholas P. Cheremisinoff is a Consultant to industry and governments on standard of care as it pertains to environmental management and worker protection. He has more than 36 years of international consulting, business, and applied research and development experiences, having worked throughout Eastern Europe, Russia, parts of the Middle East, Latin America, the Far East, Africa and domestically. He has led and participated in hundreds of pollution prevention and environmental health and safety audits of large industrial complexes ranging from refineries, to coke-chemical plants, pulp and paper mills, steel mills, pharmaceutical and pesticide manufacturing plants, wood treating facilities and other industrial operations. He has assisted several foreign governments in strengthening environmental laws and trained several thousand regulators and industry personnel on environmental auditing, pollution prevention and safe chemical management through US Agency for International Development, European Union and World Bank Organization-sponsored programs. He is the author, co-author or editor of more than 150 technical book publications.

RECOMMENDATIONS FOR INCORPORATING *WASTE: A HANDBOOK FOR MANAGEMENT* INTO SCIENCE AND ENGINEERING CURRICULA

Type of course	Year*	Chapters	As primary text	As companion text
Introduction to Waste Management	4	All	X	
Solid and Hazardous Waste Engineering	4	All	X	
Introduction to Environmental Science and Engineering	3	Preface, 1–8, 12–14, 24	X	
Green Engineering	3	Preface, 1, 2, 8–14, 16, 20–22, 29–32	X	X
Environmental Health	4	Preface, 13, 18, 23, 32		X
Water and Wastewater Treatment	4–5	Preface, 1–3 5, 6, 9, 15, 16, 19, 22–25, 27–32	X	X
Air Pollution Control	4–5	Preface, 2, 3, 5–7, 16–18, 28, 31, 32, 18		X
Hazardous Waste Management	4–5	Preface, 1–3, 5–7, 13, 17, 20–23, 27, 30–32	X	

(Continued)

Type of course	Year*	Chapters	As primary text	As companion text
Advanced Topics in Contaminant Fate and Transport	4–5	Preface, Parts 1 and 2		X
Environmental Policy Seminar	4–5	All (depending on the topic)	X	X
Advanced Environmental Engineering	4–5	All (depending on the topic)	X	X
Engineering Ethics Seminar	5	Preface, 1–3, 32		X
Risk Assessment Seminar	5	Preface, 2, Part 2, 31, 32	X	
Review Courses for Professional Exams	4	All		X

Note: Even though specific chapters are listed, those that are specifically targeted at the course topic are still valuable resources to students who may need to learn or be refreshed in technical areas where prior knowledge from prerequisite courses is assumed.
** 3 = Junior; 4 = Senior; 5 = Graduate Student.*

E-COMMUNITY

In light of the evolving nature of waste management, we will be updating the information in an electronic companion, which is available on the Internet. The companion will allow the reader to access the co-editors and chapter authors regarding queries and suggestions. It will also include any errata, related information via the internet and teaching tools, such as sample review questions, seminar guidance and bibliographies.

We hope that this project will enhance and build upon the existing waste management knowledge base.

ACKNOWLEDGEMENTS

We wish to express our gratitude to the authors who have graciously given time and effort. We also thank the Elsevier team, especially Candice Janco and Emily McCloskey for their assistance in assembling this volume. Most importantly, this work could not have been possible without the constant insights and support of our wives Janis Vallero and Valerie Letcher during the 18 months it has taken to pull this 32-chapter volume together.

Prologue

Waste - has vexed civilization for thousands of years. Most recently, however, waste concerns have grown exponentially with the industrial and petrochemical revolutions, a rapid growth in world population, and greater consumerism. Generally, engineers and scientists have done much to address previous problems long considered intractable (e.g., open dumps, lack of substitutes for dangerous products and pesticides). Advances in reduction of waste volume and hazards have encouraged a well-deserved dose of technological optimism, although the amount and hazardous nature of wastes continue to threaten society. The waste threat impinges on our public health and the integrity of ecosystems; it can compromise our aesthetic sensibilities; and it can be economically crippling. Several chapters in this book address this last point which is so crucial yet often ignored in technical handbooks.

Indeed, the economic losses posed by wastes indicate two failings. First, waste is always an indication of inefficiency. Note that every mass and energy balance includes the mass or energy exiting the control volume. The amount and type of mass or energy that exit along the waste streams are examples of inefficiency. Second, any mass or energy exiting along these pathways, which introduces costs for handling and treatment and can be staggering, must be addressed. Thus, eliminating or reducing waste helps the "bottom line" in two ways, that is, improving efficiency and avoiding the need for expensive controls.

Over the past few decades, in isolated places, waste and pollution have reached uncomfortable and, all too often, dangerous levels. From a public health point of view, one need only to think back to indelible disasters at Chernobyl (26 April 1986), Bhopal (2 December 1984), and air pollution episodes in Denora, London and other cities throughout the world. Such episodes are one of the myriad ways in which wastes pose problems.

Waste-related problems have both temporal and spatial aspects. Chernobyl provides an example of the most toxic forms of waste (radioactive carcinogens) being released rapidly (a meltdown that can be measured in hours) over vast regions of the world. It also demonstrates that the exposures and effects of a rapid release can be quite long lived, where the concentrations of isotopes from the disaster have been measured for years after, and the effects began to be detected almost immediately (acute radiation poisoning) and continue to be diagnosed (leukemia and other forms of cancer). Bhopal demonstrates a large, sudden release of a toxic gas (methyl isocyanate) with immediate and spatially confined exposures but with both short-term (death, blindness, and other acute effects) and long-term effects that are still being diagnosed. The urban episodes did not result from a single emission or release but from a cumulative release of several source types (notably power plants, refineries, steel mills, and vehicles). Open dumps had slowly led to air and water pollution that affected millions of people in developed nations and continue to be a problem in developing nations. Hazardous waste sites often affect much smaller geographic areas but with compounds so toxic they have been banned or heavily controlled worldwide over the past few decades. Indeed, all such wastes need to be managed properly, whether they are very slowly, over years and decades, affecting large areas and populations, or are being released rapidly over a confined area and a small population.

Sometimes wastes that have been "forgotten" and intentionally hidden can suddenly become an urgent problem. Corroding infrastructure of disused mines, mine dumps, and landfill sites that are not properly controlled, wastelands created by oil exploration in Russian Siberia or the Nigerian delta and very recently the Danube mud spill and BP's Deepwater Horizon disaster in the Gulf of Mexico and coal ash pit spills are examples of ecological disasters created by waste that in turn converts priceless wetlands, coasts, and other resources into vast wastelands and "no-go" areas of the most appalling kind.

Wastes manifest themselves in surprising new ways, as the BP disaster reminds us. Although the danger to workers on a rig is always present, the events that led to the explosion seem to have been similar to those in Bhopal. They are both reminders that the confluence of even unlikely events can lead to tragic results. Sometimes, the wastes that lead to problems are long forgotten. Recently, for example, the crew of clam boat encountered military munitions off the coast of Long Island, New York. The crew gathered shells that contained mustard gas (sulfur mustard), blistering a crew member [1]. This event indicates that weapons and other military wastes are present on the ocean floor around the world (United States Department of Defense estimates 17,000 tons of sulfur mustard in US waters alone). In addition to the mustard gas, these wastes include arsenic, cyanide, lewisite, and sarin gas. Interestingly, this is not necessarily a case of "improper disposal," as many would argue that these were acceptable waste handling practices at the time of disposal. After all, the vast amount of water and perceived inaccessibility made ocean dumping "acceptable." This changed in 1972 with the passage of the Marine Protection, Research, and Sanctuaries Act, which banned munitions dumping by the United States, but many orphaned waste sites remain.

As if the human health and ecological costs were not enough, as mentioned, the economic cost of cleaning up badly controlled or unmanaged industrial, mining, agricultural, and municipal waste sites has been enormous. Piles, pits, and plumes of waste are not only marks of inefficiency but are also costly to put right. Put in another way, most industrial plants involve the production of entropy and to overcome it, energy has to be expended. Each waste stream has its own unique energy and mass characteristics, each presenting an engineering challenge unique to these conditions.

This handbook can be seen as a map of a journey from exploitation to sustainability. The questions we pose at the beginning of our journey of exploring waste are

1. How serious is the global waste situation today?
2. What will our global waste situation be in 20 years' time?
3. How should waste be controlled over the next 20 years?

To answer the last two questions, first we need to decide the acceptable level of waste in an advanced society. The range is from zero level waste to doing nothing. As in most situations, the acceptable level will be somewhere between the extremes. It will need to be adapted to the particular product or system. For example, some products may be very close to zero waste (e.g., materials in newly produced automobile being 100% repurposed). Others will need to be optimized toward the level of waste that can be tolerated and managed so that an important societal need can still be met.

The answers to these three questions lie with our expert authors in the following 32 chapters. We will find that controlling waste is a fairly new concept and the constraints and controls placed on waste in our society have increased commensurately as the awareness of severity of the waste problem has heightened. This is a natural result of our expanding technologies over the past century. As always, technology is a two-edged sword. It introduces new wastes (e.g., electronic and chemical wastes) and provides solutions to the problems generated by these wastes (e.g., electronic sensors to detect waste constituents and new chemicals and microbes to clean up waste). This technological give and take is difficult to predict and, very often, we cannot foresee the problems created by waste by a new process and as a result new controls and regulations on waste appear to follow disasters.

The editors and authors have tried to be both bold and humble. We are bold when we are reasonably certain of some aspect of a particular waste or its constituents. We are simultaneously humble in pointing out our uncertainties. Knowledge about wastes has grown rapidly in recent decades and many gaps in understanding have been closed. Still, much is not known and yet to be learned. Our intent is that this handbook will not only enhance the practice of waste management but also will advance the state-of-the-science which will strengthen communications and collaborations within the waste management community.

ORGANIZATION

The handbook is organized into three sections. Part 1 considers the various ways in which wastes have been and could be addressed, beginning with a brief history. The typical transition of waste management moves from natural systems without controls to regulated and engineered systems to market-based and life cycle approaches.

Part 2 considers the specific waste streams, particularly the nature of the wastes and how they may affect human health and the environment. Note that the waste streams and waste constituents may fall into any of the categories discussed in Part 1. That is, some waste streams and chemical compounds continue to be uncontrolled, some highly regulated, and some products and systems are becoming "greener," for example, designed and based on life cycles.

Part 3 includes discussions of how to address the waste streams discussed in Part 2. These "best practices" are designed to reduce the risks posed by the wastes. These are all based on credible science and an adherence to engineering principles. Some are more precautionary, that is, actions are recommended in the absence of sufficient evidence if the problems caused by certain wastes are likely to be large and irreversible. Others draw from reliable evidence based on sound science. Part 3 should be the beginning of good waste management, because each location, waste stream, and other factors are highly variable. The waste practitioner is advised to customize the response according to the problem. For example, a highly hazardous waste that is well contained may predominantly draw upon existing designs for such waste systems, but the same waste in a less controlled environment will require additional measures and contingencies.

THE CHALLENGE

As the reader navigates through the handbook, we recommend considering the following questions, which will be revisited in the Epilogue:

1. *How much is known about each waste stream? What is the state-of-the-science?*
2. *What levels of certainty are needed to take actions?*
3. *How do the various professional and scientific disciplines vary in their approaches?*
4. *How should waste decisions be evaluated?*

Reference

[1] A. Angelle, Weapons buried at sea: Big, poorly understood problem, Live Science (2010). http://www.livescience.com/environment/underwater-military-munitions-pose-risks-100728.html?utm_source=feedburner&utm_medium=feed&utm_campaign=Feed%3A+Livesciencecom+%28LiveScience.com+Science+Headline+Feed%29; Posted July 28,2010.

[2] R.W. Emerson. Essays. First Series. Self-Reliance. 1841.

Contributors

Valerie L. Shulman (chapters 1, 21), The European Tyre Recycling Association (ETRA), 7 Rue Leroux 75116 Paris, France

Daniel A. Vallero (chapters 2, 3, 16, 18, 27, 28, 32), Pratt School of Engineering, Duke University, Durham, N. Carolina, USA

Fraser McLeod (chapter 4), Transportation Research Group, School of Civil Engineering and the Environment, University of Southampton, SO17 1BJ, United Kingdom

Tom Cherrett (chapter 4), Transportation Research Group, School of Civil Engineering and the Environment, University of Southampton, SO17 1BJ United Kingdom

Geoffrey Blight (chapters 5, 30), School of Civil and Environmental Engineering, University of the Witwatersrand, Johannesburg, South Africa

Thomas Pretz (chapter 6), Department of Processing and Recycling, RWTH Aachen University, Wüllnerstr. 2, 52062 Aachen, Germany

Jörg Julius (chapter 6), Department of Processing and Recycling, RWTH Aachen University, Wüllnerstr. 2, 52062 Aachen, Germany

John E. Marra (chapter 7), Savannah River National Laboratory, Aiken, South Carolina 29802, Untied States of America

Ronald A. Palmer (chapter 7), Institute for Clean Energy Technology, Mississippi State University, Starkville, Mississippi 39762, United States of America

Agamuthu Periathamby (chapter 8), Institute of Biological Sciences, Faculty of Science, University of Malaya, Kuala Lumpur, Malaysia

Ralf Otterpohl (chapter 9), Institute of Wastewater Management and Water Protection, TUHH-Hamburg University of Technology, D-21073 Hamburg, Germany

Christopher Buzie (chapter 9), Institute of Wastewater Management and Water Protection, TUHH-Hamburg University of Technology, D-21073 Hamburg, Germany

Gary M. Scott (chapter 10), Department of Paper and Bioprocess Engineering, State University of New York, College of Environmental Science and Forestry, One Forestry Drive, Syracuse, NY

John Butler (chapter 11), 3 The Hall, Wold Newton, Driffield, East Riding of Yorkshire YO25 3YF, United Kingdom

Paul Hooper (chapter 11), Manchester Metropolitan University, United Kingdom

Andreas Bartl (chapter 12), Institute of Chemical Engineering, Vienna University of Technology, Getreidemarkt 9/166, A-1060 Vienna, Austria

Rebecca Slack (chapter 13), School of Geography, University of Leeds, Leeds, LS2 9JT, UK

Trevor M. Letcher (chapters 3, 13), School of Chemistry, University of KwaZulu-Natal, Durban, South Africa

Marian Chertow (chapter 14), School of Forestry and Environmental Studies, Yale University, New Haven, CT 06511, USA

Jooyoung Park (chapter 14), School of Forestry and Environmental Studies, Yale University, New Haven, CT 06511, USA

Mohamed Osmani (chapter 15), Department of Civil and Building Engineering, Loughborough University LE11 ETU, United Kingdom

Paola Lettieri (chapter 17), Department of Chemical Engineering, University College London, Torrington Place, WC1E 7JE, United Kingdom

Sultan M. Al-Salem (chapter 17), Department of Chemical Engineering, University College London, Torrington Place, WC1E 7JE, United Kingdom

J. Vinicio Macías-Zamora (chapter 19), Instituto de Investigaciones Oceanológicas, Universidad Autonoma de Baja California, Ensenada, Baja California 22860, México

Jirang Cui (chapter 20), Department of Material Science and Engineering, Norwegian University of Science and Technology (NTNU), Alfred Getz vei 2, N-7491 Trondheim, Norway

Hans Jørgen Røven (chapter 20), Department of Material Science and Engineering, Norwegian University of Science and Technology (NTNU), Alfred Getz vei 2, N-7491 Trondheim, Norway

Ash Genaidy (chapter 22), WorldTek, Inc., 1776 Mentor Avenue, Suite 423, Cincinnati, OH 45212, USA

Reynold Sequeira (chapter 22), WorldTek, Inc., 1776 Mentor Avenue, Suite 423, Cincinnati, OH 45212, USA

Andrew L. Shannon (chapter 23), Michigan Department of Natural Resources and Environment, 525 West Allegan Street, Lansing, Michigan 48909, USA

Anne Woolridge (chapter 23), Independent Safety Services Limited, Dabell Avenue, Bulwell, Nottinghamshire, NG6 8WA, UK

R. Nagendran (chapter 24), Centre for Environmental Studies, Anna University, Chennai 600025, India

Victor F. Medina (chapter 25), U.S. Army Corps of Engineers, Engineers Research & Development Center, Vicksburg, MS

Scott A. Waisner (chapter 25), U.S. Army Corps of Engineers, Engineering Research & Development Center, 3909 Halls Ferry Rd., Vicksburg, MS, USA

Gene Stansbery (chapter 26), NASA Orbital Debris Program Office, NASA/Johnson Space Center/KX2, USA

Daniel J. Vallero (chapter 29), Public Works Department, Engineering Section, City of Durham, NC, USA

Nicholas P. Cheremisinoff (chapter 31), N&P Ltd., Willow Spring Road, Charles Town, West Virginia, USA

PART I

INTRODUCTION TO WASTE MANAGEMENT

Success in waste management, as in any technical enterprise, depends on three factors: awareness, decision making and action. This section gives a context for all three. Being aware of the nature of wastes and how it has been controlled through time is the first step in sound management. Good decision making depends on knowing what has happened and what has worked and failed in the past. From this knowledge flows the actions needed to solve waste problems.

For much of human history, if recognized at all, waste was considered a nuisance. Only within the past century has waste warranted its own engineering disciplines, that is, sanitary engineering and environmental engineering. The primary client of these professionals, like those of all engineers, is the public. The first canon of any engineering profession is that the public safety, health and welfare be held paramount. This begins with an awareness of the problem. From there decisions must be underpinned by credible and relevant science. Even the best decisions are worthless without action. The waste controls and management objectives can only properly be met with sound and best practices. Thus, any measure of success needs a baseline.

For wastes, the past is the key to the future. We believe that Santayana and Einstein would agree that to continue to exploit in a manner that leads to unsustainable amounts and kinds of wastes is insane. So, to begin our journey, we look at the history that has moved from frontier thinking, to an growing barrage of waste regulations and the rules that "control" waste, and more sane and sustainable ways of preventing the generation of wastes in the first place, with an introduction to green engineering and sustainable design as related to waste. Even under the most optimistic scenarios, however, wastes will be with us well into the foreseeable future, so we end this section with the most basic of waste issues—the collection of waste.

Reading between the lines, one may trace the implementation of many of the regulations to times immediately following waste disasters. For example, a new technology is implemented and we believe we have the regulations in place to cater for any eventualities. After a time of trouble-free production, some catastrophe takes place and pollution and waste occur at levels we could not have foreseen. New regulations are promulgated in an attempt to stem the tide of further disasters.

CHAPTER

1

Trends in Waste Management

Valerie L. Shulman

European Tyre Recycling Association (ETRA), Avenue de Tervueren 16, 1040 Brussels, Belgium

OUTLINE

1. INTRODUCTION

Waste management and recycling are neither new concepts nor new activities. In fact, materials had been recycled long before the term was coined in the twentieth century. People have always seen value in items cast-off by others. Witness the aphorism that 'one's trash is another man's treasure'.

Historically, waste management has been inextricably linked with the evolution of human communities, population growth and the emergence and development of commerce. During the past century, consumption and production patterns have changed radically – due in part to the greater freedom of movement of money, goods and people.

Population growth has taken precedence in terms of economic development and the creation of waste. World population trebled from approximately 2 billion in 1925–2000, when it topped 6 billion (see Box 1.1). The vast growth spurt has been attributed to the benefits of economic development, including improved healthcare, higher fertility rates, lower infant mortality and long-life expectancy, primarily in developing countries. Care must be taken when using such global data. For example, less developed nations have experienced growth without many of these benefits, that is, they continue to experience high

Waste Doi: 10.1016/B978-0-12-381475-3.10001-4

BOX 1.1

POPULATION GROWTH

Year	Estimated Population
3000 BCE	14,000,000
2000 BCE	27,000,000
1000 BCE	50,000,000
500 BCE	100,000,000
200 BCE	150,000,000
1 AD	200,000,000
1000 AD	310,000,000
1804	1,000,000,000
1925	2,000,000,000
1939	2,200,000,000
1945	2,300,000,000
1950	2,500,000,000
1960	3,000,000,000
1975	4,000,000,000
1988	5,000,000,000
2000	6,000,000,000

infant mortality because of poor nutrition and infectious diseases, whereas wealthier countries have advanced healthcare but have witnessed an overall lower fertility rate – which endures today.

The population explosion has exerted greater pressures on production and consumption and indirectly the accumulation of waste. Over time, it has become apparent that the single most important driving forces modifying the environment are population size and growth – and how man exploits available natural resources.

2. THE CATALYST FOR CHANGE

At the end of World War II, the world was in shambles from virtually every perspective: physically, economically, socially and environmentally. The war had been the most pervasive military conflict in human history – over land, on the seas and in the air. Sixty-one countries and many territories on six continents, as well as all the world's oceans, suffered devastating damage and long-term social, economic and environmental effects. Only the Western hemisphere, parts of the Near East and sub-Saharan Africa were unscathed.

Wars are most notorious for their tolls on human populations, but they also severely affect ecosystems. Rivers and lakes, jungles and forests and farmlands and deltas were obliterated – with dangerous wastes left behind. Hundreds of cities were demolished and many others rendered virtually uninhabitable. Infrastructure was decimated – bridges, roads and railroads were laid to waste – and rendered nonfunctional.

Almost 60 million civilians and military personnel were killed and tens of millions more were seriously injured and/or permanently maimed. War-induced famines took the lives of more than 2 million more in Africa and Asia [1]. Millions remained homeless throughout the war-torn world. Thousands more were captives of foreign nations – even at home.

According to the International Registry of Sunken Ships [2], more than 12,500 sunken vessels including battleships, aircraft carriers, destroyers, landing craft and more than 5000 merchant ships were scattered on ocean floors. Governments estimate that more than 335,000 aircraft were lost, primarily over Europe, Asia and Africa [3]. Thousands of tonnes of unexploded ordnance including mines, bombs and various forms of ammunition litter seabeds, fields, jungles, caves and even home gardens.

More than 60 years after the end of the war, experts estimate that it could take another 150 years to clear the detritus and neutralise the hazardous content – which continue to pose dire threats to the environment, humans and creatures in the seas, on land and in the air. In addition to military debris, every type of waste imaginable – from natural to synthetic materials – including construction rubble, plastic debris, synthetic rubber, electronic equipment and parts, transistors, microwave materials, and synthetic fuels, among hundreds of others, became the residue of the war – and had to be treated and disposed.

Many of the products created for the 'war effort' have become the most common products of today – with the same problems and issues surrounding their treatment and disposal. Pesticide formulations, such as the organophosphates, owe their basic chemical structures to chemical war agents. Petrochemical products also have grown substantially in response to war efforts. In addition, abandoned ammunition dumps, practice ranges, and other military facilities continue to be vexing hazardous waste sites.

The definition of wartime waste is complex. For example, among the most harmful and tragic wastes are abandoned land mines, which continue to cause death and inflict harm long after their initial use.

3. SUSTAINABLE DEVELOPMENT: THE CONTEXT FOR RECYCLING

As early as 1942, signatories to the Atlantic Charter had initiated discussions about an organisation that could replace the failed League of Nations. Before the final guns were silenced, world leaders had begun to prepare for the future – one without war, in which disputes could potentially be resolved through discussion and cooperation. The structure and substance of the United Nations was agreed among 50 nations – with 51 available to sign it into international law.

Signed on 26 June 1945, the UN Charter came into force on 24 October 1945, as an international organisation with the goal of providing a platform for dialogue and cooperation among nations in order 'to save succeeding generations from the scourge of war'. Inherent within the Charter is the recognition that equal rights and self-determination are imperative for each sovereign nation – large or small, wealthy or poor, and must be supported. During the next half-century, these concepts would pervade all aspects of UN undertakings – from decolonisation and economic development to environmental and waste issues.

At its inception, five interactive themes were identified: international law, international security, economic development, social progress, and human rights. The infrastructure provided for six principal organs: the Trusteeship Council,[a] the Security Council, the General Assembly, the Economic and Social Council, the International Court of Justice and the Secretariat (see Fig. 1.1). Each organisation had its own mission and objectives, which have evolved over time to reflect current issues and needs.

Actions related to the environment, and by extension to waste management, can best be described in terms of three broad periods: the post-war period (1945–1970); globalisation, scientific and environmental awareness (1970–1990) and implementation and progress (1990 to the present).

3.1. The Post-War Period

The post-war period can be described as one of far-reaching political, social and economic changes.

- Governments were responsible for assessing the war damage – and initiating the clean-up

[a] The Trusteeship Council was a bridge between the League of Nations (LoN) and the United Nations – with a limited brief. The Trust territories were former mandates under the LoN or taken from nations defeated at the end of World War II. Once its mandated responsibility for 11 colonial countries had expired in 1994, Council operations were suspended.

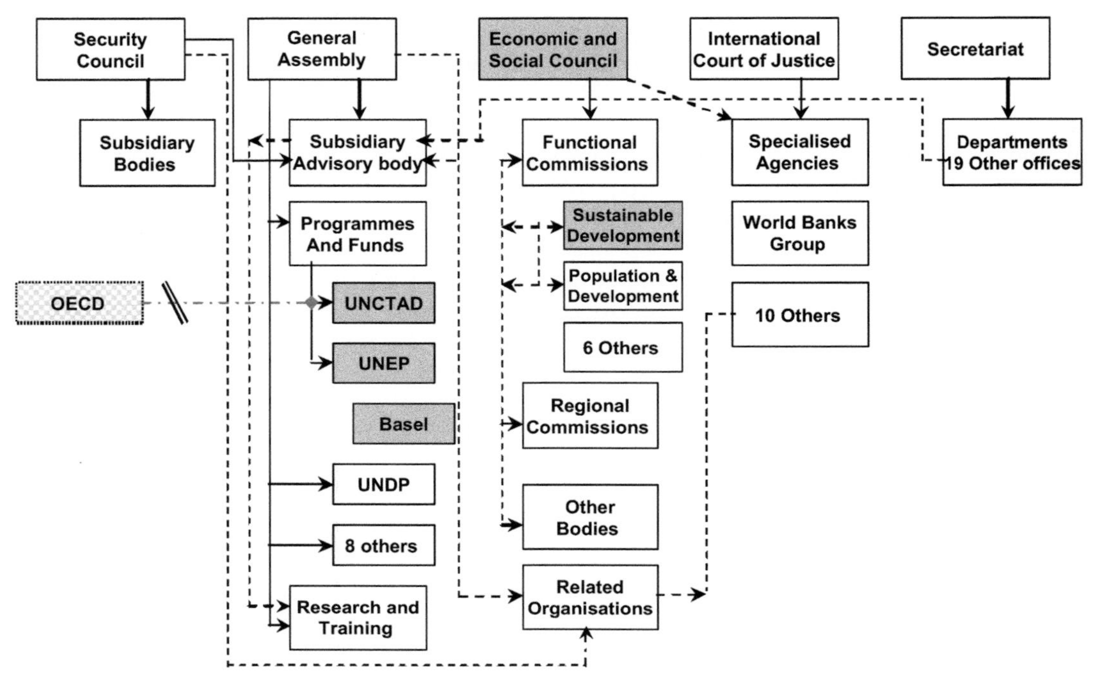

FIGURE 1.1 United Nations (UN) structure concerning the environment [4] is an adaptation of the UN organisation chart to illustrate the relationships between and among the five current organs.

and reconstruction of needed infrastructure, homes, civil institutions, business and industry.

- The Organisation for Economic Cooperation and Development (OECD) was formed in 1960 with 20 members, as an independent forum for industrialised democracies to study and formulate economic and social strategies that could involve developing nations. Today, 31 member countries focus on environmental, economic and social issues to institutionalise and integrate sustainable development concepts into national policy and strategies. Its projects are diverse, ranging from sustainable materials management to corporate responsibility and climate change.
- By mid-1961, almost 750 million people had exercised their right to self-determination, and more than 80 once-colonised territories had gained independence, including those under the Trusteeship Council.
- By the end of 1961, a Special Committee on Decolonisation was formed to aid 16 non-Trusteeship countries seeking sovereignty.
- With self-determination came new responsibilities and social commitments requiring interactions between wealthier and poorer nations (often described as 'north' and 'south'). Self-determination became increasingly important as developing countries sought a stronger role in global economics.
- The United Nations Conference on Trade and Development (UNCTAD) was formed in 1964 as a permanent body of the United Nations dealing with trade, investment and development issues. It supports the

integration of developing countries into the world economy – ensuring that domestic policy and international action toward sustainable development do not clash. It helps to assess the needs of the least developed countries in trade relationships, for example, north versus south and producers versus consumers.

The 25-year post-war period focused on clean-up and rehabilitation of affected areas. Vast quantities of wastes were collected – and often shipped from wealthier to poorer nations for disposal. The concept of self-determination came into play – and by the end of the period, poorer nations began to refuse acceptance of external wastes.

An infrastructure for debate had been created with the formation of OECD and UNCTAD. A principal outcome was the establishment of a system of organisations that had the capacity to act in unison to establish a worldwide mechanism to attain peace, as well as economic and social stability. There was a keen awareness of the relationships between policy, trade, economic development and environmental impacts.

3.2. The Period of Globalisation

The period of globalisation can be described as one of the rapid scientific and technological innovation, coinciding with the creation of the United Nations Environment Programme (UNEP) and the Basel Convention. Commercial globalisation exacerbated many environmental problems and highlighted the need for global solutions. Together, these bodies have assisted poorer nations to become a driving force in world economic development.

- UNEP was formed in 1972 to smooth a path for international agreements, with the mission of assisting poorer countries to develop and implement environmentally sound policies and practices, coordinate the development of environmental policy consensus and keep environmental impacts under review. As awareness of cross-border pollution grew, nations established agreements with neighbouring states and instituted a series of treaties, conventions and protocols for controlling pollution and similar problems that crossed national boundaries. International environment conventions promoting science and information drew great support and helped these nations to work in conjunction with policy, guidelines and treaties on international trade – particularly in terms of hazardous materials – trans-boundary air pollution and contamination of waterways, among others.
- The Basel Convention created in 1989 under UNEP, filled the gap between existing mandates that facilitate and monitor world trade, on the one hand, and those that are concerned with sound environmental practices, on the other. The mission of the Basel Convention is to monitor the trans-boundary movements and management of wastes to ensure their environmentally sound treatment and disposal and to provide support to governments by assisting them to carry out national sustainable objectives.

During the next 20 years, these organisations undertook an exhaustive awareness campaign to draw the support of national and local governments, non-government organisations (NGOs), industry and the public at large. Trans-boundary movements of wastes required the implementation of environmental management systems to evaluate the quantity and impact of emissions within the environment. New guidelines were created for the import/export of wastes for recovery with OECD and Basel support. The focus was on the prevention and minimisation of wastes – led to the study of the economics of waste – to identify barriers and failures.

4. IMPLEMENTATION AND PROGRESS

During the final years of the twentieth century, it became apparent that the unbridled economic growth of the past could not be sustained in future without irreparable damage to the environment. Discussions initiated during the 1960s culminated in a proposal for change, at the global level.

The Stockholm meeting of the United Nations Conference on Environment and Development (UNCED) in June 1972 is often marked as the critical turning point in the move towards more sustainable growth practices. It signalled a break from the past and the beginning of a new era.

The goals of the conference were limited. They were first, to introduce the concepts and practices inherent in sustainability and second, to provoke sufficient concern and interest for world leaders to make a commitment to de-link economic growth from negative environmental impacts.

Simply stated, sustainability requires policies and actions that foster economic and social growth that meet current needs without detriment to the environment. The aim is to not compromise the ability of future generations to meet their own needs.

'Environment' was defined in the broadest sense to include all of the conditions, circumstances and/or influences affecting development. The specific issue was the improved management and use of natural resources, concentrating on the prevention and control of pollution and waste.

Delegates adopted the principle and accepted the challenge of implementing the sustainable model of development for the twenty-first century. One of the most immediate results of the meeting was the creation of UNEP as the global authority on environmental issues. It was envisioned that UNEP would smooth the way for international agreements – including those between the wealthier northern and poorer southern countries.

The global economic and social nature of the plan led to the involvement of other organisations within the UN infrastructure. As described in Box 1.2, these bodies provide the international framework within which intra- and international trade occur, including the movement of wastes.

By the 1992 UNCED meeting in Rio de Janeiro, much of the groundwork had been completed. The infrastructures for both encompassing legislation and actions were in place. The goal of the conference was to propose alternative strategies and actions that could be undertaken in the short- medium- and long-term to ensure that consideration and respect for the environment would be integrated into every aspect of the development process. The Basel Convention provided the common framework for the classification, management and treatment of waste. Briefly, waste was defined as follows:

> ..substances or objects which are disposed of or intended to be disposed of or are required to be disposed of by the provisions under national law.

Both the Basel Convention and the OECD independently prepared catalogues of the substances, objects, materials, etc, which are defined as waste and separated out those defined as hazardous or dangerous. A final list contains those wastes that are not perceived to pose a risk to the environment or human health. However, it is important to note that the lists are not mutually exclusive and that under certain conditions, a 'waste' can and often does appear on more than one list. Virtually every conceivable material, product or residue is listed – those that are not specifically named fall under the rubric 'other'.

BOX 1.2

INTERNATIONAL BODIES CONCERNED WITH WASTE

UNCED formulates strategies and actions to stop and reverse the effects of environmental degradation and promote sustainable, environmentally sound development in all countries.

UNCTAD promotes trade between countries with different social and economic systems and provides a centre for harmonising the trade and development policies of governments and economic groupings.

OECD is a permanent body under UNCTAD. It was created to assist in removing restrictions and facilitating trade between and among member and non-member countries, ensuring that the substances, materials, products, and so on involved do not pose a threat to the environment or humanity in the receiving country.

UNEP is the designated authority on environmental issues at the global and regional levels. It was created to coordinate the development of environmental policy consensus and bringing emerging issues to the international community for action.

Basel Convention, under UNEP, is specifically concerned with the control of trans-boundary movements of hazardous and other wastes and their disposal, from OECD countries to non-OECD countries. Furthermore, it is concerned with the identification of those products and materials that could cause damage to the receiving countries.

The definition and annexes served as a guide for trans-boundary movements of waste, principally for their environmentally sound management. Examples of recovery and disposal operations were appended. Environmentally sound management was broadly defined as follows:

> ..taking all practicable steps to ensure that waste is managed in a manner that will protect human health and the environment against adverse effects which may result from such waste.

Within the context of the definitions of waste and its environmentally sound recovery and disposal, the OECD laid down the provisions for its trans-boundary movement and acceptance, within and outside the member countries. Each country was invited to prepare a list of those wastes that it would no longer accept for either recovery or disposal, due to lack of appropriate treatment facilities, risks to human health, among other reasons. Thus, procedures were also established for the non-acceptance of wastes and their return when they are delivered in error.

Once the framework was established, various tools were explored to assess their capacity for targeting potential environmental impacts. Life cycle analysis (LCA) was selected as the most appropriate and effective tool for determining the points at which the greatest environmental impacts occur, thus making possible the suggestion and selection of less damaging options. For example, the approach permitted the evaluation of industrial outputs from the production or extraction of raw materials through the design and manufacture of materials and products, as well as during product use (see Chapter 2).

The definitions, annexes and provisions were accepted by the delegates. However, many of the participating countries also adapted the provisions to comply with national policy and priorities.

The most hazardous wastes and the most prevalent sources of pollution were targeted for immediate attention. Five priority waste streams were distinguished. In addition to the more general category of 'household waste', post-consumer tyres, demolition waste, used cars, halogenated solvents and hospital waste were earmarked for action.

5. INTERPRETATIONS

Virtually every industry has come under scrutiny from mining to manufacturing and healthcare. A raft of legislation has been enacted, with the agreement and cooperation of the partners. A horizontal framework was established for waste management, including definitions and principles. Treatment operations were defined vertically – to include the control of landfill, incineration, etc. A body of standards is currently being prepared for treatment operations – through the International Standards Organisation, with support from national standards bodies.

During the 50 years since the initiation of the first discussions on sustainable development in the 1960s, legislation and actions have been put in place to ensure that governments work together with industry and the public at large. Today, the majority, if not all, UN member countries have enacted basic environment and waste management legislation.

Reuse and recycling are again being integrated into industrial activities. However, as they are interpreted today, the concepts of reuse and recycling are inextricably linked to the production and management of waste and by extension, to its prevention and minimisation.

Reuse and recycling have evolved into two of the four pillars that support improved resource management through the prevention of waste and the reuse, recycling and recovery of the wastes that do occur to achieve sustainable developmental goals by reducing reliance on natural resources.

References

[1] D. Reynolds, One World Divisible: A Global History Since 1945, W.W. Norton & Company, New York, NY, 2000.

[2] International Registry of Sunken Ships, Saskatchewan, Canada; <http://www.shipwreckregistry.com/>; Accessed on 26 July 2010 (2010).

[3] J. Ellis, World War II: A Statistical Survey: The Essential Facts and Figures for All the Combatants, Facts on File, New York, NY, 1993.

[4] United Nations, History of the United Nations. <http://www.un.org/aboutun/unhistory/>; Accessed on 26 July 2010 (2010).

CHAPTER

2

Green Engineering and Sustainable Design Aspects of Waste Management

Daniel A. Vallero

Pratt School of Engineering, Duke University, Durham, North Carolina, USA

OUTLINE

1. INTRODUCTION

For most of the twentieth century, wastes were viewed predominately as inevitable byproducts of modern times. Waste generation was a necessary reality associated with economic development. Thus, addressing wastes was often a matter of reacting to problems as they arose individually in a situationally dependent way. However, the processes that lead to waste can be viewed much more proactively and systematically. It is best to prevent the generation of wastes in the first place. We have decided to begin this book with this latter perspective, that is, waste streams should not only be rendered less toxic and reusable, but ultimately completely eliminated.

Engineers and other waste managers have begun to embrace waste minimization, pollution prevention, and other systematic approach, albeit incrementally. After all, these professionals are generally quite practical, so the shift to a "no-waste" paradigm has been a thoughtful one. We hope that the optimism in our view that someday the need for many of the chapters in this book will be eliminated can be balanced with the practicality that even with waste elimination, recycling, and pollution prevention,

Waste Doi: 10.1016/B978-0-12-381475-3.10002-6

there will still be vast waste management challenges in the decades to come.

2. PARADIGM EVOLUTION

Green design and sustainable approaches to waste apply scientific principles to develop objective-oriented, function-based processes. They consider every element of a product's life cycle in a way that mutually benefits the client, the public, and the environment. Waste products can decrease in volume and mass as green designs replace traditional methods of manufacturing, use, and disposal.

Let us be clear at the outset about the entrenchment of product and system design mind-sets that have relied on schemes steeped in an exploitation rather than compatibility with nature. Designs of much of the past four centuries have assumed an almost inexhaustible supply of resources. Such inertia has been and will continue to be difficult to overcome.

Wastes must be managed (and, ideally, avoided) by means of applying the laws of science. The better these principles are understood by the designer, the more likely that the products demanded by society can be produced and used predictably and sustainably. Strategic use of physical science laws must inform designs and engineering decisions.

2.1. New Thinking

New and emerging problems demand new approaches and ways of thinking. As evidence, Albert Einstein has noted:

> The significant problems we face cannot be solved at the same level of thinking we were at when we created them.[1]

Waste management is more than landfills, incinerators, infrastructure, and other "built forms." It requires a view of the totality of matter and energy, with an eye toward ways to reduce "leakage" from the system. McDonough and Braungart [1] captured quite well the need to shift the waste paradigm:

> For the engineer that has always taken—indeed has been trained his or her entire life to take—a traditional, linear, cradle to grave approach, focusing on "one-size fits-all" tools and systems, and who expects to use materials and chemicals and energy as he or she has always done, the shift to new models and more diverse input can be unsettling.

In this new paradigm, waste management begins long before any waste is generated. The function drives the product, so the product is to be considered with respect to its potential life cycles. Such a viewpoint challenges "single-purpose" thinking. For example, a detergent may be redesigned to be "phosphate free," so that it does not contain one of the nutrients that can lead to eutrophication of lakes, but this does not necessarily translate directly into an ecologically acceptable product if its life cycle includes steps that are harmful.

The phosphate waste is eliminated. However, the life cycle view does not allow the product designer to rest with this simple substitution. Could the substitute ingredient be extracted and translocated by plant life in a way that damages sensitive habitats, makes use of and releases toxic materials in manufacture, and entails persistent chemical by-products that remain hazardous in storage, treatment, and disposal?

Examples are plentiful of substitutes wreaking more havoc than the products they replace. DDT was replaced by the arguably more toxic pesticides, aldrin and dieldrin. Substituting incineration for landfills can lead to the release

[1] Attributed to A. Einstein, this quote appears in numerous publications without a source of citation.

of certain pollutants, for example, dioxins and heavy metals, in far more toxic forms than would be found in the landfill leachate.

Even substitutions that enjoy a consensus of acceptability can be associated with problems. For example, most would agree that replacing organic solvents, such as petroleum distillates, with water-soluble constituents in automobile paint has been preferable from an environmental perspective, that is, replacing an organic solvent with a water-based solution is often desirable and can rightly be called "solvent-free." However, under certain scenarios, this substitution indeed could be environmentally unacceptable. Many toxic substances, such as certain heavy metal compounds, are highly soluble in water (i.e., hydrophilic). Is it possible that the water is a better transport medium for metal pigments in the paint? Thus, our "improved" process has actually made it easier for these metals contained in the solution to enter the ecosystem and to lead to human exposures. The lesson here is to be ever mindful of the *law of unintended consequences*.

Another consideration in the new paradigm is that waste reduction and elimination must not be justified solely using a cost–benefit economics, such as those based on the return on monetary investment that can be expected over the life of a product. Often, the waste manager is presented with a list of options, but they all begin with the waste arriving at the facility. Obviously, the manager's span of control dictates the number and diversity of options. However, even if limited in options, it is incumbent on the manager to suggest upstream improvements. Products that end up in the waste stream must also be evaluated using methods beyond a comparison of the initial investment as a fraction of the total cost of manufacture, use, and disposal. Product design decisions must also include less tangible impacts on the individual, society, and ecology that may not fit neatly on a data spreadsheet.

2.2. Traditional Facility Design

The critical path from product conception to completion has changed very little over the thousands of years. The actual view of the process of design, however, varies substantially, even within the waste management community. The traditional design process from conception to completion has been sequential with distinct phases guiding the process from definition of need, drawings through technical development, fabrication, and final completion. The progression of the stepwise process from idea to realization is a sequence of events and involvement of specialized expertise. The process is direct, sequential, and linear, following a prescribed set of activities that will lead to a final solution.

This stepwise approach is often referred to as the *waterfall model*, drawing on the analogy of water flowing continuously through the phases of design. This approach is acceptable if the number of variables is manageable and a limited universe of possible solutions is predictable. An example of this approach would include a "prototype" design that is simply being adapted to a new condition. This process is often the most direct, conventional, and least costly when "first cost" is a primary consideration. For example, a reduction in the time required for design and delivery can mitigate the impact of price escalation due to inflation and other market variables. In practice, many designs are planned around schedules that appear to be linear, but the actual activity within each phase tends to be somewhat nonlinear (e.g., feedback loops are needed when unexpected events occur).

Linear progression of the process would logically begin with a clear definition of the intended use. This assumes that the product use scenarios have been clearly identified and that the variability of uses has been well understood. This would mean that the data about users and uses are ample, which is seldom the case. Once these data are collected and characterized, alternatives for meeting the use requirements are woven into a framework or *schematic* for the new product. The design process optimizes on the basis of predetermined design criteria.

2.3. Comprehensive Approach

Historically, products have been designed as an unqualified handoff, at least in terms of what to do with any waste products generated during and after the intended use. Such an approach considers only a type of contractual arrangement between the manufacturer and user, so that the product will perform according to specific criteria. Other stages in the product's life cycle (e.g., waste streams) are not part of this "contract." This has been one of the failings of the traditional design process; that is, underweighting or completely ignoring the wastes that would be generated, not only in the fabrication step but also throughout the product life. Thus, beginning in the late twentieth century, designers began to embrace *Design for Disassembly* (DfD), that is, identifying and managing the materials from the product that will be present after the useful life.

Designing without respect to disassembly was evident in a magazine advertisement in the 1970s, which showed a hand throwing away a disposable razor. The razor simply disappeared. Conversely, we can safely assume, given the biodegradation rates of the plastics used in the razor, that the handle is still intact in a landfill somewhere. Thus, DfD goes beyond evaluating the disposition of materials, vacated land, contamination of manufacturing facilities, and other remnants of the project. It is also a view of utility beyond the use phase, that is, "repurposing" the remaining materials. Certainly, this requires postuse planning, such as insisting on the use of reusable materials and considerations of obsolescence of parts and the entire system. It also addresses uses after the first stage of usage and the avoidance ("down cycling"). For example, if a neighborhood demographic were to change in the next century, is the design sufficiently adaptive to continue to be useful for this new set of users? This is not so unusual, as in the case of well-planned landfills, which may have a few decades of waste storage, followed by many decades of park facilities. How many strip malls or shopping centers were designed for but a few decades of use, followed by abandonment and desolation of neighboring communities in their wake? It is folly and professional hubris to assume that the user community will not change with respect to its social milieu. Product design must embrace the idea of "long-life/loose fit" and be sufficiently flexible to accommodate a variety of adaptive reuse scenarios.

3. LIFE CYCLE ASSESSMENT

The complexity of life cycle assessment (LCA) ranges from scant attention to inputs and outputs of materials and energy (Fig. 2.1) to multifaceted decision fields extending deeply into time and space. The latter is preferable for decisions involving large scales, such as the cumulative buildup of greenhouse gases, or those with substantially long-term implications, such as the release of genetically altered microbes into the environment. Complex LCAs are also favored over cursory models when the effects are extensive, such as externalities and artifacts resulting in geopolitical impacts. Biotechnologies may fall into any or all of these three categories.

The decision to increase the use of ethanol as a fuel additive and a reformulated fuel is such a decision. Alternative fuel standards have met with skepticism and even descent. In particular, the viability of ethanol is being challenged from scientific and policy standpoints. Corn-based ethanol is indeed a biotechnology. In fact, since the presidential proclamation, dedicated corn crops and bioreactors in these states have emerged. On the other hand, geopolitical impacts, such as food versus fuel dilemmas are being raised. Scientific challenges to any improved efficiencies and actual decreases in the demand for fossil fuels have also been

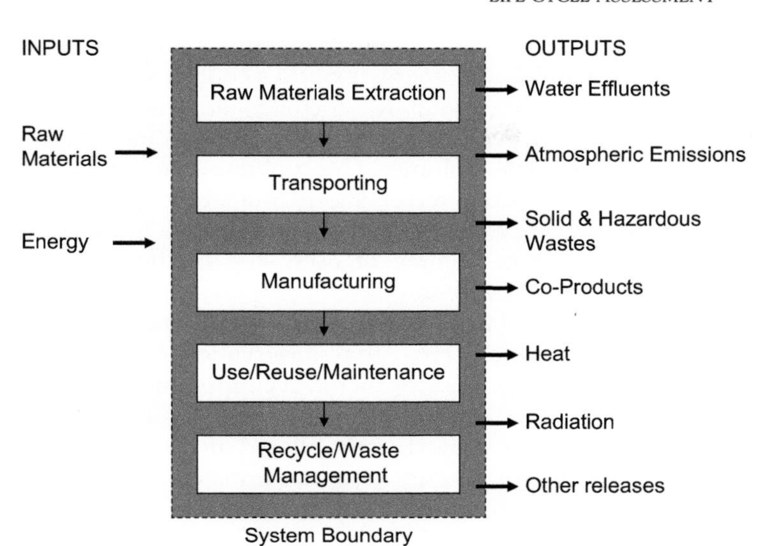

FIGURE 2.1 Life cycle stages of a process must follow the conservation law, with material and energy balances [2].

voiced. Some have accused advocates of ethanol fuels of using "junk science" to support the "sustainability" of an ethanol fuel system. Notably, some critics contend that ethanol is not even renewable, because its product life cycle includes a large number of steps that depend on fossil fuels. The metrics of success are often deceptively quantitative. For example, the two goals for increasing ethanol use include firm dates and percentages. However, the means of accountability can be quite subjective. For example, the 2017 deadline *could* be met, but if overall fossil fuel use were to increase dramatically, the percentage of total alternative use could be quite small, that is, not near the 15%. Thus, *both* absolute and fractional metrics are needed.

Another accountability challenge is how accurately energy and matter losses are included in calculations. From a thermodynamics standpoint, the nation's increased ethanol use could actually increase demands for fossil fuels, such as the need for crude oil-based infrastructures, including farm chemicals derived from oil, farm vehicle and equipment energy use (planting, cultivation, harvesting, and transport to markets) dependent on gasoline and diesel fuels, and even embedded energy needs in the ethanol processing facility (crude oil-derived chemicals needed for catalysis, purification, fuel mixing, and refining). A comprehensive LCA is a vital tool for ascertaining the actual efficiencies.

The questions surrounding life cycles of products such as ethanol can be addressed using a three-step methodology. First, the efficiency calculations must conform to the physical laws, especially those of thermodynamics and motion. Second, the "greenness," as a metric of sustainability and effectiveness can be characterized by life cycle analyses. Third, the policy and geopolitical options and outcomes can be evaluated by decision force field analyses. In fact, these three approaches are sequential. The first must be satisfied before moving to the second. Similarly, the third depends on the first two methods. No matter how politically attractive or favorable by society, an alternative fuel must comport with the conservation of mass and energy. Furthermore, each step in the life cycle (e.g., extraction of raw materials, value-added manufacturing, use, and disposal) must be considered in any benefit–cost or risk–benefit analysis. Finally, the societal benefits and risks must be viable for an alternative fuel to be accepted. Thus, even a very efficient and effective fuel may be rejected for societal reasons (e.g., religious, cultural, historical, or ethical).

The challenge is to sift through the large and diverse data sets to ascertain whether ethanol truly presents a viable alternative fuel. Of the misrepresentations being made, some clearly violate the physical laws. Many ignore or do not provide correct weights to certain factors in the life cycle. There is always the risk of mischaracterizing the social good or costs, a common problem with the use of benefit/cost relationships.

3.1. Efficiency

Fuel efficiencies are evaluated in terms of net energy production that is based on thermodynamics (first and second laws). Energy balances can be calculated from the first law of thermodynamics:

$$\text{Accumulation} = \text{creation rate} - \text{destruction rate} + \text{flow in} - \text{flow out} \quad (2.1)$$

Stated quantitatively as efficiency, Eqn (2.1) is

$$\text{Efficiency} = \frac{E_{in} - E_{out}}{E_{in}} \times 100 \quad (2.2)$$

Where E_{in} = energy entering a control volume and E_{out} = energy exiting a control volume.

The numerator includes all energy losses. However, these are dictated by the specific control volume. This volume can be of any size from molecular to planetary. To analyze energy losses related to alternative fuels, every control volume of each step of the life cycle must be quantified.

The first two laws of thermodynamics drive this step. First, the conservation of mass and energy requires that every input and output be included. Energy or mass can be neither created nor destroyed, only altered in form. For any system, energy or mass transfer is associated with mass and energy crossing the control boundary within the control volume (Fig. 2.2). If mass does not cross the boundary, but work and/or heat do, the system is a "closed" system. If mass, work, and heat do not cross the boundary, the system is an isolated system. Too often, open systems are treated as closed, or closed systems include too small of a control volume. A common error is to assume that the life cycle begins at an arbitrary point conveniently selected to support a benefit/cost ratio (BCR). For example, if a life cycle for ethanol fuels begins with the corn arriving at the ethanol processing facility, none of the fossil fuel needs on the farm or in transportation will appear.

The second law is less direct and obvious than the first. In all energy exchanges, if no energy enters or leaves the system, the potential energy of the state will always be less than that of the

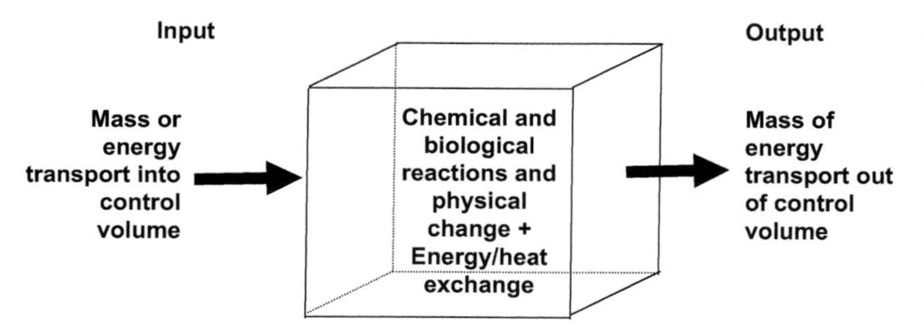

FIGURE 2.2 Control volume showing input, change, and output. The process applies to both mass and energy balances [3].

initial state. The tendency toward disorder, that is, entropy, requires that external energy is needed to maintain any energy balance in a control volume, such as a heat engine, waterfall, or an ethanol processing facility. Entropy is ever present. Losses must always occur in conversions from one type of energy (e.g., mechanical energy of farm equipment ultimately to chemical energy of the fuel). Thus, Eqn (2.2) is actually a series of efficiency equations for the entire process, with losses at every step.

These equations illustrate the importance of reliable information in the life cycle inventory (LCI). Any LCA relies on the LCI data to assess the environmental implications associated with a product, process, or service by compiling an inventory of relevant energy, material inputs, and environmental releases. From this LCI, the potential environmental impacts are evaluated. The results aid in decision-making. Thus, the LCA process is a systematic, four-component process:

1. *Goal Definition and Scoping*—Define and describe the product, process, or activity. Establish the context in which the assessment is to be made and identify the boundaries and environmental effects to be reviewed for the assessment.
2. *Inventory Analysis*—Identify and quantify energy, water, and materials usage and environmental releases (e.g., air emissions, solid waste disposal, and waste water discharges).
3. *Impact Assessment*—Assess the potential human and ecological effects of energy, water, and material usage and the environmental releases identified in the inventory analysis.
4. *Interpretation*—Evaluate the results of the inventory analysis and impact assessment to select the preferred product, process, or service with a clear understanding of the uncertainty and the assumptions used to generate the results [4].

Note that these steps, as illustrated in Fig. 2.3, track closely with the life cycle stages dictated by the laws of physics. Also, the life cycle can help to highlight possible routes and pathways of exposure to constituents during extraction, manufacture, use, and disposal (see Fig. 2.4). This can drive the availability of safer products, that is, design for the environment (DfE). The DfE process shown in Fig. 2.5 indicates that a product can be made safer at various points in the life cycle and that these improvements can greatly reduce toxicity, exposures, and risks.

This stepwise process can be used to evaluate products long before a waste is generated. First, a LCI is constructed to define the boundaries of the possible effects of an technology (e.g., microbial populations, genetically modified organisms, and toxic chemical releases). If the technology is hypothetical, this can be done by analogy with a similar conventional process. Next, experts can participate in an expert panel

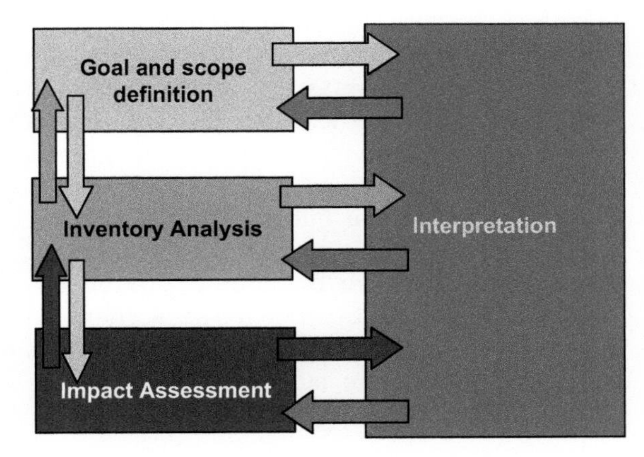

FIGURE 2.3 Life cycle assessment framework consists of (1) a specifically stated purpose of boundaries of the study (Goal and Scope Definition); (2) an estimate of the energy use and raw material inputs and environmental releases associated with each stage of the life cycle (LCI); (3) an interpretation of the results of the inventory to assess the impacts on human health and the environment (Impact Assessment); and (4) an evaluations of ways to reduce energy, material inputs, or environmental impacts along the life cycle (Interpretation) [5].

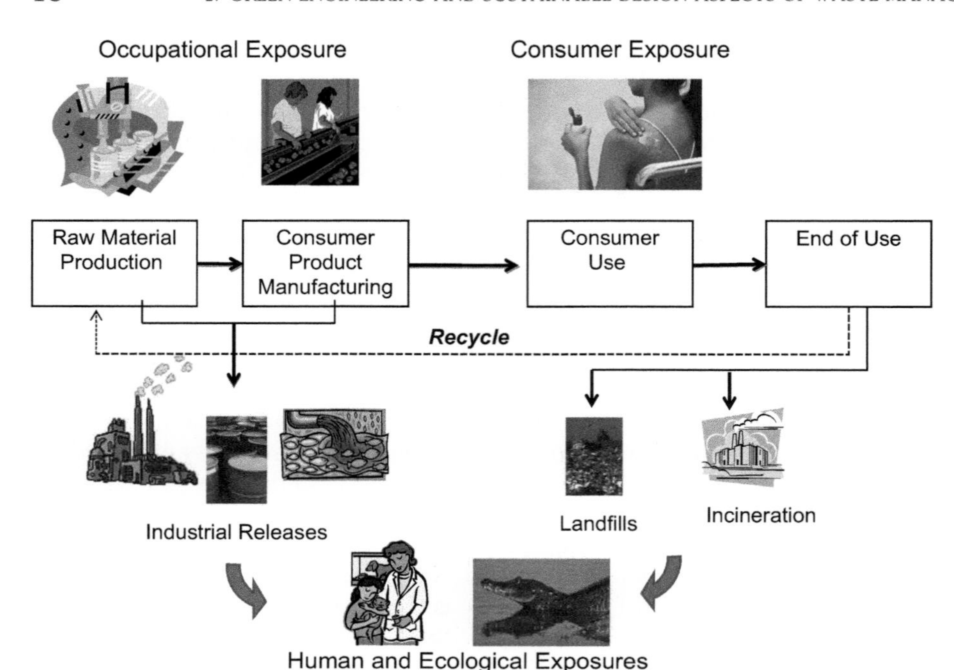

FIGURE 2.4 Life cycle of product, showing points of exposure and potential risk, including after the useful life of the product [6].

to find the driving forces involved (this is known as "expert elicitation"). Then, scenarios can be constructed from these driving forces to identify which factors are most important in leading to various outcomes. This last step is known as a sensitivity analysis. The greater the weight of the factor, the greater will be the change in the outcome. For example, if a product has a very small waste stream of a product that damages an ecosystem, but another product has 100 times the waste stream of the same chemical constituents, then the ecosystem is 100 times more sensitive to the latter product than the former in the prescribed ecosystem.

3.2. Utility and the Benefit/Cost Analysis

On the surface, the choice of whether to pursue a waste disposal option is a simple matter of benefits versus costs. Is it more or less costly to dispose of substance A using approach 1, 2, or n? Engineers make much use of the BCR, owing to a strong affinity for objective measures of successes. Thus, *usefulness* is an engineering measure of success. Such utility is indeed part of any successful engineering enterprise. After all, engineers are expected to provide reasonable and useful products. Two useful engineering definitions of utilitarianism (Latin *utilis*, useful) are imbedded in BCR and LCA:

1. The belief that the value of a thing or an action is determined by its utility.
2. The concept that effort should be directed toward achieving the most benefit for the greatest number.

The BCR is an attractive metric because of its simplicity and seeming transparency. To determine whether a project is worthwhile, one need only add up all the benefits and put them in the numerator and add up all the costs (or risks) and put them in the denominator. If the ratio is greater than 1, its benefits exceed its costs.

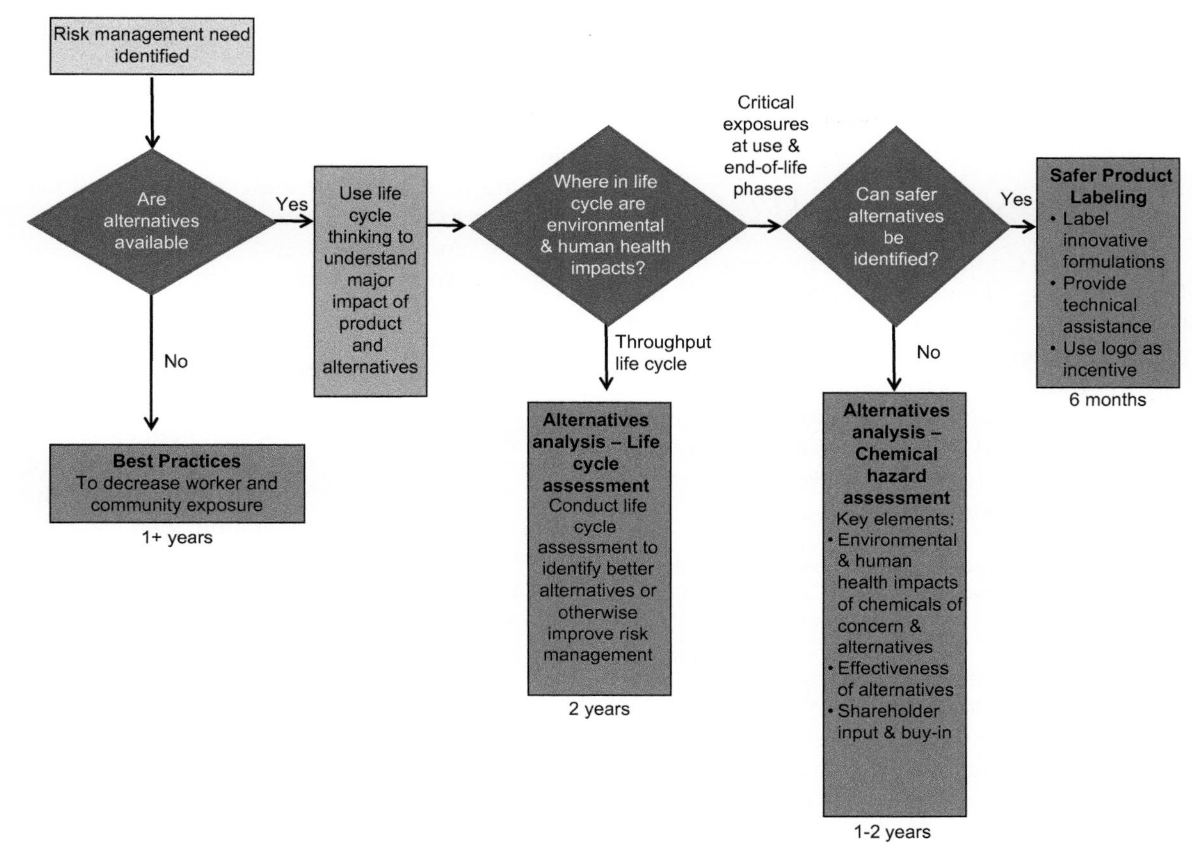

FIGURE 2.5 Decision logic for DfE approaches [7].

One obvious problem is that some costs and benefits are much easier to quantify than others. Some, like those associated with quality of life, are nearly impossible to quantify and monetize accurately. Furthermore, the comparison of action versus no-action alternatives cannot always be captured within a BCR. Opportunity costs and risks are associated with taking no action (e.g., loss of an opportunity to apply an emerging technology may mean delay or nonexistent treatment of diseases).

Simply comparing the status quo to costs and risks associated with a new product may be biased toward no action or proven technologies. Costs in time and money are not the only reasons for avoiding action. The greater availability of a product may introduce unforeseen risks that, if not managed properly, could lead to waste products that could add costs to the public (e.g., pollution or loss of ecosystem diversity) with only short-term value and little net benefit. Therefore, it is not simply a matter of benefits versus cost, it is often one risk being traded for another.

Often, selecting a critical path, including various waste streams, is a matter of optimization, which is a proven analytical tool in engineering. However, the greater the number of contravening risks that are possible, the more complicated that such optimization routines become. The product flows, critical paths, and life LCIs can become quite complicated for complex issues.

4. SUSTAINABILITY

The World Commission on Environment and Development, known as the Brundtland Commission and sponsored by the United Nations, recognized an impending global threat of environmental degradation. As such, the Commission conducted a study of the threats to the planet and issued the 1987 report, *Our Common Future*, which introduced the term *sustainable development*, defined as follows:

> ... development that meets the needs of the present without compromising the ability of future generations to meet their own needs [8].

The United Nations Conference on Environment and Development, that is, the Earth Summit held in Rio de Janeiro in 1992, communicated the idea that sustainable development is both a scientific concept and a philosophical ideal. The document, *Agenda 21*, was endorsed by 178 governments (not including the United States) and hailed as a blueprint for sustainable development.

In 2002, the World Summit on Sustainable Development identified five major areas that are considered key for moving sustainable development plans forward. Sustainable development particularly targets support to developing nations in managing their resources, such as rain forests, without depleting these resources and making them unusable for future generations. The principal objective is to prevent the collapse of the global ecosystems, tying directly with finding ways to eliminate, and to reduce the generation of wastes, which stress these ecosystems. The Brundtland report does not provide specific details on just how to achieve a sustainable global ecologic and economic system, meaning that green and sustainable actions are increasingly dependent on the advice from experts such as the authors of the chapters in this book.

From a thermodynamics standpoint, a sustainable system is one that is in equilibrium or changing at a tolerably slow rate. In the food chain, for example, plants are fed by sunlight, moisture, and nutrients and then become food for insects and herbivores, which in turn act as food for larger animals. The waste from these animals replenishes the soil, which nourishes plants, and the cycle begins again [3]. At the largest scale, manufacturing, transportation, commerce, and other human activities that promote high consumption and wastefulness of finite resources cannot be sustained indefinitely. All systems have thermodynamic inefficiencies, as articulated in the second law of thermodynamics. As such, matter and energy balances and losses from these processes must be considered over a product's entire lifetime and beyond.

4.1. The Tragedy of the Commons

The interconnectedness and need for sustainable means of reducing wastes may be illustrated by a parable presented by a biologist, Garrett Hardin, who imagines an English village with a common area where everyone's livestock may graze [9]. The common area is able to sustain the animals, making for stable village life, until one of the villagers figures out that if he raises two animals instead of one, the cost of the extra animal will be shared by everyone, whereas the profit will be his alone. Therefore, he gets two animals and prospers, but others see this and similarly want their own two animals. If two, why not three—and so on—until the village common area is no longer able to support the large number of animals, and everyone suffers when the system surpasses carrying capacity and crashes.

In some manner, is this not what we have been doing with regard to wastes? If we treat diminishing resources such as oil and minerals as capital gains, we may soon find ourselves in the "common" difficulty of scarcity of land to store the wastes and other damage from an unsustainable approach.

5. CONCLUSIONS

This book addresses wastes in its various forms and categories. The large volumes of waste generated daily indicate a lack of sustainability. Although the chapters that follow devote considerable attention to these wastes and how they can be best managed, it is worthwhile to keep in mind that many of these wastes can be minimized or even eliminated with greener approaches. This goes beyond recycling and calls for systematic and comprehensive choices in the products demanded by society. Part of the solution is in the design of these products. Another part is in making users aware of the life cycle of these products and the damage that can be avoided by wise decision-making. Such an approach requires that all decisions are rooted in sound science.

References

[1] W. McDonough, M. Braungart, Cradle to Cradle: Remaking the Way We Make Things, North Point Press, New York, 2002, p. 165.

[2] U.S. Environmental Protection Agency, Life Cycle Assessment: Inventory Guidelines and Principles. EPA/600/R-92/245, Office of Research and Development, Cincinnati, OH, 1993.

[3] D.A. Vallero, Environmental Biotechnology: A Biosystems Approach, Academic Press, Amsterdam, NV, 2010.

[4] U.S. Environmental Protection Agency, Nanotechnology White Paper. EPA 100/B-07/001. Washington, DC (2007).

[5] U.S. Environmental Protection Agency, Life Cycle Assessment: Principles and Practice. Report No. EPA/600/R-06/060 (2006).

[6] U.S. Environmental Protection Agency, Systems Analysis Research: Program Brief— Life Cycle Analysis; http://www.epa.gov/nrmrl/std/sab/lca/lca_brief.htm. Accessed on July 7, 2010 (2009).

[7] L. Sommer, U.S. EPA Design for the Environment Program. Exposure-Based Chemical Prioritization Workshop. Research Triangle Park, NC (2010).

[8] World Commission on Environment and Development, United Nations, Our Common Future, Oxford Paperbacks, Oxford, United Kingdom, 1987.

[9] G. Hardin, The tragedy of the commons, Science. 162 (1969) 1243–1248.

CHAPTER

3

Regulation of Wastes

Daniel A. Vallero[*], *Trevor M. Letcher*[†]

[*] Pratt School of Engineering, Duke University, Durham, North Carolina, USA, [†] School of Chemistry, University of KwaZulu-Natal, Durban, South Africa

OUTLINE

Waste Doi: 10.1016/B978-0-12-381475-3.10003-8

1. INTRODUCTION

Environmental problems faced today differ from those of most of the Earth's history. The difference is in both kind and degree. For example, the synthesis of chemicals, especially organic compounds, has grown exponentially since the mid-1900s. Most organisms have no mechanisms to metabolize and eliminate these new compounds. Also, stresses put on only small parts of ecosystems before the Industrial Revolutions were small in extent of damage. For example, pollutants were discharged into creeks and rivers throughout human history, but only recently these were the discharges so large and exist for long time that they have diminished the quality of entire ecosystems.

One of the reasons to regulate the generation and disposition of wastes is that they have substantial, negative impacts on society. Generation of wastes is an indication of inefficient use of resources, making scarce commodities less valuable or less fit to perform their useful purposes. For example, water pollution experts talk about a stream not meeting its "designated use," such as recreation or public water supply. Although public health is usually the principal driver for assessing and controlling environmental contaminants, ecosystems are also important *receptors* of contaminants. Contaminants also affect structures and other engineered systems, including historically and culturally important monuments and icons, such as the contaminants in rainfall (e.g., nitrates and sulfates) that render it more corrosive than would normally be expected (i.e., *acid rain*).

Society and its elected and appointed delegates increasingly respond to waste-related problems by writing and enforcing laws [1,2]. The number of laws enacted to address various aspects of wastes has grown substantially in recent decades. The number and complexity of regulations have broadened commensurately (see Fig. 3.1).

To assess and address waste management appropriately and to make sound environmental decisions, at least a basic understanding of the underlying sciences affecting those

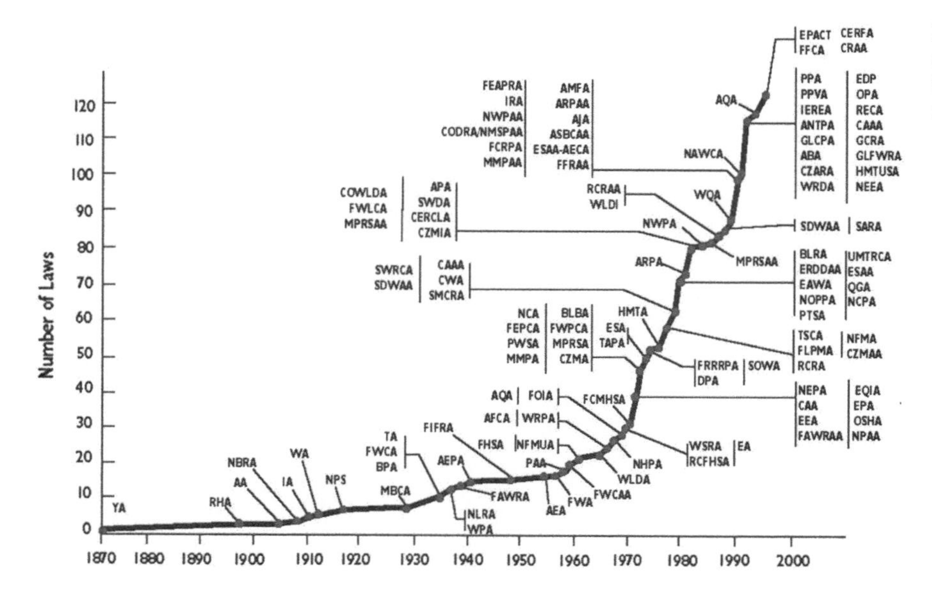

FIGURE 3.1 Growth of environmental legislation in the United States during the twentieth century [1,2].

issues, problems, and decisions is required. Although we will introduce and cover many of the scientific topics in much more detail later, it is helpful to consider how regulations and other societal mandates have come to fore in recent decades.

From a scientific perspective, waste management requires attention to the waste and the characteristics of the place where the waste is found. This place is known as the "environmental medium." The major environmental media are air, water, soil, sediment, and even biota. In fact, the regulation of wastes is even more refined. For example, water is commonly categorized into "surface water" and "groundwater." The former includes everything from puddles and rivulets, to large rivers and lakes, to the oceans. Although the various environmental disciplines apply some common language, they each have their own lexicons and systems of taxonomy. Sometimes the difference is subtle, such as different conventions in nomenclature and symbols.[1] This is more akin to different slang in the same language.

Arguably, the two media that are most important to waste management are groundwater and soil. Groundwater includes all water below the surface, but depending on the discipline, it is further differentiated from soil-bound water. Engineers commonly differentiate water in soil and groundwater because the soil water greatly affects the physical and mechanical properties of very soils. Environmental engineering publications frequently describe soil water according to the amount of void space filled, or the water filled pore space (WFPS), which is the percentage of void space containing water. The WFPS is another way of expressing the degree of saturation. Almost all environmental science professions classify the water below the soil layer based on whether the unconsolidated material (e.g., gravel and sand) is completely saturated or unsaturated. The saturated zone lies under the unsaturated zone. Hydrogeologists refer to the unsaturated zone as the "vadose zone." It is also referred to as the zone of aeration. Another type of groundwater, albeit rare, is the Karst groundwater system, which is actually made up of underground lakes and streams that flow through fractured limestone and dolomite rock strata. Small cracks in the rock erode over time to allow rapidly flowing water to move at rates usually seen only on the Earth's surface. Usually, groundwater flows quite slowly, but in these caves and caverns, water moves rapidly enough for its flow to be turbulent. We will cover all these topics in detail when we discuss pollutant transport, drawing from several different scientific disciplines.

Soil is classified into various types. This is not a soil science text, but it is important to understand that for many decades, soil scientists have struggled with uniformity in the classification and taxonomy of soil. Much of the rich history and foundation of soil scientists has been associated with agricultural productivity. The very essence of the "value" of a soil has been its capacity to support plant life, especially crops. Even forest soil knowledge owes much to the agricultural perspective, because much of the reason for investing in forests has been monetary. A stand of trees are seen by many to be a "standing crop." In the United States, for example, the National Forest Service is an

[1] Environmental handbooks and texts use different conventions in explaining concepts and providing examples. Environmental information comes in many forms. A telling example is the convention of the use of K. In hydrogeology, this means hydraulic conductivity, in chemistry it is an equilibrium constant, and in engineering it can be a number of coefficients. Likewise, units other than metric will be used on occasion, because that is the convention in some areas, and because it demonstrates the need to apply proper dimensional analysis and conversions. Many mistakes have been made in these two areas!

agency of the U.S. Department of Agriculture. The engineers have been concerned about the statics and dynamics of soil systems, improving the understanding of soil mechanics so that they may support, literally and figuratively, the built environment. The agricultural and engineering perspectives have provided valuable information about soil that environmental professionals can put to use. The information is certainly necessary, but not completely sufficient, to understand how pollutants move through soils, how the soils themselves are affected by the pollutants (e.g., loss of productivity and diversity of soil microbes), and how the soils and contaminants interact chemically (e.g., changes in soil pH will change the chemical and biochemical transformation of organic compounds). At a minimum, environmental scientists must understand and classify soils according to their texture or grain size (see Table 3.1), ion exchange capacities, ionic strength, pH, microbial populations, and soil organic matter content.

Air and water are fluids, but sediment resembles soil in that it is a matrix made up of various components, including organic matter and unconsolidated material. And, the matrix contains liquids ("substrate" to the chemist and engineer) within its interstices. Much of the substrate in this matrix is water with varying amounts of solutes. At least for most environmental conditions, air and water are solutions of very dilute amounts of compounds. For example, solutes in the air represent small percentages of the solution at the highest (e.g., water vapor) and most other solutes represent parts per million (ppm) (CO_2 is about 370 ppm). Most "contaminants" in air and water, thankfully, are found in the parts per billion range, if at all found. On the other hand, soil and sediment themselves are conglomerations of all states of matter. Soil is predominantly solid, but it frequently has large fractions of liquid (soil water) and gas (soil air, methane, and carbon dioxide) that make up the matrix. The composition of each fraction is highly variable. For example, soil and gas concentrations are different from those in the atmosphere and change profoundly with depth from the surface (Table 3.2 shows the inverse relationship between carbon dioxide and oxygen). Sediment is really an underwater soil. It is a collection of particles that have settled on the bottom of water bodies.

TABLE 3.1 Commonly Used Soil Texture Classifications [3]

Name	Size range (mm)
Gravel	>2.0
Very coarse sand	1.0–1.999
Coarse sand	0.500–0.999
Medium sand	0.250–0.499
Fine sand	0.100–0.249
Very fine sand	0.050–0.099
Silt	0.002–0.049
Clay	<0.002

Ecosystems are amalgamations of these media. For example, a wetland system consists of plants that grow in soil, sediment, and water. The water flows through living and nonliving materials. Microbial populations live in the surface water, with aerobic species congregating near the water surface and anaerobic microbes increasing with depth due to the decrease in oxygen levels, due to the reduced conditions. Air is not only important at the water and soil interfaces, but is also a vehicle for nutrients and contaminants delivered to the wetland. The groundwater is fed by the surface water during high water conditions and feeds the wetland during low water.

Thus, one of the profound challenges in regulating wastes is that the problems they cause do not fit neatly into prescribed media. Indeed, environmental media can be envisioned as compartments, each with boundary conditions, kinetics, and partitioning relationships within

TABLE 3.2 Composition of Two Important Gases in Soil Air [4]

Depth from surface (cm)	Percentage volume of air					
	Silty clay		Silty clay loam		Sandy loam	
	O_2	CO_2	O_2	CO_2	O_2	CO_2
30	18.2	1.7	19.8	1.0	19.9	0.8
61	16.7	2.8	17.9	3.2	19.4	1.3
91	15.6	3.7	16.8	4.6	19.1	1.5
122	12.3	7.9	16.0	6.2	18.3	2.1
152	8.8	10.6	15.3	7.1	17.9	2.7
183	4.6	10.3	14.8	7.0	17.5	3.0

a compartment or among other compartments. Chemicals, whether nutrients or contaminants, change as a result of the time spent in each compartment. The waste manager's challenge is to describe, characterize, and predict the behaviors of waste constituents as they move through the media. When something is amiss, the cause and cure lie within the physics, chemistry, and biology of the system. Thus, waste regulations must be underpinned by sound science.

Love Canal in New York, for example, became the waste case that, in the view of many, led to the passage of hazardous waste laws, but it is still considered to be one of the most complex cases in terms of waste management. Scholars and practitioners alike have studied its convoluted history and the critical paths of decisions. The eventual exposures of people to harmful remnant waste constituents event trees was a complicated series of events brought on by military, commercial, and civilian governmental decisions. One particularly interesting event tree is that of the public school district's decisions of accepting the donation of land and building the school on the property. As regulators and the scientific community learned more, a series of laws were passed and new court decisions and legal precedents established in the realm of toxic substances. Additional hazardous wastes sites began to be identified, which continue to be listed on the Environmental Protection Agency (EPA) Web site's National Priority Listing for sites closer to you, or even have student select sites in or near their hometowns. It would behoove managers to become familiar with each cleanup step listed in Chapter 27 for hazardous wastes.

2. THE GROWTH OF ENVIRONMENTAL REGULATIONS

Environmental awareness grew in the second half of the twentieth century. With this awareness, the public demand for environmental safeguards and remedies to environmental problems became an expectation of a greater role for government. A number of laws were on the books before the 1960s, such as early versions of federal legislation to address limited types of water and air pollution, and some solid waste issues, such as the need to eliminate open dumping. In fact, key legislation to protect waterways and riparian ecosystems was written at the end of the nineteenth century in the form of the Rivers and Harbors Act [the law that set the stage for the U.S. Army Corps of Engineers to permit proper dredging operations, later enhanced by Section 404 of the Clean Water Act (CWA)].

However, the real growth followed the tumultuous decade of the 1960s. The environment had become a social cause, akin to the civil rights and antiwar movements. Major public demonstrations on the need to protect "spaceship Earth" encouraged elected officials to address environmental problems, exemplified by air pollution "inversions" that capped polluted air in urban valleys, leading to acute diseases and increased mortality from inhalation hazards, the "death" of Erie Canal and rivers catching on fire in Ohio and Oregon.

2.1. The National Environmental Policy Act

The movement was institutionalized in the United States by a series of new laws and legislative amendments. The National Environmental Policy Act (NEPA) was in many ways symbolic of the new federal commitment to environmental stewardship. It was signed into law in 1970, after contentious hearings in the U.S. Congress. NEPA was not really a technical law. It did two main things. It created the Environmental Impact Statement (EIS) and established the Council on Environmental Quality (CEQ) in the Office of the President. Of the two, the EIS represented a sea of change in how the federal government was to conduct business. Agencies were required to prepare EISs on any major action that they were considering that could "significantly" affect the quality of the environment. From the outset, the agencies had to reconcile often-competing values, that is, their mission and the protection of the environment.

The CEQ was charged with developing guidance for all federal agencies on NEPA compliance, especially when and how to prepare an EIS. The EIS process combines scientific assessment with public review. This process is similar for most federal agencies. Agencies often strive to receive a so-called "FONSI"[2] or the finding of no significant impact, so that they may proceed unencumbered on a mission-oriented project. The Federal Highway Administration's FONSI process provides an example of the steps needed to obtain a FONSI for a project.

Whether a project either leads to a full EIS or a waiver through the FONSI process, it will have to undergo an evaluation. This step is referred as an "environmental assessment." An incomplete or inadequate assessment will lead to delays and increases the chance of an unsuccessful project, so sound science is needed from the outset of the project design.

The final step is the Record of Decision (ROD). The ROD describes the alternatives and the rationale for final selection of the best alternative. It also summarizes the comments received during the public reviews and how the comments were addressed. Many states have adopted similar requirements for their RODs.

The courts adjudicated some very important laws along the way, requiring federal agencies to take NEPA seriously. Some of the aspects of the "give and take" and evolution of federal agencies' growing commitment to environmental protection was the acceptance of the need for sound science in assessing environmental conditions and possible impacts, and the very large role of the public in deciding on the environmental worth of a highway, airport, dam, waterworks, treatment plant, or any other major project sponsored by or regulated by the federal government. This was a major impetus in the growth of the environmental disciplines since the 1970s. Experts were called on, not only to conduct credible scientific assessments, but also to meaningfully communicate the assessments to the public.

Because virtually any federal activity can have some effect on the environment, all federal agencies must follow the CEQ regulations that require agencies to meet their obligations under NEPA. Thus, environmental considerations must accompany economic, social, and other agency priorities [5].

Agencies are required to identify the major decisions called for by their principal programs and make certain that the NEPA process addresses them. This process must be set up in advance, early in the agency's planning stages.

[2] Pronounced "Fonzy" like that of the nickname for character Arthur Fonzerelli portrayed by Henry Winkler in the television show, *Happy Days*.

For example, if waste remediation or reclamation is a possible action, the NEPA process must be woven into the remedial action planning processes from the beginning, with the identification of the need for and possible kinds of actions being considered.

Noncompliance or inadequate compliance with NEPA rules regulations can lead to severe consequences, including lawsuits, increased project costs, delays, and the loss of the public's trust and confidence, even if the project is designed to improve the environment, and even if the compliance problems seem to be only "procedural."

The U.S. EPA is responsible for reviewing the environmental effects of all federal agencies' actions. This authority was written as Section 309 of the Clean Air Act (CAA). The review must be followed with the EPA's public comments on the environmental impacts of any matter related to the duties, responsibilities, and authorities of EPA's administrator, including EISs. The EPA's rating system (see Appendix 1) is designed to determine whether a proposed action by a federal agency is unsatisfactory from the standpoint of public health, environmental quality, or public welfare. This determination was published in the *Federal Register* (for significant projects) and referred to the CEQ.

Following NEPA, a number of new laws were enacted to address specific problems in various environmental media. Because waste is the main focus of this handbook, we will discuss the waste management laws first, followed by brief overviews of air, water, and consumer protection laws.

3. SOLID AND HAZARDOUS WASTES LEGISLATION

Numerous laws, including those previously mentioned, are directed at wastes, but the two principal U.S. laws governing solid wastes are the Resource Conservation and Recovery Act (RCRA) and Superfund. The RCRA law covers both hazardous and solid wastes, whereas Superfund and its amendments generally address abandoned hazardous waste sites. RCRA addresses active hazardous waste sites.

3.1. Management of Active Hazardous Waste Facilities

With RCRA, the U.S. EPA received the authority to control hazardous waste throughout the waste's entire life cycle, known as the "cradle-to-grave." This means that manifests must be prepared to keep track of the waste, including its generation, transportation, treatment, storage, and disposal. RCRA also set forth a framework for the management of nonhazardous wastes in Subtitle D.

The Federal Hazardous and Solid Waste Amendments (HSWA) to RCRA required the phase out of land disposal of hazardous waste. HSWA also increased the federal enforcement authority related to hazardous waste actions, set more stringent hazardous waste management standards, and provided for a comprehensive underground storage tank (UST) program.

The 1986 amendments to RCRA allowed the federal government to address potential environmental problems from USTs for petroleum and other hazardous substances.

3.2. Addressing Abandoned Hazardous Wastes

The Comprehensive Environmental Response, Compensation, and Liability Act (CERCLA) is commonly known as a Superfund. Congress enacted it in 1980 to create a tax on the chemical and petroleum industries and to provide extensive federal authority for responding directly to releases or threatened releases of hazardous substances that may endanger public health or the environment.

The Superfund law established prohibitions and requirements concerning closed and abandoned hazardous waste sites; established provisions for the liability of persons responsible for releases of hazardous waste at these sites; and established a trust fund to provide for cleanup when no responsible party could be identified.

The CERCLA response actions include the following:

- Short-term removals, where actions may be taken to address releases or threaten releases requiring prompt response. This is intended to eliminate or reduce exposures to possible contaminants.
- Long-term remedial response actions to reduce or eliminate the hazards and risks associated with releases or threats of releases of hazardous substances that are serious, but not immediately life threatening. These actions can be conducted only at sites listed on EPA's National Priorities List (NPL).

Superfund also revised the National Contingency Plan, which sets guidelines and procedures required when responding to releases and threatened releases of hazardous substances.

CERCLA was amended by the Superfund Amendments and Reauthorization Act (SARA) in 1986. These amendments stressed the importance of permanent remedies and innovative treatment technologies in cleaning up hazardous waste sites. SARA required that Superfund actions consider the standards and requirements found in other State and Federal environmental laws and regulations and provided revised enforcement authorities and new settlement tools. The amendments also increased State involvement in every aspect of the Superfund program, increased the focus on human health problems posed by hazardous waste sites, encouraged more extensive citizen participation in site cleanup decisions, and increased the size of the Superfund trust fund.

SARA also mandated that the Hazard Ranking System be revised to verify the adequacy of the assessment of the relative degree of risk to human health and the environment posed by uncontrolled hazardous waste sites that may be placed on the NPL.

4. CLEAN AIR LEGISLATION

The 1970 amendments to the CAA arguably ushered in the era of environmental legislation with enforceable rules. The 1970 version of the CAA was enacted to provide a comprehensive set of regulations to control air emissions from area, stationary, and mobile sources. This law authorized the EPA to establish National Ambient Air Quality Standards (NAAQS) to protect public health and the environment from the "conventional" (as opposed to "toxic") pollutants: carbon monoxide; particulate matter; oxides of nitrogen; oxides of sulfur; and photochemical oxidant smog or ozone. The metal lead (Pb) was later added as the sixth NAAQS pollutant.

The original goal was to set and to achieve NAAQS in every state by 1975. These new standards were combined with charging the 50 states to develop state implementation plans to address industrial sources in the state. The ambient atmospheric concentrations are measured at more than 4000 monitoring sites across the United States. The ambient levels have continuously decreased, as shown in Table 3.3.

The CAA Amendments of 1977 listed new dates to achieve attainment of NAAQS (many areas of the country had not met the prescribed dates set in 1970). Other amendments were targeted at air pollution, which had been insufficiently addressed, including acidic deposition (so-called "acid rain,"

TABLE 3.3 Percentage Decrease in Ambient Concentrations of National Ambient Air Quality Standard Pollutants From 1985 Through 1994 [6]

Pollutant	Decrease in concentration (%)
CO	28
Lead	86
NO_2	9
Ozone	12
PM_{10}	20
SO_2	25

tropospheric ozone pollution, depletion of the stratospheric ozone layer, and a new program for air toxins, the National Emission Standards for Hazardous Air Pollutants.

The 1990 Amendments to the CAA profoundly changed the law by adding new initiatives and imposing dates to meet the new requirements of the law. Here are some of the major provisions.

Cities that failed to achieve human health standards as required by NAAQS were required to reach attainment within 6 years of passage, although Los Angeles was given 20 years, since it was dealing with major challenges in reducing ozone concentrations.

Almost 100 cities failed to achieve ozone standards and were ranked from marginal to extreme. The more severe the pollution, the more rigorous controls required, although additional time was given to those extreme cities to achieve the standard. Measures included new or enhanced inspection/maintenance programs for automobiles; installation of vapor recovery systems at gas stations and other controls of hydrocarbon emissions from small sources; and new transportation controls to offset increases in the number of miles traveled by vehicles. Major stationary sources of nitrogen oxides would have reduced its emissions.

The 41 cities failing to meet carbon monoxide standards were ranked moderate or serious; states would have to initiate or upgrade inspection and maintenance programs and adopt transportation controls. The 72 urban areas that did not meet particulate matter (PM_{10}) standards were ranked moderate; states will have to implement Reasonably Available Control Technology; and use of wood stoves and fireplaces may have to be curtailed.

The standards promulgated from the CAA Amendments are provided in Table 3.4. Note that the new particulate standard addresses smaller particles, that is, particles with diameters $\leq 2.5\ \mu m$ ($PM_{2.5}$). Research has shown that exposure to these smaller particles is more likely to lead to health problems than do exposures to larger particles. Smaller particles are able to penetrate further into the lungs and hence are probably more bioavailable than the larger PM_{10}.

A note on units: The concentration of gases, vapors, and liquids is often expressed as volume of contaminant in a specific volume of air or water, referred to as volume per unit volume. This is most conveniently expressed as parts per million or percent by volume. Parts per million is the volume of contaminant per million volumes of air. Any volume unit can be used as long as the units for both parts are the same (e.g., liters of contaminant per million liters of air or water). Measurements in percent volume are less applicable to hazardous waste characterization because they represent very high concentrations that would not usually be found in environmental contamination situations. Percent by volume can be thought of as parts per hundred, so

$$(\% \text{ by volume}) \times (10{,}000) = \text{ppm}$$

For liquids, the volume to volume (V:V) concentration can be converted to mass per volume (M:V) concentrations if the density of

TABLE 3.4 National Ambient Air Quality Standards [6] (A Discussion on Units Is Given Below)

Pollutant	Averaging period[3]	Standard	Primary standards[4]	Secondary standards[5]
Ozone	1 h	Cannot be at or above this level on more than 3 days over 3 years.	125 ppb	125 ppb
	8 h	The average of the annual fourth highest daily 8-h maximum over a 3-yr period cannot be at or above this level.	85 ppb	85 ppb
Carbon monoxide	1 h	Cannot be at or above this level more than once per calendar year.	35.5 ppm	35.5 ppm
	8 h	Cannot be at or above this level more than once per calendar year.	9.5 ppm	9.5 ppm
Sulfur dioxide	3 h	Cannot be at or above this level more than once per calendar year.	—	550 ppb
	24 h	Cannot be at or above this level more than once per calendar year.	145 ppb	—
	Annual	Cannot be at or above this level.	35 ppb	—
Nitrogen dioxide	Annual	Cannot be at or above this level.	54 ppb	54 ppb
Respirable particulate matter (aerodynamic diameter $\leq$ 10 microns = PM_{10})	24 h	The 3 year average of the annual 99th percentile for each monitor within an area cannot be at or above this level.	155 $\mu g\ m^{-3}$	155 $\mu g\ m^{-3}$
	Annual	The 3 year average of annual arithmetic mean concentrations at each monitor within an area cannot be at or above this level.	51 $\mu g\ m^{-3}$	51 $\mu g\ m^{-3}$
Respirable particulate matter (aerodynamic diameter $\leq$ 2.5 microns = $PM_{2.5}$)	24 h	The 3 year average of the annual 98th percentile for each population-oriented monitor within an area cannot be at or above this level.	66 $\mu g\ m^{-3}$	66 $\mu g\ m^{-3}$
	Annual	The 3 year average of annual arithmetic mean concentrations from single or multiple community-oriented monitors cannot be at or above this level.	15.1 $\mu g\ m^{-3}$	15.1 $\mu g\ m^{-3}$
Lead	Quarter	Cannot be at or above this level.	1.55 $\mu g\ m^{-3}$	1.55 $\mu g\ m^{-3}$

[3] *Integrated time used to calculate the standard. For example, for particulates, the filter will collect material for 24 hours and then it is analyzed. The annual integration will be an integration of the daily values.*
[4] *Primary NAAQS are the levels of air quality that the EPA considers necessary, with an adequate margin of safety, to protect the public health.*
[5] *Secondary NAAQS are the levels of air quality that the EPA judges necessary to protect the public welfare from any known or anticipated adverse effects.*

the concentrated substance and the density of the liquid are known. Many environmental texts and models use short-hand terms of "solvent" and "solute"; however, not all substances of concern are dissolved (e.g., some contaminants are suspended as particles or in emulsions in water). The liquid in which we are usually most interested is water. For example, we want to know how much of the bad stuff (pollutants) or good stuff (e.g., dissolved oxygen) is in the water. The density of water under most environmental conditions is very nearly unity, so the $V:V$ concentration can be converted to $M:V$ concentration simply as:

$$\text{ppm} = C \times \rho \quad (3.1)$$

where C is the concentration of substance in water (mg L^{-1}) and ρ is pollutant density of the substance (g mL^{-1}).

Thus, if the $M:V$ concentration is micrograms per liter (μg L^{-1}), then the $V:V$ concentration will be in parts per billion (ppb); and if the $M:V$ concentration is nanograms per liter (ng L^{-1}), then the $V:V$ concentration will be in parts per trillion (ppt).

Converting $V:V$ to $M:V$ concentrations in air is a bit more complicated than that of water, because gas densities depend on the gas laws. The gas law states that the product of pressure (P) and the volume occupied by the gas (V) is equal to the product of the number of moles (n), the gas constant (R), and the absolute temperature (T) in degrees Kelvin. That is:

$$PV = n\text{R}T \quad (3.2)$$

4.1. Mobile Sources

Vehicular tailpipe emissions of hydrocarbons, carbon monoxide, and oxides of nitrogen were to be reduced with the 1994 models. Standards now have to be maintained over a longer vehicle life. Evaporative emission controls were mentioned as a means for reducing hydrocarbons. Beginning in 1992, "oxyfuel" gasolines blended with alcohol began to be sold during winter months in cities with severe carbon monoxide problems. In 1995, reformulated gasolines with aromatic compounds were introduced in the nine cities with the worst ozone problems, but other cities were allowed to participate. Later, a pilot program introduced 150,000 low-emitting vehicles to California that meet tighter emission limits through a combination of vehicle technology and substitutes for gasoline or blends of substitutes with gasoline. Other states are also participating in this initiative.

4.2. Toxic Air Pollutants

The number of toxic air pollutants covered by the CAA was increased to 189 compounds in 1990 (see Appendix 3). Most of these are carcinogenic, mutagenic, and/or toxic to neurological, endocrine, reproductive, and developmental systems. All 189 compound emissions must be reduced within 10 years. The EPA published a list of source categories issued Maximum Achievable Control (MACT) standards for each category over a specified timetable.

The next step beyond MACT standards is to begin to address chronic health risks that would still be expected if the sources meet these standards. This is known as residual risk reduction. The first step was to assess the health risks from air toxins emitted by stationary sources that emit air toxins after technology-based (MACT) standards are in place. The residual risk provision sets additional standards if MACT does not protect public health with an "ample margin of safety" and if they are needed to prevent adverse environmental effects.

What an "ample margin of safety" means is still up for debate, but one proposal for airborne carcinogens is shown in Fig. 3.2. That is, if a source can demonstrate that it will not contribute to more than 10^{-6} cancer risk then it meets the ample margin of safety requirements for air toxins. The ample margin needed to protect populations from noncancer toxins, such as neurotoxins, is being debated, but it will involve the application of the hazard quotient (HQ). The HQ is the ratio of the potential exposure to the substance and the level at which no adverse effects are expected. An HQ < 1 means that the exposure levels to a chemical should not lead to adverse health

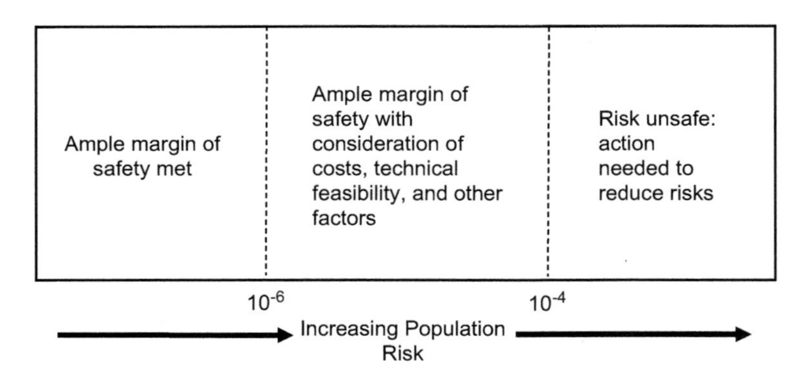

FIGURE 3.2 Ample margin of safety to protect public health and adverse environmental effects from an air pollutant. *Source: CAA Amendments of 1990, Section 112(f)(2).*

effects. Conversely, an HQ > 1 means that adverse health effects are possible. Because uncertainties and the feedback are coming from the business and scientific communities, the ample margin of safety threshold (HQ) is presently ranging from 0.2 to 1.0. So, if a source can demonstrate that it will not contribute to greater than the threshold (whether it is 0.2, 1.0, or some other level established by the federal government) for noncancer risk, it meets the ample margin of safety requirements for air toxins.

5. WATER QUALITY LEGISLATION

Environmental water laws come in two forms: those aimed at providing "clean water" to drink and use, and those cleaning up "dirty water." The Safe Drinking Water Act (SDWA) is the principal federal law designed to provide the U.S. population with potable water. The CWA addresses the many aspects of water pollution.

5.1. Drinking Water

The SDWA was passed in 1974 to protect public drinking water supplies from harmful contaminants, assuring that the concentrations of these contaminants in drinking water stay below Maximum Contaminant Levels (MCLs). The Act, as amended, authorizes a set of regulatory programs to establish standards and treatment requirements for drinking water, as well as the control underground injection of wastes that may contaminate water supplies, and the protection of groundwater resources.

The SDWA was extensively amended in 1986 to strengthen the standard-setting procedures, increase enforcement authority, and provide for additional groundwater protection programs. The U.S. EPA was mandated to issue drinking water regulations for 83 specified contaminants by 1989 and for 25 additional contaminants every 3 years thereafter (see Appendix 2). The 1986 Amendments also required that all public water systems using surface water disinfect, and possibly filter, water supplies. Thus far, the EPA has regulated 84 contaminants. The law covers all public water systems with piped water to be used for human consumption with at least 15 service connections or a system that regularly serves at least 25 people.

Most serious violations of drinking water regulations have occurred in small water systems serving populations of less than 3300. This is often the result of limited financial and technical resources that can be devoted to water monitoring and treatment. Because 87% of all community water systems are small, amendments to the SDWA have focused on increasing the abilities of these communities to meet regulatory requirements, known as "capacity building."

So new emphases have been to find how to balance risks and costs in setting standards, how and whether to discourage the formation of new drinking water systems that are unlikely to comply, and the appropriate state and federal roles in providing high-quality water supplies.

5.2. Water Pollution Abatement

The CWA, passed in 1972, represents the myriad programs aimed at surface water quality protection in the United States, using a variety of regulatory and nonregulatory approaches needed to reduce direct pollutant discharges into U.S. waterways. This has been accomplished through the issuance of effluent discharge permits, designation of water quality protection levels for water bodies, financing municipal wastewater treatment facilities, and managing nonpoint sources to allow polluted runoff to enter surface waters. The goal of these actions is to restore and to maintain the chemical, physical, and biological integrity of the nation's waters so that they can support in the words of the Act, "the protection and propagation of fish, shellfish, and wildlife and recreation in and on the water."

For years after the CWA became law, the major focus of the federal government, states, and Native American tribes was on the chemical aspects of the "integrity" goal cited above. More recently, greater attention has been paid to other provisions of the Act regarding physical and biological integrity. The implementation of the law has also moved from almost completely focusing on the regulation of pollutant discharges from "point source" facilities, such as municipal wastewater treatment plants and industrial facilities, to greater attention now being given to pollution from mining operations, roads, construction sites, farms, and other "nonpoint sources."

In many ways, nonpoint problems are more intractable than point sources, because they more heavily depend on management and comprehensive planning programs. Some successful nonpoint programs have included voluntary programs, such as cost sharing with landowners. So-called "wet weather point sources," such as urban storm sewer systems and construction sites, require regulatory actions.

The traditional "command and control" programs of enforcement and compliance programs have been evolving into a greater number of programs that consist of comprehensive watershed-based strategies in which more equity exists between the protection of healthy waters and the restoration of impaired surface waters. Such holistic approaches depend heavily on public involvement and coalition building to achieve water quality objectives that are both technically sound and publicly acceptable.

6. ENVIRONMENTAL PRODUCT AND CONSUMER PROTECTION LAWS

Although most of the authorizing legislation targeted at protecting and improving the environment is based on actions needed in specific media, such as air, water, soil, and sediment, some law has been written in an attempt to prevent environmental and public health problems while products are being developed and before their usage.

The predominant product laws designed to protect the environment are the Federal Food, Drug, and Cosmetics Act (FFDCA); the Federal Insecticide, Fungicide, and Rodenticide Act (FIFRA); and the Toxic Substances Control Act (TSCA). These three laws look at products in terms of potential risks for yet to be released products and estimated risks for products already in use. If the risks are unacceptable, new products may not be released as formulated or the uses will be strictly limited to applications that meet minimum risk standards. For products already in the marketplace, the risks are periodically reviewed. For example, pesticides have to be periodically re-registered with the government. This re-registration process consists of reviews of new research and information regarding health and environmental impacts discovered since the product's last registration.

FIFRA's major mandate is to control the distribution, sale, and applications of pesticides. This not only includes studying the health and environmental consequences of pesticide usage but also to require that those applying the pesticides register when they purchase the products. Commercial applicators must be certified by successfully passing examinations on the safe use of pesticides. FIFRA requires that the EPA license any pesticide used in the United States. The licensing and registration makes sure that pesticide is properly labeled and will not cause unreasonable environmental harm.

An important, recent product production law is the Food Quality Protection Act (FQPA), which includes new provisions to protect children and limit their risks to carcinogens and other toxic substances. The law is actually an amendment to FIFRA and FFDCA that includes new requirements for safety standard-reasonable certainty of no harm that must be applied to all pesticides used on foods. FQPA mandates a single, health-based standard for all pesticides in all foods; gives special protections for infants and children; expedites approval of pesticides likely to be safer than those in use; provides incentives for effective crop protection tools for farmers; and requires regular re-evaluation of pesticide registrations and tolerances so that the scientific data supporting pesticide registrations includes current findings.

Another product-related development in recent years is the growth in the importance of screening and prioritizing chemicals for possible harm and exposure before their appearance in the marketplace. For example, research suggests a link between exposure to certain chemicals and damage to the endocrine system in humans and wildlife. Because of the potentially serious consequences of human exposure to endocrine disrupting chemicals, the U.S. Congress added specific language on endocrine disruption in the FQPA and recent amendments to SDWA. The FQPA mandated that the EPA develop an endocrine disruptor screening program, and the SDWA authorizes EPA to screen endocrine disruptors found in drinking water systems.

The Endocrine Disruptor Screening Program focuses on methods and procedures to detect and to characterize the endocrine activity of pesticides and other chemicals (see Fig. 3.3). The EPA uses the assays in a two-tiered screening and testing process. Tier 1 identifies chemicals with the potential to interact with the endocrine system. Tier 2 characterizes the endocrine-related effects caused by each chemical and obtains information about effects at various doses [8].

The scientific data needed for the estimated 87,000 chemicals in commerce does not exist to conduct adequate assessments of potential risks. The screening program is being used by the EPA to collect this information for endocrine disruptors and to decide appropriate regulatory action by first assigning each chemical to an endocrine disruption category.

According to the existing scientifically relevant information, chemicals are sorted into four categories:

- Category 1 chemicals have sufficient, scientifically relevant information to determine that they are not likely to interact with the estrogen, androgen, or thyroid systems. This category includes some polymers and certain exempted chemicals.
- Category 2 chemicals have insufficient information to determine whether they are likely to interact with the estrogen, androgen,

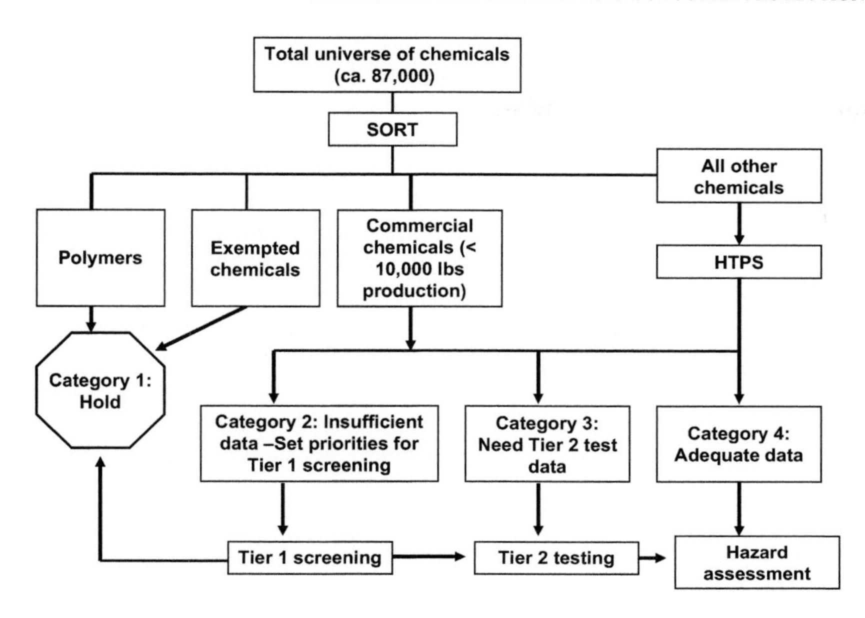

FIGURE 3.3 Endocrine Disruptor Screening Program of the U.S. Environmental Protection Agency [7]. Note: HPTS = high throughput prescreening.

or thyroid systems, thus will need screening data.

- Category 3 chemicals have sufficient screening data to indicate endocrine activity, but data to characterize actual effects that are inadequate and will need testing.
- Category 4 chemicals already have sufficient data for the EPA to perform a hazard assessment.

The TSCA gives the EPA the authority to track 75,000 industrial chemicals currently produced or imported to the United States. This is accomplished through screening of the chemicals and requiring that reporting and testing be done for any substance that presents a hazard to human health or the environment. If a chemical poses a potential or actual risk that is unreasonable, the EPA may ban the manufacture and import of that chemical.

The EPA tracks thousands of new chemicals being developed by industries each year, if those chemicals have either unknown or dangerous characteristics. This information is used to determine the type of control that would be needed to protect human health and the environment from these chemicals. Manufacturers and importers of chemical substances first submit information about chemical substances already on the market during an initial inventory. Since the initial inventory was published, commercial manufacturers or importers of substances not on the inventory have been subsequently required to submit notices to the EPA, which has developed guidance about how to identify chemical substances to assign a unique and unambiguous description of each substance for the inventory. The categories include the following:

- Polymeric substances;
- Certain chemical substances containing varying carbon chain;
- Products containing two or more substances, formulated and statutory mixtures; and
- Chemical substances of *unknown* or *variable* composition, *complex* reaction products, and *biological* materials (UVCB substance).

The product laws are likely to be changed soon. In fact, the European Commission and individual nations are strengthening their methods for screening and prioritizing chemicals based on precautions before and after these substances enter the marketplace. The Untied States is also looking to reauthorize TSCA to improve these safeguards.

7. WASTE REGULATIONS IN OTHER COUNTRIES

This chapter emphasizes U.S. legislation and rulemaking to address waste problems, but almost every nation has its own set of regulations which in each case have been drawn up in response to situations that needed controlling. For example, sparsely populated countries with wide open spaces sometimes do not have the strict landfilling regulations of more densely populated countries (see Chapter 30). What follows is a brief summary of the ways regulations have developed in the United Kingdom.

7.1. Waste Management Regulations in the United Kingdom

Waste Management Regulations have been on the statute books in the United Kingdom since 1848 with the Public Health Act. This was followed by various amendments to the Public Health Act, which were precipitated by waste problems such as burning rubbish (paper, wood, etc.) to generate electricity (1874), smoke abatement and scavenging (1875), the urgent need for refuse collection with charges levied (1907) and uncontrolled dumping (1936). Some of this related to dust and cinders from the burning of coal. In 1892, it has been estimated that more than 80% by mass of waste was dust and cinders. Today plastic is the most common waste type with 37% by mass.

During the world wars, waste regulations were not considered a great priority. In 1947, the Town and Country Planning Act gave local authorities powers to control waste and this postwar period was dominated by the construction of landfills, which in many cases developed into tips. Increased smog in cities, as a result of burning coal precipitated the CAA of 1956, which led to a decrease of open fires in homes and the acceleration of central heating fuelled by electricity, oil, or gas. This had a major effect on the composition of household waste, which was dominated by food and paper waste for the next few decades.

It took a combination of new types of chemical waste resulting from the rapid expansion of the plastics industry; the CAA of 1956, which made the public more aware of chemical waste and pollution; new Health and Safety guidelines; and the developments in the United States (e.g., the previously mentioned Love Canal site) to launch a number of new waste regulations in the 1970s. Another possible nucleus for this was the finding in 1971 of a number of metal drums of cyanide that had been dumped at a disused brick kiln at Nuneaton. This led to a huge public outcry, a report by the Royal Commission on Toxic Wastes, and the drafting of the first ever legislation on the control of hazardous waste. As a result, the Deposit of Poisonous Waste Act of 1972 was drafted in 10 days and passed through Parliament over the following month. Also in 1971, the Bristol Friends of the Earth launched their first campaign against nonreturnable bottles and returned thousands of bottles at Schweppes local headquarters. However, it was not until 1977 that the first bottle banks appeared in Britain. In 1974, the increasing concern over waste disposal led to the Control of Pollution Act of 1974, which regulated the disposal of waste and the disposal sites and began a serious tightening up of ways of disposing of waste.

In 1986, the Single European Act added new momentum to European integration and expanded Community powers in a number of fields, including environmental issues such as waste and pollution. This sea change influenced many British regulations, and in 1990 the British Government produced a white paper on the environment entitled "This Common Inheritance," setting out a strategy of waste minimization and recycling and set a target of 25% for recycling household waste by 2000 (later toned down and amended). This was followed by the Environmental Protection Act of 1990, with more regulations and controls that include the transition from the Control of Pollution Act 1974 to the new Act. Another European Union Directive (in 1994) introduced the concept of Producer Responsibility with respect to packaging and required member states to set targets on the reduction and recovery of packaging waste. This was followed by a UK regulation entitled *Producer Responsibility Obligations*, which has been amended many times since then (in 1997, the amendment required companies to recover and recycle 38% of their packaging and in 2001 this was increased to 56%). A landfill tax was introduced 2 years later (a tax of £7 per ton was levied on active waste going into landfills; this was raised every few years and by 2010 it had risen to £48 per ton). One of the aims was to encourage recycling. In 2000, as befits a millennium celebration, the "Waste Strategy 2000" was published and included revised targets for recycling or composting of household waste: 25% by 2005 and 33% by 2015.

Since then, new or amended regulations have appeared every year relating to issues such as Waste Electrical and Electronic Equipment (see Chapter 20) the Ozone and the restriction on the manufacture and use fluorocarbons, Hazardous Waste (see Chapter 13), Restricted Hazardous Waste, Low-level radiation disposal, Environmental Permits, Landfill tax, Landfill Regulations, and recently a Battery Directive that imposes new obligations and rules on the producers, suppliers, and the disposal of batteries (see Chapter 21). Other obligations included new limits on the amount of mercury and cadmium in batteries, a ban on disposing of batteries in landfills, and the introduction of a new collection regime for batteries. This surge of regulations is not unlike the situation in the United States described so well in Fig. 3.1. As always, the regulations reflect the public's interest and this, in turn, has been whetted by the ever increasing amount of data being produced by monitoring organizations using analytical tools and equipment that seem to improve their levels of sophistication, detection, and precision almost every year. The standards governing air quality reflect this.

7.1.1. Air Pollution Regulations in the United Kingdom

In the United Kingdom, air pollution regulations were first directed at the smoke produced by the burning of coal and wood in industries and households and culminated in the CAAs of 1956 and 1968. During the past 70 years, the focus has shifted to chemical pollution from factories and from vehicle emissions. In particular, the 1970s and 1980s saw legislation aimed at SO_2; NO_xs; and particulates, CO, benzene, and other hydrocarbons pollution, culminating in the National Air Quality Strategies for the United Kingdom in 1997 and 2000. The latest Air Quality Standards that came into force on 11 June 2010 are given in Table 3.5.

In many respects, the regulations in the United States and the United Kingdom (and indeed in Europe) relating to waste and pollution have followed similar paths and the laws have been drafted for similar reasons. We do seem to learn from one another!

7.2. Global Connections

Waste management must be viewed systematically. In fact, wastes tend to be transported and can cause problems well beyond the borders of nations. One of the key waste management challenges of the twenty-first century is the ease at which waste problems

TABLE 3.5 The Air Quality Standards Regulations 2010

Sulfur dioxide			
Averaging period	*Limit value*		
1 h	350 μg m^{-3} not to be exceeded more than 24 times a calendar year		
1 day	125 μg m^{-3} not to be exceeded more than 3 times a calendar year		
Nitrogen dioxide			
Averaging period	*Limit value*		
1 h	200 μg m^{-3} not to be exceeded more than 18 times a calendar year		
Calendar year	40 μg m^{-3}		
Benzene			
Averaging period	*Limit value*		
Calendar year	5 μg m^{-3}		
Lead			
Averaging period	*Limit value*		
Calendar year	0.5 μg m^{-3}		
PM_{10}			
Averaging period	*Limit value*		
1 day	50 μg m^{-3}, not to be exceeded more than 35 times a calendar year		
Calendar year	40 μg m^{-3}		
$PM_{2.5}$			
Averaging period	*Limit value*	*Margin of tolerance*	*Date by which limit value is to be met*
Calendar year	25 μg m^{-3}	20% on 11th June 2008, decreasing on the next 1st January and every 12 months thereafter by equal annual percentages to reach 0% by 1 January 2015	1 January 2015
Carbon monoxide			
Averaging period			*Limit value*
Maximum eight hour daily mean[6]			10 mg m^{-3}

[6]*The maximum daily 8 hour mean concentration of carbon monoxide must be selected by examining 8-hour running averages, calculated from hourly data and updated each hour. Each 8-hour average so calculated will be assigned to the day on which it ends, that is, the first calculation period for any 1 day will be from 17:00 on the previous day to 01:00 on that day, the last calculation period for any 1 day will be the period from 16:00 to 24:00 on that day.*

are simply transferred to countries with less-stringent controls.

An illustrative example took place in 1992, when the U.S.-based Gaston Copper Recycling Corporation transported 1000 tons of dust laden with cadmium and lead to a South Carolina fertilizer manufacturer, Stoller Chemical [9]. The fertilizer was made from dust collected from air pollution removal equipment, that is, fabric filters.[7]

The company then mixed some of the waste product with fertilizer and then exported 1.5 tons of the mixture to Australia and Bangladesh. As discussed in Chapter 27, such substances require specific manifests and must be permitted in compliance with the RCRA, in addition to notifying the EPA, as specified by TSCA. The export was even more complicated in that a broker, Hy-Tex Marketing, helped to implement the export.

Bangladesh, as is the case for many developing nations, lacks impoundment laws against waste imports. Thus, the shipment was not inspected at entry. In this case, the U.S. Department of Justice indicted the companies in 1992; however, the product remained in Bangladesh. The U.S. Embassy in Bangladesh released a warning in late 1992. Sadly, as reported by the environmental group, Greenpeace, most of the waste fertilizer sent to Bangladesh had already been distributed throughout Bangladeshi farms by October of 1992. Meanwhile, the Bangladesh government is reported to have halted the transport of the remaining Cd/Pb-laden fertilizer and set up an investigative commission. Environmental groups in both countries continued to demand that the fertilizer be returned to the U.S. and the Bangladesh Embassy has officially acknowledged the potential risks to Bangladeshis.

The interconnections of wastes can lead to a "circle of poisons," the phenomenon where nations may ban a substance, but it continues to be produced and used in other, usually less economically developed, countries. Despite the ban within the borders of certain countries, the feared substance returns. Thus, the rationale for the ban, that is reduced risks to that country's inhabitants, is forgone and people continue to be exposed when they import the products that contain or are contaminated with the banned substances.

8. CONCLUSIONS

Managing wastes is a complex challenge in the United States, as it is throughout the world. Thus, the measures discussed in this chapter combined with those in Chapter 2 are necessary. That is, risk management requires the use of numerous tools, including regulation, the marketplace, consumer education, and multinational cooperation.

[7] Air pollution controls employing fabric filters (also known as "baghouses") remove particles from the air stream by passing the air through a porous fabric. The fabric filter is efficient at removing fine particles and can exceed efficiencies of 99%. That is, the amount (mass) of particles exiting the unit is 99% less than that entering the unit. The bag filter is more than mere "sieving," but actually captures particles via diffusion, interception, and inertial impaction, whereas a cyclone generally is based on inertia.

APPENDIX 1

Summary of the Council on Environmental Quality Guidance for Compliance with the National Environmental Policy Act of 1969[1]

Title of guidance	Summary of guidance	Citation	Relevant regulation/ documentation
Forty most often asked questions concerning Council on Environmental Quality's (CEQ's) National Environmental Policy Act (NEPA) regulations	Provides answers to 40 questions most frequently asked concerning implementation of NEPA	46 FR 18026, dated March 23, 1981	40 CFR Parts 1500–1508
Implementing and explanatory documents for Executive Order (EO) 12114, Environmental Effects Abroad of Major Federal Actions	Provides implementing and explanatory information for EO 12114. Establishes categories of federal activities or programs as those that significantly harm the natural and physical environment. Defines which actions are excluded from the order and those that are not	44 FR 18672, dated March 29, 1979	EO 12114, Environmental Effects Abroad of Major Federal Actions
Publishing of three memoranda for heads of agencies on: - Analysis of impacts on prime or unique agricultural lands (Memoranda 1 and 2) - interagency consultation to avoid or mitigate adverse effects on rivers in the nationwide inventory (Memorandum 3)	1, 2. Discusses the irreversible conversion of unique agricultural lands by Federal Agency action (e.g., construction activities, developmental grants, and federal land management). Requires identification of and cooperation in retention of important agricultural lands in areas of impact of a proposed agency action. The agency must identify and summarize existing or proposed agency policies, to preserve or mitigate the effects of agency action on agricultural lands **3.** "Each Federal agency shall, as part of its normal planning and environmental review process, take care to avoid or mitigate adverse effects on rivers identified in the Nationwide Inventory prepared by the	45 FR 59189, dated September 8, 1980	1, 2. Farmland Protection Policy Act (7 U.S.C. §4201 et seq.) **3.** The Wild and Scenic Rivers Act of 1965 (16 U.S.C. §1271 et seq.)

[1] Source: National Aeronautics and Space Administration (2001). Implementing the National Environmental Policy Act and Executive Order 12114, Chapter 6.

Title of guidance	Summary of guidance	Citation	Relevant regulation/ documentation
	Heritage Conservation and Recreation Service in the Department of the Interior." Implementing regulations includes determining whether the proposed action: affects an Inventory River; adversely affects the natural, cultural, and recreation values of the Inventory river segment; forecloses options to classify any portion of the Inventory segment as a wild, scenic, or recreational river area, and incorporates avoidance/ mitigation measures into the proposed action to the maximum extent feasible within the agency's authority		
Memorandum for Heads of Agencies for Guidance on Applying Section 404(r) of the Clean Water Act at federal projects which involve the discharge of dredged or fill materials into waters of the United States, including wetlands	Requires timely agency consultation with U.S. Army Corps of Engineers (COE) and the U.S. Environmental Protection Agency (EPA) before a Federal project involves the discharge of dredged or fill material into U.S. waters, including wetlands. Proposing agency must ensure, when required, that the EIS includes written conclusions of EPA and COE (generally found in Appendix)	Council on Environmental Quality, dated November 17, 1980	Clean Water Act (33 U.S.C. §1251 et seq.) EO 12088, Federal Compliance with Pollution Control Standards
Scoping guidance	Provides a series of recommendations distilled from agency research regarding the scoping process. Requires public notice; identification of significant and insignificant issues; allocation of EIS preparation assignments; identification of related analysis requirements to avoid duplication of work; and the planning of a schedule for EIS preparation that meshes with the agency's decision-making schedule	46 FR 25461, dated May 7, 1981	40 CFR Parts 1500–1508
Guidance Regarding NEPA Regulations	Provides written guidance on scoping, CatEx's, adoption regulations, contracting provisions, selecting alternatives in licensing and permitting situations, and tiering	48 FR 34263, dated July 28, 1983	40 CFR Parts 1501, 1502, and 1508

(Continued)

Title of guidance	Summary of guidance	Citation	Relevant regulation/ documentation
National Environmental Policy Act (NEPA) Implementation Regulations, Appendices I, II, and III	Provides guidance on improving public participation, facilitating agency compliance with NEPA and CEQ implementing regulations. Appendix I updates required NEPA contacts, Appendix II compiles a list of Federal and Federal-State Agency Offices with jurisdiction by law or special expertise in environmental quality issues; and Appendix III lists the Federal and Federal-State Offices for receiving and commenting on other agencies' environmental documents	49 FR 49750, dated December 21, 1984	40 CFR Part 1500
Incorporating biodiversity considerations into environmental impact analysis under the NEPA	Provides for "acknowledging the conservation of biodiversity as national policy and incorporates its consideration in the NEPA process"; encourages seeking out opportunities to participate in efforts to develop regional ecosystem plans; actively seeks relevant information from sources both within and outside government agencies; encourages participating in efforts to improve communication, cooperation, and collaboration between and among governmental and nongovernmental entities; improves the availability of information on the status and distribution of biodiversity, and on techniques for managing and restoring it; and expands the information base on which biodiversity analyses and management decisions are based	Council on Environmental Quality, Washington, DC, dated January 1993	Not applicable
Pollution Prevention and the NEPA	Pollution-prevention techniques seek to reduce the amount and/or toxicity of pollutants being generated, promote increased efficiency of raw materials and conservation of natural resources, and can be cost-effective. Directs Federal agencies that, to the extent practicable, pollution prevention considerations should be included in the proposed action and in the reasonable alternatives to the	58 FR 6478, dated January 29, 1993	EO 12088, Federal Compliance with Pollution Control Standards

Title of guidance	Summary of guidance	Citation	Relevant regulation/ documentation
	proposal, and to address these considerations in the environmental consequences section of an EIS and environmental assessment (EA) (when appropriate)		
Considering Cumulative Effects under the NEPA	Provides a "framework for advancing environmental cumulative impacts analysis by addressing cumulative effects in either an EA or an environmental impact statement". Also provides practical methods for addressing coincident effects (adverse or beneficial) on specific resources, ecosystems, and human communities of all related activities, not just the proposed project or alternatives that initiate the assessment process	January 1997	40 CFR §1508.7
Environmental justice guidance under the NEPA	Provides guidance and general direction on Executive Order 12898 that requires each agency to identify and address, as appropriate, "disproportionately high and adverse human health or environmental effects of its programs, policies, and activities on minority populations and low-income populations"	Council on Environmental Quality, Washington, DC, dated December 10, 1997	EO 12898, Federal Actions to Address Environmental Justice in Minority Populations and Low-Income Populations

FORMAT OF AN ENVIRONMENTAL IMPACT STATEMENT

Cover Sheet (See next table for information to be included)

EXECUTIVE SUMMARY

TABLE OF CONTENTS

LIST OF ABBREVIATIONS AND ACRONYMS

MEASUREMENT CONVERSION TABLES

CHAPTERS

1. PURPOSE AND NEED FOR THE ACTION
2. DESCRIPTION AND COMPARISON OF ALTERNATIVES
   - Description of proposed action and each reasonable alternative, including No Action
   - Brief description of alternatives not considered in detail; explain why
   - Summary of environmental impacts of proposed action and reasonable alternatives, including No Action
3. DESCRIPTION OF THE AFFECTED ENVIRONMENT
   - Appropriate-level descriptions of the physical, natural, and socioeconomic aspects of the environment that will be impacted, including, but not limited to, air quality, historical/cultural resources, threatened or endangered species and habitats, wetlands, floodplains, and other sensitive/protected resources
4. ENVIRONMENTAL CONSEQUENCES
   - Impact analyses for the proposed action and reasonable alternatives, including No Action
   - Mandatory subsections
   - Relationship between short-term use of the human environment and the maintenance and enhancement of long-term productivity
   - Irreversible and irretrievable commitments of resources
5. MITIGATION AND MONITORING (optional; can be incorporated into Chapter 4, if appropriate)
6. REFERENCES
7. LIST OF PREPARERS
8. AGENCIES, ORGANIZATIONS, AND INDIVIDUALS CONSULTED
   - Consulting agencies
   - Distribution list
9. INDEX

Appendices (Final EIS must have a "Response to Comments" chapter; as either an appendix or in a separate volume)

Required Cover Sheet for an Environmental Impact Statement

POPULAR NAME of PROPOSAL INCLUDES TYPE (e.g., DRAFT or FINAL)	
Lead Agency	NASA, State name of Sponsoring Entity; name of cooperating agency if appropriate
Point of Contact for Information	Name, title, address, and phone number of NASA Point of Contact
Date	Date of Issuance (recommend using month and year)
Abstract	Succinct statement of proposed action; brief abstract of the EIS, stating proposed action, alternatives examined, and summary of key findings (the abstract maybe printed on a separate page, if necessary)

APPENDIX 2

Safe Drinking Water Act Contaminants and Maximum Contaminant Levels

Contaminant	MCLG[1] (mg L^{-1})[2]	MCL or TT[1] (mg L^{-1})[2]	Potential health effects from ingestion of water	Sources of contaminant in drinking water
Microorganisms				
Cryptosporidium	0	TT[3]	Gastrointestinal illness (e.g., diarrhea, vomiting, cramps)	Human and fecal animal waste
Giardia lamblia	0	TT[3]	Gastrointestinal illness (e.g., diarrhea, vomiting, cramps)	Human and animal fecal waste
Heterotrophic plate count (HPC)	n/a	TT[3]	HPC has no health effects; it is an analytic method used to measure the variety of bacteria that are common in water. The lower the concentration of bacteria in drinking water, the better maintained the water system	HPC measures a range of bacteria that are naturally present in the environment
Legionella	0	TT[3]	Legionnaire's Disease, a type of pneumonia	Found naturally in water and multiplies in heating systems
Total coliforms (including fecal coliform and *Escherichia coli*)	0	5.0%[4]	Not a health threat in itself; it is used to indicate whether other potentially harmful bacteria may be present[5]	Coliforms are naturally present in the environment; as well as feces; fecal coliforms and *E. coli* only come from human and animal fecal waste
Turbidity	n/a	TT[3]	Turbidity is a measure of the cloudiness of water. It is used to indicate water quality and filtration effectiveness (e.g., whether disease-causing organisms are present). Higher turbidity levels are often associated with higher levels of disease-causing microorganisms such as viruses, parasites, and some bacteria. These organisms can cause symptoms such as nausea, cramps, diarrhea, and associated headaches	Soil runoff
Viruses (enteric)	0	TT[3]	Gastrointestinal illness (e.g., diarrhea, vomiting, cramps)	Human and animal fecal waste

(*Continued*)

Contaminant	MCLG[1] $(mg\ L^{-1})^2$	MCL or TT[1] $(mg\ L^{-1})^2$	Potential health effects from ingestion of water	Sources of contaminant in drinking water
Disinfection by-products				
Bromate	0	0.010	Increased risk of cancer	By-product of drinking water disinfection
Chlorite	0.8	1.0	Anemia; nervous system effects in infants and young children	By-product of drinking water disinfection
Haloacetic acids	n/a[6]	0.060	Increased risk of cancer	By-product of drinking water disinfection
Total Trihalomethanes	None[7] n/a[6]	0.10 0.080	Liver, kidney, or central nervous system problems; increased risk of cancer	By-product of drinking water disinfection
Disinfectants				
Chloramines (as Cl_2)	MRDLG = 4[1]	MRDL = 4.0[1]	Eye/nose irritation; stomach discomfort, anemia	Water additive used to control microbes
Chlorine (as Cl_2)	MRDLG = 4[1]	MRDL = 4.0[1]	Eye/nose irritation; stomach discomfort	Water additive used to control microbes
Chlorine dioxide (as ClO_2)	MRDLG = 0.8[1]	MRDL = 0.8[1]	Anemia; nervous system effects in infants and young children	Water additive used to control microbes
Inorganic chemicals				
Antimony	0.006	0.006	Increase in blood cholesterol and decrease in blood sugar	Discharge from petroleum refineries; fire retardants; ceramics; electronics; and solder
	0[7]	0.010 as of January 23, 2006	Skin damage or problems with circulatory systems, and may have increased risk of getting cancer	Erosion of natural deposits; runoff from orchards, runoff from glass and electronics production wastes
Asbestos (fiber >10 μm)	7 million fibers/L (MFL)	7 MFL	Increased risk of developing benign intestinal polyps	Decay of asbestos cement in water mains; erosion of natural deposits
Barium	2	2	Increase in blood pressure	Discharge of drilling wastes; discharge from metal refineries; and erosion of natural deposits
Beryllium	0.004	0.004	Intestinal lesions	Discharge from metal refineries and coal-burning factories; discharge from electrical, aerospace, and defense industries

Contaminant	MCLG[1] (mg L^{-1})[2]	MCL or TT[1] (mg L^{-1})[2]	Potential health effects from ingestion of water	Sources of contaminant in drinking water
Cadmium	0.005	0.005	Kidney damage	Corrosion of galvanized pipes; erosion of natural deposits; discharge from metal refineries; runoff from waste batteries and paints
Chromium (total)	0.1	0.1	Allergic dermatitis	Discharge from steel and pulp mills; erosion of natural deposits
Copper	1.3	TT[8]; action level = 1.3	Short-term exposure: gastrointestinal distress Long-term exposure: liver or kidney damage People with Wilson's Disease should consult their personal doctor if the amount of copper in their water exceeds the action level	Corrosion of household plumbing systems; and erosion of natural deposits
Cyanide (as free cyanide)	0.2	0.2	Nerve damage or thyroid problems	Discharge from steel/ metal factories; discharge from plastic and fertilizer factories
Fluoride	4.0	4.0	Bone disease (pain and tenderness of the bones); Children may get mottled teeth	Water additive that promotes strong teeth; erosion of natural deposits; discharge from fertilizer and aluminum factories
Lead	0	TT[8]; action level = 0.015	Infants and children: delays in physical or mental development; children could show slight deficits in attention span and learning abilities Adults: kidney problems; high blood pressure	Corrosion of household plumbing systems; and erosion of natural deposits
Mercury (inorganic)	0.002	0.002	Kidney damage	Erosion of natural deposits; discharge from refineries and factories; runoff from landfills and croplands
Nitrate (measured as nitrogen)	10	10	Infants below the age of 6 months who drink water containing nitrate in excess of the MCL could become seriously ill and, if untreated, may die. Symptoms include shortness of breath and blue-baby syndrome	Runoff from fertilizer use; leaching from septic tanks, sewage; and erosion of natural deposits

(*Continued*)

Contaminant	MCLG[1] (mg L^{-1})[2]	MCL or TT[1] (mg L^{-1})[2]	Potential health effects from ingestion of water	Sources of contaminant in drinking water
Nitrite (measured as nitrogen)	1	1	Infants below the age of 6 months who drink water containing nitrite in excess of the MCL could become seriously ill and, if untreated, may die. Symptoms include shortness of breath and blue-baby syndrome	Runoff from fertilizer use; leaching from septic tanks, sewage; and erosion of natural deposits
Selenium	0.05	0.05	Hair or fingernail loss; numbness in fingers or toes; and circulatory problems	Discharge from petroleum refineries; erosion of natural deposits; and discharge from mines
Thallium	0.0005	0.002	Hair loss; changes in blood; kidney, intestine, or liver problems	Leaching from ore-processing sites; discharge from electronics, glass, and drug factories
Organic Chemicals				
Acrylamide	0	TT[9]	Nervous system or blood problems; increased risk of cancer	Added to water during sewage/wastewater treatment
Alachlor	0	0.002	Eye, liver, kidney, or spleen problems; anemia; and increased risk of cancer	Runoff from herbicide used on row crops
Atrazine	0.003	0.003	Cardiovascular system or reproductive problems	Runoff from herbicide used on row crops
Benzene	0	0.005	Anemia; decrease in blood platelets; and increased risk of cancer	Discharge from factories; and leaching from gas storage tanks and landfills
Benzo(a)pyrene	0	0.0002	Reproductive difficulties; increased risk of cancer	Leaching from linings of water storage tanks and distribution lines
Carbofuran	0.04	0.04	Problems with blood, nervous system, or reproductive system	Leaching of soil fumigant used on rice and alfalfa
Carbon tetrachloride	0	0.005	Liver problems; increased risk of cancer	Discharge from chemical plants and other industrial activities
Chlordane	0	0.002	Liver or nervous system problems; increased risk of cancer	Residue of banned termiticide
Chlorobenzene	0.1	0.1	Liver or kidney problems	Discharge from chemical and agricultural chemical factories
2,4-D	0.07	0.07	Kidney, liver, or adrenal gland problems	Runoff from herbicide used on row crops

Contaminant	MCLG[1] (mg L^{-1})[2]	MCL or TT[1] (mg L^{-1})[2]	Potential health effects from ingestion of water	Sources of contaminant in drinking water
Dalapon	0.2	0.2	Minor kidney changes	Runoff from herbicide used on rights of way
1,2-Dibromo-3-chloropropane	0	0.0002	Reproductive difficulties and increased risk of cancer	Runoff/leaching from soil fumigant used on soybeans, cotton, pineapples, and orchards
o-Dichlorobenzene	0.6	0.6	Liver, kidney, or circulatory system problems	Discharge from industrial chemical factories
p-Dichlorobenzene	0.075	0.075	Anemia; liver, kidney, or spleen damage; and changes in blood	Discharge from industrial chemical factories
1,2-Dichloroethane	0	0.005	Increased risk of cancer	Discharge from industrial chemical factories
1,1-Dichloroethylene	0.007	0.007	Liver problems	Discharge from industrial chemical factories
Cis-1,2-dichloroethylene	0.07	0.07	Liver problems	Discharge from industrial chemical factories
Trans-1,2-dichloroethylene	0.1	0.1	Liver problems	Discharge from industrial chemical factories
Dichloromethane	0	0.005	Liver problems and increased risk of cancer	Discharge from drug and chemical factories
1,2-Dichloropropane	0	0.005	Increased risk of cancer	Discharge from industrial chemical factories
Di(2-ethylhexyl) adipate	0.4	0.4	Weight loss, liver problems, or possible reproductive difficulties	Discharge from chemical factories
Di(2-ethylhexyl) phthalate	0	0.006	Reproductive difficulties; liver problems; and increased risk of cancer	Discharge from rubber and chemical factories
Dinoseb	0.007	0.007	Reproductive difficulties	Runoff from herbicide used on soybeans and vegetables
Dioxin (2,3,7,8-TCDD)	0	0.00000003	Reproductive difficulties and increased risk of cancer	Emissions from waste incineration and other combustion and discharge from chemical factories
Diquat	0.02	0.02	Cataracts	Runoff from herbicide use
Endothall	0.1	0.1	Stomach and intestinal problems	Runoff from herbicide use
Endrin	0.002	0.002	Liver problems	Residue of banned insecticide

(*Continued*)

Contaminant	MCLG[1] (mg L^{-1})[2]	MCL or TT[1] (mg L^{-1})[2]	Potential health effects from ingestion of water	Sources of contaminant in drinking water
Epichlorohydrin	0	TT[9]	Increased cancer risk, and stomach problems over a long period of time	Discharge from industrial chemical factories; an impurity of some water treatment chemicals
Ethylbenzene	0.7	0.7	Liver or kidneys problems	Discharge from petroleum refineries
Ethylene dibromide	0	0.00005	Problems with liver, stomach, reproductive system, or kidneys; increased risk of cancer	Discharge from petroleum refineries
Glyphosate	0.7	0.7	Kidney problems and reproductive difficulties	Runoff from herbicide use
Heptachlor	0	0.0004	Liver damage and increased risk of cancer	Residue of banned termiticide
Heptachlor epoxide	0	0.0002	Liver damage and increased risk of cancer	Breakdown of heptachlor
Hexachloro-benzene	0	0.001	Liver or kidney problems; reproductive difficulties; and increased risk of cancer	Discharge from metal refineries and agricultural chemical factories
Hexachloro-cyclopentadiene	0.05	0.05	Kidney or stomach problems	Discharge from chemical factories
Lindane	0.0002	0.0002	Liver or kidney problems	Runoff/leaching from insecticide used on cattle, lumber, gardens
Methoxychlor	0.04	0.04	Reproductive difficulties	Runoff/leaching from insecticide used on fruits, vegetables, alfalfa, livestock
Oxamyl (Vydate)	0.2	0.2	Slight nervous system effects	Runoff/leaching from insecticide used on apples, potatoes, and tomatoes
Polychlorinated biphenyls	0	0.0005	Skin changes; thymus gland problems; immune deficiencies; reproductive or nervous system difficulties; and increased risk of cancer	Runoff from landfills and discharge of waste chemicals
Pentachlorophenol	0	0.001	Liver or kidney problems and increased cancer risk	Discharge from wood preserving factories
Picloram	0.5	0.5	Liver problems	Herbicide runoff
Simazine	0.004	0.004	Problems with blood	Herbicide runoff
Styrene	0.1	0.1	Liver, kidney, or circulatory system problems	Discharge from rubber and plastic factories; leaching from landfills

Contaminant	MCLG[1] (mg L^{-1})[2]	MCL or TT[1] (mg L^{-1})[2]	Potential health effects from ingestion of water	Sources of contaminant in drinking water
Tetrachloroethylene	0	0.005	Liver problems and increased risk of cancer	Discharge from factories and dry cleaners
Toluene	1	1	Nervous system, kidney, or liver problems	Discharge from petroleum factories
Toxaphene	0	0.003	Kidney, liver, or thyroid problems and increased risk of cancer	Runoff/leaching from insecticide used on cotton and cattle
2,4,5-TP (Silvex)	0.05	0.05	Liver problems	Residue of banned herbicide
1,2,4-Trichloro-benzene	0.07	0.07	Changes in adrenal glands	Discharge from textile finishing factories
1,1,1-Trichloroethane	0.20	0.2	Liver, nervous system, or circulatory problems	Discharge from metal degreasing sites and other factories
1,1,2-Trichloroethane	0.003	0.005	Liver, kidney, or immune system problems	Discharge from industrial chemical factories
Trichloroethylene	0	0.005	Liver problems and increased risk of cancer	Discharge from metal degreasing sites and other factories
Vinyl chloride	0	0.002	Increased risk of cancer	Leaching from poly vinyl chloride pipes and discharge from plastic factories
Xylenes (total)	10	10	Nervous system damage	Discharge from petroleum factories and discharge from chemical factories
Radionuclides				
Alpha particles	None[7] 0	15 pCi L^{-1}	Increased risk of cancer	Erosion of natural deposits of certain minerals that are radioactive and may emit a form of radiation known as alpha radiation
Beta particles and photon emitters	None[7] 0	4 millirems per year	Increased risk of cancer	Decay of natural and man-made deposits of certain minerals that are radioactive and may emit forms of radiation known as photons and beta radiation
Radium 226 and Radium 228 (combined)	None[7] 0	5 pCi L^{-1}	Increased risk of cancer	Erosion of natural deposits

(Continued)

Contaminant	MCLG[1] (mg L^{-1})[2]	MCL or TT[1] (mg L^{-1})[2]	Potential health effects from ingestion of water	Sources of contaminant in drinking water
Uranium	0	30 μg L^{-1} as of December 8, 2003	Increased risk of cancer, kidney toxicity	Erosion of natural deposits

[1]*MCL = Maximum Contaminant Level — The highest level of a contaminant that is allowed in drinking water. MCLs are set as close to MCLGs as feasible using the best available treatment technology and taking cost into consideration. MCLs are enforceable standards. MCLG = Maximum Contaminant Level Goal — The level of a contaminant in drinking water below which there is no known or expected risk to health. MCLGs allow for a margin of safety and are nonenforceable public health goals. MRDL = Maximum Residual Disinfectant Level — The highest level of a disinfectant allowed in drinking water. There is convincing evidence that addition of a disinfectant is necessary for control of microbial contaminants. MRDLG = Maximum Residual Disinfectant Level Goal — The level of a drinking water disinfectant below which there is no known or expected risk to health. MRDLGs do not reflect the benefits of the use of disinfectants to control microbial contaminants. TT = Treatment Technique — A required process intended to reduce the level of a contaminant in drinking water.*

[2]*Units are in milligrams per liter (mg L^{-1}) unless otherwise noted. Milligrams per liter (MPL) are equivalent to parts per million under standard environmental conditions.*

[3]*Treatment Technique (TT): EPA's surface water treatment rules require systems using surface water or ground water under the direct influence of surface water to (1) disinfect their water and (2) filter their water or meet criteria for avoiding filtration so that the following contaminants are controlled at the following levels:*

- *Cryptosporidium (as of January 1, 2002, for systems serving >10,000 and January 14, 2005, for systems serving <10,000) 99% removal.*
- Giardia lamblia: *99.9% removal/inactivation.*
- *Viruses: 99.99% removal/inactivation.*
- Legionella: *No limit, but EPA believes that if* Giardia *and viruses are removed/inactivated,* Legionella *will also be controlled.*
- *Turbidity: At no time can turbidity (cloudiness of water) go above 5 nephelolometric turbidity units (NTU); systems that filter must ensure that the turbidity go no higher than 1 NTU (0.5 NTU for conventional or direct filtration) in at least 95% of the daily samples in any month. As of January 1, 2002, turbidity may never exceed 1 NTU, and must not exceed 0.3 NTU in 95% of daily samples in any month.*
- *HPC: No more than 500 bacterial colonies per milliliter.*
- *Long-term 1 enhanced surface water treatment (effective date: January 14, 2005): Surface water systems or (GWUDI) systems serving fewer than 10,000 people must comply with the applicable Long-Term 1 Enhanced Surface Water Treatment Rule provisions (e.g., turbidity standards, individual filter monitoring, Cryptosporidium removal requirements, updated watershed control requirements for unfiltered systems).*
- *Filter Backwash Recycling: The Filter Backwash Recycling Rule requires systems that recycle to return specific recycle flows through all processes of the system's existing conventional or direct filtration system or at an alternate location approved by the state.*

[4]*No more than 5.0% samples total coliform positive in a month. (For water systems that collect fewer than 40 routine samples per month, no more than one sample can be total coliform positive per month.) Every sample that has total coliform must be analyzed for either fecal coliforms or* E. coli *if two consecutive TC-positive samples, and one is also positive for* E. coli *fecal coliforms, system has an acute MCL violation.*

[5]*Fecal coliform and* E. coli *are bacteria whose presence indicates that the water may be contaminated with human or animal wastes. Disease-causing microbes (pathogens) in these wastes can cause diarrhea, cramps, nausea, headaches, or other symptoms. These pathogens may pose a special health risk for infants, young children, and people with severely compromised immune systems.*

[6]*Although there is no collective MCLG for this contaminant group, there are individual MCLGs for some of the individual contaminants:*

- *Trihalomethanes: bromodichloromethane (0); bromoform (0); dibromochloromethane (0.06 mg L^{-1}); chloroform (0.07 mg L^{-1}).*
- *Haloacetic acids: dichloroacetic acid (0); trichloroacetic acid (0.2 mg L^{-1}); monochloroacetic acid (0.07 mg L^{-1}). Bromoacetic acid and dibromoacetic acid are regulated with this group but have no MCLGs.*

[7]*Lead and copper are regulated by a Treatment Technique that requires systems to control the corrosiveness of their water. If more than 10% of tap water samples exceed the action level, water systems must take additional steps. For copper, the action level is 1.3 mg L^{-1}, and for lead is 0.015 mg L^{-1}.*

[8]*Each water system must certify, in writing, to the state (using third-party or manufacturer's certification) that when acrylamide and epichlorohydrin are used in drinking water systems, the combination (or product) of dose and monomer level does not exceed the levels specified, as follows:*

- *Acrylamide = 0.05% dosed at 1 mg L^{-1} (or equivalent)*
- *Epichlorohydrin = 0.01% dosed at 20 mg L^{-1} (or equivalent)*

NATIONAL SECONDARY DRINKING WATER REGULATIONS

National Secondary Drinking Water Regulations (or "secondary standards") are nonenforceable guidelines regulating contaminants that may cause cosmetic effects (such as skin or tooth discoloration) or aesthetic effects (such as taste, odor, or color) in drinking water. EPA recommends secondary standards to water systems but does not require systems to comply. However, states may choose to adopt them as enforceable standards. For additional information regarding nuisance chemicals, visit: http://www.ehso.com/ehso3.php?URL=http%3A%2F%2Fwww.epa.gov/safewater/consumer/2ndstandards.html (assessed on July 12, 2010).

Contaminant	Secondary standard
Aluminum	0.05–0.2 mg L^{-1}
Chloride	250 mg L^{-1}
Color	15 (color units)
Copper	1.0 mg L^{-1}
Corrosivity	Noncorrosive
Fluoride	2.0 mg L^{-1}
Foaming agents	0.5 mg L^{-1}
Iron	0.3 mg L^{-1}
Manganese	0.05 mg L^{-1}
Odor	3 Threshold odor number
pH	6.5–8.5
Silver	0.10 mg L^{-1}
Sulfate	250 mg L^{-1}
Total dissolved solids	500 mg L^{-1}
Zinc	5 mg L^{-1}

APPENDIX 3

Toxic Compounds Listed in the 1990 Clean Air Act Amendments

CAS #	Chemical or class
75070	Acetaldehyde
60355	Acetamide
75058	Acetonitrile
98862	Acetophenone
53963	2-Acetylaminofluorene
107028	Acrolein
79061	Acrylamide
79107	Acrylic acid
107131	Acrylonitrile
8107051	Allyl chloride
92671	4-Aminobiphenyl
62533	Aniline
90040	*o*-Anisidine
1332214	Asbestos
71432	Benzene (including from gasoline)
92875	Benzidine
98077	Benzotrichloride
100447	Benzyl chloride
92524	Biphenyl
117817	Bis (2-ethylhexyl) phthalate
542881	Bis(chloromethyl) ether
75252	Bromoform
106990	1,3-Butadiene
156627	Calcium cyanamide
105602	Caprolactam
133062	Captan
63252	Carbaryl

(*Continued*)

CAS #	Chemical or class
75150	Carbon disulfide
56235	Carbon tetrachloride
463581	Carbonyl sulfide
120809	Catechol
133904	Chloramben
57749	Chlordane
7782505	Chlorine
79118	Chloroacetic acid
532274	2-Chloroacetophenone
108907	Chlorobenzene
510156	Chlorobenzilate
67663	Chloroform
107302	Chloromethyl methyl ether
126998	Chloroprene
19773	Cresols/Cresylic acid (isomers and mixture)
95487	*o*-Cresol
108394	*m*-Cresol
106445	*p*-Cresol
98828	Cumene
94757	2,4-D, salts and esters
3547044	DDE
334883	Diazomethane
132649	Dibenzofurans
96128	1,2-Dibromo-3-chloropropane
84742	Dibutyl phthalate
106467	1,4-Dichlorobenzene(*p*)
91941	3,3′-Dichlorobenzidene
111444	Dichloroethyl ether (bis(2-chloroethyl) ether)
542756	1,3-Dichloropropene
62737	Dichlorvos
111422	Diethanolamine

CAS #	Chemical or class
121697	*N,N*-Diethyl aniline (*N,N*-dimethylaniline)
64675	Diethyl sulfate
119904	3,3-Dimethoxybenzidine
60117	Dimethylaminoazobenzene
119937	3,3-Dimethylbenzidine
79447	Dimethyl carbamoyl chloride
68122	Dimethyl formamide
57147	1,1 Dimethylhydrazine
131113	Dimethyl phthalate
77781	Dimethyl sulfate
534521	4,6-Dinitro-*o*-cresol, and salts
51285	2,4-Dinitrophenol
121142	2,4-Dinitrotoluene
123911	1,4-Dioxane (1,4-diethyleneoxide)
122667	1,2-Diphenylhydrazine
106898	Epichlorohydrin (L-chloro-2,3-epoxypropane)
106887	1,2-Epoxybutane
140885	Ethyl acrylate
100414	Ethyl benzene
51796	Ethyl carbamate (urethane)
75003	Ethyl chloride (chloroethane)
106934	Ethylene dibromide (dibromoethane)
107062	Ethylene dichloride (1,2-dichloroethane)
107211	Ethylene glycol
151564	Ethyleneimine (aziridine)
75218	Ethylene oxide
96457	Ethylene thiourea
75343	Ethylidene dichloride (1,1-dichloroethane)
50000	Formaldehyde
76448	Heptachlor

CAS #	Chemical or class
118741	Hexachlorobenzene
87683	Hexachlorobutadiene
77474	Hexachlorocyclopentadiene
67721	Hexachloroethane
822060	Hexamethylene-1,6-diisocyanate
680319	Hexamethylphosphoramide
110543	Hexane
302012	Hydrazine
7647010	Hydrochloric acid
7664393	Hydrogen fluoride (hydrofluoric acid)
123319	Hydroquinone
78591	Isophorone
58899	Lindane (all isomers)
108316	Maleic anhydride
67561	Methanol
72435	Methoxychlor
74839	Methyl bromide (bromomethane)
74873	Methyl chloride (chloromethane)
71556	Methyl chloroform (1,1,1-trichloroethane)
78933	Methyl ethyl ketone (2-butanone)
60344	Methyl hydrazine
74884	Methyl iodide (iodomethane)
108101	Methyl isobutyl ketone (hexone)
624839	Methyl isocyanate
80626	Methyl methacrylate
1634044	Methyl *tert*-butyl ether
101144	4,4-Methylene bis (2-chloroaniline)
75092	Methylene chloride (dichloromethane)
101688	Methylene diphenyl diisocyanate
101779	4,4′-Methylenedianiline
91203	Naphthalene
98953	Nitrobenzene

CAS #	Chemical or class
92933	4-Nitrobiphenyl
100027	4-Nitrophenol
79469	2-Nitropropane
684935	*N*-Nitroso-*N*-methylurea
62759	*N*-Nitrosodimethylamine
59892	*N*-Nitrosomorpholine
56382	Parathion
82688	Pentachloronitrobenzene (quintobenzene)
87865	Pentachlorophenol
108952	Phenol
106503	*p*-Phenylenediamine
75445	Phosgene
7803512	Phosphine
7723140	Phosphorus
85449	Phthalic anhydride
1336363	Polychlorinated biphenyls (aroclors)
1120714	1,3-Propane sultone
57578	Beta-propiolactone
123386	Propionaldehyde
114261	Propoxur (baygon)
78875	Propylene dichloride (1,2-dichloropropane)
75569	Propylene oxide
75558	1,2-Propylenimine (2-methyl aziridine)
91225	Quinoline
106514	Quinone
100425	Styrene
96093	Styrene oxide
1746016	2,3,7,8-Tetrachlorodibenzo-*p*-dioxin
79345	1,1,2,2-Tetrachloroethane
127184	Tetrachloroethylene (perchloroethylene)
7550450	Titanium tetrachloride

CAS #	Chemical or class
108883	Toluene
95807	2,4-Toluene diamine
584849	2,4-Toluene diisocyanate
95534	*o*-Toluidine
8001352	Toxaphene (chlorinated camphene)
120821	1,2,4-Trichlorobenzene
79005	1,1,2-Trichloroethane
79016	Trichloroethylene
95954	2,4,5-Trichlorophenol
88062	2,4,6-Trichlorophenol
121448	Triethylamine
1582098	Trifluralin
540841	2,2,4-Trimethylpentane
108054	Vinyl acetate
593602	Vinyl bromide
75014	Vinyl chloride
75354	Vinylidene chloride (1,1-dichloroethylene)
1330207	Xylenes (isomers and mixture)
95476	*o*-Xylenes
108383	*m*-Xylenes
106423	*p*-Xylenes
NA	Antimony compounds
NA	Arsenic compounds (inorganic, including arsine)
NA	Beryllium compounds
NA	Cadmium compounds
NA	Chromium compounds
NA	Cobalt compounds
NA	Coke oven emissions
NA	Cyanide compounds[1]
NA	Glycol ethers[2]
NA	Lead compounds
NA	Manganese compounds
NA	Mercury compounds
NA	Mineral fibers[3]
NA	Nickel compounds
NA	Polycyclic organic matter[4]
NA	Radionuclides (including radon)[5]
NA	Selenium compounds

[1] *X'CN where X = H' or any other group where a formal dissociation may occur. For example, KCN or* $Ca(CN)^2$.
[2] *Includes mono- and diethers of ethylene glycol, diethylene glycol, and triethylene glycol* $R-(OCH_2CH_2)_n-OR'$ *where n = 1, 2, or 3: R = alkyl or aryl groups; R' = R, H, or groups which, when removed, yield glycol ethers with the structure:* $R-(OCH_2CH)_n-OH$. *Polymers are excluded from the glycol category (see previous note regarding modification).*
[3] *Includes glass, rock, or slag fibers (or other mineral derived fibers) of average diameter 1 μm or less.*
[4] *Includes organic compounds with more than one benzene ring and which have a boiling point greater than or equal to 100 °C.*
[5] *A type of atom that spontaneously undergoes radioactive decay.*
Notes: For modifications to this list, visit: http://www.epa.gov/ttn/atw/pollutants/atwsmod.html (accessed on July 12, 2010). To date, the modifications have included methyl ethyl ketone, glycol esters, caporlactam, and hydrogen sulfide. A Joint Resolution to remove hydrogen sulfide, which was inadvertently included in the original Section 112(b)(1) list, was passed by the Senate on August 1, 1991 (Congressional Record page S11799), and the House of Representatives on November 25, 1991 (Congressional Record pages H11217–H11219). The Joint Resolution was approved by the President on December 4, 1991. Hydrogen Sulfide is included in Section 112(r) and is subject to the accidental release provisions.
For all listings above that contain the word "compounds" and for glycol ethers, the following applies: unless otherwise specified, these listings are defined as including any unique chemical substance that contains the named chemical (i.e., antimony, arsenic, etc.) as part of that chemical's infrastructure.

References

[1] U.S. Environmental Protection Agency, D.R. Shonnard, Chapter 3. http://www.epa.gov/oppt/greenengineering/pubs/whats_ge.html. Accessed on November 15, 2010.
[2] D.T. Allen, D.R. Shonnard, Green Engineering: Environmentally Conscious Design of Chemical Processes, Prentice Hall, Upper Saddle River, NJ, 2002.
[3] T. Loxnachar, K. Brown, T. Cooper, M. Milford, Sustaining Our Soils and Society. American Geological Institute, Soil Science Society of America, USDA Natural Resource Conservation Service, Washington, DC, 1999.

[4] V.P. Evangelou, Environmental Soil and Water Chemistry: Principles and Applications, John Wiley and Sons, Inc., New York, 1998.
[5] Council on Environmental Quality, National Environmental Policy Act. Available at: http://ceq.hss.doe.gov/, 2010. Accessed on July 15, 2010.
[6] Code of Federal Regulations, Part 40; CFR 1507.3.
[7] U.S. Environmental Protection Agency, *Report to Congress: Endocrine Disruptor Screening Program*, 2000.
[8] U.S. Environmental Protection Agency, *Endocrine Disruptor Screening Program*. Available at: http://www.epa.gov/endo/pubs/edspoverview/background.htm, 2010. Accessed on July 15, 2010.
[9] American University, The Trade & Environment Database. US-Bangladesh Waste Trade. Available at: http://www1.american.edu/TED/bengali.htm, 2010. Accessed on July 17, 2010.

CHAPTER

4

Waste Collection

Fraser McLeod, Tom Cherrett

Transportation Research Group, School of Civil Engineering and the Environment, University of Southampton, SO17 1BJ, United Kingdom

OUTLINE

1. INTRODUCTION

As will be described in this book, there are many different types and sources of waste, ranging from household garbage to debris found in outer space. This chapter will focus on the collection of municipal solid waste (MSW) which normally comprises waste from households, streets, parks, schools, hospitals and some commercial businesses. Although MSW represents only a relatively small proportion, by weight, of all waste that is generated, when compared with wastes from construction, demolition, manufacture, mining, quarrying and other industrial sources (Fig. 4.1), it is the most widespread and general waste collection problem faced by collection authorities the world over. MSW is also important from an environmental viewpoint because it contains organic materials (food waste) which may cause

Waste Doi: 10.1016/B978-0-12-381475-3.10004-X

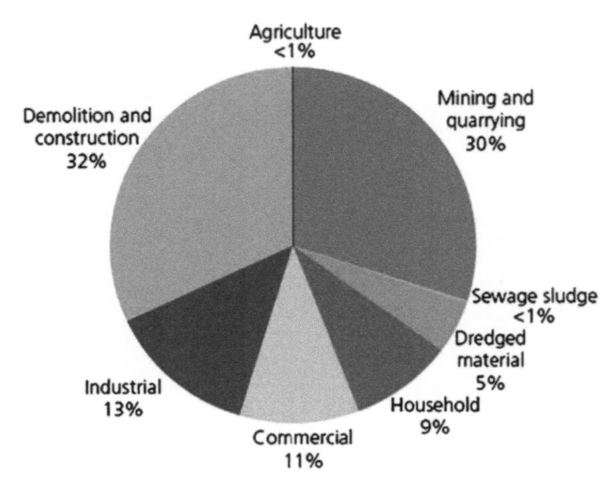

FIGURE 4.1 Waste collected in England by sector by weight. *Source*: *Ref.* [1].

pollution problems, as well as non-biodegradable materials such as glass and plastics.

The extent to which commercial waste is counted as municipal waste varies from country to country, which makes the comparison between countries difficult. Coggins and McIlveen [2] reported that in most of Europe, commercial and industrial waste from small businesses, that is similar in composition to household waste, is included in the definition of municipal waste. A current exception is the United Kingdom, although its Department for Environment, Food and Rural Affairs (Defra) [3] has announced plans to change this definition to bring the United Kingdom into line with its European neighbours. The proportion of MSW that emanates from residential properties was estimated, by the Environmental Protection Agency [4], to be 55%–65% for the United States as a whole; however, in a separate study by the Florida Department of Environmental Protection [5] this proportion was estimated to be only 33% for Florida, indicating wide variation in this statistic.

Municipal waste collection is normally the responsibility of some form of local government body. In the United Kingdom, these bodies are known as Waste Collection Authorities (WCAs) and this terminology will be used here for convenience. A WCA may have its own in-house waste collection team or, in many cases, the work is contracted out to a private company. Waste collection (and its disposal) is an expensive undertaking and can comprise a substantial proportion of a local government's expenditure. For example, in Metro Manila, Philippines, Westfall and Allen [6] estimated that over 64 million US$ was spent annually on the collection and disposal of MSW and these costs represented between 5% and 24%, with an average figure of 13%, of each local government's total expenditure. In a study produced for Germany's Federal Environment Agency (Umweltbundesamt) [7], waste collection costs were estimated to be between 60% and 80% of the total costs of waste collection and disposal. The latest statistics published by the statistical office of the European Union (EU) [8] show that an average of 524 kg of municipal waste was generated per person in 2008 across the 27 member countries of the EU, of which 40% was sent to landfill, 20% was incinerated, 23% was recycled, and 17% was composted [8]. In Florida, it was reported that around 1800 kg of MSW per resident per year was collected in 2007 [5], however, this included some construction and demolition waste that would normally be excluded from the definition of MSW. The amount of MSW generated worldwide in 2006 was estimated by Market Research.com [9] to be just over two billion tonnes (2×10^9 t, where t refers to metric tonne).

Municipal waste collection is becoming an increasingly complicated task for WCAs to undertake, with the collection requirements of the different waste streams (dry recyclables, yard waste, glass, residual waste, etc.) varying from year to year, as recycling performance improves. For example, in the United States, in 2008, some 82.9 million tons (75.2×10^6 t) of MSW was recycled (33% recycling rate) compared to 55.8 million tons (50.6×10^6 t)

(26% recycling rate) in 1995 [4]. These changing conditions mean that waste collection round designs need to be updated regularly.

Well-designed collection rounds are imperative to minimise operating costs and ensure that maximum value is gained from the collection vehicles and related infrastructure. Environmental considerations are also becoming more widely recognised: for example, current government policy for England [10] states that '... better collection and treatment of waste from households and other sources has the potential to increase England's stock of valuable resources whilst also contributing to energy policy. And achieving both of these aims helps reduce greenhouse gas emissions'. Due to the large numbers of variables, objectives and constraints, designing efficient waste collection rounds is a complex task and, consequently, optimising round structures has received a considerable amount of attention, both from the research community and from practitioners seeking to improve their performance and reduce costs. This chapter aims to draw together the research and practice to provide a summary of waste collection methods and how to use modelling methods to aid design of collection rounds and estimate costs.

2. MATERIALS COLLECTED

MSW comprises a wide range of materials. As examples of this, recent analyses of MSW in the United States [4] and the United Kingdom [11] have provided the waste compositions shown in Table 4.1. Some caution is needed in comparing the figures from the two countries as the category definitions in each case were slightly different: for example, for the United Kingdom, the 'other' category used included furniture (1.3%) and WEEE (waste, electronic and electrical equipment) (2.2%), which would clearly contribute towards wood, metal and other waste categories.

TABLE 4.1 Municipal Solid Waste Composition

	US (%)	UK (%)
Paper and card	31	23
Food	13	18
Garden/yard waste	13	14
Plastics	12	10
Metals	8	4
Textiles	8	3
Wood	7	4
Glass	5	7
Other	3	17

The above analyses give an indication of the difficulties involved in separating the various materials for subsequent disposal, recycling, or treatment and it is not surprising that there are many different methods used for collecting the materials. In countries with well-developed recycling policies and systems (e.g. Germany, Switzerland, Austria, Norway, Sweden, United States), householders are typically engaged in the recycling effort by being supplied with separate bins or containers for different materials. The number of different containers that a householder can be reasonably expected to keep is limited. Coggins and McIlveen [2] recommended that a maximum of three containers should be used: one for dry recyclable waste (paper, cardboard, plastic bottles etc.), one for food waste, and one for residual waste.

3. COLLECTION SYSTEMS

There are many different types of kerbside collection schemes, with wide variations in the types of recyclable materials targeted by individual authorities. This emanates from the different characteristics of the households within an area and the inter-relationships between the various collection systems, sorting methods, and frequency of collections. Waste collection frequencies of recyclable materials in the United Kingdom can be weekly, fortnightly or monthly. Increasingly common practice in the United Kingdom is alternate weekly collection (AWC), where each household has its recyclable waste and residual waste collected every fortnight on alternating weeks. A common alternative scheme is to have weekly collection of residual waste and fortnightly collection of recyclable waste. AWC can be justified from the viewpoints of reducing collection costs and encouraging more recycling from householders [12], although some people object to the possible health risks associated with a two-week collection frequency for residual waste. Recyclable materials can be collected together in commingled collections in one single-bodied vehicle, to be separated later, or be separated by the householder for loading into separate compartments in a vehicle (Fig. 4.2) through a 'kerbside-sort' scheme. Recyclables can also be mixed with the residual waste and collected by the same vehicle using different coloured sacks to differentiate the materials. The WCA may also offer separate collection of bulky waste (e.g. furniture, white goods), green garden waste or hazardous waste, normally at an additional charge to the customer.

FIGURE 4.2 Kerbside sorting of different recyclable materials. *Source*: http://photolibrary.recyclenowpartners.org.uk/photolibrary/index.php

In addition to collections from individual houses, WCAs may also have to collect from public litter bins and communal bins. Public litter bins located in shopping streets, parks and so on generally need to be emptied frequently, and in the United Kingdom this is normally done separately from household collections. Communal bins are relatively large bins that may be used where there is a high density of residential properties (e.g. apartment blocks) or where the collection authority decides it would be too impractical or expensive to visit properties individually. Communal bins are also commonly located at public amenity sites such as parking lots, supermarkets or at dedicated waste drop-off sites (known in the United Kingdom as 'household waste recycling centres'). Different bins can be used to collect different materials (e.g. bottle banks, clothing banks etc.).

The collection requirements for communal bins are rather different from those for household bins. The key information needed is the expected rate of fill of each bin to allow efficient collection rounds to be designed and to avoid the bin becoming full before the next collection is made, as illegal dumping of waste may be exacerbated if a communal bin becomes full. The expected fill rate may be estimated from historical data, or the bins could be equipped with sensors to detect when they are becoming full. The advantages of dynamic waste collection schedules over fixed schedules were demonstrated in Malmö, Sweden, reported by Johansson [13], where 3300 recycling containers were fitted with level sensors and wireless communication equipment, thereby giving waste collection operators access to real-time information on the status of each container.

In the United Kingdom, collections made from commercial premises are normally undertaken separately from household collections. This is mainly due to the current legislation that excludes commercial waste from being counted towards MSW recycling targets. WCAs are not obliged to offer trade waste collection services and, indeed, many in the United Kingdom do not, and the waste is collected by private waste contractors. In principle, where the waste generated by businesses is similar in composition to household waste (e.g. paper, cardboard), there may be operational benefits available through combining the collections together. However, there may be scheduling difficulties associated with the larger businesses requiring more frequent collections (e.g. daily), whereas household collections are typically on a weekly basis.

Pay-as-you-throw (PAYT) — recycling by householders can be encouraged by variable charging for residual waste collection according to the weight (or volume) collected. There are about 7000 such schemes in operation in the United States with increases in recycling between 25% and 69% in the first year being reported [5]. An alternative scheme is to provide householders with some form of incentive to recycle. For example, in Miami, Florida, a scheme was launched in March 2009 whereby householders received coupons for use at local retail stores according to the amount of waste they recycled. Although this was reported to increase recycling in areas where there was little before [5], intuitively it seems likely that the 'carrot' approach of providing incentives would be less effective than the 'stick' approach of PAYT.

3.1. Vehicles Used

The vehicles used for waste collection range from relatively small handcarts, in developing countries, to state-of-the-art refuse collection vehicles (RCVs) equipped with hydraulic bin-lifting gear. The simplest form of bin-lifting equipment requires the bin to be mounted onto a hoist by an operator, typically at the rear of the vehicle (Fig. 4.3). Vehicles that load bins from the front or from the side are also available, with the former often used to empty containers from commercial premises.

FIGURE 4.3 Rear-loading vehicle. *Source*: http://photolibrary.recyclenowpartners.org.uk/photolibrary/index.php

Multi-compartment vehicles are sometimes used to collect different materials at the same time, for example, recyclable waste alongside residual waste. While collecting different materials at the same time may bring about operational efficiencies — for example, a cost reduction of 15% was reported for a case study in Brescia, Italy [14] — their use may not be appropriate in some situations, particularly when the different materials collected have to be delivered to different locations that are far apart, or when the volumes to be collected are variable. The latter situation can lead to one vehicle compartment reaching capacity quicker than the other, requiring the vehicle to empty mid-round, which may not be optimal.

Vehicle capacities for carrying waste vary widely and are clearly a major factor affecting collection costs. Typical in the United Kingdom is a carrying capacity of between 8 and 12 tonnes for the collection of household residual waste. The corresponding weight of recyclable waste that can be collected by the same vehicle tends to be about 25% less because recyclable materials tend to be less dense than residual waste, even after compaction. From the point of view of collection efficiency, the vehicle used should be as large as possible, to reduce the number of trips made to a waste disposal site, recycling facility or transfer station; however, there are practical limitations on the size of vehicle that may be used both in urban and rural areas, associated with road widths (e.g. narrow streets), weight and other access restrictions. Where transfer stations (intermediary storage and waste consolidation areas) are used, the onwards movement of waste from the transfer station to the end disposal site tends to be undertaken by large road vehicles or, in some cases, the transfer station may have intermodal facilities to enable onwards movement of waste by rail (e.g. train, tram) or by water (e.g. canal barge, short sea shipping) [15].

4. MODELLING PROBLEMS AND METHODS

Waste collection is a multi-faceted problem that can be considered at different levels, from the high level, strategic, policy or planning viewpoints, such as where to locate facilities, how to define collection boundaries, which materials to collect and how often to collect them, to more low level, tactical or operational decisions, such as when to visit the waste disposal site on a vehicle round or which route to take to avoid traffic congestion. Some of the main issues are now discussed.

4.1. Locating Facilities

The locations of waste disposal sites, transfer stations, recycling centres and other facilities associated with waste disposal, in relation to each other and in relation to the households and businesses being served, will clearly have a significant impact on waste collection costs. Ideally, these costs should be estimated and included whenever locations for new facilities are being decided; however, this is easier said than done, as the combined facility location and vehicle routing problem is very difficult to solve, and heuristic methods tend to be based on iteration between the two separate problems [16–18]. Also, where new disposal sites are being considered, the main deciding factors will tend to be related to environmental, land use and land availability considerations rather than to transport costs, which are still relatively cheap, in comparison with real estate costs.

The location of the vehicle depot also has a significant impact on waste collection costs. Although vehicle depot location theory is usually based on placing the depot at the centre of gravity of the population being served [19], the waste collection requirements associated with visiting a waste disposal site or a recycling centre before returning to the vehicle depot changes the nature of the problem significantly. A study by the authors [12] suggested that to minimise waste collection vehicle costs, the vehicle depot should be placed as close as possible to the main waste disposal site, or transfer station, being used: a vehicle mileage savings of 13.5% was estimated when vehicles were moved from their existing depots to a hypothetical depot based at the waste disposal site for three adjoining WCAs in the county of Hampshire in England.

4.2. Districting

The term 'districting' means to sub-divide a larger region into a number of smaller districts, with the waste collection undertaken in each district being considered as separate problems. This simplifies the optimisation task, thereby reducing computational requirements, and was shown by Muyldermans et al. [20] to introduce only a small degree of sub-optimality. Districting theory is based on producing compact districts with evenly balanced workloads [21]. In a case study from Antwerp, Belgium, Muyldermans et al. [22] reported a savings of 14% in deadhead mileage through improvements made to the existing district definitions.

4.3. Defining the Collection Points

As the number of individual households in a waste collection authority area is usually large, typical modelling practice is to group households together in some way to form a reduced number of 'collection points'. Aggregation levels that have been used in waste collection modelling include (i) street level, particularly where an arc-routing model is used with arcs corresponding with streets [23]; (ii) postcode level [12]; (iii) block level [24] and (iv) macro-point level [25], defined as a cluster of collection points close to each other and sharing similar properties in terms of waste collection requirements. The volume of waste to be collected from each collection point can be estimated based on the numbers of premises at each collection point and the total volume of waste that is normally collected on a round. These estimates may also have to take the socio-demographic characteristics of the properties or areas into account, if there are wide differences between areas, for example, in the type of housing (e.g. flats, size of property) or type of household (e.g. size of family, wealth) [26].

4.4. Vehicle Routing and Scheduling

Vehicle routing and scheduling theory is vast, covering many different types of problem and situation; Cordeau et al. [27] and Golden et al. [28] provide recent reviews of the state-of-the-art. Although some of the algorithms proposed in the literature may be applied directly to the waste collection problem, there are a number of characteristics of waste collection that may limit their use or require modifications to be made. One such characteristic is the typical requirement for the collection vehicle to make more than one visit to a waste disposal site during the round. Other factors that may have to be taken into consideration include the following:

- Avoiding main roads at peak traffic periods
- Avoiding schools at opening and closing times
- Road network restrictions pertaining to vehicle weight, height, turning or other access restrictions
- Whether access is from the front or rear of the property
- Collecting from each side of a street separately where it would not be safe for the crew to walk back and forth across the street
- Avoiding awkward turning movements – for example, for countries that drive on the right hand side of the road, some left turns may be difficult due to oncoming traffic
- One-way streets – it is likely to be most efficient if any one-way streets are entered at their furthest upstream points
- Steep hills – from fuel consumption and emissions perspectives, steep hills are best climbed near the start of the round when the vehicle is not carrying much load. Also, where feasible, waste on a steep hill is best collected on both sides of the street while the vehicle is moving downhill to facilitate safety, ease and speed of collection and to reduce wear and tear on the vehicle [29].
- Time constraints include the waste disposal site opening hours and the staff working hours. Collections from commercial premises may also have to be made within stated time windows.

The 'waste collection problem' normally involves using a fleet of collection vehicles to undertake specified collections at the minimum possible cost, where the main operating costs relate to time taken and distance travelled. Related objectives may also include the need for rounds that are compact and well balanced, in terms of fair distribution of the work among the collection rounds, although Li et al. [30] reported a case in Brazil where balanced use of the waste disposal points was also desired.

Whichever waste collection problem is specified, it is normally difficult to solve. Even if constraints such as those listed above are not considered, the problem of using capacity-constrained vehicles cannot be solved optimally for more than about 100 collection points; however, the best models are able to obtain good solutions for large numbers of collection points [27]. As the problem is not generally solvable, many alternative heuristic methods have been proposed based on a variety of techniques such as branch-and-bound, tabu search, insertion, removal and clustering methods, fuzzy logic and genetic algorithms. Examples of heuristic algorithms that have been developed and applied to waste collection problems include those by De Rosa et al. [23], Viotti et al. [31], Bautista and Pereira [32], Kim et al. [33] and Nuortio et al. [34]; however, it is outside the scope of this chapter to explain the individual methods used. There are also many commercial vehicle routing and scheduling software packages available, employing such techniques, that are able to suggest new round designs and vehicle routes that may improve significantly upon existing ones, particularly where the current rounds have not been updated for some time. These packages may include parameters that affect how the algorithms operate, for example, controlling how collection points are grouped to form clusters. In some cases, there may be a trade off between run time and accuracy and some trial and error experimentation with the settings may be needed.

One area where vehicle routing and scheduling models may improve upon manually derived round structures is the choice of 'tipping' points on the round, that is, the points at which a trip to the waste disposal site (tip) is made. Intuitively, these will tend to be made when the vehicle is almost full; however, this may not be the best strategy: it may be better to tip when the vehicle is closest to the waste disposal site but not full. This will depend, naturally, on what spare capacity there is on the round as to whether or not this is feasible. Vehicle routing and scheduling models can be useful for suggesting round designs that may not occur to the waste collection manager. For example, in a study undertaken by the authors for the UK Department for Transport [35], the collection time for one round was reduced from 8 to 7 hours by improving the tipping points and reconfiguring the route (Figs 4.4a and b).

On the other hand, it should be recognised that vehicle routing and scheduling models, like all models tackling complex subjects, are simplifications of reality. Areas where models may be lacking in accuracy include the following:

1. The definition of the collection points (e.g. street, postcode), as described above, which introduces an element of coarseness in the model.
2. The specification of the available road network, including any vehicle weight, height, turning movement, access and other restrictions.
3. Assumptions made about travel speeds.
4. Assumptions made about volume of waste to be collected.
5. Assumptions made about the time needed to collect the waste.

In many cases, a lack of available detailed data and the expense of obtaining such data limit the model accuracy. This means that routes and schedules produced by models generally have to be checked and modified by an experienced manager with good local knowledge before implementation.

5. DATA REQUIREMENTS FOR MODELLING

The main data requirements for modelling waste collection relate to the volume of waste to be collected, the time needed for loading and unloading waste, and the travel time between collection points.

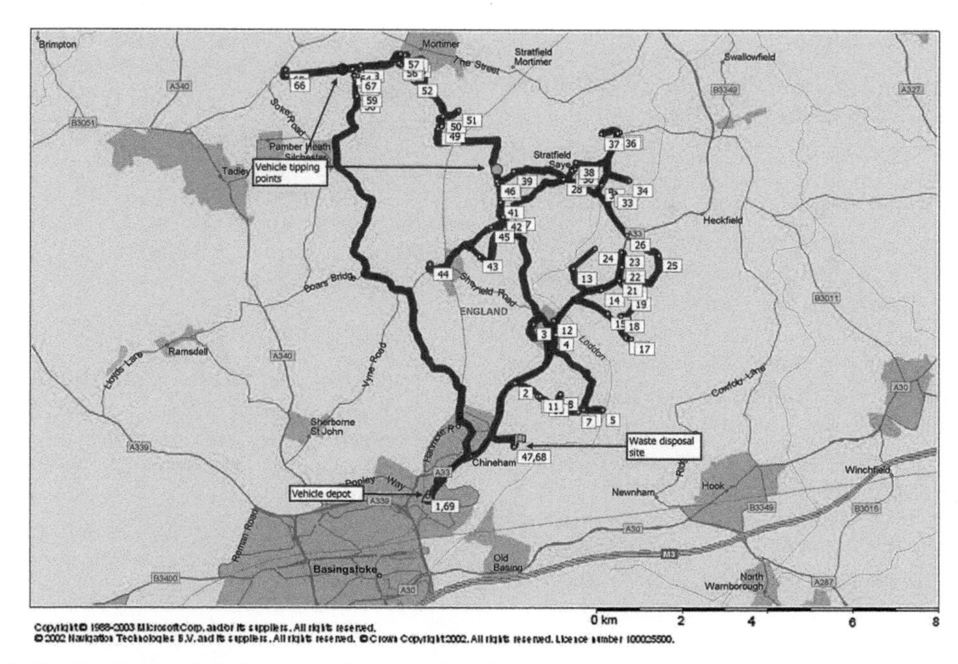

FIGURE 4.4A Original modelled route with poorly placed tipping points.

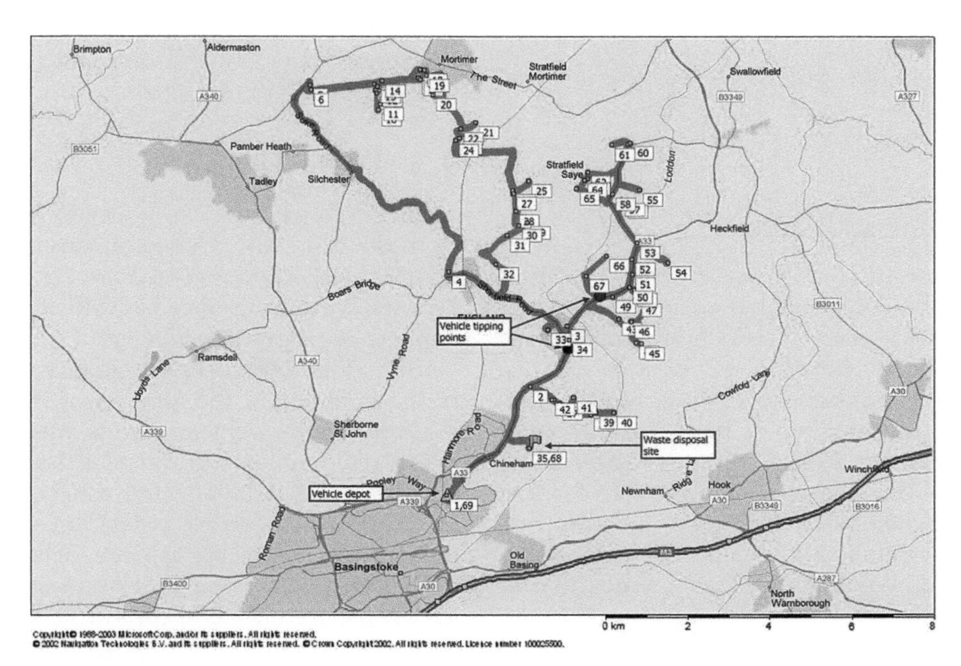

FIGURE 4.4B Redesigned route with tipping points closer to the waste disposal site.

5.1. Waste Volume

Waste volume (or weight) data are normally collected by WCAs for reporting and performance management purposes. These data show how the amount of waste collected varies over time. The amount of waste to be collected at any given time is clearly an important consideration in planning vehicle round structures and routes. The traditional approach to work with variable data has been to estimate average figures and treat the problem as a static, deterministic one. This may not be satisfactory, however, if the routes and schedules produced are infeasible on days when the amount of waste to be collected is above average. For this reason, some slack normally has to be built into the design to be able to cope in such situations. Another approach is to try to model the variability explicitly, as proposed by Albareda-Sambola et al. [18], Eisenstein and Iyer [24] and Chang and Wei [36]. The dynamic routing and scheduling approach using global positioning system (GPS) and weight sensors, reported by Johansson [13] and mentioned earlier in this chapter, is another option, however, this is a level of sophistication some way ahead of existing practice.

5.2. Loading and Unloading Times

An important parameter common to most models is the time needed to load the waste into the vehicle at each collection point. This is likely to be a function of various factors, including the number of properties at the collection point, the number of bins, bags or other units of waste set out for collection, the volume of waste and walking distances between the vehicle and the bins (or bags etc.). This information is not normally known but can be estimated from round times, taking travel times to and from the waste disposal sites into account, and from numbers of properties, bins, etc. For example, in a waste collection study undertaken by the authors for the UK Department for Transport [35], loading times of residual waste from domestic properties in Hampshire, UK were estimated as 1.2 s kg^{-1}, which equated to about 14 s per bin, whereas in a study undertaken by Everett and Shahi [37], loading times of yard waste (described as grass, leaves and brush) in Oklahoma, were estimated as (in units of seconds):

$$(\mathrm{SOR}) \times (\mathrm{NOR})\{9.23 + (11.6 \times \mathrm{AU})\}$$

where SOR = set out fraction (the proportion of households putting waste out), NOR = number of residences at the collection point, AU = average number of units (bin, bag, bundle) at a residence.

The unloading time at a waste disposal site can either be modelled as a fixed average time: for example, a survey by Mazzotti [38] of 628 vehicle trips at a landfill site near Southampton, UK gave an average turnaround time of 11 minutes; or variability may be explicitly modelled by taking any waiting factors into account: for example, Bhat [39] considered the variability due to the number of vehicles queuing to use the facilities.

5.3. Travel Times

Travel times between collection points and from the collection area to and from the waste disposal site, can be estimated via observation of typical running speeds. Vehicle speed will depend on the size and engine power of the vehicle, the road network (e.g. road widths, traffic lights), and the amount of traffic on the road. Normally, vehicle rounds are designed to avoid main roads during peak traffic periods both to avoid delays to the collection vehicles and to avoid receiving complaints about their slow-moving vehicles causing delays to other road users. This can be done by assuming significantly reduced travel speeds on certain roads at certain times of day, thereby making them unattractive for use. An even more sophisticated approach is to incorporate real-time traffic data into the waste collection model, as done by Taniguchi and Shimamoto [40].

6. EXAMPLE STUDIES

Some example case studies from around the world are described to give an indication of the benefits that may be possible through the use of waste collection routing and scheduling models. Comparisons between studies are difficult to make as the operating conditions tend to be unique to each and as the results are very much dependent on the quality of the original routes that were used in the base cases. It also seems likely that in many cases, reported results may present an overly optimistic picture, as various practical restrictions or modelling limitations may not have been fully considered.

6.1. Hampshire, UK

Commercially available vehicle routing and scheduling software was used to assess benefits associated with joint working between three neighbouring WCAs in Hampshire, UK (Basingstoke and Deane, Hart and Rushmoor), comprising around 130,000 households. This joint working effectively removed the existing collection boundaries between the authorities and allowed vehicles to be moved to a neighbour's depot. A total of 25 rounds (=25 vehicle days) were selected as being most suitable for joint working, as they lay closest to the existing boundaries between the authorities. Redesigning these rounds resulted in one fewer vehicle days' work; a time savings of 1.4% (approximately 6 minutes per vehicle day); a vehicle mileage savings of 5.9% (approximately 3.3 km per vehicle day), and an annual distance savings of around 4300 km, which was estimated to be equivalent to an annual cost savings of around £35,000 and a carbon savings of around 2.1 t [35].

6.2. Taipei City, Taiwan

Joint working between Taipei City (12 districts in the centre of the region) and Taipei County (29 districts surrounding the city), Taiwan, with a total population of around 6 million and served by a waste collection fleet of just over 1000 vehicles, was estimated to achieve annual savings of around US$ 4×10^6 over the existing system where all 41 districts operate independently. This estimate was based on a model devised by Chang et al. [41] which employed specific goal constraints in an integer programming model.

6.3. Porto Alegre, Brazil

Li et al. [30] reported an interesting slant on the typical waste collection problem where, for social reasons, one of the objectives was to balance the rounds to make equal use, as far as possible, of recycling centres run by cooperatives whose members are poor and who gain financially through use of their facilities. Their heuristic approach, which incorporated an auction algorithm and a dynamic penalty method, was shown to reduce the number of truck days per week required from 103 to 77, with the total distance travelled in a week reducing from 2418 to 1760 km (27% savings), when compared with the existing rounds. The stated objective of balancing the use of the recycling centres was also achieved.

6.4. Finland

Nuortio et al. [34] presented waste collection vehicle routing and scheduling results, based on a 'guided variable neighbourhood thresholding metaheuristic', for three real-life waste collection problems in eastern Finland. Vehicle distance savings were 4% in one of their case studies and 44% in the other two, although they admitted that the latter savings were partly due to some clearly inefficient operating practices in the 'before' case.

7. CONCLUSION

Waste collection is a multi-faceted and highly complex problem with no easy solutions. Common solutions are also difficult to identify due to the huge diversity of operating conditions throughout the world. It is very much in the interests of WCAs worldwide to devise efficient working methods to keep their operating costs as low as possible while providing a good level of service to householders, trade customers and other interested parties. This chapter has provided an introduction to the waste materials that are collected, the collection systems that are typically used and the types of vehicles and other equipment used. It has also described vehicle routing and scheduling, in terms of its capabilities, limitations, and data requirements and has provided some case study examples from around the world to give an indication of the range of waste collection problems and of what benefits may be achievable through the use of waste collection vehicle routing and scheduling methods.

References

[1] UK Department of Environment, Food, and Rural Affairs (Defra). Commercial and industrial waste in England. Statement of aims and actions. <http://www.defra.gov.uk/environment/waste/topics/documents/commercial-industrial-waste-aims-actions-091013.pdf> (2009)

[2] C. Coggins, R. McIlveen (ed. B. Caldecott). A wasted opportunity? How to get the most out of Britain's bins. Policy exchange <http://www.policyexchange.org.uk/images/publications/pdfs/A_wasted_opportunity_1.pdf> (2009)

[3] UK Department of Environment, Food and Rural Affairs (Defra). Changing the UK approach to the EU landfill diversion targets. <http://www.defra.gov.uk/environment/waste/strategy/legislation/landfill/targets.htm> (2009)

[4] US Environmental Protection Agency. Municipal solid waste generation, recycling and disposal in the United States: facts and figures for. <http://www.epa.gov/osw/nonhaz/municipal/pubs/msw2008rpt.pdf> (2008).

[5] Florida Department of Environmental Protection. 75% Recycling goal report to the Legislature. <http://www.dep.state.fl.us/waste/quick_topics/ publications/shw/recycling/75percent/75_recycling_report.pdf> (2010)

[6] M. Westfall, M. N. Allen. The Garbage Book. Asian Development Bank. <http://www.adb.org/documents/books/garbage-book/> (2004)

[7] Umweltbundesamt. Best practice municipal waste management. <http://www.umweltbundesamt.de/abfallwirtschaft-e/best-practice-mwm.htm> (2009)

[8] European Commission Eurostat. Municipal waste generated, kg per capita (Structural Indicator). <http://epp.eurostat.ec.europa.eu/portal/page/portal/ waste/data/sectors/municipal_waste> (2010)

[9] Market Research.com. Global waste management market assessment. <http://www.marketresearch.com/product/display.asp?productid=1470786> (2007)

[10] UK Department of Environment, Food and Rural Affairs (Defra). Waste strategy for England. <http://www.defra.gov.uk/environment/waste/strategy/strategy07/index.htm> (2007)

[11] UK Department of Environment, Food and Rural Affairs (Defra). Municipal waste composition: review of municipal waste component analyses – WR0119. <http://randd.defra.gov.uk/Default.aspx?Menu=Menu&Module=More&Location=None&Completed=0&ProjectID=15133> (2008)

[12] F.N. McLeod, T.J. Cherrett, Waste Management 28 (2008) 2271–2278.

[13] O.M. Johansson, Waste Management 26 (2006) 875–885.

[14] R. Mansini, M.G. Speranza, Computers and Operations Research 25 (1998) 659–673.

[15] STRAW (Sustainable Transport Resources and Waste). Intermodal infrastructure maps. <http://www.straw.org.uk/intermodal/> (2005)

[16] H. Min, V. Jayaraman, R. Srivastava, European Journal of Operational Research 108 (1998) 1–15.

[17] G. Nagy, S. Salhi, European Journal of Operational Research 177 (2007) 649–672.

[18] M. Albareda-Sambola, E. Fernández, G. Laporte, European Journal of Operational Research 179 (2007) 940–955.

[19] J. Hayford, Journal of the American Statistical Association 8 (1902) 47–58.

[20] L. Muyldermans, D. Cattrysse, D.V. Oudheusden, Journal of Operational Research Studies, 54 (2003) 1209–1221.

[21] S. Hanafi, A. Freville, P. Vaca, Information Systems and Operational Research (INFOR) 37 (1999) 236–254.

[22] L. Muyldermans, D. Cattrysse, D. Van Oudheusden, T. Lotan, European Journal of Operational Research 139 (2002) 521–532.

[23] B. De Rosa, G. Improta, G. Ghiani, R. Musmanno, Transportation Science 36 (2002) 301–313.
[24] D.D. Eisenstein, A.V. Iyer, Management Science 43 (1997) 922–933.
[25] E. Angelelli, M.G. Speranza, Journal of Operational Research Studies 53 (2002) 944–952.
[26] M. Purcell, W.L. Magette, Waste Management 29 (2009) 1237–1250.
[27] J.F. Cordeau, G. Laporte, M.W. Savelsbergh, D. Vigo, B. Cynthia, L. Gilbert, Handbooks in operations research and management science, Chapter 6, Vehicle routing, Elsevier, 2007.
[28] B. Golden, S. Raghavan, E. Wasil (Eds.), The vehicle routing problem: latest advances and new challenges, Springer, 2008.
[29] The Asia Foundation. Solid waste collection and transport: service delivery training module 1 of 4. <http://asiafoundation.org/publications/pdf/499> (2008).
[30] J.-Q. Li, D. Borenstein, P.B. Mirchandani, Omega 36 (2008) 1133–1149.
[31] P. Viotti, A. Polettini, R. Porni, C. Innocenti, Waste Management and Research 21 (2003) 292–298.
[32] J. Bautista, J. Pereira, Ant algorithms for urban waste collection routing. In Ant colony, optimisation and swarm intelligence, Berlin, Springer, 2004.
[33] B.-I. Kim, S. Kim, S. Sahoo, Computers and Operations Research 33 (2006) 3624–3642.
[34] T. Nuortio, J. Kytöjoki, H. Niska, O. Bräysy, Expert Systems with Applications 30 (2006) 223–232.
[35] Department for Transport. Optimising Vehicles Undertaking Waste Collection. Research study PPAD 9/142/024, London (2006).
[36] N.B. Chang, Y.L. Wei, Fuzzy Sets and Systems 114 (2000) 133–149.
[37] J.W. Everett, S. Shahi, Waste Management and Research 15 (1997) 627–640.
[38] A. Mazzotti, A critical assessment of a routing and scheduling application for optimising domestic waste collections in Hampshire, M.Sc. dissertation, Transportation Research Group, School of Civil Engineering and the Environment, University of Southampton, 2004.
[39] V.N. Bhat, Waste Management and Research 14 (1996) 87–96.
[40] E. Taniguchi, H. Shimamoto, Transportation Research C: Emerging Technologies, 12 (2004) 235–250.
[41] N.B. Chang, Y.H. Chang, Y.L. Chen, Journal of Environmental Engineering 123 (1997) 178–190.

PART II

WASTE STREAMS

Waste can be characterized in many ways. Most books on waste tend to look at the sources of the waste as the primary means of classification. Certain sectors indeed generate common waste types. Examples include slag and gob piles from mining and extraction, high organic matter content wastes from rendering and agricultural sectors, and low-level and high-level nuclear wastes from nuclear electric generating and defense-related sectors. Waste may also be subdivided by the problems they cause, such as those that require large expanses and depths of soil (solid waste) and those that may lead to specific types of disease (hazardous waste). Other wastes are classified based on what they contain, such heavy metals, volatile organics, isotopes and pathogens.

No matter how they are classified, wastes come in many forms and here our journey breaks up into many streams, each related to a type of waste with its unique mode of generation, recycling possibilities, magnitude and life history.

CHAPTER

5

Mine Waste: A Brief Overview of Origins, Quantities, and Methods of Storage

Geoffrey Blight

School of Civil and Environmental Engineering,
University of the Witwatersrand, Johannesburg, South Africa

OUTLINE

1. ORIGINS AND QUANTITIES OF MINE WASTE

With an expanding world market for mineral commodities such as chrome, coal, copper, diamonds, fluorspar, gold, iron, manganese and zinc, so necessary for the functioning of the modern world, mining companies are exploiting ever low-grade ore bodies on an ever-increasing scale. Mining on a vast scale is usually necessary for profitability of a lower-grade mine, and volumes of waste are commensurately large.

Waste Doi: 10.1016/B978-0-12-381475-3.10005-1

The actual volume of mine waste that has to be disposed of in dumps and tailings storage facilities, worldwide, is difficult to assess. In 1996, the International Commission on Large Dams (ICOLD) [1] gave an estimate of "almost certainly exceeds 5 thousand million tonnes per annum" (5×10^9 t a^{-1}). Considering that some valuable commodities occur in their ores in concentrations of grams or carats per ton (1 carat = 0.2 g, 1 tonne = 1000 kg), and that many individual mines extract in excess of 50 million tons of ore per year (5×10^7 t a^{-1}), even the estimate of ICOLD is probably too low. For example, a single platinum tailings storage at Rustenburg, South Africa, has a storage capacity of almost 1×10^9 t over a life of 50 years [2]. The mine currently sends 0.5×10^6 t of tailings to storage every month, and plans to increase this to 2×10^6 t every month.

More recent estimated quantities of mine waste are as follows:

- The world's iron, copper, gold, lead, and bauxite (aluminum) mines together generated 35×10^9 t of waste in 1995 alone [3].
- The South African gold mining industry produced 7.4×10^5 t of gold tailings in the decade from 1997 to 2006, that is, 7.4×10^4 t a^{-1} [4].
- All gold mining waste produced in the past century in South Africa amounts to 6×10^9 t, which covers a total area of 400 to 500 km^2, and contains 4.30×10^5 t of uranium and 3.0×10^4 t of sulfur, both of which, and especially the sulfur, have a high pollution potential [5].

The term "mine waste storage" is preferred to the more common "mine waste disposal," because advances in extractive metallurgy and increased demand and price for a commodity periodically coincide to allow a particular mine waste deposit to be reworked and further resources to be extracted from it at a profit. As examples, some gold mine waste storages in South Africa have been re-mined and reprocessed three times in the past 100 years. Some of these deposits started out as waste rock which was unprofitable to process at the time, but was necessary to remove to access richer ores. Some platinum mines are now considering reprocessing their older tailings storages for platinum and other minerals, and coal mines are reprocessing their old coal discard storages. Thus, there is a realization that mine waste deposits are really storages of low-grade ore. They do not consist of the waste they were formerly considered to be. Even if the grade of mineral they contain remains too low for economical extraction, there may be a present or future economic value for other minerals in what is presently waste, or as a construction material.

It is for this reason that there is a great reluctance in the mining industry to "dispose" of waste by placing it in locations that render the waste inaccessible for future reprocessing, for example in "worked out" parts of a mine, except where it can be used for strata support. In time to come, not only may the waste become profitable to reprocess but the stopes themselves may also be worth re-mining to remove seams or reefs of ore previously regarded as uneconomical to mine.

Ore bodies usually contain more than one type of mineralization. As examples, ore mined primarily for its platinum content, usually also contains chrome, used in stainless steel manufacture, and apatite from which phosphate fertilizers can be produced. At least one platinum mine in South Africa also extracts chrome as a by-product and supports a phosphoric acid plant for fertilizer manufacture. Another example relating to by-products is the South African gold mining industry which produced uranium as a by-product, much of which contributed to the Hiroshima and Nagasaki bombs during World War II. This activity stopped in the mid-1960s when the price of uranium fell, but it is now starting up again as it is being realized that nuclear power will have to be used in the future to replace dwindling coal and oil supplies. Figure 5.1 shows a dump of waste sand, from early mining operations (1885–1905), in the

FIGURE 5.1 A 100-year-old dump of gold tailings sand being re-mined for processing to extract the uranium and residual gold content.

process of being removed for reprocessing to extract gold and uranium left in the sand after the initial extraction process, 100 years ago.

Some mine waste can be recycled for other purposes. For example, waste rock can be used as a fill material in civil engineering works, or if the rock is sound, durable, and unweathered and has a satisfactory mineralogy, as aggregate for concrete and in asphalt and road layer works. Gypsum, a by-product of fertilizer manufacture from apatite rock, can be used to make building boards or can be reprocessed to produce sulfuric acid and building cement.

2. WASTE CHARACTERISTICS

Mine waste may arise in a number of forms: as stripped soil and coarse, broken, partly weathered rock overburden in open cast or strip-mining operations; as unweathered development waste rock in underground mining; and as fine-grained tailings, the residuum of the process of comminution and mineral-extraction from ores. The various wastes are usually stored separately. The top-soil is stock-piled for eventual use in environmentally rehabilitating the dumps of coarse wastes, or the surfaces of backfilled surface-mining voids. The coarse broken rock is usually stored in dumps, either with or without compaction, or is used to progressively backfill opencast or strip-mining voids.

Figure 5.2 shows typical particle size analyses for a range of tailings from various sources. In the diagram, the horizontal axis represents the particle size in millimeters and the vertical axis represents the proportion of material finer than a specific size. For example, for the vanadium tailings, 63% is finer than 0.06 mm, that is, the tailings contain 63% of silt and clay-sized particles.

In addition to its physical characteristics, mine waste and especially tailings may have characteristics or contain substances that may be prejudicial to the health of those living near the waste storage or the local natural environment. Examples of these are:

- combustible substances, usually carbon,
- free asbestos fibre,
- metallic sulfides, sulfuric acid, and metal sulfates,
- radon gas, and
- soluble salts of heavy metals, for example arsenic, cadmium, copper, lead, or nickel.

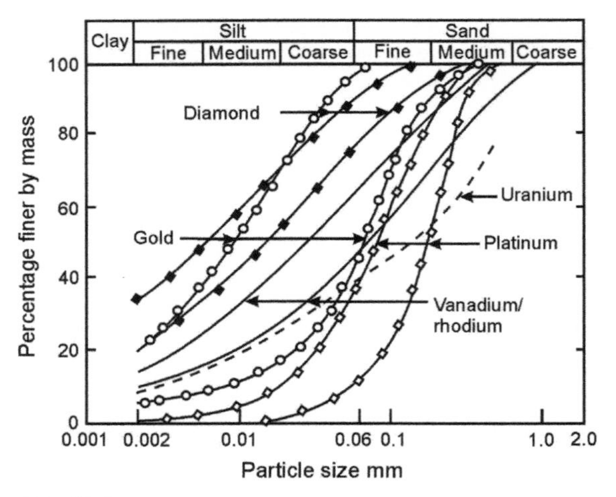

FIGURE 5.2 Particle size analyses for typical tailings from mineral extraction of various ores. (e.g., Vanadium tailings contain 63% of particles finer than 0.06 mm.)

The first and third examples are particular problems with coal wastes, which often catch fire as a result of spontaneous heating caused by oxidation of metallic sulfides contained in the coal. Oxidation of sulfides also results in the production of seepage containing sulfuric acid and soluble metal salts such as iron and magnesium sulfates. Figure 5.3 shows a burning dump of coal waste. Once on fire, a large dump like this, often containing several million tons of combustible material, is a major source of air pollution and very difficult, dangerous, and costly to extinguish.

3. STORAGE OF FINE-GRAINED WASTES

Fine-grained wastes (tailings) containing a high proportion of silt-sized particles are the most difficult wastes to store. Some fine-grained wastes can be "dry-dumped" or "stacked" either by truck or belt conveyor, although they must always contain some water to prevent dust pollution arising during transport and deposition. However, economics dictate that most of the tailings be transported hydraulically either as a slurry or as a "thickened tailings" or "paste" and be deposited or "beached" into hydraulic fill tailings storages where the tailings flow under their own weight, settle, and consolidate to form fine-grained silty deposits. Alternative methods of deposition are to discharge a thickened tailings slurry from a single, or a series of, point discharges around each of which the tailings form

FIGURE 5.3 A burning dump of coal waste.

FIGURE 5.4 A typical hydraulic fill tailings storage constructed by "beaching" toward a decant shaft.

a flat-sloping conical deposit; or to transport a tailings paste by conveyor belt and discharge it from a preconstructed earthen ramp to form a wedge-shaped deposit by viscous flow.

Figure 5.4 shows a hydraulic fill storage of fine tailings formed by "beaching", that is, by depositing a slurry of tailings around the outer perimeter of the tailings storage and allowing it to run down a "beach" toward the pool of water formed in the arms of the Y-shaped causeway or pool training wall. The darker (wet) surfaces show where slurry has recently been deposited. Water that collects in the pool is decanted through a vertical decant shaft that is visible at the centre of the Y. The decant shaft leads to a sub-horizontal outfall pipe or conduit that leads the decanted water to a return water reservoir, before being returned to the mineral extraction plant.

4. WATER BALANCES FOR MINE WASTE STORAGES

A water balance is a statement that sets out how much water enters, leaves, and is stored in tailings storage. Water enters as a component of the waste slurry and as rain, and it can be recovered as water decanted from the pool and collected from the drains around the storage. A large proportion of the water remains in the void spaces between the solid waste particles and cannot be recovered, and a proportion is lost by evaporation from the outer surface of the storage and by seepage into the ground strata under the storage. Water balances are used to estimate and control the quantities of water involved in the waste storage process and also for environmental control by checking on the quantities of water lost in seepage to the natural ground water.

Every hydraulic fill waste storage has a water or seepage surface in it, called the phreatic surface, where the pressure of the water is the same as the atmospheric pressure. Below the phreatic surface, the water pressure is greater than atmospheric pressure and acts to destabilize the outer containing slopes of the storage. Above the phreatic surface, the pressure of the water is less than atmospheric pressure and helps to stabilize the slopes.

Figure 5.5 shows the components of the water balance for a hydraulic fill tailings storage, including the phreatic surface and the drains

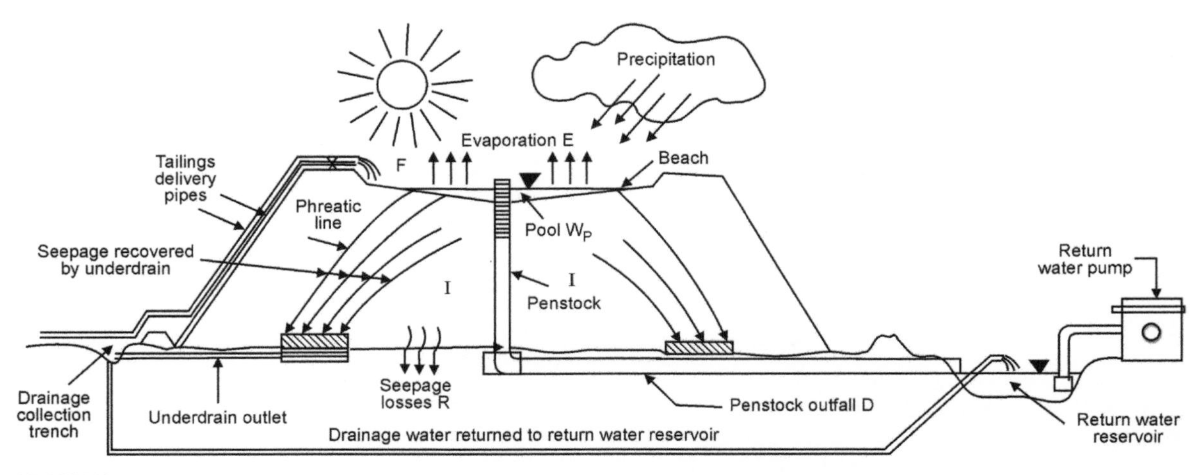

FIGURE 5.5 Diagrammatic representation of components of a water balance for a beached unthickened tailings storage.

designed to release water from the tailings and enable it to be returned to the mineral extraction plant, via the return water reservoir.

When a mine waste storage ceases operation, the pool is drained and thereafter remains dry except for rainfall that runs off from the surface. The phreatic surface slowly subsides once the supply of slurry water ceases and the outer slopes of the storage become more stable. At this stage, the surface of the waste storage needs to be "rehabilitated" which means that the waste surface is protected from erosion by water and wind and is usually planted with indigenous vegetation to enable it to blend, as much as possible, into the natural landscape.

5. SAFETY OF MINE WASTE STORAGE STRUCTURES DURING THEIR OPERATIONAL LIFETIME

Mine waste storages are very large structures, easily visible from space, that have very long operating lives (often more than 50 years) and often were not properly planned in the first place or carefully operated. They are under construction for the whole of their operating lives and are operated by a succession of people, not all of whom are dedicated to carrying out their assigned tasks to the best of their abilities, not all of whom are properly trained, and not all of whom understand why they have to undertake certain tasks and what the consequences of negligence may be. Not all of the workers can recognize that a dangerous situation may be developing, and not all of them know the correct course of action to be taken in an emergency. The foregoing statement is not flattering, but it is a reality. Adding to the unknown dangers are natural hazards; severe rain storms; earthquakes; undetected adverse geological or ground water conditions; human errors such as well-intentioned, but faulty design, theft, or lack of maintenance of vital components; warning systems that fail at the crucial time; and finally, so-called "Acts of God".

Failures of tailings and coarse waste storages can take many forms, the most dangerous and destructive of which are those in which the waste loses strength, becomes mobile and flows as a viscous liquid in which the supporting fluid can either be air in deposits of dry waste, but more commonly, water. Nineteen major flow failures of tailings storages occurred between 1928 and 2000, which together caused at least 1080 deaths of which 1065 occurred between 1965 and 1996, an average of more than 34 deaths per year. Deaths from failures of other forms of waste storage do not lag behind. Deaths caused by only three failures of municipal solid waste dumps between 1993 and 2005 alone, totaled 464, an average of more than 38 deaths per year (see Chapter 30) [6]. However, to keep these numbers in perspective, it must be remembered that they are miniscule in comparison with the yearly death toll on the roads of the world and with other preventable causes of death, such as HIV-AIDS.

It is also interesting to note that many of the 19 failures occurred in small waste storages at operations that were being run on a "shoe-string," or at mines threatened by closure. The failures at Stava, in 1985, which killed 268 people and Star Diamonds (see Fig. 5.9) were of this type. Also, by no means all mine waste "incidents or accidents" are reported in the widely read or viewed news media nor do all lead to disasters.

Figure 5.6 summarizes the causes of 185 failures of hydraulic fill tailings storages, collected by the U.S. Commission on Large Dams [7].

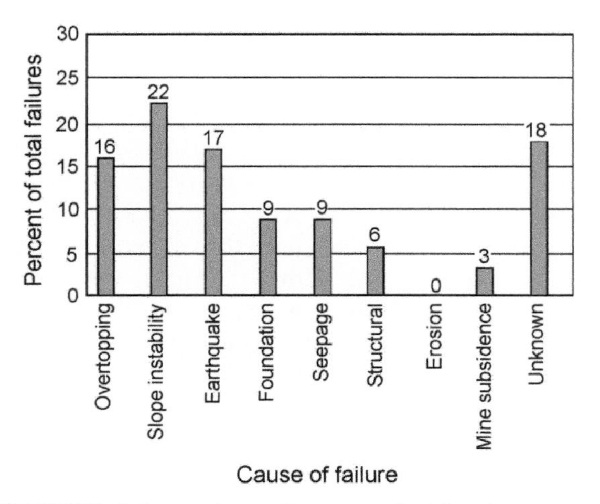

FIGURE 5.6 Analysis of causes of 185 tailings dam failures.

Explanations of the listed causes of failure are given below.

5.1. Overtopping

Overtopping occurs when the pool of water around the decant shaft (see Fig. 5.4) either grows so large that it reaches and overtops the impoundment wall, or the pool moves away from the decant shaft due to uneven distribution of the tailings around the pool. If this happens, the water cannot be decanted and if the problem is not noticed and corrected, the pool may overtop the outer wall. The usual cause of overtopping is holding too much water on the top of the storage, resulting in an insufficient height margin or freeboard to contain water suddenly deposited by a large rain storm. Overtopping usually results in the eroding of a breach in the storage outer wall, possibly causing the contents of the storage to liquefy and flow out of the breach. (It is not quite correct to rule out erosion as a cause of failure, as in Fig. 5.6. Erosion may not be a primary cause of failure, but it often forms part of the process of failure.)

5.2. Slope Instability

Slope instability occurs when the strength of the material forming the outer wall of the storage is insufficient to carry the weight of the wall, with the result that a segment of the wall slides out on a failure surface that is often cylindrical in shape. Figure 5.7 shows a typical slope instability failure or slide in the outer wall of a tailings storage. (The moment of the weight *W* about the centre of rotation *O* became too large for the moment of the shear forces *S* about *O* to resist.) In this real example, the rotational movement on the failure surface was relatively small and the freeboard of the storage surface was reduced by height AA^{I} of just more than 2 m. However, because tailings storages are sometimes operated with freeboards as low as 1.0 m, even a loss of freeboard of 2 m could have resulted in

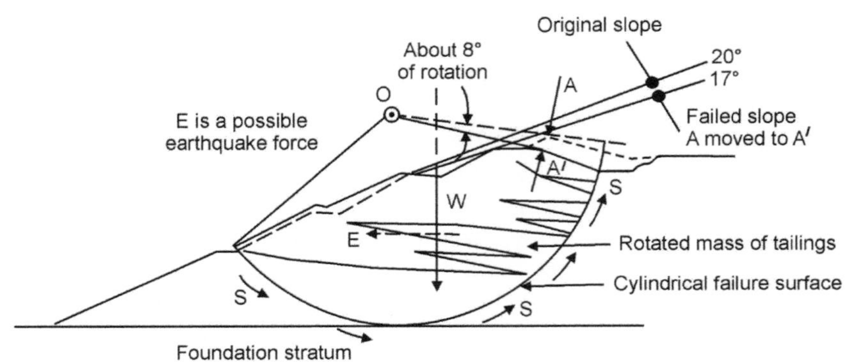

FIGURE 5.7 Postfailure profile of outer slope of a tailings storage. Failure was caused by insufficient strength of the tailings forming the wall.

overtopping, followed by erosion and a flood of escaping liquefied tailings.

5.3. Earthquakes

Earthquakes can cause failure of structures, including slopes by imposing alternating vertical and horizontal accelerations (and therefore forces) that may be appreciable fractions (e.g., 0.2–0.3) of normal gravitational acceleration. Referring to Fig. 5.7, vertically downward accelerations would effectively increase the disturbing force of the weight *W* and horizontal accelerations would impose a lateral force *E*, both of which would tend to destabilize the mass of tailings above the potential failure surface. Because earthquake forces are cyclical and periodically reverse, usually with a frequency of 1 to 2 applications and reversals per second, water in the pool of the tailings storage may begin to slop backward and forward with a magnitude that increases with every cycle until the pool overtops the outer wall, with all the consequences of an overtopping.

5.4. Foundation Failures

Foundation failures occur when the strength of the foundation strata of the tailings storage becomes inadequate, as the storage increases in height. This usually results in instability of the entire outer wall along a failure surface, such as the failure surface shown in Fig. 5.7, but which cuts into the weak foundation stratum.

5.5. Seepage

Seepage from the pond of a storage through the outer wall can cause slope instability because it weakens the tailings by increasing water pressures in the slope (i.e., by a rise in the phreatic surface). The same effect can occur if the storage is built too rapidly, so that the "rate of rise" in metres per year is too quick for the tailings to settle, consolidate, and gain strength to the extent assumed in the design. Figure 5.8 is an aerial photograph of a 28-m high tailings storage that was built at an excessive rate of rise and as a consequence suffered three rotational failures in the course of 3 days. The first failure took place on a Friday night, the second, next to the first, on Saturday night and the third on Sunday night. The safe rate of rise had been determined as 1.5 m a^{-1}, but because of an increase in production at the mine, this had been increased to 2.57 m a^{-1}, and in the month prior to the failure, it increased to 2.83 m a^{-1}. Fortunately, nobody was injured and damage was confined to the mine property.

5.6. Structural Failure

Structural failure is a poorly defined term, but might include the result of damage from

FIGURE 5.8 Failures of a gold tailings storage at Saaiplaas mine that was being built at an excessive rate of rise. (Two failures are visible, side-by-side in the centre of the photograph, and one in the centre-right.)

FIGURE 5.9 Aftermath of the collapse of a tailings storage at a diamond mine into shallow underground workings. The level of the tailings before the collapse can be seen as a level surface in the background. The small pump-barge is located over the site of the collapse.

burst tailings delivery pipes, resulting in the cutting of an erosion gully, or the collapse of the decant control structure, either the outlet shaft or the outlet pipe, also resulting in erosion damage.

5.7. Mine Subsidence

Failure caused by mine subsidence usually occurs when tailings are stored on surface over shallow mine workings, and the strata between the underground workings and the ground surface collapse, allowing tailings to flood into the underground workings. Figure 5.9 shows the result of one such occurrence in which four miners were drowned underground. In a similar accident in Zambia in 1970, 89 miners were drowned underground.

6. DECOMMISSIONING, CLOSING AND REHABILITATING TAILINGS, AND OTHER MINE WASTE STORAGES

The decommissioning and closure processes consist of removing all the installations associated with operating the storage, such as tailings delivery pipes, conveyors, and so on. A permanent surface drainage system is installed, sealing the decant system and replacing it with a permanent spillway, and the surface of the storage is reshaped so that surplus surface water can enter the surface drainage system and either be held on the top surface of the deposit, to evaporate, or be conducted to natural ground level without causing any erosion or flooding. At natural ground level, the surplus surface drainage should be purified, if necessary, and channeled into a natural water course.

The biggest problem in rehabilitating the surface of a closed waste storage is to permanently protect the surface from water and wind erosion, thus maintaining a stable surface in which vegetation can be established.

It is becoming more and more common for regulatory agencies to require that the rehabilitation of a tailings dam or other waste deposit be designed to be maintenance free for 500 to 1000 years. This seems quite unrealistic when one thinks of the changes only 100 years can cause to a landscape (e.g., the areas now occupied by the cities, freeways, and waste deposits of many cities in the "new world" were untouched countryside just over a century ago). However, very ancient man-made earth

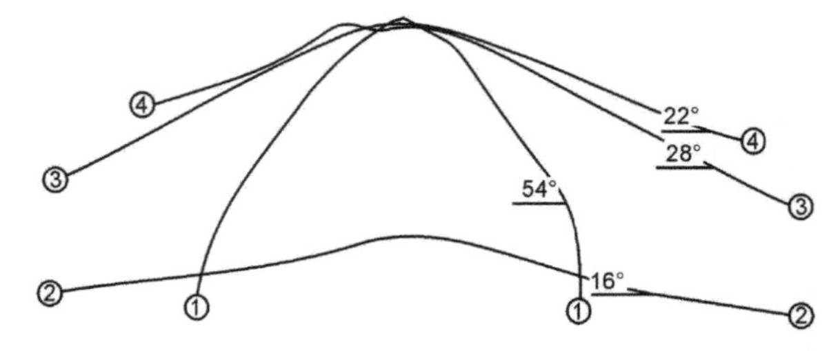

FIGURE 5.10 Profiles of ancient earth mounds in China.
1. Xia Kingdom, 99-700 a BP (desert climate) near Yinchuan (a BP = years before present).
2. Qin Shi Huan, 2200 a BP, said originally to have been 116-m high, now 52-m high (near X'ian).
3. Liu Che, 2100 a BP, now 46-m high.
4. Gaozong, 1320 a BP (2, 3, and 4 – semi-arid climate, near Xi'an).

mounds exist in many parts of the world with slopes that have been subject to 1000 or more years of erosion and still exist in good condition. Figure 5.10 shows the profiles of several ancient man-made monumental mounds in China [8]. All these are accurately dated and are from 700 to more than 2000 years old. The mounds around Xi'an are all constructed of fine sandy silts from river alluvium and loess, which are intrinsically fairly erodible. The climate has distinct wet and dry seasons, rainfall is between 500 and 1000 mm a^{-1}, and there is an annual water deficit, that is, potential evaporation from the soil exceeds rainfall. The mound near Yinchuan is also of loess, but the area has a desert climate. It might have been expected that over a period of 1000 years, the slopes would all have eroded to similar and very flat slope angles. But this has not occurred. As shown in Figure 5.10, the slopes vary from 16° to 28°. Hence, one cannot conclude that if a slope has less than a certain limiting angle, it will not erode. The contrast between the mounds near Xi'an and those near Yinchuan shows, however, that climate plays a role in determining an erosion-resistant slope. Observations and measurements of the erosion of the slopes of gold tailings dams in South Africa [9] have shown that an "erosion rate surface" exists in a (slope angle–slope length–erosion loss) space. This surface, illustrated in Fig. 5.11, shows that erosion rates increase with slope length but are low both at very flat and very steep slope angles. The "belly" of the erosion rate "sail" at intermediate slope angles represents the range of slope angles often used for tailings dams (25° to 35°), which thus usually have the worst possible slope angles for erosion losses.

The larger the surface strength of a slope, the lower is the rate of erosion. Hence, the effect of a varying surface strength would be like the wind on a spinnaker sail; it would move the erosion rate surface inward (as the shear strength increases) or outward (as the shear strength reduces) relative to that of the erosion loss axis (or "mast"). Protecting the surface of a slope, by armouring it or covering it with vegetation, has a similar effect to increasing the surface shear strength, and therefore moves the "sail" in, that is, back toward the "mast."

Figure 5.11 illustrates the following very important principles:

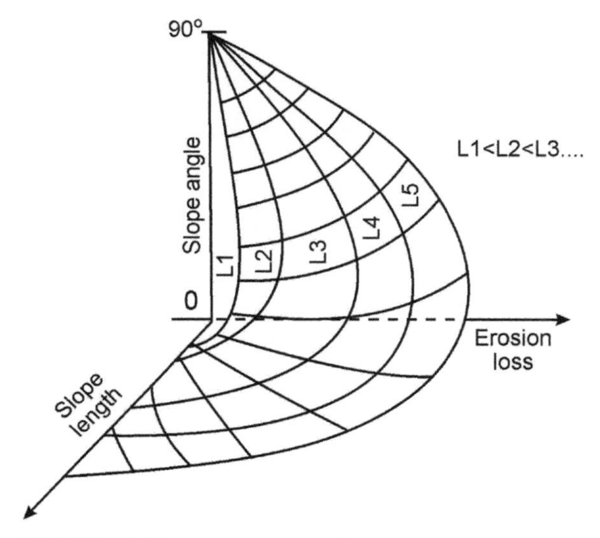

FIGURE 5.11 Erosion rate surface.

- *Slope angles* should be made as flat as possible. Around 10° to 15° is practical and approaches the ideal angle.
- *Slope lengths* should likewise be limited. The maximum length should not exceed 25 to 30 m.

To achieve these two requirements, long slopes should be interrupted at height intervals of about 7 to 8 m by horizontal berms or steps equipped with lined, erosion-proof surface drains designed to conduct the surface run-off from the slope above to ground level, without causing any erosion of the lower slope surfaces.

- *Very flat slopes and horizontal surfaces* (e.g., the flat top surface of a tailings storage) do not erode to any significant extent and hence do not require protection against erosion.

So far, the term "erosion" has not differentiated between wind erosion and water erosion. Both can be very significant. In wetter climates, water erosion predominates, and in drier climates, there is more wind erosion. In South Africa, for example, wind erosion accounts for two-thirds of total erosion and (perhaps surprisingly) in Iceland, wind erosion also predominates over water erosion.

Concerning *rates of erosion*, unprotected tailings slopes may erode at rates of up to 1000 t ha^{-1} a^{-1}, whereas natural slopes usually erode at 50 to 100 t ha^{-1} a^{-1} or less. The aim of protecting tailings slopes should be such that their rate of erosion is reduced to less than 100 t ha^{-1} a^{-1}.

There is a controversy [10] as to whether the Qin mound (No. 2) was ever actually 116 m high, or whether this was the planned height, but the mound was only built to 52 m, its present height. If one calculates back 2200 years from the present height and plan dimensions of 330 m × 330 m, assuming a rate of erosion of 100 t $ha^{-1}a^{-1}$ (appropriate for the soil from which it is built) and maintaining the present average slope of 16°, the calculated original height works out as 115 m. Hence, 115 m is a very likely original height. Because it was a memorial mound to the first emperor of a united China, it is likely that the mound has been maintained over the 2000 years since it was built. Nevertheless, its height has halved in that time. This illustrates the futility of legislating for a design period of even 100 years as a maintenance-free life. Closed tailings storages, like the Chinese mounds, cannot be made erosion free.

7. SUMMARY

Mining activities produce larger quantities of waste and have more adverse environmental impacts than waste from any other human activity. It is difficult to do more than briefly outline the subject of mine waste storage in a single chapter, and the reader is referred to a recently published full-length book on the subject [6] for further details.

Mine waste can be divided into coarse-grained wastes that are usually stored in surface dumps, and fine-grained wastes, usually stored in hydraulic-fill structures. Both coarse and fine mine wastes could contain toxic substances, emit radioactive radon gas, or be combustible. Therefore, mine waste storages need to be constructed and protected in such a way that their adverse effects on human health and the natural environment are minimized on a long-term, continuing basis. Fortunately, controls and statutory regulations for the storage of mine waste are being tightened in most countries, and at the same time, the engineering skills needed to construct and maintain storages safely and for long periods of time are also improving.

It is very important to realize that once a mine waste storage has been created, it constitutes a hazard that requires ongoing maintenance and care for millennia to come.

References

[1] International Commission on Large Dams (ICOLD), A Guide to Tailings Dams and Impoundments, Bulletin 106, Paris, France, ISBN 92-807-1590-7, 1996.

[2] R.J. Stuart, T.C. Holmes, M. Jurcevici, I.B. Watt, G.D.J. van Rensburg, Tailings and Mine Waste '03, Swets & Zeitlinger, Lisse, Netherlands, 2003. 95–100.

[3] D. van Zyl, M. Sasoon, C. Digby, A.M. Fleurly, K. Kyeyune, Mining for the Future, International Institute for Environment and Development, 2002. Available at: http://www.iied.org/mmsd/retrieved 25 October 2010.

[4] Chamber of Mines of South Africa (COMSA), Facts and Figures, COMSA, Johannesburg, South Africa, 2006.

[5] M.W. Sutton, L.M. Weiersbye, J.S. Galpin, D. Heller, Mine Closure 2006, Perth, Australia, 2006.

[6] G. Blight, Geotechnical Engineering for Mine Waste Storage Facilities, CRC Press/Balkema, Leiden, Netherlands, 2009. ISBN: 9780415468282.

[7] R.C. Lo, E.J. Klohn, Proceedings of International Symposium on Seismic and Environmental Aspects of Dams Design, Santiago, Chile, 1996. 35–50.

[8] G.E. Blight, F. Amponsah-Da Costa, Ground and Water Bioengineering for Erosion Control and Slope Stabilization, in: D.H. Barker, A.J. Watson, S. Sombatpanit, B. Northcutt, A.R. Maglinao (Eds.), Science Publishers, Plymouth, U.K., 2004, pp. 365–377.

[9] G.E. Blight, S Afr Inst Min Metal 89 (1989) 23–29.

[10] J. Man, The Terracotta Army, Bantam Books, London, UK, 2007, 233–236.

CHAPTER

6

Metal Waste

Thomas Pretz, Jörg Julius

Department of Processing and Recycling, RWTH Aachen University, Wüllnerstr. 2, 52062 Aachen, Germany

OUTLINE

1. INTRODUCTION

The reuse of secondary metals in the production cycle probably constitutes the oldest and most important kind of recycling activity in the history of mankind. The recycling of metal dates back to approximately 10,000 years ago, and the main reason why metals have always been recycled is because of the large amount of energy that must be expended to extract a metal from its ores in the first place. This has not changed over the millennia. Steel and nonferrous (NF) metal industries are among those sectors with the highest energy intensities and recycling saves up to 95% of the energy consumption expended in the production of the primary metal. Thus, scrap and metal-bearing residues are of great importance when considering the purchase of raw materials to produce virgin metal. This is especially important for those countries that have little or no ore deposits. For instance, in Europe, in the field of end of life vehicles (ELVs) and waste from electric and electronic equipment (WEEE)

Waste Doi: 10.1016/B978-0-12-381475-3.10006-3

recycling, the EU directives created additional incentives concerning metal reclamation.

In this chapter, the fundamentals of metal recycling are described. Only the mechanical processing of scrap is considered, because these processes are all that is required to produce metals that can be used without any difficulties and without further treatment in metallurgical plants. These metallurgical processes include the machinery that is actually a part of the downstream equipment involved in smelting and refining and thus these processes complete the recycling loop for metals.

2. SCRAP METALS

Metals for recycling include those that are recovered from large bulky equipment and often consist of both ferrous (Fe) and NF materials. The metals involved are mainly iron, steel, stainless (inox-) steel, and several NF metals. In this section, the advantages and approximate quantities of metals recovered for recycling are discussed.

2.1. Ferrous Metals

In 2008, the total world steel production amounted to 1.3 billion tonnes (1.3×10^9 t), of which 0.50×10^9 t was fabricated from discarded metal waste [1]. Thus, roughly 40% of the world's steel output originates from scrap iron and steel. The energy saved from using recycled steel scrap amounts to approximately 75% of the energy that would have been spent to generate the steel from primary mineral raw materials, and the CO_2 emissions are reduced by 58%. Furthermore, the recycling process results in an 86% reduction in air pollutants and a 97% reduction in mining waste [1].

Stainless steel is an iron alloy containing nickel, chromium and other elements that are present to protect the metal against corrosion and other unwanted chemical reactions. The market demand for this metal has doubled over the past decade, with an annual production amounting to more than 25×10^6 t [2]. The recycling of stainless steel is especially important as it saves natural resources such as the different alloying metals that are beginning to show signs of running out in the future. Today, an ordinary stainless steel product is composed of about 60% recycled material [2].

2.2. NF Metals

The most commonly used NF metals in the world today are aluminum, copper, brass, zinc, and lead. Because of their limited availability, their high value and the considerable energy saved if recycled, large quantities of these NF metals are reclaimed and recycled in smelters, refineries, foundries, and other producers of these metals and metal products in most countries of the world. This is not only ecologically sound (up to 99% reduction of CO_2 emissions) but also energy efficient (80%–95% savings). The following list of percentages of recycled metals in new products gives some idea of the scale of recycling and its importance: aluminum $> 33\%$, copper $> 32\%$, zinc $> 30\%$ and lead $> 35\%$ [3].

3. MANAGEMENT OF METAL WASTE

The essential demands of modern waste management are to reduce the total amount of waste arising and to reuse and recycle as much of the waste as possible. In many fields of recycling, it is difficult to meet these requirements because the recovered products are often of reduced quality and hence value. As a result, these materials can only be used in a downgraded form. A typical example is the recycling of waste plastics. In most cases, the collected material is of mixed types of plastic that are not fit for easy recycling. The mixed plastics can however be used in the production of components with inferior properties.

However, for metal recycling, the situation is very different. The most important reasons for this is that metal can be recycled indefinitely and that subsequent metallurgical treatments do not substantially change the physical and chemical properties of the metal. Furthermore, metals (including recycled metals) have a high market value, which is very rare in the recycling industry. Here, with metals, supply and demand are the determining factors that affect the market prices. Moreover, the prices can be influenced to a certain degree by operators of recycling plants as they have the chance to sell their products at convenient points of time, that is when adequate revenues can be achieved and the "price is right."

In general, the recycling industry for metals is structured in an organizational form of a pyramid: at the bottom of the pyramid are many small companies that purchase and collect scrap metal. This scrap is then sold to larger business establishments that process and separate the metals and finally it is sold to companies that have metallurgical plants for further treatment or it is sold to big multinational trading companies at the top of the pyramid.

4. METAL CONTAINING RAW MATERIALS FOR RECYCLING

Metal scrap is made up of a mixture of metals originating from a variety of sources that include commerce and industry, municipalities, and households. This obsolete scrap is collected, stored, processed, and sold from scrap yards or other specialized facilities. In this context, "home" scrap and "in-house" industrial scrap are not discussed as these materials can be considered as "works scrap," which never reaches the consumer. As a rule, this type of scrap does not contain foreign material, and does not need any treatment or separation process and can be reused directly. However, obsolete scrap requires target-oriented mechanical processing to meet the quality demands of the customers, which commonly are metallurgical plants. The following goals are aimed in this stage:

- modifications of physical properties, for example, dimensions and bulk densities,
- modifications of chemical properties, for example, metal content and unmixed grades of final products,
- modifications of safety properties, for example, removal of explosives and hollow bodies,
- modifications of other properties, for example, liberation and separation of unwanted adherences.

The collected scrap appears in a great variety of forms and properties such as different lump sizes and shapes; materials with differing bulk densities; different kinds of metals, each with their own properties (e.g., hardness, abrasiveness, etc.); and materials made of composite substances that could include plastics and other nonmetals. Consequently, it is necessary to adapt the treatment processes to the special characteristics of the most frequently found complex-composed feed mixtures. Mechanical processing of scrap is carried out predominantly with mechanical equipment, such as shears, compaction units, shredders, and other types of machines. Table 6.1 shows an overview of the different kinds of scrap and the mechanical processing methods used in recovering the different metals.

5. MACHINERY AND PROCESSES FOR SCRAP TREATMENT

In the following section, typical machinery and processes used for scrap treatment are described. Moreover, the most relevant machines that are used in practice are listed in Table 6.2.

TABLE 6.1 Examples of Different Scrap Types and the Possible Processing Methods Used

Type of scrap (Ferrous [Fe] and Nonferrous [NF] metals)	Potential processing method
Light-mixed consumer scrap, End-of-Live-Vehicles (ELVs)	Comminution with shredder, subsequent separation with air classifier, magnetic separator, and handpicking
Waste of Electric and Electronic Equipment (WEEE)	Comminution with hammer mill, subsequent separation with air classifier, magnetic separator, and other equipment
Cable scrap	Comminution with rotor shears and granulators, subsequent separation with air tables and other equipment
Ash from waste incineration including Fe and NF metals	Classifying with screens, comminution with shredder, subsequent separation of metals
Mixed stainless steel scrap	Comminution with hydraulic shear or shredder, subsequent separation of metals
Intermediate metal-products from waste sorting plants	Comminution with hammer mill, subsequent separation with magnetic and eddy current separators and other equipment
Sheet metals and residues of stamping	Compaction with scrap baling press
Fe and NF turnings	Comminution with turnings crusher, subsequent separation of cutting fluids with centrifuge, magnetic separator
Heavy scrap with wall thicknesses up to 150 mm	Comminution with hydraulic scrap shear, where necessary with subsequent screening of fines
Heavy scrap with wall thicknesses more than 150 mm	Comminution with flame cutting or blasting
Cast iron scrap	Comminution with vertical drop work

5.1. Scrap Shears

Guillotine shears are used in most industrialized countries for the cutting up of lumpy, thick-walled steel and NF metal scrap, the majority of which is reasonably ductile. In the process, the material is usually stressed by compressing and shearing forces. The desired goal of this cutting process, using the hydraulic shears, is to achieve a single-stage reduction in the scrap to dimensions with a final length of approximately 400 to 1000 mm. The process meets the demands of steel mills, foundries, and NF metal smelters in respect to the physical properties and condition of the metals. Thus, all types of scrap from the more compacted steel plates, pipes, and beams, as well as the bulky scrap in the form of tanks, casings, and structural elements, can be cut into similar-sized pieces.

The functioning operation sequence of guillotine shears is characterized by spatial precompression with a side press, a press lid, and a feeding pusher forming a compact cord that is pushed underneath the cutting blade. This allows the scrap to be cut into pieces of variable length [4].

However, it is not possible to obtain liberation of complex and compounded material such as metal/metal or metal/nonmetal composites from the pieces cut with these scrap shears. Therefore, the application of guillotine shears is limited to the processing of materials containing a minor quantity of foreign substances. Such shears are predominantly operated in steady

TABLE 6.2 A Selection of Machinery and Equipment Used in Metal Processing

Machines for comminution	Typical applications
Hydraulic guillotine shears	Comminution of heavy, bulky, and long scrap constituents
Alligator shears	Comminution of bulky and long scrap constituents
Rotor shears	Comminution of cable and Waste of Electric and Electronic Equipment (WEEE) scrap
Granulators	Comminution of cable scrap
Rotor impact mills	Comminution of WEEE and shredder residues
Grinder mills	Comminution of light mixed scrap, compaction of cooling scrap
Pre-shredders	Pre-crushing of End-of-Life-Vehicles (ELVs), subsequent processing with shredder
Shredders	Comminution of light mixed scrap and ELVs
Hammer mills	Comminution of turnings, WEEE and metal cans
Cross-flow crushers	Comminution of WEEE and metal cans
Single shaft crushers	Comminution of WEEE and metal cans
FOR SCREENING	
Drum screens	Classification of cooling scrap, turnings, and shredder residues
Vibrating screens	Classification of shredder residues
Flip-flow screens	Classification of shredder residues and WEEE
FOR SEPARATION	
Magnetic separators	Separation of magnetizable waste items as a rule Fe metals
Eddy current separators	Separation of electrically conductive waste items like NF metals
Corona roll separators	Separation of metals from nonmetals in the size range < 4 mm
Air classifiers	Separation of lights such as shredder fluff
Cyclones	Separation of air classifier lights from the carrier air
Air tables	Separation of metals from nonmetals
Sensor-based sorting devices	Separation of nonferrous metals and stainless steel mixtures
Heavy media separators	Separation of NF metals such as Mg and Al from shredder heavy fraction
Wet shaking tables	Separation of NF metals < 2 mm from WEEE processing
FOR COMPACTION	
Baling presses	Compaction of metal sheets and cans
Briquetting presses	Compaction of turnings

continuous operation. The required electrical power ranges from about 300 to 900 kW and the shearing forces can be as much as 14,000 kN.

5.2. Shredders and Shredder Plants

There are about 700 shredder plants installed worldwide, and these plants are used mainly for the processing of light-mixed scrap and ELVs. The first step in these plants is the shredder itself, which has the task of crushing and liberating the feed material by intensive impact and shearing stresses. As can be seen in Fig. 6.1, the input material slides via an inclined chute into the shredder. The scrap is then seized and compacted by the feed rollers (1). To avoid current peaks, the feed rollers are automatically controlled. Consequently, a load-dependent shredder operation can be guaranteed. At the edge of the anvil (2), the in-feed material is grasped by the fast rotating hammers that have a circumferential velocity of up to 70 m s^{-1} and the scrap is stressed and disintegrated by the shearing forces. The hammers, which are suspended on the carrying shafts of the rotor, force the disintegrated scrap pieces into the interior of the shredder where they are further stressed by impact and flexing.

The baffle plate edge (3) and the baffle plate (4) are structurally formed in such a way that ductile metals are further crushed and compressed until they are small enough to pass through the grid (5). The position of the grid (11) can be adjusted hydraulically; this has an important role in influencing the final grain size and the compression ratio. As a rule, the steel scrap produced in shredder plants is compressed so that a bulk density of 1 t m^{-3} can be achieved. The discharge of the crushed material from the shredder is carried out via a vibrating conveyor (8) and a belt conveyor. If heavy scrap lumps are found in the shredder and cannot be further crushed, an ejection flap (7), operated manually or automatically, can be opened to discharge these bulky pieces. A branch connection (6) to the de-dusting system of the shredder plant fulfils the demand of a largely emission-free operation.

The material that is discharged through the shredder grid exhibits a very broad grain size distribution ranging from coarse (~150 mm) to very fine particles (<1 mm). As can be seen in the simplified flow-sheet of Fig. 6.2, the crushed material is then fed into an air classifier. This device separates light constituents from the remaining heavy particles (heavies) in a rising air current. The proper adjustment of the air

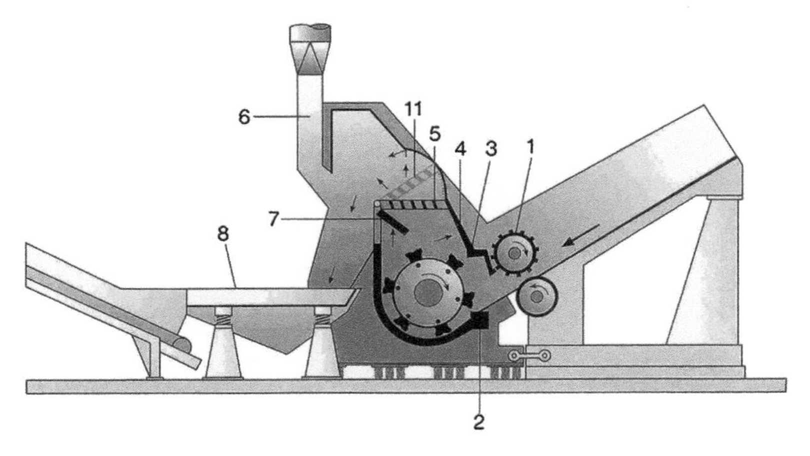

FIGURE 6.1 Schematic view of a shredder [5].

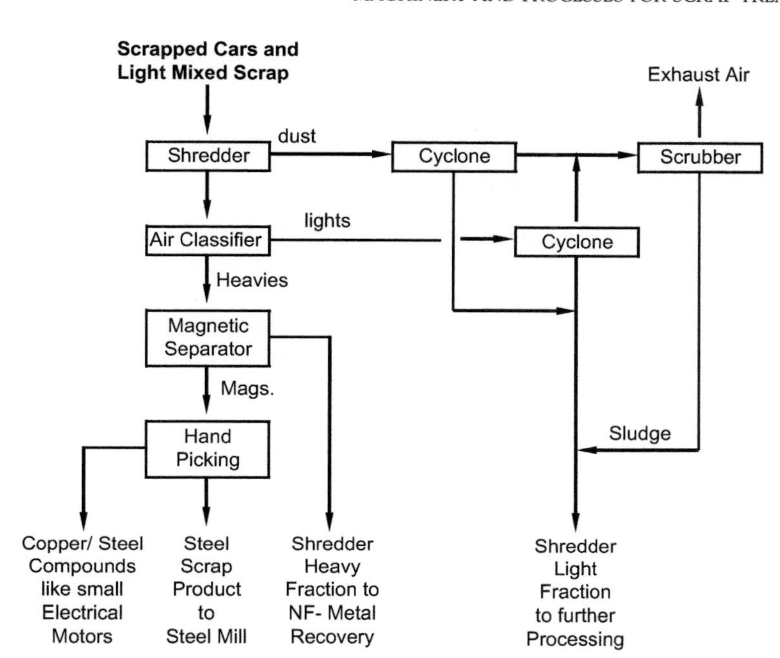

FIGURE 6.2 Example of a simplified flow sheet for shredder plants.

classifier velocity is the determining factor in the amount of misplaced heavies, especially metals in the light fraction. There is always a compromise necessary as the heavies need to be free, to a certain extent, of the bothersome fluffy materials. Fines that cannot be separated by the cyclones are segregated from the carrier air very effectively with a subsequently arranged scrubber. The separated shredder light fraction amounts to approximately 16% to 22% by weight of the shredder input.

Within the heavy fraction of the air classifier, the metals are enriched. In most cases, magnetic drum separators are used for the selective separation of the iron and steel scrap—this accounts for approximately 60% to 70% by weight of the shredder in-feed material. After hand picking, for the removal of nonliberated copper-containing items, for example parts of electric motors, the final ferrous product fulfils the quality demands in respect of grade and bulk density so that it can be recycled in steel mills without further treatment. However, it may be expected that new developments in the field of sensor-based sorting will replace manual picking in the future. The material that is not attracted by the magnetic separator—the so-called shredder heavy fraction—is an intermediate product (5%–8% by weight of shredder input) containing NF metals. The processing of this fraction allows recovering of valuable metals such as magnesium, aluminum, copper, zinc, bronze, lead, and stainless steel. In total, approximately 50% to 60% by weight of metals are to be found in the heavy fraction.

5.3. Processing of the Shredder Heavy Fraction

Further processing of the shredder heavy fraction with the main aim to recover NF metals mainly occurs in specialized plants, which, however, very rarely belong to the downstream equipment of shredder plants. A simplified flow sheet of such a plant illustrating the state of the art for the processing of NF–metal-rich mixtures is shown in Fig. 6.3. As the first step, a suspended magnetic belt separator is used to achieve a further reduction in the ferrous metal content. The following drum screen divides the feed material into four fractions with the intention of reducing the width of the grain size distribution in the screen products, thus improving their sorting capabilities. The grain size ranges used proved to be advantageous in practice. However, these separating sizes can be modified. The screen overflow is directed to a hand-picking station at which NF metals and stainless steel are separated.

The screen underflow fractions are then processed on three eddy current separators. These devices are very effective in segregating a good portion of the NF metals contained in the mixture [6]. The NF metals from eddy current separator 1 in the grain size range of <15 mm are subsequently treated on a jig that provides a separation of aluminum and magnesium from the heavier metals according to different densities.

The segregation efficiency of eddy current separators depends to a great extent on the physical properties of the feed material. These are mainly the grain size range, the grain shape, and the ratio of electrical conductivity to density of the metallic components. With these systems, some metals such as wires, metal/nonmetal compounds, lead, and stainless steel cannot be deflected and remain in the so-called tailings fraction for further processing. The sensor sorters 1 and 2 are used for separating the NF metal remainders. Additionally, two more sensor sorters are arranged for the removal of stainless steel items. For this purpose, sensor-based sorters with inductive metal detectors are applied (q.v. Section 5.4). Following this is another hand-picking station that, as far as is possible, provides a separation of NF metals and stainless steel.

Finally, two heavy media separators are applied for the density separation of the NF metal mixture from the preceding process stages. A separation into three products (magnesium, aluminum, and a mixture of heavy metals) can be conducted based on the different densities using heavy liquids of densities of approx. 1.9 and 2.9 g cm^{-3}. The recovered metals from such plants are final products that can be sent directly to metallurgical plants without further treatment. At the moment, most of the nonmetallic residues are disposed in landfills. However, co-processing together with the shredder light fraction is conceivable either to recover plastics and rubber or refuse-derived fuel.

5.4. Sensor-Based Sorting for Metal Recycling

Single grain separation using externally identifiable properties, which can be determined with suitable detectors, is defined as sensor-based sorting. This technology has revolutionized the design of dry working processes in the field of waste treatment over the past 20 years [7]. In principle, most of modern sensor-based sorting machines consist of a conveyance system for dissemination and singling of the feed material, a sensor system for the recognition of specific material properties, a software-controlled electronic device for the interpretation of data received from the sensor system and a pneumatically working discharge device for the separation of positively detected components. Today, newly developed machines are often equipped with a combination of different sensors to recognize multiple material properties in a single step whereas the sensor data are evaluated in real time by a computer system. As a consequence, these machines guarantee a better separation efficiency in comparison with single sensor devices, especially for sorting of complex material mixtures with a large range of qualitative characteristics. A distinct advantage of modern machines is their learning aptitude, which is based on software-controlled

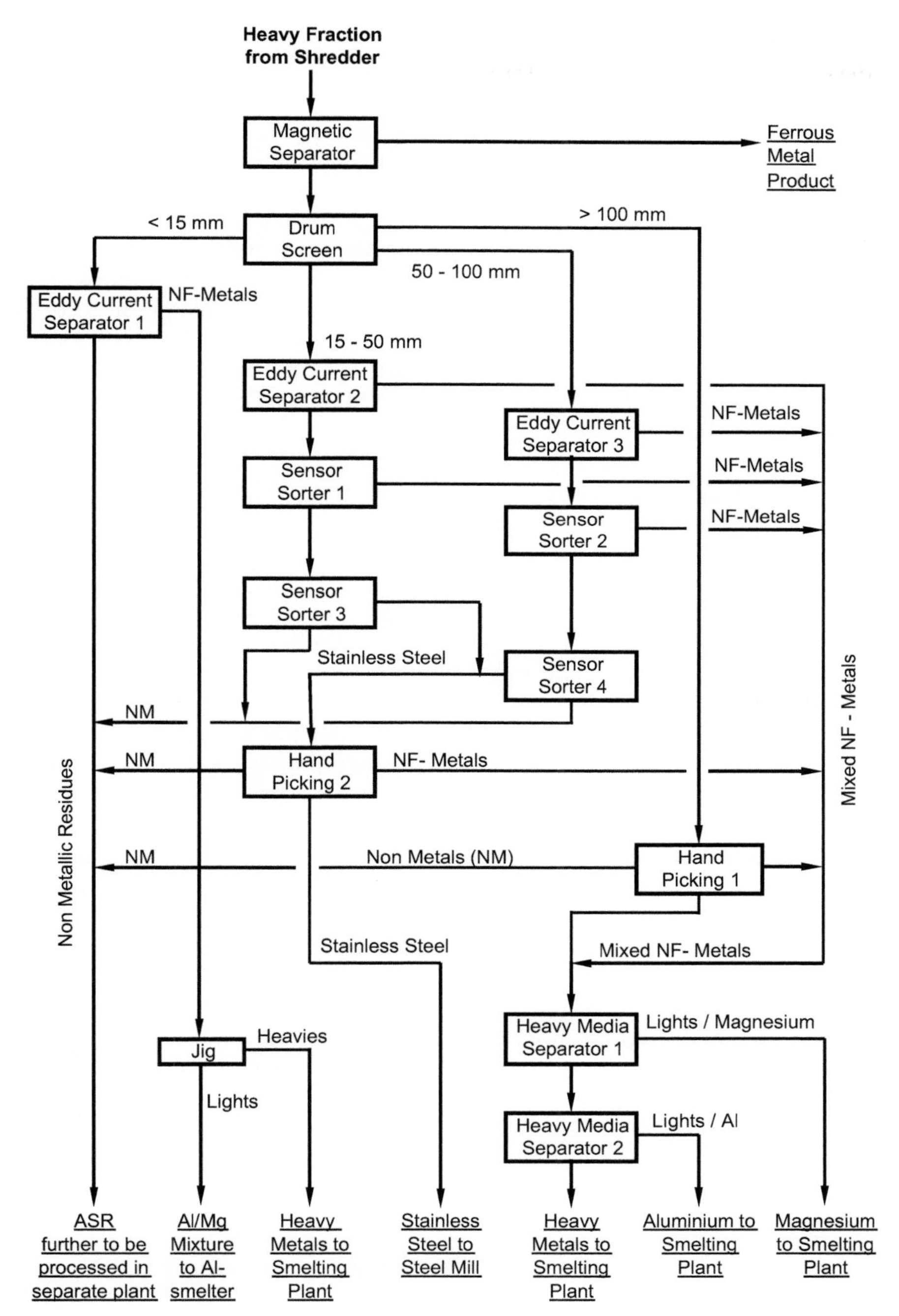

FIGURE 6.3 Example of a simplified process for the treatment of the shredder heavy fraction.

data processing. This feature allows an increased adaptability especially with regard to changes in the process flow sheet, the input material composition, or altered demands concerning the product quality of the recovered materials.

To achieve effective separation with sensor-based sorters, it is very important to condition the feed material before entering the detection zone so that the different components of the feed material can be recognized by the sensors. Thus, single particles should not touch or overlap each other. The most common method of achieving this is to use a cascade of different belt conveyors and/or inclined sliding chutes with increasing transport velocities. The application of sensor-based sorters, in recycling processes, has great advantages in comparison with conventional methods. This is becoming increasingly more beneficial as new separating criteria for separation are made accessible. In addition, the continuous development of new sensor systems continues to open up innovative fields related to separation and recycling, and remarkable progress concerning novel sensors and sensor combinations has been achieved over the past few years.

A number of different types of sensor sorters have entered the market in the last decade especially designed for the separation of metals. Table 6.3 gives an overview of the different types of sorters, the separating attributes used, and typical applications.

Today, sensor-based sorters with inductive detectors are frequently used for metal separation. The assembly and functioning principle of such devices are given in Fig. 6.4 [8]. This type of automatic picking is a combined opto-electronic system that can operate with a belt width of 600 mm or 1200 mm. It combines the special characteristics of an optical system incorporating a high-speed camera with recognition of 1 billion colors and a special conductivity sensor permitting the identification of all metals practically. The machine can handle mass streams of up to 10 ton h^{-1}, for instance, in the size class 15 to 50 mm. With a belt width of 1200 mm, 160 air nozzles are used to blow out the detected materials.

A vibrating feeder (1) and a belt conveyor (2) running with a speed of 3 m s^{-1} are used to line up the particles on the belt into a single layer. The metal sensor (3) that is located under the conveyor belt delivers the initial characteristics of each single metal particle to the computer. In addition, the camera (4) analyses the size, shape, color, and position of particles on the belt. The information from both sensors is then transferred to a special electronic system (5) that evaluates the data and generates the impulses and instructing the nozzles (7) to blow out single particles or allow them to pass. Both, the accepted (9) and the

TABLE 6.3 Sensor-based Sorters for Metal Separation and Typical Applications

Separating attribute	Sensor technology	Examples of separation
Color, brightness	Line scan color camera	Copper and brass from nonferrous (NF) metal mixtures, printed circuit boards from Waste of Electric and Electronic Equipment (WEEE)
Electrical conductivity	Inductive detector	Metals from diverse waste mixtures, stainless steel from waste mixtures
Density	X-ray transmission sensor	Aluminum and magnesium from metal mixtures, WEEE from waste mixtures
Chemical composition	Laser induced break-down spectroscopy	Different aluminum and steel alloys from metal mixtures

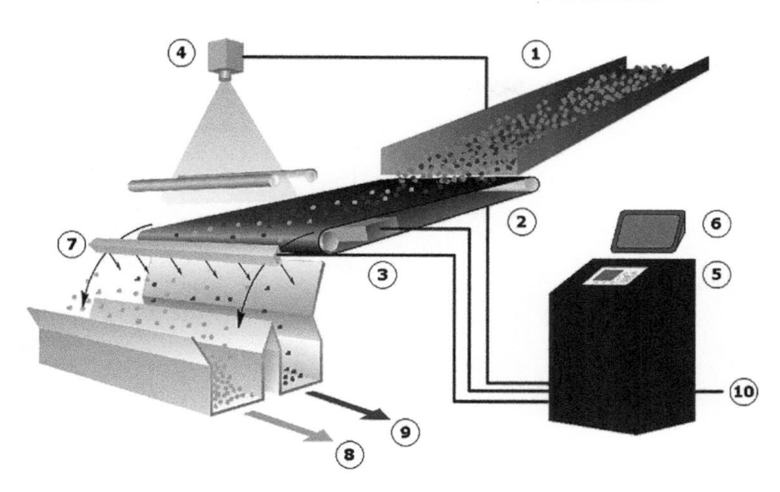

FIGURE 6.4 Principle of inductive sensor sorter with additional line scan camera [8].

rejected (8) products are transported by belt conveyors for further treatment or storage.

6. CONCLUSION

Mechanical treatment of different waste streams containing metals involves the use of well-established methods, which, as a rule, is profitable because of the high market value of the recovered metals and also because metals do not change their properties with use and hence can be recycled an unlimited number of times. The main aim of the processing methods is to achieve high recovery values as well as the best possible grades of the final metallic products. Newly developed systems such as sensor-based sorters are increasingly being implemented to improve separation efficiencies. Finally, re-smelting of the reclaimed products constitutes the closing of the complete recycling loop for metals.

References

[1] http://www.bir.org/ferrous-metals/- accessed on 21.10.2010

[2] http://www.bir.org/stainless-steel/- accessed on 21.10.2010

[3] http://www.bir.org/non-ferrous-metals/- accessed on 21.10.2010

[4] http://www.metso.com/corporation/home_eng.nsf/WebWID/WTB-090526-2256F-0D901?OpenDocument - accessed on 21.10.2010

[5] http://www.metso.com/recycling/mm_recy.nsf/WebWID/WTB-050615-2256F-984C1/$File/END_GB_121009.pdf - accessed on 21.10.2010

[6] P.C. Rem, Eddy Current Separation, Uitgeverij Eburon, ISBN 90-5166-702-8.

[7] Th. Pretz, J. Julius, Stand der Technik und Entwicklung bei der berührungslosen Sortierung von Abfaellen, Österreichische Wasser- und Abfallwirtschaft, No. 07-08/2008, 105–112.

[8] J. Julius, J. Müller, Entwicklung und Erprobung eines Sortierverfahrens für die Rückgewinnung der Edelstahlfraktion Abschlussbericht über ein Entwicklungsprojekt, gefördert unter dem Aktenzeichen 15926 von der Deutschen Bundesstiftung Umwelt, Juli 2002.

CHAPTER

7

Radioactive Waste Management

John E. Marra[†], *Ronald A. Palmer*[‡]

[†] Savannah River National Laboratory, Aiken, South Carolina 29802, United States of America,
[‡] Institute for Clean Energy Technology, Mississippi State University, Starkville, Mississippi 39762, United States of America

OUTLINE

1. INTRODUCTION

The ore pitchblende was discovered in the 1750s near Joachimstal in what is now the Czech Republic. Used as a colorant in glazes, uranium was identified in 1789 as the active ingredient by chemist Martin Klaproth. In 1896, French physicist Henri Becquerel studied uranium minerals as part of his investigations into the phenomenon of fluorescence. He discovered a strange energy emanating from the material, which he dubbed "rayons uranique." Unable to explain the origins of this energy, he set the problem aside.

About 2 years later, a young Polish graduate student was looking for a project for her dissertation. Marie Sklodowska Curie, working with her husband Pierre, picked up on Becquerel's work and, in the course of seeking out more information on uranium, discovered two new elements (polonium and radium) that exhibited the same phenomenon, but were even more powerful. The Curies recognized the energy, which they now called "radioactivity," as something very new, requiring a new interpretation, a new science. This discovery led to what some view as the "golden age of nuclear science" (1895–1945) when countries throughout Europe devoted large resources to understand the properties and potential of this material.

By World War II, the potential to harness this energy for a destructive device had been recognized and by 1939, Otto Hahn and Fritz

Waste Doi: 10.1016/B978-0-12-381475-3.10007-5

Strassman showed that fission not only released a lot of energy but that it also released additional neutrons that could cause fission in other uranium nuclei leading to a self-sustaining chain reaction and an enormous release of energy. This suggestion was soon confirmed experimentally by other scientists and the race to develop an atomic bomb was on. The rest of the development history that lead to the bombing of Hiroshima and Nagasaki in 1945 is well chronicled. After World War II, the development of more powerful weapons systems by the United States and the Soviet Union continued to advance nuclear science. It was this defense application that formed the basis for the commercial nuclear power industry.

1.1. The Dawn of the Commercial Nuclear Power Industry

Both the Soviet Union and the West realized that the tremendous heat produced in the process could be tapped either for direct use or for generating electricity. It was also clear that this new form of energy would allow development of compact long-lasting power sources that could have various applications. The first nuclear reactor to produce electricity was the Experimental Breeder Reactor (EBR-1) in Idaho, USA, December 1951. In 1953, President Eisenhower proposed the "Atoms for Peace" program, which reoriented significant research effort toward electricity generation and set the course for civil nuclear energy development in the United States. The main US effort was under Admiral Hyman Rickover, which developed the Pressurized Water Reactor (PWR) for naval (particularly submarine) use. The PWR used enriched uranium oxide fuel and was moderated and cooled by ordinary (light) water. The Mark 1 prototype naval reactor started up in March 1953 in Idaho, and led to the US Atomic Energy Commission building the 60 MWe Shippingport demonstration reactor in Pennsylvania, which started up in 1957 and operated until 1982. The Shippingport reactor spurred the commercial nuclear power industry in the United States.

Similar development occurred across the globe and today there are more than 400 reactors of varying configurations operating throughout the world with a total installed capacity of more than 370 GWe or about 15% of the world's electricity (vs. coal, 40%; oil, 10%; natural gas, 15%; and hydro and other, 19%) [1]. Today, the nuclear industry is at the eye of a "perfect storm." Fuel oil and natural gas prices are at near record highs, worldwide energy demands are increasing at an alarming rate, and increased concerns about greenhouse gas (GHG) emissions have caused many to look negatively at long-term use of fossil fuels. This convergence of factors has led to a growing interest in the revitalization of the nuclear power industry within the United States and across the globe. The International Atomic Energy Agency (IAEA) upwardly revised its projections for 2030 [2]. Its low projection shows an increase from 372 GWe today to 511 GWe in 2030, the high one gives a projection of 807 GWe, in line with a higher forecast growth in power generation.

1.2. Radioactive Waste Generation

As with any industrial process, commercial nuclear power results in the generation of process wastes. To facilitate communication and information exchange regarding treatment and handling of radioactive wastes, the IAEA instituted a revised waste classification system in 1994 that takes into account both qualitative and quantitative criteria [3]. As defined below, the IAEA developed a system to classify these wastes in three principal classes including exempt waste (EW), low- and intermediate-level waste (LILW), and high-level waste (HLW). (Note: The quantitative cutoff points for each class of waste vary from country to country, typically associated with radiation risk to the public, which depends on storage and/or disposal characteristics. The IAEA has provided general guidelines and these are noted below.)

Exempt Waste: EW contains such a low concentration of radionuclides that it can be excluded from nuclear regulatory control because radiological hazards are considered negligible. As per the IAEA, the definition of EW is based on an annual dose to members of the public of less than 0.01 mSv.

Low- and Intermediate-Level Waste: LILW contains enough radioactive material that it requires actions to ensure the protection of workers and the public for short or extended periods of time. This class includes a range of materials from just above exempt levels to those with sufficiently high levels of radioactivity to require use of shielding containers and in some cases periods for cooling off. LILW may be subdivided into categories according to the half-lives of the radionuclides it contains, with "short-lived" being less than 30 years and "long-lived" being greater than 30 years. The IAEA suggests limiting concentrations of short-lived isotopes to less than 4000 Bq g^{-1} in an individual waste package and concentrations of long-lived isotopes to less than 400 Bq g^{-1} in a waste package.

High-Level Waste: HLW contains sufficiently high levels of radioactive materials that a high degree of isolation from the biosphere, normally in a geologic repository, is required for long periods of time. Such wastes normally require both special shielding and cooling periods. In the IAEA guidelines, HLW contains thermal power above 2 kW m^{-3} and concentrations exceeding those for LILW.

Substantial amounts of radioactive waste are generated through civilian applications of radionuclides in medicine, research, and industry. A typical 1000 MWe nuclear power station produces approximately 300 m^3 of LILW per year and some 27 tonnes (27 t) of high-level solid packed waste per year. By comparison, a 1000 MWe coal plant produces some 270,000 tons of ash alone per year containing radioactive material and heavy metals which end up in landfill sites and in the atmosphere [4]. Worldwide, nuclear power generation facilities produce about 200,000 m^3 of LILW and 10,000 m^3 of HLW (including spent fuel designated as waste) each year.

2. NUCLEAR WASTE TREATMENT AND PROCESSING

Research studies in the management of radioactive waste began in the 1930s. As the commercial nuclear industry evolved through the 1960s and 1970s, additional emphasis was placed on developing long-term solutions for radioactive wastes. Exempt and LILW from commercial nuclear power facilities are handled much like ordinary municipal wastes, although most LILW is disposed of in stable near-surface disposal sites, or as is the case with transuranic (TRU) waste from the United States defense program in stable salt-based repositories (such as the Waste Isolation Pilot Plant, WIPP in New Mexico, USA).

In 2007, in response to growing concerns about management of LILW in the United States, the US Government Accountability Office (GAO) performed a comprehensive review of worldwide practices associated with LILW handling [5]. This report provides a comprehensive analysis of management approaches for LILW, including soil, debris, rubble, process materials, and clothing that have been exposed to radioactivity or contaminated with radioactive material. The report also looked at disposition of excess sealed radiological sources that are no longer useful for industrial or medical applications. The GAO found that most countries maintain waste inventory databases that include information on waste generators (nuclear utilities, hospitals, universities, and research laboratories), waste types, storage locations, and present and future waste generation predictions and disposal capacity needs. The report also found that disposal practices vary according to the hazard presented by the LILW in question. As discussed earlier, lower-activity LILW is handled much similar to municipal waste and is typically disposed in near-surface burial sites that are monitored over time. Depending on the level of activity being treated, most of this disposal is handled as EW per the IAEA definitions (see above) and no review is required from the nuclear regulatory authority. For higher-activity LILW, most countries have centralized storage and disposal options that are licensed by the appropriate nuclear regulatory authority. Funding for operations of these facilities is either provided by the central government or by collecting disposal fees at the time of disposal. For sealed sources used in industrial and medical applications, the disposal fee is often collected at the time of purchase.

The discussion in the remainder of this chapter will focus on handling of spent nuclear fuel and HLW. A typical 1000 MWe nuclear reactor generates about 23 t of used fuel each year. In the United States and Canada, this used fuel is regarded as waste and is slated for direct disposal. In most of Europe and Japan, the used fuel is reprocessed to recover unused uranium and to efficiently manage TRU and fission products. France is widely considered to be the world leader in commercial nuclear fuel recycling, although Russia also has several decades' worth of reprocessing experience. The French program uses state-of-the-art processing facilities at La Hague in northern France and uses both the recovered uranium and plutonium to produce mixed oxide (MOX) fuel (see discussion below). The Japanese have worked extensively with the French and are currently constructing a reprocessing facility similar to that at La Hague. Countries such as India and China have emerging reprocessing programs associated with the increase in nuclear power generation in those countries.

In either reprocessing or direct disposal, the used fuel is first stored for several years under water in cooling basins at the reactor site (see Fig. 7.1). The water covering the fuel assemblies provides radiation protection, while removing the heat generated during radioactive decay.

If the used fuel is reprocessed, the nuclear fuel assemblies are disassembled and chopped into small pieces in a highly secure, remote processing environment. The fuel core is typically dissolved in nitric acid and separated chemically into uranium, plutonium, and HLW solutions. About 97% of the used fuel can be recycled leaving only 3% as HLW. The resulting hulls from the fuel assemblies are treated as LILW.

For a typical 1000 MWe nuclear reactor, about 230 kg of plutonium (1% of the spent fuel) is separated in reprocessing annually. The separated Pu can be used in fresh MOX fuel or stored for later handling and disposal. MOX fuel fabrication has been ongoing in Europe, with some 25 years of operating experience. A similar plant is scheduled to start up in Japan in 2012. A MOX fuel plant is also under construction in the United States (at the Savannah River Site in Aiken, SC) for disposition of excess Pu from the United States nuclear weapons program.

FIGURE 7.1 Typical cooling basin for used nuclear fuel.

Because of the presence of long-lived radionuclides, the separated HLWs (about 3% of the typical reactor's used fuel) need to be isolated from the environment. Research into the processing and disposal of HLW has been active for many years. Isolation of the HLW has typically been in large underground storage tanks. In the United States, about 75 million gallons (284 m^3) have been stored in 177 tanks at Hanford, 53 tanks at Savannah River, and two tanks at the West Valley Demonstration Project. The HLW is in the form of a liquid or slurry and needs to be solidified into a waste form before transport to a final storage/disposal facility.

Development of appropriate waste forms began in the United States in the late 1940s at the Brookhaven National Laboratory [6]. The materials studied were based on montmorillonite clay and phosphate glass. France began making borosilicate glass in 1963 [7]. Over the years, a wide variety of materials have been developed and studied for the ultimate isolation and disposal of HLW. The list includes:

- Borosilicate glass
- SYNROC
- Porous glass matrix
- Tailored ceramic
- Pyrolytic carbon and SiC-coated particles
- FUETAP (and other) concretes
- Glass marbles in a lead matrix
- Plasma spray coatings
- Phosphate glass
- Titanate ceramic
- Various calcines

In the late 1970s, the US Department of Energy (DOE) formed an alternative waste form peer review panel consisting of prominent, independent engineers and scientists with expertise in materials science, ceramics, glass, metallurgy, and geology [8]. Using a rating process to evaluate the relative merit of all the various proposed waste forms, the panel-selected borosilicate glass was used as the reference waste form.

Borosilicate glass is now the material of choice for incorporating and immobilizing the potentially hazardous radionuclides in HLW. Factors that contribute to the suitability of glass waste forms fall into two main categories. First, glass waste forms possess good product durability. Various glass systems are able to incorporate a variety of waste compositions into durable waste forms. These forms have demonstrated good chemical and mechanical performance as well as good radiation and thermal stability. Second, waste-glass forms possess good processing characteristics. The technology for making waste-glass forms is both well developed and well demonstrated. Waste-glass forms ranging in size from bench- and laboratory-scale products to multiton canisters have been successfully produced by using ceramic melters as well as in-can melting techniques. The vitrification process provides a substantial volume reduction; a piece, the size of a hockey puck (25 mm thick, 76 mm in diameter) would contain the total HLW arising from nuclear electricity generation for one person throughout a normal lifetime (see Fig. 7.2) [9]. Vitrification has been used for nuclear waste immobilization for more than 40 years across Europe, Japan, and the United States.

In a typical processing facility, the HLW solution is combined with borosilicate glass-forming materials and sent to a Joule-heated ceramic

FIGURE 7.2 Approximate size of borosilicate HLW glass produced from nuclear electricity generation over a human lifespan.

melter. Operating at ~1200°C, the melter turns the waste slurry into molten glass that exits the melter into large stainless steel canisters. An international reference stainless steel canister holds approximately 400 kg of waste glass. After filling, the canisters are decontaminated and sealed by welding a steel plug into the neck of the canister. In most cases, the canisters are temporarily stored at the vitrification facility before eventual transport to a geological repository.

Cement and cement-based materials are used to contain by-products from reprocessing operations that are categorized as LILW. Common advantages of cement stabilization include continuous or batch processing at ambient temperatures, low-cost raw materials, suitability for large or small volumes of many different waste types, and ability to use modular equipment.

Waste stabilization/solidification is most commonly accomplished by mixing aqueous-based wastes with hydraulic or pozzolanic materials such as Portland cements, calcium aluminate cements, calcium sulfoaluminate (CSA) cements, magnesium (aluminum) phosphate cements, kiln dusts, fly ashes, and reactive slags. These materials react with water to form insoluble binders. Composite cement systems using several of these phases are commonly used by the nuclear industry. The hydrated binder phases encapsulate solid particles in the waste, coprecipitate selected contaminant species, and adsorb excess water and soluble contaminants. In addition, the aqueous chemistry of the cement–waste mixture can be adjusted so that the soluble contaminants are precipitated from solution simultaneously with the formation of the matrix phases. Mixtures of the cementitious ingredients plus other additives such as sodium silicate (hardening agent), set accelerators, and retarders are commonly used. As a result, a monolithic waste form can be produced at ambient temperatures. The waste forms can also be designed to have a wide range of properties. Compressive strengths typically range from 50 to 3000 psi (350 kPa to 20×10^3 kPa). Viscosity and set time can also be adjusted to meet mixing and placement requirements dictated by the production process. Composite cements are typically associated with a highly alkali environment that is not suitable for all wastes; Al metal for example will corrode in such an environment. As a result, a toolbox of cement systems is being developed including geopolymers, CSA cements, and alkali-activated systems with at least one suitable for all waste types [10].

Hydrated waste forms are typically used for stabilizing aqueous wastes, such as, condensed-off gas wastes, electroplating sludges, salt solutions, incinerator ash, electrostatic precipitator and bag house wastes, and process residues, such as, metal chloride and hydroxide bottoms from ore refining processes. Cementitious materials are also used in a variety of environmental remediation actions to stabilize seepage basin sludges, contaminated soils, and waste disposal sites. In addition, cement-based materials are also used for underground waste tank and pipeline closures. The standard requirement for this application is subsidence prevention. Portland cement-based grouts or pumpable, self-leveling, self-compacting backfills containing Portland cement are typically used for tank stabilization. Special grout or backfill formulations are also being designed to stabilize residual contaminants that may not have been removed from these tanks.

3. GEOLOGIC DISPOSAL

As discussed earlier, the HLW fraction of the radioactive waste will require geological isolation for extended periods of time (up to 1,000,000 years). When dealing with such extended time periods, the fate of the stabilized waste is often questioned. In fact, there is a natural example that suggests that final disposal of HLW underground is safe. Over 2 billion years ago (10^9 years) at Oklo in Gabon, West Africa, chain reactions started spontaneously in concentrated deposits of uranium ore. Scientists estimate that these natural nuclear reactors continued operating for hundreds of thousands of years, forming plutonium and the other by-products created today in a nuclear power reactor. This same area remained highly saturated following the end of the nuclear reaction. Evidence shows that the materials that would be classified as HLW using the current IAEA guidelines remained where they were formed and eventually decayed into nonradioactive elements; they were not mobile in the environment. It is this natural analog that provides the basis for geological isolation of HLW and/or spent nuclear fuel.

Worldwide, countries are working to develop deep geologic disposal facilities for a variety of more radioactive LILW and all forms of HLW. Although there is certainly international cooperation in developing programs for geologic disposal, each country has taken a slightly different approach due to differences in geological formation and their domestic nuclear policies. These repository programs have been ongoing for several decades and as of this writing no country has fully licensed and opened a geologic repository of long-term disposition of radioactive wastes. Sweden will likely have the first licensed repository open sometime after 2015.

Initial discussions regarding a permanent solution to the disposal of high-level radioactive waste in the United States culminated in a report by the National Academy of Sciences—National Research Council in 1957 [6]. The primary recommendation was that "disposal in salt is the most promising method for the near future." Other report recommendations included stabilization of the waste in a "slag or ceramic material forming a relatively insoluble product," the potential of disposing of the waste in a deep repository, the separation of Cs^{137} and Sr^{90}, and consideration of the transportation costs.

A generation later, a similar report [11] expanded on the original as a result of the research completed in the interim. This report concluded that the technology for geologic disposal "is predicted to be more than adequate for isolating radioactive wastes." Other recommendations examined the criterion for system performance, repository design and construction, waste package design, prediction of the performance of the system, and expanded the list of potential host-rock candidates to basalt, granite, salt, and tuff.

Following this report, the United States enacted a policy to dispose of used nuclear fuel and HLW from defense applications at Yucca Mountain, Nevada. The US maintained this policy for more than 20 years and submitted a License Application for the Yucca Mountain Repository to the US Nuclear Regulatory Commission (NRC) in 2008. Recently, however, there has been a Presidential decision to cancel the Yucca Mountain project and re-evaluate disposal alternatives in the United States. A recent article [12] compared the repository programs of the United States and Sweden. In addition to examining the technological aspects of geologic disposal, this report discusses the regulatory and bureaucratic aspects of the two programs and how those issues affect the success or failure of overall nuclear waste management systems.

The experience in the United States proves that an essential aspect of the waste isolation strategy is that long-term safety of geologic disposal must be convincingly presented, and accepted, long before a repository can be opened [13]. As discussed above, this requires safety assessments that consider timescales far beyond the normal horizon of societal thinking. For example, it must be acknowledged that the most robust and passively safe system that can be devised by current generations may ultimately be compromised by the actions of a future society, through inadvertent intrusion

[13]. These probabilities must be taken into account in assessing the performance of the repository throughout its operational lifetime. Finally, the scientific analysis and decision-making process must involve stakeholders (i.e., regulators, political leaders, public interest groups, and other non-government entities) at local, regional, and national levels.

4. CONCLUSIONS

Expansion of nuclear energy worldwide is necessary to meet increased power demands and GHG reduction targets. Virtually any international energy forecast predicts large growth in nuclear power generation over the next 30 to 50 years. The expansion of commercial nuclear power production will result in increased radioactive waste generation. Managing this waste effectively is a critical component of a worldwide "nuclear renaissance" and will continue to be a primary consideration for future nuclear fuel cycles. Past practice has proven that the wastes generated from nuclear power generation can be effectively managed using technologies that are well demonstrated at industrial scales, with decades of safe operating practices. As the worldwide community continues to investigate advanced nuclear fuel cycles, radioactive waste management considerations are being given increased emphasis and are, in fact, being considered at the advent of proposed fuel cycles. This emphasis is necessary to ensure that future nuclear fuel cycles remain economically and *environmentally* competitive (as compared with other forms of energy production) and do not produce legacy waste management issues for future generations.

References

[1] World Nuclear Association, World Energy Needs and Nuclear Power. Available at: http://www.world-nuclear.org/info/inf16.html. Accessed 2009.

[2] International Atomic Energy Agency, Energy, Electricity and Nuclear Power Estimates for the Period to 2030, IAEA Reference Data Series No. 1, 2009 Edition, Vienna, Austria, 2009.

[3] International Atomic Energy Agency, Classification of Radioactive Waste: A Safety Guides, SAFETY SERIES No. 111-G-1.1, Vienna, Austria, 1994.

[4] Alex Gabbard, Coal Production: Nuclear Resource or Danger, ORNL Review, Volume 26, 3/4, Oak Ridge National Laboratory, Oak Ridge, Tennessee USA, 1993.

[5] United States Government Accountability Office, Low-Level Radioactive Waste Management: Approaches Used by Foreign Countries May Provide Useful Lessons for Managing U.S. Radioactive Waste, GAO-07−221, US Government Accountability Office, Washington, DC, USA, 2007.

[6] United States National Academy of Sciences − National Research Council, The Disposal of Radioactive Waste on Land, Report of the NRC Committee on Waste Disposal of the Division of Earth Sciences, September 1957.

[7] Yves Sousselier, Use of Glasses and Ceramics in the French Waste Management Program, in Ceramics in Nuclear Waste Management, CONF-7990420, T.D. Chikalla and J.E. Mendel, editors, 1979, p. 9.

[8] L.L. Hench, The 70's: From Selection of Alternative Waste Forms to Evaluation Storage System Variables, in Environmental Issues and Waste Management Technologies in the Ceramic and Nuclear Industries, Ceramic Transactions, Volume 61, V. Jain and R. Palmer, editors, 1995, p. 129.

[9] World Nuclear Association, Waste Management. Available at: http://www.world-nuclear.org/education/wast.htm. Accessed 2007.

[10] N.B. Milestone, Reactions in Cement Encapsulated Nuclear Wastes: Need for a Toolbox of Different Cement Types Adv. Appl. Ceramics 105(1), 2006, 13−20.

[11] T.H. Pigford, The National Research Council Study of the Isolation System for Geologic Disposal of Radioactive Wastes, in: Peter L. Hofmann (Ed.), The Technology of High-Level Nuclear Waste Disposal, 1987.

[12] L.G. Eriksson, Spent Fuel Disposal—Success vs. Failure, Radwaste Solutions, January/February, 2010, 22.

[13] Organization for Economic Cooperation and Development, Nuclear Energy Association, The Environmental and Ethical Basis of Geological Disposal of Long-Lived Radioactive Wastes: The Geological Disposal Strategy for Radioactive Wastes, OECD Publications, 2, rue André-Pascal, 75775 Paris Cedex 16, France, 1995.

CHAPTER

8

Municipal Waste Management

Agamuthu Periathamby

Institute of Biological Sciences, Faculty of Science, University of Malaya, Kuala Lumpur, Malaysia

OUTLINE

1. INTRODUCTION

Municipal solid waste (MSW) is an inevitable by-product of human activity. In nature, there is no such thing as waste, because nature is able to "recycle" the elements in the ecosystem. However, urbanization and rapid population increase has generated waste. The generation of waste, which has exceeded the earth's carrying capacity by more than 30%, has resulted in an accumulation of waste in specific sites. MSW has unique features compared with other waste types, because it involves the public, where the generator frequently meets the waste management representative. As such, MSW management is highly influenced by the socio-economic and political driver in society [1].

2. DEFINITION OF MSW

Broadly speaking, MSW includes all wastes generated within a municipality. However, the definitions differ from country to country and

Waste Doi: 10.1016/B978-0-12-381475-3.10008-7

individuals, authors or researchers. In some developing nations, industrial waste and fecal material, though normally not considered as part of MSW, are often found in MSW and, thus, disposed together in normal landfills. MSW generally refers to all wastes generated, collected, transported, and disposed of within the jurisdiction of a municipal authority. In most cases, it comprises mainly food waste, and rubbish from residential areas, street sweepings, commercial and institutional nonhazardous wastes as well as (in some countries) construction and demolition waste.

3. MSW MANAGEMENT

MSW management incorporates several interrelated aspects, which needs complete cooperation and collaboration for efficient delivery. It comprises aspects of waste generation, waste composition, collection, recycling (if any), pretreatment and treatment, and finally disposal. These management aspects thus require input from legal, economic, governmental, political, administrative, and environmental players. Thus, it requires the involvement of multiprofessional drivers, and at times, the failure of one component is sufficient to cause the whole management to collapse. The management structure and function is site-specific and depends on socioeconomic, behavioral, cultural, institutional, and political frameworks. These stakeholders need to interact and cooperate for the management system to achieve its target.

3.1. Challenges in MSW Management

Management of MSW is most challenging compared with other waste types. The challenge is even more glaring in the developing world. Issues that are often associated with poor management of MSW are as follows:

- Inadequate waste collection system,
- Low recycling rate,
- Poor treatment or no treatment,
- Uncontrolled disposal,
- Inadequate technology, and
- Low awareness of health risks

These situations are serious in low-income and middle-income countries where there is uncontrolled rural–urban migration, urbanization without proper planning, and rapid industrialization with poor infrastructure. Rapid population increases are also the cause of some of the ill effects.

4. MSW GENERATION

Daily global MSW generation is estimated to be more than 2×10^9 t [2], where t refers to metric tonne. The generation volume is influenced by several factors such as family income level, education, season, type of residence, waste collection system and frequency, consumption pattern, and socioeconomic practices. Economic factors directly influence the waste generation per capita, and higher economic status results in an increase in MSW volume (Fig. 8.1) [3].

With few exceptions, there is a strong correlation between gross national income (GNI) and waste generation per capita. For example, the United States, Germany, and Switzerland, which each record a high GNI, generate MSW at a rate of 700 kg per capita per annum (700 kg ca^{-1} a^{-1}). Japan, too, has high GNI, but it generates only 400 kg ca^{-1} a^{-1}. On the other hand, low-income countries generate 100 kg ca^{-1} a^{-1}, and this is seen in India.

MSW generation among world cities too shows a strong correlation between the gross domestic product and MSW generation, irrespective of the size of the cities (Fig. 8.2).

5. MSW COMPOSITION

Composition of MSW is dynamic and changes with factors such as income level,

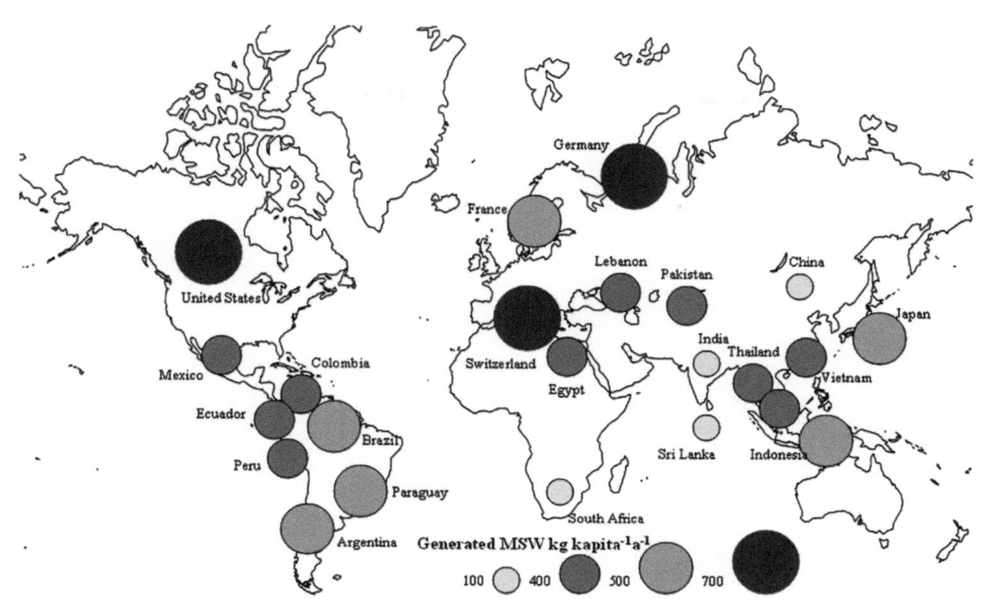

FIGURE 8.1 Municipal solid waste generation (kg capita^{-1} a^{-1}) in 25 countries grouped according to their gross national income (GNI). *Source: eawag: Swiss Federal Institute of Aquatic Science and Technology.*

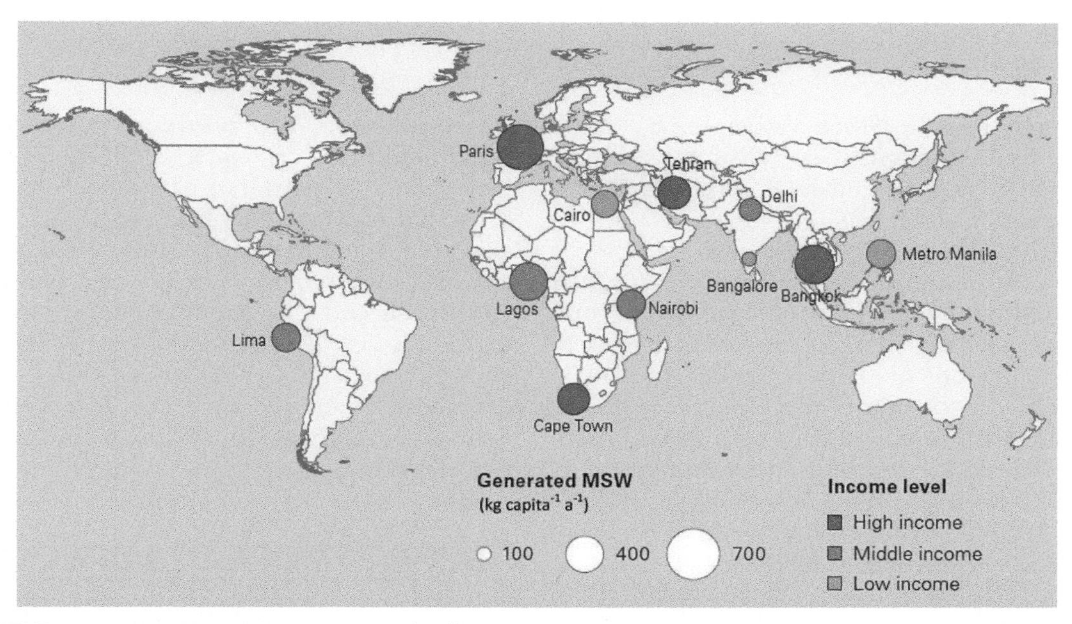

FIGURE 8.2 Generation of MSW (kg capita^{-1} a^{-1}) in 11 cities and their gross domestic product in 2005 (in US$, using *purchasing power parity* exchange rates) per capita according to the World Bank's income classification of 2006. Source: eawag: Swiss Federal Institute of Aquatic Science and Technology.

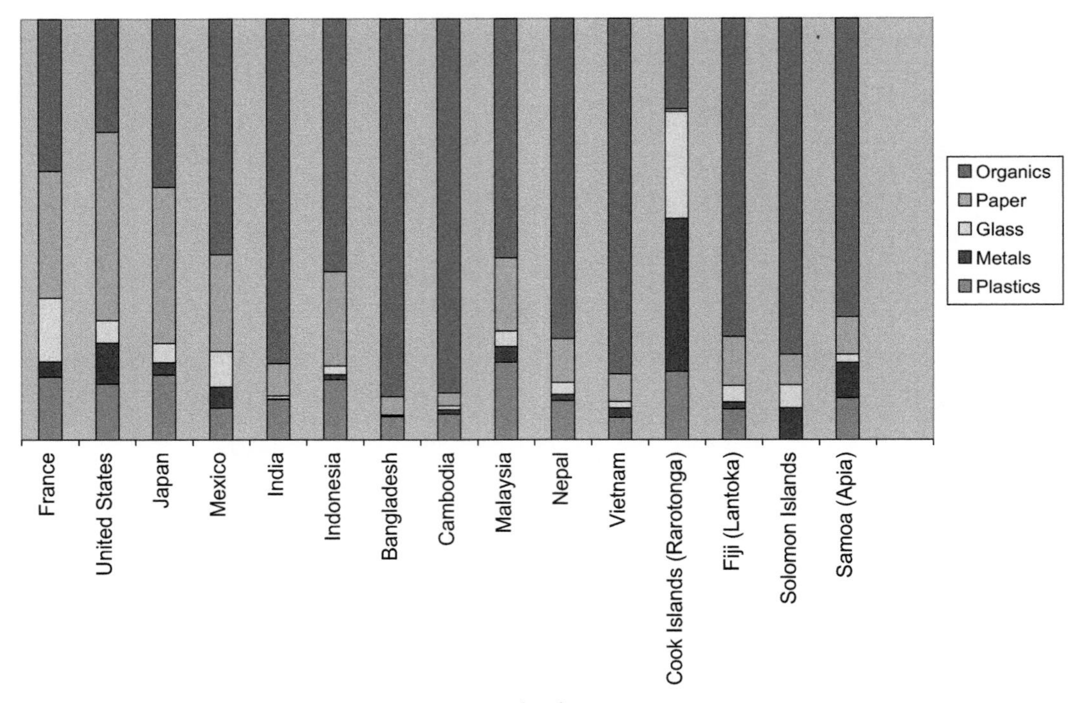

FIGURE 8.3 Top: Composition of the MSW (kg capita^{-1} a^{-1}) in 12 countries grouped according to their gross national income (GNI). Bottom: Composition of the MSW (kg capita^{-1} a^{-1}) in 23 cities. *Source: eawag: Swiss Federal Institute of Aquatic Science and Technology.*

changing lifestyle, season, residence type, affluence, and location. Generally, the organic component is predominant, especially in developing nations (Fig. 8.3).

It is also observed that the poorer households, with lower income, generate more organic food waste. A similar trend was observed in the rural areas where more organic waste is recorded. Higher amount of metals, plastics, and glass are typically the output of high-income households because it reflects the consumption of processed food. Opportunities to recycle depend on the MSW composition. For example, organic waste could be composted to produce cheap fertilizer for low-income nations.

The efficiency of waste management is determined by waste collection coverage which in turn is dependent on the wealth of the community. In most developing nations, waste collection and transport takes up the major portion of waste collection costs. However, this does not guarantee complete collection of waste. For example, in Sri Lanka and Philippines, only 40% of the waste generated is collected; in Vietnam and Paraguay, waste collection is about 50%, whereas in India, it is 70%. For Malaysia, waste collection covers almost 100%. This is also seen in most developed nations, such as in Germany, Switzerland, Japan, United States, and Argentina (Fig. 8.4).

6. TREATMENT AND DISPOSAL

MSW treatment and disposal depends very much on the waste quantity, the composition, and the available funds to pay for it. Rich

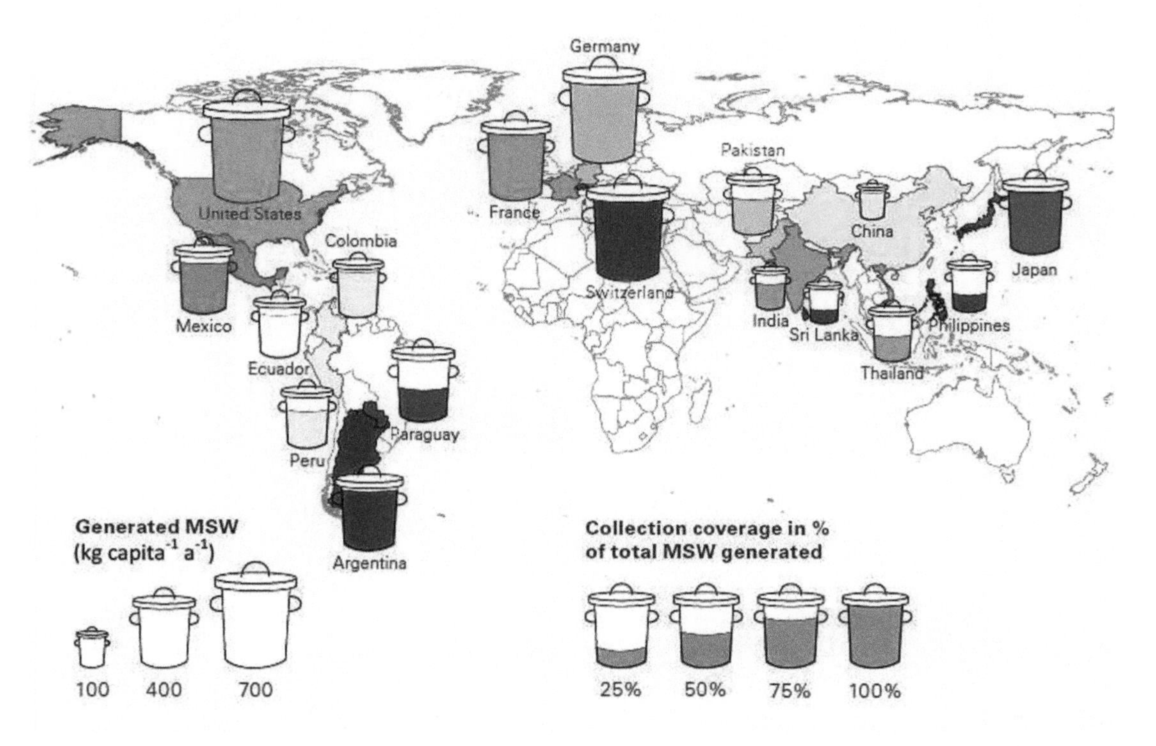

FIGURE 8.4 Total MSW generated (kg $capita^{-1}$ a^{-1}) and collection coverage in percentage in 17 countries. *Source: eawag: Swiss Federal Institute of Aquatic Science and Technology.*

nations can afford high-end technology such as incineration or pyrolysis, whereas most developing nations still depend on landfill disposal. Waste dumps are still prevalent in most Asian and African nations (Fig. 8.5). These are uncontrolled dumping sites, which do not use any geomembrane to prevent the flow of leachate. This results in frequent soil and water contamination problems.

Sanitary landfills are found in transitory nations, such as Malaysia, Bolivia, Brazil, Peru, and Mexico. These are engineered landfills where waste is frequently packed in layers with a geomembrane liner and a proper landfill cover [4]. Levels 3 and 4 landfills are modern ones where landfill leachate and gas are collected and treated before disposal or reuse.

7. WASTE MANAGEMENT AND CLIMATE CHANGE

MSW management requires major attention from the authorities, because it has been identified as one of the three main sources of environmental degradation in Asian countries by the World Bank. Emissions from landfills including leachate and landfill gas (LFG) require appropriate treatment technologies to curb environmental contaminations. It can easily be tackled in a sanitary landfill where apposite methods have been planned to accommodate the treatment of the emissions. However, it is an issue of concern when the emissions are released by nonsanitary landfills that lack lining systems and treatment facilities. In many nonsanitary

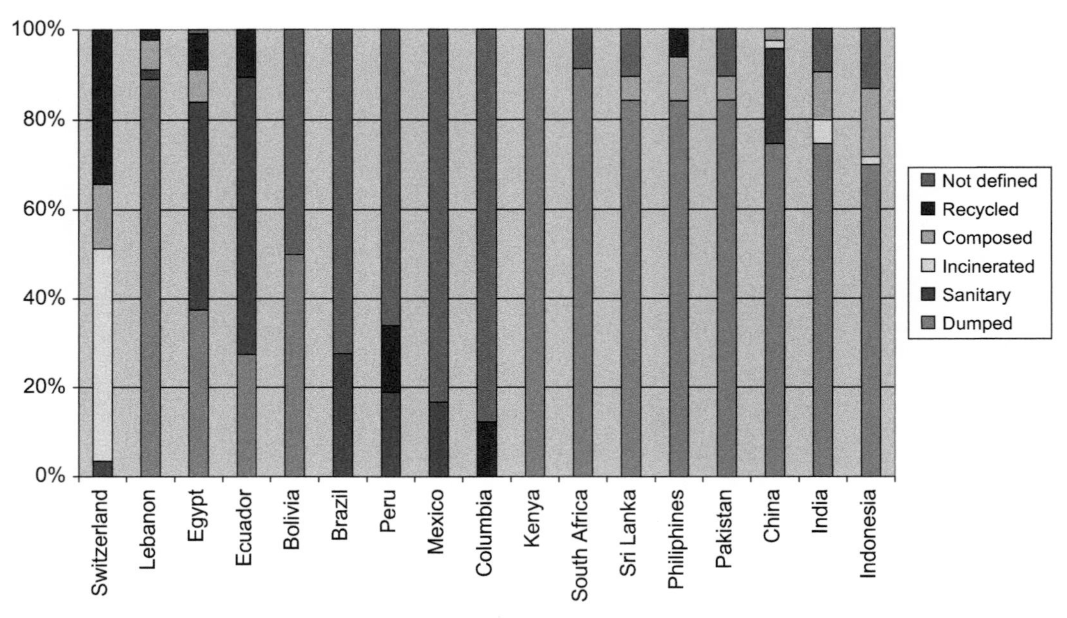

FIGURE 8.5 Percentage of the commonly used MSW treatment and disposal technologies in 21 countries. *Source: eawag: Swiss Federal Institute of Aquatic Science and Technology.*

landfills in Asia, the emissions are released directly into the environment. Although methane (CH_4) can be used to generate energy, the improper designs in nonsanitary landfills make it economically impossible for collection. Thus, passive release of LFG to the atmosphere is most widely practiced. LFG that contains greenhouse gases (GHG), including carbon dioxide (CO_2), CH_4, nitrous oxide (N_2O), perflourocarbons, sulphurhexaflouride (SF_6), and hydrofluorocarbon, contributes negatively to global warming [5]. It has been reported that solid waste and wastewater contribute 3% of the total GHG emission [6].

Approximately 18% of global anthropogenic CH_4 is sourced from landfill and wastewater, contributing 90% of the total emissions from the waste segment (Fig. 8.6). The daily disposal of MSW into landfill ensures continuous emission of LFG, which makes it necessary to take measures to reduce the volume of LFG generation. Table 8.1 details the quantitative assessment of GHG by common waste treatment technologies.

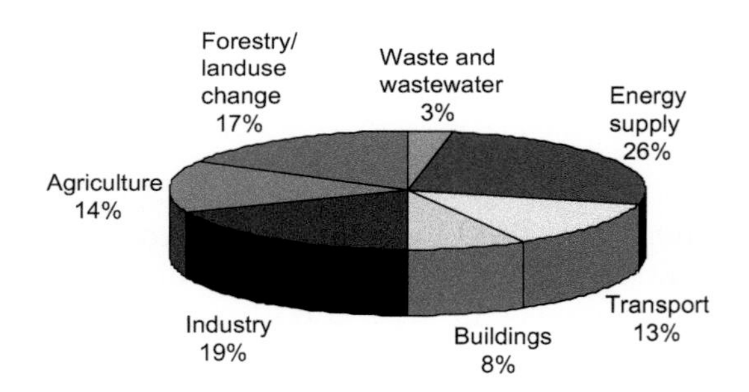

FIGURE 8.6 Distribution of anthropogenic GHG emissions by sectors in 2004 *(Adapted from Ref. [7]).*

For comparison, each unit of volume of CH_4 has 25 times more global warming potential than that of a unit of CO_2, whereas N_2O has global warming potential of 310 [1, 5]. It was reported that approximately 40–60 Mt CH_4 are contributed by LFG per year representing 11%–12% of global CH_4

TABLE 8.1 Qualitative Assessment of GHG Emissions by Common Waste Treatment Techniques [5]

	GHG generation					
	Pyrogenic			Biogenic		
	CO_2	CH_4	N_2O	CO_2	CH_4	N_2O
Landfilling	−	−	−	+	+	+
Landfill with gas utilization	+	−	+	+	+	+
Incineration	+	−	+	−	−	−
Mechanical biological	+	−	+	+	+	+

Treatment (with combustion of the light fraction). +, present; −, absent.

emission [7]. As a result, it is important that such gases are to be treated appropriately so as to reduce the global warming potential from landfill sites.

Developing countries produced more than 4.57×10^6 t of CO_2 Equivalent [where Equivalent (Eq) refers to all GHG including CH_4, perfluorocarbons, nitrous oxides, etc.] in 2000, and this is expected to increase by 84% in 2025 [8]. In a developing country such as Malaysia, daily MSW generation in 2000 of approximately 16×10^3 t resulted in a total amount of 26×10^6 t CO_2 Eq from CH_4 emission alone [9]. This is generally because of the lack of appropriate technology to mitigate LFG emissions [10]. On the other hand, a developed country such as Austria has managed to reduce the total GHG emission to 23×10^6 t by 2005 with the implementation of appropriate mitigation measures [5]. Thus, the recent Copenhagen meeting, COP15, had been a strengthening point to enable developed nations to assist the developing and underdeveloped countries in mitigating the release of GHG. The implementations of the Cleaner Development Mechanisms have been identified as a practical strategy toward the reduction of GHG emissions. The technologies should integrate various factors, including flexible strategies and financial incentives in the waste management options, to make it applicable to local conditions and more practical to be implemented.

8. MSW MANAGEMENT IN ISLANDS AND MARINE POLLUTION

As urbanization continues to take place, the management of solid waste is becoming a major environmental problem around the world. One of the most vulnerable ecosystems that needs attention in this respect are the islands. Very small settlements have historically required little or no waste management. When islands become the hub for tourist activities, the entire scenario changes. The human activities start producing more waste as the economy generators try to constantly meet the demands of the tourists so as to attract them in larger numbers [2]. This change often comes about so gradually that it is hardly noticed until the problem is serious. The rapid increase in the density of human population, in previously Virgin Islands, because of leisure or tourism activities is making the collection, treatment, and disposal of MSW an insurmountable problem [11]. The cornerstone of successful planning for a MSW management program is the availability of reliable information about the quantity and the type of material being generated and an understanding of how much of that material the collection program manager can expect to prevent or capture [12].

MSW management on islands includes waste generation, composition, collection, treatment, and disposal. MSW generation rates and composition on islands vary from country to country depending on the economic situation, industrial structure, waste management regulations, and lifestyle. Generally, the daily MSW

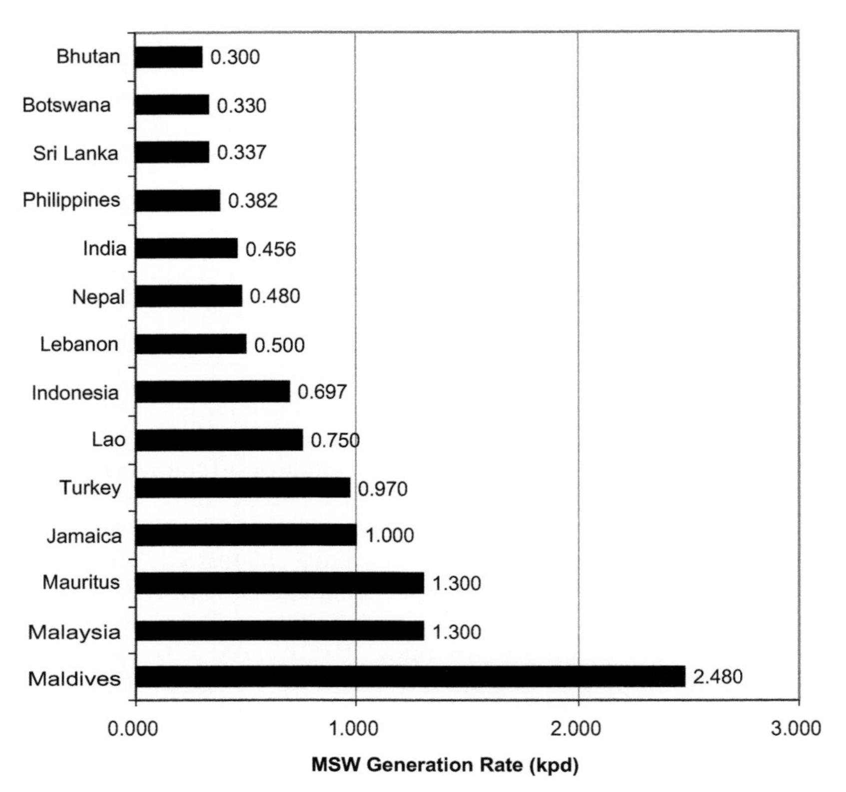

FIGURE 8.7 MSW generation rate in various countries/islands. Garden waste (14%–44%), food waste (2%–45%), and plastic (4%–16%) are the main components that form the MSW generated on islands. The generation rates are given in units of kg d^{-1} [44].

generation rate for islands is within the range of 0.3–1.4 kg ca^{-1} d^{-1} (Fig. 8.7), except for Maldives that records the highest waste generation rate of 2.5 kg ca^{-1} d^{-1} [13]. MSW composition in different countries, islands, and areas is summarized in Table 8.2 [14, 15].

Increasing trends in waste generation were commonly observed on islands. Most of the small islands have poor infrastructure such as roads and are only accessible by boats. Therefore, collection of waste is very limited. In some islands, almost 40–80% of wastes are not collected [11]. In the case of Sri Lanka, Gunawardana et al. [16] reported that only about 24% of households have access to waste collection. The low collection rate is primarily due to resource constraints such as lack of collection vehicles, labor, or appropriate disposal sites. For the disposal of waste, the most common methods were dumping into a backyard pit (57%) and open backyard dumping (37%). Few incidences of disposing waste into waterways were observed (1%). Waste disposal activities can cause a problem when trash is lost during collection or transportation or when trash blows off or is washed away from disposal facilities. In some islands such as Hawaii [17] and Hong Kong [18] due to limited disposal options in the island, the MSW has to be transported to the mainland via large boats/barges. Waste disposal has been a major concern for Hawaii. A total of 4.56×10^5 t of MSW has to be transported to landfills annually, using baling facilities in Washington, Idaho, and Oregon [17]. If a barge can carry 4535 t of MSW, this would result in

TABLE 8.2 MSW Composition (% on Wet Weight) in Different Countries, Islands, and Areas [14], [15]

Component	Corfu Island[a]	Manila[b]	Green Island[c]	Singapore[d]	Phuket[e]	Penang Island[f]
Organics	45.0	53.7	39	38.8	49.3	63
Paper	22	12.9	27	20.6	14.7	5
Glass	4.0	3.5	6	1.1	9.7	NA
Plastics	11.0	1.6	5	5.8	15.1	17
Metals	4.5	5.8	2	2.7	3.4	4
Leather	5.0	1.8	20	9.8	2.3	3
Inerts	3.0	20.7	1	4.5	1.4	NA
Others	5.5	NA	NA	16.7	4.1	8

NA, not available. Source: [a] Skordilis, 2004; [b] USEPA, 1998; [c] EPA Taiwan, 2002; [d] Renbi and Mardina, 2002; [e] Liamsanguan and Gheewala, 2008; [f] UNDP Malaysia, 2008.

approximately 100 barge trips per year, slightly more than 1 per week. The trip across the Pacific Ocean will take at least 12–18 days. It is very common for few bags of garbage to drop into the ocean due to full barge/boat loads or rough sea conditions.

Landfill is the most common MSW disposal method on islands. Green Island is a 15-km^2 island, 33 km east of Taiwan, all wastes are landfilled on the island, and the site will be full by 2010 if current trends of waste generation and handling continue [19]. Alternative landfill sites are not available [20]. As an alternative, incineration has been used as a disposal method [21]. Taiwan has built 22 incinerators over a short span of time, combusting large amounts of MSW as much as 2.33×10^4 t d^{-1}. Approximately 53% of MSW was treated by incineration [22]. The same scenario has been observed in Hong Kong [18] and Singapore Islands [23]. In Singapore, about 90% of wastes are incinerated and 10% is landfilled [24]. Because there is no other site on the mainland suitable to develop a landfill, the Singapore government has resorted to the more costly option of developing an offshore landfill – Pulau Semakau [23]. Other treatment options such as recovery, recycling, and reuse (three Rs) have been successfully practiced in some urban islands. However, the effort for three Rs is driven by government incentives or regulations [25]. The Guam Integrated Solid Waste Management Plan provides incentives to recycling companies [25]. The benefits include a 100% corporate income tax rebate. Materials reported to be recycled have increased dramatically over the past 5 years, from just 250 t of nonferrous metals in 2002 to more than 11.1×10^3 t of the same in 2005 [25]. While in Singapore, upon implementation of Public Cleansing and General Waste Collection Regulations, in 2007, about 3.03×10^6 t of waste was recycled, and an overall recycling rate of 54% was achieved [23]. The major issues and challenges on waste management on islands are summarized in Table 8.3.

Ocean dumping has been a common method of waste disposal around islands in the world. About 6.4×10^6 t of litter normally ends up in the ocean annually [1]. Almost 80% of ocean debris can be traced to land-based sources such as inadequately treated municipal waste, storm water runoff, beach use, and littering [26]. Barges from US coastal cities routinely carried trash out into the open ocean and dumped it. It was not until 1988 that the US banned the dumping of industrial and sewage wastes into the ocean

TABLE 8.3 Issues and Challenges of Waste Management on Islands [2]

No.	Issues and Challenges
1	Pollution of groundwater, surface, and marine pollution from land-based sources such as domestic sewage, industrial effluents, and agricultural runoff: they carry risks for human health and can degrade habitats such as coral reefs and tourist attractions such as beaches; many islands receive bad publicity related to disease outbreaks and the destruction of fisheries, which can have major adverse economic impacts
2	The management of toxic substances such as pesticides, waste oil, heavy metals: most islands do not have the systems or physical capacity to isolate and dispose off such substances
3	Sewage treatment facilities: on many islands such facilities are inadequate either because they are overloaded or because of a shortage of trained manpower; as a result, poorly treated effluent is often discharged into the environment
4	Ineffective regulations: some islands have spent a considerable amount of time and financial resources on developing regulations; however, regulations have not been very effective in many cases because of inadequate institutional and human resource capacities to enforce them
5	Lack of waste disposal sites: gullies and the marine environment are still used as disposal sites by some islands because of the shortage of land and inadequate capacity to collect garbage; the inability to manage solid waste disposal facilities is a common problem for islands, and disposal sites can easily become foci of disease transmission
6	Lack of facilities for storage and disposal of hazardous wastes

[27]. Alarming quantities of rubbish thrown out to sea continue to endanger people's safety and health, entrap wildlife, damage nautical equipment, and deface coastal areas around the world [28]. Plastic – especially plastic bags and PET bottles – is the most pervasive type of marine litter around the world, accounting for more than 80% of all rubbish collected in several of the regional seas assessed [1]. In the Ocean Conservancy's 2007, International Coastal Cleanup (ICC), 378,000 volunteers cleaned 53,000 km of shoreline worldwide and removed 2.7×10^6 kg of debris in 1 day [29]. Table 8.4 summarizes the top 10 debris items collected worldwide from 1989 to 2008 during the ICC. About 57% of the debris was related to tourism/shoreline recreational activities.

The tourism and recreation sector has a significant impact on the state of seas and coastlines around the world. In some tourist areas in the Mediterranean, more than 75% of the annual waste production is generated during the summer season [29]. In Thailand, it is recognized that marine litter affects tourism, a high-value industry, for the entire region [30]. Shoreline activities account for 58% of the marine litter in the Baltic Sea region and almost half in Japan and the Republic of Korea [30]. In Jordan, the major source of marine litter is recreational and leisure usage contributing up to 67% of the total discharge, whereas shipping and port activities

TABLE 8.4 Top Ten Debris Items Collected Worldwide From 1989 to 2008 [29]

	Debris items	Counts	Percent
1	Cigarettes/cigarette filters	28.4×10^6	25.2
2	Caps, lids	10.3×10^6	9.2
3	Food wrappers/containers	10.1×10^6	9.0
4	Bags (paper and plastic)	8.0×10^6	7.1
5	Cups, plates, forks, knives, spoons	7.8×10^6	7.0
6	Beverage bottles (plastic)	6.3×10^6	5.7
7	Beverage bottles (glass)	5.4×10^6	4.8
8	Beverage cans	5.2×10^6	4.6
9	Straws, stirrers	5.0×10^6	4.5
10	Rope	2.4×10^6	2.1
	Top ten totals	89.1×10^6	79.1
	Global debris totals	112.6×10^6	100.0

contribute around 30% and the fishing industry 3% only [30].

The problem of marine litter is likely to be particularly severe in the East Asian Seas region—home to 1.8×10^9 people, 60% of whom live in coastal areas. The lack of adequate solid waste management facilities results in MSW entering the waters of the Western Indian Ocean, South Asian Seas, and southern Black Sea, among others [31]. This trash is washed, blown, or dumped on land, eventually ending up on beaches or floating out to sea, where ocean currents may take it hundreds of miles from its launching point. The Great Pacific Garbage Patch, a massive garbage pile floating in the ocean about 1600 km north of Hawaii, consists of an estimated 3.5×10^6 t of trash and is scattered over an area roughly twice the size of Texas, United States [32]. The garbage comes from countries all over the world and is trapped there by the Pacific gyre (a rotating system of the ocean currents) and wreaked havoc on fish and seabirds. Seagulls dragging a piece of fishing line, pelicans with six-pack rings around their necks, or sea lions struggling to remove a piece of discarded fishnet are some common examples of the problems marine debris causes to wildlife [32]. In addition, sea birds, sea turtles, and whales have been known to mistake floating plastic pellets and plastic bags for natural prey, such as fish eggs, jellyfish, and squid. Ingesting plastic can cause internal injury, blockage of the digestive tract, and starvation in these animals [26]. The United Nations Environment Programme estimates that more than 1 million seabirds and 100,000 marine animals die every year from ingesting plastics [1].

Unsightly and unsafe marine litter can also cause serious economic losses through damaged boats, fishing gear, contamination of tourism, and agriculture facilities. For example, the cost of cleaning the beaches in Bohuslän on the west coast of Sweden in just 1 year was at least 10 million SEK (US$ 1.55×10^6) [30]. In the United Kingdom, Shetland fishermen had reported that 92% of them had recurring problems with debris in nets, and it has been estimated that each boat could lose between US$ 10,500 and US$ 53,300 per year because of the presence of marine litter. The cost to the local industry could then be as high as US$ 4.3×10^6 [30]. According to the US National Oceanographic and Atmospheric Administration's Office of Response and Restoration, in 2005, the US Coast Guard found that floating and submerged objects caused 269 boating accidents resulting in 15 deaths, 116 injuries, and US$ 3×10^6 in property damage [27]. Although policies on ocean dumping in the recent past took an "out of sight- out of mind" approach, it is now known that accumulation of waste in the ocean is detrimental to marine and human health [30]. Another unwanted effect is eutrophication [33]. A biological process where dissolved nutrients cause oxygen-depleting bacteria and plants to proliferate creating a hypoxic, or oxygen poor, environment that kills marine life. In addition to eutrophication, ocean dumping can destroy entire habitats and ecosystems when excess sediment builds up and toxins are released. Although ocean dumping is now managed to some degree, and dumping in critical habitats and at critical times is regulated, toxins are still spread by ocean currents [33]. Alternatives to ocean dumping include recycling, producing less wasteful products, saving energy, and changing the dangerous material into more beneficial waste [34]. Marine pollution is closely linked to the wider problem of waste management and cannot be resolved independently.

9. INTRODUCTION TO MSW POLICY AND LEGISLATION

This section provides an overview of policies and legislation in MSW management in terms of its trends in an international context, its key features, and potential impacts on MSW management. This is expected to provide a brief and selected temporal and spatial framework of the evolution of MSW policies and legislation.

9.1. MSW Policies and Legislation in an International Context

Generally, MSW policies and legislation around the world have evolved from simple and informal policies on waste management to specialized and complex waste management policies and legislation. Initially, MSW policies and legislation were focused on waste collection, transport, and disposal but gradually have advanced to incorporate elements of source reduction, waste minimization, and sustainable patterns of production and consumption. Today, polices and legislation on MSW management are no longer just about cleansing and sanitation but encompass a larger global perspective of environmental protection and sustainable development.

Specifically, one of the most significant policies on MSW management is found in Agenda 21, which is a comprehensive global program on sustainable development adopted at the United Nations Conference on Environment and Development in 1992, and is a reflection of a global consensus and commitment at the highest level by governments on development and environmental cooperation. Chapter 21 of Agenda 21 addresses the environmentally sound management of solid waste and emphasizes that MSW management must go beyond the mere safe disposal or recovery of wastes and seek to address the root cause of the problem by attempting to change unsustainable patterns of production and consumption. Agenda 21 also requires a national program on MSW reuse and recycling for industrialized countries by the year 2000 and for developing countries by 2010 [35]. The commitments of Agenda 21 on addressing unsustainable patterns of consumption and production and waste management were reaffirmed in the World Summit on Sustainable Development at Johannesburg in 2002. A concise review of MSW policies and legislation in the European Union, Japan, and Malaysia is presented below.

European Union: The main MSW legislation in the European Union is the EU Directive 2008/98/EC that was enacted in 1975 and amended in 1991 and 2008. The EU Directive 2008/98/EC aims to protect the environment and human health through the prevention of the harmful effects of waste generation and waste management [36]. The EU Directive 2008/98/EC has set the targets for selected waste streams and is complemented by the following legislation:

Directive 94/62/EC on packaging and packaging waste.
Directive 96/61/EC concerning integrated pollution prevention and control.
Directive 1999/31/EC on the landfill of waste.
Directive 2000/76/EC on the incineration of waste.
Directive 2000/53/EC on end-of-life vehicles
Directive 2002/96/EC on waste electrical and electronic equipment
Directive 2005/64/EC on reusability, recyclability, and recoverability of motor vehicles
Directive 2006/66/EC on batteries and accumulators.

MSW policy in the European Union was initiated by its EU 1st Waste Strategy in 1989 and EU 2nd Waste Strategy in 1996 [37]. The current main MSW policy in the European Union is the EU Strategy on the Prevention and Recycling of Waste 2005 that sets out guidelines and measures to reduce the environmental impacts of waste production and management. The main thrust of the strategy is on preventing waste and promoting effective recycling (Table 8.5).

Japan: The main MSW legislation in Japan is the Waste Management and Public Cleansing Law (WMPC) enacted in 1970 and amended in

TABLE 8.5 European Union's MSW Recycling Targets

Reuse/recycling target	2020
Reuse and recycling of waste materials such as paper, metal glass from households, and similar waste streams	50%
Reuse and recycling of nonhazardous construction and demolition waste	70%

2001 [38]. The WMPC aims to protect the environment and improve public health and is complemented by the following legislation:

Law for Promotion of Effective Utilization of Resources 1991
Containers and Packaging Recycling Law 1995
Home Appliances Recycling Law 1998
Basic Law for Establishing the Recycling-Based Society 2000
Construction Materials Recycling Law 2000
Food Recycling Law 2000
Law on Promoting Green Purchasing 2000
End-of-life Vehicle Recycling Law 2002

The main MSW policy in Japan is the Basic Policy for Comprehensive and Systematic Promotion of Measures on Waste Reduction and Other Proper Waste Management 2001 [39]. The basic waste policy sets the direction and targets for waste generation, reduction, and disposal as well as outlines the responsibilities of citizens, businesses, local governments, and the national government. The Japan Waste Policy 2001 has set the targets for MSW management in Japan (Table 8.6) [40].

Malaysia: The main MSW legislation in Malaysia is the Solid Waste and Public Cleansing Management Act 2007 [41]. The MSW Act aims to maintain proper sanitation and matters relating to sanitation and has provisions to prescribe additional legislation for the following:

Reduction, reuse, and recycling of MSW including the use of environmentally friendly or recycled material.

TABLE 8.6 Japan's MSW Targets

Target criteria	1997	2005	2010
Volume generated	53	51	49
Volume recycled	5.9 (11%)	10 (20%)	12 (24%)
Volume by treatment	35 (66%)	34 (67%)	31 (63%)
Volume for disposal	12 (23%)	7.7 (15%)	6.4 (13%)

Unit: 907,184 t/yr.

Take back and deposit refund systems including the responsibilities for manufacturers, assembler, importer, or dealers.

Methods, levels of recycling, and separation and storage of MSW.

The main MSW policy in Malaysia is the National Strategic Plan for Solid Waste Management 2005 (NSP) and the Master Plan for Waste Minimization 2006 (MWM). The NSP aims to achieve sustainable waste management through the reduction, reuse, recycling, and use of appropriate technologies, facilities, and equipment and has set the MSW management targets for Malaysia. The MWM aims to provide the strategic direction for stakeholders to minimize the amount of MSW disposed in Malaysia (Table 8.7) [42].

9.2. Key Trends in MSW Policies and Legislation

Key trends in MSW policies and legislation observed from an overview of the European

TABLE 8.7 Malaysia's MSW Targets

Level of service	2002	2003–2009	2010–2014	2015–2020
Extend collection service	75%	80%	85%	90%
Reduction and recovery	3%–4%	10%	15%	17%
Closure of dump sites (112 sites)	0%	50%	70%	100%
Source separation (urban areas)	None	20%	80%	100%

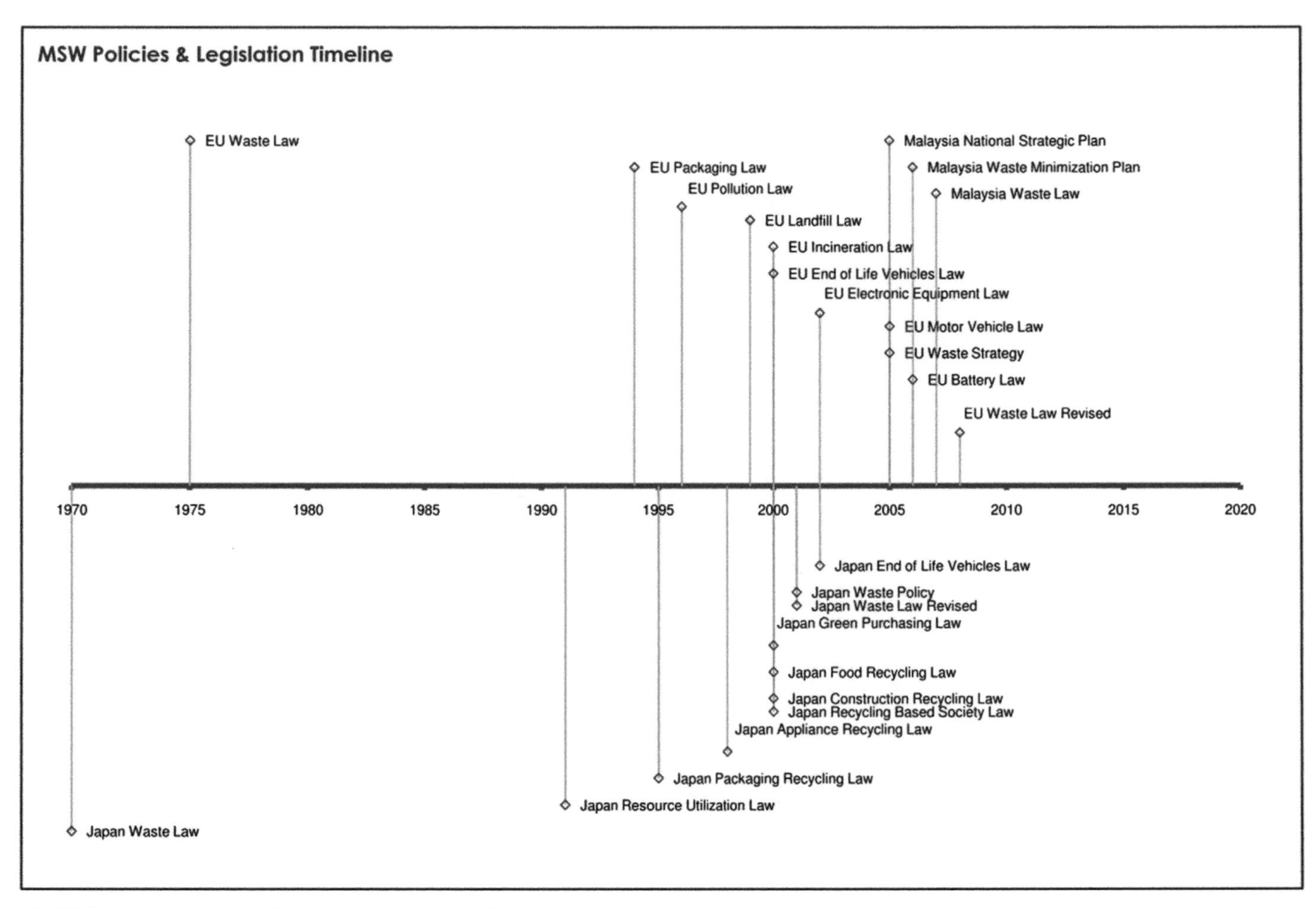

FIGURE 8.8 Municipal Policies & Legislation Timeline

Union, Japan, and Malaysia indicate the following:

The basic waste legislation for both European Union and Japan were formulated in the 1970s and expanded to other waste-related legislation in the 1990s and early 2000. Malaysia, representative of a developing country, only enacted its waste legislation in 2007.

Initial MSW legislation formulation focused on basic aspects of MSW management such as collection, transport and disposal, and sanitation, which were the focus of European Union and Japan in the early 1970s. Subsequently, MSW legislation expanded in the 1990s to include MSW reduction, reuse, and recycling including environmental protection measures. Finally, MSW legislation began to incorporate elements of sustainable forms of consumption and production in the 2000s.

Policies and strategies of MSW management initially seem to be embedded within the European Union and Japan's waste legislation and only recently have been formulated into a holistic waste policy and strategy in the 2000s. Interestingly, Malaysia that is a developing country first formulated its policy and strategy before enacting its waste legislation that may be indicative of lessons learned from the more developed countries on MSW management.

Common elements addressed in MSW policies and legislation are content restriction, source separation, producers responsibility, recycling targets, minimum recycled material content standards, landfill restriction/diversion targets, and environmentally sound treatment standards [43] (Fig. 8.8).

Both the European Union and Japan as developed countries have enacted specific legislation on packaging recycling, appliance recycling, and end-of-life vehicle recycling, whereas Japan has enacted additional legislation pertaining to recycling-based society, construction waste recycling, and green purchasing. Malaysia has provisions in its basic waste legislation for enacting additional and specific legislation pertaining to most of the above but has not yet implemented it, including the requirement for source separation or packaging recycling.

The potential impacts of MSW policies and legislation varied as these were evolving over time. Generally, impacts of MSW policies and legislation on basic waste management and sanitation can be considered positive, whereas impacts of MSW policies and legislation on waste recycling and environmental pollution control are varied, and finally, impacts of MSW policies and legislation on decoupling MSW generation with economic growth have not been positive [4].

Experience in the European Union indicates that MSW policies and legislation may have unintended effects in directing MSW in the wrong direction when market forces are not integrated in the planning [37].

Revisions of European Union and Japan's basic waste legislation in 2000s may indicate that MSW management is attempting to move toward sustainable patterns of consumption and production, where MSW policies and legislation will soon integrate life cycle thinking.

References

[1] P. Agamuthu, Waste Management and Research 27 (2009) 551–552.

[2] UNEP, Marine Litter in the Wider Caribbean Region: A Regional Overview and Action plan, 2008. Online: <http://www.cep.unep.org/about-cep/amep/marine-litter-in-the-wider-caribbean-a-regional-overview-action-plan> (Retrieved 11.10.2009).

[3] eawag, Swiss Federal Institute of Aquatic Science and Technology, Collaborative Working Group on Solid Waste Management in Low- and Middle-Income Countries, 2008, Eawag, Dubendorf, Switzerland, 2008.

[4] P. Agamuthu, S.H. Fauziah, V. Navarani, L. Khairudin, Proceedings of the international symposium of environmental science and technology. Vol. II Part A, Science Press, Shanghai, China, 2009 823–829.

[5] A. Hackl, G. Mauschitz, Waste Management and Research 26 (2008) 5–10.

[6] H.H. Rogner, D. Zhou, R. Bradley, P. Crabbé, O. Edenhofer, B. Hare, L. Kuijpers, M. Yamaguchi, Introduction, in: B. Metz, O.R. Davidson, P.R. Bosch, R. Dave, L.A. Meyer (Eds.), Climate Change 2007, Cambridge University Press, Cambridge, United Kingdom, 2007.

[7] M. Ritzkowski, R. Stegmann, International Journal of Greenhouse Gas Control 1 (2007) 281–288.

[8] N.L. Bindu, ENSEARCH Workshop, Asian Development Bank, Climate Change: Challenges and Business Opportunities, 2009, Environmental Management and Research Association of Malaysia, Kuala Lumpur, 2009.

[9] P. Agamuthu, S.H. Fauziah, International Conference on Climate Change and Bioresources, Tiruchirapalli, India, 2010.

[10] USEPA, Guidance for Landfilling Waste in Economically Developing Countries, USEPA., 1998.

[11] World Health Organization, Guides for Municipal Solid Waste Management in Pacific Island Countries. Western Pacific Regional Environmental Health Centre (EHC), Kuala Lumpur, Malaysia, 1996, pp. 12.

[12] E. Gidarakos, G. Havas, P. Ntzamilis, Waste Management 26 (2006) 668–679.

[13] A.M. Troschinetz, J.R. Mihelcic, Waste Management 29 (2008) 915–923.

[14] C. Liamsanguan, S.H. Gheewala, Journal of Cleaner Production 16 (2008) 1865–1871.

[15] UNDP Malaysia, Developing Solid Waste Management Model for Penang. United Nations Development Programme (UNDP), Malaysia, 2008. Online: <www.undp.org.my/uploads/SWM-2008_final.pdf> (Retrieved 11.06.2009).

[16] E.G.W. Gunawardana, S. Shimada, B.F.A. Basnayake, T. Iwata, Influence of biological pre-treatment of municipal solid waste on landfill behaviour in Sri Lanka, Waste Management & Research 27 (2009) 456–462.

[17] R.H. Shannon, Regional Movement of Plastic-baled Municipal Solid Waste from Hawaii to Washington, Oregon, and Idaho, 2008, Online: <http://www.regulations.gov/search/Regs/contentStreamer> (Retrieved 24.01.2009).

[18] Z. Keyuan, Ocean and Coastal Management 5 (2009) 383–389.

[19] M.C. Chen, A. Ruijs, J. Wesseler, Resources, Conservation and Recycling 45 (2005) 31–47.

[20] A. Skordilis, Resources, Conservation and Recycling 41 (2004) 243–254.

[21] EPA Taiwan, Solid Waste: Year Book of Environmental Protection Statistics, Taiwan Area, Environmental Protection Agency, Taiwan, 2002, 126–150.

[22] J.H. Kuo, H.H. Tseng, R.P. Srinivasa, M.Y. Wey, Applied Thermal Engineering, Article in Press, 2008.

[23] Eugene, Waste Management and Recycling in Singapore, 2008, Online: <http://www.asiaisgreen.com.> (Retrieved 11.10.2009).

[24] B. Renbi, S. Mardina, Waste Management 22 (2002) 557–567.

[25] Green Island Alliance, Approaches to Pacific Island Waste-Stream Reduction Through Scrap Recycling, 2005, Online: <http://www.greenislandalliance.org/main2/EPAReport_fnl.pdf> (Retrieved 11.06.2009).

[26] C. Gillian, Island Studies Journal 1 (2006) 125–142.

[27] GOJ, Waste Management and Public Cleansing, Law, Government of Japan (GOJ) (2001).

[28] Earth Force, Littering and Illegal Dumping: A Historical Perspective, 2006, Online: <http://www.earthforce.org/content/article/detail/1568> (Retrieved 11.10.2009).

[29] A. Michelle, W. Adam, S. David, J. Paul, Plastic Debris in the World's Oceans, Greenpeace International, Amsterdam, Netherlands, 2009, Online: <www.oceans.greenpeace.org> (Retrieved 11.10.2009).

[30] Ocean Conservancy, Guide to Marine Debris and the International Coastal Cleanup, 2009, Online: <http://www.oceanconservancy.org/site/PageServer?pagename=icc_home> (Retrieved 11.10.2009).

[31] J. Ljubomir, S. Seba, A. Ellik, Marine Litter: A Global Challenge, 2009, Online: <http://www.unep.org/regionalseas/marinelitter/publications/docs/Marine_Litter_A_Global_Challenge.pdf> (Retrieved 11.10.2009).

[32] UNEP, Global Environment Outlook 3 (GEO 3), 2009, Online: <http://www.unep.org/GEO/geo3/english/index.htm> (Retrieved 26.04.2010).

[33] E. Jesse, Pacific Garbage Patch Plastic Takes Toll on Birds, 2009, Online: <http://www.newsweek.com/id/226075> (Retrieved 11.10.2009).

[34] The Marine Bio Conservation Society, Ocean Dumping, 2009, Online: <http://marinebio.org/Oceans/ocean-dumping.asp> (Retrieved 11.10.2009).

[35] UNEP Island, Waste Management in Small Island Developing States, Progress in the Implementation of the Programme of Action for the Sustainable Development of Small Island Developing States, United Nations Commission on Sustainable Development, 2009.

[36] EU, Directive 2008/98/EC of the European Parliament and of the Council, Official Journal of the European Union, 2008.

[37] UN, Conference on Environment & Development Rio de Janerio, Brazil, Agenda 21, United Nations, 1992.

[38] C. Monkhouse, A. Farmer, Applying Integrated Environmental Assessment to EU Waste Policy, European Forum on Integrated Environmental Assessment, 2003, Online <http://www.ieep.eu/publications/pdfs/2003/efieafinalreport.pdf> (Retrieved 26.04.2010).

[39] GOM, Solid Waste and Public Cleansing Management Bill 2007, Government of Malaysia (GOM), 2007.

[40] MOEJ, Basic Policy for Comprehensive and Systematic Promotion of Measures on Waste Reduction and Other Proper Waste Management, Announcement No. 34 of the Ministry of the Environment Japan (MOEJ), 2001.

[41] N. Tojo, Waste Management Policies and Policy Instruments in Europe, International Institute for Industrial Environmental Economics (IIIEE) at Lund University, Sweden, 2008.

[42] MHLG, National Strategic Plan for Solid Waste Management, Local Government Department, Ministry of Housing and Local Government (MHLG), Malaysia, 2005.

[43] EEA, Taking Sustainable Use of Resources Forward: A Thematic Strategy on the Prevention and Recycling of Waste, Commission of the European Communities, Brussels, 2005.

[44] K. Sakurai, T. Hoo, A Practitioner's Guide for Municipal Solid Waste Management in Pacific Island Countries, WHO Environmental Health Centre, Kuala Lumpur, 1996, pp. 60–74.

CHAPTER

9

Wastewater: Reuse-Oriented Wastewater Systems—Low- and High-Tech Approaches for Urban Areas

Ralf Otterpohl, Christopher Buzie

Institute of Wastewater Management and Water Protection, TUHH-Hamburg University of Technology, D-21073 Hamburg, Germany

OUTLINE

1. INTRODUCTION

In biowaste management, as in solid waste management, the priority is to reuse rather than dispose and there are several options available. The key issue as far as the reuse of wastewater systems is concerned is to collect and treat toilet water (blackwater or dry faecal matter and urine) and wastewater (greywater) separately. There are three main lines of approaching the problem: high-tech systems based mainly on vacuum technology, membrane bioreactors, and biogas plants; urine diverting flush systems; and, finally, the very promising modern dry sanitation systems that can be very cost efficient and furthermore produce black soil and water that can be reused [terra preta sanitation (TPS)]. This chapter gives an overview to some

Waste Doi: 10.1016/B978-0-12-381475-3.10009-9

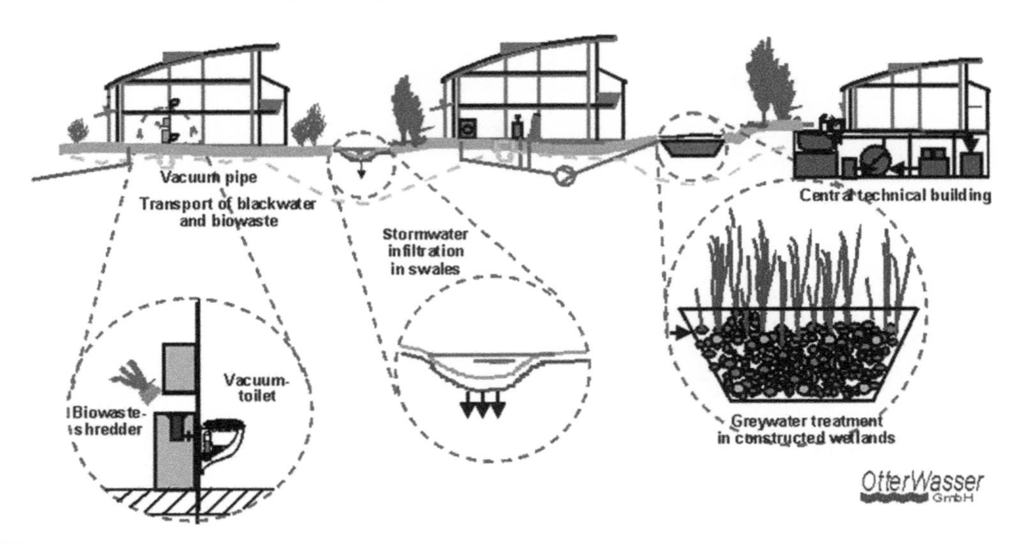

FIGURE 9.1 The vacuum–biogas system Lubeck Flintenbreite, Germany.

of the different solutions; as there are many installations and systems operating, this chapter will focus on just a few examples.

2. HIGH-TECH SYSTEMS FOR REUSE OF WASTEWATER AND NUTRIENTS

2.1. The Vacuum–Biogas Concept

The vacuum–biogas system was developed by Otterwasser, Lübeck, and first published by Otterpohl in 1993 [1]. The first system was installed in Lübeck Flintenbreite in 1999 and was designed for a population of 250 people. The houses were equipped with vacuum toilets, small diameter blackwater pipes (40–50 mm) from toilets to vacuum pump. The waste was then mixed with gridded biowaste, sanitised, and treated in a digestor to produce biogas and liquid fertiliser that could either be dried or transported for agricultural use [2].

The system is sketched in Fig. 9.1, and the toilet is photographed in Fig. 9.2.

FIGURE 9.2 The toilet bowl of the vacuum–biogas system.

The summary of specifications involved in the vacuum toilet system (Roediger, Germany) is:

- 0.7 L/flush
- small-diameter pipes
- evacuation pump station needed
- pneumatic control of the valves
- can lift water up to 4.5 m
- technically complex, maintenance needed
- rather for groups of houses or hotels, office buildings
- scaling in pipes: acid around every 5 years (depending on water hardness)
- there is a need to explain the functioning of the process to users, to avoid clogging

Another vacuum—biogas system was installed in Freiburg Vauban by Jörg Lange and Arne Panesar for a combined flat/office house at around the same time [3]. Although the cost of the system in Lübeck is similar to the cost of a conventional sanitation (taking into account both investment and operation), the cost of the system in Freiburg is relatively high. This highlights the need to consider the size and extent of each project. It turns out that these systems require a minimum size of installation for around several hundred people with the houses not too far apart, if the costs are to be reasonably low. However, prices today reflect a niche market and can drop drastically with large numbers of installations.

At about the same time in the late 1990s, Petter Jenssen independently installed a vacuum—blackwater system in Norway. In this case, the treatment was done using a standard aerobic thermophilic treatment system. This type of system has also been built in Berlin by Berliner Wasserbetriebe (Berlin Water Utility) [4] with the further development of a gravity urine diversion in the vacuum toilets, making the brownwater (flush toilet water containing faeces only, urine excluded) a lot more concentrated [5]. See Fig. 9.3.

In an EU-Project related to the Centre of Competence for Water Berlin Water Works/ Veolia Water [6] gravity separation toilets have

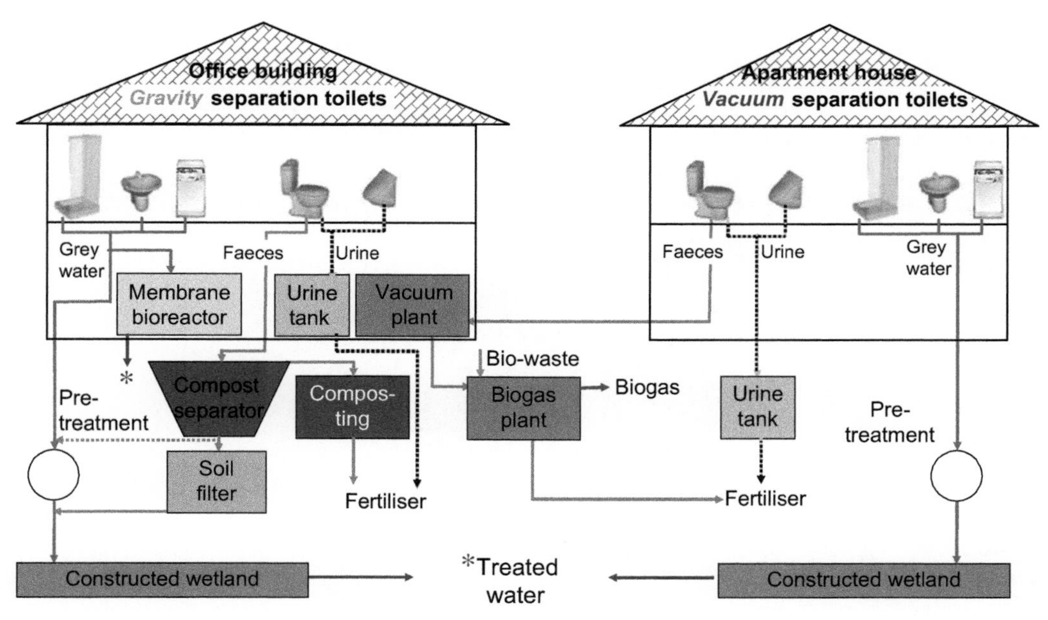

FIGURE 9.3 Research installation by Berliner Wasserbetriebe, Germany.

been installed in an office building and vacuum separation toilets have been installed in an apartment house of the wastewater treatment plant Stahnsdorf in Berlin.

Recently (2006), another project based on the vacuum–biogas concept was built in Sneek, in the Netherlands, for about 30 houses [7] with plans for expansion. Yet another one is in the final design phase by the Hamburg water utility for a new housing estate for about 2000 people [8]. There have also been installations in China [9].

Historically, there was the Liernur vacuum system [10] in Amsterdam and Leiden, in the Netherlands, where blackwater was collected by a vacuum truck. It served several thousand people for more than a decade but suffered from technical limitations.

2.2. Black-/Brownwater Loop Concept

A very promising concept is the blackwater loop involving the conversion of toilet wastewater to toilet flushwater [11]. This system was invented and patented by Ulrich Braun, Hamburg [12] and makes water for toilets independent of freshwater supplies and produces a treated liquid with concentrations of nutrients not unlike urine. With this system, the toilet wastewater is not wasted but is treated for reuse as toilet flushwater. The treatment with a membrane bioreactor plus ozonisation, including nitrification for nitrogen removal, assures a high quality of water for reuse. Pharmaceutical residues and hormones can be eliminated by ozonolysis after the biological treatment but this is rather energy intensive

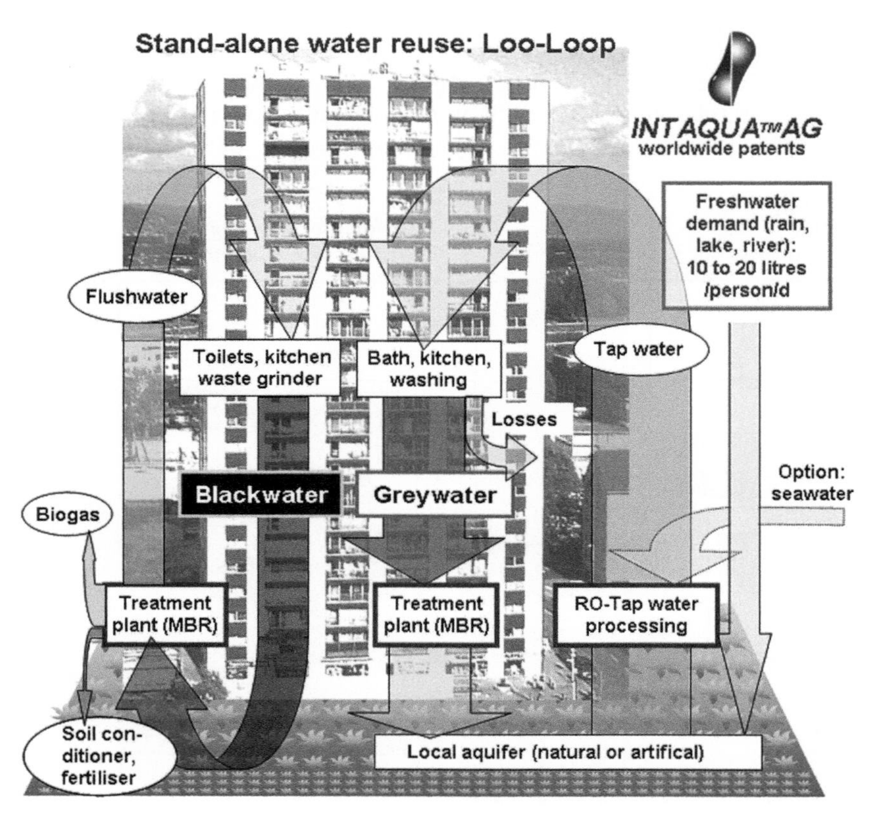

FIGURE 9.4 The blackwater loop combined with greywater reuse for dry areas.

and is costly. The closed-loop system produces a clear liquid with about the same volume and concentrations of nutrients as was put into the toilet in the first place. This system can solve the hygienic and pollution problems that beset hospital effluent, which usually enters the local sewerage system and can be exposed to the public through possible stormwater overflows. Furthermore, it is possible that there is inadequate treatment for such waste at conventional sewerage plants. The size of these installations would start at systems for 100 people/hotel beds. The cost of the construction of a new system of this type can be significantly lower than the cost of a conventional system and furthermore it reduces the freshwater demand to 10 L per person per day. For larger systems, anaerobic treatment of faecal matter with biowaste would be incorporated into the system. Technically, the system is complex, and the operation and maintenance would need special care and organization. Figure 9.4 illustrates the system in combination with local water reuse.

The first worldwide installation of the blackwater loop was done at Hamburg University of Technology in 2005 with a design capacity of around 20 people. The brownwater urine diversion loop will involve an additional installation and has not yet been built. The inventor Ulrich Braun holds several patents on these systems and plans to develop several projects in the near future [12].

3. URINE DIVERSION WITH FLUSH SANITATION

Urine diversion was "rediscovered" and developed in Sweden around 1990 and has been the focus of many projects. In a project for Wupperverband, Wuppertal, Germany, started in 1996, Otterwasser added brownwater treatment with the "Rottebehälter" system (alternate filterbags for sieving and precomposting of the solids) that is currently being marketed by companies mainly in German-speaking countries. Otterwasser designed the "Lambertsmühle concept," which involves the combination of urine diversion and the Rottebehälter treatment unit for faecal matter while the Hamburg University of Technology developed vermicomposting that dramatically increased the composting performance [13]. However, this vermicomposting process does require temperatures above 20°C. It is thus well suited for warmer climates, for indoor installations, or for systems involving a simple solar heating unit during the warmer seasons of regions with temperate climates. The Lambertsmühle is actually an ancient water mill, which has been refurbished as a museum displaying the two concepts: "from grain to bread" and "from bread to grain" [14]. Figure 9.5 shows a urine tank with constructed wetland for treatment of greywater including leachate, from the sieving unit at Lambertsmühle.

Otterwasser has used this concept in a design for 100 flats plus a school, to be built by Linz AG, the water utility company for the city of Linz in Austria (www.otterwasser.de) [2]. Huber Technology, a company in Berching, Germany, that supplies wastewater treatment units for the international market, has installed a urine diverting system for 200 employees in its new office building (see Fig. 9.6).

Furthermore, Gesellschaft für Technische Zusammenarbeit (German Technical Co-operation) has equipped its new office building in Eschborn, Frankfurt, with urine diverting toilets [15].

4. DRY SANITATION/LOW-COST SOLUTIONS

The concept of sustainable sanitation and ecological living is gradually spreading to many parts of the world and dry sanitation—with urine separation—is becoming more and more important. This increased tendency

FIGURE 9.5 The Urine tank for 10 people together with the constructed wetland for treatment of greywater plus leachate, from the sieving unit at Lambertsmühle.

FIGURE 9.6 Huber Technology office building with urine and brownwater sorting.

toward dry sanitation is being driven by a growing public consciousness of impending water shortages and the limitations of the water closet. Although many homes, schools, multistorey buildings, residential areas, and public buildings are equipped with dry sanitation, the global uptake rate remains low and this can be blamed on the absence of attractive innovative technologies that should be included in national long-term development schemes and implemented on large scales.

Although recent sources state that the pace of population growth is slowing down, estimates indicate that with the current trend, 25–40% of people will face a shortage of fresh water in the near future [16]. Furthermore, these figures will increase significantly if waterborne sanitation systems continue to be installed at the present rate, particularly in the fast expanding cities of the developing world. An innovative alternative in dry sanitation, that has the potential for mass production and installation on a large scale, is TPS [17]. This option is the focus of this section. Before discussing this process, it is necessary to discuss the other dry sanitation options in existence.

Dry sanitation systems range from single-pit systems to waterless systems with single or alternating pits and to waterless systems with urine diversion.

4.1. The Single-Pit System

The single-pit system, be it a simple single pit or a single ventilated improved pit (VIP) system consists of a user interface and a pit for excreta collection and storage [18–20]. The user interface can either be a raised pedestal on which the user sits or a squat pan that the user squats over. There are two ways of managing a filled pit: (1) when space is available, a movable superstructure can be built; the full pit is then filled with soil and planted with a tree while the superstructure is relocated to a new pit; (2) in densely populated areas where there is no space, the pit can be emptied and the faecal sludge treated off-site. Where pit emptying is not feasible, decommissioning of the pit is practiced. For example, the "Arboloo," [21] which is a shallow pit that is filled with excreta and soil/ash, then covered with soil and tree-planted, operates on this principle. The main advantages of this system are as follows: the pit can be built and repaired with locally available materials; it does not require a constant supply of water; and it has low capital and operating costs. However, this option has major drawbacks: odours and insect vectors (flies, cockroaches, and insect breeding in stagnant pit water) are frequently a serious problem even when the system is equipped with a ventilation pipe; the costs of desludging may be high compared with capital costs and the sludge often requires secondary treatment and appropriate disposal; in most cases, Biological oxygen demand (BOD) and pathogen reduction are low as the treatment processes (anaerobic, aerobic, dehydration, or composting) are usually very basic; and most importantly, for health reasons, groundwater contamination by leachate and failure to overcome/over-flowing during floods makes this option unsuitable even for periurban or rural areas where there is space.

4.2. Waterless Systems with Composting Chambers

In the waterless sanitation systems with single or alternating pits such as the composting chamber, fossa alterna or double VIP [18–21], the user interface connects directly to the collection and storage unit, and the pedestal is designed such that it can be lifted and moved from one pit to another when two pits are used. The purpose of alternating the pits is to allow the contents of one pit to drain, decompose, and reduce in volume while the other is in use. Despite its numerous advantages, notably significant odour reduction due to ventilation, there is a major concern of low pathogen reduction and groundwater contamination by latrine leachate.

4.3. Waterless System with Urine Diversion

The waterless system with urine diversion—otherwise known as the urine diverting dry toilet (UDDT)—is designed so that the urine is collected and drained away from the front area of the toilet, while faeces falls through a larger hole at back. Like other dry toilets, drying material such as lime, ash, or earth is added into the same hole after defecation. Significant nutrient recovery, limited odours and vectors, and low-capital costs are some of the advantages associated with the system. However, there are drawbacks: people perceive this system as being difficult to maintain; there is the likelihood of depositing faeces in the urine section causing blockages and cleaning problems; the urine pipes are liable to blockage over time, requiring the services of skilled maintenance staff. In short, the system has not been well received.

In general, most professionals recommend the implementation of urine diversion systems over the simple pit or VIP systems for the reasons (in addition to those already mentioned) that the UDDT systems protect groundwater and allow recycling of materials in agriculture. But, from an environmental and public health point of view, even the UDDT is not a practical option for large-scale implementation. The main reasons are that the faeces often only undergo partial decomposition, the degradation process is notably slow as a result of dehydration, and the decomposition does depend on ambient weather conditions. A temperature of greater than 50°C is required to ensure pathogen die-off and this is rarely achieved, except in hot climates. Even with the addition of lime or ash (creating a high pH environment), the fate of pathogens still remains unclear.

Much has been said and written about the potential of dry sanitation systems. From what has been discussed in this chapter, the importance and urgency to develop a unique system that is neither waterborne nor a pit latrine cannot be overemphasized.

4.4. The TPS System

The breakthrough appears to be with TPS. TPS is a dry sanitation system based on a two-phase treatment of biowaste: first by lactic-acid fermentation and then by vermicomposting. No external inputs of water for moisture regulation, ventilation, or external energy are required. It includes urine diversion, addition of a charcoal mixture, and subsequent degradation processes that transform faecal matter into highly fertile humus-like organic matter that can be used in urban agriculture [17]. This development owes its origin to the ancient culture of the Amazon, nowadays Brazil. An analysis of the cultures of the Amazonians has provided hints that might enable the highly efficient handling of organic wastes. Terra Preta do Indio is the anthropogenic black soil that was produced by ancient cultures through the conversion of biowaste and faecal matter into long-term fertile soils. These soils have maintained high amounts of organic carbon over the several thousand years since the fields were abandoned [22]. It was recently discovered that around 10% of the originally infertile soils in the Amazon region was converted this way from around 500 to 7000 years ago [23].

Figure 9.7 shows two soils; one (left image) having been left fallow for many years without any addition of charcoal and having a thin dark soil layer and the other soil (right image), located in the same region, having been left fallow for presumably a similar time period. The latter soil, because of the accumulation of

FIGURE 9.7 Photograph showing terra preta soils from [24] Gunther, modified.

charred biomass and other organic residues, shows the deep, distinctly dark, and highly fertile soil layer that is the terra preta.

Factura et al. [17] reported on these excavations and have evidence to show that this early South American culture had a superior sanitation and biowaste system that was based on the separation of faecal matter and urine at source. This together with cleverly thought out additives, particularly charcoal dust, and treatment steps for the solids, produced superior composting which resulted in high-yielding agriculture. The initial silage process makes the two-stage process superior to other combined processes such as composting and vermicomposting.

Preliminary qualitative experimental evidence by Factura et al. [17] suggest that it is not only lactic acid fermentation but also some other organisms such as *Bacillus subtilis*, in combination with the addition of some thin wood particles, that has produced the high quality of compost. The authors have pointed out that one of the main advantages of the TPS system is that the lacto-fermentation process works efficiently and is stable without air exchange and produces no offensive odours. Also, the vermicomposting step proceeds quite efficiently after adapting the earthworms to the special compost. Other researchers [25–28] have shown that vermicomposting is an efficient way of treating faecal matter. Their findings also suggest that vermicomposting may be a feasible option for eliminating pathogens.

Research into this ancient process is at the very beginning but the initial results are most promising. There is the possibility for immediate application although there are a lot of open questions to be answered before the whole process is fully understood.

Adaptation of existing dry toilets is feasible where space requirements for the common double vault systems can be reduced by half. The anaerobic, but oxygen-tolerating TPS offers odour-free operation without ventilation pipes.

Besides the modification of existing urine-diverted dry toilets, there are three further development lines for implementation to make an impact round the world:

1. Upgrade the pit latrines as used by about two billion people;
2. Install simple bucket toilets with some kind of urine diversion mechanism that ensures a closed air tight system after defecation and involves the addition of charcoal and lactic-acid bacteria mixture; and
3. Design-effective and comfortable UDDTs with optional automatic addition of the conditioning/inoculation material.

People with inadequate or no sanitation should have access to one of the three types of installations [16].

5. CONCLUSIONS

Integration of the anaerobic dry toilet and vermicomposting thus promises to be an ideal approach for managing wastes; even wastes generated by urban households. The product, terra preta, can address the problems of soil degradation and food insecurity, common in many areas across the world. TPS could become important in the design of highly efficient eco-friendly houses and housing areas with the added value of improved urban agriculture, which can be combined with local greywater reuse. Thus, TPS can close regional cycles and improve hygienic conditions and soil fertility in a sustainable manner with the creation of local added value. There are still questions to be answered such as the issue of macronutrients and also micronutrients.

References

[1] R. Otterpohl, J. Naumann, Kritische Betrachtung der Wassersituation in Deutschland, in: K. Gutke (Ed.), Wieviel Umweltschutz braucht das Trinkwasser?, 1993, pp. S.217–S.233. Kirsten Gutke Verlag.

[2] See www.otterwasser.de Accessed 13 May 2010
[3] See www.vauban.de Accessed 13 May 2010
[4] See www.kompetenzwasser.de/SCST.22.0.html?&L=1&type=title%3D Accessed 17 May 2010
[5] M. Oldenburg, A. Peter-Fröhlich, C. Cloaks, L. Pawlowski, A. Bonhomme, Water Science and Technology 56, 2007, 239–249.
[6] A. Peter-Fröhlich, Isabelle Kraume, André Lesouëf, Martin Oldenburg. Separate Ableitung und Behandlung von Urin, Fäkalien und Grauwasser – ein Pilotprojekt – Korrespondenz Abwasser, Abfall, 51/1, 2004, 38–42.
[7] See www.wetsus.nl. Accessed 7 May 2010
[8] Kim Augustin, Hamburg Water, Germany, personal Information 2010-07-1,.
[9] Xiaochang C. Wang, Xi'an University of Architecture and Technology, China personal information 2010.
[10] See http://enviren.gr/content/view/14/35/lang, en/ Accessed 19 May 2010
[11] See www.intaqua.com Accessed 17 May 2010
[12] U. Braun, patent: PCT/EP98/03316.
[13] See www.tuhh.de/aww. Accessed 5 June 2010
[14] See www.otterwasser.de/english/concepts/lande.htm Accessed 27 May 2010
[15] GTZ Basic Overview of Urine Diversion Components (Waterless Urinals, UD Toilet Bowls and Pans, Piping and Storage). Technical Datasheets, Ecosan program - Deutsche Gesellschaft für Technische Zusammenarbeit (GTZ) GmbH. Available at: www.gtz.de/en/themen/umwelt infrastruktur/wasser/9397.htm. 2008.
[16] World Population Prospects: The 2008 Revision. Population Division of the Department of Economic and Social Affairs of the United Nations Secretariat. June 2009.
[17] H. Factura, T. Bettendorf, C. Buzie, H. Pieplow, J. Reckin, R. Otterpohl, Water Science and Technology 61, 2010, 2673–2679.
[18] P. Morgan, D. Mara, Ventilated Improved Pit Latrines: Zimbabwean Brick Designs, The World Bank, Washington, D.C., 1985.
[19] P. Morgan, Zimbabwe's Rural Sanitation Programme. Updated version of paper, 2002. 1998
[20] B. Brandberg, SanPlat System: Lowest Cost Environmental Sanitation. Proceedings of 17th WEDC Conference, Nairobi, Loughborough, 1991, pp. 193–196.
[21] P. Morgan and Sei. An ecological approach to Sanitation in Africa: A compilation of experiences (Aquamor) Harare, Zimbabwe, Chapter 4: Arborloo; Chapter 5: Fossa Alterna, 2004.
[22] J. Lehmann, M. Rondon, Biological Approaches to Sustainable Soil Systems 36, 2003, 517–530.
[23] B. Glaser, Philos Trans R Soc Lond B Biol Sci 362, 2007, 187–196.
[24] F. Günther, Personal communication, also submitted to, Energy and Environment 2007,.
[25] N. Basja, O. Nair, J. Mathew, G.E. Ho, Vermiculture as a Tool for Domestic Wastewater Management. In: Proceedings of the International water association fifth specialized conference on small water and wastewater systems. Instanbul. (2002) ISBN 9755612254.
[26] M. Shalabi, Vermicomposting of Faecal Matter as a Component of Source Control Sanitation, PhD thesis, TUHH, Hamburg University of Technology, Germany, 2006.
[27] Singh Yadav K.D.V., Vermiculture and Vermicomposting Using Human Faeces as Feed and Eisenia foetida as earthworm species, Doctoral thesis: Department of civil engineering. Indian Institute of Technology, Kampur, India, 2008.
[28] C. Buzie, Development of a Continuous Flow Vermicomposting Urine Diversion Toilet for On-site Application, PhD thesis, Hamburg University of Technology, Germany, 2010.

CHAPTER

10

Recovered Paper

Gary M. Scott

Department of Paper and Bioprocess Engineering, State University of New York, College of Environmental Science and Forestry, One Forestry Drive, Syracuse, NY 13210

OUTLINE

1. INTRODUCTION

Paper is probably the most commonly recycled material in use today. Over the past several years, over 60% of all paper produced in the United States has been recycled and similar amounts are used worldwide. Recovered paper, as a raw material for papermaking, is as important to the paper industry as is chemically pulped virgin fiber. Within the paper industry, the term *waste paper* is rarely used, the préferred term being *recovered paper*, which better reflects the value and importance of it as a raw material. The terms *recycled paper* and *secondary fiber* are also often used to refer to this waste stream.

Recovered paper, or waste paper, can come from a number of different sources, including internally at the paper mill. A number of different terms are used to indicate when in the life cycle of the paper that it is returned for recycling (Fig. 10.1).

- *Post-consumer recycled fiber:* Paper that is recycled after its final use by the consumer and recovered from homes, offices, and retail locations. This type of waste paper is often the most contaminated and difficult to collect

Waste Doi: 10.1016/B978-0-12-381475-3.10010-5

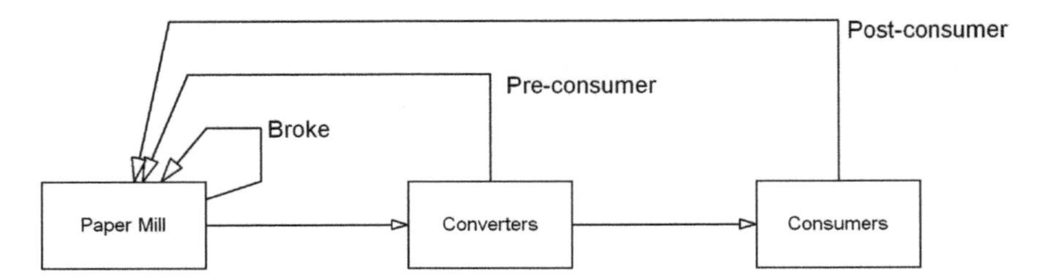

FIGURE 10.1 Overview of the life cycle of waste paper. Paper can be recycled internally within the paper bill (broke), returned to the mill from converting operations (preconsumer), or collected after its final use (postconsumer).

and recycle as it often must be separated from municipal waste [1].

- *Pre-consumer recycled fiber:* Paper that is recycled during the manufacturing or conversion process of paper into consumer products such as envelopes. This recycled paper consists primarily of the trimmings and rejected or obsolete products and represents a relatively clean source of fiber for papermaking [1].
- *Broke:* Paper recycled internally within a single mill and is typically trimmings and paper that does not meet the specification of the grade being made. Broke can represent from 5 to 25% of the production of a mill depending on the grade being produced and the efficiency of the machine. The broke may be recycled into the same grade of paper or into a different grade of paper being made at the same mill.

In general, only the first two categories (post-consumer recycled fiber and preconsumer recycled fiber) are considered recovered waste paper; broke is not typically considered recovered waste paper as it had never left the manufacturing site. More information about papermaking can be found in a number of textbooks on the subject [2,3] and general references [4,5].

1.1. History of Use

Papermakers have been reusing their raw material to create new paper from the earliest history of papermaking. Although the making of paper dates back to ancient China, when the first sheets were made from the bark of the mulberry tree in the first century C.E., it was not until the invention of the printing press and the industrialization of the papermaking process in the late eighteenth century that demand for paper necessitated a wider range of raw materials to be used for papermaking. Before this time, most paper was made from recycled textiles, primarily cotton, linen, and hemp. The use of wood as a raw material for papermaking was prompted by the shortage of textile-based raw materials in the eighteenth and nineteenth centuries, resulting in the development of technology for the mechanical and chemical pulping of wood into a fiber suitable for papermaking.

The modern papermaking age (since the early nineteenth century) and the conversion to wood as the primary raw material for papermaking brought in the recycling of this raw material for papermaking. The use of recovered paper as a raw material has especially grown in the latter half of the twentieth century due to a number of reasons including increased demand, environmental concerns, government regulations, and improving technology for the use of waste paper as a raw material. The increased use over the past 30 years have also seen a number of challenges in the use of the raw material that needs to be addressed including the degradation of the raw materials and increased and varied contamination that needs to be removed before reuse.

1.2. Reasons for Use

Paper recycling is important for a number of reasons. Paper is the largest fraction of the municipal waste stream (Fig. 10.2); recycling can significantly reduce the amount of material heading to the ever-decreasing landfill space. In fact, for every tonne of paper recycled about 2 m^3 of landfill space are saved. Although it is often stated that recycling saves trees, at least in the United States both wood and recovered paper are plentiful. The decision to use recovered paper is often one of the economics and the desired properties of the final paper sheet. In some cases, recycled fiber may be more economical to produce the desired sheet, whereas in other cases, virgin fiber may be the best. Elsewhere in the world, however, the economics and supply may dictate different choices. Each manufacturer must decide what is best to produce their product under the market conditions at any particular time [6]. Government regulations and recycled content mandates have also been a significant driving force in the increased use of recovered paper in the industry.

1.3. Industrial Statistics

The amount of paper recovered has been significantly increasing over the past two decades. In the United States, the percentage of paper recovered for reuse (as a percentage of the amount produced in the United States) has been steadily rising since the mid-nineties (see Fig. 10.3) [7]. Although recovery rate has been increasing, the absolute amount recovered has decreased due to a reduction in the production of paper in the United States and the reduction of the amount being landfilled (Fig. 10.4) [7]. In addition, the per capita consumption of paper has decreased over the past several years; for example, in the United States, per capita consumption has decreased from 299.7 kg ca^{-1} in 2007 to 286.7 kg ca^{-1} in 2008. Globally, the per capita consumption has decreased from 59.7 to 58.0 kg ca^{-1} over the same time frame [8].

Globally, the recovery rate of paper in 2008 was roughly on par with that in the United States. Figure 10.5 shows the worldwide recovery rate as a fraction of the regions production. As can be seen, all regions of the world are

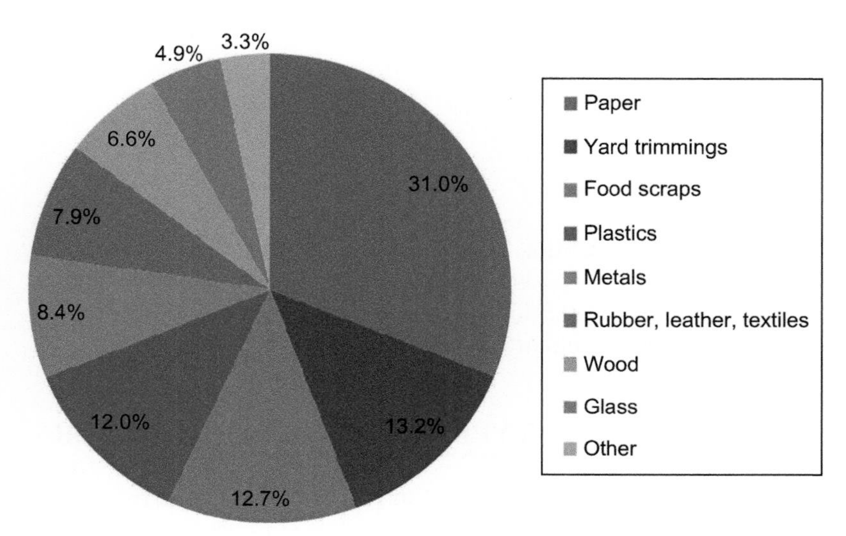

FIGURE 10.2 Composition of municipal solid waste in the United States before recycling in 2008 [15].

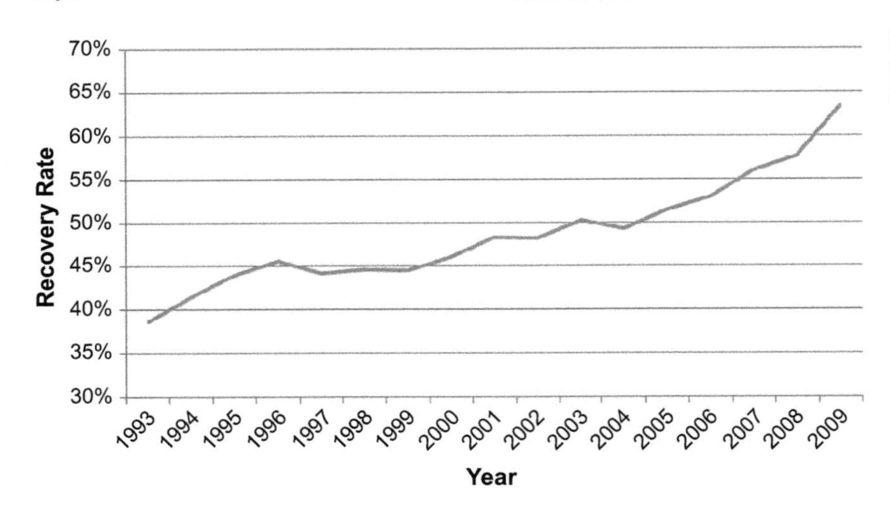

FIGURE 10.3 Waste paper recovery rate in the United States from 1993 to 2009 [7].

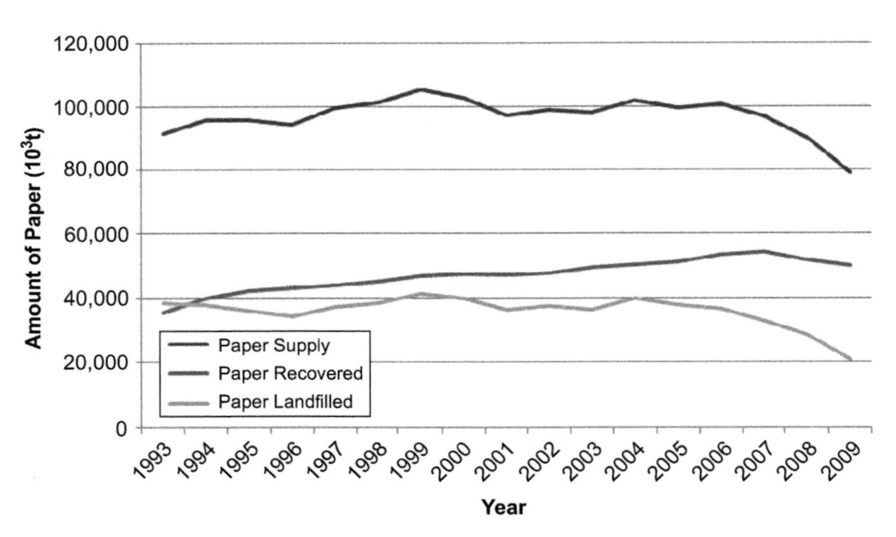

FIGURE 10.4 The paper supply compared with the amount of paper recovered and the amount of paper landfilled in the United States [7].

recovering paper at a rate greater than 50%. In fact, at these recovery rates, recovered paper will soon, if not already, account for more than virgin pulp in terms of fiber content of the worldwide paper and board production [9], The demand for recovered paper will continue to be strong, with increasing demand being primarily driven by the fiber needs of China and other countries with limited resources for the production of virgin pulp (Table 10.1). For China, the majority of the recovered paper used is in the form of old corrugated containers (49.5%), mixed paper (32.1%), and old newspapers and magazines (16.5%) [8].

2. TYPES OF RECOVERED PAPER

As depicted in Fig. 10.1, recovered paper can come from a number of sources, including the

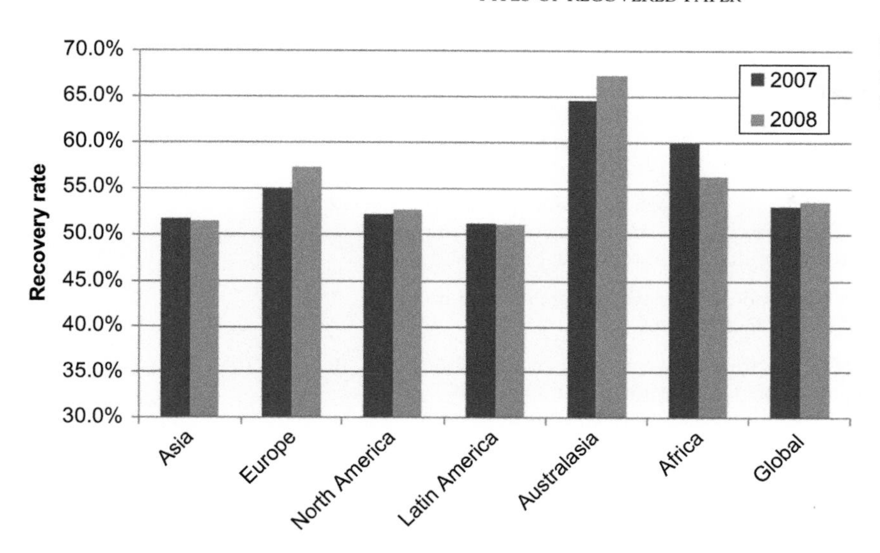

FIGURE 10.5 Paper recovery rates (as a fraction of production) regionally around the world [8].

TABLE 10.1 Countries Importing the Most Waste Paper in 2008 [8]

Country	Recovered paper imported (10^3 t a^{-1})
China	24,205
Indonesia	2,080
Canada	1,765
India	1,670
Mexico	1,435
South Korea	1,307
Thailand	1,217
Taiwan	834

internally recycled broke, preconsumer recycled paper from converters, and postconsumer recycled paper from various collection methods. Generally, as the paper gets further into the life cycle (from leaving the mill to being used by the consumer), the paper becomes further contaminated, harder to recycle, and worth less in the recovered paper market. To help with the marketing of the recovered paper, a number of organizations have defined various grades of paper and determined their characteristics.

2.1. Waste Paper Grades

In the United States, recovered paper grades are defined by the Institute of Scrap Recycling Industries (ISRI), Inc. [10]. These grades determine the characteristics and quality of the recovered paper, allowing comparisons between different suppliers. Table 10.2 provides some examples of the grades defined by ISRI, categorized here into preconsumer recovered paper and postconsumer recycled paper. In this table, prohibitive materials are defined as any material which, by their presence in a packing of paper stock in excess of the amount allowed, will make the packaging unusable as the grade specified or any material that may be damaging to equipment. In addition, sorted recovered paper must be free of food, medical and hazardous wastes, and poisonous or harmful substances. Out-throws are defined as "all papers that are so manufactured or treated or are in such a form as to be unsuitable for consumption as the grade specified" [10].

TABLE 10.2 Selected Recovered Paper Grades as Defined by ISRI [10]

Grade	Definition	Prohibitive materials (%)	Out-throws allowed (%)
Postconsumer grades			
(1) Residential mixed paper	Consists of a mixture of various qualities of paper not limited as to the type of fiber content, normally generated from residential, multimaterial collection program	2	5
(6) News	Consists of newspaper as typically generated from news drives and curbside collections	1	5
(7) News, de-ink quality (#7 ONP)	Consists of sorted, fresh newspapers, not sunburned, containing not more than the normal percentage of rotogravure and colored sections. May contain magazines	0	0.25
(11) Corrugated containers (OCC)	Consists of corrugated containers having liners of test liner, jute, or kraft	1	5
(12) Double-sorted corrugated (DS OCC)	Consists of double-sorted corrugated containers, generated from supermarkets and/or industrial or commercial facilities, having liners of test liner, jute, or kraft. Material has been specially sorted to be free of boxboard, off-shore corrugated, plastic, and wax	0.5	2
(36) Unsorted office paper (UOP)	Consists of printed or unprinted paper typically generated in an office environment that may include a document destruction process. This grade may contain white, colored, coated and uncoated papers, manila and pastel-colored file folders	2	10
(37) Sorted office paper (SOP)	Consists of paper, as typically generated by offices, containing primarily white and colored groundwood-free paper, free of unbleached fiber. May include a small percentage of groundwood computer printout and facsimile paper	1	5
(40) Sorted white ledger (SWL)	Consists of uncoated, printed or unprinted sheets, shavings, guillotined books, and cuttings of white groundwood-free ledger, bond, writing, and other paper that has similar fiber and filler content.	0.5	2
Preconsumer grades			
(4) Boxboard cuttings	Consists of new cuttings of paperboard used in the manufacture of folding cartons, set-up boxes, and similar boxboard products	0.5	2
(16) Mixed kraft cuttings	Consists of new brown kraft cuttings, sheets, and bag scrap free to stitched paper	0	0.1
(24) White blank news	Consists of unprinted cuttings and sheets of white newsprint or other uncoated white groundwood paper of similar quality	0	1
(30) Hard white shavings (HWS)	Consists of shavings or sheets of unprinted, untreated white groundwood-free paper	0	0.5

As can be seen from the table, the postconsumer recovered papers have a much higher tolerance for contaminating materials than the preconsumer grades. The price that each grade demands is strongly dependent on the amount of contamination that is present, which usually must be removed in the recycling process.

In Europe, a similar classification system defining the various types and qualities of recovered paper was created by the European Recovered Paper Association and the Confederation of European Paper Industries [11]. In this system (see Table 10.3), recovered paper is first classified into a number of broad groups, depending on the quality and homogeneity of the paper collected. In general, groups 1 and 2 tend to be the postconsumer grades such as various mixed papers that are collected from residential and commercial collection systems. Groups 3 and 4 tend to be of a higher quality and represent primarily those preconsumer grades collected from paper converters. Group 5 represents a variety of specialty grades that may be of interest only to the producers of specific grades of paper. As can be seen from these definitions, recovered paper is a well-specified commodity that is regularly traded both within various countries and globally.

2.2. Recovered Paper Collection

Recovered paper collection is done in a number of different ways, depending on the type of paper being collected and the source of the paper. In general, preconsumer recycled paper is easier to collect as it tends to be concentrated in specific manufacturing locations and also tends to be much more homogeneous. These collections, often of the form of cuttings, trimmings and over issues, are typically baled and packaged directly at the collection site with little additional processing needed [12].

The recovery of postconsumer recycled paper is more difficult as the sources tend to be less concentrated and the paper is often mixed in

TABLE 10.3 Selected Paper Grades as Defined by CEPI and ERPA [11]

Group	Grade	Definition
Group 1—Ordinary grades		
1.01	Mixed paper and board, unsorted, but unusable materials removed	A mixture of various grades of paper and board, without restriction on short fiber content
1.02	Mixed papers and boards (sorted)	A mixture of various qualities of paper and board, containing a maximum of 40% of newspapers and magazines
1.05	Old corrugated containers	Used boxes and sheets of corrugated board of various qualities
1.08	Mixed newspapers and magazines 1	A mixture of newspapers and magazines, containing a minimum of 50% of newspapers, with or without glue
Group 2—Medium grades		
2.01	Newspapers	Newspapers, containing a maximum of 5% of newspapers or advertisements colored in the mass
2.05	Sorted office paper	Sorted office paper
Group 3—High grades		
3.04	Tear white shavings	White wood-free lightly printed shavings without glue, free from wet-strength paper and paper colored in the mass
3.05	White wood-free letters	Sorted white wood-free writing papers, originating from office records, free from cash books, carbon paper and non-water soluble adhesives
3.14	White newsprint	Shavings and sheets of white unprinted newsprint, free from magazine paper

(*Continued*)

TABLE 10.3 Selected Paper Grades as Defined by CEPI and ERPA [11] — *cont'd*

Group	Grade	Definition
Group 4—Kraft grades		
4.01	New shavings of corrugated board	Shavings of corrugated board, with liners of kraft or testliner
4.02	Used corrugated kraft 1	Used boxes of corrugated board, with kraft liners only, the fluting made from chemical or thermo-chemical pulp
4.07	New kraft	Shavings and other new kraft paper and board of a natural shade
Group 5—Special grades		
5.01	Mixed recovered paper and board	Unsorted paper and board, separated at source
5.02	Mixed packaging	A mixture of various qualities of used paper and board packaging, free from newspapers and magazines

terms of the types of paper as well as being mixed with other waste. Some postconsumer waste paper grades, such as sorted office paper (see Table 10.2), can be relatively clean as they are collected in dedicated locations, baled, and transported. Often, these collections can be sold directly to the recycled fiber users directly with little additional processing.

Additional postconsumer recycled paper is collected through municipal collection systems. These papers tend to be the least homogeneous and require the greatest amount of processing and handling to be usable as recovered paper. The collection of these types of paper can be done through a number of different mechanisms including pick-up containers, drop-off containers, curbside collection, and other systems. Although curbside collection is the most convenient for the consumer, it results in an extremely diffuse source. The effectiveness of curbside recycling in relation to population density can be easily seen with the variation of access across the United States. Heavily populated states, such as New York and California, have curbside recycling access rates more than 85%. However, less densely populated states, such as Montana and South Dakota, have minimal curbside programs [13]. Overall in the United States in 2005, 56% of the population has access to curbside recycling, 65% has access to drop-off recycling programs, and 86% has access to some type of recycling program [14].

These collection methods depend heavily on the final user separating the paper from the municipal solid waste (MSW) stream to be effective. This can be done fairly easily as paper is by far the largest part of the MSW stream before recycling (Fig. 10.2) provided homeowners and businesses separate the recyclable material before collection [15]. The recovery of paper after being mixed with other waste is a much more difficult prospect. In this case, the type and amount of contamination can significantly increase the processing costs of using the recovered paper, perhaps beyond economic feasibility. In addition, consideration must also be given to the fate of the balance of the MSW stream. Although there is an advantage to remove the paper if the collected MSW is destined for a landfill, separation of paper from MSW that will be incinerated for energy production will reduce the energy value of the material, as paper is a significant portion of the combustible material.

3. PROCESSING OF RECOVERED PAPER

Depending on the source of the recovered paper and the collection method, the recovered paper must be processed before being sent to a mill for recycling. As mentioned above, some grades collected can be sent immediately to

TABLE 10.4 Uses of Recovered Paper in the United States [7]

Recovered paper	Fraction (%)
Net exports	41.3
Containerboard	29.3
Boxboard	11.5
Tissue	8.0
Newsprint	4.4
Other	5.4

the mill for processing, while other grades require more extensive processing before reaching the mill. Even after reaching the mill, the paper often requires extensive processing to remove contaminants and restore the paper-making ability of the fibers. The extent of this processing heavily depends on the grade of paper being recycled and the product that the recovered paper will be used in. Although much of the paper recovered in the United States is converted back into paper products domestically, a good fraction of it is exported to other countries for use and a small fraction of it is used in non-paper products or uses (see Table 10.4) [7].

3.1. Paper to paper

In the processing of recovered paper into new paper products, a number of different unit operations are necessary to make the fibers suitable for papermaking. The number and variety of processing steps stem mainly from the large range of contaminants that must be removed from the recovered paper to make it usable; the various operations removing contaminants with differing characteristics. Depending on the cleanliness of the fiber (e.g., whether it is pre- or postconsumer and the homogeneity of the recovered paper) more or fewer processing steps would be necessary to recover the paper.

The objective of a recycling mill is to take a material that can vary considerably in properties, and convert it into fiber that can consistently be made into a final product. Unlike virgin fiber mills producing pulp from wood, the quality of the fiber and the cleanliness of the fiber can vary greatly from day-to-day or even hour-to-hour. By comparison, wood as a raw material for producing pulp is a much more consistent raw material. Recovered paper can vary greatly in its composition (hardwood content vs. softwood content, mechanical pulp vs. chemical pulp), the types of contaminants present (adhesives, plastics, metals, etc.), and the cleanliness of the stock (inks, dyes, and other foreign materials). Although the grading and sorting of this material ameliorates these variations to some degree, recycled mills must still deal with a significant variation in their raw material.

It is impossible to describe a "typical" process used in the recycling of recovered paper, as each grade requires different operations to handle the contaminants that may be present. In fact, the same grade of recovered paper may be treated with significantly different operations depending on the grade of paper used. Table 10.5 summarizes the primary operations that are used to separate the fiber fraction of the recovered paper from the contaminants.

TABLE 10.5 Separation Operations Used in the Processing of Recovered Paper to Remove Contaminants from the Feedstock

Unit operation	Removal mechanism	Contaminants removed
Washing	Particle size and shape	Water soluble inks, fine particles
Cleaning	Specific gravity, size, and shape	Dirt, sand, metal, plastics
Screening	Particle size, shape, and deformability	Dirt, metal, large particles
Flotation	Surface characteristics and particle size	Laser and photocopy inks

The various operations separate by a number of different principles and relative characteristics between the fibers and the contaminants. Washing, which is essentially retaining the fibers on a fine screen while letting the liquid pass through, is primarily used for removing very fine particles and water-soluble inks including those typically used in newspapers. Centrifugal cleaning separates primarily by differences in the specific gravity between the fibers and the contaminants. Cleaners can be used to separate material that is both more dense than fiber (e.g., dirt, metal, etc.) or less dense than fiber (e.g., plastics, most adhesives, etc.) depending on the operation of the cleaner. Screening uses a barrier to prevent the passage of contaminants through the screen while allowing the fibers to pass through. The shape and size of the opening in the screen determines which contaminants are prevented from entering the accepted pulp in the operation. Finally, flotation uses a surfactant to attach the contaminants to air bubbles that are then removed from the surface of the flotation cell as dirty foam. It is the primary mechanism for the removal of laser and photocopy toner from office waste. Greater details of these separation operations can be found in many references and proceedings [16–19].

Several other unit operations are also used in the recycling of paper into new paper products as listed in Table 10.6. Although some of these are basic operations needed to process the paper into pulp suitable for papermaking, some of them are used to address a specific need in the processing of the recovered paper. Repulping is the first necessary step to get the fibers into an aqueous slurry for further processing and transport. Fractionation is sometimes used to sort the fibers, typically by length, into those that are more suited for papermaking. Refining is used to improve the surface characteristics of the fibers themselves to promote greater fiber-to-fiber bonding. Dispersion is a mechanical action used to further detach any ink particles or other contaminants from the fiber so that it can be removed in one of the separation processes. Because many of the processes must be done at fairly dilute conditions to be effective, it is often necessary to remove or add water from the pulp from one operation to the next. Finally, in cases in which the final brightness of the pulp is important, bleaching may be necessary. Again, details of these operations can be found in many references [16–19].

TABLE 10.6 Other Unit Operations Used in the Processing of Recovered Paper

Unit operation	Effect on recovered paper
Repulping	Breaks down the dry paper into individual fibers by mixing vigorously with water and chemicals if needed
Fractionation	Separates a fiber mixture into two separate streams typically by fiber length. The long fibers are usually used for papermaking
Refining	Imparts mechanical action on the fibers to change the surface characteristics to enable the bonding of the fibers to each other
Dispersion	Detaches ink particles from the fiber and reduces the particle size of contaminants
Dewatering	Removes excess water from the pulp stock so that it can be further processed
Bleaching	Enhances the brightness of the resulting pulp using chemical agents such as hydrogen peroxide

3.2. Paper to Other Products

The various processes used to recycled paper into other products are too numerous to describe in detail in a survey article such as this. However, the varieties of products that can be produced are summarized in Table 10.7. In addition, the processing of recovered paper into usable fiber for papermaking often results in a secondary stream typically termed sludge.

TABLE 10.7 Non-Paper Products from Recovered Paper and Recovered Paper Sludge

Products from recovered paper	Products from recovered paper sludge
Insulation	Fertilizer/soil amender
Animal bedding	Landfill caps
Egg cartons	Construction materials
Planting pots for seedlings	Animal bedding
Lamp shades	Ethanol
Ethanol	Energy
Energy	
Dust masks	

Although considered a waste product of the recycling process, this stream often can also be used to produce a number of different products [20]. Paper can also be converted to energy through the use of various combustion technologies that are available. Paper, being an organic material, has a relatively high energy value and can make an excellent fuel. In many cases, the paper (or the sludge from the recycling operation) can be co-fired with other fuels in power boilers and can also be processed as fuel for small-scale (e.g., residential) burners.

4. BARRIERS TO RECOVERED PAPER USE

Although there are many benefits to the use of recovered paper, there are a number of issues that will limit its use. Most importantly, it is the fact that paper is an organic material, and it is subject to wear and degradation each time it is recycled. The drying of paper, one of the last steps in the papermaking process, causes irreversible changes in the structure and morphology of the fibers. The subsequent repulping and processing of the fiber causes damage to the fibers that reduce their strength and bonding ability, causing a degradation of the paper properties made from recycled fiber. This degradation limits the amount of recycled fiber that can be used, especially in papers that have high strength requirements. Fractionation, the separation of fibers by length and using only the longer and stronger fibers, is one method of ameliorating this degradation. However, a certain fraction of virgin fibers will always be needed to make up the difference [21]. As more paper is recovered to be used in recycled paper, the quality of the paper recovered will continue to degrade. This can already be seen in the quality differences in old corrugated containers, where boxes from Asia are considered to be of significantly lower quality and a greater percentage of recovered paper is used in their manufacture.

Most of the easier paper to collect and recycle is already being recovered. For old corrugated containers, 81% was recovered and newsprint reached 70% in 2009 [7]. To increase the amount of paper recycling further, collections will need to increase primarily in the postconsumer grades of paper. The incremental paper collected will suffer from greater contamination, less homogeneity, and lower fiber quality. Most of the growth in recovered papers will likely be in the mixed paper categories, which also suffer from greater contamination.

Digging deeper into the recovered paper streams will also have economic and environmental consequences. As the paper recovered becomes more contaminated, greater costs will occur to make it suitable for papermaking. The disposal of the separated contaminants will need to be done in an environmentally friendly manner, which will become more costly as landfill space becomes more scarce. Environmental concerns with the use of forests may drive paper producers to use a greater fraction of recycled fiber. At the end, paper producers will usually look to the economic bottom line, using whatever raw material that will produce the needed paper at the least cost.

5. CONCLUSIONS

Paper recycling will continue long into the future. It is a key raw material for the papermaking process, representing a significant portion of the fiber used. However, because of the degradation of the fibers through the papermaking and recycling process, the limit to the amount of paper being recycled is slowly being approached. In addition, there are a number of grades of paper that are currently not recycled and will not be in the foreseeable future. This is particularly true for such personal care products such as toweling and tissue.

Future improvements in recycling will need to deal mainly with the fiber quality issues. If recovered paper is to be used in the higher and more demanding grades of paper, technology will need to be developed to "upgrade" the fibers to meet these needs. Although fiber fractionation can improve a portion of the fibers by separating out the longer, stronger fibers, it is at the cost of a portion of the recovered paper. Technology will also be needed to handle greater levels of contamination in the recovered paper stream. As we delve further into the recycled stream, the contaminant level of the recovered paper will increase. Additionally, as the percentage of recovered paper as a raw material increases, we will see a greater quantity of contaminants carrying over from one cycle to the next.

Research continues at a number of universities, government research centers, and industrial research locations. There are also a number of organizations that deal with the issues of recovered paper. Table 10.8 summarizes a number of these; additional information about recovered paper can be found on their websites.

TABLE 10.8 Organizations Dealing with Waste Paper and Paper Recycling

Organization	Website
Bureau of International Recycling	www.bir.org
Technical Association of the Pulp and Paper Industry	www.tappi.org
American Forest and Paper Association	www.afandpa.org
Institute of Scrap Recycling Industries, Inc.	www.isri.org
European Recovered Paper Association	www.erpa.info
Confederation of European Paper Industries	www.cepi.org www.paperonline.org
Environmental Protection Agency	www.epa.gov
British Recovered Paper Association	www.recycledpaper.org.uk
Canadian Association of Recycling Industries	www.cari-acir.org
National Recycling Coalition	nrc-recycle.org
The Paper and Paperboard Packing Environmental Council	ppec-paper.com

References

[1] D.B. Mulligan, Sourcing and grading of secondary fiber, in: R.J. Spangenberg (Ed.), Secondary Fiber Recycling, Tappi Press, Atlanta, 1993.

[2] G.A. Smook, Handbook for Pulp and Paper Technologists, Angus Wilde Publications, Vancouver, 1992.

[3] C.J. Biermann, Essentials of Pulping and Papermaking, Academic Press, Inc., San Diego, 1993.

[4] J. Evans, J. Burley, J. Youngquist (Eds.), Encyclopedia of Forest Sciences, Elsevier, London, 2004.

[5] H. Holik, Handbook of Paper and Board. Weinheim, Germany, Wiley-VCH, 2006.

[6] Tappi. Earth answers—why recycle? Paper University—All About Paper. [Online] TAPPI,. http://www.tappi.org/paperu/all_about_paper/earth_answers/Whyrec1.htm. Accessed June 12, 2010 2001.

[7] AF&PA announces increase in paper recovery, meets goal ahead of schedule. Paper Industry Association Council. [Online] AF&PA, http://paperrecycles.org/news/press_releases/2009_recovery_stats_released.html. Accessed June 8, 2010. March 22, 2010.

[8] G. Magnighi, The World Recovered Paper Market in 2008, Bureau of International Recycling, Brussels, 2008.

[9] W. Moore, Global recovered paper markets outlook, Progress in Paper Recycling 15 (2006) 4.
[10] Scrap Specifications Circular, Institute of Scrap Recycling Industries, Inc., Washington, DC, 2009.
[11] C.E.P.I. ERPA, European List of Standard Grades of Recovered Paper and Board, Confederation of European Paper Industries, Brussels, 2002.
[12] P. Hans-Joachim, Collection systems, sources, and sorting of recovered paper, in: L. Göttsching, H. Pakarinen (Eds.), Recycled Fiber and Deinking, Tappi Press, Atlanta, 2000.
[13] AF&PA. Paper Industry Association Council. Paper/Paperboard Collection by State, Population (Millions). [Online] AF&PA, http://stats.paperrecycles.org/maps/. Accessed June 11, 2010. 2010.
[14] EPA. Paper grades and collection. Paper Recycling. [Online] Environmental Protection Agency, May 12, 2010. http://www.epa.gov/wastes/conserve/materials/paper/basics/grade.htm#collection. Accessed June 11, 2010.
[15] Environmental Protection Agency, Municipal Solid Waste Generation, Recycling, and Disposal in the United States: Facts and Figures for 2008, Environmental Protection Agency, Washington, DC, 2008. EPA-530-F-009-021.
[16] H. Holik, Unit operations and equipment in recycled fiber processing, in: L. Göttsching, H. Pakarinen (Eds.), Recycled Fiber and Deinking, Tappi Press, Atlanta, 2000.
[17] R.J. Spangenberg (Ed.), Secondary Fiber Recycling, Tappi Press, Atlanta, 1993.
[18] S. Abubakr (Ed.), Recycling, Tappi Press, Atlanta, 1997.
[19] M.R. Doshi, Recycled Paper Technology: An Anthology of Published Papers, Tappi Press, Atlanta, 1994.
[20] G.M. Scott, S. Abubakr, A Smith, Sludge characteristics and disposal alternatives for recycled fiber plants. Proceedings of the 1995 Recycling Symposium. Atlanta: Tappi Press, 1995.
[21] S. Abubakr, G.M. Scott, J.H. Klungness, TAPPI Journal 78 (1995) 123–126.

CHAPTER

11

Glass Waste

John H. Butler, Paul Hooper

Department of Environmental and Geographical Sciences, Manchester Metropolitan University, UK

OUTLINE

1. THE GLASS INDUSTRY

Glass is in the background of the daily lives of most people. It is manufactured from plentiful raw materials and can be readily reused as feedstock in glass production. Between 80% and 85% of the mass output from the worldwide glass industry is either in the form of containers for the food, beverage, and pharmaceutical industries, or flat glass for building construction or for motor vehicle manufacture [1]. Other product segments, whilst only constituting about 15% of the mass output, produce high-value technical and consumer products by comparison with container and flat glass. However, the potential for glass recycling comes largely from the container and flat glass sectors, because of their dominance in terms of mass, and their relatively uniform chemical composition, with soda lime—silica glass accounting

Doi: 10.1016/B978-0-12-381475-3.10011-7

for virtually all the container and flat glass produced. Hence, this chapter will focus on these categories of glass when discussing the environmental issues arising from glass production and consumption.

1.1. Glass Production

The demand for glass containers, being dependent on sales of beverages and food, does not fluctuate greatly with business cycles, by contrast with flat glass demand. Annual production has increased over the past decade, but not at the same rate as growth in demand for packaged beverages and food, because of competition from other packaging materials. We have calculated that global production of glass containers in 2007 was 72 million tonnes (72×10^6 t), based on data in ISO Business Plan TC/63 for Glass Containers in 2004 [2], the EU 2008 Non-metallic Mineral Products Report for Glass, and *Official Journal of the European Union C* 317/7 on competitiveness of the European glass industry [3,4].

Table 11.1 summarises production for 2007 by regional groupings, citing source data and assumptions. Although data for the European Union (EU) and the United States can be quoted

TABLE 11.1 Global Container Glass Production

		Reported data		Conversion	Production	Assumed
	Reference year	Unit of measure	Total	rate	tonnes	Prodn 2007
Europe (EU27)[1]	2007	Tonnes	22,429,390	1.00	22,429,390	22,430,000
Russian Federation[2]	2006	Units million	9,800	0.0025	2,450,000	2,570,000
USA	2006[3]	Tonnes	9,638,516	0.907184	8,743,907	9,180,000
	2008[4]	Tonnes				
South America[5]	2007	Tonnes	12,500,000	1.00	12,500,000	12,500,000
Japan[6]	2006	US$ Million	1,864.62	0.00080	2,119,708	2,230,000
China[7]	2004	Tonnes	7,852,200	0.907184	7,123,390	10,010,000
India[8]	2008	Tonnes	1,400,000	1.00	1,400,000	1,400,000
Subtotal						60,320,000
Rest of world						11,680,000
Total[9]	2001	Tonnes	57,000,000	1.00	57,000,000	72,000,000

[1]*FEVE European Container Glass Federation 2009 for container glass statistics. European Commission 2009, Draft Reference Document on Best Available Techniques in the Glass Manufacturing Industry.*
[2]*Conversion units to tonnes based on Faraday Packaging and Glass Technology Services 2006. Light-weight Glass Containers—The Route to Effective. Waste Minimisation, WRAP, Banbury, UK Unipack RU 2007, Russian Market of Food Glass Containers. http://article.unipack.ru/eng/20538/.*
[3]*Container Recycling Institute 2008, Beverage Market Data Analysis for 2006, CRI, Culver City, California. Beverage containers only.*
[4]*US Environmental Protection Agency 2009. http://www.epa.gov/osw*
[5]*Undata, Industrial Statistics Commodity Database, April 2009 update, United Nations Statistics Division Argentine National Wine Institute, Rigolleau S.A. 2009, Buenos Aires, Argentina.*
[6]*UN Statistics Database—Container Glass Production. Conversion at reported tonnes/US$ ratio for year 2001, then uplifted for yen appreciation to the US $ to 2008.*
[7]*Chinese Ceramic Society, 2006, Review and Prospect of the Glass Industry in China.*
[8]*T. E. Narasimhan, 2009, Business Standard 2 April 2009, Glass Container Sector Gallops Ahead. http://www.business-standard.com/india/index2.php.*
[9]*ISO 2004, Business Plan ISO TC/63, Glass Containers, (forecasting 14% p.a. growth to 2005) Assumed Growth = 5% p.a.*

with some confidence, those for other regions are subject to varying degrees of accuracy through incomplete data collection, and the use of reported data to justify commercial and tariff cases.

By contrast with container glass, the demand for flat glass can be very cyclical, depending on the level of activity in the building construction and automotive industries. Global flat glass production for 2006 is presented in Table 11.2. As in the case of container glass, we believe that the figures for the EU and United States are more reliable than some of the data for other regional groupings.

1.2. Environmental Issues

The main environmental impacts in glass making are the high-energy use in batch melting, and the resultant gaseous emissions from fuel combustion and the heat reaction of components of the batch mix. The usual way of providing heat to melt glass is by burning fossil fuels above a bath of batch material, which is continuously fed into, and then withdrawn from the furnace in a molten condition. Heat is provided mainly by radiative transmission from the furnace crown, which is heated by the flames to up to 1650 °C, with some heat coming also from the flames themselves. The molten glass in the furnace is held at a constant temperature for approximately 24 h for production of containers and 72 h for float glass [5]. In general, the energy necessary for melting and mixing the batch components accounts for more than 75% of the total energy requirements of glass manufacture [5], with the raw material procurement and formation of life cycle stages accounting for the other 25%.

Although pure silica can be made into high-quality glass, this requires the batch to be heated to a temperature of around 2300 °C, at which point its viscosity is reduced to a liquid state suitable for the subsequent formulation stage, the 'melting point'. By adding sodium oxide (Na_2O) obtained from the addition of soda ash (Na_2CO_3), the melting point is lowered to about 1500 °C [6]. However, the soda makes any glass produced water soluble. To overcome this, calcium oxide (CaO) obtained from limestone ($CaCO_3$) is added to the batch to render the glass chemically durable. Magnesium oxide (MgO) and aluminium oxide (Al_2O_3) may also be used to enhance the chemical durability, whereas other materials are added to provide colour. The resulting glass contains about 70% to 74% silica, 12% to 15% sodium oxide, and 10% to 15% calcium oxide by weight, plus a small amount of colouring and other material, and is called silica–soda lime glass. It accounts for about 90% of manufactured glass.

TABLE 11.2 Global Flat Glass Production

	Million tonnes								
	Europe	Former Soviet Union	North America	South America	Japan	China	South East Asia	Rest of world	Total
Flat glass capacity (2006)	11.00	2.70	6.90	1.00	1.25	19.00	4.80	1.35	48.00
Flat glass production (2006)	9.70	2.70	6.00	1.60	1.05	16.00	3.80	3.15	44.00

Sources: Pilkington Group Ltd., St. Helens, U.K. 2007, Pilkington and the Flat Glass Industry; Haley C.V.U., 2009, Through China's Looking Glass; Subsidies to the Chinese Glass Industry 2004—2008, Economic Research Institute; European Commission 2009, Draft Reference Document on Best Available Techniques in the Glass Manufacturing Industry; Ecorys Research and Consulting 2008, FWC Sector Competitive Studies—Competitiveness of the Glass Sector, Rotterdam, Netherlands. http://ec.europa.eu/enterprise/newsroom/cf/document.cfm?action=display&doc_id=4044&userservice_id=1.

Fuel oil and natural gas are the predominant energy sources for melting, a small amount of electricity is also used. The theoretical energy requirements for soda–silica lime glass are given in Table 11.3 [7]. The calculation assumes that all available heat is fully used and has three components:

- The heat required to raise the temperature of the raw materials from 20 °C to 1500 °C
- The latent heat required to enable the reactions between the batch components to form the glass
- The heat content of the gases (principally CO_2) released from the batch during melting

The delivered process energy actually needed is higher than the theoretical figures [8] due to waste gas and structural heat losses, and it depends on the furnace efficiency. Large modern cross-fired regenerative furnaces [capacity > 500 tonnes per day (500 t d^{-1})] operating with a typical energy efficiency of 50% would result in energy use of approximately 5.5 GJ t^{-1} for a container batch containing virgin feedstock only.

The principal emissions to air from the batch-melting process result from the combustion of fuel and decomposition of the soda ash and limestone as they heat up. Once limestone is heated above 850 °C, it will start to decompose as in the reaction:

$$\text{Heat} + CaCO_3 \rightarrow CaO + CO_2$$

TABLE 11.3 Theoretical Secondary Energy Requirements in Batch Melting Using Virgin Feedstock

Soda lime–silica glass	Energy/GJ t^{-1}
Endothermic melting heat	1.89
Latent heat of fusion of materials (HChem)	0.49
Heat of gases emitted (HChem)	0.30
Total energy use	2.68

Similarly, soda ash decomposes to produce sodium oxide (Na_2O) as in:

$$\text{Heat} + Na_2Co_3 \rightarrow Na_2O + CO_2$$

Emissions of gaseous outputs from other additions to the batch produce oxides of sulphur and nitrogen. Based on an input of 150 and 190 kg of limestone and soda ash into the batch mix, 145 kg of process CO_2 per tonne of glass would be produced.

Emissions from combustion per unit of energy will vary depending on the energy source, the most common of which for batch melting are methane processed from natural gas and fuel oil. Emissions from methane combustion will follow the reaction:

$$CH_4 + 2O_2 \rightarrow CO_2 2H_2O$$

and that for fuel oil:

$$C_{14}H_{20} + 21.50_2 \rightarrow 14CO_2 + 15H_2O$$

Based on a delivered energy use of 5.5 GJ t^{-1} in the batch melt, and converting this to the mass of methane and fuel oil consumed, CO_2 emissions would be 280 and 415 kg t^{-1} of glass, respectively, because of the different combustion carbon outputs for given masses of methane and fuel oil with the same energy content. As all the inputs are in powder or granular form, there may also be releases of particulates into the atmosphere. The principal emissions to air are summarised in Table 11.4.

There are three broad approaches to reduce the environmental impacts of glass production: first, reductions in energy use; second, 'end of pipe' emission abatement measures; and third, for glass containers, product 'lightweighting'. Energy intensity efficiencies are achieved through more energy efficient furnace design and substituting recycled glass cullet for virgin raw materials. The use of cullet avoids the use of heat in thermal reactions between batch

TABLE 11.4 Principal Emissions to Air from Soda Lime–Silica Glass Batch Melting Per Tonne of Container Glass Produced

CO_2	Emissions/kg
Using natural gas (CH_4)	430
Using fuel oil	560
NO_x	2.4
SO_x	2.5
Dust (without secondary abatement)	0.4
Dust (with secondary abatement)	0.024
HCl (without secondary abatement)	0.041
HCl (with secondary abatement)	0.028
HF (without secondary abatement)	0.008
HF (with secondary abatement)	0.003
H_2O (evaporation and combustion)	1800

Source: European Commission. Integrated Pollution Prevention and Control - Reference Document on Best Available Techniques in the Glass Making Industry. Brussels: European Union, 2001, Table 3.8.

components and loss of heat in gaseous emissions, and it provides additional liquidity at lower temperatures in the batch, thereby reducing the energy used to heat the components. Compared with the theoretical energy requirements of 2.7 GJ t^{-1} for batch melting primary of raw materials, the energy required to simply melt glass is 1.9 GJ t^{-1}, and it is commonly estimated that substituting 10% of cullet for a similar weight of requisite virgin raw material mix can save 2.5% of energy.

Container glass typically has a short life cycle, being primarily used to package beverages and food. Production and use are often within the same country or region, although they may be distant from one another in the case of specialist products, for example, estate bottled wines, or pharmaceutical products. By contrast, flat glass, principally used in motor vehicle manufacture and the construction industries, has a long in-use life span. In the case of buildings, production of the glass used may be distant from its point of use, although usually within the same national or regional boundaries, whereas the glass used for motor vehicle windows may well be shipped across national and regional boundaries to its point of use. However, the two types of glass each present their own set of recycling challenges, which will be reviewed separately.

2. GLASS REUSE

Within the waste management hierarchy, reuse is considered before moving to the next option down, recycling. In the case of flat glass, because of the dispersed nature of its use, lack of homogeneity, and its long life span, reuse is not a viable option financially or environmentally. By contrast, the glass used for manufacturing containers has a similar raw material mix, apart from colouring agents, and has a short in-use life. Furthermore, production and use of the containers often take place within the same country or administrative region. For glass containers, the reuse option is therefore considered before recycling.

Supermarket retailers are the most influential decision makers in determining the viability of reuse of primary packaging, including glass. Since 1970, supermarket chains have accelerated the development of distribution systems on the basis of a one-way packaging flow from producer to consumer, and this trend has been further stimulated by the increasing globalisation of retail supply chains. Even in those countries with retake systems, underpinned by container deposit legislation, recycling rather than refill is becoming the norm. In effect, packaging recovery costs have very largely been externalised into the recycling route, where the burden is picked up by consumers, city and regional government recycling infrastructure, and the waste management industry.

For refill systems, the environmental and financial transport burdens of the collect and return system and the cleaning and sterilisation process before refill have to be measured against the burdens of cullet collection, processing, and batch melting. This has been the subject of numerous studies demonstrating the significant environmental benefits of reusing rather than recycling glass packaging [9–13]. A comprehensive study, placing economic valuations on internal costs and social and environmental externalities, presented data demonstrating that refillable glass containers are preferable to the recycling of single trip bottles from a purely environmental standpoint, subject to a combination of distances involved between filler and distribution centre and number of times the bottles are refilled [14].

Even given globalised markets and the current dominance of supermarket and hypermarket chains in retail distribution, there may still be opportunities for smaller scale glass container reuse, for example, in the rapidly growing microbrewery sector serving localised markets in the United Kingdom and New England. There are also a number of countries and market segments where the refillable glass bottle is used extensively, in many cases supported by container deposit legislation.

Canada's brewers maintain a 'closed-loop' container return system. The Brewers Association of Canada reports that returnable and reusable bottles make up almost 73% of packaged Canadian-brewed beer and 97% of them are returned by consumers [15]. The Environmental register of Packaging PVR, Ltd., Helsinki, shows that in 2007, 76% of all used glass packaging in Finland was refilled [16]. The 2008 Annual Report of Dansk Retursystem A/S shows that while in Denmark the one-way packaging share of the beverage market is increasing year on year, refillables still accounted for 53% with a 100% return rate [17]. Data for Germany [18] show that refillable bottles accounted for 47% of all types of beverage consumed. It is claimed that in Russia some 60% to 70% of all glass bottles used are returned for refill [19]. In other countries, the refillable glass container lives on for locally produced beverages or for niche markets, for example in the United Kingdom, the refillable milk bottle delivered to and collected from the doorstep. Finally, in some developing nations such as India and Brazil, the cost of new bottles often stimulates the collection and refill of glass bottles for selling carbonated and other drinks.

2.1. Container Glass Recycling

In theory, container glass can be made from 100% cullet, and there is no limitation on the number of times that used container glass can be fed back into the raw material input cycle. Consequently, the total potential for recycling is all the container glass used in a given period, which, given its short life cycle, is for all practical purposes the amount produced. However, there are practical limitations to this theoretical 100% use of cullet. First, given the dispersed nature of the waste stream, the marginal environmental and financial burdens of collecting increasing fractions of the post-consumer waste (PCW) container glass waste stream may increase to the point where they exceed the marginal benefits. Second, production waste cullet normally contributes about 10% of the batch mix, which limits the amount of PCW cullet that can be used. Having said that, a review of the current status of PCW glass recycling across the world reveals that any limits on using PCW-sourced cullet in container glass production are far from being reached.

On the basis of public domain information, we have estimated worldwide glass container consumption, reuse and recycling, the results of which are given in Table 11.5. Principal data sources are shown in Annex A. Although numerical data have not been available, we have made an assessment of the waste disposal options being used in a region or country, based on reviews of the municipal waste management practices obtained from press and industry

TABLE 11.5 Global Glass Container Consumption and Recycling

Glass waste management hierarchy	Australasia/Oceania			North America		Latin America and Caribbean		Europe			Mena		Sub Saharan Africa		Asia			
	Australia	New Zealand	Other (excl Hawaii)	U.S.A.	Canada	Brazil	Other	EU 27 + NOR, CH, Turkey	Russian Federation	Other	Egypt	Other	South Africa	Other	India	China (Urban Areas)	Japan	Other
Re-use (%)					68	45		≈ 5	60–70						≥ 70	≈ 50		
Recycle (%)	37	50		28	11		3–10	64			≤ 25		31				14	
Sanitary landfill (%)	63	50		72	20	55	90–97	31	30–40		≥ 75		79		≤ 30	≈ 50	86	
Disposal sites																		
Year (tonnes mission)	2005	2008	2007	2008	2008	2007	No data available	2008	2006	No data available	2007	No data available	2008	No data available	2007	2004	2007	2007
Consumption	0.95	0.30	0.30	13.46	1.37	1.57		17.82	3.04		1.90		0.80		1.69	7.60	2.28	4.05
Recycle/re-use	0.35	0.15	No data	3.78	1.09	0.71		11.47	1.98		≤ 0.48		0.25		1.18	≈ 3.8	0.32	No data

articles and Web sites. These appear as shaded areas in Table 11.5. As there is no national or local government infrastructure for municipal waste management and recycling for many underdeveloped and developing countries, there is a consequent lack of reliable data in these regions. However, this does not mean that no recycling or reuse takes place. On the contrary, high rates of urban recycling in many countries are tied in with poverty, so that the very poor, such as the Kabari in India, find a source of income by picking recyclable material from waste left in streets or on municipal dumps.

Calculated consumption amounts to 57×10^6 t for those regions and countries for which data are available and for the years quoted. This compares with the global production of 72×10^6 t given in Table 11.1. The difference between the two figures is largely due to the lack of consumption data for some regions and the limitation of some data to beverage containers only. In developing countries in particular, there is a huge annual increase in beverage container use, up to 15% per annum in the case of China, but glass takes an increasingly lower share of the total, with predictions for the growth in global container glass consumption in the region of 3% to 5% [20].

2.2. Flat Glass Recycling

The recycling of flat glass largely depends on the way in which construction and demolition wastes (C&DW) end-of-life vehicles (ELVs) and are treated. Approximately 70% of global flat glass production is used in the building and construction industries, 10% in motor vehicle manufacture [21], and the remainder for other uses. Recycling of such glass is largely dependent on the management of the C&DW and ELV waste streams.

2.2.1. Flat Glass Construction and Demolition Waste

Despite C&DW waste being one of the largest waste flows in the world, there is a significant lack of consistent data about the total waste stream and its management. One estimate for China is that urban C&DW has reached 30% to 40% of the total urban waste generation because of the large-scale construction and demolition activities resulting from the accelerated urbanization and city rebuilding [22]. Applying these percentages to the total reported industrial waste for China in 2004 of 1.089×10^9 t would give estimated C&DW of between 325 and 425 million tonnes per annum ($325–425 \times 10^6$ t a^{-1}) [23]. With the rapid growth in construction projects in China, these estimates are likely to significantly understate the amount of C&DW for later years. Estimates for some other countries/regions are shown in Table 11.5, which also includes the source references. Two studies characterising the composition C&DW estimated the proportion of flat glass in C&DW to be 0.4% ± 0.2% and 0.2%, respectively [24,25]. These ratios are also applied to the estimated C&DW to indicate the size of flat glass waste arisings from that source.

Currently, there is little recycling of glass by demolition companies due to financial viability. Much brick and concrete C&DW is reused on the construction sites in the form of hard core once it has been crushed to an acceptable particle size. Given that glass is an insignificant part of total C&DW, it is frequently absorbed into the hard core, or any material removed from site for further processing. The exception to this occurs in buildings where glass is a significant part of the external or internal construction. The reader is referred to Chapter 15 for a fuller description of C&DW processing.

If flat glass can be collected without contamination, it can be recycled to be incorporated in new flat glass production. St. Gobain Glass, United Kingdom, claims that it uses 30% flat glass cullet in the manufacture of its float glass, amounting to 3.6×10^4 t a^{-1} [26], which would include production waste. Nevertheless, the incorporation of flat glass C&DW into building aggregates for substrate is likely to remain the main recycling option.

2.2.2. *Flat Glass End-of-Life Motor Vehicle Waste*

Within the EU, the management of ELV waste is regulated by Directive 2000/53/EC, which aims *inter alia*, to increase the reuse, recycling, and recovery of materials from ELVs. Total ELV waste for 2007 was calculated as 6.12×10^6 t with 5.02×10^6 t being reused or recycled [27]. According to a report submitted by GHK to the EU DG XI [28], the average weight of glass per ELV was calculated as 21.2 kg, which applied to the 2007 ELV waste data would give a figure of 1.3×10^5 t of ELV glass waste in that year.

In the United States, the objectives of the Automotive Recyclers Association include 'to promote automotive recycling'. A report published in the United States [29] calculated that the number of vehicles taken out of use in the period 1989–1998 was 11.374×10^6 (Tables 1 and 2 of the report) with an average glass weight of 39 kg (86 lbs) out of a total average vehicle weight of 1.44 t (3165 lbs). This would amount to 4.45×10^5 t a^{-1} of waste glass.

In Japan, roughly five million cars are disposed every year, with around a million of these exported as second-hand vehicles [30]. Of the four million remaining, it is claimed that nearly 100% are subject to recycling, with a recycling rate of 75% by vehicle weight [31]. Based on the EU average of 21.2 kg of glass per vehicle, flat glass waste from ELVs would amount to approximately 8.5×10^4 t a^{-1}.

By the end of 2007, there were 43 million vehicles with an average life of 15 years on the roads of China. It is estimated that in each year up to 2010, 4.8 million vehicles will be scrapped [32]. Using an average of 21.2 kg of glass per vehicle, ELV flat glass waste would amount to around 1.00×10^5 t a^{-1}.

There is little, if any, collection of ELV glass for feeding back into the flat glass production loop, and the glass is generally treated as a waste product from the metal and other materials recovered from ELVs (see Chapters 6 and 20 of this handbook for a description of ELV recycling).

2.3. Summary of Glass Waste Streams

Based on data presented in the preceding sections, the relative importance of the three

TABLE 11.6 Flat Glass Construction and Demolition Waste (C&DW) in Selected Countries Where C&DW Refers to Construction and Demolition, Respectively

Tonnes million	Europe[1]	USA[2]	China[3]	Japan[4]	Brazil[5]	India[6]
Year	2004	pre 2006	2005	2005	2001	2005
C&DW	450	295	425	76	69	≥30
Glass proportion[6] (%)	0.2	0.2	0.2	0.2	0.2	0.2
Flat glass C&DW	0.90	0.59	0.85	0.15	0.14	0.06

[1]*EU Taskforce on Sustainable Construction 2007, Accelerating the Development of the Sustainable Construction Market in Europe, EU Commission, Brussels.*
[2]*Construction Materials Recycling Association USA, Amount of C&DW Annually CMRA, Eola, Illinois.*
[3]*National Bureau of Statistics of China 2005, Industrial Solid Wastes Produced, Table 21-1, Beijing.*
[4]*Ministry of Land, Infrastructure and Transport of Japan White Paper on Land, Infrastructure, Transport, and Tourism in Japan, 2008, 66.*
[5]*John V.M., Angulo S.C., Miranda F.R., Agopyan V., Vasconcellos F. 2004, Strategies for Innovation in Construction and Demolition Waste Management in Brazil, Department of Civil Construction, University of Sãn Paulo, Sãn Paulo Brazil.*
[6]*Technology Information Forecasting and Assessment Department 2001. Utilisation of Waste from Construction Industry. Department of Science and technology, Government of India 12% p.a. growth rate applied to reported 15 million tonnes for 2000 to 2005. http://www.docstoc.com/docs/17050867/ Indian-Construction-industry.*

TABLE 11.7 Relative Mass of Four Glass Waste Streams

	EU27	USA	Japan	China
Container glass				
Year	2008	2008	2007	2007
Tonnes million	17.82	13.46	3.28	7.6
Flat Glass				
C&DW				
Year	2007	Pre 2006	Pre 2005	2005
Tonnes million	0.9	0.59	0.15	0.85
ELVs				
Year	2007	Pre 2005	Pre 2005	2007
Tonnes million	0.01	0.04	0.08	0.10
Proportion flat glass of total (%)	5	4	7	11

principal glass waste streams in terms of mass is summarised in Table 11.7. Despite the variation in the source years and assumptions for the data presented, the many orders of magnitude difference between the three waste streams illustrate the dominant position that recycling container glass, compared with flat glass, can play in the recycling challenge to reduce the environmental and resource impacts of glass production.

3. CONTAINER GLASS RECYCLING PROCESSES

Although waste recycling has been regarded as the waste management option of choice, it has to be recognised that it carries its own environmental and financial burdens [33]. Conceptually, converting post-consumer glass into cullet is a straightforward process of collecting material and removing contaminants, followed by colour separation and crushing to feedstock size ready for inclusion in the batch melt, but in practice, this is often difficult to achieve. Furthermore, in the drive to achieve high levels of recycling, sight is often lost of the aim of optimising the environmental gains, or at least this becomes of secondary importance.

In Table 11.8, some key characteristics of container glass recycling have been classified according to the end use of the cullet. Using cullet to produce containers is the most environmentally benign option, not only because of the energy saved in the batch melt but also because the used glass containers can be fed back into the product loop continuously. The ability to do so depends on there being sufficient demand, which in turn requires that the cullet supplied meets the manufacturer's specification for colour mix and purity, for example, in the United Kingdom, the WRAP PAS 101 specification, and in the United States, the Glass Packaging Institute's 'High Quality Cullet' guide.

A key dependency for optimising environmental benefit is the achievement of a balanced flow of material through the system. For this to happen, it is essential to have the necessary capacity at each stage, without an over or under supply of material. In practice, the different motivations of the actors in the system can, and often do, prevent this system balance being achieved, and may result in open-loop recycling, for example, using cullet as a substrate in road construction. This situation arises when there is insufficient demand for cullet of a specific colour and grade for glass container production. One reason for this may result from there being an imbalance between regulatory recycling targets and commercial demand.

Replacing virgin feedstock with cullet avoids the Hchem and Hgas energy use shown in Table 11.3. Based on a furnace thermal efficiency of 50%, a theoretical saving of the energy used in the glass container batch melt from 100% cullet rather than 100% virgin feedstock would be in excess of 1.5 GJ t^{-1} of delivered (secondary) energy. Table 11.8 shows that, as a result of the reduction in energy use and avoidance of heat reactions with soda ash and limestone, CO_2

TABLE 11.8 Characteristics of Container Glass Recycling and Cullet End Use

	Glass container cullet		
	Closed loop	→	**Open loop**
	Market demand	→	**Regulatory driven**
Type of recycling	Product to product	Material to material	Material substitution
Type of use	Used in container glass production	Used in other glass production	Used in nonglass applications
Typical end products	Glass bottles	Fibre glass insulation	Aggregates and substrate
Secondary energy saving/ tonne (batch melt)	≥1.5 GJ	≥1.5 GJ	0
CO_2 emissions avoidance kg/tonne (batch melt)	215–250	200–230	0
Maximum cullet proportion (%)	90	50	10–20
Continual loop recycling	Yes	No	No

Sources: Enviros Consulting Ltd. 2003, Glass Recycling—Lifetime Carbon Dioxide Emissions, British Glass Manufacturers Confederation, Sheffield; Butier and Hooper, Dilemmas in optimising the environmental benefit from recycling: a case study of glass container waste management in the UK, Resources Conservation and Recycling 45 (2005) 331–355.

emissions from the batch melt are reduced by 215 to 250 kg t^{-1} if a theoretical 100% cullet is used in place of virgin feedstock in glass container production. As there is no limit to the number of times glass can be recycled, these savings can be repeated, depending on the efficiency of the recycling regime in keeping waste container glass within the loop [34]. Although a similar one of energy saving is obtainable from using cullet as raw material for producing fibreglass, it is not possible to then recycle fibreglass as feedstock into further production cycles [35]. At the other end of the spectrum, it has been shown that reductions in the energy burden through the use of cullet in aggregate production are largely dependent on the reduced transport resulting from using locally produced cullet rather than more distant virgin raw materials. In future cases, using cullet may actually increase the energy burden compared with using virgin feedstock [34].

4. THE FUTURE OF GLASS RECYCLING

There are clear environmental and financial benefits accruing from using glass cullet rather than virgin feedstock in glass production. Based on the data in Tables 11.1 and 11.5, we estimate that 30% to 35% of the container glass consumed globally enters the recycling loop, leaving room for significant enhancements in recycling rates to meet the demand of glass container manufacturers for quality cullet. Some initiatives to enhance container glass recycling are considered below.

4.1. Introduction of Container Deposit Schemes

Container deposit schemes have been shown to be very effective in motivating householders to recycle food and beverage containers. Thus, the 11 U.S. states with container deposit legislation consistently return glass container recycling rates of between 66% and 96%, compared with the 35% average for those states without container deposit legislation [36]. Similar differences between those member states with and without container deposit legislation are found in the EU.

4.2. Bottle Deposit Return Machines

An enhancement to deposit schemes is the provision of conveniently located reverse vending machines, often in supermarket stores, where bottle deposits are returned once the empty has been deposited in the machine.

4.3. Regulatory Systems

Regulatory systems where the target is to maximise the amount collected for recycling without the need to take into account maximising environmental benefit may encourage the easy option of open-loop recycling. In the United Kingdom, the Department of the Environment, Food, and Rural Affairs (DEFRA) has taken a lead on this issue by proposing to set differentiated glass packaging recycling targets for businesses by 2011, based on whether material is recycled into open- or closed-loop processes [37].

4.4. Coloured Glass

In some countries, to minimise city and local government collection costs and maximise recycling, there is a move away from collecting colour-separated glass containers at source, to collecting mixed colour glass containers or even mixed material recyclate. This passes the sorting and cleaning burden on to cullet processors. One development in overcoming the problem of mixed coloured cullet is the introduction of colour separation systems to identify and remove glass cullet of different colours.

4.5. Infrastructure Maintenance and Change

In the case of rapidly developing countries and regions, there will be an increasing movement from unregulated to city and local government regulated systems, financed by local taxes and other financial stimuli. During this transition, it will be important to ensure that regulatory systems take over from market-driven ones, without there being a void created by lack of financial motivation for those at the picking and sorting end of the cycle, due to the availability of better employment opportunities.

The challenge for the regulatory systems found in the developed countries and regions is to ensure that all the links in the cycle from household to glass producer are in balance in terms of the flows of material through the system.

5. CONCLUSION

The core challenge for environmentally and cost-effective recycling of container glass arises from the dispersed nature of its sources, principally households, and the consequent need for an environmentally and cost-effective infrastructure providing for its colour separation, collection, and transportation to processors to produce furnace ready feedstock. In assessing the scope for increasing the amount of glass recycled, there is an overall need to quantify the resultant energy and other environmental burdens to allow valid comparisons to be made with the burdens of using virgin feedstock. Nevertheless, based on the data in Table 11.8, overall significant environmental and cost benefits can result from substituting cullet for virgin feedstock in container glass production.

Assuming a global glass container recycling rate in the region of 30%, there is a huge potential for energy savings and a resultant reduction in carbon emissions by increasing the proportion of PCW sourced cullet used in container production. Based on the 2007 figure of 72×10^6 t of container glass produced globally and assuming an overall 50% thermal efficiency in the batch melt, an 80% use of cullet would result in about 54×10^6 GJ of energy saved per annum compared with the energy consumed using 30% cullet. Based on the same assumptions, reductions in CO_2 emissions using the same cullet/virgin feedstock ratios would be about 8.3×10^6 t a^{-1}. These potential reductions represent a very strong case for striving to enhance the proportion of PCW cullet used in container glass production in the future.

ANNEX A—SOURCES FOR DATA PRESENTED IN TABLE 8.5

Sources for Glass Packaging Consumption, Reuse, and Recycling Data

1. UN Data, Industrial Commodity Statistics Database, http://data.un.org/Data.aspx?q=glass+&d=ICS&f=cmID:37191-0&c=2,3,5,6&s=_crEngNameOrderBy:asc,yr:desc,_utEngNameOrderBy:asc&v=8
2. OECD 2009, Environmental Data: Compendium 2006–2008—Waste, OECD, Washington
3. Hassan W. 2006, Glass Recycling, O-I Australia, http://www.acor.org.au/presentations/Glass.pdf

   Packaging Council of New Zealand 2009, "Five years on: recycling up by 26%," http://www.packaging.org.nz/packaging_info/packaging_consum.php
4. Miller C. 2007, "Glass Containers," Waste Age August 1 2007, http://wasteage.com/Recycling_And_Processing/waste_glass_containers_4/index.html
5. The Glass Recycling Company, "Glass Recycling Facts and Figures," http://www.theglassrecyclingcompany.co.za/main%20pages/FactsAndFigures/GlassRecycling.html
6. Container Recycling Institute 2008, Beverage Market Data Analysis for 2006, CRI, Culver City, California Beverage containers only, http://www.container-recycling.org/media/newsrelease/general/2008-12-BMDA.htm
7. US Environmental Protection Agency 2008, http://www.epa.gov/osw

Sources for Glass Packaging Consumption, Reuse, and Recycling Data

8. Brewers Association of Canada 2009, "Annual Statistical Bulletin 2008," http://www.brewers.ca/default_e.asp?id=98
9. Glassworks 2001, "Glass Container Recycling in Canada," http://www.glassworks.org/statistics/default.html
10. Brenda Platt and Doug Rowe 2002, "Reduce, Reuse, Refill," Institute for Local Self Reliance, Washington, D.C., produced under a joint project with Grass Roots Recycling Network, http://www.grrn.org/beverage/refillables/refill_report.pdf
11. The World Bank 2008, "Solid Waste Management in LAC; Actual and Future CH4 Emissions and Reductions," http://siteresources.worldbank.org/INTUWM/Resources/340232-1221149646707/Solid_Waste_Management_in_LAC.pdf
12. Borzino M. A., "Promotion of 3Rs at the National Level," Ministry of the Environment, Brazil
13. Look M. 2009, "Trash Planet: Brazil," www.Earth911.com
14. Rodriguez C. M., Zanetto E. D., "Glass Industry and Research in Brazil," http://www.lamav.ufscar.br/artpdf/glint25.pdf
15. FEVE European Container Glass Federation 2009, "Recycling Statistics for 2008," http://www.feve.org/index.php?option=com_content&view=article&id=10&Itemid=11
16. PRO Europe, "Packaging waste legislation in Denmark". http://www.pro-e.org/Denmark
17. Ecorys Research and Consulting 2008, "FWC Sector Competitive Studies—Competitiveness of the Glass Sector," Rotterdam, The Netherlands. http://ec.europa.eu/enterprise/newsroom/cf/document.cfm?action=display&doc_id=4044&userservice_id=1
18. Tatiana Vyugina, Elena Nikonova 2008, Review of Russian Market of Glass Containers, http://article.unipack.ru/eng/22036/
19. Gonopolsky Adam M. 2007, "The Waste Recycling Industry in Russia: Challenges and Prospects," WasteTech-2007
20. Glass International reports. "Egypt's glassmaking sector: a 2000 year old industry continues to grow in this fast-developing country," January 1, 2006, http://www.allbusiness.com/nonmetallic-mineral/glass-glass-manufacturing/874638-1.html.

(*Continued*)

Sources for Glass Packaging Consumption, Reuse, and Recycling Data

21. Hania Moheeb. 2006, "Talking Trash," Business Today, December 2006, http://www.businesstodayegypt.com/article.aspx?ArticleID=7054

    The Glass Recycling Company, "Glass Recycling Up 65%," http://www.bizcommunity.com/Article/196/457/43247.html

22. Issues and Recommendations, East Asia Infrastructure Department, World Bank, http://siteresources.worldbank.org/INTUSWM/Resources/463617-1144078790304/Hoornweg.pdf

23. The World Bank 1999, "What a Waste: Solid Waste Management in Asia"; figures presented projected on growth rate of 4% p.a. for glass containers, http://siteresources.worldbank.org/INTEAPREGTOPURBDEV/Resources/whatawaste.pdf

24. David Hanrahan, Sanjay Srivastava, A. Sita Ramakrishna 2006, Improving the Management of Municipal Solid Waste in India—Overview and Challenges, Published by Environment and Social Development Unit, South Asia Region. The World Bank (India Country Office), 70 Lodi Estate, New Delhi 110003, Internet: www.worldbank.org/in

25. T. E. Narasimhan/Chennai, April 02, 2009, Glass container sector gallops ahead, Business Standard http://www.business-standard.com/india/index2.php

26. Japanese Container Packaging Recycling Association 2007, "Recycling Statistics," http://www.jcpra.or.jp/eng/statistics.html

27. D. Hoornweg, P. Lam, M. Chaudhry, 2005, Waste Management in China

28. J. Lin, N. Lin, L. Qiao, J. Zheng, Tsao C-C 2007, Municipal Solid Waste Management in China. http://www.docstoc.com/docs/19960465/Municipal-Solid-Waste-Management-in-China

References

[1] Edificio EXPO, Sevilla, Spain, Integration Pollution Prevention and Control; Draft Reference Document on Best Available Techniques in the Glass Manufacturing Industry, European Union, Brussels, 2009.

[2] ISO. Business Plan ISO TC/63, Glass Containers, ISO, Geneva, 2004.<http://isotc.iso.org/livelink/livelink/fetch/2000/2122/687806/ISO_TC_063__Glass_containers_.pdf?nodeid=1267119&vernum=0>, 2004. Accessed March 2010.

[3] European Commission Enterprise and Industry, Non-Metallic Mineral Products Report—Glass, European Commission, Brussels, 2008. <http://ec.europa.eu/enterprise/sectors/metals-minerals/non-metallic-mineral-products/index_en.htm>. Accessed March 2010.

[4] Official Journal of the European Union C 317/7, Opinion of the European Economic and Social Committee on 'The competitiveness of the European glass and ceramics industry, with particular reference to the EU climate and energy package', European Union, Brussels, 2009. <http://eur-lex.europa.eu/LexUriServ/LexUriServ.do?uri=OJ:C:2009:317:0007:0014:EN:PDF>. Accessed March 2010.

[5] European Commission. Integrated Pollution Prevention and Control - Reference Document on Best Available Techniques in the Glass Making Industry, European Union, Brussels, 2001.

[6] J.E. Shelby, in: Introduction to Glass Science and Technology, second ed., The Royal Society of Chemistry, Cambridge, UK, 2005.

[7] ETSU, Energy Use in the Glass Industry Sector, AEA Environment and Energy, Abingdon, Oxfordshire, UK, 1992.

[8] R.K. Dhir, M.C. Limbachiya, T.D. Dyer, Recycling and Reuse of Glass Cullet, Thomas Telford Publishing, London, 2001.

[9] J. Hancock, G.F. Bousted, Report on the Energy and Raw Material Requirements of Liquid Food Container Systems in the United Kingdom, HMSO, London, 1986.

[10] E.P.A. Danish, Life Cycle Assessment of Packaging Systems for Beer and Soft Drinks, Danish Ministry of Environment and Energy, Copenhagen, 1998.

[11] V.R. Sellers, J.D. Sellers, Comparative Energy and Environmental Impacts for Soft Drink Delivery Systems, Franklin Associates, Kansas, 1989.

[12] G. Hartmann, F. Coffey, Moving Up the Ladder: The Place of Re-Use and Refill in Canadian Waste Management Strategies, Toronto Environmental Alliance, Toronto, 1994.

[13] R. Lanoie, P. Lachance, Refillable and Disposable Beer Containers—An Analysis of the Environmental Impacts, Ecole des Hautes Etudes Commerciales, Montreal, 1999.

[14] RDC Environment and PIRA International, Evaluation of Costs and Benefits for the Achievement of Reuse and Recycling Targets for the Different Packaging Materials in the Frame of the Packaging and Packaging Waste Directive 94/62/EC—Final Consolidated Report, EU Commission DG XI, Brussels, 2003, Annex 12, pp. 167–173.

[15] Brewers Association of Canada, 2007, Bottling and Packaging Recovery in 2006. <http://www.brewers.ca/default_e.asp?id=31>. Accessed March 2010.

[16] The Environmental Register of Packaging PVR Ltd, Statistics on the Reuse of Packaging for 2007, PVR Ltd, Helsinki, 2008. <http://www.pyr.fi/eng/statistics/reuse.html>, 2008. Accessed March 2010.

[17] Dansk Retur System, Annual Report of the Dansk Retur System for 2008, Dansk Retur System, Copenhagen, 2009. <http://www.dansk-retursystem.dk/content/us/news/>, 2009. Accessed March 2010.

[18] Gesellschaft für Verpackungsmarktforschung mbH (GVM), Share of Reusable Packaging in Drinks Consumption by Type of Drink from 1991 to 2007 (in %), Federal Ministry for the Environment, Nature Conservation and Nuclear Safety, Berlin, 2009. <http://www.bmu.de/files/pdfs/allgemein/application/pdf/mehrweganteil_zeitverlauf_en.pdf>, 2009. Accessed March 2010.

[19] A.M. Gonopolsky, Market Survey the Waste Recycling Industry in Russia: Challenges and Prospects, Waste Tech-2007, Moscow, 2007. <http://w2007.sibico.com/?content=list§ion_id=12>, 2007. Accessed March 2010.

[20] Glasstech, *Successful Year for Container Glass Manufacturers*. 2005. <http://www.glassteconline.com/cipp/md_glasstec/custom/pub/content, lang,2>. Accessed March 2010.

[21] Pilkington Group, *Pilkington and the Flat Glass Industry*. 2007. <http://www.pilkington.com/pilkington-information/downloads/pilkington+and+the+flat+glass+industry+2009.htm>. Accessed March 2010.

[22] W. Zhao, R.B. Leeftink, V.S. Rotter, Evaluation of the Economic Feasibility for the Recycling of Construction and Demolition Waste in China—The Case of Chongqing, Resources, Conservation, and Recycling, 54, April 2010, pp. 377–389.

[23] National Bureau of Statistics of China. Industrial Solid Wastes Produced, Table 21-1, Beijing, 2005.

[24] Cascadia Consulting Group, Detailed Characterization of Construction and Demolition Waste, California Environmental Protection Agency Integrated Waste Management Board, Sacramento, 2006.

[25] B. Kourmpanis, A. Papadopoulos, K. Moustakas, F. Kourmousis, M. Stylianou, M. Loizidou, An integrated approach for the management of Demolition Waste in Cyprus, Waste Management and Research 26 (2008) 573–581.

[26] St Gobain Glass UK. *'More-is-less' with Unique Cullet Recycling Scheme*. 2009. <http://uk.saint-gobain-glass.com/b2b/default.asp?nav1=act&id=23833 and http://in.saint-gobain-glass.com/b2c/default.asp?nav1=act&id=24145>. Accessed March 2010.

[27] Eurostat, End of Life Vehicles, Data 2007 (2009).

[28] GHK in association with Bio Intelligence Service, A study to examine the benefits of the End of Life Vehicles Directive and the costs and benefits of a revision of the 2015 targets for recycling, re-use and recovery under the ELV Directive, GHK, Birmingham, U.K, 2006.

[29] J. Staudinger, G.A. Keoleian, Management of End of Life Vehicles (ELVs) in the US. Michigan: Center for Sustainable Systems, University of Michigan, 2001.

[30] Kiyoshi Koshiba, The Recycling of End-of-Life Vehicles in Japan; Newsletter No.50 (October 2006). s.l.: Japan for Sustainability Mail Magazine (2006). <http://www.japanfs.org/en/mailmagazine/newsletter/pages/027816.html>. Accessed March 2010.

[31] Japan Automobile Manufacturers Association, Vehicle Recycling and Waste Reduction in The Motor Industry of Japan 2010 http://www.jama-english.jp/publications/MIJ2010.pdf. Accessed March 2010.

[32] M. Zhang, C. Fan, End-of-Life Vehicle Recovery in China: Consideration and Innovation following the EU ELV Directive, Journal of the Minerals, Metals and Materials Society, 61, 2009, 3.

[33] J.H. Butler, P.D. Hooper, Factors determining the post-consumer waste recycling burden, Journal of Environmental Planning and Management 43 (2000) 407–432.

[34] J.H. Butler, P.D. Hooper, Dilemmas in optimising the environmental benefit from recycling: A case study of glass container waste management in the UK, Resources Conservation and Recycling 45 (2005) 331–355.

[35] Enviros Consulting Ltd, Glass Recycling—Life Cycle Carbon Dioxide Emissions, British Glass Manufacturers Confederation, Sheffield, 2003.

[36] Container Recycling Institute, Beverage Market Data Analysis, Container Recycling Institute, Culver City, California, 2008.

[37] DEFRA Advisory Packaging Committee, Note of the Meeting held on 22 October 2009, DEFRA, London, 2009.

CHAPTER

12

Textile Waste

Andreas Bartl

Institute of Chemical Engineering, Vienna University of Technology, Getreidemarkt 9/166, A-1060 Vienna, Austria

OUTLINE

1. INTRODUCTION

Over the past few decades, the attitude towards waste has dramatically changed. Up to the 1960s, waste treatment was basically a sanitary activity. Garbage was simply transported out of the cities to prevent epidemic diseases. Since the 1970s, waste management has progressed in most parts of the world and especially in the developed counties such as the European Union (EU), the United States, and Japan. Initially, technologies were developed to prevent environmental problems at landfill sites. Later on, recycling strategies for special types of wastes such as glass or paper were introduced. Parallel to this, waste incinerations with highly efficient gas cleaning devices have been extensively installed. Today, waste is increasingly seen as (secondary) raw material, and 'zero waste' represents the current ideal in waste management.

Waste Doi: 10.1016/B978-0-12-381475-3.10012-9

Fibres and fibre containing products have followed the same trend towards sustainability. Because fibre production and textile processing are sophisticated and demands large amounts of energy, reuse and recycling are highly recommended, not only for ecological reasons but also for economic reasons. Unfortunately, because fibre reprocessing is difficult and complex, a large portion of fibrous waste still ends up in landfill sites or, largely based on legislative forces, in waste incinerators.

2. TECHNOLOGICAL, ECONOMICAL, AND ECOLOGICAL BACKGROUND

2.1. Fibre Technology

Although fibres are omnipresent in our society, it is striking that there exists no clear definition of a fibre. According to the Bureau International pour la Standardisation des Fibres Artificielles (BISFA), fibre is *a morphological term for substances characterised by their flexibility, fineness, and high ratio of length to cross-sectional area* [1]. Thus, fibres are explicitly distinguishable from rods and wires that are either too stiff or too coarse. A similar requirement is given in a DIN standard (German Standard) [2] but additionally the capability of textile processing is a part of the definition. Although no definite threshold values for length and width exist, BISFA [1] defines some fibre-related terms according to the unit length as given in Table 12.1.

TABLE 12.1 Definitions of Fibre-Related Terms Ranked with Decreasing Unit Length [1]

Term	Definition
Filament	A fibre of very great length; considered as continuous.
Staple fibre	A textile fibre of limited but spinnable length.
Flock	Very short fibres intentionally produced for other purposes than spinning.
Fibre fly	Airborne fibres or parts of fibres (light enough to fly), visible as fibres to the human eye.
Fibril	A subdivision of a fibre; can be attached to the fibre or loose.

The cross-section of fibres is not necessarily circular but frequently exhibits profiles such as angular (e.g. triangular), lobal (e.g. trilobal), serrated, oval (e.g. bean-shaped), ribbon-like, or even hollow. It is thus clear that the diameter is not a universally applicable property for defining the fineness of a fibre. To circumvent this problem, it is well established in the textile industry to use the fibre denier (i.e. the linear density) [3]. The mass of a certain length of a fibre or a yarn is usually given in terms of a 'tex' unit (1 tex = 1 g per 1000 m), a decitex (dtex = 1 g per 10,000 m), or a denier (1 den = 1 g per 9000 m [4]).

Fibres are composed of a variety of materials but are usually categorised into natural and man-made fibres. Natural fibres comprise vegetable as well as animal fibres. Within this group, cotton (a crop fibre) is the most important. Man-made fibres are classified according to their chemical composition, with the most fundamental differentiation being between organic and inorganic materials. Within organic man-made fibres, there is a distinction between polymers from natural resources, mainly cellulose (i.e. cellulosics) and from synthetic polymers originating from petroleum (i.e. synthetics). Figure 12.1 shows the major fibre categories including representative examples.

Although fibres represent an intermediate product, they can be considered as being semi-finished. Depending on the final end-use, fibre processing can take many forms. Basically, one can distinguish between conventional textiles manufactured from yarns and nonwovens. The classical route comprises spinning (yarn making) and weaving or knitting resulting in woven or knitted fabrics. Alternatively, nonwovens are formed directly

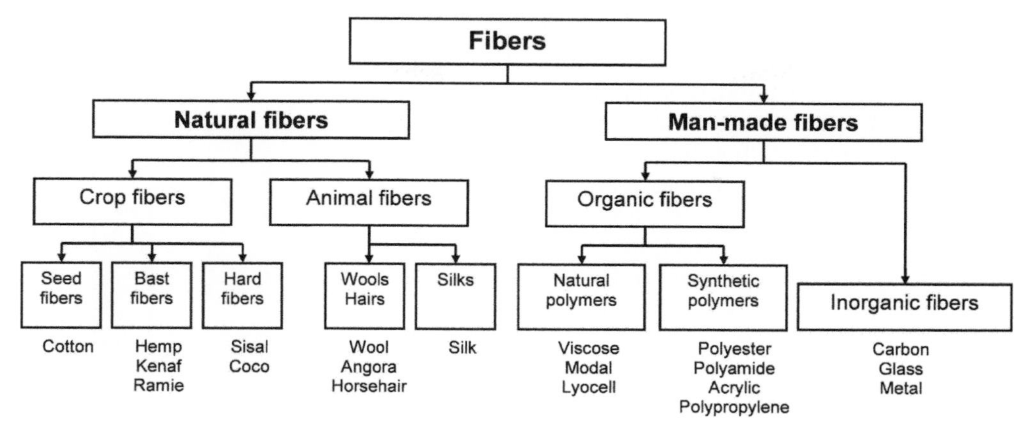

FIGURE 12.1 Categorisation of the most important fibres [1,2,5].

from individual fibres. However, certain applications do not require fabrics but yarns, rovings, or fibres.

The use of fibres and fibre containing goods is widespread. The main end-applications comprise apparel as well as home furnishing (curtains, carpets, soft furnishings, etc.) and several other uses as shown in Table 12.2. Because fibres can be found in a great variety of products, it is to be expected that fibres end up in diverse waste streams.

TABLE 12.2 Main End-Users for Fibres Including Their Respective Share (Volume of Fibre Usage) and Typical Examples [6]

Main end-user	Share (%)	Subcategories
Apparel	44	• Outerwear (e.g. trousers, coats, dresses) • Underwear (e.g. briefs, stockings, undershirts)
Interior and home textiles	33	• Carpets • Home textiles (e.g. curtains, blankets, table clothes)
Industrial and technical textiles	24	• Transport • Protective end uses • Building and construction • Medical, pharma, and health • Filters and membranes

2.2. Fibre Market

Worldwide fibre production reached a peak of 73.7 million tonnes (73.7×10^6 t) in 2007 [7]. In following year, 2008, there was a slight decrease due to the economic crisis and the world production was 69.1×10^6 t [8]. Synthetic fibres make up 60% (i.e. 41.4×10^6 t) of all fibre production and within this group polyester fibre is the most important, amounting to about 76% (30.6×10^6 t) of all synthetic fibres and showing a significant increase over the past few years. These figures are for 2008. Cotton is the second most important fibre – amounting to 34% (i.e. 23.3×10^6 t) of all types of fibre, whereas other natural fibres are of less commercial importance (e.g. wool: 1.7% or 1.2×10^6 t). The trend for cotton shows a slight increase over the past 10 years. The percentage of cellulosic fibres amounts to about 4.7% (i.e. 3.2×10^6 t) and its production is relatively constant [8].

In the United Kingdom, the textile consumption is approximately 2.15×10^6 t and, as is typical for many industrialised countries, this corresponding to 35 kg per capita, with about half being made up of wearing apparel [9]. With regard to the total amount of worldwide waste, the fibre fraction is rather small. The portion of textiles in municipal solid waste ranges between 4% and 5% [10].

It is expected that fibre production will increase significantly in the future. The two main reasons are, on the one hand, world population is steadily growing creating an increase in fibre demand, and on the other hand, overall worldwide prosperity is predicted to increase, especially in many emerging marks, and this is closely linked to fibre consumption (i.e. increasing consumption per capita). It can further be assumed that, with a certain time delay, a comparable amount of fibres will end up as waste. Thus, we can expect the fibre portion in waste to increase over the next few years.

2.3. Saving Potentials

Natural fibres, among which cotton is of major importance, are renewable products. During their growing stage, cotton plants absorb carbon dioxide from the atmosphere to form natural polymers, in particular, cellulose. However, cotton cannot be seen as a sustainable product as demonstrated by the data shown in Table 12.3.

Synthetic fibres are based on petroleum and are *a priori* not sustainable. However, petroleum is not only used as feedstock but also as source for energy in polymer production and fibre manufacture. The overall energy consumption is, thus, higher when compared with cotton or cellulosic fibre production as demonstrated in Fig. 12.2. However, the actual energy consumption and the ratio between renewable and non-renewable energy are strongly dependent on the particular situation of the production site [11].

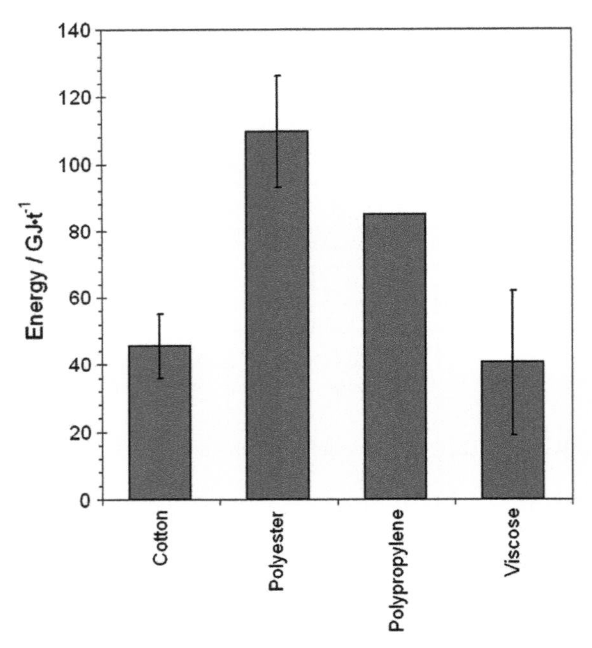

FIGURE 12.2 Energy required to produce 1 t of fibres [10,11].

Table 12.4 shows the energy requirement for all production steps necessary to manufacture 1 t of apparel that corresponds to approximately 4000 T-shirts (cotton) or 5000 blouses (Viscose) [9]. The total consumption ranges between 160

TABLE 12.3 Consumption of Resources for Cotton Cultivation

Resource	Demand for 1 t raw cotton	Remark	References
Crop land	$8 \times 10^3 - 18 \times 10^3$ km^2	2.4% of worldwide crop land for cotton (i.e. 340×10^3 km^2)	[11,12]
Agricultural chemicals	8.3 – 13.8 kg	Based on petroleum 18% of agricultural chemicals for cotton (only 2.4% of crop land)	[12]
Water	Average: 5700 m^3 up to 29,000 m^3	Desiccation of Aral Lake associated with excessive cotton irrigation Frequently insufficient irrigation systems	[11,12]
Energy	36–55 GJ	Fuel for harvesters, etc.	[11]

340 m^2

TABLE 12.4 Energy Demand ($GJ \cdot t^{-1}$) for the Production of Garment Made of Cotton, Polyester, and Viscose

		Cotton		Polyester	Viscose
	Reference	[10]	[9]	[10]	[9]
Production step					
Material	Fibre production	55	64	126	173
	Preparation/ blending	29		29	
	Spinning	89		90	
Production	Knitting	32	96	49	68
	Dying/finishing	27		27	
	Making up	9		9	
Total		241	160	330	241

and 330 GJ t^{-1}, depending on the type of the fibre and sources.

TABLE 12.5 Definitions of Recycling Related Terms Valid for This Communication

Term	Description [13]	Definition valid for this communication
Reuse	Any operation by which products or components that are not waste are used again for the same purpose for which they were conceived.	*Reuse*: Complete products (e.g. deposit bottle). *Component reuse*: Only parts of products (e.g. reusing an alternator from an end-of-life vehicle).
Recycling	Any recovery operation by which waste materials are reprocessed into products, materials, or substances whether for the original or other purposes.	*Product recycling*: Processes that require a considerable effort and result in distinct lower product qualities (e.g. rubber powder from end-of-life tyres). *Material recycling*: Processes in which the physical but not the chemical constitution is changed (e.g. melting and reprocessing of metals). *Feedstock recycling*: Also 'raw material recycling' or 'chemical recycling'; the physical as well as the chemical constitution of the product is destroyed (e.g. repolymerisation, pyrolysis, gasification).

3. TEXTILE WASTE TREATMENT SCENARIOS

3.1. Definitions and Overview

In waste management, a set of terms is frequently used for waste treatment methods but there is no mandatory regulation about their exact meanings. In Table 12.5, some terms used in this chapter are proposed to avoid confusion.

As mentioned in Section 2 (see Table 12.4), the production procedure from raw materials to end-consumer textiles is rather long and includes several intermediate stages. Figure 12.3 shows a processing chain for semi-manufactured products, including possible recycling processes and energy consumptions. Possible recycling paths are indicated according to the definitions given above. Because energy consumption increases significantly with increasing number of intermediate steps, it is thus favourable to develop recycling routes at processing stages that involve large energy inputs. However, it is not *a priori* clear that the saving potential can actually be used and, furthermore, whether the recycling process is technically and economically viable or not. In the next sections, some examples of textile recycling are described.

3.2. Reuse

According to the EU waste hierarchy [13], reuse is a more favourable option than recovery or recycling. Multiple utilisations of apparel and textiles are quite common with a washing or cleaning step after each cycle. In these cases,

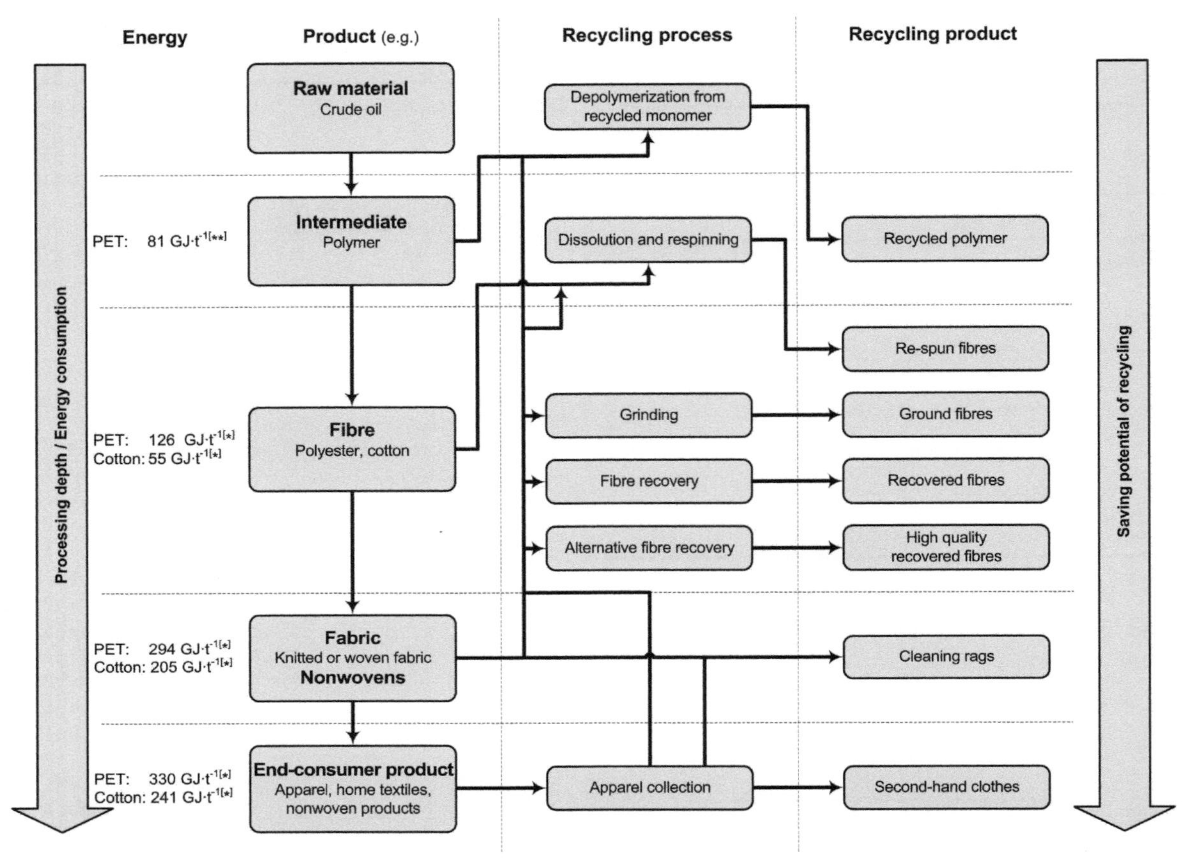

FIGURE 12.3 Scheme of textile processing chain including possible recycling processes; References for energy consumption: *[10]; **[14].

reuse means a further utilisation after the primary customers have sold or given away the product. In most industrialized countries, end-of-life apparel as well as home textiles are collected separately from other types of waste and subsequently sorted by hand into the different fractions specified in Table 12.6.

Even if only about half of the collected textiles can be reused as second-hand products, its collection and sorting represents the most favourable solution in textile recycling. Energy consumption of waste collection is mainly dependent on the population density. For household waste, a consumption of diesel of up to 10 L t^{-1} [17] is reported, which corresponds to about 0.35 GJ t^{-1}. The total energy demand (including transport, sorting, packing, etc.) is of the order of 6 GJ t^{-1} [10]. The energy requirement for second-hand clothes is, thus, negligible compared with the effort in its production (up to 330 GJ t^{-1}).

The sale of second-hand clothes in industrialised countries does give a reasonably good financial return. Typically, a second-hand blouse can be sold for £ 2 in the United Kingdom [9], corresponding to roughly 10,000 € per tonne. The market price of second-hand quality apparel in emerging and developing countries ranges between 730 and 1100 € per tonne [15].

TABLE 12.6 Mass Flow of End-of-Life Apparel During the Sorting Process [15,16] According to Definition Shown in Table 12.5

Category (compare Table 12.5)	Use	Portion (%)	Economics
Reuse First quality	Second-hand clothes for second-hand shops in industrialised countries	1–3	Excellent
Reuse Second quality	Second-hand clothes for export to emerging and developing countries	40–48	Good
Component reuse	Cleaning and wiping rags	29–38	Low
Recycling	Fibre recovery		
Incineration	Energy recovery	7–12	Deficit
Landfill	None		

Thus, reuse of apparel and home textiles is not only an ecological sound solution but also pays for itself and can fund the treating of the residual fractions that are unsuitable for direct reuse or recycling [16].

In a broader sense, the reuse of fabric can be seen as component reuse. In contrast to second-hand clothes, it is also convenient for worn and damaged apparel as well as for production rejects that arise in the course of garment making. The saving potential is relatively high as the production of the fabric would have required a considerable input of energy of up to 294 GJ t^{-1}. In practice, when fabric or non-woven textiles are reused, they are used as cleaning rags.

The requirements for cleaning rags are reasonably high. Ideally, they should consist of cotton, linen or viscose with minor portions of synthetic material and should be either white or coloured [18]. However, textiles with a high fraction of synthetic material can be used as oil absorbent rags [19]. Heavy and hard constituents such as carpet residues, fastenings, eyelets, or zippers must not be present. In practice, a minimum size of 20 × 30 cm is required [18]. The market price of cleaning rags is relatively high and range between 1200 and 1600 € per tonne [15]. However, the manufacture of cleaning rags demands a high input of man power and due to the high personnel cost in developed countries such as in Europe, the overall economic situation for cleaning rags is rather poor.

3.3. Recycling

3.3.1. *Respun Fibres*

The respinning of fibres refers to the process involved when a polymer, for example polyester, is melted (or dissolved) and new fibres or filaments are produced. Even if the fibre quality is similar to the fibres from the original material, this procedure shows two main disadvantages. On the one hand, the fibrous structure is destroyed, and as a result, this recycling route corresponds to *material recycling* (see Section 1) and the amount of energy saved is low since the energy required for fibre formation has to be repeated. For polyester, only 81 GJ t^{-1} instead of 126 GJ t^{-1} can be saved. On the other hand, very often the secondary polymer, such as polyester, does not originate from fibres but from other products. In practice, a so-called plastic 'bottle to fibre' principle is realized. It means that secondary polyester from bottles is used to produce filaments or staple fibres [20, 21]. Although recycled polyester shows reduced material properties such as molecular weight, it can be blended with virgin material, resulting in comparable fibre qualities [22]. In practice, however, a 'fibre to fibre' recycling procedure is not usually viable, because the recycled fibre material commonly consists of a blend. A selective dissolution of certain polymers out of a mixture is possible but is only used for analytical purposes [23]. In conclusion, the production of respun fibres is uneconomical.

3.3.2. Recovered Fibres

Fibre recovery refers to fibres that are derived from end-of-life apparel, fabrics, or rejects arising in the course of the textile processing chain. Fibre recovery is a rather complicated process and product quality is significantly reduced, thus, the procedure is classified as *product recycling* (see Section 1). Because the fibrous structure is maintained and reused, the saving potential is higher than that for respun fibres and ranges between 55 GJ t^{-1} (cotton) and 126 GJ t^{-1} (polyester).

The conventional process, to disintegrate textile structures and yarns, uses equipment such as cylinder raising or fearnought opener machines. In the process, unwanted and damaged fibres are separated and finally the recovered fibres are obtained. The process is, however, damaging to fibres. In particular, fibre length is drastically reduced, and only 25% to 55% of the fibres are longer than 10 mm [24]. Furthermore, the material contains a considerable portion of dust as well as residual yarns. Fibre length distribution is wide and commonly recovered fibres are a blend of various fibre types. As a result, the recovered fibres exhibit a significant lower level of quality than does the virgin products. To a certain extent, the recovered fibres can be used for technical textiles as well as for non-woven materials [24,25]. Generally, fibre recovery in industrial countries, such as the EU, is in competition with low-cost imported textiles from Asia, and as a result, textiles from recycled fibres are becoming increasingly less competitive and less important.

Apart from the classical textile disintegration process discussed above, it is reported that an alternative method can result in fibres showing properties similar to new fibres [26,27]. In particular, fibres up to 30- to 40-mm length can be obtained. However, the processing costs are high and the process is only convenient for high-value fibres such as aramid polymers.

3.3.3. Ground Fibres

The potential sources for ground fibres are widespread. Feedstock for ground fibres includes apparel that can be torn or damaged, rejects from the textile processing chain, or even alternative fibre-containing waste such as fluff from end-of-life tyres. In regard to material originating from apparel sorting (Table 12.6), fibre grinding results in an almost complete spectrum of fractions and thus the portion of waste is significantly minimised.

Ground fibres are similar to ground flock and, thus, are not convenient for spinning (Table 12.1). One can distinguish between cut and ground flock. On the one hand, cut flock requires a guillotine cutter, exhibits a uniform fibre length, and is exclusively manufactured from filaments. On the other hand, ground flock shows a rather broad distribution of fibre length and can be produced from any staple fibre by a grinding process. It is well established that ground flock can be used as a reinforcement or viscosity modifiers. For instance, ground cellulose is frequently applied as additive to bitumen to increase the load capacity and temperature resistance of asphalt pavement material [28].

Because ground flock allows a rather broad fibre length distribution, it is usually made from end-of-life fibre products. Because the grinding machinery is very sensitive towards metallic and hard components, it has to be ensured that all unwanted and potentially damaging bits and pieces are separated and removed. The feasibility of the process has already been shown for fibres derived from end-of-life tyres. And such fibres have been successfully used as an additive in bitumen instead of the well-established ground cellulose [29]. Apart from tyre-derived fibres, it is also

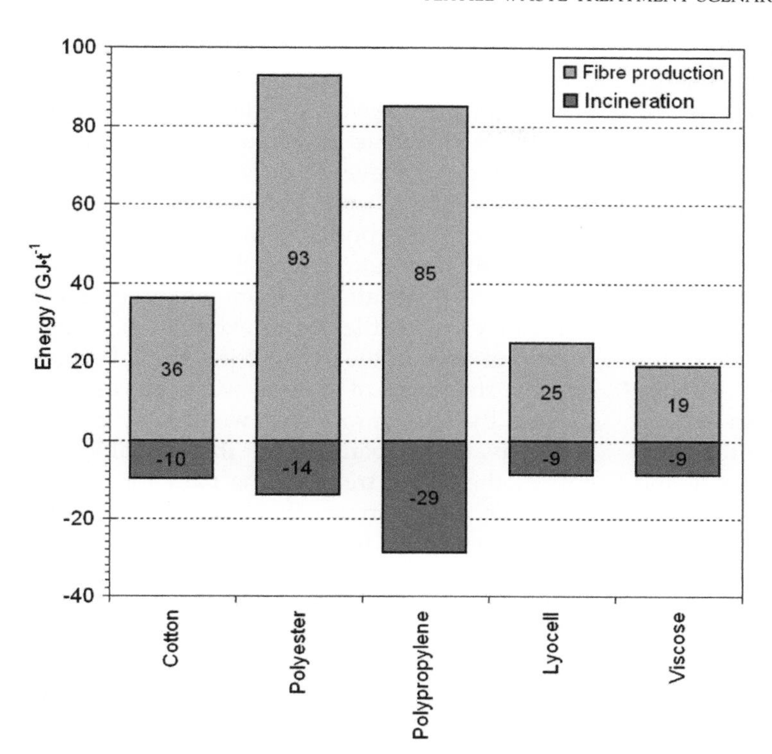

FIGURE 12.4 Comparison of energy required for fibre production and energy recovery (negative values) of incineration (assumption: energetic efficiency of 60 %) [11].

reported that the non-recoverable fraction from apparel collection (see Table 12.6) can be processed to a short-fibre product to be used in bituminous and cementitous construction materials [30,31].Ground fibres derived from textiles and other fibre-containing waste seems an interesting recycling option not only from an ecological but also from an economical point of view. Ground fibres and recycled fibres have been successfully introduced into many markets.

3.3.4. *Recycled Polymers*

As sketched in Fig. 12.3, polymer waste such as polyester [32] can be used as chemical feed stock. It is reported that polyester fibres can be converted to bis(2-hydroxyethylene terephthalate) monomer by glycolysis [33] to be used for fibre production or, alternatively, for the manufacture of textile dyestuffs [34]. Even if procedures are technically viable, it has to be questioned whether it is ecologically and economically worthwhile. On the one hand, the waste polymer competes with intermediates from petroleum and, thus, the saving potential is fairly low. On the other hand, waste, in particular fibrous waste, is commonly contaminated with a variety of extraneous materials (e.g. other polymers, textile auxiliaries) that complicates the processing schedule. Fibre recovery and not polymer recycling seems to be the more favourable option from both an ecological as well as from an economical point of view.

3.4. Incineration and Landfill

Incineration of waste textile materials can be used as an energy source. It is the least favourable of all the textile recycling options such as reuse and recycling. According to EU legislation [13], incineration is not recycling but recovery or disposal. A classification as 'recovery' (referring here to energy) demands a highly efficient incineration plant. With incineration, it is possible to only partially reclaiming the energy used in the developing and processing of the fibre material. Figure 12.4 compares the reclaimable energy (energy efficiency: 60%) with the energy used for the fibre production. It is clear that incineration exhibits some advantages over landfill, but the other options as discussed above, are more favourable from an ecological as well as from an economical point of view.

Landfill represents the least favourable option of waste treatment. However, despite this, a large portion of waste and also end-of-life fibres are presently disposed of in this way. Because most fibres are not biodegradable and also because of the high energy that has gone into producing textile fibres, landfilling of textiles and fibres should be avoided whenever a viable alternative exists.

4. DISCUSSION

4.1. Driving Forces

As shown above, a variety of methods does exist for treating waste and end-of-life textiles. It is of interest to investigate the driving forces that exist to avoid landfilling or incineration in favour of recycling and reusing textile and fibre waste.

Legislation plays an important role in the treatment of waste. For instance, the minimum recycling quotas as required by the EU directive on end-of-life vehicles [35] have initialised a great push to develop new methods and procedures. Of course, the treatment of textile waste is also significantly influenced by regulations such as the deposition ban for untreated waste in Germany and Austria.

However, economics is also an important driving force in determining the direction followed by waste streams. If the economics of the recovery process is advantageous, the processes will run without intervention – such is the case with the recovery of noble metals from end-of-life catalytic converters. On the other hand, even if legislation dictates recycling quotas, waste can be directed to illegal routes if no economically viable procedures for recycling and recovery are available. For instance, it has been reported that up to 75% of end-of-life vehicles are illegally exported from the EU countries [36] and, thus, circumventing the stringent European waste legislations. From an economical point of view, textile waste exhibits an exceptional position. In the EU systems, the collection and the treatment of general waste are usually funded by communities as well as by public and private companies and thus, finally, the cost of the waste treatment is paid for by the customer. By contrast, the collection, sorting and disposal costs for end-of-life textiles do not require funding but pays for itself by the profit obtained by the sale of second-hand textiles.

Finally, environmental and ethical attitudes can significantly influence waste treatment. Textiles represent a basic need for all human beings either as simple covering or as protected against cold. However, in industrialised countries, textiles are frequently discarded and changed due to fashion reasons even if the products are almost new and fully functional. For ethical reasons, many people in the industrialised countries do support the idea of second-hand textiles and, thus, in the field of apparel collection, charity organisations as well as public and commercial organisations are well established to deal with the reuse of textiles and second-hand clothing. This is unique in the area of waste disposal and waste management.

4.2. Life Cycle Thinking

It has been demonstrated above that textile production requires a complex manufacturing process with the need of significant energy demands and resources. Thus, for end-of-life textiles, reuse is favoured over recycling and recovery. However, when taking into account the total life cycle of the textile or clothing, the

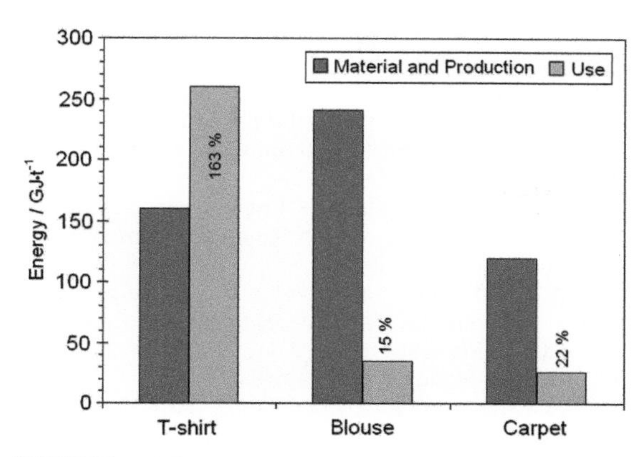

FIGURE 12.5 Energy consumption for material and production in comparison with use phase for T-shirts, blouses, and carpets (compare Table 12.7) [9].

utilisation phase must not be neglected. Most textiles are reuseable products that are commonly cleaned between each cycle. As shown in Fig. 12.5, energy consumption during the use phase can significantly exceed the energy demand for the material and the production of the product. However, the impact of the use phase is strongly dependent on the material and the cleaning behaviour of the customer (Table 12.7).

4.3. Global Markets

Today, regional markets are steadily loosing their importance and global markets are influencing world economy. This development is particularly true for textiles. Asia plays a predominant position in the textile industry and industrialised countries such as the EU can only keep up with speciality products. The EU stringent regulations for waste also valid for textile wastes only affect a small share of the total textile volume. Production residues or rejects primarily arise in Asia and are thus not affected by EU legislation. A high degree of reuse and recycling will only affect end-consumer products imported to the EU but not the majority of the textile industry. This is not only valid for textile waste but also for the by-products of the textile-processing industry such as waste water [37] and textile chemicals [38].

4.4. Moving Towards 'Zero Waste'

Zero waste is an extensively used keyword today. It refers to the reuse and recycling of waste and the avoidance of landfilling and combustion. Within the last few years, many countries have moved towards zero waste [39], but it is clear that from an economical or from a technical point of view, a recycling rate of 100% is not possible for textile waste.

Regarding the treatment of textile waste in the EU, it is clear that a relatively high standard of reuse and recycling has already been achieved. The latest development in textile recycling aims at reducing the fraction which is not reused and recovered (Table 12.6) and to realize

TABLE 12.7 Detail of Textile Products Chosen for Evaluating the Impact of the Use Phase (see Fig. 12.5) [9]

Textile product	Material	Use (i.e. cleaning)
T-shirt	100% Cotton 250 g per piece (i.e. 4000 t^{-1})	25 lifecycles washing: 60°C/Tumble drying ironing
Blouse	100 % Viscose 200 g per piece (i.e. 5000 t^{-1})	25 lifecycles washing: 40°C/hang drying no ironing
Carpet	1100 $g \cdot m^{-2}$ Polyamide pile fibre 133 $g \cdot m^{-2}$ primary backing 1400 $g \cdot m^2$ secondary backing (i.e. 380 $m^2 \cdot t^{-1}$)	Lifetime: 1 year Vacuum cleaning

a closed-loop system [40]. However, even if zero waste can (almost) be realised for textile waste, it is clear that it is, strictly speaking, 'recycling' but more a case of 'downcycling'. This term refers to a reuse or a recycling at a lower level of quality that can be thought of as a consequence of the second law of thermodynamics. For textiles, a set of cascade utilisation processes makes sense and can help to achieve significant savings of energy and resources.

References

[1] Terminology of Man-Made Fibers, The International Bureau for the Standardization of Man-Made Fibres, Brussels, 2009.

[2] German Standard DIN 60 000, 1969.

[3] German Standard DIN 60 905-1, 1985.

[4] German Standard DIN 60 910, 1985.

[5] German Standard DIN 60 001-1, 2001.

[6] A Vision for 2020, The European Apparel and textile Organisation (EURATEX), Brussels, Belgium, 2004.

[7] Information on Man-Made Fibres – 44th volume, The International Rayon and Synthetic Fibres Committee, Brussels, Belgium, 2008.

[8] Information on Man-Made Fibres – 45th volume, The International Rayon and Synthetic Fibres Committee, Brussels, Belgium, 2009.

[9] J.M. Allwood, S.E. Laursen, C.M. Rodriguez, N.M.P. Bocken, Well dressed? University of Cambridge, Cambridge, UK, 2007.

[10] A.C. Woolridge, G.D. Ward, P.S. Phillips, M. Collins, S. Gandy, Life cycle assessment for reuse/recycling of donated waste textiles compared to use of virgin material: An UK energy saving perspective, Resources, Conservation, and Recycling 46 (2006) 94–103.

[11] L. Shen, M.K. Patel, Life cycle assessment of man-made cellulose fibres, Commissioned by the European Polysaccharide Network of Excellence (EPNOE) and Lenzing AG, Austria (2008).

[12] K. Paulitsch, Am Beispiel Baumwolle: Flaechennutzungskonkurrenz durch exportorientierte Landwirtschaft, Wuppertal Institut für Klima, Umwelt, Energie GmbH, paper 148, Wuppertal, Germany, 2004.

[13] Directive 2008/98/EC of the European Parliament and of the Council of 19 November 2008 on waste and repealing certain Directives.

[14] I. Boustead, Eco-profiles of the European Plastics Industry – Polyethylene Terephtalate (PET), Brussels, Belgium, 2005.

[15] J. Hawley, in: Y. Wang (Ed.), Recycling in Textiles, Woodhead Publishing Limited, Cambridge, 2006, pp. 7–24.

[16] Textilrecycling Zahlen – Daten – Fakten, Bundesverband Sekundaerrohstoffe und Entsorgung, E.V., Bonn, Germany, 2001.

[17] A.W. Larsen, M. Vrgoc, T.H. Christensen, P. Lieberknecht, Waste Management and Research 27 (2009) 652–659.

[18] German Standard DIN 61 650, 2005.

[19] K. Langley, Y. Kim, in: Y. Wang (Ed.), Recycling in Textiles, Woodhead Publishing Limited, Cambridge, 2006, pp. 137–164.

[20] K. Gurudatt, P. De, A. Rakshit, M. Bardhan, Spinning Fibers from Poly(ethylene terephthalate) Bottle-Grade Waste, Journal of Applied Polymer Science 90 (2003) 3536–3545.

[21] G.M. Bhatt, Adding value to recycled PET flakes, Chemical Fibers International 58 (2008) 223–226.

[22] A. Elamri, A. Lallam, O. Harzallah, L. Bencheikh, Mechanical characterization of melt spun fibers from recycled and virgin PET blends, Journal of Materials Science 42 (2007) 8271–8278.

[23] Y. Shao, M. Filteau, A Systematic Analysis of Fiber Contents in Textiles, Textile Technology Center, Canada (2004).

[24] B. Gulich, in: Y. Wang (Ed.), Recycling in textiles, Woodhead Publishing Limited, Cambridge, 2006, pp. 117–136.

[25] P. Boettchber, W. Schilde, Zum Einsatz von Reissfasern in Vliesstoffen, ITB Vliesstoffe Technische Textilien (1994) 26–27.

[26] G. Ortlepp, R. Luetzkendorf, R. Bauer, Recycling fibers of waste textiles without tearer, Melliand Textilber 86 (2005) 704–707.

[27] G. Ortlepp, R. Luetzkendorf, Long carbon fibers from textile wastes, Chemical Fibers International 56 (2006) 363–365.

[28] S. Rettenmaier, European Patent EP288863 (1988).

[29] A. Bartl, A. Hackl, B. Mihalyi, M. Wistuba, I. Marini, Recycling of fibre materials, Journal on Process Safety and Environmental Protection 83 (2005) 351–358.

[30] A. Bartl, I. Marini (Eds.), Abfallwirtschaft, Abfalltechnik, Deponietechnik und Altlasten - DepoTech 2008, VGE-Verlag, K.E. Lorber, Essen, Germany, 2008, pp. 195–200.

[31] A. Bartl, I. Marini, Recycling of Apparel, Chemical Engineering Transactions 13 (2008) 327–333.

[32] S.R. Shukla, K.S. Kulkarni, Depolymerization of Poly (ethylene terephthalate) Waste, Journal of Applied Polymer Science 85 (2002) 1765–1770.

[33] S.R. Shukla, A.M. Harad, Glycolysis of Polyethylene Terephthalate Waste Fibers, Journal of Applied Polymer Science 97 (2005) 513–517.

[34] S. Shukla, A.M. Harad, L.S. Jawale, Chemical recycling of PET waste into hydrophobic textile dyestuffs, Polymer Degradation and Stability, journal 94 (2009) 604–609.

[35] Commission Decision of replacing Decision 94/3/EC establishing a list of wastes pursuant to Article 1(a) of Council Directive 75/442/EEC on waste and Council Decision 94/904/EC establishing a list of hazardous waste pursuant to Article 1(4) of Council Directive 91/689/EEC on hazardous waste. 3 May 2000.

[36] S. Scherhaufer, P. Beigl (Eds.), Abfallwirtschaft, Abfalltechnik, Deponietechnik und Altlasten - DepoTech 2008, VGE-Verlag, K.E. Lorber, Essen, Germany, 2008, pp. 201–206.

[37] R. Schneider, in: Y. Wang (Ed.), Recycling in textiles, Woodhead Publishing Limited, Cambridge, 2006, pp. 73–94.

[38] G. Buschle-Diller, in: Y. Wang (Ed.), Recycling in textiles, Woodhead Publishing Limited, Cambridge, 2006, pp. 95–113.

[39] R. Cossu, Driving forces in national waste management strategies, Waste Management 29 (2009) 2799–2800.

[40] A. Bartl, A.S. Haner, Fiber Recovery from End-of-Life Apparel, Chemical Engineering Transactions 18 (2009) 875–880.

CHAPTER

13

Chemicals in Waste: Household Hazardous Waste

Rebecca Slack[†], *Trevor M. Letcher*[‡]

[†] School of Geography, University of Leeds, Leeds, LS2 9JT, UK,

[‡] School of Chemistry, University of KwaZulu-Natal, Durban, Private Bag X54001, Durban 4000, South Africa

OUTLINE

1. INTRODUCTION

'Household hazardous waste' (HHW) is a term used to describe hazardous wastes entering the municipal waste stream. It represents a variety of waste types classified together based on the possession of hazardous properties. Although various national classification systems exist, either legislatively imposed or as voluntary schemes, that categorise the relevant household products in a hazardous subcategory of municipal waste, separate collection is rare in many countries, with the result that such products are generally discarded alongside non-hazardous household waste. This chapter looks at the problems related to the disposal of chemicals and other hazardous substances used in the home. Examples of chemical HHW are shown in Fig. 13.1.

2. SOURCES OF HHW

Hazardous products at home are a diverse and variable group categorised together due to possession of one or more hazardous properties (see Boxes 13.1a and 13.1b). Common items

Waste Doi: 10.1016/B978-0-12-381475-3.10013-0

FIGURE 13.1 Typical household products with potentially hazardous properties.

considered to possess such hazards are household cleaning products (oven cleaners, household bleach etc.; Table 13.1); garden pesticides (insecticides, fungicides, and herbicides); paint and wood preservatives; mineral oils and oily substances, including motor fuels and lubricants; inks and dyes; pharmaceuticals; photographic chemicals and processing materials; swimming pool cleaners; heavy metal containing products (fluorescent lamps that contain mercury; end-of-life vehicles containing oils, asbestos, antifreeze etc.; electrical and electronic equipment with refrigerants); and batteries. Lists of household hazardous products vary from country to country and region to region depending upon chemicals permitted for use in products. For instance, the REACH regulations in the European Union (EU) require that each chemical produced or used in the EU in amounts over 10,000 tonnes

BOX 13.1A

PROPERTIES OF PRODUCTS/WASTES THAT RENDER THEM HAZARDOUS TO HUMAN HEALTH AND THE ENVIRONMENT [46]

- Explosive – substances and preparations that may explode under the effect of flame or are more sensitive to shocks or friction than dinitrobenzene
- Oxidising – substances and preparations that exhibit highly exothermic reactions when in contact with other substances, particularly flammable substances
- Highly flammable – liquid substances and preparations having a flash point below 21 °C (including extremely flammable liquids); or substances and preparations that may become hot and finally catch fire when in contact with air at ambient temperature without any application of energy; or solid substances and preparations that may readily catch fire after brief contact with a source of ignition and may continue to burn or to be consumed after removal of the source of ignition; or gaseous substances and preparations that are flammable in air at normal pressure; or substances and preparations that, in contact with water or damp air, evolve highly flammable gases in dangerous quantities
- Flammable – liquid substances and preparations that have a flash point equal to or greater than 21 °C and less than or equal to 55 °C
- Irritant – non-corrosive substances and preparations that, through immediate, prolonged, or repeated contact with the skin or mucous membrane, can cause inflammation

BOX 13.1A (cont'd)

- Harmful – substances and preparations that if inhaled or ingested or penetrate the skin may involve limited health risks
- Toxic – substances and preparations (including very toxic substances and preparations) that if inhaled or ingested or penetrate the skin may involve serious, acute, or chronic health risks and even death
- Carcinogenic – substances and preparations that if inhaled or ingested or penetrate the skin may induce cancer or increase its incidence
- Corrosive – substances and preparations that may destroy living tissue on contacts
- Infectious – substances that contain viable micro-organisms or toxins that are known or reliably believed to cause disease in man or other living organisms
- Teratogenic – substances and preparations that if inhaled, ingested, or penetrate the skin may induce non-hereditary congenital malformations or increase their incidence
- Mutagenic – substances and preparations that if inhaled or ingested or penetrate the skin may induce hereditary genetic defects or increase their incidence.
- Substances/preparation that release toxic or very toxic gases in contact with water, air, or an acid.
- Substances/preparations capable by any means, after disposal, of yielding another substance, for example, a leachate, which possesses any of the characteristics listed above
- Ecotoxic/dangerous for the environment – substances and preparations that present or may present immediate or delayed risks for one or more sectors of the environment

(1×10^4 t) are thoroughly examined for effects on health and the environment, helping to flag potentially harmful substances. Nevertheless, it has been suggested that the average household in many parts of the world (e.g. United States or Europe) contains about 50 kg of toxic material [1]. A list of potentially hazardous household chemicals is given in Table 13.1.

Not all household hazardous products such as a desktop computer or energy-saving light bulb are immediately dangerous to human health and the environment. However, a number of chemical products such as household cleaners can be dangerous to particular population groups and/or environmental compartments. Children and pets are vulnerable to poisoning from direct exposure to such products. A number of reports indicate that toilet cleaners, chlorine bleach, glass/window cleaners, roach killers, oven cleaners, drain openers, other cleaning products, personal care products, and pharmaceuticals are responsible for most cases of household poisoning incidents around the world – some of which prove to be fatal [2–4]. Localised environmental pollution incidences can also result from the inappropriate use of pesticides, wood preservatives, and other chemicals, although it is usually the disposal of these chemicals that realise the greatest risk. Labelling on packaging will help householders to identify potentially hazardous products. However, if householders are not aware of the reasons behind the hazardous status, it is possible that they may not take steps to limit their exposure to any dangerous properties possessed by the product or waste, and may not be fully aware of the environmental consequences arising from inappropriate use and disposal.

BOX 13.1B

EUROPEAN CHEMICALS BUREAU (ECB) HAZARD SYMBOLS ACCORDING TO EUROPEAN UNION'S DANGEROUS SUBSTANCES DIRECTIVE 67/548/EEC AND USED ON PACKAGING FOR CONSUMER PRODUCTS WITH POTENTIALLY HAZARDOUS PROPERTIES

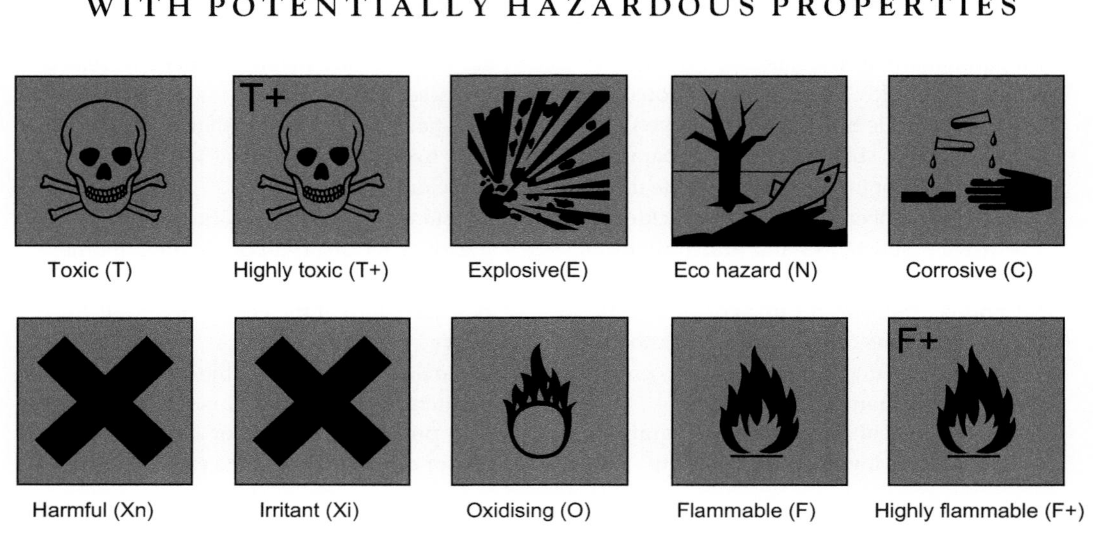

TABLE 13.1 Examples of Potentially Hazardous Chemicals Found in Households

Product	Ingredient	Hazardous property
Oven and cleaners	Sodium hydroxide and/or ammonium hydroxide (ammonia), and possibly methylene chloride	Very corrosive and can cause death if ingested, burns skin and eyes
Drain cleaners	Concentrated sodium hydroxide (can be solid or in aqueous solution of 50% m/m), or hydrochloric acid or sulphuric acid (up to 70% m/m)	Very corrosive and can cause death if ingested, burns skin and eyes
Household bleach	Sodium or calcium hyperchlorite in concentrations up to 10% m/m, or hydrogen peroxide	Can burn skin and eyes
Toilet bowl cleaners	Hydrochloric acid or sodium hyperchlorite and if coloured blue can contain chromium compounds	Very corrosive and can cause death if ingested, burns skin and eyes
Mould and mildew cleaners	Sodium hyperchlorite and formaldehyde	Very corrosive and can cause death if ingested, burns skin and eyes
Other cleaning products	Ammonium hydroxide (ammonia) and ethanol, chlorinated phenols and complex phosphates	Harmful if ingested and can cause burns to the skin and eyes

TABLE 13.1 Examples of Potentially Hazardous Chemicals Found in Households—*cont'd*

Product	Ingredient	Hazardous property
Air fresheners	Formaldehyde or phenol	Harmful if ingested and can cause burns to the skin and eyes
Pesticides	Organophosphates and chlorinated compounds such as chlorinated pyrethrums	Death on ingestion
Pharmaceuticals	Cytotoxic/cytostatic, antibiotics	Carcinogenic, teratogenic, risk of overdose, etc.
Paints and related products	Alcohol, glycols, ethers, ketones, toluene, xylene, acetone, esters, ketones, petroleum distillates, and solvents	Organic compounds can cause lung and kidney damage
Anti-freeze	Ethylene glycol	Fatal if swallowed. Affects the central nervous system
Pool chemicals	HCl and NaOCl	Very corrosive and can cause death if ingested, burns skin and eyes

3. TYPES AND QUANTITIES OF HHW

When a potentially hazardous household product is discarded, it becomes household hazardous waste (HHW). Attempts to quantify and characterise HHW have proved to be difficult. The complexity of separating each hazardous product from the general waste stream for individual group quantification has restricted many researchers to comparisons of cumulative hazardous content to non-hazardous municipal solid waste (MSW). Characterisation is also fraught with complications as workers use different classification systems for HHW. The European Commission provides one example of a hazardous waste list that includes some household products. This is provided by the European Waste Catalogue (EWC), Decision 2000/532/EC as amended by Council Decisions 2001/118/EC, 2001/119/EC, and 2001/573/EC, which is a list of waste descriptions, divided into 20 chapters, the last of which concerns municipal waste [5]. In the final chapter, 14 waste descriptions are marked as hazardous, but only 11 describe a particular waste product (Table 13.2). Solvents, acids, and alkalis are listed without further clarification; many household cleaning agents will fall within these categories. A number of household products are not specifically included in this list, namely personal care products (including cosmetics), non-heavy metal containing batteries, medicines and veterinary products beyond the cytostatic and cytotoxic categories, aerosols, and bottled gas products.

As HHW is an extremely variable waste stream compared to other MSW streams, attempts at quantification have met with mixed success. Each of the HHW sub-categories is produced in very different amounts, by mass, due to very variable composition, initial mass, and product usage patterns. Thus, by mass, waste electrical and electronic equipment (WEEE) is the greatest component of HHW and hence disguises the true scope of HHW. Although WEEE has been estimated to be produced in quantities ranging from 200,000 (2×10^5) to over 900,000 tonnes per annum (9×10^5 t a^{-1}) in the United Kingdom and chlorofluorocarbon (CFC)-containing waste at over 1.6×10^5 t combined estimates for the remaining HHW categories

TABLE 13.2 Municipal Hazardous Waste Listed in Chapter 20 of the European Waste Catalogue [5], notified in the list by an asterisk. The four-digit prefix, 20 01, indicates separately collected fractions

Waste category	Six-digit waste specific code
Solvents	20 01 13
Acids	20 01 14
Alkalines*	20 01 15
Photochemicals	20 01 17
Pesticides	20 01 19
Fluorescent tubes and other mercury-containing waste	20 01 21
Discarded equipment containing chlorofluorocarbons	20 01 23
Oil and fat other than those mentioned in 20 04 25	20 01 26
Paint, inks, adhesives, and resins containing dangerous substances	20 01 27
Detergents containing dangerous substances	20 01 29
Cytotoxic and cytostatic medicines	20 01 31
Mixed batteries and accumulators containing batteries or accumulators included in 16 06 01, 16 06 02 or 16 06 03	20 01 33

* *'Alkalines' are referred to as a category within the EWC but the more correct terminology should be 'alkalis'.*

varies from 15–2 $\times$ 10^4 t a^{-1} to 2.5 $\times$ 10^5 t a^{-1} depending upon the waste types considered [6–10]. One estimate assumes that each household in developed countries disposes 3.5 kg of HHW per annum [11]. Slack et al. [12] report a study that included nine HHW categories (as defined in EU legislation) and estimated that these nine categories contribute 1 $\times$ 10^5 t a^{-1} to the MSW stream in the United Kingdom; the study notes that some of the categories contain products with negligible hazardous content such as water-based paints (compared to the more hazardous solvent-based paints). Generally as a proportion of MSW, estimates vary from >0% to 4% with inclusion of WEEE or 0% to 1% for all other HHW categories [13–17]; this value is fairly consistent across North America and Europe. If adequate collection facilities and disposal mechanisms are to be provided, more accurate estimates will be needed.

4. COLLECTION AND DISPOSAL ROUTES

Recent quantification studies have also looked at HHW disposal pathways. It has been estimated that almost 50% of the HHW stream is likely to be co-discarded with general, non-hazardous household waste in the general household bin, so forming part of MSW [12,18]. Internationally, almost 70% of MSW is disposed of to landfill [19,20]; the remaining fractions are disposed of through incineration, energy-from-waste incineration, recycling, and reuse. Much of the remaining HHW (about 45%) is discarded at household waste reception centres (HWRCs) or collected from homes. Discarding HHW items at HWRCs or having a separate collection from households need not imply that the items are separately collected; householders may not necessarily be aware of the operation of oil, paint, and battery collection banks and hazardous waste safes, simply placing the items in the general household waste receptacle. In the EU, WEEE is also separately collected as is equipment containing ozone-depleting substances as required by legislation. Increasingly, however, improved accessibility and signage at HWRCs is making it simpler for householders to identify the alternative disposal options for their waste and hence will make it increasingly likely that those discarding HHW may do so via a separate collection facility. Separate collection means that HHW can be treated as hazardous waste and hence treated appropriately prior to disposal at, for instance, hazardous waste landfills.

Unlike other waste types, in many countries general MSW (including co-disposed HHW) sent to landfill is relatively unregulated, with no waste separation or proof of contents. Consequently, hazardous materials can be disposed of alongside non-hazardous waste, in such a way that the presence of hazardous organic contaminants can be identified in MSW landfill leachate (Fig. 13.2 [21]). As markets open up, consumers may be exposed to a greater range of products and manufacturers are likely to produce a wider variety of merchandise to compete effectively. This could result in the utilisation of a greater diversity, and amount, of chemicals, while ongoing assessments throw more and more substances into the domain of 'potential hazard', such as phthalates, monosodium glutamate, and anti-bacterial agents. Hence, waste streams are not only constantly growing but are also becoming more varied than at any time in the past. Although HHW represents a small fraction of total household waste, and an even smaller amount of total MSW, the range and types of hazardous substances used in households have the potential to cause problems on

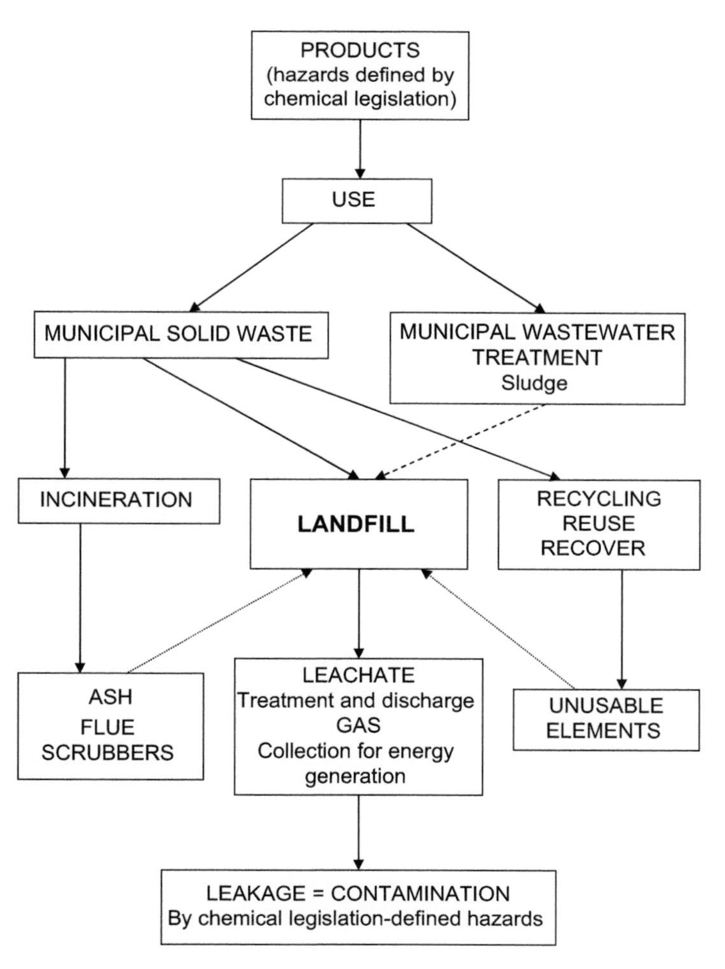

FIGURE 13.2 Hazardous household waste is predominantly disposed of to landfill. Alternative means of disposal, such as to sewer, incineration, or through recycling, contribute fractions of waste to landfill [21].

disposal, particularly to the aquatic environment. Because of the predominance of landfill disposal around the world, through landfill the environmental consequences of HHW disposal will be most apparent (Fig. 13.2). Alternative disposal practices, from recycling to incineration, generate waste that is ultimately landfilled.

5. ENVIRONMENTAL AND HEALTH RISKS

The level of risk posed to the environment and human health through the use and disposal of HHW is associated with the amounts of wastes produced (hazard potential) and the disposal routes adopted (probability that exposure pathways leading to a pollution event). As discussed in the preceding sections, data are scarce regarding the amount of individual waste categories and even HHW as a whole, but it is generally acknowledged that landfilling of HHW is currently the most common disposal method adopted around the world. Emissions from landfill take two forms: gaseous emissions of volatile organic compounds (VOCs) and leachate, and both have been the subject of much research as they have the potential to cause localised air quality issues and groundwater contamination [15,22–33].

The leakage potential of a landfill is mitigated by a number of factors. Locating landfills in appropriate strata that act to attenuate releases resulting from waste disposal to land is common practice for older 'dilute-and-disperse' landfills as well as newer engineered landfills. In modern landfills, the use of barrier material, most usually a combination of geomembrane and clay liner, at the base of the landfill slows the migration of landfill leachate to the surrounding environment whilst leachate collection systems transfer leachate for recirculation within the landfill or for removal from the landfill for treatment prior to release. Capping of filled waste cells (used to separate the landfill site into manageable sections) and closed sites, combined with daily covering of disposed waste, reduce rainwater ingress and limit leachate production. A cap barrier restricts gaseous emissions, particularly if a gas collection system is incorporated. Failure of any of the control measures or insufficient treatment of collected emissions may result in the release of a cocktail of chemicals, as reported by Schwarzbauer and co-workers [29]. For older landfills (largely completed before the 1990s), the implementation of measures to prevent release to the environment is less well defined with the result that aquifer contamination has been far more common alongside high levels of localised VOCs [28]. As water extraction from surface water and groundwater sources increases, the knowledge of the geochemistry of leachate plumes becomes an important consideration.

Leachate is inherently variable due to the heterogeneity of waste composition, water infiltration rate and amount, residual refuse moisture content plus factors relating to the landfill: design, operation, management, and age [15]. Leachate components result from the types of waste disposed and the processes occurring within the landfill. The presence of inorganic compounds including heavy metals and hazardous organic contaminants, such as halogenated aliphatic compounds, aromatic hydrocarbons, phenolic compounds, and pesticides, in MSW landfill leachate is a direct indicator of the disposal of hazardous wastes in MSW [22,23,34]. However, care must be taken with MSW leachate analyses that reveal the presence of harmful substances due to the co-disposal of industrial liquid wastes and manufacturing wastes with MSW, a practice no longer permitted in the EU. The consequences of HHW disposal are therefore obscured in many leachate studies. Where it is possible to differentiate waste sources, leachate composition has the potential to act as a useful tool in HHW risk evaluation.

A simulation of the landfill environment, using the LandSim modelling software, demonstrated that contaminants linked to HHW may enter the environment around the landfill as a consequence of the degradation of the engineered barrier system (EBS) and cessation of leachate management [35]. A worst-case scenario found that certain inorganic substances (arsenic, chromium, cadmium, and mercury) might exceed groundwater-reporting thresholds several decades or even centuries after waste was first deposited and after management measures have long ceased. The concentrations of most contaminants, however, were lower than most statutory guidelines/regulations and demonstrated significant levels of attenuation within the unsaturated zone beneath the landfill and above the aquifer. These risks, though small, are likely to be reduced further in the future with improved management of HHW.

Other disposal routes also have environmental and health risks with which they are associated. Energy-from-waste (EfW) incineration provides a more rapid, less space intensive disposal alternative to landfill with the benefit of energy (heat and power) production. The production of dioxins from chlorinated organic hydrocarbons is one major disadvantage [36], but the use of flue scrubbing agents helps reduce emissions to the atmosphere. The disposal of resulting scrubbers and ash is another problem. EfW incineration is the preferred means of disposal for a number of European countries including the Netherlands, Denmark, and Germany where emissions from incinerators are heavily regulated by EU legislation. Disposal to sewer rather than as solid waste may also be used as a means of discarding HHW, particularly household cleaners/detergents (and during use), pharmaceuticals, pesticides, paint-related products, and so on. Sewage treatment processes can remove some of these hazardous substances but some substances may affect the sewage treatment process and/or pass into receiving waters unchanged. Reported risks associated with sewer disposal include localised pollution events (e.g. heavy metal, solvents, etc.), leakage to groundwater from septic tanks, release of endocrine-disrupting substances, antibiotic resistance and even eutrophication arising from phosphate elevation associated with detergent use (conventionally a non-hazardous product) [37—43]. Illegal disposal of HHW may also be practiced, for example, side-of-the-road dumping or fly tipping, and this can lead to large-scale pollution events. However, it is rare and criminal charges can result from such activities. Other risks can result from events happening during collection, treatment, and storage prior to disposal, including accidental spills, toxic/flammable gas release, explosion, or fire [44]. Risks are greatest at these times to waste handlers and other workers such as firefighters.

6. WASTE LEGISLATION

In the EU, the Landfill Directive 1999/31/EC is having a considerable effect on the composition of the waste being disposed of to landfill as Europe attempts to reduce the amount of waste landfilled and improves recycling rates to improve resource efficiency [45]. Diversion of putrescible waste (garden and kitchen waste) to composting facilities and separation at source of paper/card, glass, and metal cans will result in a different composition of household waste. A proportional increase of HHW in MSW may then result which may lead to changes in the composition of leachate. It is currently unclear whether these changes will be beneficial or disadvantageous to HHW co-disposal to MSW landfills. Diversion of HHW from MSW landfill would, however, reduce the likelihood of HHW-associated environmental contamination scenarios from occurring. The disposal of certain substances, not only asbestos but also some waste types that can be considered part of HHW such as car tyres, WEEE, and end-of-life vehicles, are specifically controlled by legislation. These are exceptions and despite inclusion as a hazardous waste in the appropriate lists [5], HHW is generally excluded from the requirements of hazardous waste legislation in the EU unless it is separately collected at, for instance, HWRCs [46]. This exclusion from the Hazardous Waste Directive 91/689/EC, therefore, permits householders to avoid being classed as hazardous waste producers but allows hazardous waste co-disposal with MSW.

HHW in the United States is exempt from the definition of hazardous waste provided in Subtitle C (hazardous waste management) of the Resource Conservation and Recovery Act (RCRA) but is still subject to Subtitle D (solid waste management) regulations with other municipal wastes [47,48]. However, if HHW is mixed with small quantity or large generator waste, full Subtitle C regulations may be applicable. European and US regulation of HHW is therefore very similar. In Australia, the inappropriate disposal of HHW to sewer, storm water drain or landfill is a breach of a person's general environmental duty as described in the Environment Protection Act 1993 [49]. However, separate collection is not mandatory and a variety of schemes are used to encourage householders to remove HHW from general MSW, many of which include extended producer responsibility and voluntary product stewardship schemes. The Canadian province of Manitoba has developed draft regulations along similar lines, aimed at introducing compulsory product stewardship of 11 categories forming HHW. Most Canadian provinces have adopted a non-regulated, voluntary approach to the management of HHW [50], an approach that is adopted to some extent elsewhere around the world. In South Africa and Japan, for example, most HHW disposal guidelines are restricted to advice offered to householders [51].

7. MANAGEMENT

Current management of HHW is extremely variable within, as well as between, countries, often with neighbouring regions operating very different strategies for the collection and disposal of HHW. In Europe, the United States, and other countries, the central governing bodies have deferred the control and disposal of HHW to local authorities. Most local authorities have Web sites focussing on HHW (see references to HHW from Indiana (USA) [52], New Mexico (USA) [53], Somerset County (UK) [54], Durban (South Africa) [51], Tarrant County (USA) [55], Lewisville (USA) [56]). South Africa, like many developing countries, is yet to establish mechanisms for special HHW collections. Many of the local authorities do offer collection advice and some even offer to assist in the collection of household hazardous material. However, the lack of consistency in treatment of HHW can result in confusion among householders and increase the possibility of a failure to adapt to changes in waste management.

In many areas of the United States, local community programs have been implemented to inform and educate the public of HHW. This has resulted in a well-publicised collection of programs that can take the form of one day a month or week collection at a central convenient site. Here the HHW is identified, packaged, and then transported to a permitted hazardous waste facility. In spite of the apparent popularity of the program, only a small proportion of the public is currently involved and a more sustained effort is required to raise awareness of the services offered [57,58]. The local community programs also place a stress on informing the public of the hazardous nature of many of the materials used within the home, promoting not only waste reduction, but also product replacement with less potentially hazardous substances, for example, use of lemon juice as a household cleaner [39]. Educational programs have also sought to change householders' behaviour regarding the use and disposal of potentially hazardous substances [59].

Changes in the management of all municipal wastes will also have consequences for the future collection and disposal of HHW. Source segregation of MSW is increasing as many recyclable or reusable materials are diverted from disposal to encourage the sustainable use of natural resources, for example, paper, plastics, and metals. As a consequence, the proportion of HHW in the residual waste stream may also increase, resulting in potentially more concentrated hazardous waste being disposed of to landfill. However, increased provision of separate collection for many HHW streams will lead to a diversion from MSW co-disposal and so may prevent a proportional increase of HHW in MSW. Separate collection of HHW will also require that the wastes are treated as hazardous wastes rather than as a MSW stream. Based on current practice in Europe, United States, and elsewhere, it is likely that, despite the absence of any legal obligation to do so, waste disposal authorities will seek to enlarge and encourage the separate collection of HHW and diversion from MSW co-disposal. The cost of improving HHW collection services must, however, be considered alongside the pattern of declining hazardous content of consumer products, principally through the implementation of legislation including the EU's REACH Regulation (EC) No. 1907/2006, Restriction of the Use of Certain Hazardous Substances Directive 2002/95/EC and Batteries Directive 91/157/EEC [60–62]. For instance, flat-screen televisions will eventually make CRT recycling facilities redundant although many household chemicals, principally cleaners, become more benign. HHW management strategy needs further clarification at national levels to ensure

BOX 13.2

STRATEGY FOR HHW MANAGEMENT

HHW management may take the form of the following points:

- The public must be involved and must appreciate the problems related to HHW;
- Publicising local disposal and collection schemes and guidelines, providing adequate explanation for reasoning behind the schemes adopted;
- There should be both national and local initiatives, for example, flyers, Web sites, public notices, news stories, and so on, for publicising the management of HHW, giving the same unambiguous message;
- Regular door-to-door collections to include householders without their own transport;
- Easily accessible internet and telephone help lines should be available;
- Access to clearly signposted household waste recycling centres;
- Take back schemes at retail centres for WEEE, CFC-containing household appliances, and so on;
- Voluntary schemes/legislation to encourage/ensure manufacturers disclose the contents of all potentially hazardous products and use appropriate labelling schemes;
- Ensure householders and waste handlers are educated of the dangers inherent in collecting and sorting of HHW.

BOX 13.3

GOOD HOUSEKEEPING CHECKLIST FOR THE PURCHASE, USE, STORAGE, AND DISPOSAL OF HHW

- Always buy just enough for the job in hand and try not to keep/store hazardous chemicals – use up, recycle, or dispose in the proper manner.
- Seek a less hazardous alternative if possible.
- Use hazardous chemicals in well-ventilated areas using the appropriate protective clothing/equipment.
- Always follow the manufacturers' instructions for use, storage, and disposal as provided on the product label, the instruction manual, or safety data sheet.
- If you have to store a chemical, ensure that it is out of the way of children and pets, is properly sealed, and labelled.
- Do not store flammable liquids or gas canisters in the house, storing instead in well-ventilated sheds outside.
- Always know how and/or where to dispose of your hazardous chemicals – including medicines, engine oil, WEEE, pesticides, and so on. Most householders in Europe and North America have access to a household waste recycling centre, many of which cater for the disposal of HHW.
- Discard any unlabelled products, particularly if not stored in original containers.
- Never dispose of a hazardous chemical by burning, mixing with other chemicals, pouring/burying in gardens, disposing to storm water drains or directly to surface water bodies, flushing to septic tanks, or pouring to sewer drains, unless explicitly permitted in manufacturers' instructions.
- Do not add HHW to the normal household refuse collection if an alternative collection facility (e.g. HWRC) is available.
- Contact your local environmental health, solid waste management, or public works department for information on correct disposal of HHW and collection/disposal facilities.

that practical and cost-effective measures for collection and disposal are adopted with the communication of clear and unambiguous instructions to householders. Such a strategy may consider the points provided in Box 13.2 and communication to householders may consider the checklist provided in Box 13.3.

8. CONCLUSIONS

HHW is a small proportion of the municipal waste stream, but the potential risks to the environment and health are disproportionate to its size. Although estimates of the amount of HHW vary from region to region and across national boundaries, it generally comprises 1–4% of MSW. Estimates vary partly due to different use patterns around the world and partly to poor definition of what is considered as HHW with some definitions including bulky items such as waste electrical and electronic equipment and CFC-containing household appliances. Internationally, most HHW is co-disposed with MSW to municipal waste landfills. Co-disposal of potentially hazardous wastes can lead to an increase in hazard status; not only are these substances potentially dangerous to the environment and health, but they can also induce changes in other waste streams by reacting directly with the waste or by altering the redox environment (particularly acids, alkalis, and solvents). Certain substances may also adversely affect the EBS in landfills and the scrubbers used in incineration flues, increasing likelihood of environmental release. Landfill simulations, based on current MSW disposal patterns, reveal the risk to the environment from leakage of potentially hazardous materials from landfills to be small but existent. There is a growing tendency for waste disposal authorities to make provision for the separate collection of many items of HHW. The increased cost of separate HHW collection and disposal must be considered alongside the new environmental and health risks posed by changes to disposal practice and the general trend of reduced hazardous content of most consumer goods.

References

[1] Rhode Island Department of Health. <http://www.uri.edu/ce/wq/has/PDFs/WQP.hazardous.pdf>

[2] P.A. Koushki, J.M. Al-Humoud, Practice Periodical of Hazardous, Toxic, and Radioactive Waste Management 6 (2002) 250–255.

[3] R.L. Franklin, G.B. Rodgers, Pediatrics 122 (2008) 1244–1251.

[4] M. McParland, K. Kennedy, N. Sutton, Z. Tizzard, J.N. Edwards, S. Wyke, R. Duarte-Davidson, J. Tempowski, Description of the nature of the accidental misuse of chemicals and chemical products (DeNaMiC) in Chemical Hazards & Poisons Report, Health Protection Agency, http://www.hpa.org.uk/web/HPAwebFile/HPAweb_C/1211266315123, 2008, p. 42.

[5] European Commission Commission Decision 2000/532/EC of 3 May 2000 replacing Decision 94/3/EC establishing a list of wastes pursuant to Article 1(a) of Council Directive 75/442/EEC on waste and Council Decision 94/904/EC establishing a list of hazardous waste pursuant to Article 1(4) of Council Directive 91/689/EEC on hazardous waste. Official Journal of the European Communities, L 226, p. 003–024 (as amended by Commission Decisions 2001/118/EC (OJ L 47, 1-31), 2001/119/EC (OJ L 47, 32) and 2001/573/EC (OJ L 203, 18-19)) (2000).

[6] Department for Environment, Food and Rural Affairs, Waste strategy 2000: England & Wales (Part 2), Her Majesty's Stationary Office, London, 2000, p. 63.

[7] A. Gendebien, A. Leavens, K. Blackmore, A. Godley, K. Lewin, B. Franke, et al., Study on hazardous household waste (HHW) with a main emphasis on hazardous household chemicals (HHC), European Commission – General Environment Directorate, Brussels, 2002, p. 150.

[8] W. Pendle, A.J. Poll, Common household products: a review of their potential environmental impacts and waste management options. Warren Spring Laboratory and Dept. of Enterprise report LR927 (RAU), Warren Spring Laboratory, Stevenage, 1993, p. 55.

[9] A.J. Poll, The composition of municipal solid waste in Wales, Cardiff, Welsh Assembly Government and AEA Technology (2003) p. 84.

[10] P. Stevens, Priority waste stream project – household hazardous waste: Stage 1, Save Waste & Prosper Ltd (SWAP) and Recycling Advisory Group Scotland (RAGS), Leeds, 2003, p. 62.

[11] B.D. Otoniel, M.-B. Liliana, P.G. Francelia, Waste Management 28 (2008) 52–56.

[12] R.J. Slack, M. Bonin, J.R. Gronow, A. Van Santen, N. Voulvoulis, Environmental Science and Technology 41 (2007) 2566–2571.

[13] S.J. Burnley, J.C. Ellis, R. Flowerdew, A.J. Poll, H. Prosser, Resources, Conservation and Recycling 49 (2007) 264–283.

[14] A.J. Poll, Variations in the composition of household collected waste. Didcot, AEAT for EB Nationwide, 2004, p. 42.

[15] D.R. Reinhart, Waste Management and Research 11 (1993) 257–268.

[16] E.J. Stanek, R.W. Tuthill, C. Willis, G.S. Moore, Archives of Environmental Health 42 (1987) 83–86.

[17] T.M. Letcher, R. Schutte, Journal of Energy Research and Development in Southern Africa 3 (1992) 26–28.

[18] R.J. Slack, P. Zerva, J.R. Gronow, N. Voulvoulis, title, Environmental Science and Technology 39 (2005) 1912–1919.

[19] Organization for Economic Cooperation and Development Household energy and water consumption and waste generation: trends, environmental impacts and policy responses. Report: ENV/EPOC/WPNEP(2001)15/FINAL, Sector Case Studies Series, Programme on Sustainable Development 1999–2001. Paris, Organization for Economic Cooperation and Development Environment Directorate. p. 97, 2001.

[20] A. Zacarias-Farah, E. Geyer-Allely, Journal of Cleaner Production 11 (2003) 819–827.

[21] R. Slack, J. Gronow, N. Voulvoulis, Critical Reviews in Environmental Science and Technology 34 (2004) 419–445.

[22] T.H. Christensen, P. Kjeldsen, P.L. Bjerg, D.L. Jensen, J.B. Christensen, A. Baun, et al., Applied Geochemistry 16 (2001) 659–718.
[23] J.V. Holm, K. Rugge, P.L. Bjerg, T.H. Christensen, Environmental Science and Technology 29 (1995) 1415–1420.
[24] P. Kjeldsen, A. Grundtvig, P. Winther, J.S. Andersen, Waste Management and Research 16 (1998) 3–13.
[25] P. Kjeldsen, M.A. Barlaz, A.P. Rooker, A. Baun, A. Ledin, T.H. Christensen, Critical Reviews in Environmental Science and Technology 32 (2002) 297–336.
[26] N. Mikac, B. Cosovic, M. Ahel, S. Andreis, Z. Toncic, Water Science and Technology 37 (1998) 37–44.
[27] E. Noaksson, M. Linderoth, A.T.C. Bosveld, L. Norrgren, Y. Zebuhr, L. Balk, The Science of the Total Environment 305 (2003) 87–103.
[28] M. Reinhard, N.L. Goodman, J.F. Barker, Environmental Science and Technology 18 (1984) 953–961.
[29] J. Schwarzbauer, S. Heim, S. Brinker, R. Littke, Water Research 36 (2002) 2275–2287.
[30] M.R. Allen, A. Braithwaite, C.C. Hills, Environmental Science and Technology 31 (1997) 1054–1061.
[31] T. Assmuth, K. Kalevi, Chemosphere 24 (1992) 1207–1216.
[32] K.J. James, M.A. Stack, Chemosphere 34 (1997) 1713–1721.
[33] S.C. Zou, S.C. Lee, C.Y. Chan, K.F. Ho, X.M. Wang, L.Y. Chan, Z.X. Zhang, Chemosphere 51 (2003) 1015–1022.
[34] R.J. Slack, J.R. Gronow, N. Voulvoulis, Science of the Total Environment 337 (2005) 119–137.
[35] R.J. Slack, J.R. Gronow, D.H. Hall, N. Voulvoulis, Environmental Pollution 146 (2007) 501–509.
[36] J. Wienecke, H. Kruse, U. Huckfelct, W. Eickhoff, O. Wassermann, Chemosphere 30 (1995) 907–913.
[37] F. Baquero, J.-L. Martinez, R. Cantón, Current Opinion in Biotechnology 19 (2008) 260–265.
[38] A.J. Watkinson, E.J. Murby, D.W. Kolpin, S.D. Costanzo, Science of the Total Environment 407 (2009) 2711–272.
[39] M.J. Focazio, D.W. Kolpin, K.K. Barnes, E.T. Furlong, M.T. Meyer, S.D. Zaugg, et al., Science of the, Total Environment 402 (2008) 201–216.
[40] H. Palmquist, J. Hanæus, Science of the Total Environment 348 (2005) 151–163.
[41] S. Mompelat, B. Le Bot, O. Thomas, Environmental International 35 (2009) 803–814.
[42] P.J.A. Withers, H.P. Jarvie, Science of the Total Environment 400 (2008) 379–395.
[43] X.-C. Jin, S.-R. Wang, J.-Z. Chu, F.C. Wu, Pedosphere 18 (2008) 394–400.
[44] S. Vecchio, Thermochemica Acta, in press.
[45] European Council Council Directive 99/31/EC of 6 April 1999 on the landfill of waste. Official Journal of the European Communities, L 182, pp. 001–019. 1999.
[46] European Council, Council Directive 91/689/EEC of 12 December 1991 on hazardous waste. Official Journal of the European Communities, L 377, p. 020–027. (Amended by Council Directive 94/31/EC). Official Journal of the European Communities, L 168, p. 028.
[47] United States Federal Code Resource Conservation and Recovery Act (RCRA) (Public Law 94-580), amending Solid Waste Disposal Act 1965: Subtitle C (Hazardous Waste Management). US Code (Acts of Congress) Title 42, Chapter 82, Subchapter III (Section 6921-6939). (1976).
[48] United States Federal Code Resource Conservation and Recovery Act (RCRA) (Public Law 94-580), amending Solid Waste Disposal Act 1965: Subtitle D (Solid Waste Program), amended 1984. US Code (Acts of Congress) Title 42, Chapter 82, Subchapter I (Section 6901-6908a). (1976).
[49] Parliament of South Australia, Environment Protection Act 1993. Part 4, Section 25 – General Environmental Duty, Parliament of South Australia, Adelaide, 1993.
[50] D. Smith Discussion paper: household hazardous waste/HHW. <(http://www.greenmanitoba.ca/cim/dbf/HHW_DiscPaper.pdf?im_id=16&si_id=1000). Winnipeg, Green Manitoba Eco Solutions. p. 21 (2005)>.
[51] Google search – South Africa, household hazardous waste, chemical cocktails
[52] Indiana Web site: www.idem.IN.gov
[53] Google search – New Mexico State University safe use and disposal of household chemicals guide G-312, dated 11/07.
[54] http://www.somersetwaste.gov.uk
[55] Tarrant County: <http://www.tarrantcounty.com/ehealth/cwp/view.asp>
[56] Lewisville HHW site <http://www.utrwd.com/HHW.htm>
[57] S. Cassel, Environmental Impact Assessment Review, Journal 8 (1988) 307–322.
[58] E.P.A. Report, A survey of householders hazardous wastes and related collection programs, EPA Office of Solid Waste and Emergency Response, Washington, DC, 1986.
[59] C.M. Werner, Journal of Environmental Psychology 23 (2003) 33–45.
[60] European Council Council Directive 91/157/EEC of 18 March 1991 on batteries and accumulators containing certain dangerous substances. Official Journal of the European Communities, L 78, pp. 038–041. (1991)

[61] European Parliament and Council Council Directive 2002/95/EC of the European Parliament and of the Council of 27 January 2003 on the restriction of the use of certain hazardous substances in electrical and electronic equipment. Official Journal of the European Communities, L 37, pp. 019–023. (2002).

[62] European Parliament and Council Proposal for a Directive of the European Parliament and of the Council amending Council Directive 67/548/EEC in order to adapt it to Regulation (EC) of the European Parliament and of the Council concerning the registration, evaluation, authorisation and restriction of chemicals {SEC(2003) 1171}. COM/2003/0644 final – COD 2003/0257. (2003).

CHAPTER

14

Reusing Nonhazardous Industrial Waste Across Business Clusters

Marian Chertow, Jooyoung Park

School of Forestry and Environmental Studies, Yale University, New Haven, CT 06511, USA

OUTLINE

1. INTRODUCTION

Nonhazardous industrial waste (NHIW) is distinct from both municipal solid waste (MSW), the more familiar mix from homes and businesses, and hazardous waste, materials that are more highly regulated owing to their toxicity and related public health concerns. Both definitions and regulations of NHIW vary greatly from jurisdiction to jurisdiction and country to country. The US Environmental Protection Agency reports that "industrial non-hazardous waste" or "industrial solid waste" [1]:

> consists primarily of manufacturing process wastes from sectors such as organic and inorganic chemicals, primary iron and steel, plastics and resin manufacturing, stone, clay, glass and concrete, pulp and paper, and food and kindred products, including wastewater and non-wastewater sludges and solids, and construction and demolition materials.[1]

2. STATUS OF NHIW

The amount of NHIW is vast. Yet, quantifying NHIW generation, reuse, and disposal, has been difficult and uncertain, because there are many diverse sources of waste with few means to track them. As for the global trend of NHIW generation, one of the few global estimates was completed for the OECD Environmental Data Compendium [2]. The waste statistics reported by the OECD for selected countries as available include waste generated from some or all these categories: agriculture, mining, manufacturing,

[1] This chapter does not cover construction and demolition waste that is handled separately in Chapter 15.

Waste Doi: 10.1016/B978-0-12-381475-3.10014-2

energy production, water purification and distribution, construction industries, municipal waste, and other sources (Fig. 14.1). Although there is no one standard classification of NHIW across countries or even within units of the same country, of the categories compiled by the OECD, definitions usually include at least manufacturing, energy production, and water purification and distribution wastes. As shown in Fig. 14.1, for most of the countries selected, total MSW is far less than the NHIW categories. However, agriculture and mining wastes may or may not be counted as NHIW as some systems aggregate them separately.

The mid-1980s was the last time when a comprehensive estimate was made in the United States. At that time, the amount of NHIW, including water and wastewater components, totaled seven billion metric tonnes (7×10^9 t) when compared with only 0.20×10^6 t of MSW. Of the 7×10^9 t, pulp and paper comprised the largest portion, approximately 30%, followed by primary iron and steel (17%), electric power generation waste (14%), and inorganic chemicals (12%), with these four industries representing more than 70% of NHIW [3]. Disposal ranged from onsite pits, ponds, and lagoons to special impoundments or landfills located on or off site. It is important to differentiate between NHIW quantities offered by various sources that count water and wastewater in the estimates as they can be an order of magnitude higher than if only the solid portions of NHIW are counted.

To develop a better sense of quantities, it is instructive to examine one state in the United States that has more extensive regulation of NHIW than do most other jurisdictions. Since 1994, the State of Pennsylvania has tracked 110 material categories of what they term "residual waste" on a biennial basis from all companies that discard at least 12 t a^{-1} (13 short t a^{-1}) of NHIW. Not only does the state report where the waste is generated but it also reports each waste destination across 13 classifications, some of which represent disposal, such as landfills and impoundments, and others that represent recovery, such as composting or recycling. From these data, we can see both broad and specific patterns. With respect to the overall trend, total NHIW declined from 635×10^6 t (700 million short tons) in 1994 to 476×10^6 t in 2004 when wastewater is included. For solid residues only, total generation declined from 34×10^6 t (37 million short tons) in 1994 to 18×10^6 t in 2004. The 18×10^6 t of solid residues still exceeds the total MSW generated in the state. The data also permit spatial mapping of specific wastes by quantity and geography.

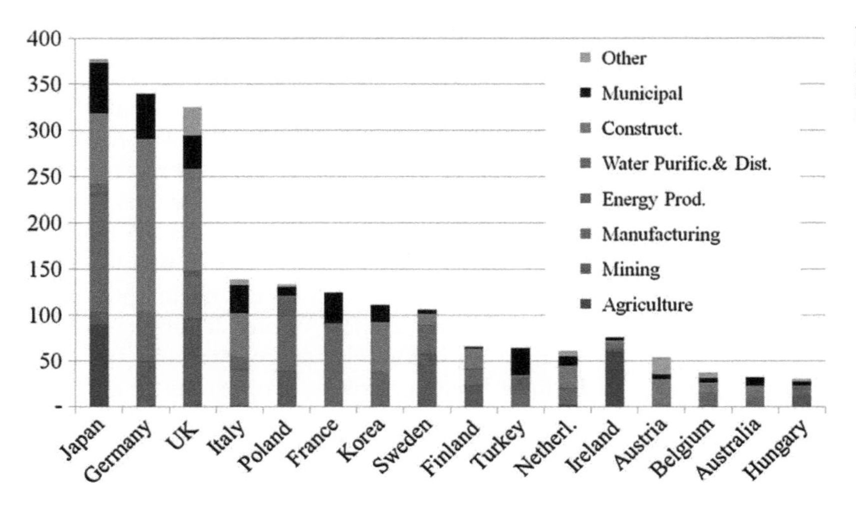

FIGURE 14.1 Total waste generation and its composition in selected countries [2] (OECD 2008) (unit: million metric tonnes).

Although waste and the bartering of by-products go back to ancient times, as industry has expanded, waste, too, has increased greatly especially since World War II. The sheer quantities of NHIW raise the question of how best to manage it. The first defense is cleaner production: generating less waste by increasing industrial efficiency and effectiveness. A broad range of companies now produce "annual sustainability reports" detailing the amount of waste reduced in particular categories, year to year. For wastes that continue to be generated, many large volume industrial streams are amenable to separate handling such as foundry sand, coal ash, and paper mill sludge. Indeed, some of these streams have an excellent track record for reuse. In 2006, for example, 43% of all coal combustion by-products were beneficially reused in the United States, 93%, were beneficially reused in 15 countries in the European Union, and 97% in Japan [4–6].

3. INDUSTRIAL SYMBIOSIS

A critical strategy discussed in this chapter is reuse of NHIW materials within industrial clusters such that one company's waste can become another company's raw material. There is a long history of these types of reuse networks especially within single-industry-dominated clusters such as pulp and paper operations, integrated petroleum complexes, and sugarcane complexes [7]. Less attention has been paid to multiple-industry clusters where a great deal of waste can be exchanged safely among companies in quite different industries with significant environmental benefits. The idea of creating economic and environmental benefits through interfirm waste exchange has taken on the name "industrial symbiosis." The expression "symbiosis" builds on the notion of biological symbiotic relationships in nature in which at least two otherwise unrelated species exchange materials, energy, or information in a mutually beneficial manner—the specific type of symbiosis known as mutualism [8].

The model of industrial symbiosis in a multiple-industry cluster was first fully realized in the small city of Kalundborg, Denmark. The primary partners in Kalundborg, an oil refinery, power station, gypsum board facility, pharmaceutical plant, and the City of Kalundborg, share water, steam, and electricity, and also exchange a variety of wastes that become feedstocks in other processes. High levels of environmental and economic efficiency have been achieved and have led to many other less-tangible benefits involving personnel, equipment, and information sharing.

Specifically, industrial symbiosis is part of a new field called industrial ecology. Industrial ecology is principally concerned with the flow of materials and energy through systems at different scales, from products to factories and up to national and global levels. Industrial symbiosis focuses on these flows through networks of businesses and other organizations in local and regional economies as a means of approaching ecologically sustainable industrial development. Industrial symbiosis engages traditionally separate industries in a collective approach to competitive advantage involving physical exchange of materials, energy, water, and/or by-products. The keys to industrial symbiosis are collaboration and the synergistic possibilities offered by geographic proximity [8]. Benefits achieved by industrial symbiosis have been shown to reduce environmental impacts, increase material use efficiencies and ultimately, present a more sustainable means of industrial production.

4. THE PATTERN OF INDUSTRIAL SYMBIOSIS

In this chapter, approximately 200 instances of interfirm exchanges of industrial waste were collected at 13 industrial clusters in different parts of the world. As shown in Fig. 14.2, the geographical location of these sites spans Europe (Kalundborg, Styria, Forth Valley, and Humberside), North America (Guayama, Campbell, and Sarnia-Lambton), Asia (Kawasaki, TEDA, Guigang and Ulsan), and Oceania (Kwinana and Gladstone). The selection of the sample sites is determined by data availability: waste recovery in these sites has been studied in depth and is sufficiently documented to serve as a basis for tracking material flows among waste generators and users. There is great variability among these clusters as shown in Table 14.1. Kwinana, for example, is a 120-km^2 mineral processing region in Western Australia [22]. In Japan, Kawasaki's industrial cluster features waste reuse among steel, cement, chemical, and paper industries as a part of the national "Eco-town" Program that helps cities reuse secondary materials [16]. In China, the Tianjin Economic-Technological Development Area spans 45 km^2 and has thousands of companies. Guigang is a sugar-refining operation that uses its fiber by-products from sugar cane and the organic by-products from molasses [19,20]. In the United States, both Guayama and Campbell have fewer than 12 actors and are anchored by coal-fired power plants that reuse wastewater and provide steam for other industrial operations as well as beneficially reuse coal ash [14,15]. Similar exchanges also compose part of the cluster in Kalundborg, Denmark [24].

Based on information from academic journals and reports, we analyzed industries and industrial sectors involved in waste reuse, either as generators or as users, as well as the types of industrial wastes that are exchanged. Nineteen industrial sectors, from the letter "A" to the letter "S," are listed in Table 14.2 based on "The North American Industry Classification System" (NAICS) codes. To present the maximum number of waste reuse possibilities encountered for each exchange, all waste reuse cases are compiled, whether they are currently operating, proposed, or discontinued.

Numerous industrial waste materials were identified at the 13 industrial cluster sites and then grouped into 10 categories: chemical waste, metallic waste, ash, organic waste, sludge, paper and wood waste, nonmetallic waste, waste plastics and rubber, waste oil, and others (Table 14.3). Among a total of 199 industrial waste exchanges across companies, the top five waste categories in terms of the number of linkages occur in these sectors: chemical waste, metallic waste, ash,

FIGURE 14.2 Thirteen industrial sites selected in this study to analyze industrial waste reuse.

TABLE 14.1 Descriptions of 13 Industrial Clusters Worldwide Chosen for Review

Location (country)	Main industries	Number of firms and linkages[2]	Base year	Reference
Kalundborg (Denmark)	Coal-fired power station, oil refinery, pharmaceutical/ biotech industry, wallboard plant, cement, soil remediation, farms	12 19	2008	[9,10]
Styria (Austria)	Paper, saw mills, cement, construction materials, plastic, stone and ceramic, wastewater treatment, power plant	19 38	1994	[11]
Forth Valley (UK)	Coal-fired, biomass, combined heat and power generation, cement, oil refinery, plastic, chemical, fertilizer industry, wastewater and sludge treatment	14 10	2004	[12]
Humberside (UK)	Oil refinery, chemical, food, cement, furniture, steel, wastewater treatment, combined heat and power generation	16 22	2004	[12,13]
Guayama (US)	Chemical refining, coal-fired power station, wastewater treatment, aggregate, pharmaceuticals	6 5	2005	[14]
Campbell (US)	Coal-fired power station, cement, oil refining, water supply	11 11	2006	[15]
Sarnia-Lambton (Canada)	Chemicals, fertilizer, polystyrene, rubber, wallboard, wastewater treatment, and hydro-power station	7 4	1997	[12]
Kawasaki (Japan)	Steel, stainless steel, cement, chemical, paper waste, wastewater treatment	7 15	2009	[16]
TEDA (China)	Construction materials, glass, paper, pharmaceuticals, chemicals, food, electric/electrical equipment, tires, metal smelting, automobiles, rubber, wood, furniture, waste industry, cogeneration plant, water supply, wastewater	68 60	2008	[17,18]
Guigang (China)	Sugar, alcohol, cement, fertilizer, and paper industry	10 11	2004	[19,20]
Ulsan (South Korea)	Chemicals, oil refining, petrochemicals, textiles, metal processing, paper, water supply, wastewater treatment	12 10	2004	[21]
Kwinana (Australia)	Alumina, nickel and oil refining, iron and titanium oxide, cement, chemicals, fertilizer, coal- and gas-fired power generation, water supply and treatment	32 71	~2008	[22,23]
Gladstone (Australia)	Alumina refining, aluminium smelting, cement, coal-fired power generation, sewage treatment	6 5	2005	[22]

[2]*When visualizing a web of waste exchanges as a network, a firm participating in a waste exchange represents a node in the network. One linkage is defined as a single instance of waste reuse between a generator and a user.*

organic waste, and sludge. These are the waste materials that have been reused the most frequently and, generally, where there is likely to be more recognition, knowledge, technology, and experience in reusing these materials.

Understanding how actors reuse different waste materials is critical to planning and facilitating industrial waste reuse linkages. Industrial waste reuse can occur within a facility, but the opportunity is largely limited. In this

TABLE 14.2 The Classification of Industries

	NAICS Code	Description
A	11	Agriculture, forestry, fishing, and hunting
B	21	Mining, quarrying, and oil and gas extraction
C	2211	Energy utilities (electric power generation, transmission and distribution)
D	2213	Water utilities (water, sewage, and other systems)
E	23	Construction
F	311	Food manufacturing
G	313/314	Textile mills and textile product mills
H	321	Wood product manufacturing
I	322	Paper manufacturing
J	324	Petroleum and coal product manufacturing
K	325	Chemical manufacturing
L	326	Plastic and rubber product manufacturing
M	327	Nonmetallic mineral product manufacturing
N	331	Primary metal manufacturing
O	334/335	Computer and electronic/electrical equipment manufacturing
P	336	Transportation equipment manufacturing
Q	337	Furniture and related product manufacturing
R	562	Waste industries (waste collection, treatment/disposal, remediation, and other management services)
S	-	Other industries

analysis, most waste reuse occurs across industries and industrial sectors. Interindustry reuse can happen in at least two different patterns: one pattern of reuse is more concentrated in a single industrial sector whereas the other pattern is more distributed across diverse industries. As shown in Fig. 14.3, the largest supplier and user of metallic waste is the primary metal manufacturing industry, although five different industries generate and five industries reuse the metallic waste. In contrast, sludge is more dispersed, as it is generated from eight industrial sectors and reused by seven different sectors. One implication of the degree of dispersion across sectors is the opportunity provided for reuse within and around multiple industry clusters that foster cooperation among the participants in the cluster. By offering more market channels across industries, there is likely to be less dependence on any one industry should that industry become unavailable as a reuse channel.

Industrial sectors vary in their potential either to accept or reuse waste. For example, companies in the nonmetallic mineral manufacturing sector (Industry "M"), including cement/concrete producers, lime/gypsum manufacturers, and glass and clay products manufacturers, accept waste materials from 59 other industries, but send their waste to only five other industries (Fig. 14.4). Companies in this industrial sector primarily act as waste users and have a high capability to incorporate waste materials in their production. However, energy utilities (C) and food manufacturers (F) supply more waste materials than they receive. In the case of chemical manufacturing (K) and primary metal manufacturing (N), companies supply and use a similar quantity of waste materials in aggregate as most reuse happens within a single industrial sector.

Industries exchange not only solid waste materials (by-products) but also residual heat and secondary effluent. Often, power plants and public wastewater treatment works provide energy, residual steam, and treated wastewater to industries nearby, a practice known as "utility sharing." Figure 14.5 shows that in addition to the 199 instances of solid material reuse, there are 82 utility sharing linkages observed in these industrial clusters. While energy utilities (C) and water utilities (D) dominate in supplying energy and water, petroleum and coal product

TABLE 14.3 Summary of Waste Linkages by Industrial Waste Category

Waste Category	Material Description	Number of Observations	Industries that Generate This Waste	Industries that Use This Waste
Chemical waste	Spent solvents, residual acids/alkali, sulfur, industrial gases (CO_2, H_2), activated carbon, spent catalyst	54	C, D, F, I, J, K, N, O, R	B, C, D, F, I, J, K, L, M, N, O, R
Metallic waste	Metal scraps (iron, steel, stainless steel, copper, lead, zinc), slag (blast furnace, steel, lead), solder materials, bauxite residue, spent lead acid batteries	28	I, N, O, P, R	D, K, M, N, R
Ash	Fly ash, bottom ash, mixed ash, burnt residue	26	C, I, R	A, B, E, K, M, R, S
Organic waste	Food waste, biomass, fertilizer, other organic waste	23	A, F, K, N, R, S	A, C, F, I, K, M, R, S
Sludge	Sewage sludge, refinery sludge, paper sludge, fiber muds, filter cakes	19	D, F, I, J, M, N, P, R	A, B, C, D, K, M, R
Paper and wood waste	Cardboard, mixed paper, wood dust, chips, trimmings	15	H, I, M, Q, R, S	A, H, I, M, S
Non-metallic waste	Synthetic gypsum, construction and concrete waste, glass scrap, coal mine overburden, lime kiln dust, silica fume	14	B, C, K, M, O, R	E, M, N
Waste plastics and rubber	Polystyrene, waste plastics, off-spec plastics, rubber scrap	7	F, I, L, R	J, L, M, N, R, S
Waste oil	Used oil from chemical processes, edible oil from food manufacturing	7	F, K, R	J, K, M
Others	Textile waste, fine materials, biogas, excess gas	6	D, G, I, J, R	J, M

manufacturing (J), chemical manufacturing (K), and primary metal manufacturing (N) also provide some utilities as resources to other industries.

It is not the intent of this analysis to suggest that all industrial wastes could or should be reused. Many wastes pose numerous risks to human health and the environment and the first task of any program is safety. Neither can we presume, however, that many common practices today such as storing waste in piles or impoundments is protective or useful. The enormous flood of coal ash escaping from a compromised impoundment in Tennessee in 2008 ended up covering 1.2 km^2 (300 acres) of land with 4.1×10^6 m^3 (1.1 billion gallons) of coal ash flowing into the tributaries of the Tennessee River [25]. By the logic of symbiosis, it is far preferable to find safe, beneficial uses for that coal ash in the first instance. In this way of thinking, the products of the 13 industrial clusters discussed in this chapter need not be considered wastes, but should be thought of as secondary materials that could find economic and environmentally sound reuse opportunities.

Some argue that the solution to the NHIW problem is to regulate it much more closely, or, at a minimum, to increase information flows about it. Establishing more uniform definitions of NHIW across broader geographies would

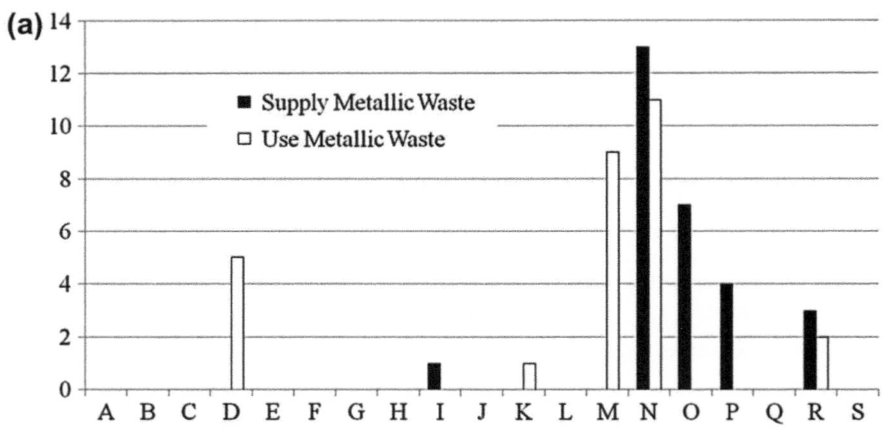

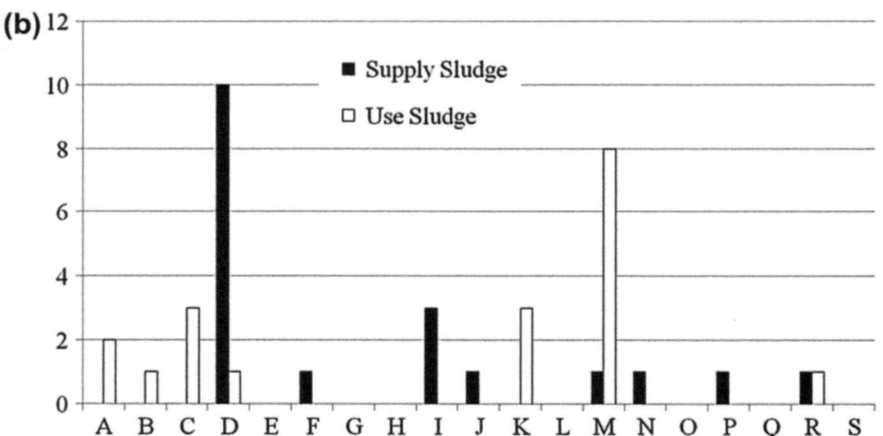

FIGURE 14.3 The number of linkages for supply and reuse of waste in different industrial sectors: for metallic waste (a, upper) and for sludge (b, lower).

help clarify available market opportunities. Routine collection and distribution of data on the type and amount of NHIW that is available, along with clear standards for NHIW management including legal liability issues, would greatly facilitate reuse [26]. Interested jurisdictions could experiment in a systematic way with different incentives and requirements to achieve decreases in generation and disposal and increases in beneficial reuse.

In this chapter, we have analyzed industrial waste exchange networks in business clusters around the globe; discussed what kinds of industrial by-products and utility resources have been reused; determined how much or how often each waste reuse stream has been observed in different places; and also identified the industry sectors that generate and reuse them. Still, we have not dealt with the question of how waste reuse becomes organized. Some reuse networks, particularly in Asian regions including Kawasaki, TEDA, and Ulsan in this study, are primarily planned from the top—down as a way of conserving resources and increasing efficiency. However, many exchanges have been self-organized by companies such as in Kalundborg, Campbell, and Styria, through bilateral discussion, communication, and

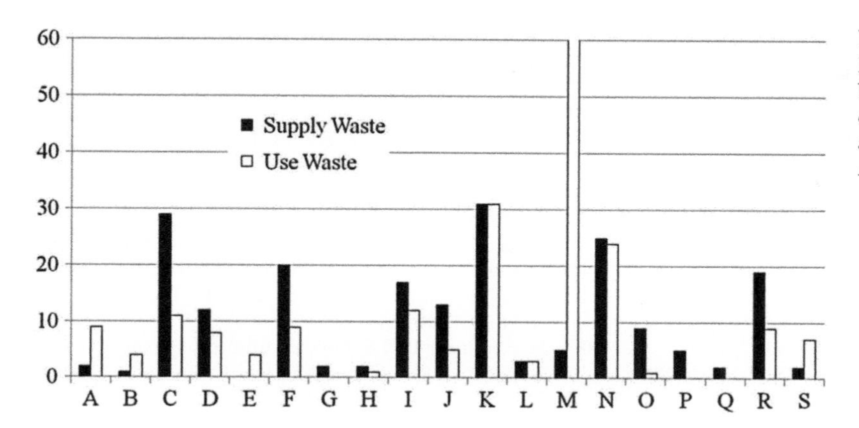

FIGURE 14.4 The number of linkages for supplying waste (black bars) or using waste (white bars) observed in 13 industrial clusters across different industrial sectors from A to S.

negotiation across industries [27]. Some networks are facilitated by third parties including Forth Valley and Humberside in this study organized by dedicated entities such as the National Industrial Symbiosis Program of the United Kingdom.

A business rationale for symbiotic activity is that by acting together, proximate firms can find collective economic and environmental benefits that are greater than the sum of the individual benefits each would realize if it acted alone [28]. Specifically, firms benefit from avoiding disposal or substituting input materials at a lower total cost, and may also be driven by a long-term interest in supply security. A less visible attribute of industrial symbiosis lies in the necessary condition of inter-firm cooperation: industries that recognize the possibility that the whole can be greater than the sum of its parts, find an approach that extends beyond their own boundaries that is good for business and often good for the environment at the same time. The 13 industrial areas described here only capture a tiny percent of the symbiotic activity involving industrial waste that is already occurring in the field. When NHIW can be used in the ways cited in these examples, however, there is the double benefit of landfill avoidance and reuse that contributes to the productive economy. By learning about and understanding these phenomena, we can facilitate and expand the structures that enable such

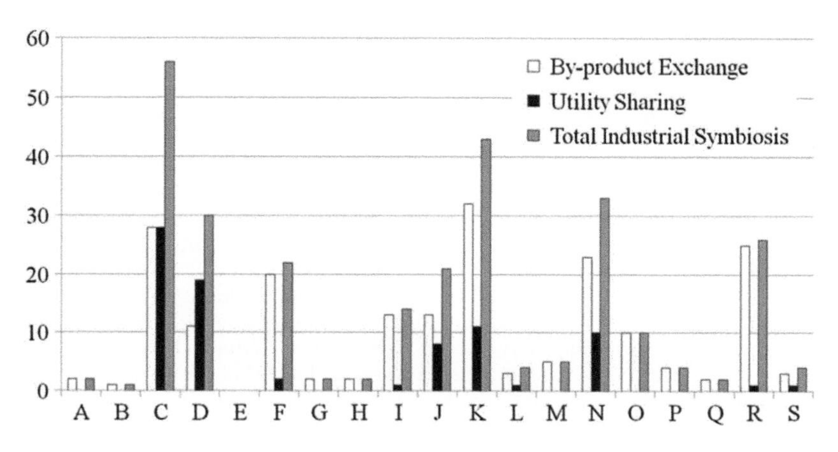

FIGURE 14.5 The number of linkages for supplying by-products, utilities, and total industrial symbiosis linkages observed in 13 industrial sites across different industrial sectors from A to S.

linkages to occur in service to a more efficient and sustainable industrial system.

References

[1] USEPA. 2010. Waste—Non-Hazardous Waste—Industrial Waste. <http://www.epa.gov/waste/nonhaz/industrial/index.htm>. Accessed May 28, 2010.

[2] OECD. 2008. OECD Environmental Data: Compendium 2006–2008. OECD.

[3] USEPA. Report to Congress: Soild Waste Disposal in the United States, USEPA, Washington DC, 1988.

[4] ACAA. 2006 Coal Combustion Product (CCP) Production and Use Survey, ACAA, 2007.

[5] ECOBA, Production and Utilisation of CCPs in 2006 in Europe (EU15). <http://www.ecoba.com> Accessed September 15, 2009.

[6] JCOAL. 2009. Utilization of Coal Ash. <http://www.jcoal.or.jp/coaltech_en/coalash/ash02e.html>. Accessed September 15 2009.

[7] N. Nemerow, Zero Pollution for Industry: Waste Minimization through Industrial Complexes, John Wiley & Sons, New York, 1995.

[8] M.R. Chertow, Industrial Symbiosis: Literature and Taxonomy, Annual Review of Energy and the Environment 25 (2000) 313–337.

[9] J. Kryger, 2009. Kalundborg Industrial Symbiosis, Paper presented at 6th Industrial Symbiosis Research Symposium, Kalundborg, Denmark, June 19, 2009.

[10] N.B. Jacobsen, Industrial Symbiosis in Kalundborg, Denmark: A Quantitative Assessment of Economic and Environmental Aspects, Journal of Industrial Ecology 10 (2006) 239–255.

[11] E.J. Schwarz, K.W. Steininger, Implementing Nature's Lesson: The Industrial Recycling Network Enhancing Regional Development, Journal of Cleaner Production 5 (1997) 47–56.

[12] R. van Berkel, Regional Resource Synergies for Sustainable Development in Heavy Industrial Areas: An Overview of Opportunities and Experiences, Centre of Excellence in Cleaner Production, Curtin University of Technology, Perth, 2006.

[13] M. Mirata, Experiences from Early Stages of a National Industrial Symbiosis Programme in the UK: Determinants and Coordination Challenges, Journal of Cleaner Production 12 (2004) 967–983.

[14] M.R. Chertow, D.R. Lombardi, Quantifying Economic and Environmental Benefits of Co-located Firms, Environmental Science & Technology 39 (2005) 6535–6541.

[15] M. Chertow, Y. Miyata, Assessing Collective Firm Behavior: Comparing Industrial Symbiosis with Possible Alternatives for Individual Companies in Oahu, Hawaii, Business Strategy and the Environment, Published online DOI:10.1002/bse.694, 2010.

[16] R. van Berkel, T. Fujita, S. Hashimoto, M. Fujii, Quantitative Assessment of Urban and Industrial Symbiosis in Kawasaki, Japan, Environmental Science & Technology 43 (2009) 1271–1281.

[17] H.M. Shi, M. Chertow, Y. Song, Developing Country Experience with Eco-Industrial Parks: A Case Study of the Tianjin Economic-Technological Development Area in China, Journal of Cleaner Production 18 (2010) 191–199.

[18] H. Shi, Industrial Symbiosis from the Perspectives of Transaction Cost Economics and Institutional Theory Thesis, Yale University, 2009.

[19] Q.H. Zhu, E.A. Lowe, Y.A. Wei, D. Barnes, Industrial Symbiosis in China: A Case Study of the Guitang Group, Journal of Industrial Ecology 11 (2007) 31–42.

[20] Q.H. Zhu, R.P. Cote, Integration Green Supply Chain Management into an Embryonic Eco-Industrial Development: A Case Study of the Guitang Group, Journal of Cleaner Production 12 (2004) 1025–1035.

[21] H.S. Park, E.R. Rene, S.M. Choi, A.S.F. Chiu, Strategies for Sustainable Development of Industrial Park in Ulsan, South Korea: From Spontaneous Evolution to Systematic Expansion of Industrial Symbiosis, Journal of Environmental Management 87 (2008) 1–13.

[22] D. van Beers, G. Corder, A. Bossilkov, R. van Berkel, Industrial Symbiosis in the Australian Minerals Industry: The Cases of Kwinana and Gladstone, Journal of Industrial Ecology 11 (2007) 55–72.

[23] D. van Beers, Capturing Regional Synergies in the Kwinana Industrial Area: 2008 Status Report, Centre of Excellence in Cleaner Production, Curtin University of Technology, Perth, 2008.

[24] J.R. Ehrenfeld, N. Gertler, Industrial Ecology in Practice: The Evolution of Interdependence at Kalundborg, Journal of Industrial Ecology 1 (1997) 67–79.

[25] USEPA. Action Memoremdum Fact Sheet on Selected Engineering Evaluation/Cost Analysis Alternative Kingston Fossil Fuel Plant Release Site Harriman, Roane County, Tennessee: USEPA, 2010.

[26] T. Chase, K. Twaite, Non-Hazardous Industrial Wastes, Yale School of Forestry and Environmental Studies, New Haven, 2010.

[27] M.R. Chertow, "Uncovering" Industrial Symbiosis, Journal of Industrial Ecology 11 (2007) 11–30.

[28] E.A. Lowe, L.K. Evans, Industrial Ecology and Industrial Ecosystems, Journal of Cleaner Production 3 (1995) 47–53.

CHAPTER

15

Construction Waste

Mohamed Osmani

Department of Civil and Building Engineering, Loughborough University LE11 3TU, United Kingdom

OUTLINE

1. INTRODUCTION

The built environment consumes more natural resources than necessary and therefore generates a large amount of waste. A study by the World Resource Institute of material flows in a number of industrialised countries showed that one half to three quarters of the annual material input to these societies was returned to the environment as waste within 1 year [1]. The international community started realising that resources are finite and that nature can no longer absorb the vast quantities of waste continually released to it. Achieving 'zero waste' will be a breakthrough strategy for a world in an environmental crisis; however, this is a highly challenging target in construction, but involving and committing all stakeholders to reduce waste at source and developing efficient waste management strategies by reusing and recycling materials and components can take the industry closer to the 'zero waste' vision, hence, moving

Waste Doi: 10.1016/B978-0-12-381475-3.10015-4

from myth to reality. The aim of this chapter is to rethink construction waste management by re-engineering processes and practices to reduce construction waste at source. The chapter examines the concept of waste and definitions, discusses construction waste quantification and source evaluation, explores current thinking on construction waste research and appraises the current construction waste management and minimisation status in the United Kingdom (UK) in terms of drivers and pressures for change, design and onsite practices, and challenges and enablers.

2. CONCEPTS AND DEFINITIONS

Emerging sustainable thinking is redefining the concept of waste from a 'by-product' of processes to missed opportunities to cut costs and improve performance. Koskela [2] went further to argue that waste adds costs but does not add value. Similarly, Formoso et al. [3] classified waste as 'unavoidable', for which the costs to reduce it are higher than the economy produced, and 'avoidable', when the necessary investment to manage the produced waste is higher than the costs to prevent or reduce it. Therefore, the concept of waste should be looked at in terms of activities that increase costs directly or indirectly but do not add value to the project.

There is no generally accepted definition of waste. As a result, the European Council revised the Waste Framework Directive (WFD) in October 2008, which must be fully implemented within all European Union (EU) member states by December 2010. The changes to the WFD can be broadly separated into major and 'sorting out' measures. The major changes are aimed at encouraging the greater reuse and recycling of waste, whereas the sorting out measures are aimed at simplifying the fragmented legal framework that has regulated the waste sector to date. Significantly, the definition of 'waste' has been clarified in the revised WFD through specific articles that formally introduce the concepts of 'by-products' and 'end-of-waste'. The introduction of a definition of by-products in Article 5 (1) formally recognises the circumstances in which materials may fall outside the definition of waste. This change is intended to reflect the reality that many by-products are reused before entering the waste stream. In the United Kingdom, a consultation process on draft guidance on the legal definition of waste and its application was launched in January 2010, and a report summarising the consultation responses and their guidance on the interpretation of the definition of waste is scheduled for publication in July 2010 [4]. For the scope of this chapter, the following definitions are adopted:

- Waste is 'any substance or object which the holder discards or intends or is required to discard' [5]. This definition applies to all waste irrespective of whether it is destined for disposal or recovery operations.
- 'Construction waste' is a material or product which needs 'to be transported elsewhere from the construction site or used on the site itself other than the intended specific purpose of the project due to damage, excess or non-use or which cannot be used due to non-compliance with the specifications, or which is a by-product of the construction process' [6].
- 'Design waste' is the waste arising from construction sites owing directly or indirectly to the design process.
- 'Waste minimisation' is the reduction of waste at source, (i.e. designing out waste) by understanding its root causes and re-engineering current processes and practices to alleviate its generation.
- 'Waste management' is the process involved in dealing with waste once it has arisen, including site planning, transportation, storage, material handling, onsite operation, segregation, reuse and recycling and final disposal.

3. CONSTRUCTION WASTE COMPOSITION AND QUANTIFICATION

It is difficult to give exact figures of construction waste produced on a typical construction site, but it is estimated that it is as much as 30% of the total weight of building materials delivered to a building site [7]. In the United States, around 170 million tonnes of construction and demolition waste was generated during 2003, of which 48% was estimated to be recovered [8]. Chun Li et al. [9] related the production of construction waste to the designed facilities' floor areas by stating that most buildings in the United States generate between 20 to 30 kg m^2. In the EU, more than 450 million tonnes of construction and demolition waste is generated every year, which makes it the largest waste stream in quantitative terms, with the exception of mining and farm wastes [10]. At present, 75% of construction and demolition waste in the EU is being landfilled, although over 80% recycling rates have been exceptionally achieved in countries such as Germany and the Netherlands [11]. In the United Kingdom, the disposal of construction waste accounts for more than 50% of overall landfill volumes [12]. Furthermore, Guthrie et al. [13] reported that at least 10% of all materials delivered to UK construction sites are wasted due to damage, loss and over-ordering. However, Fishbein [7] estimated this amount to be as much as 30% of the total weight of building materials delivered to a building site. Equally, 38% of solid waste in Hong Kong comes from the construction industry [14], and in 2006, about 40% of the available landfill capacity was used to manage construction waste [15]. In addition, Bossink and Brouwers [16] revealed that in the Netherlands, each building material generates between 1 to 10% waste of the amount purchased resulting in an overall average of 9% of purchased materials becoming waste. Pinto and Agopyan [17] went further to report that, in Brazil, the construction project waste rate is 20 to 30% of the weight of the total site building materials.

In terms of weight, brick masonry and concrete present by far the largest potential for recycling in the building sector [18]. This has been supported by the findings of comprehensive research conducted across the United States, the United Kingdom, China, Brazil, Korea and Hong Kong, which compared the types and volumes of construction waste in these countries [19]. However, the types and composition of onsite wastes are highly variable, depending on the construction techniques used. For example, 'there will be very little waste concrete and timber forms for disposal if pre-cast concrete elements are adopted' [20]. Guthrie and Mallett [21] split construction and demolition waste into three categories as follows: materials which are (1) potentially valuable in construction and easily reused/recycled, including concrete, stone masonry, bricks, tiles/pipes, asphalt and soil; (2) not capable of being directly recycled but may be recycled elsewhere, including timber, glass, paper, plastic, oils and metal and (3) not easily recycled or which present particular disposal issues, including chemicals (i.e. paint, solvents), asbestos, plaster, water and aqueous solutions. Coventry et al. [22] identified seven different types of waste: bricks, blocks and mortar (33%); timber (27%); packaging (18%), dry lining (10%); metals (3%); special waste (1%) and other waste 10%.

4. CONSTRUCTION WASTE SOURCE EVALUATION

There are a variety of different approaches to the evaluation of the main origins, sources and causes of construction waste. The extant of literature reveals a number of construction waste generation sources, which can be broadly categorised into 11 clusters. Table 15.1 shows that construction waste is generated throughout the

TABLE 15.1 Origins and causes of construction waste (compiled from the main sources within the literature)

Origins of Waste	Causes of Waste
Contractual	• Waste client-driven/enforced. • Errors in contract documents. • Contract documents incomplete at commencement of construction.
Procurement	• Lack of early stakeholders' involvement. • Poor communication and coordination among parties and trades. • Lack of allocated responsibility for decision making. • Incomplete or insufficient procurement documentation.
Design	• Design changes. • Design and detailing complexity. • Design and construction detail errors. • Inadequate/incoherent/incorrect specification. • Poor coordination and communication (late information, last minute client requirements, slow drawing revision and distribution).
On-site Management and Planning	• Lack of on-site waste management plans. • Improper planning for required quantities. • Delays in passing information on types and sizes of materials and components to be used. • Lack of on-site material control. • Lack of supervision.
Site Operation	• Accidents due to negligence. • Unused materials and products. • Equipment malfunction. • Poor craftsmanship. • Use of wrong materials resulting in their disposal. • Time pressure. • Poor work ethics.
Transportation	• Damage during transportation. • Difficulties for delivery vehicles accessing construction sites. • Insufficient protection during unloading. • Methods of unloading.
Material ordering	• Ordering errors (i.e. ordering items not in compliance with specification). • Over allowances (i.e. difficulties to order small quantities). • Shipping and suppliers' errors.
Material Storage	• Inappropriate site storage space leading to damage or deterioration. • Improper storing methods. • Materials stored far away from point of application.
Materali Handling	• Materials supplied in loose form. • On-site transportation methods from storage to the point of application. • Inadequate material handling.
Residual	• Waste from application processes (i.e. over-preparation of mortar). • Off-cuts from cutting materials to length. • Waste from cutting uneconomical shapes. • Packaging.
Other	• Weather. • Vandalism. • Theft.

project from inception to completion and the pre-construction stage has its considerable share. A recent research on construction procurement systems-related waste sources showed that these fall under four main themes: uncoordinated early involvement of project stakeholders, ineffective project communication and coordination, unclear allocation of responsibilities and inconsistent procurement documentation [23]. Furthermore, it has been estimated that 33% of wasted materials is due to architects failing to design out waste [24]. However, construction waste minimisation through design is complex because buildings embody a large number of materials and processes. Equally, Osmani et al. [25] reported that 'waste accepted as inevitable', 'poor defined responsibilities' and 'lack of training' are major challenges facing architects to design waste reduction measures in their projects. This is made more complex when further waste is created directly or indirectly by other projects' stakeholders, namely, clients, contractors, sub-contractors and suppliers. Nonetheless, there is a general consensus that design changes during operation activities are one of the key origins of construction waste [16, 26]. The main drivers for design variations during construction are lack of understanding the underlying origins and causes, design changes to meet client's changing requirements, complex designs, lack of communication between design and construction teams, lack of design information, unforeseen ground conditions and long project duration.

5. CONSTRUCTION WASTE MANAGEMENT AND MINIMISATION APPROACHES

Despite international governmental, industrial and academic efforts to develop waste reduction thinking in construction, uptake globally is piecemeal. The current and ongoing research in the field of construction waste management and minimisation can be broadly categorised into the following 13 clusters:

1. construction waste quantification and source evaluation;
2. procurement waste minimisation strategies;
3. designing out waste;
4. onsite construction waste sorting methods and techniques;
5. development of waste data collection models, including flows of wastes and waste management mapping, to help with the handling of onsite waste;
6. development of onsite waste auditing and assessment tools;
7. impact of legislation on waste management practices;
8. improvements of onsite waste management practices;
9. reuse and recycle in construction;
10. benefits of waste minimisation;
11. waste minimisation manuals, including guides for designers;
12. attitudes towards construction waste minimisation and
13. comparative waste management studies.

Research reports, such as the work of Coventry et al. [22], aim to promote awareness in the building construction industry about the benefits of waste minimisation, including cost savings, and environmental issues and use of recycled and reclaimed materials. The 'three Rs' principle of waste (reduction, reuse and recycle), otherwise known as the waste hierarchy, has been widely adopted. Similarly, the impact of legislation, particularly the Landfill Tax, and its effects on the behaviour and practices of the construction industry has resulted in a number of research studies. Furthermore, in the last few years, many waste minimisation and recycling guides have been produced such as Waste and Resources Action Programme (WRAP) [27]. These documents give broad guidance for designers to adopt a waste minimisation approach in their projects; however, the recommendations in these guides do not realistically relate waste to all parameters of the designers' environment, including the complex design and construction process and the supply chain. In addition, they do not specifically identify waste-stream components in relation to their occurrence during the architectural design stages. In addition, tools, models and techniques, such as SMARTWaste in the United Kingdom and WasteSpec in the United States, have been developed to help handle and better manage onsite waste generation and assess the associated cost implications. These tools, which facilitate onsite auditing, waste management and cost analysis, deal with waste that has already been produced. Consequently, there is insufficient effort and no structured approach to address waste at source, that is, 'design waste', to prevent it from being generated at the first place.

6. CONSTRUCTION WASTE MANAGEMENT AND MINIMISATION: THE UK CONTEXT

There are a number of existing data sources in the United Kingdom that quote the amount of materials and products used in construction activities, wasted, managed, recovered and landfilled; however, the resulting statistics vary in terms of scope, methodology, reliability, accessibility and frequency of updatability. As a result, available data is not robust enough to provide benchmarks and baselines on waste generation which should inform the setting up of realistic targets for waste reduction and to measure improvements.

6.1. Construction Waste Minimisation Drivers

The key drivers for waste reduction in the UK construction industry could be broadly categorised into four main groups which are environmental, legislative, economic and business.

6.1.1. Environmental Drivers

As shown in Fig. 15.1, the construction and demolition activities account for 32% of all waste arisings in England, which makes it the largest waste stream. This figure is substantially higher if additional construction-related wastes from other sectors are added, namely, through construction material product manufacturing processes in the industrial sector, and during raw material excavation and production in the mining and quarrying sector.

The UK construction, demolition, refurbishment and excavation activities produce around 120 million tonnes of waste each year, including an estimated 13 million tonnes of unused materials [28]. Furthermore, it is responsible for

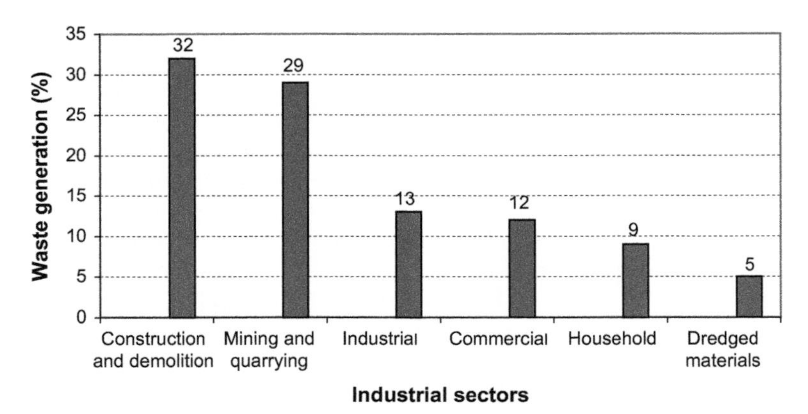

FIGURE 15.1 Estimated waste arisings in England by industrial sector (combined data from a number of UK government sources, 2003–2007).

generating over 36 million tonnes of landfill waste every year. It has been estimated that UK landfill sites will be filled in as little as 6 years, making it imperative to reduce and manage waste. Public opinion in the United Kingdom has emphasised the difficulties of minimising construction waste, but with Germany recycling over 80% of its construction waste and Denmark over 90%, this is clearly a misperception. Fortunately, the situation is changing in the United Kingdom, and there are a growing number of sustainable waste management solutions that can be used as examples of best practice. Indeed, the proportion of construction and demolition waste recycled by crushers and screeners from 2001 to 2005 has increased from 49 to 52%; however, the amount of construction and demolition waste sent to landfill has increased from 26 to 31% [29]. The latter indicates that there is a pressing need to reduce waste at all stages of construction and divert it from landfill by considering the long-term impacts of design, build and waste management.

6.1.2. Legislative Drivers

The UK government has been using a combination of regulation, economic instruments and voluntary agreements to meet targets of ethical, social and environmental performance in driving the waste management agenda. The government's *Strategy for Sustainable Construction*, published in June 2008, calls for a step change in the sustainability of procurement, design and operation of all built assets to be driven by innovation [28]. The aim of the strategy is to improve the built environment performance with a focus on reducing carbon emissions and resource consumption in new buildings. In encouraging the construction industry to drive its own resource efficiency programme, the strategy calls for zero construction waste to landfill by 2020 [30]. It also set a target to halve the amount of construction, demolition and excavation wastes going to landfill by 2012 in comparison with 2008 levels, as a result of waste reduction, reuse and recycling. This is a significant challenge for the industry. In addition, existing waste-related legislation – especially the Landfill Tax (£48 per tonne in 2010, which will make the current waste disposal methods too costly for construction firms), the Aggregates Levy (£2 per tonne for on the extraction of aggregates) and Site Waste Management Plans (SWMPs) – should contribute to a transition away from land-filling towards waste reduction, reuse and recycling. However, as yet this does not appear to have seriously reduced the amount of waste production, the UK government is likely to introduce other fiscal measures and legislation in the future, which will push the construction industry towards a closed loop production system.

6.1.3. Economic Drivers

The construction industry in the United Kingdom spends more than £200 million on Landfill Tax each year. Waste typically costs companies 4% of turnover with potential savings of 1% through the implementation of a comprehensive waste minimisation programme. Furthermore, WRAP [27] estimates that £1.5 billion is wasted in materials that are delivered to the site but unused. Construction-related businesses can take advantage of government funding to implement waste minimisation practices. Indeed, from April 2005 to March 2008, the government granted £284 million of Landfill Taxes to the Business Resource Efficiency and Waste (BREW) programme. More than 65% of this funding was approved for waste management initiatives.

Waste minimisation financial benefits are related to the direct costs of both waste disposal and raw material purchase. However, the true cost of waste is estimated to be around 20 times the disposal of waste. A study by a major UK contracting company revealed that that a typical construction skip costs around £1343. This figure is broken into £85 for skip hire (6.4% of cost), £163 for labour (12.1% of cost) and £1095 of cost of wasted materials (81.5% of cost). Therefore, the financial cost of waste for a generic house (5 skips) is around £6715, of which £5439 is attributed to the cost of discarded materials.

6.1.4. Business Drivers

For construction to improve its performance in this competitive age, it has become essential that sustainable practices, including waste minimisation, are adopted and implemented. Indeed, clients are increasingly demanding for enhanced sustainable project performance and are exerting more influence on the industry to reduce onsite waste and cut costs. This is gradually becoming a necessary requirement for procurement across the entire supply chain. In response to such pressures, businesses are abandoning their narrow theory of value in favour of a broader approach, which not only seeks increased economic value but also considers corporate social responsibilities and stakeholders' engagement and commitment.

6.2. Construction Waste Minimisation Practices

With increasing waste legislation and fiscal measures in the United Kingdom, research was thus undertaken by the author to explore current waste minimisation practices and associated barriers in the UK construction industry. A questionnaire survey and follow up interviews were used in this research as a method of collecting data. The sampling frame was confined to the top 100 architectural practices and contracting firms in the United Kingdom.

Architects were asked to rate the waste minimisation practices that they employed during design; their answers are shown in Fig. 15.2.

It is evident that very few attempts were being made to minimise waste during the design process; for example, more than 92% of architects reported that they did not conduct a feasibility study of waste estimation. However, around a third of the firms claimed that they did use standard materials and prefabricated units frequently, to avoid cutting onsite. Most of the participating architects acknowledged that designing out waste is not being implemented at present; as one respondent put it, 'waste reduction is rarely considered during daily life in an architect's office'. However, respondents reported that lack of interest from clients and 'waste accepted as inevitable' were their major concerns.

Similarly, contractors were asked to rate onsite waste management strategies; their answers are shown in Fig. 15.3. It is interesting

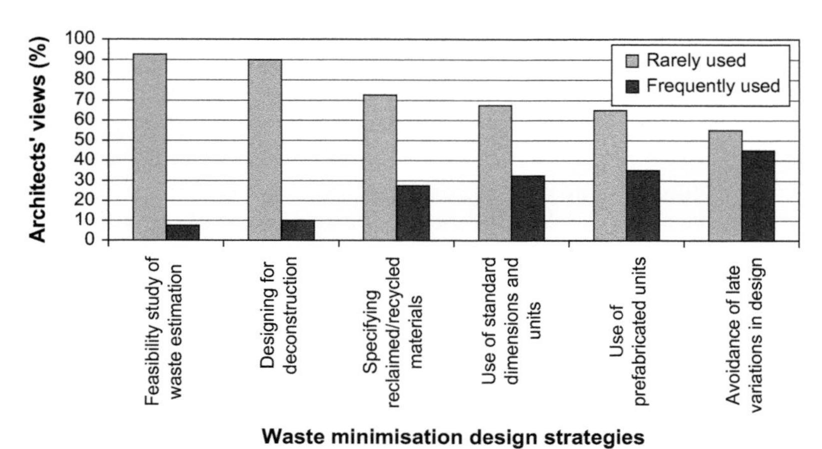

FIGURE 15.2 Current waste minimisation design strategies (architects' views).

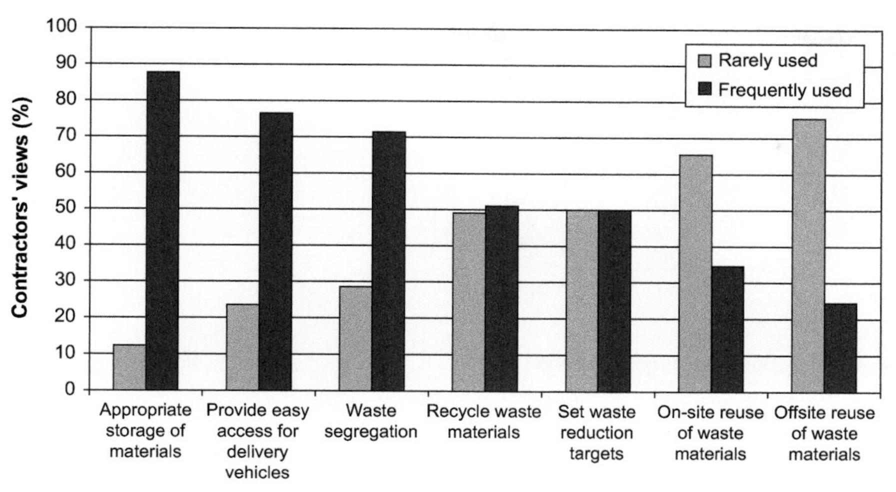

FIGURE 15.3 Current onsite construction waste management strategies (contactors' views).

to note that, contrary to expectations, the majority of contractors indicated that they used 'appropriate storage of materials' (88%) and 'provided easy access for delivery vehicles' (77%) in most or all their projects. However, few efforts were made to segregate and reuse materials. Indeed, over 26% implemented onsite segregation of non-hazardous waste, and about 12 to 6%, respectively, reused onsite and offsite waste materials in all their projects. However, half of the responding companies said they did set waste reduction targets, which appears somewhat contradictory.

6.3. Construction Waste Minimisation Barriers and Incentives

Architects and contractors were asked to identify the most influential barriers and incentives relating to waste management, using

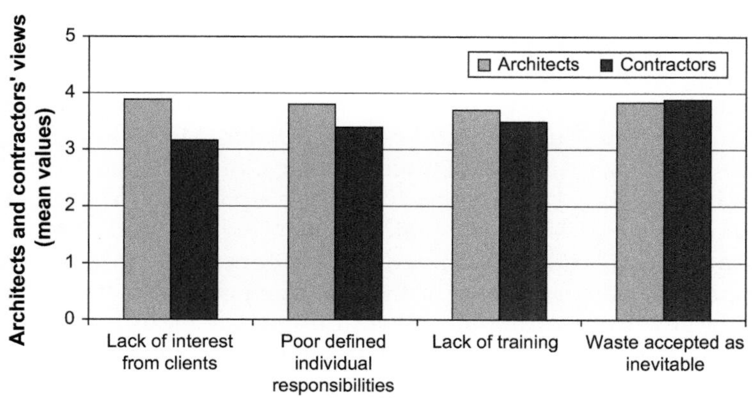

FIGURE 15.4 Barriers to construction waste minimisation (architects and contactors' views).

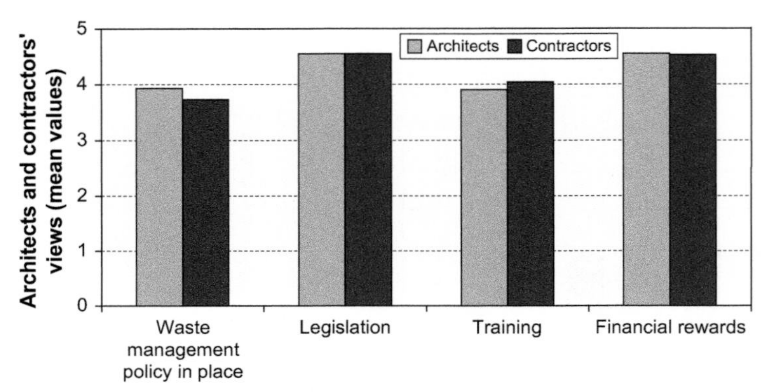

FIGURE 15.5 Incentives to construction waste minimisation (architects and contactors' views).

a Likert scale of 1 to 5. Their responses are shown in Figs. 15.4 and 15.5, respectively. Figure 15.4 shows that the barrier of 'waste accepted as inevitable' was rated the highest mean importance rating by contractors, while architects considered 'lack of interest from clients' as the major constraint, followed closely by 'waste accepted as inevitable' and 'poor defined individual responsibilities'.

However, there was a greater degree of consistency in respondents' views on major incentives to waste minimisation practices, which is shown in Fig. 15.5. Both architects and contractors ranked 'financial rewards' and 'legislation' equally as the main incentives that could drive waste minimisation in the construction industry. Although there is a consensus that legislation can be effective in maintaining the pressure in improving waste minimisation, it was suggested that financial drivers at project level — that is, allocated fees for architects and reward performance against agreed targets for contractors and through government initiatives, that is, tax incentives — will have a far-reaching impact on waste reduction practices. The latter was further emphasised by one respondent who argued that 'the government uses a penal system when a reward system would help clients address the issue with more enthusiasm'.

7. DISCUSSION AND CONCLUSIONS

The current thinking of waste minimisation practices is heavily focussed on the physical minimisation of construction waste and identification of site waste streams. Tools, models and techniques have been developed to help handle and better manage onsite waste generation. Although these tools facilitate auditing, assessment and benchmarking, their waste source evaluation approach is curtailed and piecemeal, as it fails to effectively address the causative issues of waste production throughout all stages of a construction project. The challenge now is to provide a novel platform for the next generation of tools and techniques that will identify and resolve the fundamental causes and origins of construction waste. The basis for such an approach could utilise Building Information Modelling (BIM) and related technologies, in particular Virtual Prototyping, to provide a platform for 'virtual' waste evaluation which reviews and assesses the severity of waste generation throughout all stages of the construction project life cycle. Although BIM design methods are not currently as fully utilised in the construction industry as in other industries, there is general recognition that BIM adoption will become more pronounced to demonstrate not only the entire building lifecycle but also assess and evaluate the environmental performance and impacts of buildings. In addition, 'Lean design thinking' is emerging as a fundamentally and holistic new approach to create value by eliminating 'design waste'. That said, a successful mapping process of Lean principles could only be applied fully and effectively in construction by focusing on improving the whole production, flow of materials and life cycle information processes, hence moving from traditional practices and 'going for lean'.

Construction waste minimisation can be viewed as a threat requiring ever-increasing expenditure on end-of-pipe recycling tools and technologies to meet ever-increasing legislation, or as an opportunity to cut costs and improve performance. The choice should be obvious, but there is a need for a culture change. Rethinking waste management in construction requires adopting 'cyclic' rather than 'linear' approach to design and construction. This requires re-engineering current practice to contribute to a cleaner environment through efficient and cost effective sustainable waste minimisation strategies. However, for waste minimisation to be effective and self-sustaining, it is important that all stakeholders along the construction supply chain embrace a more proactive approach in dealing with waste. In recognition of the responsibility of the architectural profession, through its leading role in project management and a key player in the construction industry, architects should move beyond the concept of 'eco-efficiency' through bolt-on environmental strategies and strive to adopt 'eco-effective' practices by implementing a holistic approach to design out waste, which will be reinforced in tender documents and implemented during the construction stage, in addition to the capture and dissemination of lessons learnt to inform construction waste reduction baselines and benchmarking in future projects.

References

[1] C. Hutter, The Weight of Nations – Material Outflows From Industrial Economies, World Resources Institute, Washington, DC, 2000.

[2] L. Koskela, Application of New Production Theory in Construction, Technical Report No.72, CIFE, Stanford University, Stanford, CA, 1992.

[3] C.T. Formoso, E.L. Isatto, E.H. Hirota, Method for Waste Control in the Building Industry, In: Tommelein, I.D, Ballard G. eds., Proceedings of the Seventh Annual Conference of the International Group for Lean Construction (IGLC-7), Berkeley, CA, July 1999, 325–334.

[4] DEFRA (Department for Environment, Food and Rural Affairs), Consultation on the Legal Definition of Waste and Its Application, DEFRA, London, 2010.

[5] European Council, Waste Framework Directive (2008/98/EC), November 2008.

[6] E.R. Skoyles, J.R. Skoyles, Waste Prevention on Site, Mitchell, London, 1987.

[7] B.K. Fishbein, Building for the Future: Strategies to Reduce Construction and Demolition Waste in Municipal Projects, INFORM Publications, New York, 1998.

[8] EPA, Construction and Demolition materials Amounts, Office of Resource Conservation and Recovery (March 2009) 29. EPA530-R-09-002.

[9] P. Chun-Li, D.E. Scorpio, C.J. Kibert, Strategies for successful construction and demolition waste recycling operations, Construction Management and Economics 15 (1997) 49–58.

[10] EC (European Commission), Directorate-General Environment, Directorate Industry and Environment, ENV.E.3-Waste Management, Management of Construction and Demolition Waste Working Document No. 1 (2000).

[11] M. Erlandsson, P. Levin, Environmental assessment of rebuilding and possible performance improvements effect on a national scale, Building and Environment 40 (2005) 1459–1471.

[12] J. Ferguson, N. Kermode, C.L. Nash, W.A.J. Sketch, R.P. Huxford, Managing and Minimizing Construction Waste: A Practical Guide, Institution of Civil Engineers, London, 1995.

[13] P. Guthrie, C. Woolveridge, S. Coventry, Managing Materials and Components on Site, The Construction Industry Research and Information Association – CIRIA, London, 1998.

[14] Hong Kong Government Environmental Report 2006, Environmental Protection Department, Hong Kong, 2006.

[15] C.S. Poon, Reducing construction waste, Waste Management 27 (2007) 1715–1716.

[16] B.A.G. Bossink, H.J.H. Brouwers, Construction waste: quantification and source evaluation. Journal of Construction Engineering and Management, ASCE 122 (1996) 55–60.

[17] T.P. Pinto, V. Agopyan, Construction Waste as Row Materials for Low-Cost Construction Products, In: Kibert, C.J. (Ed.), Proceedings of the First Conference of CIB TG 16 on Sustainable Construction, Tampa, FL, 1994, 335–342.

[18] R. Emmanuel, Estimating the environmental suitability of wall materials: preliminary results from Sri Lanka, Building and Environment 39 (2004) 1253–1261.

[19] Z. Chen, H. Li, C.T.C. Wong, An application of bar-code system for reducing construction wastes, Automation in Construction 11 (2002) 521–533.

[20] C.S. Poon, A.T.W. Yu, L.H. Ng, On-site sorting of construction and demolition waste in Hong Kong, Resources, Conservation and Recycling 32 (2001) 157–172.

[21] P. Guthrie, H. Mallett, Waste Minimisation and Recycling in Construction, CIRIA Special Publication 122, CIRIA, London, 1995.

[22] S. Coventry, B. Shorter, M.M. Kingsley, Demonstrating Waste Minimisation Benefits in Construction, CIRIA C536, CIRIA, London, 2001.

[23] I.S.W. Gamage, M. Osmani, J. Glass, An Investigation Into the Impact of Procurement Systems on Waste Generation: The Contractors' Perspective, Association of Researchers in Construction Management (ARCOM), Nottingham, UK, September 2009. 103–104.

[24] S. Innes, Developing tools for designing out waste pre-site and onsite. In: Proceedings of Minimising Construction Waste Conference: Developing Resource Efficiency and Waste Minimisation in Design and Construction, New Civil Engineer, London, UK, October 2004.

[25] M. Osmani, J. Glass, A.D. Price, Architect and contractor attitudes towards waste minimisation, Waste and Resource Management 59 (2006) 65–72.

[26] M. Osmani, J. Glass, A.D. Price, Architects perspectives on construction waste minimisation by design, Waste Management 28 (2008) 1147–1158.

[27] WRAP (Waste and Resources Action Programme), Designing Out Waste, A Design Team Guide for Building, WRAP, Banbury, Oxon, 2009.

[28] WRAP (Waste and Resources Action Programme), Halving Construction Waste to Landfill by 2012, WRAP, Banbury, Oxon, 2007.

[29] DEFRA (Department for Environment, Food and Rural Affairs), Key Facts About: Waste and Recycling, DEFRA, London, 2006.

[30] DEFRA (Department for Environment, Food and Rural Affairs), Strategy for Sustainable Construction, DEFRA, London, 2008.

CHAPTER

16

Thermal Waste Treatment

Daniel A. Vallero

Pratt School of Engineering, Duke University, Durham, N. Carolina, USA

OUTLINE

1. INTRODUCTION

Chemical reactions at elevated temperatures can yield products that are either good or bad from an environmental perspective. This depends mainly on whether toxic and otherwise harmful molecules react to become less or more toxic or harmful molecules, or are transformed into molecules that are more or less mobile in the environment, which could mean greater exposures and risks. Worse yet, relatively nontoxic and innocuous waste constituents can speciate into compounds that are toxic and bioavailable.

Many pollutants are formed thermochemically, including the six National Ambient Air Quality Standards (carbon monoxide [CO], oxides of sulfur, oxides of nitrogen, ozone, lead compounds, and particulate matter) as well as many so-called air toxics, for example, benzene, 1,3-butadiene, and coke oven emissions (see Chapter 17). Thermally mediated reactions also lead to myriad air and soil pollutants, including those from all the waste streams discussed throughout Section 2. This chapter focuses, conversely, on the thermal reactions that can be used to treat numerous wastes. The organic portion of wastes has heat value. This

Waste Doi: 10.1016/B978-0-12-381475-3.10016-6

means that this portion of any waste can be completely destroyed using principles based on thermodynamics, ultimately yielding carbon dioxide and water:

$$\text{Hydrocarbons} + O_2 \rightarrow CO_2 + H_2O + \text{energy} \quad (16.1)$$

Complete combustion may also result in the production of molecular nitrogen (N_2) when nitrogen-containing organics are burned, such as in the combustion of methylamine. Incomplete combustion can produce a variety of compounds. Some are more toxic than the original compounds being oxidized, such as polycyclic aromatic hydrocarbons (PAHs), dioxins, furans, and CO.

Thermal destruction begins with the mixing of wastes with oxygen, sometimes in the presence of an external energy source, and within several seconds, the byproducts of gaseous carbon dioxide and water are produced and exit the top of the reaction vessel, while a solid ash that is produced exits the bottom of the reaction vessel. Energy may also be produced during the reaction and the heat may be recovered.

If the wastes contain other chemical constituents, particularly chlorine and/or heavy metals, the situation becomes complex. For such wastes, the potential exists for destroying the initial contaminant as well as exacerbating the problem by generating hazardous off-gases containing chlorinated hydrocarbons or ashes containing heavy metals. For example, the improper incineration of certain chlorinated hydrocarbons can lead to the formation of the highly toxic chlorinated dioxins, furans, and hexachlorobenzene.

The benefits of thermal systems include the following:

(1) the potential for energy recovery;
(2) volume reduction of the contaminant;
(3) detoxification as selected molecules are reformulated;
(4) the basic scientific principles, engineering designs, and technologies are well understood from a wide range of other applications, including electric generation and municipal solid-waste incineration;
(5) application to most organic contaminants that compose a large percentage of the total contaminants generated worldwide;
(6) the possibility to scale the technologies to handle a single gallon per pound (liter per kilogram) of waste or millions of gallons per pound (megaliter per kilogram) of waste; and
(7) the land-area requirements are small relative to other hazardous-waste management facilities such as landfills.

The common drawbacks of thermal destruction technologies include:

(1) the capital intensity of equipment, particularly the refractory material lining the inside walls of each combustion chamber that must be replaced as cracks form whenever a combustion system is cooled and heated;
(2) the need for very skilled operators, and thermal systems are more costly to operate when fuel must be added;
(3) the buildup of ash that will need to be handled, often as a new hazardous waste, which is particularly costly and difficult if heavy metals and/or chlorinated compounds are present; and
(4) the emission of air pollutants (particularly products of incomplete combustion [PICs] and heavy metals), which must be monitored for chemical constituents and controlled.

In general, the same reaction applies to most thermal processes of gasification, pyrolysis, hydrolysis, and combustion:

$$C_{20}H_{32}O_{10} + x_1O_2 + x_2H_2O_{\Delta} \rightarrow y_1C + y_2CO_2 + y_3CO + y_4H_2 + y_5CH_4 + y_6H_2O + y_7C_nH_m \quad (16.2)$$

The coefficients x and y balance the compounds on either side of the equation.

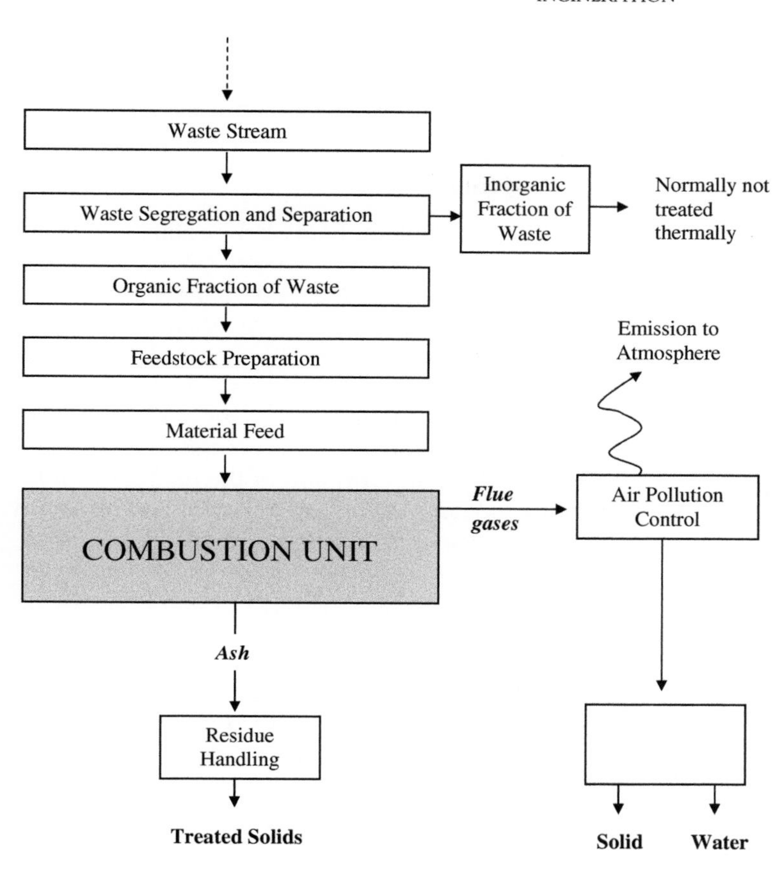

FIGURE 16.1 Steps in the thermal destruction of wastes. *(Adapted from U.S. Environmental Protection Agency, 2003, Remediation Guidance Document, EPA-905-B94-003, Chapter 7).*

Stoichiometrically, x_1 will be much lower than combustion reactions (near 0) for pyrolytic reactions. In many thermal reactions, C_nH_m includes the hydrocarbons, C_2H_2, C_2H_4, C_2H_6, C_3H_8, C_4H_{10}, C_5H_{12}, and benzene, C_6H_6. Of all of the thermal processes, incineration is the most common process for destroying organic contaminants in industrial wastes. Incineration simply is heating wastes in the presence of oxygen to oxidize organic compounds (both toxic and nontoxic). The principal incineration steps are shown in Fig. 16.1.

2. INCINERATION

Incineration is the most common thermal treatment process for organic contaminants in industrial and municipal wastes. Incineration simply is heating wastes in the presence of oxygen to oxidize organic compounds. The principal reactions from test burns for commonly incinerated compounds are provided in Table 16.1. Note that these generally follow the stoichiometry of Equation (16.2), but vary by the chemical composition of the fuel.

Wood is considered to have the composition of $C_{6.9}H_{10.6}O_{3.5}$. Therefore, the combustion consists simply of the oxidation of carbon and hydrogen:

$$C + O_2 \rightarrow CO_2, H_2 + 0.25\ O_2 \rightarrow 0.5\ H_2O$$

There are a number of stages in the incineration process where new compounds may need

TABLE 16.1 Balanced Combustion Reactions for Selected Organic Compounds [1]

Chlorobenzene	$C_6H_5Cl + 7O_2 \rightarrow 6CO_2 + HCl + 2H_2O$
Tetrachloroethene (TCE)	$C_2Cl_4 + O_2 + 2H_2O \rightarrow 2CO_2 + HCl$
Hexachloroethane (HCE)	$C_2Cl_6 + ½O_2 + 3H_2O \rightarrow 2O_2 + 6HCl$
Post-chlorinated polyvinyl chloride (CPVC)	$C_4H_5Cl_3 + 4.5O_2 \rightarrow 4CO_2 + 3HCl + H_2O$
Natural gas fuel (methane)	$CH_4 + 2O_2 \rightarrow CO_2 + 2\ H_2O$
PTFE Teflon	$C_2F_4 + O_2 \rightarrow CO_2 + 4HF$
Butyl rubber	$C_9H_{16} + 13O_2 \rightarrow 9CO_2 + 8H_2O$
Polyethylene	$C_2H_4 + 3O_2 \rightarrow 2CO_2 + 2H_2O$

Source: D.A. Vallero (2008). Fundamentals of Air Pollution. U.S. Environmental Protection Agency/Elsevier Academic Press, Burlington, MA.

to be addressed. Ash and other residues may contain high levels of metals, at least higher than the original feed. The flue gases are likely to include organic and inorganic compounds resulting from temperature-induced volatilization or newly transformed PICs with higher vapor pressures than the original contaminants.

3. TYPES OF THERMAL UNITS

Although they are often called "combustion" units, they are more accurately "thermal destruction" units, as they may or may not have excess oxygen. Thermal units range in size; characteristics; and varying time, temperature, and turbulence, the "3Ts," to achieve optimal destruction of harmful constituents and to reduce the volume of wastes. Five general categories are available to destroy contaminants: (1) rotary kiln, (2) multiple hearth, (3) liquid injection, (4) fluidized bed, and (5) multiple chamber.

It is important to note that unlike many chemical engineering reactor thermal waste treatment systems must handle feedstocks with highly variable loading volumes with widely ranging chemical compositions.

3.1. Rotary Kiln

The combustion chamber in a rotary kiln incinerator (Fig. 16.2) is a heated rotating cylinder mounted at an angle with baffles sometimes added to the inner face to provide the turbulence necessary for the target 3Ts. Engineering design options from specific laboratory results from tests of a specific contaminant, including: (1) angle of the drum, (2) diameter and length of the drum, (3) presence and location of the baffles, (4) rotational speed of the drum, and (5) use of added fuel to increase the temperature of the combustion chamber as the specific contaminant requires. The liquid, sludge, or solid hazardous waste is input into the upper end of the rotating cylinder, rotates with the cylinder-baffle system, and falls with gravity to the lower end of the cylinder. The heated, upward-moving off-gases are collected, monitored for chemical constituents, and subsequently treated as appropriate prior to release, whereas the ash that falls with gravity is collected, monitored for chemical constituents, and also treated as needed before ultimate disposal. The newer rotary kiln systems [2] consist of a primary combustion chamber, a transition volume, and a fired afterburner chamber. After exiting the afterburner, the flue gas is passed through a quench section followed by a primary air pollution control system (APCS). The primary APCS can be a venture scrubber followed by a packed-column scrubber. Downstream of the primary APCS, a backup secondary APCS, with a demister; an activated carbon adsorber; and a high-efficiency particulate air (HEPA) filter can collect contaminants not destroyed by the incineration.

The rotary kiln can be used to incinerate most organic contaminants; it is well suited for solids and sludges, and in special cases, liquids and gases can be injected through auxiliary nozzles

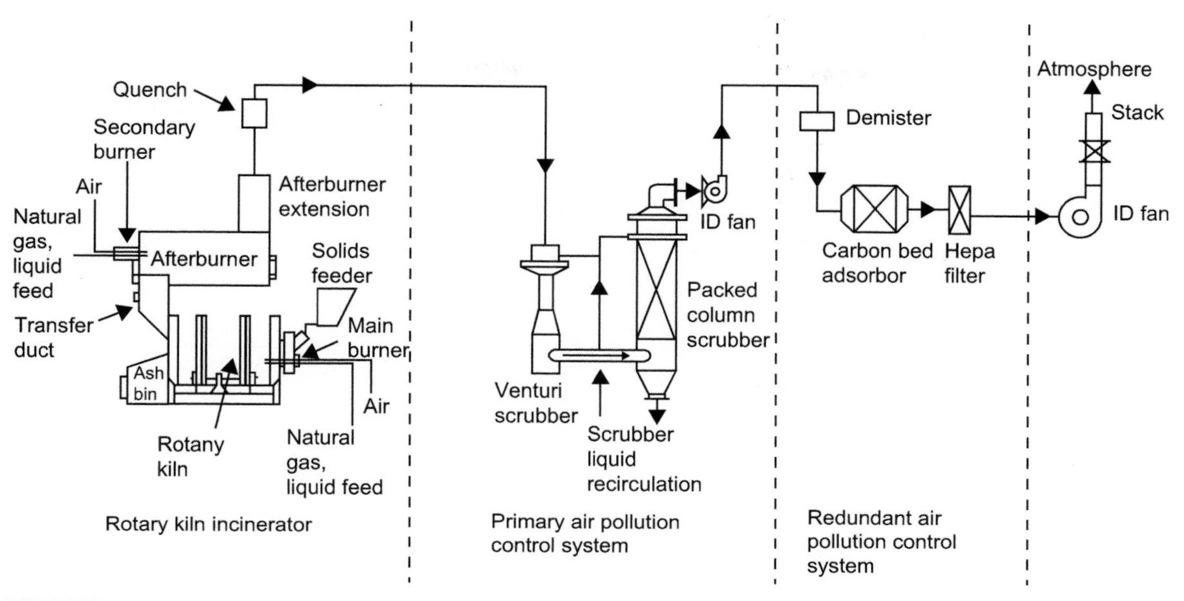

FIGURE 16.2 Rotary kiln system. *(Adapted from U.S. Environmental Protection Agency, 1997. In: J. Lee, D. Fournier Jr., C. King, S. Venkatesh, C. Goldman, Project Summary: Evaluation of Rotary Kiln Incinerator Operation at Low-to-Moderate Temperature Conditions).*

in the side of the combustion chamber. Operating temperatures generally vary from 800 °C to 1650 °C. Engineers use laboratory experiments to design residence times of seconds for gases and minutes or possibly hours for the incineration of solid material.

3.2. Multiple Hearths

Multiple hearths (Fig. 16.3) usually are used for contaminants in solid or sludge form and is fed slowly through the top vertically stacked hearth; in special configurations, hazardous gases and liquids can be injected through side nozzles. They were first used to burn municipal wastewater treatment biosolids (i.e., sludges), relying on gravity and scrapers working the upper edges of each hearth to transport the waste through holes from upper, hotter hearths to lower, cooler hearths. Heated, upward-moving off-gases are collected, monitored for chemical constituents, and treated as appropriate prior to release; the falling ash is collected, monitored for chemical constituents, and subsequently treated prior to ultimate disposal.

Numerous waste types with varying organic content can be incinerated using a multiple-hearth configuration. Operating temperatures generally vary from 300 °C to 980 °C. These systems are designed with residence times of seconds if gases are fed into the chambers to several hours if solid materials are placed on the top hearth and allowed to eventually drop to the bottom hearth exiting as ash.

3.3. Liquid Injection

Nozzles mount horizontally and/or vertically spray liquid hazardous wastes into liquid injection incinerators. That is, the wastes are atomized to match the waste being handled with the combustion chamber as determined in laboratory testing. The application obviously is limited to liquids that do not clog these nozzles, though some success has been

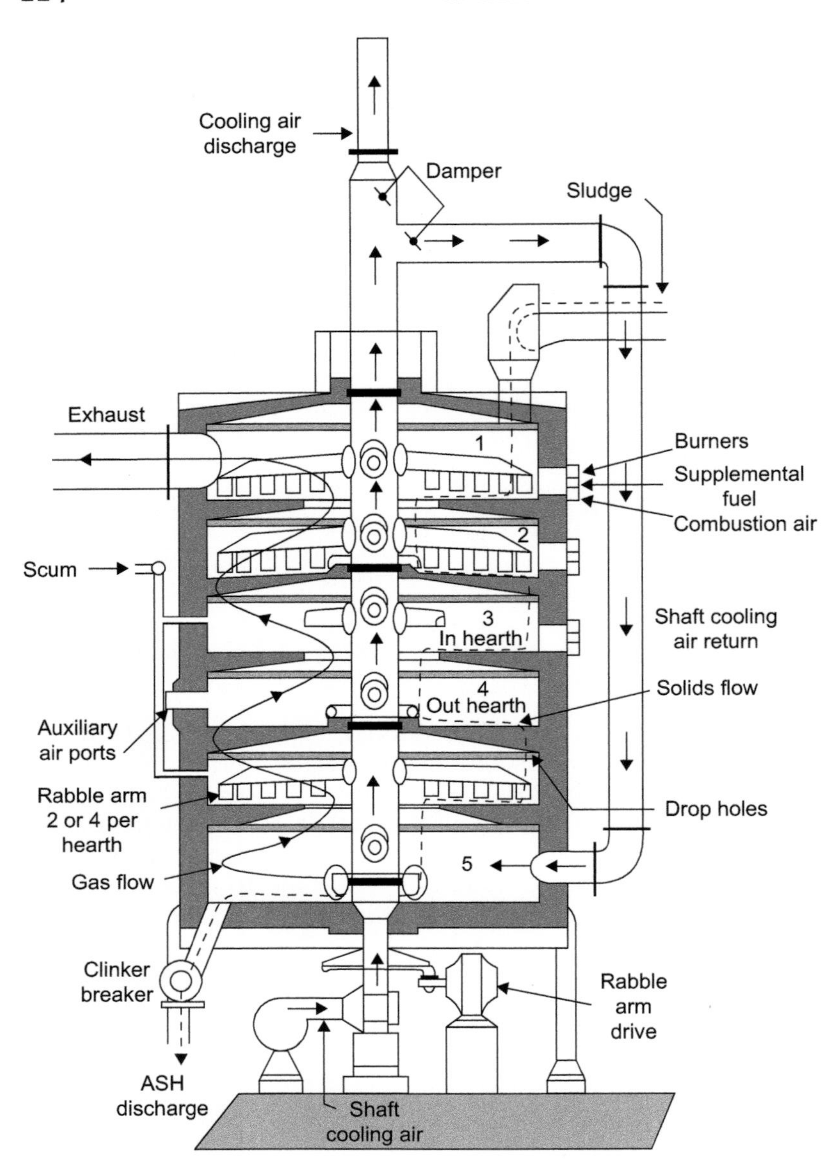

FIGURE 16.3 Multiple hearth incineration system. *Source: U.S. Environmental Protection Agency, 1998, Locating and Estimating Air Emissions from Sources of Benzene, EPA-454/R-98-011, Research Triangle Park, NC.*

experienced with hazardous waste slurries. Operating temperatures generally vary from 650 °C to 1650 °C (1200 °F–3000 °F). Liquid injection systems (Fig. 16.4) are designed with residence times of fractions of seconds as the upward-moving off-gases are collected, monitored for chemical constituents, and treated as appropriate prior to release to the lower troposphere of the atmosphere.

3.4. Fluidized Bed

Contaminated feedstock is injected under pressure into a heated bed of agitated, inert

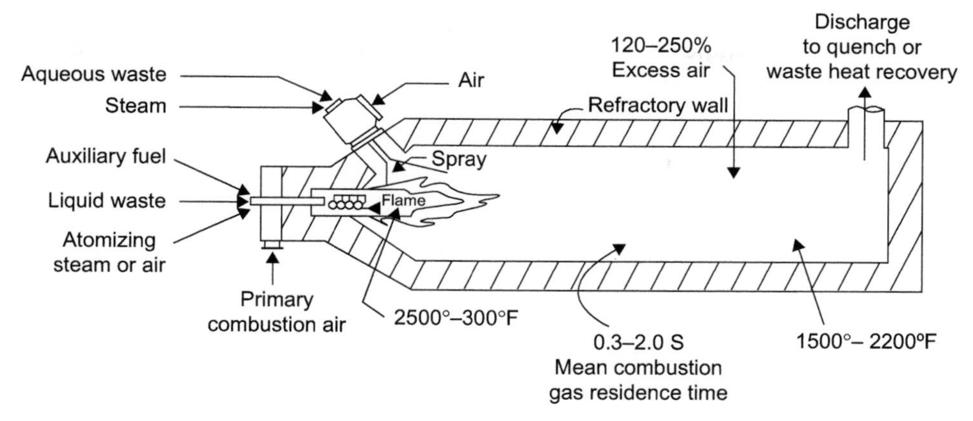

FIGURE 16.4 Prototype of liquid injection system. *Source: US Environmental Protection Agency, Locating and Estimating Air Emissions from Sources of Benzene, EPA-454/R-98-011, Research Triangle Park, NC, 1998.*

granular particles, usually sand, as the heat is transferred from the particles to the waste, and the combustion process proceeds as shown in Fig. 16.5. External heat is applied to the bed prior to the injection of the waste. Heat is also continually applied throughout the combustion operation as the situation dictates. Heated air is forced into the bottom of the particle bed and the particles become suspended among themselves during this continuous fluidizing process. The openings created within the bed permit the introduction and transport of the waste into and through the bed. The process enables the contaminant to come into contact with particles that maintain their heat better than, for example, the gases inside a rotary kiln. The heat maintained in the particles increases the time the contaminant is in contact with a heated element, and thus, the combustion process could become more complete with

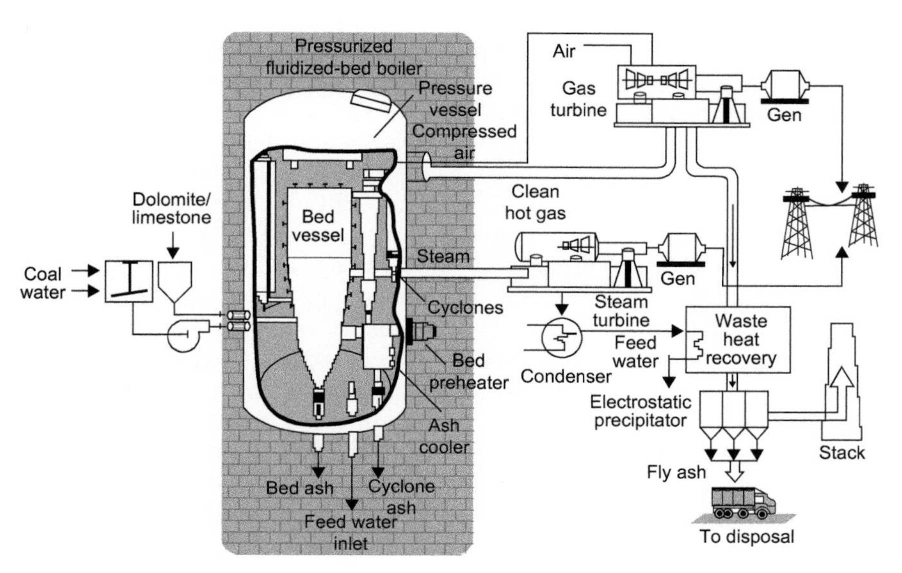

FIGURE 16.5 Pressurized fluidized bed system. *Source: U.S. Department of Energy, 1999, TIDD PFBC Demonstration Project.*

fewer harmful byproducts. Off-gases are collected, monitored for chemical constituents, treated as appropriate prior to release, and the falling ash is collected, monitored for chemical constituents, and subsequently treated prior to ultimate disposal.

Most organic wastes can be incinerated in a fluidized bed, although the system is best suited for liquids. Operating temperatures generally vary from 750 °C to 900 °C. Liquid injection systems are designed with residence times of fractions of seconds as the upward-moving off-gases are collected, monitored for chemical constituents, and treated as appropriate prior to release to the lower troposphere.

3.5. Multiple Chambers

Wastes are first vaporized on a grate in the ignition chamber of a multiple chamber system. The gases created in this ignition chamber travel through baffles to a secondary chamber where the actual combustion process takes place. Often the secondary chamber is located above the ignition chamber to promote natural advection of the hot gases through the system. Heat may be added to the system in either the ignition chamber or the secondary chamber as required for specific burns.

The application of multiple chamber incinerators is mainly restricted to solid wastes, wherein the waste enters the ignition chamber through an opened charging door in batch, not continuous, loading. Combustion temperatures are typically near 540 °C for most applications. Multiple chamber units normally are designed with residence times of minutes to hours for solid hazardous wastes as off-gases are collected, monitored for chemical constituents, and treated as appropriate prior to release to the lower troposphere. At the end of each burn period, the system must be cooled so that the ash can be removed prior to monitoring for chemical constituents and subsequent treatment prior to ultimate disposal.

Incineration alone does not "destroy" some wastes. Some elements, like the metals, are actually concentrated in the ash and other incineration residues. Thermal treatment changes the oxidation state (valence) of the metals. In fact, incineration can actually replace one problem with another, such as increasing the potential movement of metals via oxidation and allowing the metals to migrate into the environment. Conversely, processes like slagging (operating at sufficiently high temperatures to melt and remove incombustible materials) or vitrification (producing nonleachable, basalt-like residue) can substantially reduce the mobility of many metals.

4. THERMAL OPTIMIZATION

The chemical and physical characteristics of the waste determine how a waste is to be thermally treated. The key is to make the waste less harmful. This often means making a waste less mobile and less toxic.

4.1. Leachability

The measure of the ease with which compounds in the waste can move into the accessible environment is known as leachability. The increased leachability of metals would be problematic if the ash and other residues are to be buried in landfills or stored in piles. The leachability of metals is generally measured using the toxicity characteristic leaching procedure (TCLP) test, discussed earlier. Incinerator ash that fails the TCLP must be disposed of in a waste facility approved for hazardous wastes. Enhanced leachability would be advantageous only if the residues are engineered to undergo an additional treatment step of metals. Again, the engineer must see incineration as one component within a systematic approach for any contaminant treatment process.

New compounds can be created at various steps as the contaminant moves through the incineration process. As mentioned, ash and other residues may contain high levels of metals, at least higher than the original feed. The flue gases are likely to include both organic and inorganic compounds that have been released as a result of temperature-induced volatilization and/or newly transformed PICs with higher vapor pressures than the original contaminants.

Based on these underlying principles, seven general incineration guidelines can be highlighted:

(1) Liquid phase, nearly pure organic contaminants are best candidates for combustion;
(2) Halogen-containing organic materials deserve special consideration if in fact they are to be incinerated at all, that is, special materials used in the construction of the incinerator, long (many seconds) combustion time, and high temperatures (>1600 °C), with continuous mixing if the contaminant is in the solid or sludge form;
(3) Feedstock containing heavy metals generally should not be incinerated;
(4) Sulfur-containing organic material will emit sulfur oxides that must be controlled;
(5) The formation of nitrogen oxides can be minimized if the combustion chamber is maintained above 1100 °C;
(6) Destruction depends on the interaction of a combustion chamber's temperature, dwell time, and turbulence; and
(7) Off-gases and ash must be monitored for chemical constituents; each residual must be treated as appropriate so the entire combustion system operates within the requirements of the local, state, and federal environmental regulators; and hazardous components of the off-gases, off-gas treatment [3] processes, and the ash must reach ultimate disposal in a permitted facility.

4.2. Destruction and Removal Efficiency

In the United States, the Emergency Planning and Community Right-To-Know Act (EPCRA) requires that certain emergency planning activities be conducted for extremely hazardous substances (EHSs) in quantities at or above the threshold planning quantity (TPQ). The same chemicals may also have to meet requirements under the Comprehensive Environmental Response, Compensation, and Liability Act (CERCLA) of the Clean Air Act. In fact, the Environmental Protection Agency (EPA) updates the so-called "list of lists" for these substances [4].

In the United States, any incineration of wastes containing hazardous substances requires a permit. The owner or operator of an incinerator is only allowed to burn wastes as specified in the permit and only under operating conditions specified for those wastes. Incinerating new wastes may have to be based on either trial burn results or alternative data to establish appropriate conditions for each of the applicable requirements related to allowable waste feedstocks and operating conditions [5]. In particular, an incinerator burning a substance that contains principal organic hazardous constituents (POHCs) must be designed, constructed, and maintained so that it will achieve a destruction and removal efficiency (DRE) of 99.99% for each POHC designated in its permit for each waste feed. For each such waste feed, the permit will specify acceptable operating limits, including the following conditions:

(1) CO level in the stack exhaust gas;
(2) Waste feed rate;
(3) Combustion temperature;
(4) An appropriate indicator of combustion gas velocity;
(5) Allowable variations in incinerator system design or operating procedures;
(6) Operating requirements as are necessary to ensure compliance with the hazardous

waste incineration performance standards (Section 264.343 of EPA regulations);

(7) During start up and shut down of an incinerator, hazardous waste (except wastes exempted in accordance with Section 264.340 of EPA regulations) must not be fed into the incinerator unless the incinerator is operating within the conditions of operation (temperature, air feed rate, etc.) specified in the permit;

(8) Fugitive emissions from the combustion zone must be controlled by the following:
  (a) Keeping the combustion zone totally sealed against fugitive emissions,
  (b) Maintaining a combustion zone pressure lower than atmospheric pressure, or
  (c) An alternate means of control demonstrated (with Part B of the permit application) to provide fugitive emissions control equivalent to maintenance of combustion zone pressure lower than atmospheric pressure;

(9) An incinerator must be operated with a functioning system to automatically cut off waste feed to the incinerator when operating conditions deviate from limits established in 8(a) of Section 264.343

(10) An incinerator must cease operation when changes in waste feed, incinerator design, or operating conditions exceed limits designated in its permit.

The DRE is calculated as follows:

$$DRE = \frac{W_{in} - W_{out}}{W_{in}} \times 100 \qquad (16.3)$$

where W_{in} is the rate of mass of waste flowing into the incinerator and W_{out} is rate of mass of waste flowing out of the incinerator.

Extremely hazardous waste constituents require at least 99.9999% DRE.

For example, is an incinerator meeting the federal DRE if, during a stack test, the mass of Chemical X (POHC and EHS) that is loaded into the incinerator is measured at $W_{in} = 10\ \text{mg min}^{-1}$ and the mass flow rate of the compound measured downstream in the stack is $W_{out} = 200\ \text{pg min}^{-1}$ (where p refers to pico, 10^{-12})?

$$DRE = \frac{10\ \text{mg min}^{-1} - 200 \times 10^{-9}\ \text{mg min}^{-1}}{10\ \text{mg min}^{-1}} \times 100$$

$$= 999999.98\%\ \text{removal}$$

Even though Chemical X is considered to be "extremely hazardous," this DRE is better than the "rule of six nines," so the incinerator is operating up to code.

If during the same stack test the mass of Chemical Y loaded into incinerator was measured at the rate of 100 L min^{-1} and the mass flow rate of the compound measured downstream was 1 ml min^{-1}, is the incinerator up to code for this POHC? This is a lower removal rate as 100 L are in and 0.001 are leaving, so the DRE = 99.999. This is acceptable, that is, better removal efficiency than 99.99% by an order of magnitude, so long as Chemical Y is not considered an extremely hazardous compound. If it were, then the incinerator would not have met the rule of six nines (it only has five).

From a qualitative perspective, if either of these compounds is chlorinated or otherwise halogenated (e.g. Br and Fl substitutions), special precautions must be taken when dealing with such halogenated compounds, as more toxic compounds than those being treated can end up being generated. Incomplete reactions are very important sources of environmental contaminants. As mentioned, these reactions generate PICs, such as dioxins, furans, CO, PAHs, and hexachlorobenzene.

5. OTHER THERMAL TECHNOLOGIES

High-temperature incineration may not be needed to treat many wastes, including most volatile organic compounds (VOCs). Also, in soils with heavy metals, the high temperatures will lead to an increase in the vapor pressure of metals, so they will find their way into the combustion flue gas. High concentrations of volatile trace-metal compounds in the flue gas poses complicate pollution control strategies.

When successful in decontaminating soils to the necessary treatment levels, thermally desorbing contaminants from substrates has the benefits over incineration, including lower fuel consumption, no formation of slag, less volatilization of metal compounds, and less complicated air pollution control demands. Beyond monetary costs and ease of operation, a less energy (heat) intensive system can be more advantageous in terms of actual pollutant removal efficiency.

5.1. Pyrolysis

Chemical decomposition induced in organic materials by heat in the absence of oxygen is known as pyrolysis, that is, it is the thermal conversion of organic matter in the total absence of oxygen at relatively low temperatures of between 500 °C and 800 °C and short vapor residence times of 3 to 1500 s. It is practically impossible to achieve a completely oxygen-free atmosphere, so pyrolytic systems run with less than stoichiometric quantities of oxygen. Because some oxygen will be present in any pyrolytic system, there will always be a small amount of oxidation. Also, desorption will occur when volatile (i.e., vapor pressure $> 10^{-2}$ kPa) or semivolatile (i.e., vapor pressure between 10^{-2} and 10^{-5} kPa) compounds are present in the feed.

During pyrolysis [6], organic compounds are converted to gaseous components, along with some liquids, as coke, that is, the solid residue of fixed carbon and ash. CO, H_2, and CH_4 and other hydrocarbons are also produced during pyrolysis. If these gases cool and condense, liquids will form and leave oily tar residues and water with high concentrations of total organic carbon (TOC). The secondary gases need their own treatment, such as by a secondary combustion chamber, by flaring, and partial condensation. Particulates must be removed by additional air pollution controls, for example, fabric filters or wet scrubbers.

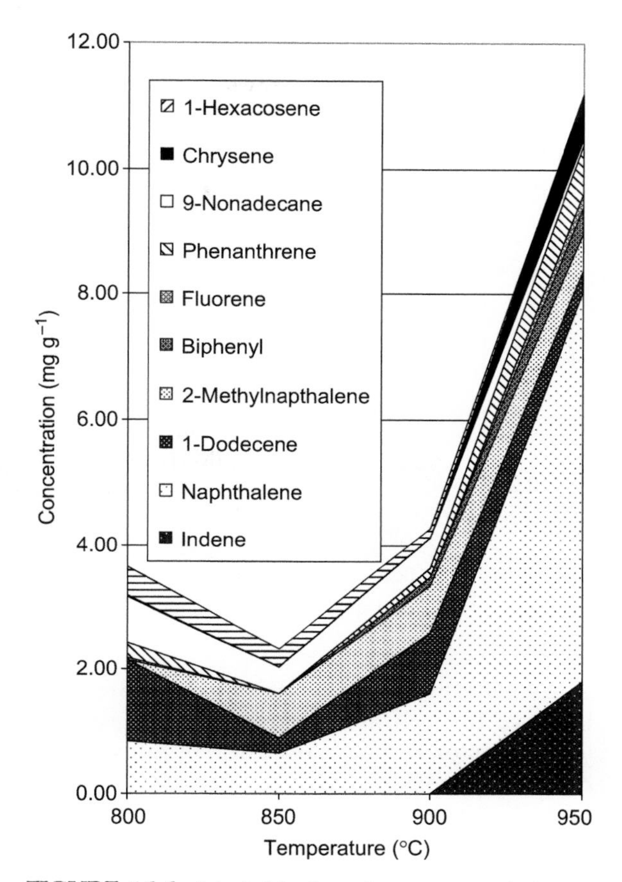

FIGURE 16.6 Selected hydrocarbon compounds generated in a low-density polyethylene fire (pyrolysis) in four temperature regions. *Source: D.A. Vallero (2007). Biomedical Ethics for Engineers: Ethics and Decision Making in Biomedical and Biosystems Engineer, Chapter 9, Elsevier Academic Press, Amsterdam, NV; data from R.A. Hawley-Fedder, M.L. Parsons, and F.W. Karasek, 1984, "Products Obtained During Combustion of Polymers under Simulated Incinerator Conditions," Journal of Chromatography, 314, 263-272.*

Many of the same conventional thermal treatment methods used for incineration, such

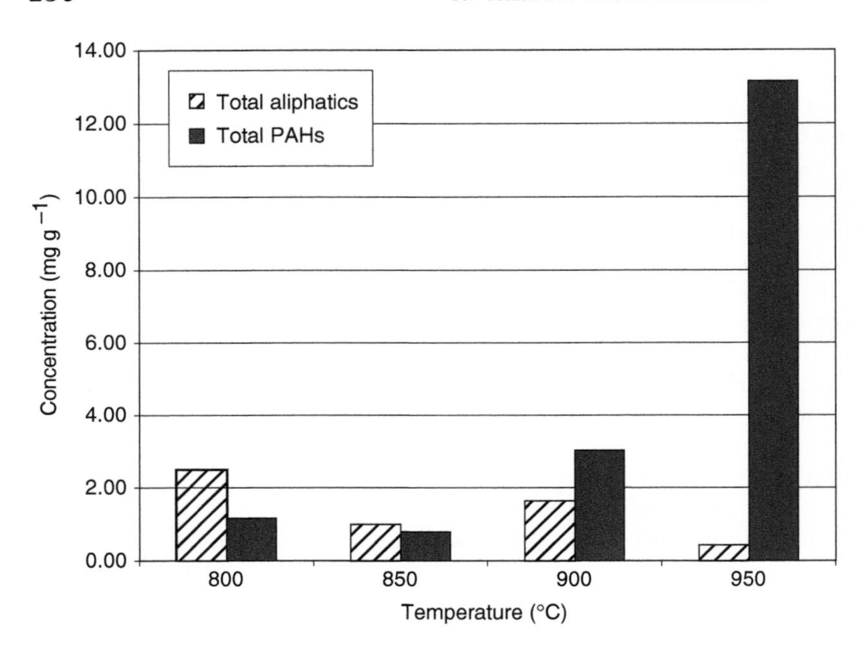

FIGURE 16.7 Total aliphatic (chain) hydrocarbons versus polycyclic aromatic hydrocarbons (PAHs) generated in a low-density polyethylene fire (pyrolysis) in four temperature regions. *Source: D.A. Vallero (2007). Biomedical Ethics for Engineers: Ethics and Decision Making in Biomedical and Biosystems Engineer, Chapter 9, Elsevier Academic Press, Amsterdam, NV; data from R.A. Hawley-Fedder, M.L. Parsons, and F.W. Karasek, 1984, "Products Obtained during Combustion of Polymers under Simulated Incinerator Conditions," Journal of Chromatography, 314, 263-272.*

as rotary kiln, rotary hearth furnace, or fluidized bed furnace, are also used for waste pyrolysis. Kilns or furnaces used for pyrolysis may be of the same design as those used for combustion discussed earlier but operate at lower temperatures and with less air than in combustion.

Pyrolysis allows for separating organic contaminants from various wastes and may be used to treat a variety of organic contaminants that chemically decompose when heated (i.e., "cracking"). Pyrolysis is not effective in either destroying or physically separating inorganic compounds that coexist with the organics in the contaminated medium.

5.2. Specialized Thermal Systems

Other promising thermal processes include high-pressure oxidation and vitrification [7]. High-pressure oxidation combines two related technologies, that is, wet air oxidation and supercritical water oxidation, which combine high temperature and pressure to destroy organics. Wet air oxidation can operate at pressures of about 10% of those used during supercritical water oxidation, an emerging technology that has shown some promise in the treatment of polychlorinated biphenyls (PCBs) and other stable compounds that resist chemical reaction. Wet air oxidation has generally been limited to conditioning of municipal wastewater sludges but can degrade hydrocarbons (including PAHs), certain pesticides, phenolic compounds, cyanides, and other organic compounds. Oxidation may benefit from catalysts.

Vitrification uses electricity to heat and destroy organic compounds and immobilize inert contaminants. A vitrification unit has a reaction chamber divided into two sections: the upper section to introduce the feed material containing gases and pyrolysis products, and the lower section consisting of a two-layer molten zone for the metal and siliceous components of the waste. Electrodes are inserted into the waste solids and graphite is applied to the surface to enhance its electrical conductivity. A large current is applied, resulting in rapid heating of the solids and causing the siliceous components of the material to melt as temperatures reach about 1600 °C. The end product is a solid, glass-like material that is very resistant to leaching.

The next chapter stresses the potential value of thermochemical reactions in handling plastic wastes, and caution is in order. Thermal processing of plastics must be managed very carefully to prevent the release of highly toxic compounds. As those who have had to fight fires in municipal landfills know all too well, plastic fires can produce very noxious and toxic compounds. In fact, burning plastics can release over 450 different organic compounds [8]. The role of oxygen, that is, the relative amount of combustion and pyrolysis, is a major determinant as to the actual amounts and types of compounds released. Temperature is also important, but a direct relationship between temperature and pollutants released has not been established. For example, Fig. 16.6 shows that in low-density polyethylene pyrolysis of plastics, some compounds are generated at lower temperatures, whereas for others the optimal range is at higher temperatures. However, the aliphatic compounds (i.e., 1-dodecene, 9-nonadecane, and 1-hexacosene) are generated in higher concentrations at lower temperatures (about 800 °C), whereas the aromatics need higher temperatures (see Fig. 16.7).

6. CONCLUSIONS

This chapter has addressed the benefits and drawbacks of various thermal destruction and removal technologies. Recalcitrant waste constituents can be degraded using these techniques but caution should always be applied as improper applications of thermal technologies can lead to harm caused by PICs and other pollutants.

References

[1] D.A. Vallero, Fundamentals of Air Pollution, U.S. Environmental Protection Agency/Elsevier Academic Press, Burlington, MA, 2008.

[2] U.S. Environmental Protection Agency, in: J. Lee, D. Fournier Jr., C. King, S. Venkatesh, C. Goldman (Eds.), Project Summary: Evaluation of Rotary Kiln Incinerator Operation at Low-to-Moderate Temperature Conditions, EPA/600/SR-96/105, U.S. Environmental Protection Agency, Cincinnati, OH, 1997.

[3] Title 40: Protection of Environment. Part 262: Standards for Owners and Operators of Hazardous Waste Treatment, Storage, and Disposal Facilities. [46 FR 7678, Jan. 23, 1981, as amended at 47 FR 27533, June 24, 1982; 50 FR 4514, Jan. 31, 1985; 71 FR 16907, Apr. 4, 2006].

[4] U.S. Environmental Protection Agency, List of Lists: Consolidated List of Chemicals Subject to the Emergency Planning and Community Right-to-Know Act (EPCRA), Comprehensive Environmental Response, Compensation and Liability Act (CERCLA) and Section 112(r) of the Clean Air Act. EPA 550-B-10–001. <http://www.epa.gov/emergencies/docs/chem/list_of_lists_05_07_10.pdf>. Accessed June 7, 2010. (2010).

[5] Code of Federal Regulations, *Part 40:* Protection of the Environment. Chapter I (7–1–05 Edition) section 264.344 of EPA regulations.

[6] Federal Remediation Technologies Roundtable, Remediation Technologies Screening Matrix and Reference Guide. fourth ed. (2002).

[7] A Principal Source for All of the Thermal Discussions is the U.S. Environmental Protection Agency. Remediation Guidance Document, EPA-905-B94-003 Chapter 7. Washington, DC (2003).

[8] M. Paabo, B.C. Levin, A literature review of the chemical nature and toxicity of the decomposition products of polyethylenes, Fire and Materials 11 (1987) 55–70.

CHAPTER

17

Thermochemical Treatment of Plastic Solid Waste

Paola Lettieri, Sultan M. Al-Salem

Department of Chemical Engineering, University College London, Torrington Place, London WC1E 7JE, United Kingdom

OUTLINE

1. INTRODUCTION

From 1995 to 2000, the world's urban population grew at a rate of 2.2% per year, and by 2000, 75% of the population in the developed world lived in urban areas. This latter figure is projected to rise to nearly 83% by 2030, whereas in the developing world, the rate of urbanisation is even faster [1]. Along with this comes increased demand for energy and natural resources, fuelled by increasing consumption levels per capita in rich countries and rapid rise in consumption in developing countries, in particular China and India. If we are to meet this increased demand and, at the same time, stabilise climate change, then by the year 2050 we must triple the planet's current energy-producing capacity, with all new additions being carbon neutral.

If we are to satisfy our basic needs and 'enjoy a better quality of life without compromising the quality of life of future generations' [2], we must,

Waste Doi: 10.1016/B978-0-12-381475-3.10017-8

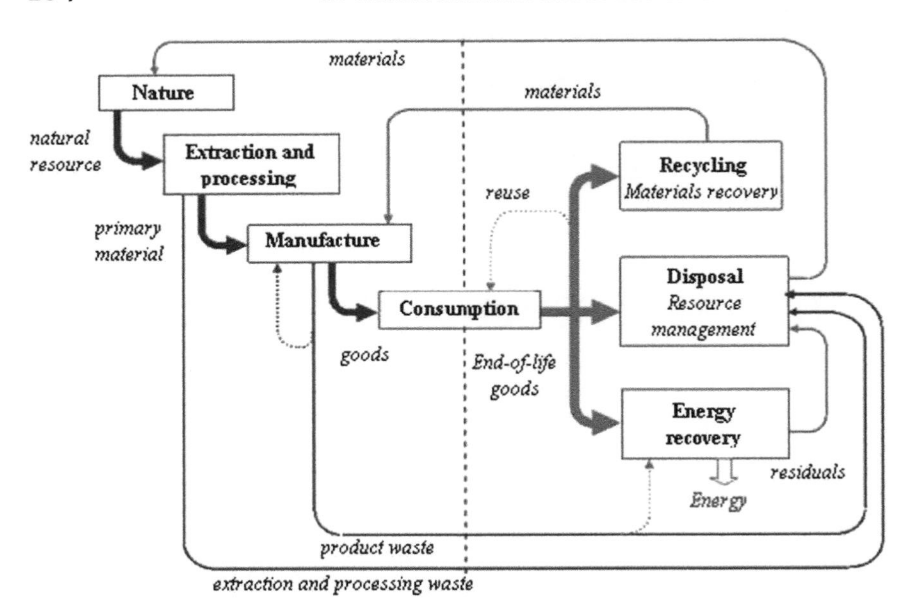

FIGURE 17.1 The resource cycle [1].

among other things, improve our resource efficiency and reduce our impacts on the environment. This involves getting the most out of our finite resources and minimising waste. Ultimately, we need to shift processes from linear and 'open loop' systems, where natural resources and capital investments move through the system to become waste, to 'closed loop' systems, where waste outputs from one process may be used as resource inputs to another and where any residual wastes are returned to the environment in a way that enables them to be extracted and used again. This 'resource cycle' is depicted in Fig. 17.1.

1.1. Plastic Solid Waste, Generation and Trends

The amount of municipal waste produced on average by each European citizen is projected to increase from 520 kg in 2004 to 680 kg by 2020 [3]), an increase of 25%. Furthermore, waste production in the new European Union 12 (EU12) countries is projected to reach the current levels of those in the EU15 countries by 2020.

Plastics contribute to our daily life functions in many aspects. Household goods nowadays are mainly composed of plastic or plastic-reinforced materials, from packaging, clothing, appliances and electrical and vehicle equipments, to insulations, industrial applications, greenhouses, automotive parts, aerospace and mulches. Thus, plastic materials represent an important component of the municipal solid waste produced. Production of plastics has reached global maximum capacities levelling at 260 million tonnes (260×10^6 t) in 2007, whereas in 1990, the global production capacity was estimated at 80×10^6 t. It is estimated that production of plastics worldwide is growing at a rate of about 5% per year [4].

Over the past 70 years, the plastic industry has witnessed a drastic growth in the production of synthetic polymers represented by polyethylene (PE), polypropylene (PP), polystyrene (PS), polyethylene terephthalate (PET), polyvinyl alcohol (PVA) and polyvinyl chloride (PVC). In 1996, the total plastics consumption in Western Europe was estimated at 33.4 × 10^6 t, whereas in 2004, it was estimated at 48.3 × 10^6 t, increasing by an average of 4% per year. This brings a high estimate of almost 60% of plastic solid waste (PSW) being discarded in open space or landfilled worldwide [5]. In the United Kingdom, average consumption of plastic is currently in excess of 5 million tonnes per year (5 × 10^6 t a^{-1}), with a distribution by market sector and plastic type as shown in Fig. 17.2 [6].

Traditionally, landfilling has been used as the major waste management method. However, European and national waste management policies have successfully resulted in an increase in recycling and incineration and a subsequent decrease in the amount of waste sent to landfill. Examples of legislations include the EU Packaging Directive (94/62/EC), introduced in 1994, aimed at increasing recycling and recovery of packaging, and the Landfill Directive (99/31/EC), introduced in 1999, aimed at diverting biodegradable municipal waste from landfill. These targets were subsequently revised by the EU, following which the U.K. government published the national packaging recycling and recovery targets for 2006 and beyond which required 23% of plastics waste to be recovered by 2006, rising to 25.5% by 2010.

Predictions for municipal waste management in the EU project an overall increase in recycling of up to 34%, an increase in energy recovery of up to 27% and a reduction in landfill disposal of up to 34% by 2020 [3]. While the increased waste quantities are expected to increase direct emissions, a net positive effect will result from increasing recycling and energy recovery which will represent savings on greenhouse gas emissions avoided, where recycling is expected to contribute to 75% of total avoided emissions by 2020.

1.2. Chapter Outline

The aim of this chapter is to provide a brief introductory guide to the challenge of managing PSW to its use as a material and an energy resource. The chapter will review briefly the current situation in the UK, EU and worldwide in relation to solid waste arising and strategies

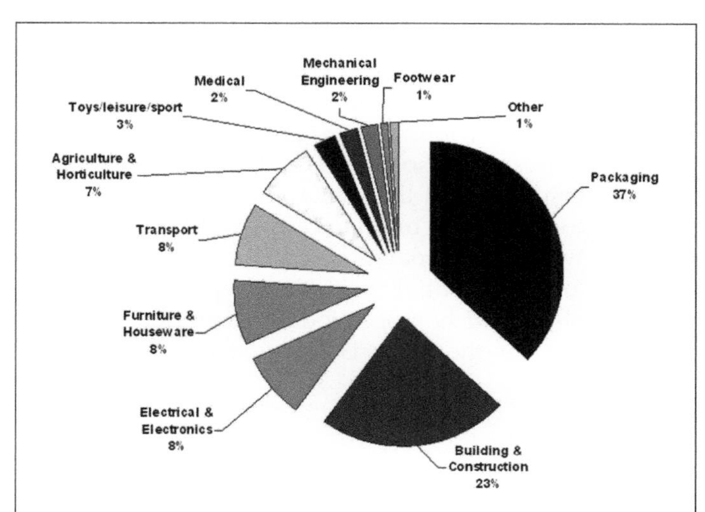

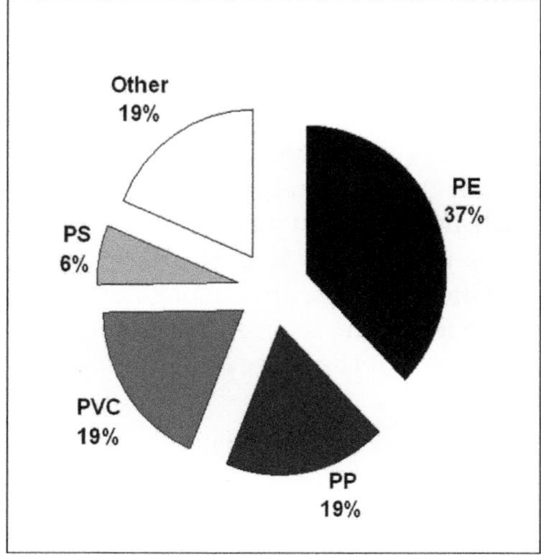

FIGURE 17.2 U.K. average plastic consumption by market sector [6] (left); U.K. average plastic consumption by polymer type. Other include PC, PBT, PA, PMMA, POM, PUR, PC/ABS, UPR) [6] (right).

to address waste management. It will briefly introduce the different routes for PSW management, namely, re-extrusion, mechanical recycling and energy recovery. Petrochemical recovery via thermochemical treatment will be discussed and in particular some of the thermal treatment options available, namely, gasification, hydrogenation and pyrolysis, will be presented. Finally, a case study based on pyrolysis will be briefly discussed.

2. TECHNOLOGIES FOR PSW MANAGEMENT

Increasing recovery of waste and diverting waste away from landfill play a key role in tackling the environmental impacts of increasing waste volumes. PSW not only represents a significant proportion of the total waste produced but also has the highest potential for energy recovery, having the highest calorific value when compared with other waste fractions, see Table 17.1. Energy can be recovered as electricity and/or thermal heat via either incineration or gasification.

Reuse and recycling of plastics have also become a key component of the material life cycle, as shown in Fig. 17.3.

PSW recycling processes can be classified in four major categories [7], re-extrusion (primary), mechanical (secondary), chemical (tertiary) and energy recovery (quaternary). Each method provides a unique set of advantages that makes it particularly beneficial for specific locations, applications or requirements. In the following sections, we briefly review from primary to tertiary recycling processes.

2.1. Re-extrusion and Mechanical Recycling

Re-extrusion, or primary recycling, is the reintroduction of scrap, industrial or single-polymer plastic edges and parts to the extrusion cycle to produce products of the similar material. This process utilizes semi-clean scrap plastics that have similar features to the original products [9]. Currently, most of the recycled PSW is process scrap from industry which is recycled via primary recycling techniques. In the UK, process scrap represents 0.25×10^6 t of the plastic waste and approximately 95% of it is primary recycled [9]. Primary recycling can also involve the re-extrusion of post-consumer plastics; generally, households are the main source of such waste stream. This type of recycling however is not the most cost effective as it involves collecting relatively small quantities of mixed PSW from a large number of sources.

Mechanical recycling (i.e. secondary or material recycling) is the process in which PSW is used in the manufacturing of plastic products via mechanical means, using recyclates, fillers and/or virgin polymers [7,10]. Mechanical recycling of PSW can only be performed on single-polymer plastic. Examples of products found in our daily lives that come from mechanical recycling processes are grocery bags, pipes, gutters, window and door profiles, shutters, blinds and so on. The more complex and contaminated the waste, the more difficult it is to recycle it mechanically [11,12]. Separation, washing, preparation and granulation of PSW are all essential to produce high quality, clear, clean and homogeneous end products [12,13]. One of the main issues that face mechanical

TABLE 17.1 Calorific Value of Waste Fractions [3]

	Food	Garden	Paper	Wood	Textile	Plastics	Inert
Calorific value ($GJ\ t^{-1}$)	2	5	15	15	16	30	0

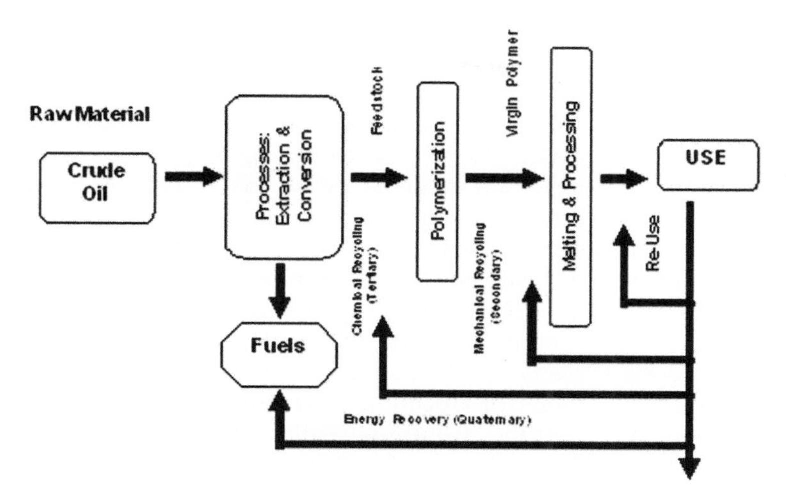

FIGURE 17.3 Treatment methods related to the production cycle of polymers [8].

recyclers is the degradation and heterogeneity of PSW.

2.2. Petrochemicals Recovery via Thermochemical Treatment

Chemical recycling (tertiary) refers to advanced technology processes that convert polymers to either monomers (monomer recycling) or useful petrochemicals (feedstock recycling) which are suitable for use as a feedstock or fuel [7], replacing traditional refinery cuts as end products. The term chemical is used because an alteration is bound to occur to the chemical structure of the polymer. The main advantage of chemical recycling is the possibility of treating heterogeneous and contaminated polymers with limited use of pre-treatment. PE, for example, can be used as a potential source for the production of fuels (gasoline) and synthetic lubricants. Other materials such as PET and nylon can be used instead to produce monomer units [13], whereas vinyl polymers such as polyolefins can produce a mixture containing numerous components for use as a fuel.

Advanced thermochemical treatments of PSW, such as thermolysis, provide a high product yield and minimum waste. Such thermal treatments enable the recovery of monomer fractions up to 60% [14] and produce valuable petrochemicals, including gases (rich with low-cut refinery products and hydrocarbons), tars (waxes and liquids very high in aromatic content) and char (carbon black and/or activated carbon) [15]. Thermolysis processes can be divided into gasification (in the sub-stoichiometric presence of air, usually leading to CO, CO_2 and H_2 production), hydrogenation (or hydrocracking) and advanced thermochemical or pyrolysis (thermal cracking in an inert atmosphere) [16].

2.2.1. Gasification and Hydrogenation

Gasification is the thermal conversion of organic matter by partial oxidation into a gaseous product called syngas. This consists mainly of H_2, CO and small amounts of methane (CH_4), water vapour, CO_2, N_2 and tar. The oxidant used for the gasification process can be air, pure oxygen, steam or a mixture of these gases. Air-based gasifiers typically produce a product gas containing a relatively high concentration of nitrogen with a low heating value. Oxygen and steam-based gasifiers produce a product gas containing a relatively high concentration of hydrogen and carbon monoxide, with a higher heating value [17]. Several types of gasification processes have already been developed and reported. Their practical performance data, however, have not necessarily been satisfactory for universal application. A significant amount of char is always produced in gasification which needs to be further processed and/or burnt. An ideal gasification process for PSW should produce a gas with high calorific value, completely combusted char, produce an easy metal product to separate ash from and should not require any additional installations for air/water pollution abatement. Examples of gasification studies of solid waste have been reported in the literature since the early 1970s [18,19], with more recent work on the gasification of plastic and in particular PVC [20], PP [21] and PET [22].

Hydrogenation by definition means the addition of hydrogen by chemical reaction through unit operation. Many technologies employing PSW hydrogenation have failed or disintegrated in pilot stages. Examples of these are the Entsorgungs (RWE) process, which employed hydrogenation after depolymerisation of plastic waste, and the Hiedrierwerke and Freiberg hydrogenation processes, both terminated for financial reasons. These processes employed a hydrocracking reactor above 400 °C to produce rich oils [23]. The Veba process is the main technology applied in PSW recycling via hydrogenation. The treatment technology employs a depolymerization section, where the agglomerated plastic waste is kept between 350 and 400 °C to effect depolymerisation and dechlorination (in the case of PVC rich waste). The main outputs of the process are (i) hydrochloric acid, (ii) syncrude from the Veba-Combi-Cracking (VCC) section (chlorine free), (iii) hydrogenated solid residue and (iv) off-gas.

2.2.2. *Pyrolysis*

Pyrolysis is the thermal conversion of organic matter in the total absence of oxygen at relatively low temperatures of between 500 and 800 °C and short vapour residence times of 3 to 1500 s. It produces a liquid fuel, a solid char and some combustible gas, which is usually used within the process to provide heat requirements or utilized offsite in other thermal processes. The liquid fuel, or bio-oil, can be used directly as a substitute for fuel oil in heat and power applications or to produce a wide range of speciality and commodity chemicals. Pyrolysis enables the treatment of many different solid hydrocarbon-based wastes while producing a clean fuel gas with a high calorific value, typically 22 to 30 MJ m^{-3} depending on the waste material being processed. Gases can be produced with higher calorific values when the waste contains significant quantities of synthetic materials such as rubber and plastics. Solid char is also produced from the process, which contains both carbon and the mineral content of the original feed material.

A number of studies have been carried out on polyolefins thermal cracking in inert (pyrolysis) and/or partially oxidized atmospheres (e.g. step pyrolysis, gasification). Thermogravimetry is the most commonly used technique for the determination of kinetic parameters, although the experimental conditions utilized are very different, involving broad ranges of temperature, sample amount, heating rates (in the case of dynamic runs), reaction atmospheres and pressures [24]. In this chapter, a novel approach for the analysis of depolymerization reactions of high-density polyethylene (HDPE) during isothermal pyrolysis is briefly presented.

2.2.3. *Case Study: Analysis of the Kinetics of Polymers in Isothermal Pyrolysis*

One of the main materials present in significant quantities in MSW streams is PE. Numerous kinetic studies have been conducted in the past, focusing on reaction mechanism but rarely considering product formation and interaction analysis. In this section, we present a novel kinetic approach of the thermal cracking mechanism of HDPE. The product formation analysis was considered from an engineering perspective using lumped product analysis, and kinetics parameters were evaluated accordingly. This case study considers pure HDPE (T_m = 133 °C) Belgian commercial grade supplied by Ravago Plastics (Arendonk, Belgium). Pyrolysis products were lumped into gases (C_1–C_4), liquids (non-aromatic C_5–C_{10}), *single ring* aromatics (C_5–C_{10}) and waxes (> C_{11}). Figure 17.4 illustrates the product distribution of HDPE thermal degradation as a function of operating temperature (see Ref. [4] for the full details). With increasing temperature, both waxes and liquids were observed to decrease; the fractions of gases and aromatics increased instead.

Products at 500, 550 and 600 °C were used to derive the thermal cracking model of polymer degradation fraction (x_p) with time. The thermal degradation scheme adopted to describe the reactions taking place is illustrated below:

$$(P) \xrightarrow{k_1} (W) \begin{bmatrix} \xrightarrow{k_2} (G) \\ \xrightarrow{k_3} (L) \\ \xrightarrow{k_4} (A) \end{bmatrix} (L) \xrightarrow{k_5} G \qquad (17.1)$$

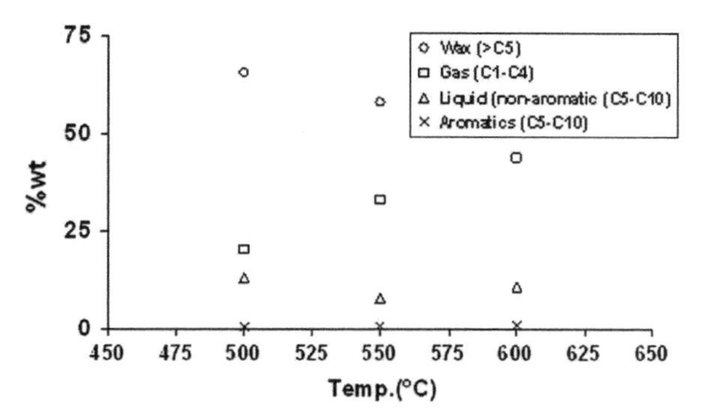

FIGURE 17.4 Lumped products (% wt) at different operating temperatures.

where P, G, L, W and A stand for the fractions of HDPE, gases, liquids, waxes and aromatics. k_1, k_2, k_3, k_4 and k_5 represent the kinetic rate constant (s^{-1}) of the polymer thermal degradation to waxes (primary reactions); waxes to gases, liquids and aromatics (secondary reactions) and liquid to gases (tertiary reactions). The mathematical model of the mechanism proposed was based on mass balances and rate equation analysis; all reactions were assumed to be irreversible. The model given by Equations (2) to (6) was solved by fourth-order Runge–Kutta method:

$$\frac{dx_p}{dt} = -k_1\left[x_p^n\right] \qquad (17.2)$$

$$\frac{dx_w}{dt} = k_1\left[x_p^n\right] - \left[k_2x_g + k_3x_l + k_4x_a\right] \qquad (17.3)$$

$$\frac{dx_g}{dt} = x_w[k_2] + x_l[k_5] \qquad (17.4)$$

$$\frac{dx_l}{dt} = x_w[k_3] - x_l[k_5] \qquad (17.5)$$

$$\frac{dx_a}{dt} = x_w[k_4] \qquad (17.6)$$

Figure 17.5 shows the exported modelled results for the predictions of polymer loss against experimental results obtained at 500, 550 and 600 °C. It was observed that time needed for total conversion decreases with higher operating temperatures. At 500 °C, the reaction time was 510 s, whereas at 550 and 600 °C, it decreased to 91.7 and 66.3 s, respectively. To obtain the kinetic parameters, the differential equations [Eqns. (2)–(6)] were solved and kinetic rate constants were calculated. Table 17.2 shows the kinetic parameters estimated at

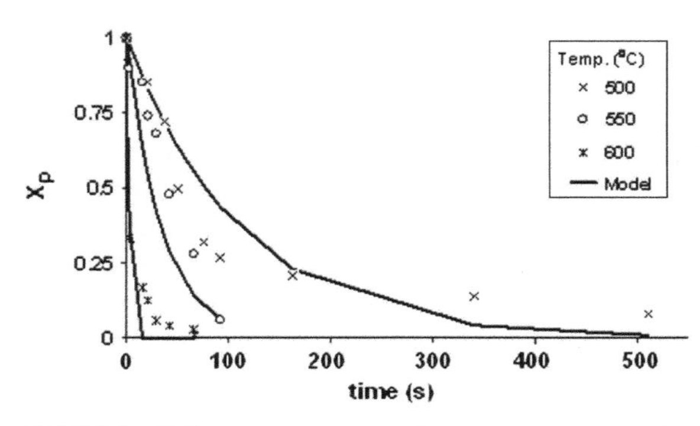

FIGURE 17.5 Experimental and model results of polymer weight loss obtained at 500, 550 and 600 °C as a function of time (s).

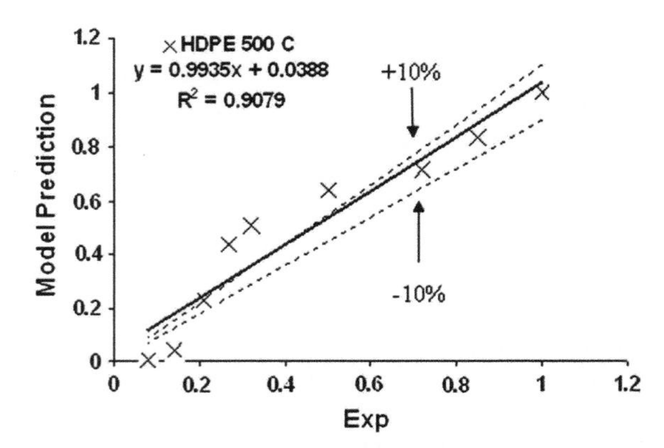

FIGURE 17.6 Model versus experimental results of polymer weight fraction (x_p) at different times between 0 to 510 s (at 500 °C).

the three different temperatures. The activation energies (E_a) and pre-exponential factors (P_0) were estimated for every primary product by first-order Arrhenius equation fitting. The overall activation energy (E_o) and pre-exponential factors (P_0) were found to be equal to 216.85 kJ mol^{-1} and 2.51 $\times$ 10^{12} s^{-1}, respectively, with a reaction order equal to 0.97. These results are in agreement with previous results reported in the literature [25], which estimated the reaction order between 0.81 and 1, values of E_o between 220 and 259 kJ mol^{-1} and P_0 = 1.9–7.2 $\times$ 10^{13} s^{-1}. Earlier work by Mastellone [7] reported values of E_o for PE between 192 and 301 kJ mol^{-1}, whereas [26] reported E_o to be between 237 and 279 kJ mol^{-1}, for temperatures between 450 and 470 °C.

The polymer fraction (x_p) resulting from the model prediction were plotted against the experimental values with respect to the same reaction time (see Fig. 17.6). The scatter of the data reveals that the model predicts more accurate polymer weight fractions at earlier stages of the reaction.

2.3. Summary

Based on the previous findings, pyrolysis and gasification are the most promising techniques for the processing of PSW. Till now, however, most pyrolysis and gasification processes applied on an industrial scale lack a designed end-product manner. Thermal decomposition schemes on the end-product (employing lumped product analysis) are an essential step in the development of pyrolysis and gasification reactors.

TABLE 17.2 Results Summary of the Depolymerization Reactions of HDPE in Isothermal Pyrolysis (500–600 °C), Showing Kinetic Rate Constant of Single Path Reactions (k_i), Overall Rate Constant (k_o) and Reaction Order (n)

T (°C)	Reaction order (n)	$k_1 \approx k_o$	k_2	k_3	k_4	k_5
500	0.97	0.9 $\times$ 10^{-2}	0.5 $\times$ 10^{-3}	2.7 $\times$ 10^{-4}	2.1 $\times$ 10^{-5}	1 $\times$ 10^{-5}
550		2.9 $\times$ 10^{-2}	0.3 $\times$ 10^{-2}	1.9 $\times$ 10^{-3}	5.4 $\times$ 10^{-5}	1 $\times$ 10^{-5}
600		0.4	0.4 $\times$ 10^{-2}	3.8 $\times$ 10^{-3}	3.8 $\times$ 10^{-4}	1 $\times$ 10^{-5}

3. CONCLUDING REMARKS

In this chapter, we have introduced the problem of waste management and plastic generation and trends. We have briefly introduced the legislations at United Kingdom and EU level that regulate the treatment of waste and have shown how their implementation may lead to reduced greenhouse gas emissions from the treatment of waste. We have looked at the different ways of treating PSW *viz.*, re-extrusion and mechanical recycling, petrochemical recovery via advanced thermochemical treatment (gasification, hydrogenetion and pyrolysis). We have shown through a case study on isothermal pyrolysis of HDPE the usefulness of Thermogravimetric analysis (TGA) to determine the product distribution obtained from thermal degradation as a function of operating temperature and the resulting kinetics of the reaction.

References

[1] P. Lettieri, L. Yassin, S.R.J. Simons, Management, recycling and reuse of waste composites, in: V. Goodship (Ed.), Woodhead Publishing, Cambridge, 2009, pp. 152–191.
[2] DEFRA. Securing the Future: Delivering UK Sustainable Development Strategy, London (2006), UK.
[3] EEA Briefing 2008/01 Better management of municipal waste will improve greenhouse gas emissions, ISSN 1830–2246.
[4] S.M. Al-Salem, P. Lettieri, Baeyens, The valorization of plastic solid waste (PSW) by primary to quaternary routes: From re-use to energy and chemicals, Prog Energy Combust Sci 36 (1) 103–129.
[5] A. Valavanidis, N. Iliopoulos, G. Gotsis, K. Fiotakis, Persistent free radicals, heavy metals and PAHs generated in particulate soot emissions and residue ash from controlled combustion of common types of plastic, J Hazard Mater 156 (1-3) (2008) 277–284.
[6] Waste watch. Plastics in the UK economy: a guide to polymer use and the opportunities for recycling. Final report of the Programme of sustainable use (UK). Waste Watch Group (2008).
[7] M.L. Mastellone, Thermal treatments of plastic wastes by means of fluidized bed reactors. PhD thesis, Department of Chemical Engineering, Second University of Naples, Italy, 1999.
[8] S.M. Al-Salem, Recycling and recovery routes of plastic solid waste (PSW): A review, Waste Manage 29 (10) (2009) 2625–2643.
[9] Waste Online. Plastic recycling information sheet. http://www.wasteonline.org.uk/resources/InformationSheets/Plastics.htm
[10] C.N. Kartalis, C.D. Papaspyrides, R. Pfaendner, K. Hoffmann, H. Herbst, Mechanical recycling of postused high-density polyethylene crates using the restabilization technique. I. Influence of artificial weathering, J App Polym Sci 77 (5) (2000) 1118–1127.
[11] K.M. Zia, H.N. Bhatti, I.A. Bhatti, Methods for polyurethane and polyurethane composites, recycling and recovery: A review, React Funct Polym 67 (8) (2007) 675–692.
[12] M.P. Aznar, M.A. Caballero, J.A. Sancho, E. Francs, Plastic waste elimination by co-gasification with coal and biomass in fluidized bed with air in pilot plant, Fuel Process Tech 87 (5) (2006) 409–420.
[13] T. Yoshioka, G. Gause, C. Eger, W. Kaminsky, A. Okuwaki, Pyrolysis of poly (ethylene terephthalate) in a fluidized bed plant, Polym Degrad Stab 86 (3) (2004) 499–504.
[14] K. Smolders, J. Baeyens, Thermal degradation of PMMA in fluidised beds. "http://www.sciencedirect.com/science?_ob=ArticleURL&_udi=B6VFR-4D981F7-2&_user=125795&_coverDate=12%2F31%2F2004&_alid=830030874&_rdoc=1&_fmt=high&_orig=search&_cdi=6017&_sort=d&_docanchor=&view=c&_ct=1&_acct=C000010182&_version=1&_urlVersion=0&_userid=125795&md5=b5fc4b115ffd2029cb4e52619c2c57f8", Waste Manage 24 (8) (2004) 849–857.
[15] S.M. Al-Salem, P. Lettieri, J. Baeyens, Thermal pyrolysis of High Density Polyethylene (HDPE). Proc. of the ninth European gasification conference: clean energy and chemicals 2009, Dusseldorf, Germany, 23rd-25th March.
[16] A.V. Bridgwater, Renewable fuels and chemicals by thermal processing of biomass. "http://www.sciencedirect.com/science?_ob=ArticleURL&_udi=B6TFJ-4728F81-1&_user=125795&_coverDate=03%2F15%2F2003&_alid=1527361298&_rdoc=1&_fmt=high&_orig=search&_origin=search&_zone=rslt_list_item&_cdi=5228&_sort=r&_st=13&_docanchor=&view=c&_0&_userid=125795&md5=156f45dec408f4b92d171a93d4888058&searchtype=a", Chemical Engineering Journal 91 (2-3) (2003) 87–102.
[17] A.G. Buekens, Resource recovery and waste treatment in Japan, Resour Recovr Conserv 3 (3) (1978) 275–306.
[18] M. Hasegawa, X. Fukuda, D. Kunii, Gasification of solid waste in a fluidized with circulating sand, Conserv Recy 3 (2) (1974) 143–153.

[19] C. Borgianni, P.D. Filippis, F. Pochetti, M. Paolucci, Gasification process of wastes containing PVC, Fuel 14 (81) (2002) 1827–1833.

[20] G. Xiao, M. Ni, Y. Chi, B. Jin, R. Xiao, Z. Zhong, et al., Gasification characteristics of MSW and an ANN prediction model. "http://www.sciencedirect.com/science?_ob=ArticleURL&_udi=B6VFR-4S97J91-1&_user=125795&_coverDate=01%2F31%2F2009&_alid=869832724&_rdoc=1&_fmt=high&_orig=search&_cdi=6017&_sort=d&_docanchor=&view=c&_ct=3&_acct=C000010182&_version=1&_urlVersion=0&_userid=125795&md5=d41307dae086f2bd672c4a8c24b2298d", Waste Manage 29 (1) (2009) 240–244.

[21] J. Matsunami, S. Yoshida, O. Yokota, M. Nezuka, M. Tsuji, Y. Tamaura, Gasification of waste tyre and plastic (PET) by solar thermochemical process for solar energy utilization. "http://www.sciencedirect.com/science?_ob=ArticleURL&_udi=B6V50-3V93771-3&_user=125795&_coverDate=01%2F01%2F1999&_alid=869676442&_rdoc=4&_fmt=high&_orig=search&_cdi=5772&_sort=d&_docanchor=&Version=0&_userid=125795&md5=f61d2874a846e3c9b8fa521c4f388bc7", Solar Energy 65 (1) (1999) 21–23.

[22] J. March, Advanced Organic Chemistry: Reactions, Mechanisms, and Structure, fourth ed., John Wiley & Sons (US), 1992.

[23] J. Nishino, M. Itoh, Y. Fujiyoshi, Y. Matsumoto, R. Takahashi, Y. Uemichi, Proceedings of the third international symposium on feedstock recycle of plastics & other innovative plastics recycling techniques 25–29th, Karlsruhe, Germany; pp. 325–332. 2005.

[24] W.C. McCaffrey, M.R. Kamal, D.G. Cooper, Polymer Degradation and Stability 47 (1995) 133–139.

[25] R.W.J. Westerhout, J. Waanders, J.A.M. Kuipers, W.P.M. van Swaaij, Industrial Engineering Chemistry Research 36 (1997) 1955–1964.

[26] N. Horvat, F.T.T. Ng, Fuel 78 (1999) 459–470.

CHAPTER

18

Air Pollution: Atmospheric Wastes

Daniel A. Vallero

Pratt School of Engineering, Duke University, Durham, N. Carolina, USA

OUTLINE

1. INTRODUCTION

Air pollution is the presence of contaminants or substances in the air that interfere with human health or welfare, or produce other harmful environmental effects.[1] An air pollutant is interfering with a desired condition that, at a minimum, provides for air quality that supports human and other life. Usually, the pollutant is the result of inefficiency. In trying to produce something of value, a waste is released to the atmosphere. In environmental connotations, the term "waste" generally applies to materials that find their way to the land and soil. Materials found in water and air are referred to as pollution. This is a distinction of convenience and regulation, not one of science. Thus, this chapter focuses on the waste streams that affect the atmosphere.

[1]Based on the definition in United States Environmental Protection Agency (2007). "Terms of Environment: Glossary, Abbreviations and Acronyms"; http://www.epa.gov/OCEPAterms/aterms.html; accessed on May 26, 2010.

Waste Doi: 10.1016/B978-0-12-381475-3.10018-X

2. AIR POLLUTION

For centuries, people have known intuitively that something was amiss when their air was filled with smoke or when they smelled an unpleasant odor. But, for most pollutants, those that were not readily sensed without the aid of sensitive equipment, a baseline had to be set to begin to take action. One way to look at the interferences mentioned in the definition is to put them in the context of "harm." The objects of the harm have received varying levels of interests. In the 1960s, harm to ecosystems, including threats to the very survival of certain biological species, was paramount. This concern was coupled with harm to humans, especially in terms of diseases directly associated with obvious episodes, such as respiratory diseases and even death associated with combinations of weather and pollutant releases.

Myriad atmospheric wastes have accompanied modern life. Nuclear power plants are associated with the possibilities of meltdown and the release of airborne radioactive materials. Burning fossil fuels releases products of incomplete combustion, for example, polycyclic aromatic hydrocarbons. Even complete combustion releases the greenhouse gas (GHG), carbon dioxide (CO_2). Leaks from chemical, pesticide, and other manufacturing facilities may cause exposures to toxic substances, such as the release of methyl isocyanate in Bhopal, India. In the last quarter of twentieth century, these apprehensions led to the public's growing wariness about "toxic" chemicals added to the more familiar "conventional" pollutants like soot, carbon monoxide, and oxides of nitrogen and sulfur. The major new concern involving toxic chemicals relates to cancer and also threats to hormonal systems in humans and wildlife, neurotoxicity (notably from lead and mercury exposure in children), and immune system disorders. Growing numbers of studies have provided evidence linking disease and adverse effects to extremely low levels of certain particularly toxic substances. For example, exposure to tetrachlorodibenzo-*para*-dioxin (See Chapter 28) at almost any currently detectable concentration could be associated with numerous adverse effects in humans.

At the threshold of the new millennium, new atmospheric waste problems were identified, including the loss of aquatic diversity in lakes due to deposition of acidic precipitation, so-called "acid rain." Acid deposition was also being associated with the corrosion of materials, including some of the most important human-made structures, such as the pyramids in Egypt and monuments throughout the world. Presently, global pollutants have become the source of public concern, such as those that seemed to be destroying the stratospheric ozone (O_3) layer or those that appeared to be affecting the global climate. This escalation of awareness of the multitude of pollutants complicated matters.

A principal challenge in addressing air pollution, as other wastes discussed in this book, is that most pollutants under other circumstances would be "resources," such as compounds of nitrogen. In the air, these compounds can cause respiratory problems directly or in combination with hydrocarbons and sunlight indirectly can form O_3 and smog. But, in the soil, nitrogen compounds are essential nutrients. So, it is not always simply a matter of "removing" pollutants but one of managing systems to ensure that optimal conditions for health and environmental quality exist. Impurities are common, but in excessive quantities and in the wrong places they become harmful.

Recently, in the transportation sector, the term "zero emission" has been applied to vehicles, as the logical next step following low-emission vehicles (LEVs) and ultra-low-emission vehicles (ULEVs). However, zero emissions of pollutants will not be likely for the foreseeable future, especially if one considers that even electric cars are not emission free but actually are a type of emission trading from a mobile source to a stationary source. That is, the electricity is generated at a power plant that is emitting pollutants as it burns fossil fuels or has the problem of radioactive wastes if it is a nuclear power plant. Even hydrogen, solar, and wind systems are not completely pollution free as manufacture, transport, and installation of the parts and assemblages require energy and materials, some of which find their way to the atmosphere (see the discussion of life cycles in Chapter 2).

2.1. Atmospheric Contamination

The predominant concern with atmospheric waste is chemical contamination, which presents a hazard to human health. Thus, public health is usually the principal driver for assessing and controlling air contaminants. However, air pollution abatement laws and programs have also recognized that effects beyond health are also important, especially *welfare* protection. One of the main welfare considerations is that ecosystems are important *receptors* of contamination. Another welfare concern is that contaminants impact structures and other engineered systems by corrosion. Thus, from an air pollution perspective, there is a cascade of hazards from human health to ecosystems to abiotic (i.e., nonliving) systems.

Certainly, not all pollutants are chemical hazards. They can also be physical, such as the energy from ultraviolet (UV) light. Often, even though our exposure is to the physical contamination, this exposure is indirectly brought about by chemical contamination. For example, the release of chemicals into the atmosphere in turn reacts with O_3 in the stratosphere, decreasing the O_3 concentration and increasing the amount of UV radiation at the Earth's surface. This means that the mean UV dose in the temperate zones of the world has increased. This has been associated with an increase in the incidence of skin cancer, especially the most virulent form, melanoma.

Air pollutants may also be biological, as when bacteria and viruses are released to the atmosphere from medical facilities. Other biological air pollutants include irritants and allergens, such as pollen and molds (i.e., bioaerosols).

2.2. Legislation

The U.S. Congress enacted the 1970 amendments to the Clean Air Act (CAA) to provide a comprehensive set of regulations to control air emissions from area, stationary, and mobile sources. This law authorized the U.S. Environmental Protection Agency (EPA) to establish National Ambient Air Quality Standards (NAAQS) to protect public health and the environment from the "conventional" (in contrast to "toxic") pollutants: carbon monoxide, particulate matter (PM), oxides of nitrogen, oxides of sulfur, and photochemical oxidant smog or O_3. The metal lead (Pb) was later added as the sixth NAAQS pollutant. The original goal was to set and to achieve NAAQS in every state by 1975.These new standards were combined with charging the 50 states to develop state implementation plans (SIPs) to address industrial sources in the state.

The ambient atmospheric concentrations are measured at over 4000 monitoring sites across the United States. The ambient levels have continuously decreased, as shown in Table 18.1. The Clean Air Act Amendments (CAAA) of 1977 mandated new dates to achieve attainment of NAAQS (many areas of the country had not met the prescribed dates set in 1970). Other amendments were targeted at types of air pollution that had not previously been addressed sufficiently, including acidic deposition, tropospheric O_3 pollution, depletion of the stratospheric O_3 layer, and a new program for air toxins, known as the National Emission Standards for Hazardous Air Pollutants (NESHAPS).

TABLE 18.1 Percentage Decrease in Ambient Concentrations of National Ambient Air Quality Standard Pollutants From 1985 Through 1994

Pollutant	Decrease in concentration (%)
Lead	86
Carbon monoxide	28
Sulfur dioxide	25
Particulate matter (PM_{10}) – aerodynamic diameter = 10 micron	20
Ozone	12
Nitrogen dioxide	9

Source: U.S. Environmental Protection Agency.

3. SCALE OF THE PROBLEM

Air is polluted at all scales, from extremely local[2] to global. These can be subdivided into five ascending scales: local, urban, regional, continental, and global. The spheres of influence of the air pollutants themselves range from molecular (e.g., gases and nanoparticles) to planetary (e.g., diffusion of GHGs throughout the troposphere). The local scale is up to about 5 km of the Earth's surface. The urban scale extends to the order of 50 km. The regional scale is from 50 to 500 km. Continental scales range from several 1000 km. Of course, the global scale is planetwide [1].

3.1. Local Scale

Local air pollution problems are usually characterized by one or several large emitters or a large number of relatively small emitters. Lowering the release height of a source worsens the potential local impact for a given release. Examples include carbon monoxide and hydrocarbon released from tailpipes or oxides of sulfur released from power plants. In fact, this phenomenon is the reason that certain controls have led to larger-scale problems; that is, solving a local air pollution problem by increasing the height of the point of release (i.e., tall stacks) allowed pollutants to travel long distances from the source. Any ground-level source, such as evaporation of volatile organic compounds from a gasoline station, will produce the highest concentrations near the source, with concentrations generally diminishing with distance. This phenomenon is known as a concentration gradient.

Large sources that emit high above the ground through stacks can also cause local problems, especially under unstable meteorological conditions that cause portions of the plume to reach the ground in high concentrations.

There are many releases of pollutants from relatively short stacks or vents on the top of one- or two-story buildings. Under most conditions, such releases are caught within the turbulent downwash downwind of the building. This allows high concentrations to be brought to the ground surface. Many different pollutants can be released in this manner, including compounds and mixtures that can cause odors. Usually, the effects of accidental releases are confined to the local scale.

3.2. Urban Scale

Air pollution problems in urban areas generally result from either primary or secondary pollutants. Primary pollutants are released directly from sources, whereas secondary pollutants form through chemical reactions of the primary pollutants in the atmosphere after release. Air pollution problems can be caused by individual sources on the urban scale as well as the local scale. For pollutants that are

[2] At an even smaller scale are *micro-environmental* (e.g., indoor air in a home, office, or garage) and *personal* (i.e., at the individual's breathing zone).

relatively nonreactive, such as carbon monoxide and PM,[3] or relatively slowly reactive, such as sulfur dioxide (SO_2), the contributions from individual sources combine to yield high concentrations. As a major source of carbon monoxide is motor vehicles, "hot spots" of high concentration can occur especially near intersections with high traffic. The emissions are especially high from idling vehicles. The hot spots are exacerbated if high buildings surround the intersection, as the volume of air in which the pollution is contained is severely restricted. The combination of these factors results in high concentrations.

Tropospheric O_3 is the dominant urban problem resulting from the formation of secondary pollutants. Many large, metropolitan areas experience the formation of O_3 from photochemical reactions of oxides of nitrogen and various species of hydrocarbons. These reactions are catalyzed by the ultraviolet light in sunlight and are therefore called photochemical reactions. Many metropolitan areas are in nonattainment for O_3; that is, they are not meeting the air quality standards. In the United States, the CAAA of 1990 classifies the various metropolitan areas to be in nonattainment according to the severity of the problem for that attainment with the NAAQS.

Oxides of nitrogen, principally nitric oxide (NO) but also nitrogen dioxide (NO_2), are emitted from automobiles and from combustion processes. Hydrocarbons are emitted from many different sources. The various species have widely varying reactivities. Determining the emissions of these chemical species from myriad sources as a basis for pollution control programs can be difficult but methods continue to improve.

3.3. Regional Scale

Three particular types of problems contribute to air pollution problems on the regional scale. The first is the blend of urban oxidant problems at the regional scale. Many major metropolitan are in close proximity to one another and continue to grow. Urban geographers refer to some of the larger urban aggregations as "megalopolises." As a result, the air from one metropolitan area, containing both secondary pollutants formed through reactions and primary pollutants, flows on to the adjacent metropolitan area. The pollutants from the second area are then added on top of the "background" from the first.

The second problem involves the release of relatively slow-reacting primary air pollutants that undergo reactions and transformations during lengthy transport times. Protracted transport times, result over regional scales, not only mean long residence times for the parent compounds but also numerous transformation byproducts that form during the transport. The gas, SO_2, released primarily through combustion of fossil fuels (especially from coal and oil) is oxidized during long-distance transport to sulfur trioxide (SO_3):

$$2SO_2 + O_2 \rightarrow 2SO_3 \tag{18.1}$$

Although SO_2 is a gas, both gas phase and liquid phase oxidation of SO_2 occurs in the troposphere. The SO_3 in turn reacts with water vapor to form sulfuric acid:

$$SO_3 + H_2O \rightarrow H_2SO_4 \tag{18.2}$$

Sulfuric acid reacts with numerous compounds to form sulfates. These are fine (submicrometer) particulates. NO results from high-temperature combustion, both in stationary sources—such as power plants or industrial plants in the production of process heat—and in internal combustion engines in vehicles. The NO is oxidized in the atmosphere, usually

[3]Particulate matter is highly variable in its composition. Most of the particle may be nonreactive, but aerosols often contain chemically reactive substances, especially those adsorbed onto particle surfaces.

rather slowly but more rapidly if there is O_3 present, to NO_2, which also reacts further with other constituents forming nitrates, which is also in fine particulate form.

The sulfates and nitrates existing in the atmosphere as fine particulates, generally in the size range less than 1μm (aerodynamic diameter), can be removed from the atmosphere by several processes. "Rain out" occurs when the particles serve as condensation nuclei that lead to the formation of clouds. If the droplets grow to a sufficient size, they fall as raindrops with the particles in suspension or in solution. Another mechanism, known as "washout," also involves rain, but the particles in the air are captured by raindrops falling through the air. Both mechanisms contribute to "acid rain," which results in the sulfate and nitrate particles reaching lakes and streams, and increasing their acidity. As such, acid rain is both a regional and continental problem [1].

A third type of regional problem is visibility, which may be reduced by specific plumes or by the regional levels of PM that produce various intensities of haze. The fine sulfate and nitrate particulates just discussed are largely responsible for reduction of visibility. This is especially problematic in locations of natural beauty, where it is desirable to keep scenic vistas as free of obstructions to the view as possible. Regional haze is a type of visibility impairment that is caused by the emissions of air pollutants from numerous sources across a broad region. The CAAA provides special protections for such areas—the most restrictive denoted as mandatory Federal Class I areas that cover over 150 national parks and wilderness areas in the United States. Decreased visibility can also impair safety, especially concerning aviation.

3.4. Continental Scale

In Europe, there is little difference between what would be considered regional scale and continental scale. However, on larger continents, there would be a substantial difference. Perhaps of greatest concern on the continental scale is that the air pollution policies of a nation are likely to create impacts on neighboring nations. Acid rain in Scandinavia has been considered to have had impacts from Great Britain and Western Europe. Japan has considered that part of their air pollution problem, especially in the western part of the country, has origins in China and Korea. For decades, Canada and the United States have cooperated in studying and addressing the North American acid rain problem. Likewise, tall stacks in Great Britain and the lowland countries of continental Europe have contributed to acid rain in Scandinavia. For some years, British industries simply built increasingly tall stacks as a method of air pollution control, reducing the immediate ground-level concentration but emitting the same pollutants into the higher atmosphere. The local air quality improved but at the expense of acid rain in other parts of Europe.

3.5. Planetary Scale

The release of radioactivity from the accident at Chernobyl would be considered primarily a regional or continental problem. However, elevated levels of radioactivity were detected in the Pacific Northwest part of the United States soon after the accident. Likewise, persistent organic pollutants, such as polychlorinated biphenyls (PCBs), have been observed in Arctic mammals, thousands of miles from their sources. These observations demonstrate the effects of long-range transport.

In this and other instances, the gases are released from numerous locations on the Earth's surface but when added together can change the temperature and other climatological features of the troposphere. In the case of chlorofluorocarbons (CFCs), their release results in free chlorine (Cl) atoms that attack O_3 in the stratosphere, which is ordinarily quite stable and resists vertical air exchange between layers. However, halogen compounds can be transferred from the troposphere into the stratosphere by injection through the tops of thunderstorms that occasionally penetrate the tropopause, the boundary between the troposphere and the stratosphere. Some transfer of stratospheric air downward also occurs through occasional gaps in the tropopause. As the O_3 layer is considerably above the troposphere, the transfer of CFCs upward to the O_3 layer is expected to occur gradually. Thus, there is a lag from the first release of these gases until an effect is seen, that is, the so-called thinning of the O_3 layer at the poles. Similarly, with the cessation of use of these materials worldwide, there has been a commensurate lag in restoration of the O_3 layer.

Other important planetary scale pollutant is the attendant global problem of climate change that is generated by excessive amounts of radiant gases (commonly known as GHGs), especially methane (CH_4) and CO_2, which is not normally considered an air pollutant. A portion of radiation from the Earth's surface is intercepted by the CO_2 in the air and is reradiated both upward and downward. That which is radiated downward keeps the ground from cooling rapidly. As the CO_2 concentration continues to increase, the Earth's temperature is expected to increase [2].

4. AIR QUALITY

The main aim in controlling air pollution is to prevent adverse responses to receptor categories exposed to the atmosphere: human, animal, plant, and material. These adverse responses have characteristic response times: short term (i.e., seconds or minutes), intermediate term (i.e., hours or days), and long term (i.e., months or years) (Table 18.2). To elicit no adverse responses, the pollutant concentration in the air must be lower than the concentration level at which these responses occur. Figure 18.1 illustrates the relationship between these concentration levels. This figure displays response curves, which remain on the concentration duration axes because they are characteristic of the receptors, not of the actual air quality to which the receptors are exposed. The odor response curve, for example, to hydrogen sulfide, shows that a single inhalation requiring approximately 1 s can establish the presence of the odor but that, due to odor fatigue, the ability to continue to recognize that odor can be lost in a matter of minutes. Nasopharyngeal and eye irritation (e.g., by O_3) is similarly subject to acclimatization due to tear and mucus production. The three visibility lines correlate with the concentration of suspended PM in the air.

TABLE 18.2 Examples of Receptor Category Characteristic Response Times for Exposures to Air Pollutants [3]

	Characteristic response times		
Receptor category	**Short term (seconds–minutes)**	**Intermediate term (hours–days)**	**Long term (month–years)**
Human	Odor, nasopharyngeal and visibility, eye irritation	Acute respiratory disease	Chronic respiratory disease and lung cancer
Animal, vegetation	Field crop loss and ornamental plant damage	Field crop loss and ornamental plant damage	Fluorosis of livestock, decreased fruit and forest yield
Material	Acid droplet pitting and nylon hose destruction	Rubber cracking, silver tarnishing, and paint blackening	Corrosion, soiling, and materials deterioration

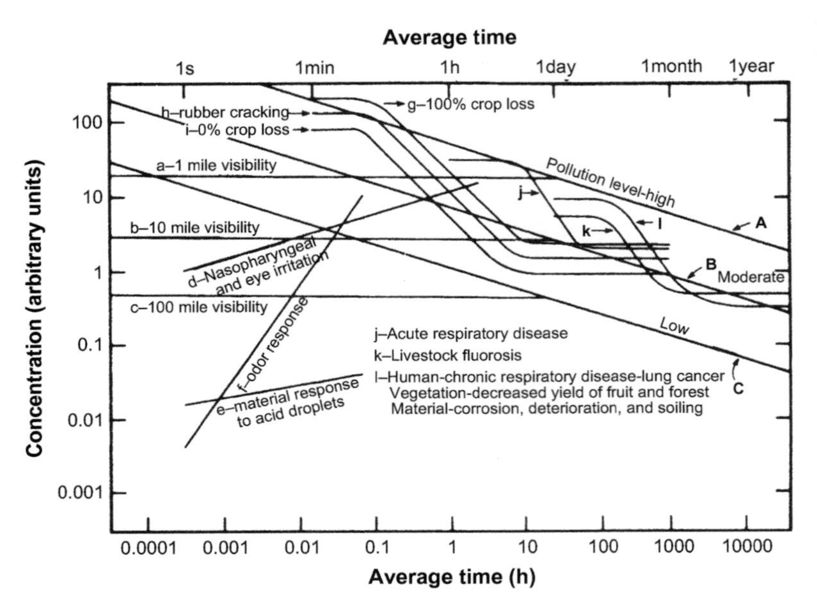

FIGURE 18.1 Adverse responses to various pollution levels [3].

Attack of metal, painted surfaces, or nylon hose is shown by a line starting at 1 s and terminating in a matter of minutes (when the acidity of the droplet is depleted by the material attacked).

Plant damage can be measured biologically or socioeconomically. Biological damage can be measured as stress on growth and survival. Socioeconomic damage can be measured as 0% loss when there is no loss of the sale value of the crops or ornamental plants but a 100% loss if the crop is damaged to the extent that it cannot be sold. Such responses are related to dose, that is, concentration times duration of exposure, as shown by the percent loss curves on the chart. A number of manifestations of material damage, for example, rubber cracking by O_3, require an exposure duration long enough for the adverse effects to be economically significant. That is, attack for just a few seconds or minutes will not affect the utility of the material for its intended use, but attack for a number of days will.

The biological response line for acute respiratory disease is a dose–response curve, which for a constant concentration becomes a duration–response curve.

The shape of such a curve reflects the ability of the human body to cope with short-term, ambient concentration respiratory exposures and the overwhelming of the body's defenses by continued exposure. For example, fluorosis of livestock is not induced until there has been a long-enough period of deposition of a high-enough ambient concentration of fluoride to increase the level of fluoride in the forage. As the forage is either consumed by livestock or cut for hay at least once during the growing season, the duration of deposition ends after the growing season. The greater the duration of the season, the greater the time for deposition, hence the shape of the line labeled "fluorosis." Long-term vegetation responses—decreased yield of fruit and forest—and long-term material responses—corrosion, soiling, and material deterioration—are shown on the chart as having essentially the same response characteristics as human chronic respiratory disease and lung cancer.

The relationship of these response curves to ambient air quality is shown by Lines A, B, and C, which represent the maximum or any other chosen percentile line from a display such as Fig. 18.1, which shows actual air quality. Where the air quality is poor (Line A), essentially all the adverse effects displayed will occur. Where the air quality is good (Line C), most of the intermediate and long-term adverse effects displayed will not occur. Where the air quality is between good and poor, some of the intermediate and long-term adverse effects will occur but in an attenuated form compared with those of poor air quality.

4.1. Air Quality Indices

Air quality indexes (AQIs) categorize the air quality measurements of several individual pollutants by one composite number. The index described in Table 18.3 is used by the U.S. EPA, as authorized by Section 319 of the CAA. Comparable values for international standards are shown in Table 18.4.

The AQI is designed to convey information to the public regarding daily air quality and its associated health risks. The scientific understanding of pollutants, especially PM and O_3, has increased steadily. This is reflected in the NAAQS promulgated in 1997. In particular, much more is now known about the nature of the relationships between exposure to ambient concentrations of these pollutants and the health effects likely to be experienced, especially near the level of the NAAQS. Sensitivity varies by pollutant. Thus, when the AQI exceeds 100, the sensitive groups of concern to be at greatest risk include:

- *O_3*: Children and anyone with asthma.
- *$PM_{2.5}$*: Those with respiratory or heart disease, the elderly and children.
- *PM_{10}*: Those with respiratory disease.
- *Carbon monoxide*: Those with heart disease.
- *SO_2*: Those with asthma.
- *NO_2*: Children and anyone with respiratory disease.

The precautionary statements for the NAAQS pollutants are shown in Table 18.5.

5. AIR POLLUTION CONTROL

Air pollution control must be strategic and tactical. The former is the long-term reduction of pollution levels at all scales of the problem from local to global. Goals can be set for air quality improvement 5, 10, or 15 years ahead and plans can be made to achieve these improvements.

5.1. Strategies

One notable example is the requirement of the CAAA of 1990 to reduce the emissions of hazardous air pollutants (so-called "HAPs"). The law requires the EPA to establish an inventory to track progress in reducing HAPs in ambient air. The first step of this strategy was to promulgate technology-based emission standards (maximum achievable control technology, that is, MACT). For major sources of HAPs, Section 112(f) requires standards to address risks remaining after implementation of MACT standards, known as residual risks standards. Section 112(c)(3) and Section 112(k) of the CAAA requires EPA to address the emissions and risks of HAPs from area sources and to show a 75% reduction in cancer incidence of emissions from stationary sources of HAPs since 1990. To monitor the success in meeting these strategies in reducing emissions and human health, EPA compiles the National Emissions Inventory (NEI) for HAPs. The EPA previously compiled a baseline 1990 and 1996 National Toxics Inventory (NTI) and 1999 NEI for HAPs and has recently completed version 3 of the 2002 NEI [1].

The NEI includes major point and area sources, nonpoint area and other sources, and mobile source estimates of emissions. Stationary major sources of HAPs are defined as sources that have the potential to emit 9 tonnes per annum (9 t a^{-1} or 10 U.S. tons per annum) or more of any single HAP or 22.7 t a^{-1} (25 U.S. tons per year) or more of any combination of HAPs [7].

TABLE 18.3 Air Quality Index [1,4]

Category	Good	Moderate	Unhealthy for sensitive groups[a]	Unhealthy	Very unhealthy	Hazardous	
Index value	0–50	51–100	101–150	151–200	201–300	301–400	401–500
Pollutant			Concentration ranges				
Carbon monoxide (ppm)	0–4.4	4.5–9.4	9.5–12.4	12.5–15.4	15.5–30.4	30.5–40.4	40.5–50.4
Nitrogen dioxide (ppm)	–	–	–	–	0.65–1.24	1.25–1.64	1.65–2.04
Ozone (average concentration for 1 h in ppm)	–	–	0.125–0.164	0.165–0.204	0.205–0.404	0.405–0.504	0.505–0.604
Ozone (average concentration for 8 h ppm)	0–0.0604	0.0605–0.084	0.085–0.104	0.105–0.124	0.125–0.374	–	–
$PM_{2.5}$ (μg m^{-3})	0–15.4	15.5–40.4	40.5–65.4	65.5–150.4	150.5–250.4	250.5–350.4	350.5–500.4
PM_{10} (μg m^{-3})	0–54	55–154	155–254	255–354	355–424	425–504	505–604
Sulfur dioxide (ppm)	0–0.034	0.035–0.144	0.145–0.224	0.225–0.304	0.305–0.604	0.605–0.804	0.805–1.004

[a]*Each category corresponds to a different level of health concern. The six levels of health concern and what they mean are as follows:*

"Good": The AQI value for your community is between 5 and 50. Air quality is considered satisfactory and air pollution poses little or no risk.

"Moderate": The AQI for your community is between 51 and 100. Air quality is acceptable; however, for some pollutants there may be a moderate health concern for a very small number of people. For example, those who are unusually sensitive to ozone may experience respiratory symptoms.

"Unhealthy for sensitive groups": When AQI values are between 101 and 150, members of sensitive groups may experience health effects. This means they are likely to be affected at lower levels than the general public. For example, people with lung disease are at greater risk from exposure to ozone, whereas people with either lung disease or heart disease are at greater risk from exposure to particle pollution. The general public is not likely to be affected when the AQI is in this range.

"Unhealthy": Everyone may begin to experience health effects when AQI values are between 151 and 200. Members of sensitive groups may experience more serious health effects.

"Very unhealthy": AQI values between 201 and 300 trigger a health alert, meaning everyone may experience more serious health effects.

"Hazardous": AQI values over 300 trigger health warnings of emergency conditions. The entire population is more likely to be affected.

TABLE 18.4 International Air Quality Indices [1,5]

WHO Air Quality Guidelines (AQGs) for Particulate Matter, Ozone, Nitrogen Dioxide, and Sulfur Dioxide

	PM_{10} ($\mu g\ m^{-3}$)	PM_{25} ($\mu g\ m^{-3}$)	Basis for the selected level
Interim target-1 (IT-1)	70	35	These levels are associated with about a 15% higher long-term mortality risk relative to the AQG level
Interim target-2 (IT-2)	50	25	In addition to other health benefits, these levels approximately 6% (2–11%) relative to the IT-1 level
Interim target-3 (IT-3)	30	15	In addition to other health benefits, these levels reduce the mortality risk by approximately 6% (2–11%) relative to the IT-2 level
AQG	20	10	These are the lowest levels at which total cardiopulmonary and lung cancer mortality have been shown to increase with more than 95% confidence in response to long-term exposure to $PM_{2.5}$

WHO AQGs and interim targets for particulate matter[a]: 24-h concentrations[b]

	PM_{10} ($\mu g\ m^{-3}$)	PM_{25} ($\mu g\ m^{-3}$)	Basis for the selected level
Interim target-1 (IT-1)	150	75	Based on published risk coefficients from multicentre studies and meta-analyses (about 5% increase of short-term mortality over the AQG value)
Interim target-2 (IT-2)	100	50	Based on published risk coefficients from multicentre studies and meta-analyzes (about 2.5% increase of short-term mortality over the AQG value)
Interim target-3 (IT-3)[c]	75	37.5	Based on published risk coefficients from multicentre studies and meta-analyzes (about 1.2% increase of short-term mortality over the AQG value)
AQG	50	25	Based on relationship between 24-h and annual PM levels

WHO AQG and interim target for ozone: 8-h concentrations

	Daily maximum 8-h mean ($\mu g\ m^{-3}$)	Basis for the selected level
High levels	240	Significant health effects; substantial proportion of vulnerable populations affected
Interim target-1 (IT-1)	160	Important health effects; does not provide adequate protection of public health. Exposure to this level of ozone is associated with the following: • Physiological and inflammatory lung effects in healthy exercising young adults exposed for periods of 6.6 h • Health effects in children (based on various summer camp studies in which children were exposed to ambient ozone levels) • An estimated 3–5% increase in daily mortality[d] (based on findings of daily time-series studies)
AQG	100	Provide adequate protection of public health, though some health effects may occur below this level. Exposure to this level of ozone is associated with the following: • An estimated 1–2% increase in daily mortality[d] (based on findings of daily time-series studies) • Extrapolation from chamber studies exclude highly sensitive or clinically compromised subjects or children • Likelihood that ambient ozone is a marker for related oxidants

(*Continued*)

TABLE 18.4 International Air Quality Indices [1,5]—*cont'd*

	WHO AQGs for nitrogen dioxide: annual mean	
AQG	40	Recent indoor studies have provided evidence of effects on respiratory symptoms among infants at NO_2 concentrations below 40 μg m^{-3}. These associations cannot be completely explained by co-exposure to PM, but it has been suggested that other components in the mixture (such as organic carbon and nitrous acid vapor) might explain part of the observed association

	WHO AQGs for nitrogen dioxide: 1-h mean	
	1-h mean (μg m^{-3})	
AQG	200	Epidemiological studies have shown that bronchitic symptoms of asthmatic children increase in association with annual NO_2 concentration and that reduced lung function growth in children is linked to elevated NO_2 concentrations within communities already at current North American and European urban ambient air levels. A number of recently published studies have demonstrated that NO_2 can have a higher spatial variation than other traffic-related air pollutants, for example, particle mass. These studies also found adverse effects on the health of children living in metropolitan areas characterized by higher levels of NO_2 even in cases where the overall city-wide NO_2 level was fairly low. Since the existing WHO AQG short-term NO_2 guideline value of 200 m^{-3} (1-h) has not been challenged by more recent studies, it is retained

	WHO AQGs and interim targets for SO_2: 24-h and 10-min concentrations		
	24-h average (μg m^{-3})	**10-min average (μg m^{-3})**	**Basis for the selected level**
Interim target-1 (IT-1)[e]	125	–	–
Interim target-1 (IT-1)	50	–	Intermediate goal based on controlling motor vehicle emissions, industrial emissions, and/or emissions from power production. This would be a reasonable and feasible goal for some developing countries (it could be achieved within a few years) that would lead to significant health improvements that, in turn, would justify further improvements (such as aiming for the AQG value)
AQG	20	500	–

[a] *The use of $PM_{2.5}$ guideline value is preferred.*
[b] *99th percentile (3 days per year).*
[c] *For management purposes. Based on annual average guideline values, precise number to be determined on the basis of local frequency distribution of daily means. The frequency distribution of daily $PM_{2.5}$ or PM_{10} values usually approximates to a log-normal distribution.*
[d] *Deaths attributable to ozone. Time-series studies indicate an increase in daily mortality in the range of 0.3–0.5% for every 10 μg m^{-3} increment in 8-h ozone concentrations above an estimated baseline level of 70 μg m^{-3}.*
[e] *Formerly the WHO Air Quality Guideline (WHO, 2000).*

TABLE 18.5 Cautionary statements for criteria air pollutants in the United States [6]

Index values	Levels of health concern	Cautionary statements[a]			
		Ozone	Particulate matter	Carbon monoxide	Sulfur dioxide
0–50	Good	None	None	None	None
51–100[a]	Moderate	Unusually sensitive people should consider reducing prolonged or heavy exertion outdoors	Unusually sensitive people should consider reducing prolonged or heavy exertion	None	People with asthma should consider reducing exertion outdoors
101–150	Unhealthy for sensitive groups	Active children and adults, and people with lung disease such as asthma, should reduce prolonged or heavy exertion outdoors	People with heart or lung disease, older adults, and children should reduce prolonged or heavy exertion	People with heart disease, such as angina, should reduce heavy exertion and avoid sources of CO, such as heavy traffic	Children, asthmatics, and people with heart or lung disease should reduce exertion outdoors
151–200	Unhealthy	Active children and adults, and people with lung disease such as asthma, should avoid prolonged or heavy exertion outdoors. Everyone else, especially children, should reduce prolonged or heavy exertion outdoors	People with heart or lung disease, older adults, and children should avoid prolonged or heavy exertion. Everyone else should reduce prolonged or heavy exertion	People with heart disease, such as angina, should reduce moderate exertion and avoid sources of CO, such as heavy traffic	Children, asthmatics, and people with heart or lung disease should avoid outdoor exertion. Everyone else should reduce exertion outdoors
201–300	Very unhealthy	Active children and adults, and people with lung disease such as asthma, should avoid all outdoor exertion. Everyone else, especially children, should avoid prolonged or heavy exertion outdoors	People with heart or lung disease, older adults, and children should avoid all physical activity outdoors. Everyone else should avoid prolonged or heavy exertion	People with heart disease, such as angina, should avoid exertion and sources of CO, such as heavy traffic	Children, asthmatics, and people with heart or lung disease should remain indoors. Everyone else, should avoid exertion outdoors

(Continued)

TABLE 18.5 Cautionary statements for criteria air pollutants in the United States [6]—*cont'd*

Index values	Levels of health concern	Cautionary statements[a]			
		Ozone	**Particulate matter**	**Carbon monoxide**	**Sulfur dioxide**
301–500	Hazardous	Everyone should avoid all physical activity outdoors	People with heart or lung disease, older adults, and children should remain indoors and keep activity levels low. Everyone else should avoid all physical activity outdoors	People with heart disease, such as angina, should avoid exertion and sources of CO, such as heavy traffic. Everyone else should reduce heavy exertion	

[a]*Nitrogen dioxide can cause respiratory problems in children and adults who have respiratory disease, such as asthma. The AQI for nitrogen dioxide is not included here because ambient nitrogen dioxide concentrations in the United States have been below the national air quality standard for the past several years. These concentrations are sufficiently low so as to pose little direct threat to human health. Nitrogen dioxide, however, is a concern because it plays a significant role in the formation of tropospheric ozone, particulate matter, haze, and acid rain.*

As concentrations increase, the proportion of people prone to experience health effects and the seriousness of these effects are expected to increase. Thus, the 1997 standards were intended to include an ample margin of safety as required by Section 109(b) of the Clean Air Amendments. The margin includes special concern about production the health of sensitive individuals. However, they were not considered risk free and exposures to ambient concentrations just below the numerical level of the standards may be problematic for the most sensitive individuals. However, exposures to levels just above the NAAQS are not expected to be associated with health concerns for most healthy individuals. Such is the complicated nature of individual response to air pollutants, and one of the objectives of the revised index is to provide sufficient information to allow sensitive people to avoid unhealthy exposures.

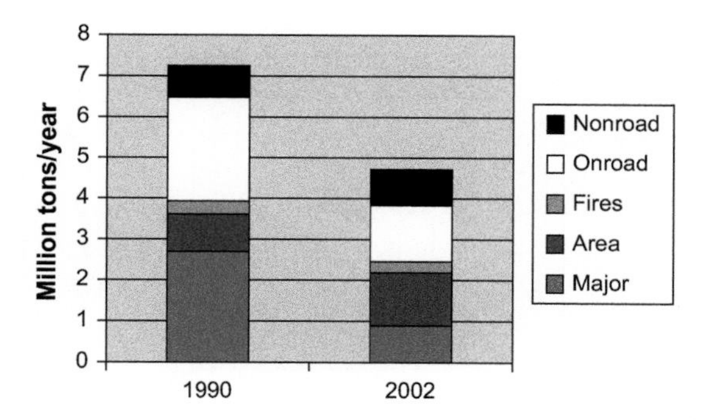

FIGURE 18.2 Trends in U.S. emissions of hazardous air pollutants [7].

Stationary area sources of HAPs are defined as sources that have the potential to emit less than 9 t a^{-1} or more of any single HAP or less than 22.7 t a^{-1} or more of any combination of HAPs. Mobile sources include on-road vehicles, nonroad equipment, and aircraft/locomotive/commercial marine vessels (ALM). EPA has developed the National Air Toxics Assessment (NATA) to estimate the magnitude of HAP emissions reductions and demonstrate reduced public risk from HAP emissions attributable to CAA toxics programs [7].

Emissions of HAPs fell through the 1990s in response to the MACT standards and mobile source regulations. However, area and other source emissions have increased because the EPA has not yet fully implemented its area source program as required by Section 112c(3) and 112(k) of the CAAA. Figure 18.2 and Table 18.6 present emissions trends for the sum of 188 HAPs by source sectors. Toxicity-weighted emissions have also declined between 1990 and 2002 for cancer and noncancer respiratory and neurological effects. Figures 18.3 to 18.5 and Tables 18.7 to 18.9 provide the toxicity-weighted emissions scaled to the sum of 6.57×10^6 t (7.24 million U.S. tons) for 1990 total emissions.

In 2002, benzene accounts for 28% of cancer risks in the toxicity-weighted NEI, whereas manganese accounts for 77% of noncancer neurological effects in the toxicity-weighted NEI and acrolein accounts for 90% of noncancer respiratory effects in the toxicity-weighted NEI [7].

A regional strategy can affect planned reductions at the urban and local scales, as well as a state or provincial strategy to achieve reductions at the state, provincial, urban, and local scales, and a national strategy to achieve them at national and lesser scales. The continental and global scales require an international strategy for which an effective instrumentality is being developed.

TABLE 18.6 Sum of Emissions of the 188 Hazardous Air Pollutants Listed in the Clean Air Act Amendments of 1990 [7]

Sector	1990 Emissions	2002 Emissions	% Reduction
Total	7.24 million tons	4.7 million tons	35
Major	2.69 million tons	0.89 million tons	67
Area	0.91 million tons	1.29 million tons	−42
Fires (wildfires and prescribed burns)	0.34 million tons	0.28 million tons	18
Onroad mobile	2.55 million tons	1.36 million tons	47
Nonroad mobile	0.75 million tons[b]	0.86 million tons	−15
ALM mobile[a]		0.02 million tons	

[a] *ALM-Aircraft Locomotive and Commercial Marine Vessels.*
[b] *1990 nonroad mobile include ALM.*

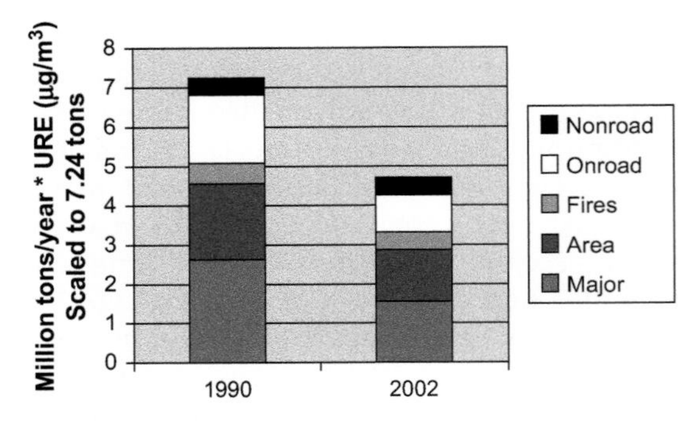

FIGURE 18.3 Trends in scaled cancer toxicity-weighted U.S. emissions of hazardous air pollutants. *Note*: Unit risk estimate (URE) is upper bound risk estimate of an individual's probability of contracting cancer over a lifetime of exposure to a concentration of 1 μg of the pollutant per cubic meter of air. For example, if an URE is 1.5×10^{-6} μg^{-1} m^{-3}, 1.5 excess tumors are expected to develop per 1,000,000 people if they are exposed daily for a lifetime to 1 μg of the pollutant in 1 m^3 of air [7].

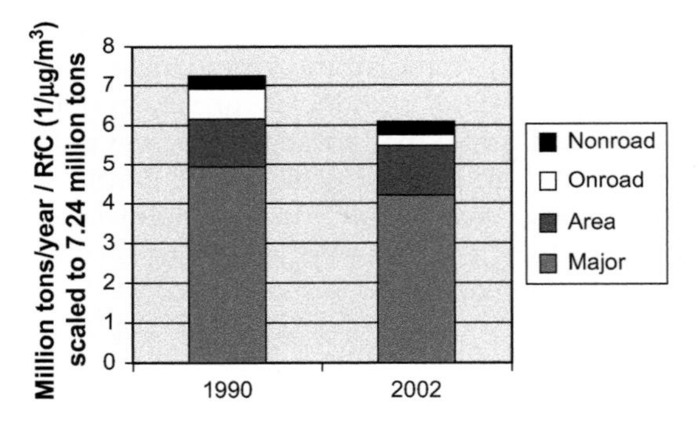

FIGURE 18.4 Trends in scaled noncancer neurological toxicity-weighted U.S. emissions of hazardous air pollutants. *Note*: RfC is the reference concentration, that is, the level below which no adverse effect is expected [7].

5.2. Tactics

The other major aspect of air pollution reduction is the control of short-term episodes on the urban scale. Prior to an episode, a scenario of tactical maneuvers must be developed for application on very short notice to prevent an impending episode from becoming a disaster. As an episode usually varies from a minimum

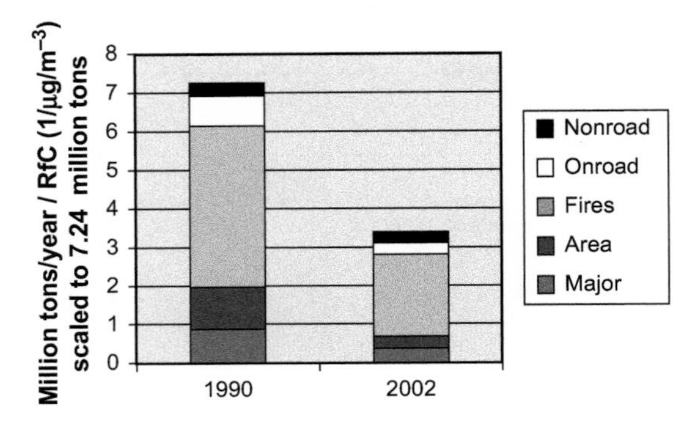

FIGURE 18.5 Trends in scaled noncancer respiratory toxicity-weighted U.S. emissions of hazardous air pollutants [7].

TABLE 18.7 Trends in Scaled Cancer Toxicity-Weighted U.S. Emissions of Hazardous Air Pollutants [7]

Sector	% Reduction from 1990 to 2002
Total	36
Major	44
Area	31
Fires (wildfires and prescribed burns)	−4
Onroad mobile	49
Nonroad mobile	7.5

TABLE 18.8 Trends in Scaled Noncancer Neurological Toxicity-Weighted U.S. Emissions of HAPs [7]

Sector	% Reduction from 1990 to 2002
Total	16
Major	15
Area	−7
Fires (wildfires and prescribed burns)	0
Onroad mobile	62
Nonroad mobile	12

TABLE 18.9 Trends in Scaled Noncancer Respiratory Toxicity-Weighted U.S. Emissions of Hazardous Air Pollutants [7]

Sector	% Reduction from 1990 to 2002
Total	54
Major	69
Area	64
Fires (wildfires and prescribed burns)	50
Onroad mobile	58
Nonroad mobile	35

of about 36 h to a maximum of 3 or 4 days, temporary controls on emissions much more severe than are called for by the long-term strategic control scenario must be implemented rapidly and maintained for the duration of the episode. After the weather conditions that gave rise to the episode have passed, these temporary episode controls can be relaxed and controls can revert to those required for long-term strategic control [3].

The mechanisms by which a jurisdiction develops its air pollution control strategies and episode control tactics are outlined in Figure 18.6. Most of the boxes in the figure have already been discussed—sources, pollutant emitted, transport and diffusion, atmospheric chemistry, pollutant half-life, air quality, and air pollution effects. To complete an analysis of the elements of the air pollution system, it is necessary to explain the several boxes not yet discussed.

5.2.1. *Episodes*

The distinguishing feature of an air pollution episode is its persistence for several days, allowing continued buildup of pollution levels. Consider the situation of the air pollution control officer who is expected to decide when to use the stringent control restrictions required by the episode control tactics scenario (Fig. 18.7 and Table 18.10). If these restrictions are imposed and the episode does not mature—that is, the weather improves and blows away the pollution without allowing it to accumulate for another 24 h or more—the officer will have required a very large expenditure by the community and a serious disruption of the community's normal activities. Also, part of the officer's credibility in the community will be diminished and the urgency when a real episode occurs will be weakened. If, however, the reverse situation occurs—that is, the restrictions are not invoked and an episode does occur—there can be illness or possibly deaths in the community that could have been averted.

In deciding whether to initiate episode emergency plans, the control officer cannot rely solely on measurements from air quality monitoring stations, because even if pollutant concentrations rise toward acute levels over the preceding hours, these readings give no information on whether they will rise or fall during the succeeding hours [3].

The only way to avert this dilemma is for the community to develop and utilize its capability of forecasting the advent and persistence of the stagnation conditions during which an episode occurs and its capability of computing pollution concentration buildup under stagnation conditions. Social and political considerations provide a point in the flow chart for the public debates, hearings, and action processes necessary to decide, well in advance of an episode, what control tactics to use and when to call an end to the emergency. The public needs to be involved because alternatives have to be written into the scenario concerning where, when, and in what order to impose restrictions on sources. This should be done in advance and should be well publicized because during the episode there is no time for public debate. In any systems analysis, the system must form a closed loop with feedback to keep the system under control. It will be noted that the system for tactical episode control is closed by the line connecting episode control tactics to sources, which

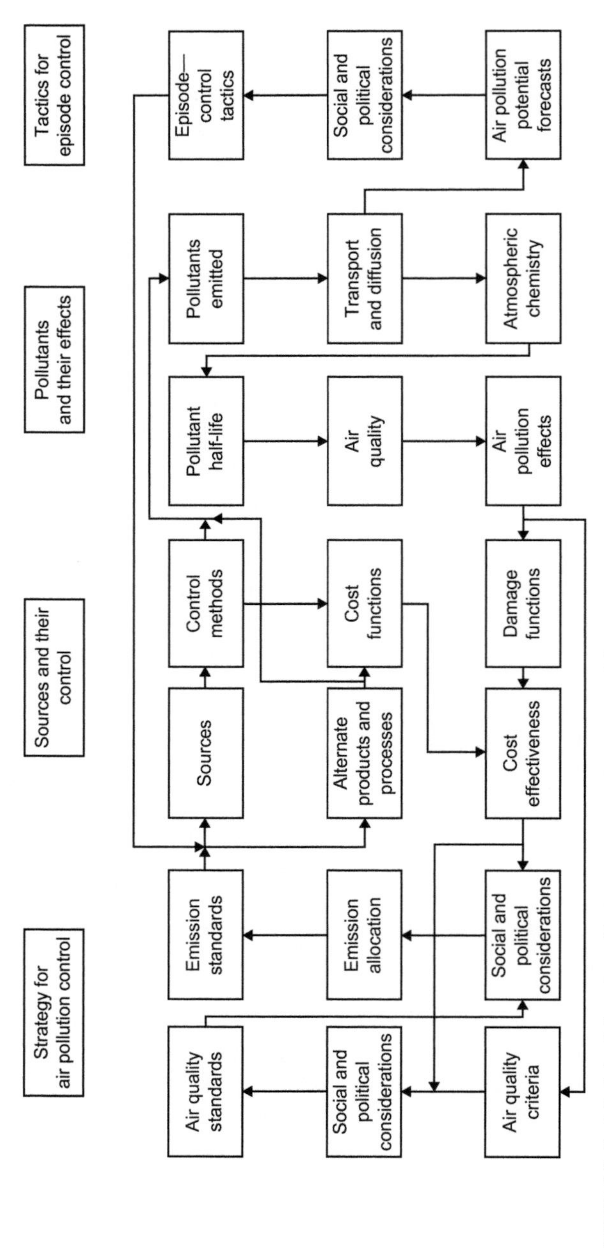

FIGURE 18.6 Model of the air pollution management system [3].

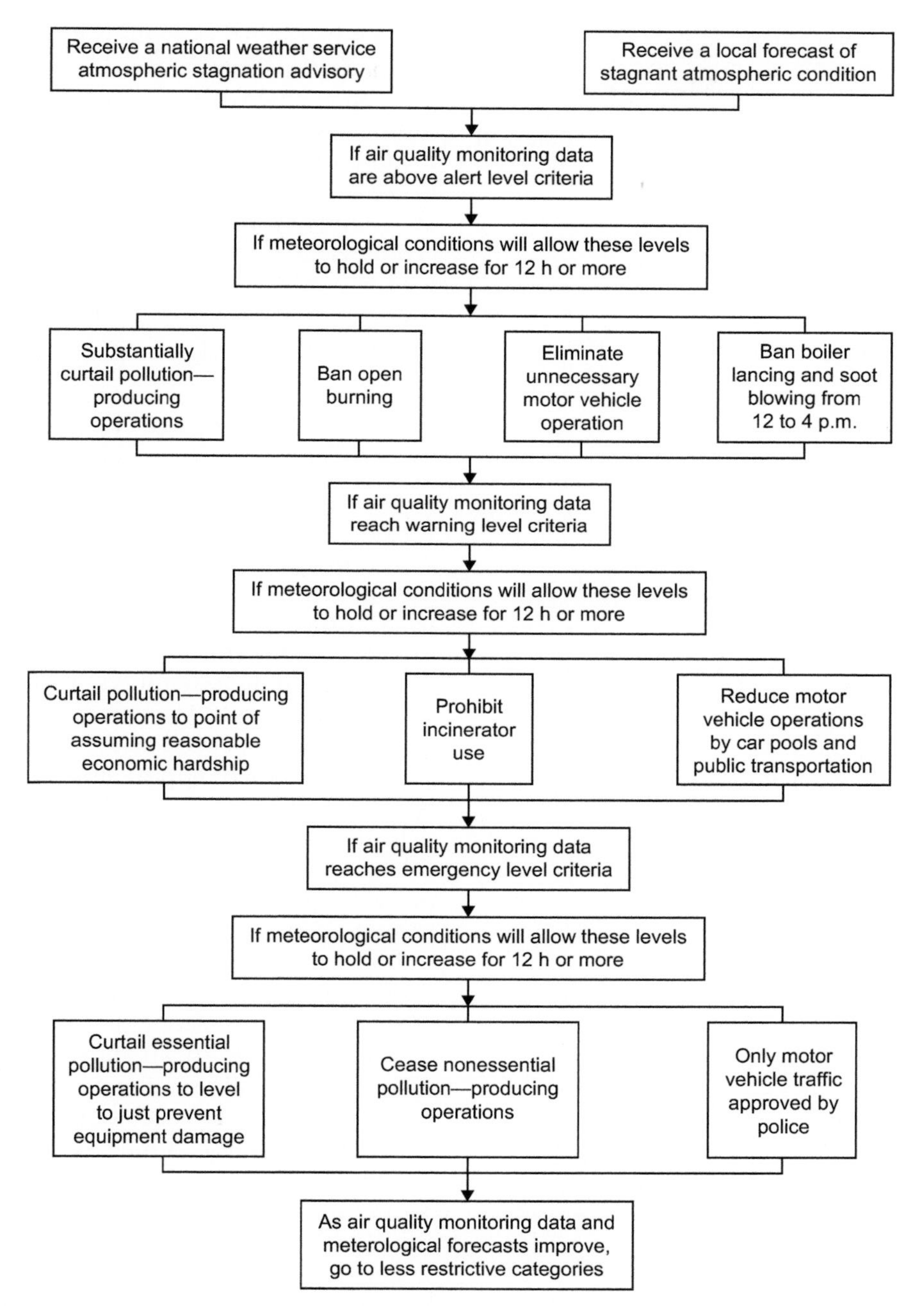

FIGURE 18.7 Air pollution episode control scenario. See Table 18.10 [3].

TABLE 18.10 United States Alert, Warning, and Emergency Level Criteria[a,b,c] [1]

	Concentration ($\mu g\ m^{-3}$)	Concentration (ppm)	Averaging Time (hrs)
Alert-level criteria			
SO_2	800	0.3	24
PM_{10}	350		24
CO	17	15	8
Ozone	400	0.2	1
NO_2	1130	0.6	1
	282	0.15	24
Warning-level criteria			
SO_2	1600	0.6	24
PM_{10}	420		24
CO	34	30	8
Ozone	800	0.4	1
NO_2	2260	1.2	1
	565	0.15	24
Emergency-level criteria			
SO_2	2100	0.8	24
PM_{10}	500		24
CO	46	15	8
Ozone	1000	0.1	1
NO_2	3000	0.6	1
	750	0.15	24

[a] *2003 Code of Federal Regulations, Title 40-Protection of Environment, Chapter 1—Environment Protection Agency, Appendix L, Example Regulations for Prevention of Air Pollution Emergency Episodes, 1.1 Episode Criteria, pp. 351–352.*
[b] *There is no criterion for lead, due to the chronic nature of the health effects of concern.*
[c] *Note: Append to each entry: Meterological conditions are such that pollutant concentrations can be expected to remain at the above levels for 12 or more hours or increase; or in the case of ozone situations is likely to reoccur within the next 24 h unless control actions are taken.*

means that the episode control tactics are to limit sources severely during the episode. As it takes hours before emergency plans can be put into effect and their impact on pollution levels felt, it is possible that by the time the community responds, the situation has disappeared. Experience has shown that the time for community response is slowed by the need to write orders to close down sources and to respond in court to requests for judicial relief from such orders. To circumvent the former type of delay, orders can be written in blank in advance. To circumvent the latter type of delay, the agency's legal counsel must move as rapidly as the counsel seeking such relief [3].

6. AIR QUALITY MANAGEMENT

Means of controlling air pollution sources include: (1) using devices to remove all or part of the pollutant from the gases discharged to the atmosphere, (2) changing the raw materials used in the pollution-producing process, or (3) modifying the operation of the process so as to decrease pollutants emitted. These are control methods (Table 18.11). Such control methods have a cost associated with them and are the cost functions that appear in the system. There is always the option of seeking alternate products or processes that will provide the same utility to the public but with less quantity of pollutants emitted. Such products and processes have their own cost functions [1].

Both controlling and not controlling air pollution have costs. All the adverse air pollution effects represent economic burdens on the public for which an attempt can be made to assign dollar values, that is, the cost to the public of damage to vegetation, materials, structures, animals, the atmosphere, and human health. These costs are called damage functions. To the extent that there is knowledge of cost functions and damage functions, the cost effectiveness of control methods and strategies can be determined by their interrelationship. Cost effectiveness is an estimate of how many dollars worth of damage can be averted per dollar expended for control. It gives information on

TABLE 18.11 Examples of Air Pollution Control Methods [3]

- **I.** Applicable to all emissions
  - A. Decrease or eliminate production of emission
    1. Change specification of product
    2. Change design of product
    3. Change process temperature, pressure, or cycle
    4. Change specification of materials
    5. Change the product
  - B. Confine the emissions
    1. Enclose the source of emissions
    2. Capture the emissions in an industrial exhaust system
    3. Prevent drafts
  - C. Separate the contaminant from effluent gas stream
    - Scrub with liquid
- **II.** Applicable specifically to particulate matter emissions
  - A. Decrease or eliminate particulate matter production
    1. Change to process that does not require blasting, blending, buffing, calcining[a], chipping, crushing, drilling, drying, grinding, milling, polishing, pulverizing, sanding, sawing, spraying, tumbling, etc.
    2. Change from solid to liquid or gaseous material
    3. Change from dry to wet solid material
    4. Change particle size of solid material
    5. Change to process that does not require particulate material
  - B. Separate the contaminant from effluent gas stream
    1. Gravity separator
    2. Centrifugal separator
    3. Filter
    4. Electrostatic precipitator
- **III.** Applicable specifically to gaseous emissions
  - A. Decrease or eliminate gas or vapor production
    1. Change to process that does not require annealing, baking, boiling, burning, casting, coating, cooking, dehydrating, dipping, distilling, expelling, galvanizing, melting, pickling, plating, quenching, reducing, rendering, roasting, smelting, etc.
    2. Change from liquid or gaseous to solid material
    3. Change to process that does not require gaseous or liquid material
  - B. Burn the contaminant to CO_2 and H_2O
    1. Incinerator
    2. Catalytic burner
  - C. Adsorb the contaminant
    1. Activated carbon
    2. Other sorbents

[a]*Calcining is the process of heating a substance to a temperature less than melting or fusing point, but sufficiently high to cause moisture loss, redox, and decomposition of carbonates and other compounds.*

how to economically optimize an attack on pollution, but it gives no information on the reduction in pollution required to achieve acceptable public health and well-being. However, when these goals can be achieved by different control alternatives, it behooves us to utilize the alternatives that show the greatest cost effectiveness [1,3].

To determine what pollution concentrations in air are compatible with acceptable public health and well-being, air quality criteria is used, which are statements of the air pollution effects associated with various air quality levels. It is inconceivable that any jurisdiction would accept levels of pollution it recognizes as damaging to health. However, the question of what constitutes damage to health is judgmental and therefore debatable. The question of what damage to well-being is acceptable is even more judgmental and debatable. Because they are debatable, the same social and political considerations come into the decision-making process as in the previously discussed case of arriving at episode control tactics. Cost effectiveness is not a factor in the acceptability of damage to health, but it is a factor in determining acceptable damage to public well-being. Some jurisdictions may opt for a pollution level that allows some damage to vegetation, animals, materials, structures, and the atmosphere as long as they are assured that there will be no damage to their constituents' health. The concentration level the jurisdiction selects by this process is called an air quality standard. This is the level the jurisdiction says it wishes to maintain [3].

7. CONCLUSIONS

The atmosphere can be seen as a receptor of wastes. Substantial progress has been made in improving air quality since the 1970s. All of the criteria pollutants in many countries have been reduced dramatically. The science, engineering, and technology needed to control the release of air pollutants have progressed, but still new ways of thinking about air pollution are needed. Economic progress has unfortunately been accompanied by seemingly intractable problems. Global GHG emissions continue to rise. Numerous toxic compounds continue to be released. Risks to human health persist. Sensitive ecosystems are threatened. Certainly, improved technologies and better regulations have resulted in cleaner air, but these emergent problems are not readily resolved by the old "command and control" approaches. Innovations and market forces must also be part of the solution, as must life cycle, preventive, and sustainable processes. Indeed, improved waste management approaches will translate into improved air quality.

References

[1] D.A. Vallero, Fundamentals of Air Pollution, fourth ed., Academic Press, Burlington, MA, 2008.

[2] Intergovernmental Panel on Climate Change, IPCC Special Report on Carbon Dioxide Capture and Storage. Prepared by Working Group III of the Intergovernmental Panel on Climate Change, in: B. Metz, O. Davidson, H.C. de Coninck, M. Loos, L.A. Meyer (Eds.), Cambridge University Press, Cambridge, UK and New York, NY, 2005.

[3] R.W. Boubel, D.L. Fox, D.B. Turner, A.C. Stern, in: Fundamentals of Air Pollution, third ed., Academic Press, Burlington, MA, 1984.

[4] U.S. Environmental Protection Agency, Air Quality Index (AQI) – A Guide to Air Quality and Your Health. http://www.airnow.gov/index.cfm?action=aqibasics.aqi, 2010 (accessed on July 5, 2010.)

[5] World Health Organization (2006). Report No. WHO/SDE/PHE/OEH/06.02.

[6] U.S. Environmental Protection Agency (2006). Guidelines for the Reporting of Daily Air Quality – The Air Quality Index (AQI). Report No. EPA-454/B-06-001. Research Triangle Park, NC.

[7] A. Pope, M. Strum (2007). 1990–2002 NEI HAP Trends: Success of CAA Air Toxic Programs in Reducing HAP Emissions and Risk. 16th Annual International Emission Inventory Conference, Emission Inventories: "Integration, Analysis, and Communications", Raleigh, NC, May 14–17.

CHAPTER

19

Ocean Pollution

J. Vinicio Macías-Zamora

Instituto de Investigaciones Oceanológicas, Universidad Autonoma de Baja California, Ensenada, Baja California 22860, México

OUTLINE

1. INTRODUCTION

Ocean pollution has become one of the largest and sometimes even conflicting subjects which very often result in threats to marine ecosystems, human health, and diminishing food sources. The pollution, as a result of oil spills and many other discharges, very often produces large "dead zones" due to nutrient enrichment and organic matter overload in coastal areas as well as extensive damage to amenities and resources. Human development has resulted in an acceleration of these effects due to the need for more resources.

To understand these changes to our ocean environment pollution, marine pollution studies have evolved and developed over the past few decades. It's been a long time since pollutant concentration measurement was at the center of the research, but today there is a great variety of focused research design to explain the characteristics of a particular concentration value measured in some compartment of the environment for a particular pollutant. For example, if a particular pollutant measurement is recorded at say 250 ng L^{-1}, one would like to know just what factors have contributed to this figure. It could be that the

Waste Doi: 10.1016/B978-0-12-381475-3.10019-1

number is made up of 150 ng L^{-1} from waste water, 57 ng L^{-1} from the atmospheric contribution, and another 82 ng L^{-1} from lateral advection and so forth. If the pollutant is located in the sediment, one may wonder, for example, as to the origin of the pollutant, what will be its final destination, how long will it take to decompose, what are its main metabolites, are these metabolite molecules more or less toxic than the parent molecule, and will it be available for incorporation into living organisms? Society in particular would like to know whether that pollutant concentration represents a threat to human and animal health, will it endanger some organisms, will it affect the quality of food collected locally, and will it alter part of the local ecosystem? Furthermore, managers of industries concerned about the environment would like to know where the quantities of pollutants are coming from and where they are going to, but mostly, they would like to identify the sources, to use the information to mitigate any potential effects of the pollutant in question by reducing its release from the source. In this chapter, we will first briefly identify the types of pollutants and then review a partial list of pollutants, with an emphasis on the most recent trends in research.

It has been well established and recognized that coastal marine pollution and marine pollution in general is mainly derived from land (as much as 80%) [1]. Consequently, the type of pollution found in coastal environments, including lagoons, bays, and estuaries, is largely a result of human activities being conducted on the coast and on land adjacent to the ocean environment.

1.1. Types of Waste

There are many classifications of pollutants in terms of the type, origin, and form of entry into the ocean. The most frequently used classification is the one that relates the pollutants to its source, these being a point source or a nonpoint source (or diffuse source). Clearly, pollutants coming from point sources are easier to pinpoint and consequently are also easier to control, reduce, mitigate, or eliminate. On the contrary, diffuse or nonpoint sources are intrinsically more difficult to handle.

However, we will try to avoid the classification problem and will start by listing those pollutants that have been important from the point of view of their impact on the coastal zones:

(i) Particle bound or insoluble pollutants (hydrophobic compounds)
- Organic matter
- Bacteria
- The "legacy" persistent organic pollutants (POPs)
  - Dichlorodiphenyltrichloroethanes (DDTs) and related compounds dichlorodiphenyldichloroethane (DDD) and dichlorodiphenyldichloroethylene (DDE)
  - Polychlorinatedbiphenyls (PCBs)
  - The Drins (aldrin, dieldrin, and endrin)
  - Mirex
  - Dioxins and furans
  - Hexachlorocyclohexane (HCH) (alpha, beta, gama(lindane) isomers mainly)
  - Hexachlorobenzene (HCB)
  - Endosulfanes
    Toxaphene
  - Chirality of POPs
- Oil
- Trace metals (soluble but particle bound)

(ii) New "POPs"
- Plasticizers (phthalates)
- Flame retardants polybrominateddiphenylethers (PBDEs)
- Perfluoroctanesulphonates (PFOSs)
- Herbicides

(iii) Garbage
- Plastic
- Batteries, paper, and cardboard

(iv) Soluble pollutants
- Nutrients (phosphates and nitrates)
- Hormones
- Antidepressants
- Opioids
- Penicillin
- Steroids
- Other pharmaceuticals endocrine disruptors (ECDs)

(v) Radioactive waste

The two fundamental reasons for studying the concentrations of some of the above chemicals in the environment (including the oceans and the sediments of the oceans) are the following: they are toxic (lethal or chronic) to organisms and they persist in the environment with half lives measure in years and tens of years.

2. SOURCES OF POLLUTANTS TO COASTAL ENVIRONMENTS

2.1. Organic Matter and Nutrients

The four most important sources of waste in the marine environment are, not necessarily in order of importance, as follows: discharges from wastewater treatment plants, rivers, coastal run-off, and atmospheric transport.

In many parts of the world, the main mechanism for waste introduction from land into the marine environment is the discharge of wastewater. There are many substances involved in this discharge (Fig. 19.1), including organic matter (both dissolved and particulate), nutrients (particularly nitrates and phosphates), POPs, trace metals, linear alkyl benzenes (LABs), and linear alkyl benzene sulfonates (LAS), oil-related compounds, and other hydrocarbons and what are a very broad conglomerate of substances designated as new pollutants. One type of material entering into the ocean, which is of particular concern, is particulate organic matter (POM) along with nutrients. A consequence of the dramatic impact of these materials in coastal waters is the promotion of large hypoxic ($<2\ mg\ L^{-1}$) O_2 concentration compared to the average O_2 concentration in the ocean of >8 mg/L in a well-saturated surface water and even, anoxic environments [2, 3]. As can be expected, large areas of hypoxic environments are often associated with river discharges because of the large volumes that are required to make a noticeable impact in the oxygen concentrations of the ocean in open waters. However, in some shallow coastal marine areas, zones of hypoxia and anoxia have developed without the aid of river pollution [4].

The hypoxic–anoxic dead zones are, in addition, also associated with global change in more ways than one. Warmer temperatures in some parts of the world will result in an increase of rain, which in turn will result in the transporting of larger quantities of nutrients into the ocean. At the same time, these zones of below normal oxygen concentrations, that also include large organic material (OM) loads, represent adequate environments for methane generation, which when entering the atmosphere will contribute to global warming [5].

The characteristics of the allochthonous OM introduced into the sea are not trivial. These materials may include refractory materials in addition to nutrients that promote, on the one hand, oxygen consumption directly via the oxidation of the OM, and, on the other hand, they also help to promote the explosion or phytoplankton growth with their subsequent death and decomposition followed by oxidation of this authocthonous OM. According to Zonneveld et al. [6], the preservation of OM in the marine sediment is not a well-understood process. In any case, as a result of anthropogenic discharges of wastewater into restricted marine coastal zones, local events of hypoxia and anoxia may result as well as an increased release of methane to the atmosphere.

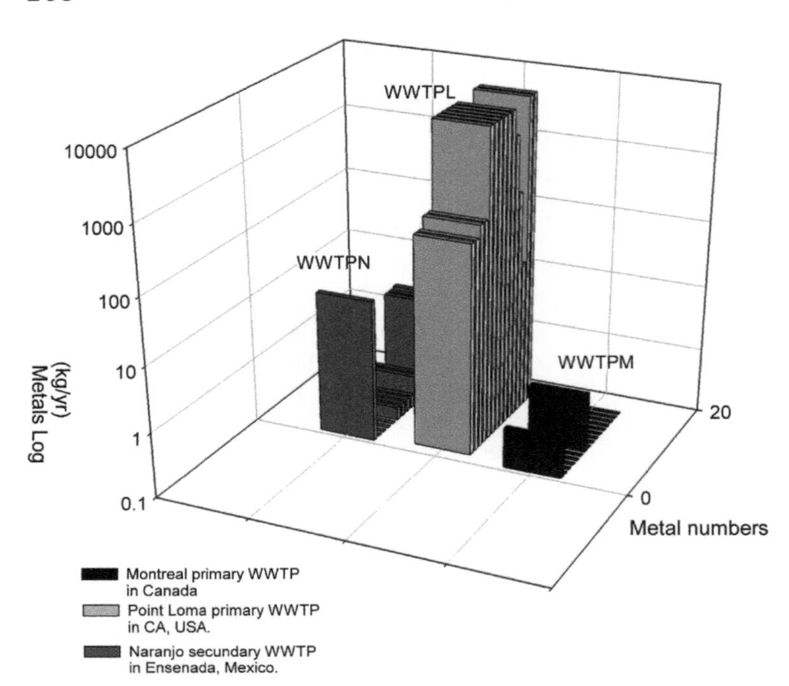

FIGURE 19.1 Typical discharges for three different wastewater treatment plants in three different countries. Differences are more related to volumes than efficiency of plants.

In the sediment, the excess of OM from allochthonous origin can give rise to other effects such as changes in the benthic communities [7]. In addition, the presence of OM in the sediment will diminish the oxygen concentration in interstitial water, which in turn will produce changes in redox potential and changes in the concentration of trace metals such as Fe, Mn, and Co, for instance, which are affected by redox processes.

The OM plays an important role in the feeding habits of coastal organisms. For example, Ramírez-Álvarez et al. [8] have shown that organic matter from wastewater sewage derived particle organic matter (SDPOM) released from coastal point sources is being used by bottom feeders to obtain a percentage of their carbon intake. In their research, Ramírez-Álvarez et al. have explained that using stable isotopes for carbon and nitrogen, as well as the C/N ratio as an indicator, shows that as much as 40% of the carbon incorporated in tissue of the polichaeta *Sphiophanes duplex* may come from wastewater. It is important to realize that sometimes the pollutant concentration itself may yield only partial information in elucidating the origin of pollutants. In several instances, the information provided by the interpretation of the isotopic signatures has proved to be very helpful in improving our knowledge of material movement between compartments in ecosystems.

3. TRACE METALS

3.1. Dissolved

One of the most often associated pollutants into the marine environment is the group of elements known as trace metals. These elements can enter from rivers, atmosphere (dust particles), and wastewater discharges. They are ubiquitous chemicals into the marine environment as they are mainly the result of erosion processes of rocks, minerals, and sediments in land and ocean. Trace metals in the ocean are often in the form of ions (cations) and to a lesser extent attached to particles bound to silicates, aluminates and/or to organic matter. These charged ions, dissolved in the water, are often trapped by solid particles (clay and other materials) and consequently end up at the bottom in the sediments.

Generally speaking, the concentration of most trace elements in seawater are extremely low, usually in the parts per billion (ppb or ng mL^{-1}) to the parts per trillion (ppt or pg mL^{-1}). The concentration of the ions of a trace metal in the marine environment has an upper limit defined by its solubility product, K_{sp}. Particulates (clays and OM) in the marine environment are responsible for scavenging trace metals, removing them from the water, and decreasing their concentrations below the equilibrium limits defined by the solubility product.

There are a number of issues related to the dissolved trace metals: their measurement has only been possible by the advent of ultra-clean collection techniques, by the understanding of their behavior in the marine environment (oceanographic consistency), and by the better sensitivities obtained by using the latest analytical instrumentation. Measuring trace metals at such low levels usually requires concentrating and/or eliminating the saltwater matrix. This is carried out in three major schemes: (a) using organic chelating agents such as dithiocarbamate (for example, work by Boyle and Edmond [9] and by Sawatari et al. [10]), (b) concentration of trace metals by ion exchange chromatography in a chelating resin such as Chelex 100 (for example, work by Delgadillo-Hinojosa et al. [11] and Dwinna et al. [12]), and (c) by co-precipitation with aluminum (see, for example, work by Doner and Ege [13]).

However, even at such low dissolved concentrations, there are very clear concentration differences in many trace metals in coastal waters compared to oceanic waters. For example, silver concentration has been reported as large as 307 pM (pico-molar) inside San Diego Bay and as low as ~3 pM at the North Pacific Ocean [14] and references therein. They also found that largest coastal concentrations for silver were located near wastewater discharges along the coast, underlining the role as point sources for this and other trace metals. They further argued that most of the coastal concentration was accounted for by wastewater discharges.

Other trace metals such as Co or even Cu and Ni have also shown to have gradients initiating in coastal areas associated to point as well as diffuse sources. For example, the gradient concentration reported for Co in the coastal waters of Baja California ranged from 0.93 to as much as 3.6 pM [15]. That is about four times the concentration found in oceanic waters. In other parts of the world, a similar pattern has been identified and reported (see, for example, work by Girbin et al. [16] and Stauber, et al. [17]).

3.2. In Sediments

The presence of trace metals in marine sediments are primarily, even in coastal environments, the result of normal biogeochemical processes such as dissolution of rocks by rain, wind erosion, and transport of soil, combined with riverine transport or water run-offs. Their concentration in the sediment depends on several parameters that include sediment composition, texture, physicals sorting, oxidation/reduction reactions, and also anthropogenic inputs. The main purpose of measuring trace metals in sediments is to measure both the contribution from normal biogeochemical processes and the contribution from processes of anthropogenic origin. To do this, and because of the need to compare sites with very different granulometric compositions, methods have been devised to correct for the textural differences between the sites. In the past, corrections between sites were mostly accounted for by using physical methods. Sieving of samples followed by collection and analysis of a particular range of sizes (say >63 μm) was the frequently used method. There were limitations to this method; one was to ignore the possible contributions of larger sediment grain sizes. As a consequence, geochemical normalization methods were developed.

In using geochemical normalization procedures, it is important to consider what characteristics or conditions should be used in selecting a normalizer. First to consider are the two physical processes that bring the sedimentary materials into the marine environment. One such process involves the breakdown of the rocks into smaller particles such as clays and silts. The other process involves the increase in the surface area as the particle size decreases, with the resultant capacity to capture charged particles and ions as well as incorporating lipophyllic organic molecules. Due to their characteristic hydrophobicity and their tendency to be excluded from the water phase, organic molecules also tend to attach to these particles. As a result, sites made up of deposited material will often contain large quantities of fine sediments and hence more trace metals and organic matter than sites which do not contain such depositions.

Luoma [18] has suggested that if a normalized procedure is to be used then this normalizer must be highly correlated with the surface area of the sediment. This implies that the normalizer can be considered as a proxy for surface area of the sediments. Hence, for a particular site, any metal that covaries with the size distribution of the sediments can also be a good normalizer. There are certain things that have to be complied with for a normalization procedure. The preferred process is as follows:

a) In the first place, a good normalizer is one that is abundant and hence is not the trace metal itself. Also, due to its large size (usually expressed as a percentage) there is little chance of it being affected by anthropogenic activities.

b) It is necessary to generate a large database. This means a relatively large number of sites in a relatively large area. This will insure that the required variability is captured in the database.

c) It is a good idea if the same digestion method is used to determine both the trace metal concentrations and the normalizer concentration.

d) Aluminum and iron are two elements that are often considered as normalizers (see Fig. 19.2 and Fig. 19.3). But because of the difficulties in extracting all of the available aluminum, iron is often the preferred choice. If Al is chosen, then it requires HF to dissolve the matrix, together with the strong acid digestion of nitric + hydrochloric acids ($HNO_3 + HCl$). If one only uses the strong acid procedure, then Fe would be the better alternative because Al recoveries are poor without HF.

e) Once the normalizer is selected, then linear regressions are developed for all samples. If possible, data near point sources are excluded a priori from the database. Then, from the regression, those data points that are clearly outside one standard deviation are eliminated from the database and the linear regression is recalculated. The process of elimination is repeated until no more data points are outside one standard deviation.

f) This final regressions is expected to represent the normal geochemical variation characteristic for the area due to erosion, wind transport, and so on., of material into the coastal area.

g) Points above the 95% prediction lines [19] represent enriched sites and could be explained in terms of anthropogenic influences.

Sometimes, gradients in trace metal concentrations determined from surface sediments starting from well-known point sources are not clear. This often arise from the combined effects of many factors, including the influence of other nearby point sources, effects of diffuse sources, variability in water currents and winds, topography, and so on.

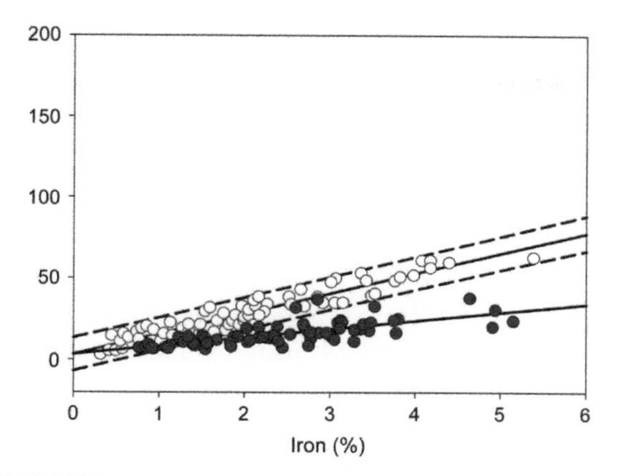

FIGURE 19.2 Normalization for trace metals in marine sediments. The open circles are part of the data showing the normal geochemical variations for Chromium at the Southern California Bight 1998. Open circles are from point conception to the Mexican border. The filled circles are data for the Southern California Bight from the Mexican border to Punta Banda Baja California.

Once the trace metal concentration has been determined to be outside the normal variation expected by natural processes, then and only then may we suspect enrichment due to waste from a nearby source.

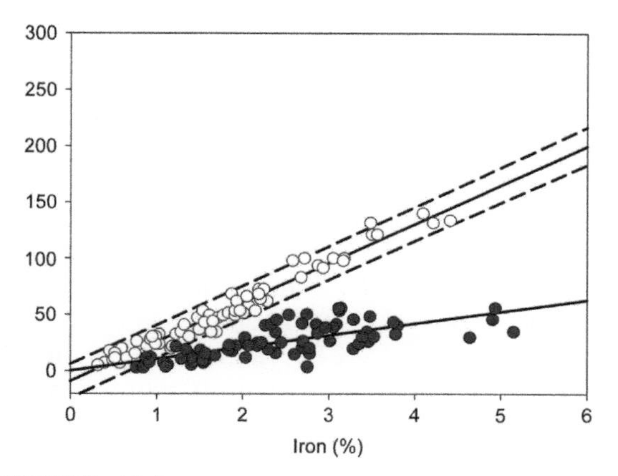

FIGURE 19.3 Normal variability for trace metals in marine sediments. The open circles represent part of the database for the normal geochemical variations for Zinc in the Southern California Bight 1998 from point conception to the Mexican border. The dark red data points are the concentrations found in the Mexican portion of the SCB from the border to Punta Banda in Ensenada, B.C. México.

Sediment quality guidelines (SQG) is an area of intense research with the specific purpose of finding concentration levels of trace metals below which no effects on biota, ecosystems and or humans is expected. For example, the use of Effect Range Low (ERL) and Effect Range Medium (ERM) [20] were first used to relate concentration levels of pollutants, including trace metals, to toxicity effects on organisms. This approach was followed by the development of more refined guidelines; these were more specifically oriented to determine threshold levels below which biological adverse effects have been noticed either rarely, occasionally, or frequently. Such approaches include using the Threshold Effect Level (TEL) and Probable Effect Level (PEL) proposed by MacDonald et al. [21]. A better alternative to understand the possible effects of human intervention in sediments can be obtained with the multiple evidence such as that of the triad approach.

4. ORGANIC WASTE

4.1. Oil Pollution

Oil pollution is one of the most predominant forms of ocean pollution causing severe damages to amenities, ecosystems, and resources. There have been over 25 major oil spills in the oceans of the world since 1967 when the Torrey Canyon ran aground off Cornwall on March 18 discharging 38 million gallons (0.14×10^6 L) of oil. The most recent disaster in the Gulf of Mexico is a result of the Deepwater Horizon drilling rig explosion of April 20, 2010, which precipitated a sea floor gusher of over 60,000 barrels of oil per day (9.5×10^6 L d^{-1}). This is the largest oil spill disaster in U.S. history. Before the Deepwater Horizon accident, oil spill disasters were more or less "tolerated" because the pollution was not always very visible and after all, over a period of months rather than years, the oil, or at least most of it, appeared to be decomposed by bacteria. As a result, up to recent times, it was not always included in monitoring programs in spite of the large number of oil spills, which appear to be on the increase.

The hydrocarbons in crude oil, gasoline, and lubricating oils encompass a large number of chemicals, and these also find their way into the coastal zone from wastewater discharges [22]. Waste oil (from spent oil from cars) is frequently discharged into municipal water works in particular in underdeveloped countries. This, together with oil washed out from roads and parking lots by rain, often finds its way to the coast. It has been estimated that as much as 2.67×10^9 L of oil find their way to the sea each year. Till recently, the main source of oil in the oceans was from land sources as opposed to oil spills and marine accidents from ocean operations. This is changing, and in 1993, 50% of the ocean's oil pollution was from land sources and 50% was from marine accidents. In 2002, land sources contributed only 37% (see work by Gesamp [23] and oils.gpa.unep.org/facts/sources.htm). A third source is the occurrence of natural seepage in coastal environments.

The composition of the wasted oil and crude oil are variable. One of the main components is the family of saturated or aliphatic hydrocarbons. Their molecular size range from the one-carbon atom compound (methane) and other gaseous hydrocarbons to larger molecules that include liquid phase n-hydrocarbon components, waxes, and solid hydrocarbons. Other components include naphtenes, aromatics, and asphaltenes among many [24].

However, to complicate matters, there are also other sources of hydrocarbons that reach the marine environment. They can originate in cuticle waxes from superior plants (see, for example, the work done by Eglinton and Hamilton, [25], Prahl et al. [26], and Page et al. [27]), hydrocarbons from bacteria (see the work by Nishimura and Baker [28]), and phytoplankton (work done by Hayakawa et al. [29]). As in the case of trace metals, it can be important to know whether a particular hydrocarbon in the sediment or in the water has a natural origin or whether it comes from municipal wastewater discharges or run-off.

To identify the origin, special indexes are used. The most frequently used index is the carbon preference index (CPI). To identify the presence or absence of species, the Unresolved Complex Matter (UCM) index is used. The CPI is an index that identifies and differentiates between biological sources and crude oil sources of hydrocarbons. It is usually calculated as a ratio of the sums of odd/even chain lengths of n-hydrocarbons.

$$\mathrm{CPI} = \frac{\sum \mathrm{odd}}{\sum \mathrm{even}} \tag{19.1}$$

This index takes into account n-hydrocarbons of differing molecular sizes. When the CPI value for a sample of hydrocarbons is close to unity, it is most likely that the hydrocarbons originate from crude oil or from an oil spill. However, if the index is very different from unity (either larger or smaller than one), then it is most likely that the hydrocarbons originated from some biological process [30].

The UCM is often hidden in a broad peak of a chromatogram that represents a large number of compounds that are not separated by the normal chromatographic process (Fig. 19.4). It includes heavily branched and often cyclic or aromatic compounds. Many of these molecules are not taken up by bacteria. Very often, compounds with a CPI value close to unity together with at least one chromatograph peak representing the UCM are indicative of a nonbiological origin [31–33]. There are other indexes that have been used in conjunction with the CPI and the UCM indexes. The presence and relative abundance of pristane and phytane and their respective ratios associated to n-C_{17} and n-C_{18} have also been used [29]. Unfortunately, the index technique is not perfect and the indices

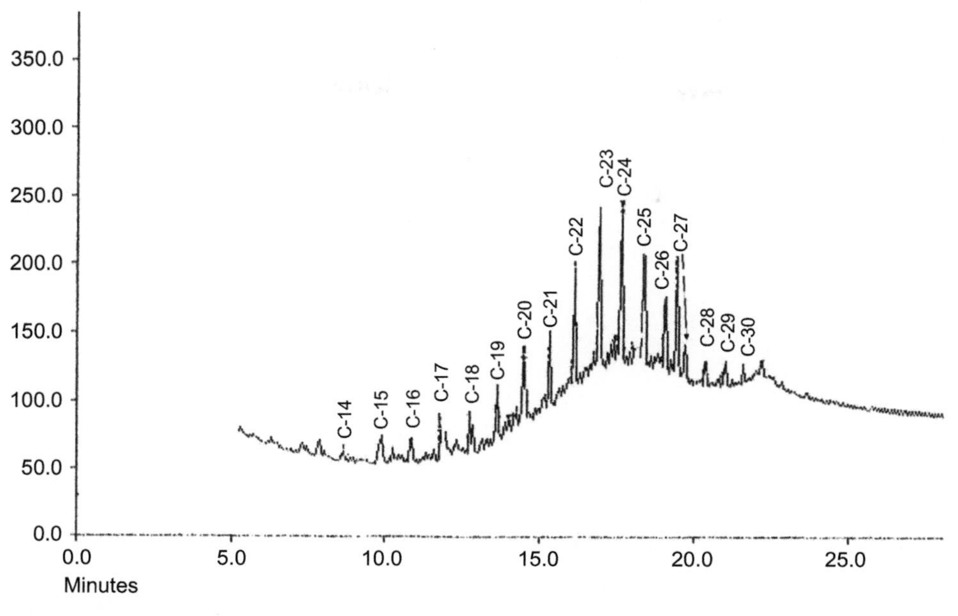

FIGURE 19.4 Typical gas chromatogram showing a sequence of n-hydrocarbons from a sediment sample at SCB near the Mexican-USA border. The Unresolved Complex Mixture is also evident.

may give contradictory information; additional sources of information are then required to resolve the apparent contradiction.

One important component of wasted oils and petroleum is the presence, in variable quantities, of aromatic hydrocarbons and in particular the polyaromatic hydrocarbons or PAHs. This is a very important family of compounds because they represent some of the most toxic components found in crude oil.

PAHs can have several origins. They are present in crude oil as mentioned above, smoke from grass or forest fires (a very important source; see Fig. 19.5), exhaust fumes of internal combustion and diesel engines, and in coke plants [33]. Sometimes, it is important to know the origin of hydrocarbon. Determining the source can help in the decision of how important anthropogenic sources are in relation to natural sources. One of the criteria used for source identification is the presence or absence of alkyl substituted PAHs. The temperature of formation determines the degree of substitution on PAH molecules. At high temperatures, one can expect mostly unsubstituted PAHs. At intermediate temperatures, a mixture of substituted and unsubstituted PAHs is usually obtained and at low temperatures, typical of those of crude oil formation, a large number of substituted PAHs are formed [34–36]. Frequently, in addition, the stable isotopic signature of carbon in the molecules has to be determined to help in the decision of the origin of the PAHs [37]. Other approaches have also been used. For example, if the ratio of methyl phenanthrenes to phenanthrene PAHs is >2 then its origin is petrogenic; if the ratio is <0.5 then it is pyrogenic [38]. Finally, principal component analysis may also help in determining the origin and hence source of the PAHs.

The recent accident in the Gulf of Mexico, with the consequent introduction of large

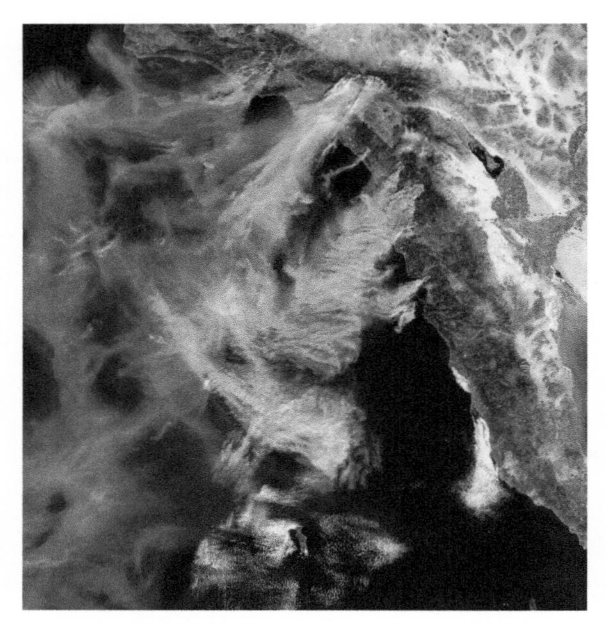

FIGURE 19.5 Satellite image from summer brush fires in California-Baja California region. Large quantities of PAHs are released into the atmosphere and later they reach coastal zones.

amounts of crude oil, has made the need to understand the physicochemical behavior of such a complex substance more pressing.

According to news releases, somewhere between 5000 and 60,000 barrels of oil per day (0.8×10^6 L d^{-1} to 9.5×10^6 L d^{-1}) are being released in the ocean, and it has become today the worst oil spill accident in the U.S. history. Several approaches have been used to diminish the impact of this catastrophic event. First, skimmers have been used to collect a small but significant amount of oil from the surface of the sea. They have also used collectors called booms that can resist fire and part of the oil has been burned. This is controversial as it really means changing one form of pollution (marine) for a different one (atmospheric).

In addition, the oil has also been dispersed in two different forms. Large amounts of oil dispersant called Corexit 9500 has been used at depth, mixing it with the crude oil released under the water. This is certainly a novel approach but it might indeed be ill conceived. For one thing, there is little evidence that the oil will magically disappear. The only (dubious) "advantage" is that amounts of dispersed oil will not be seen on the ocean surface and public outrage can be temporarily subdued. The disadvantage is that the degradation of the oil at the lower temperatures of the deeper water may be slowed down and the dispersed oil may be transported to unsuspected locations in a less predictive manner, with unknown consequences. In fact, the advantage of surface oil is that the oil is very visible and can be dealt with in a more appropriate manner. The immediate tragedy is that birds, fish turtles, and sea grasses among other organisms will die and the fishermen will loose their livelihood. The other tragedy, through more subtle effects, involves the biota that will be victims of the destroyed habitat.

We recently conducted an experiment in laboratory to monitor the degradation of oil in marine sediments in the presence and in the absence of the surfactant and dispersant Corexit 9500. One of the conclusions was that the dispersant did not help the rate of degradation in the 4 months long study [39].

Finally, the dispersant has also been used at the surface. Although oil spill engineers have had a lot of experience with Corexit 9500 on the ocean surface, its use is still controversial. It might be rated as being as much as 10 times less toxic than other dispersants, but it is not classified as an innocuous substance. Removal of the oil from the surface does not mean that the oil has been completely removed—much remains in the mile long column of sea water, and organisms living in the sea are still exposed to this oil. In spite of the severity of this accident, it is almost certain that oil drilling in the Gulf will continue as it will in many other deep water sites around the world, and we can expect further oil spill disasters. The severity of the Gulf oil spill is slowly being revealed as more research establishments in the United States and around the world are studying the possible impacts on life in the Gulf, and there is also an incipient program in Mexico because of the strong possibility that part of the oil, either on the surface mainly wind driven or as dispersed oil below the surface via currents, may reach its shores in a near future.

4.2. Persistent Organic Pollutants

The pollutants known as POPs include the organo-chlorine compounds, some of which can affect the nervous system and reproductive system and some are known carcinogens, and all have been shown to degrade very slowly. In this group of substances are (a) DDT and the *ortho* and *para* isomers of DDT, DDE, and DDD; (b) polychlorinated biphenyl or PCB; (c) aldrin, dieldrin, and endrin; (d) HCH (alpha, beta, gamma (lindane) isomers); (e) HCB; (f) dioxins and furans; (g) toxaphene; (h) mirex; (i) endosulfanes; and (j) metoxychlor.

Another characteristic that makes POPs a threat to organisms, ecosystems, and human health is their hydrophobicity and strongly lipophilic properties. These slow decomposing compounds persist in the environment and many have long half-lives measured in years. Because they last a long time in the environment before degrading, there is a good chance that they could be incorporated into the human food chain through uptake by fishes and readily bioaccumulate and biomagnify. Because of their hydrophobicity, they have a tendency to adhere and dissolve in organic fatty tissues (e.g., fish) and POM when released into water bodies. This tendency is quantified by the octanol-water partition coefficient which for most POPs is large (e.g., for DDTs, chlordanes, and many other POPs, Log $K_{ow} > 5$). This partitioning process involving uptake by POM results in much of the POP compounds settling in the sediments. It is often important to know when a particular pollutant was discharged or applied. Very often, simple concentration distributions of pollutant and degradation products can answer such questions. For example, in an area where DDT has been used, if the ratio is

$$\frac{[\mathrm{DDE}] + [\mathrm{DDD}]}{[\mathrm{DDT}]} > 1 \tag{19.2}$$

it suggests an old residue as the concentration of the degradation products is greater than that of the parent compound, DDT (see Fig. 19.6). Put in another way, if the ratio is

$$\frac{[\mathrm{DDT}]}{[\mathrm{DDE}]} > 1 \tag{19.3}$$

it suggests a relative recent DDT application as the concentration of the degradation product is small.

The same principle applies in the case of aldrin. If the ratio is

$$\frac{[\mathrm{Aldrin}]}{[\mathrm{Dieldrin}]} > 1 \tag{19.4}$$

it suggests a recent application of aldrin, as dieldin is a degradation product of aldrin.

As far as these simple measurements are concerned, we can only obtain a certain level of information regarding pollutant histories. Recently however, many authors have relied on enantiomeric separation of racemers of organochlorine compounds with chiral centers. The following are some of the POPs molecules that are chiral; α-HCH, cis- and trans-chlordane (CC, TC), heptachlor (HEPT), and o,p′-DDT. Using chirality as an indicator requires a detailed separation of chiral species using gas chromatography or liquid chromatography with chiral columns.

The industrial processes involved in manufacturing chiral compounds are usually nonselective and racemic mixtures (a 50% Levo and 50% Dextro) are produced. This is true o,p-DDT.

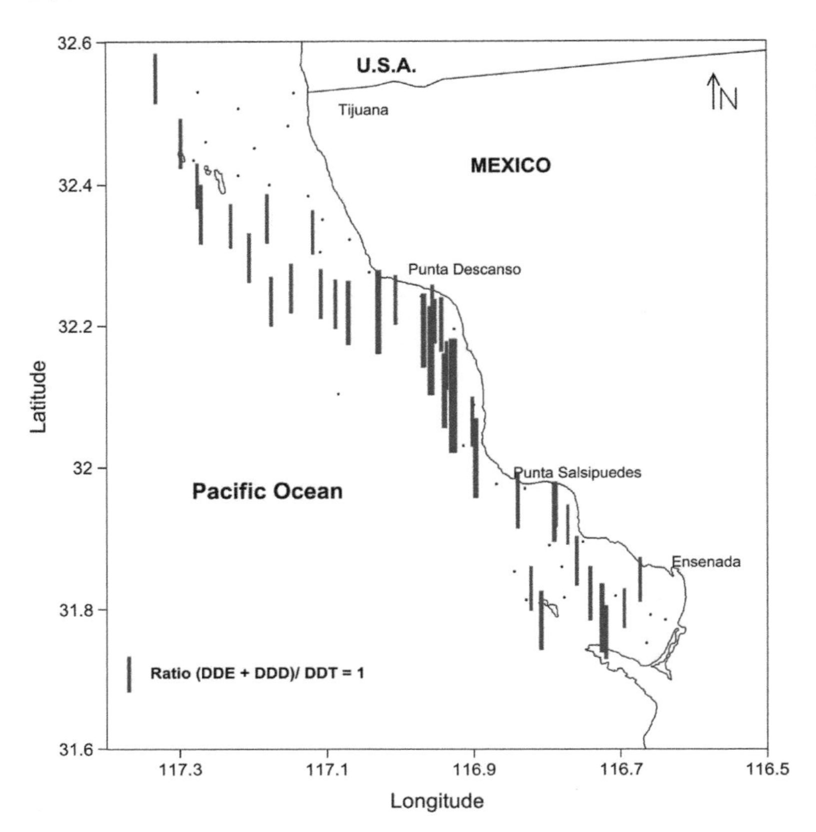

FIGURE 19.6 Ratio of DDT metabolites to DDT for the Southern part of the Southern California Bight. Values smaller than one would suggest recent introduction of DDT into the marine environment. Most ratios in figure are >1.

In a racemic mixture (Fig. 19.7), half of the mixture has the ability to rotate polarized light to the right (clockwise) and thus are known as *dextro* or simply D or (+). The other half has the ability to rotate polarized light anticlockwise and are labeled *levo* or simply L or (−). The rotation of polarized light is however not a parameter that is determined in chiral separation. What is important is that there are two distinct molecules. As it happens, all physical processes

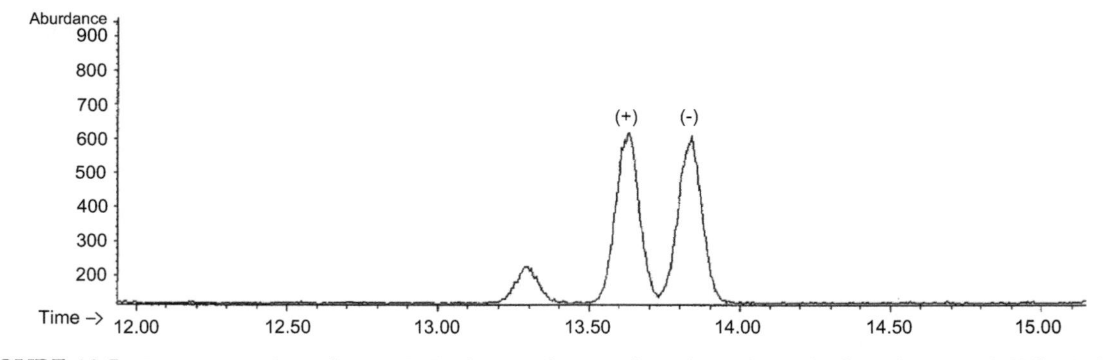

FIGURE 19.7 Enantiomeric forms for o, p-DDT. The sample was collected in soil samples from the Mexicali Valley in 2008.

such as evaporation, dissolution, adsorption, and so on, will not select one form of the molecule over the other. However, once the mixture reaches the soil or sediment for example, it will be subjected to biochemical processes that are selective. The selectivity is due to the presence of biomolecules such as enzymes that are able to select and degrade only one of the two enantiomeric forms. Consequently, one of the two racemics will diminish in concentration and the other one will remain mostly unchanged. In this sense, this chiral property may have several implications. For example, it means that, very frequently, half of the amount used to kill plagues may have been useless for the intended purpose and in fact may have had unintentional toxic effects for other nontargeted organisms.

For research purposes, the degradation of one enantiomer may serve for several purposes based on the simple enantiomeric ratio or ER = (+)/(−). This ratio is still not well understood. Sometimes, for example in soils, you may find a ratio >1 and in some instances a ratio <1 [40]. Maybe, both enantiomers are being degraded to some degree. However, it is not clear what role physical factors play in producing clear cut trends.

More research directed to better understand how the enantio-selectivity works will go a long way in understanding the ER as a clock for enzymatic degradation reactions. In any case, these ratios have been helpful in identifying the connection between soil and air composition for these compounds. It has been also used to identify sources and destinations. Still, much remains to be done to fully comprehend the functioning of these selective degradation pathways of toxic molecules in the environment.

Finally, with some of these compounds trapped in remote areas in ice and snow, and with the onset of global warming, we may see the release of "old" enantiomeric signatures. These need to be investigated and may further complicate the understanding of this topic.

4.3. Polychlorinated Biphenyls

These organochlorine compounds are ubiquitous in nature. They are found in most environmental compartments—air, water, biota, soils, and sediments and are considered to be toxic substances. The compounds, that make up the group known as PCBs, are differentiated by different numbers of chlorine atoms in the biphenyl rings. The larger the number of chlorine atoms in the molecule, the less soluble is the compound in water and the less volatile and more persistent (or less prone to degradation) it is in the environment. PCBs are, because of their nature (high K_{ow} values, hydrophobic and lipophilic), frequently attached to particles and removed to the bottom of water bodies. Concentrations are then larger in marine sediments than in seawater. These compounds, as in the case of POPs, will bioconcentrate in fish [41] and biomagnify up the food web.

Distribution studies in the marine environment have proved again the large influence that a source has in determining the gradient distributions. Very recently, Zhang and Lohman [42] have shown gradient concentration of PCBs in oceanic waters of the central Pacific. Furthermore, they have also shown that PCBs have a tendency to transfer from seawater to the atmosphere—indicating a nonequilibrium state.

Although these and other organo-chlorine compounds are called the "legacy" pollutants as they are no longer being manufactured, and hence are no longer playing the preponderant role that they once did, they are still very much present in the environment. Furthermore, their presence is expected to increase in the atmosphere and in general circulation due to the global warming process. In addition to these compounds, new pollutants such as the poly brominated diphenyl ethers (PBDE), a flame retardant, and perfluoroctanesulphonate compounds (PFOS), a fire fighting foam, among others, have been observed in several environmental compartments. They have similar characteristics, including persistency, as do the PCBs and POPs, and their levels are on the increase although many have been banned and are no longer being manufactured.

4.4. Ocean Waste Management

Ocean waste includes all material that is discarded accidentally or purposefully from vessels, ships, and other ocean transport. It also includes waste that enters waterways and rivers which flows into the sea and also waste from human habitation living in coastal areas close to the sea. Ocean waste can also result from ocean storms waves, tsunamis, and other extreme marine events. Some of these are the result of poor handling of cargo on ships and vessels. Ocean debris and waste are not only found along coast lines, in bays, and in beaches and ports but also in the middle of oceans. In particular, plastic (floats on water) in many forms is found in the center of the ocean, for example, in the pacific gyre [43].

In some countries such as the United States, the Environmental Protection Agency (EPA) and the National Oceanic and Atmospheric Administration are responsible, along with the U.S. Navy and the Coast Guard and other agencies, for the management and control of ocean waste. The Inter-agency of the Marine Debris Coordinating Committee are responsible for the rules and regulation and recommendations for the handling this form of waste. In less developed nations such as Mexico, there are also incipient programs to handle ocean waste.

It is important that progress be made in controlling waste discharges into the ocean. There is strong evidence that when correct measures are taken, not only the ocean but also land and many organisms are spared from the consequences of chemical pollutants. The following three cases illustrate this point:

1. After the gasoline additive, tetraethyl lead was banned, there was a marked decrease in the lead content in the atmosphere [44] and consequently there was a reduction of lead in the oceans.
2. With recent improvements to industrial processes, stringent requirements to discharges municipal water along with improvements in wastewater treatment procedures, there is evidence that trace metals and organic pollutant discharges that eventually enter the oceans have decreased substantially [45].
3. There is also good evidence that the improvements and changes to industrial processes have decreased the amount of dioxins and furans that find their way into the atmosphere and into water bodies [46].

References

[1] United Nations Environmental Program. (2005). Intergovernmental Conference to Adopt a Global Programme of Action for the Protection of the Marine Environment From Land-Based Activities, Washington, DC. UNEP(OCA)/LBA/IG.2/7

[2] H.W. Paerl, J.L. Pinckney, J.M. Fear, B.L. Peierls, Marine Ecology Progress Series 166 (1998) 17–25.

[3] R.J. Diaz, Journal of Environmental Quality 30 (2001) 275–281.

[4] N.N. Rabalais, R.J. Dıaz, L.A. Levin, R.E. Turner, D. Gilbert, J. Zhang, Biogeosciences 7 (2010) 585–619.

[5] K. Castro-Morales, J.V. Macías-Zamora, S. Raúl Canino-Herrera, R.A. Burke. (2010). In preparation.

[6] K.A.F. Zonneveld, G.J.M. Versteegh, S. Kasten, T.I. Eglinton, K.C. Emeis, C. Huguet, et al., Biogeosciences 7 (2010) 483–511.

[7] V. Rodríguez-Villanueva, R. Martinez-Lara, V. Macías-Zamora, Hydrobiologia 496 (2003) 385–399.

[8] N. Ramirez-Álvarez, J.V. Macias-Zamora, R.A. Burke, L.V. Rodriguez-Villanueva, Environmental Toxicology and Chemistry 26 (2007) 2332–2338.

[9] E.A. Boyle, J.M. Edmond. Determination of Trace Metals in Aqueous Solution by APDC Chelate Co-precipitation. Advances in Chemistry. Vol. 147, chapter 6, pp 44–55. in: T.R.P. Gibb (Ed.), Analytical Methods in Oceanography, (1975). DOI: 10.1.021/ba-1975-0147.vh006. ISBN 13:9780841202450 eISBN:9780841223288, Publication date June 01,1975. American Chemical Society, Washington DC.

[10] H. Sawatari, T. Hayashi, E. Fujimori, A. Hirose, H. Haragushi, Bulletin of the Chemical Society of Japan 69 (2005) 1925–1931.

[11] F. Delgadillo-Hinojosa, J.V. Macías-Zamora, J.A. Segovia-Zavala, S. Torres-Valdés, Marine Chemistry 75 (2001) 109–122.

[12] R. Dwinna, Y. Zhu, F. Eiji, U. Tomonari, H. Hiroki, Talanta 72 (2007) 600–606.

[13] G. Doner, A. Ege, Analytica Chimica Acta 547 (2005) 14–17.

[14] S. Sañudo-Whilelmy, R.A. Flegal, Environmental Science & Technology 26 (1992) 2147–2151.

[15] S. Sañudo-Whilelmy, R.A. Flegal, Environmental Science & Technology 30 (1996) 1575–1580.

[16] D.C. Girvin, A.T. Hodgson, M.E. Tatro, R.N. Anaclerio Jr., Spatial and seasonal variations of silver, cadmium, copper, nickel, lead and zinc in South San Francisco Bay water during two consecutive draught years. Regional Water Quality Control Board, University of California, Berkeley, CA, 1978, p. 117.

[17] J.L. Stauber, S. Andrade, M. Ramirez, M. Adams, J.A. Correa, Marine Pollution Bulletin 50 (2005) 1363–1372.

[18] S.N. Luoma, in: R.F. Rainbow (Ed.), Processes affecting metal concentrations in estuarine and coastal marine sediments. Heavy Metals in Marine Environment, CRC Press, Boca Raton, FL, 1990, pp. 51–66.

[19] H.L. Windown, S.J. Schropp, F.D. Calder, J.D. Ryan, R.G. Smith Jr., L.C. Burney, et al., Environmental Science & Technology 23 (1989) 314–320.

[20] E.R. Long, L.G. Morgan, NOAA Technical Memorandum NOS OMA 52, National Oceanic and Atmospheric Administration, Seattle, WA, 1990. 175 pp and appendices.

[21] D.D. Macdonald, R. Scott Carr, F.D. Calder, E.R. Long, C.G. Ingersoll, Ecotoxicology 5 (2004) 253–278.

[22] J.V. Hunter, T. Sabatino, R. Gomperts, M.J. Mackenzie, Water Pollution Control Federation 51 (1979) 2129–2138.

[23] GESAMP (IMO/FAO/UNESCO/WMO/WHO/IAEA/UN/UNEP Joint Group of Experts on the Scientific Aspects of Marine Pollution). (1999). GESAMP working group 32. Proceedings of 1999 International Oil Spill Conference, London, UK.

[24] S.M. Libes, Introduction to Marine Biogeochemistry, Wiley, New York, 1992. ISBN 0-471-50946-9.

[25] G. Eglinton, R.J. Hamilton, Science 156 (1967) 1322–1355.

[26] F.G. Prahl, J.T. Bennett, R. Carpenter, Geochimica et Cosmochimica Acta 44 (1980) 1967–1976.

[27] M. Page, L.J. Nelson, M.I. Haverty, G.J. Blomquist, Journal of Chemical Ecology 16 (1990) 1178–1193.

[28] M. Nishimura, E.W. Baker, Geochimica et Cosmochimica Acta 50 (1986) 299–305.

[29] K. Hayakawa, N. Handa, N. Ikuta, M. Fukushi, Organic Geochemistry 24 (1996) 511–521.

[30] N. Harada, N. Handa, M. Fukushi, R. Ishiwata. Proceedings of the NIPR Symposium on Polar Biology 8 (1995) 163–176.

[31] J.V. Macías-Zamora, Environmental Pollution 92 (1996) 49–53.

[32] S. Ou, J. Zheng, B.J. Richardson, P.K.S. Lam, Chemosphere 56 (2004) 107–112.

[33] X. Gao, S. Chen, X. Xie, A. Long, F. Ma, Environmental Pollution 148 (2007) 40–47.

[34] M.F. Simsick, S.J. Eisenreich, P.J. Lloy, Atmospheric Environment 33 (1999) 5071–5079.

[35] R.A. Hites, R.E. Laflamme, J.G. Windsor, in: L. Petrakis, F.T. Weiss (Eds.), Polycyclic aromatic hydrocarbons in marine/aquatic sediments: their ubiquity. Petroleum in the Marine Environment, Advances in chemistry Series 185, American Chemical Society, Washington D.C, 1980, pp. 289–311.

[36] C.D. Simpson, A.A. Mosi, W.R. Cullen, K.J. Reimer, Science of the Total Environment 181 (1996) 265–278.

[37] L. Mazeas, H. Budzinski, Analusis 27 (1999) 200–2002.

[38] R. Boonyatumanond, G. Wattayakom, A. Togo, H. Takada, Marine Pollution Bulletin 52 (2006) 942–956.

[39] A. de la, L. Meléndez Sánchez, Efecto del dispersante corexit 9500© durante el proceso de degradacion de los hidrocarburos en ambiente marino. 2009, Master Thesis 57 (2009).

[40] E.J. Aigner, A.D. Leone, R.L. Falconer, Environmental Science and Technology 32 (1998) 1162–1168.

[41] C.J. Moore, S.L. Moore, M.K. Leecaster, S.B. Weisberg, Marine Pollution Bulletin 42 (2001) 1297–1300.

[42] G.A. Leblanc, Environmental Science and Technology 29 (1995) 154–160.

[43] L. Zhang, R. Lohman, Cycling of PCBs and HCB in the Surface Ocean-Lower Atmosphere of the Open Pacific. Environmental Science and Technology (2010). 10.1021/es9039852.

[44] H. von Storch, M. Costa-Cabral, C. Hagner, F. Feser, J. Pacyna, E. Pacyna, et al., The Science of the Total Environment 311 (2003) 151–176.

[45] G.S. Lyon, E.D. Stein. Effluent discharges to the Southern California Bight form Industrial facilities in 2005. SCCWRP. Annual Reports. 2005.

[46] K.H. Chai, M.B.Chang, S.J. Kao. Historical Trends of PCDDFs and dioxin like PCBs in the sediment buried in a reservoir in Northern Taiwan. Chemosphere 68 (2007) 1733–1740.

CHAPTER

20

Electronic Waste

Jirang Cui, Hans Jørgen Roven

Department of Materials Science and Engineering, Norwegian University of Science and Technology (NTNU), Alfred Getz vei 2, NO-7491 Trondheim, Norway

OUTLINE

1. INTRODUCTION

The production of electronic equipment is increasing worldwide. Both technological innovation and market expansion continue to accelerate the replacement of equipment resulting in a significant increase in electronic waste (e-waste). In 2008, the number of personal computers in use around the world has surpassed 1 billion and will become obsolete in the next 5 years. An investigation has estimated that the volume of household e-waste was up to 20 million tonnes (20×10^6 t) globally in 2005 [1].

Waste Doi: 10.1016/B978-0-12-381475-3.10020-8

Because of their hazardous material contents, electronic waste may cause environmental problems during the waste management phase if it is not properly pretreated. Growing attention is being given to the impacts of the hazardous components in electronic waste on the environment. Many countries have drafted legislation to improve the reuse, recycling, and other forms of recovery of such wastes so as to reduce disposal problems [2]. For example, the Directive 2002/96/EC of the European Union (EU) on Waste Electrical and Electronic Equipment (WEEE) had to be implemented into national legislation by the 13 August 2004 [3].

The purpose of this chapter is to give an overview of electronic waste, current status of the management of electronic waste, and recycling technologies for the recovery of metals from end-of-life electronic equipment.

2. MANAGEMENT OF ELECTRONIC WASTE

2.1. Definition of Electronic Waste

According to the definition from the European Directive on the WEEE [3], WEEE or e-waste refers to waste equipment that is dependent on electric currents or electromagnetic fields to work properly and equip for the generation, transfer, and measurement of such currents. The major electric and electronic equipment that fall into the EU electric and electronic equipment (EEE) categories are listed in Table 20.1.

2.2. Quantity and Global Electronic Waste Flow

Different methods have been used for estimating e-waste generation [4], such as:

1. The market supply method uses past domestic sales data coupled with the average life of products for a certain region;
2. The consumption and use method is based on extrapolation from the average amount of electronic equipment in a typical household;
3. The saturated market method, applied in Switzerland, assumes that for each new appliance bought, the old one reaches its end-of-life.

It should be pointed out that in an unsaturated market, assumptions need to be made about the average life time of EEE products as well as their average weight.

Estimation of actual electronic waste streams generated by the EU, North America, and Asia for the year 2005 was investigated by Bastiaan et al. [1]. In the EU, 7×10^6 t of electronic waste was the estimated total based on a value of 15 kg per capita per annum (15 kg ca^{-1} a^{-1}) for the 457 million inhabitants, of which

- 50% was large household appliances (fridges and washing machines)
- 10% was small household appliance (vacuum cleaners and toasters)
- 20% was office and communication waste (computers and cell phones)
- 20% was entertainment electronic equipment (radios, TVs, and stereos)

An investigation [5] from the Industry Council for Electronic Equipment Recycling in United Kingdom (UK) showed that every year 1×10^6 t (16 kg ca^{-1} a^{-1}) e-waste are produced in the UK of which 88% of large household appliances is recycled, 26% of office and communication waste is recycled, but only 4% of entertainment electronic equipment is recycled. About 10% of e-waste was shipped illegally to non-OECD (The Organisation for Economic Co-operation and Development) countries in Asia, Africa and Eastern Europe.

The study for the year 2005 by Bastiaan et al. [1] showed that the total supply of e-waste in the United States amounted to 6.6×10^6 t (22 kg ca^{-1} a^{-1}) of which 20% was estimated

TABLE 20.1 EEE Categories According to the EU Directive [3]

No.	Category	Major electric and electronic equipments
1	Large household appliances	Refrigerators Washing machines Cooking Electric fans, air conditioner appliances
2	Small household appliances	Vacuum cleaners Irons Toasters, grinders, coffee machines Electric knives, hair-cutting, tooth brushing Clocks, watches, scales, etc.
3	IT and telecommunications equipment	Mainframes Data processing, personal computers, laptops, notepads, calculators Printer units, copying equipment, facsimile, telephones
4	Consumer	Television sets Radio sets Video cameras, recorders, amplifiers, etc.
5	Light equipment	Luminaires for fluorescent lamps Straight or compact fluorescent lamps High-intensity discharge lamps Low-pressure sodium lamps
6	Electrical and electronic tools	Drills Saws Sewing machines
7	Toys, leisure, and sports equipment	Electric trains or car racing sets Video games or consoles Computers for biking, diving, running, rowing, etc. Coin slot machines
8	Medical devices	Radiotherapy equipment Cardiology Dialysis Pulmonary ventilators
9	Monitoring and control instruments	Smoke detectors Heating regulators Thermostats
10	Automatic dispensers	Automatic dispensers for hot or cold drinks, solid products, etc.

to be exported to Asia; in Japan, 3.1×10^6 t (24 kg ca^{-1} a^{-1}) e-waste is produced; and in China, 0.5×10^6 t. The latter number is believed to increase in an accelerated manner, which is similar in the case of the rapid developing countries such as India and Brazil.

2.3. Characteristics of Electronic Waste

2.3.1. Hazardous Components in e-Waste

Electronic waste consists of a large number of components of various sizes and shapes, some of which contain hazardous components that need be removed for separate treatment. Major categories of hazardous materials and components of WEEE that have to be selectively treated are shown in Table 20.2 [6].

Flame retardants are widely used in plastics to prevent or delay a developing fire in electronic equipment. According to the report from the Association of Plastics Manufacturers in Europe [7], about 12% of all plastics used in the electric and electronic equipment contains flame retardants, mainly television housing, computer monitors, and cases. In the consumer electronic equipment sector, up to 55% of plastics is treated with flame retardants.

2.3.2. *Material Composition of e-Waste*

From the point of material composition, electronic waste can be defined as a mixture of various metals, particularly copper, aluminium, and steel, attached to, covered with, or mixed with various types of plastics and ceramics [8]. Precious metals have a wide application in the manufacture of electronic appliances, serving as contact materials due

TABLE 20.2 Major Hazardous Components in Waste Electric and Electronic Equipment [6]

Materials and Components	Description
Batteries	A large proportion of heavy metals such as lead, mercury, and cadmium is present in batteries
Cathode ray tubes (CRTs)	Lead in the cone glass and fluorescent coating over the inside of panel glass
Mercury containing components, such as switches	Mercury is basically used in thermostats, sensors, relays, and switches (e.g. on Printed Circuit Boards and in measuring equipment and discharge lamps). Furthermore, it is used in medical equipment, data transmission, telecommunication, and mobile phones
Asbestos waste	Asbestos waste has to be treated selectively
Toner cartridges, liquid and pasty, as well as colour toner	Toner and toner cartridges may contain hazardous chemicals that are moderately toxic if acute exposure occurs
Printed Circuit Boards	In Printed Circuit Boards, lead, and cadmium are common in solder; cadmium occurs in certain components, such as SMD chip resistors, infrared detectors, and semiconductors
Polychlorinated biphenyl (PCB) containing capacitors	PCB-containing capacitors have to be removed for safe destruction
Liquid Crystal Displays (LCDs)	Lead and mercury occur in LCDs, especially for old designs. LCDs of a surface greater than 100 cm^2 have to be removed from WEEE
Plastics containing halogenated flame retardants	During incineration/combustion of the plastics, halogenated flame retardants can produce toxic components
Equipment containing CRC HCFC or HFCs	The CFC present in the foam and the refrigerating circuit must be properly extracted and destroyed; HCFC or CFCs present in the foam and refrigerating circuit must be properly extracted and destroyed or recycled
Gas discharge lamps	The mercury has to be removed

to their high chemical stability and their good conducting properties. Platinum group metals are used among other things in switching contacts (relays, switches) or as sensors to ascertain the electrical measure and as a function of the temperature [9].

Table 20.3 gives examples of the metal composition of different electronic scraps from the literature. It is clear that electronic waste varies considerably in age, origin, and between the different manufacturers. There is no average scrap composition, even the values given as typical averages actually only represent scrap of a certain age and manufacturer.

The value distributions V_i for different electronic scrap samples (as shown in Table 20.4) were calculated by using the following equation:

$$V_i = \frac{100 Wt_i Pr_i}{\sum Wt_i Pr_i} \tag{20.1}$$

where Wt_i is the mass percent of metal i in the electronic scrap sample (as shown in Table 20.3), Pr_i is the current price of metal i. The metal price data are from London Metal Exchange (LME) official prices for cash seller and settlement or London Fix Prices (precious metals) on the 4th of May 2010.

TABLE 20.3 Mass Composition of Metals for Different Electronic Scrap Samples from the Literature

	Mass (%)					Mass (mg kg^{-1})			
Electronic waste	**Fe**	**Cu**	**Al**	**Pb**	**Ni**	**Ag**	**Au**	**Pd**	**References**
TV board scrap	28	10	10	1.0	0.3	280	20	10	[10]
PC board scrap	7	20	5	1.5	1	1000	250	110	[10]
Mobile phone scrap	5	13	1	0.3	0.1	1380	350	210	[10]
Portable audio scrap	23	21	1	0.14	0.03	150	10	4	[10]
DVD player scrap	62	5	2	0.3	0.05	115	15	4	[10]
Calculator scrap	4	3	5	0.1	0.5	260	50	5	[10]
PC mainboard scrap	4.5	14.3	2.8	2.2	1.1	639	566	124	[11]
Printed circuit boards scrap	12	10	7	1.2	0.85	280	110	–	[12]
TV scrap (CRTs removed)	–	3.4	1.2	0.2	0.038	20	<10	<10	[13]
Electronic scrap	8.3	8.5	0.71	3.15	2.0	29	12	–	[14]
PC scrap	20	7	14	6	0.85	189	16	3	[15]
Typical electronic scrap	8	20	2	2	2	2000	1000	50	[16]
E-scrap sample 1	37.4	18.2	19	1.6	–	6	12	–	[17]
E-scrap sample 2	27.3	16.4	11.0	1.4	–	210	150	20	[17]
Printed circuit boards	5.3	26.8	1.9	–	0.47	3300	80	–	[18]
E-scrap (1972 sample)	26.2	18.6	–	–	–	1800	220	30	[8]
Nokia cell phones	8	19	9	0.9	1	9000	–	–	[19]

Note that "–" denotes not reported.

TABLE 20.4 Calculated Value Distribution for Different Electronic Scrap Samples [Eq. (20.1)]

	Value-share (%)							
Electronic waste	**Fe**	**Cu**	**Al**	**Pb**	**Ni**	**Ag**	**Au**	**Pd**
Prices* ($/ton)	500	7175	2164	2129	25650	601500	38,100,000	17,000,000
TV board scrap	6	32	10	1	3	8	33	7
PC board scrap	0	10	1	0	2	4	69	13
Mobile phone scrap	0	5	0	0	0	4	71	19
Portable audio scrap	5	69	1	0	0	4	17	3
DVD player scrap	22	25	3	0	1	5	40	5
Calculator scrap	1	8	4	0	5	6	73	3
PC mainboard scrap	0	4	0	0	1	2	85	8
Printed circuit boards scrap	1	13	3	0	4	3	76	–
TV scrap (CRTs removed)	–	82	9	1	3	4	–	–
Electronic scrap	2	35	1	4	30	1	27	–
PC scrap	5	25	15	6	11	6	30	3
Typical electronic scrap	0	3	0	0	1	3	90	2
E-scrap sample 1	8	54	17	1	–	–	19	–
E-scrap sample 2	2	15	3	0	–	2	74	4
Printed circuit boards	0	27	1	–	2	28	43	–
E-scrap (1972 sample)	1	12	–	–	–	9	73	4
Nokia cell phones	1	19	3	0	4	74	–	–

** The metal price data are from LME official prices for cash seller and settlement (base metals) or London Fix Prices (precious metals) on the 4th of May 2010: Ag 18.71 $ per Troy oz; Au 1185 $ per Troy oz; Pd 529 $ per Troy oz. Here ton refers to a US short ton which is equivalent to 0.9072 metric tonnes.*

It is obvious that the copper and precious metals make up more than 80% of the value for most of the e-waste samples. It indicates that the recovery of precious metals and copper may remain as the major economic driver for a long time. However, it should be pointed out that the precious metals content have gradually decreased in concentration in scrap. This is due to the falling power consumption of modern switching circuits and the rising clock frequency (surface conduction). Although the contact layer thickness in the 1980s was in the region of 1 to 2.5 μm, in modern appliances today this is between 300 and 600 nm (gold wafer) [9].

2.4. Hierarchy in the Treatment of Electronic Waste

Currently, the main options for the treatment of electronic waste are reuse, remanufacturing, and recycling, as well as incineration and landfilling. In many cases, electronic equipment that is no longer useful to the original purchaser still has value for other people. In this case, equipment can be resold or donated to schools or charities without any modification. Reuse of end-of-life electronic equipment has first priority on the management of electronic waste because the usable lifespan of equipment is extended to a secondary market, resulting in a reduced volume of waste stream encompassing treatment. Remanufacturing is a production-batch process in which used products or cores are disassembled, cleaned, repaired or refurbished, reassembled and qualified for new or like-new equipments [20]. Recycling means the reprocessing of the waste materials for the original purposed products or for other purposes. Recycling of electronic waste involves disassembly and/or destruction of the 'end-of-life' equipment to recover materials.

Incineration of electronic waste by traditional methods for municipal solid waste is dangerous. For example, copper is a catalyst for dioxin formation when flame-retardants are incinerated. This is of particular concern for the incineration of brominated flame retardants at low temperature. It was estimated that emissions from waste incineration accounts for 36 tonnes per year (36 t a^{-1}) of mercury and 16 t a^{-1} of cadmium in the EU Community [6]. Additionally, in the United States, the vast majority of e-waste currently ends up in landfills. According to the US Environmental Protection Agency (EPA), in year 2000 more than 4.6 million tonnes of e-waste ended up in landfills nationally [21].

The hierarchy in the treatment of e-waste encourages the reuse of original equipment as a first priority, second remanufacturing, then recovery of materials by recycling techniques, and as a last resort, disposal by incineration and landfilling.

Recycling of electronic waste is an important subject not only from the point of waste treatment but also from the recovery aspect of valuable materials. The US EPA has identified seven major benefits when scrap iron and steel are used instead of virgin ore resources. Using recycled materials in place of virgin materials results in significant energy savings [22].

Recycling of e-waste can be broadly divided into three major steps: (a) disassembly: selective disassembly, targeting by singling out hazardous or valuable components for special treatment, is an indispensable process in the recycling of e-waste; (b) upgrading: using mechanical processing and/or metallurgical processing to up-grade desirable materials content, that is preparing materials for refining processes; (c) refining: in the last step, recovered materials are retreated or purified by using chemical (metallurgical) processing so as to be acceptable for their original use. Mechanical processing and disassembly are mainly used for the pretreatment of e-waste for upgrading the valuable materials content. However, mechanical recycling cannot efficiently recover precious metals. In the last refining step, recovered metals are melted or dissolved and finally separated by using metallurgical and chemical techniques, including pyrometallurgical and hydrometallurgical processing.

3. DISASSEMBLY OF ELECTRONIC WASTE

Selective disassembly involving the singling out of hazardous or valuable components is an indispensable process in electronic waste recycling. It is a systematic approach that allows removal of a component or a part; or a group of parts or a sub-assembly from a product (i.e., partial disassembly); or separating a product into its parts (i.e., complete disassembly) for a given purpose.

The areas of disassembly that are being pursued by researchers are focused on disassembly process planning (DPP) and innovation of disassembly facilities. The objective of DPP is to develop procedures and software tools for forming disassembly strategies and configuring disassembly systems. The following phases for developing a disassembly process plan have been proposed [23]:

- Input and output product analysis: In this phase, reusable, valuable, and hazardous components and materials are defined. After preliminary cost analysis, optimal disassembly is identified.
- Assembly analysis: In the second phase, joining elements, component hierarchy, and former assembly sequences are analyzed.
- Uncertainty issues analysis: Uncertainty of disassembly comes from defective parts or joints in the incoming product, upgrading/downgrading of the product during consumer use, and disassembly damage.
- Determination of dismantling strategy: In the final phase, it is decided whether to use nondestructive or destructive disassembly.

In addition to generating a good disassembly process plan, the implementation of disassembly needs highly efficient and flexible tools. Although automated assembly of electronic equipment is well advanced, the full (semi-) application of automation disassembly for recycling of electronic equipment is unfortunately full of frustrations. The main obstacles preventing automated disassembly from becoming a commercially successful activity are (1) too many different types of products; (2) the amount of products of the same type is too small; (3) general disassembly-unfriendly product design; (4) general problems in return logistics; and (5) variations in returned amounts of products to be disassembled.

In the practice of recycling of waste electric and electronic equipment, selective disassembly (dismantling) is an indispensable process because: (1) the reuse of components is the first priority; (2) the dismantling of hazardous components is essential; (3) it is also common to dismantle highly valuable components and high-grade materials such as printed circuit boards, cables, and engineering plastics so as to simplify the subsequent recovery of materials.

Most of the recycling plants use manual dismantling. Ragn-Sells Elektronikåtervinning AB in Sweden is a typical electronics-recycling corporation. Figure 20.1 illustrates the current disassembly process that they use [2]. A variety of tools are involved in the dismantling process for removing hazardous components and in the

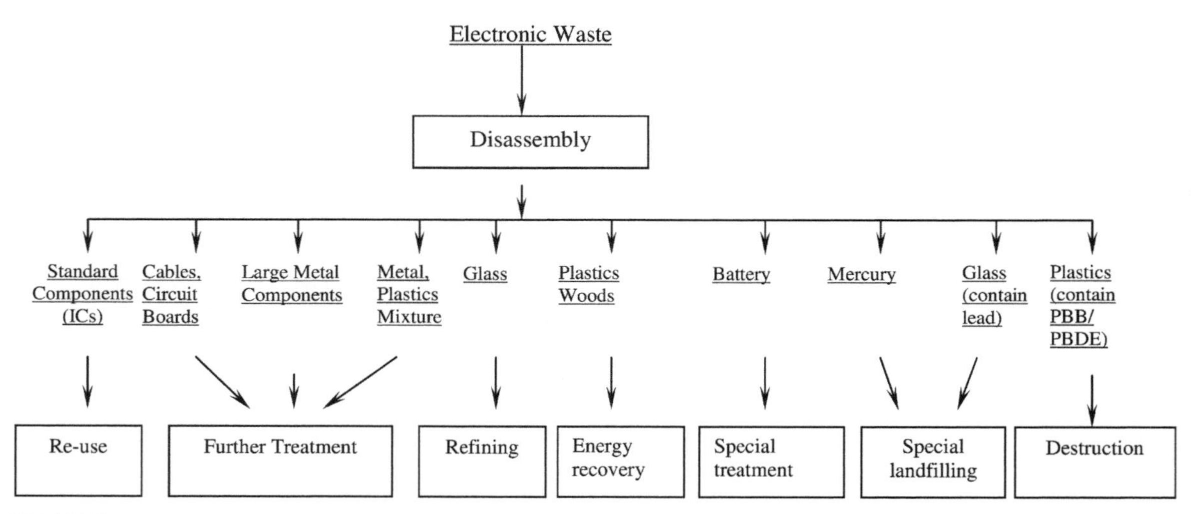

FIGURE 20.1 The recycling process developed by Ragn–Sells Elektronikåtervinning AB in Sweden.

recovery of reusable or valuable components and materials.

4. RECYCLING TECHNOLOGY OF ELECTRONIC WASTE

4.1. Magnetic Separation

Magnetic separators, in particular, low-intensity drum separators, are widely used for the separation of ferromagnetic metals from non-ferrous metals and other nonmagnetic wastes. Over the past decade, there have been many advances in the design and operation of high-intensity magnetic separators, mainly due to the introduction of rare earth alloy permanent magnets capable of providing very high-field strengths and gradients.

4.2. Eddy Current Separation

In the past decades, one of the most significant developments in recycling industry was the introduction of eddy current separators based on the use of rare earth permanent magnets. The principal of the eddy current separator is illustrated in Figure 20.2. Eddy currents can be induced in an electrical conductive particle by a time-dependent magnetic field

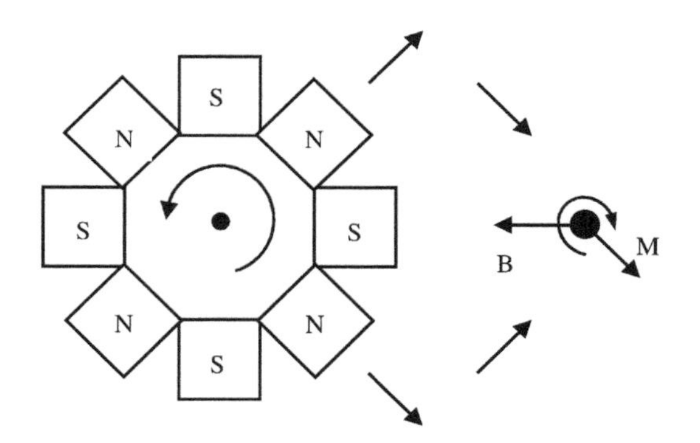

FIGURE 20.2 Illustration of Eddy current separation: magnet rotor (left) produces a time-dependent magnetic field B including Eddy currents in a particle (right) resulting in a particle magnetic moment *M*.

(magnet rotor in the figure). Further, the eddy currents will in return result in magnetic fields that oppose the inducing fields, giving rise to a so-called Lorentz force, a repulsive force.

4.3. Density-Based Separation

Gravity concentration separates materials of different specific gravity by their relative movement in response to the force of gravity and one or more other forces, the latter often being the resistance to motion offered by a fluid, such as water or air [24]. The motion of a particle in a fluid is dependent not only on the density of the particle but also on its size and shape, large particles will be affected more than smaller ones. In practice, close size control of feeds to gravity processes is required to reduce the effect on the size and make the relative motion of the particle specific gravity dependent.

4.4. Optical Sorting

With the fast development of the Charge-Coupled Device (CCD) sensor, computing, and software technology, optical sorting processes have been developed in both recycling and mineral processing industry. In addition, recording more and better data with sensors improves the separation performance of automated sorting equipment. The measuring of particle properties such as colour, texture, morphology, conductivity, and others allows high-quality sorting of mixed materials into almost pure fractions. Hence, multisensor systems involving the use of two or more different sensors have been developed over the past few years.

4.5. Pyrometallurgical Process

Pyrometallurgical processing, including incineration, smelting in a plasma arc furnace or blast furnace, dross formation, sintering, melting, and reactions in a gas phase at high temperatures

TABLE 20.5 Summary of Typical Pyrometallurgical Methods for Recovery of Metals from Electronic Waste

Techniques	Metals recovered	Main process features	Main results	References
Noranda process at Quebec, Canada	Cu, Au, Ag, Pt, Pd, Se, Te, Ni	Feeding to copper smelter with copper concentration (14% of the total throughput); Upgrading in converter and anode furnaces; Electrorefining for metal recovery	High recovery for both copper and precious metals	[28]
Boliden Rönnskär Smelter, Sweden	Cu, Ag, Au, Pd, Ni, Se, Zn, Pb	Feeding to a Kaldo reactor with lead concentrates (totally 100,000 tonnes every year); Upgrading in copper converter and refining; Precious metals refining for recovery	High recovery of copper and precious metals	[18,29]
Test at Rönnskär Smelter	Copper and precious metals	PC scrap feeding to a Zinc Fuming process (1:1 mixture with crushed revert slag); Plastics were tested as reducing agent and fuel; Copper and precious metals following the copper collector to be recovered to the copper smelter	Almost complete recovery of copper and precious metals in the Zinc Fuming process	[30]
Umicore's Precious metal refining process at Hoboken, Belgium	Base metals, Precious metals, Platinum group metals and Selenium, Tellurium, Indium	IsaSmelt, copper leaching and electro-winning, and precious metal refinery for Precious Metals Operation (PMO); E-waste cover up to 10% of the feed (250,000 tonnes of different wastes per annum); Plastics partially substitute the coke as reducing agent and fuel in IsaSmelt; Base Metals Operations process by-products from the PMO, including Lead blast furnace, lead refinery, and special metals plant; Offgas emission control system is installed at the IsaSmelt furnace	Recovering both base metals, precious metals and special metals such as Sb, Bi, Sn, Se, Te, In	[10,31,32]
Full-scale trial at Umicore's smelter	Metals in electronic scrap	Plastics-rich materials from WEEE were tested to replace coke as a reducing agent and energy source for the IsaSmelt	The smelter operation and metal recovery are not negatively affected by using 6% WEEE plastics and 1% of coke to replace 4.5% coke	[33]
Dunn's patent for gold refining	Gold	Gold scrap reacted with chlorine at 300°C to 700°C; Hydrochloric acid washing to dissolve the impurity-metal chlorides; Ammonium hydroxide and nitric acid washing respectively to dissolve the silver chloride; Samples should contain more than 80% of gold	Gold with 99.9% purity was recovered from gold scraps	[34]

(*Continued*)

TABLE 20.5 Summary of Typical Pyrometallurgical Methods for Recovery of Metals from Electronic Waste—*cont'd*

Techniques	Metals recovered	Main process features	Main results	References
Day's patent for refractory ceramic precious metals scraps	Precious metals such as platinum and palladium	The scrap was charged to a plasma arc furnace at a temperature of at least 1400°C; A molten metallic phase containing precious metals, and collector metal was produced; Ceramic residues went into a slag phase; Silver and copper are suitable collector metals in the process	Recoveries of 80.3% and 94.2% were obtained for platinum and palladium, respectively	[35]
Aleksandrovich's Patent for recovery of platinum group metal and gold from electronic scraps	Platinum group metal and gold	Fusing together scraps based on chalcogenides of base metals with carbon reducer; After the settling and cooling of melted materials, solidification and separation of solidified products are facilitated by formed phase boundaries	Platinum group metal and gold were recovered	[36]

[8,16,25] has become a traditional method to recover non-ferrous metals as well as precious metals from electronic waste over the past two decades. In the process, the crushed scraps are burned in a furnace or in a molten bath to remove plastics, and the refractory oxides form a slag phase together with some metal oxides.

Table 20.5 gives a summary of typical pyrometallurgical methods for the recovery of metals from electronic waste. The traditional technology of pyrometallurgy has been used for the recovery of copper and precious metals from waste electronic equipment for years. However, most methods involving pyrometallurgical processing of electronic waste have the following limitations [10,16,26,27]:

- Integrated smelters cannot recover aluminum and iron as metals; they are transferred into the slag component. Unfortunately, aluminum has influences on the slag properties that are undesirable in most cases.
- The presence of halogenated flame retardants in the smelter feed can lead to the formation of dioxins unless special installations and measures are present. Traditional smelters designed for the treatment of mining concentrates or simple copper scrap encounter some challenges for electronic waste treatment. However, state-of-the-art smelters are highly dependent on investments.
- Ceramic components and glass in the e-waste increase the amount of slag from blast furnaces, which results in a loss of precious metals and base metals from the scrap.
- Energy recovery and the using of organic constituents as reducing agents are beginning to be used only now.
- Only partial separation of metals can be achieved using pyrometallurgy, resulting in a limited upgrading of the metal value. Subsequent hydrometallurgical techniques and/or electrochemical processing are therefore necessary.
- Precious metals take a long time to separate in pyrometallurgical processes and are thus only obtained at the very end of the process.

4.6. Hydrometallurgical Process

Precious metals contribute the highest value to electronic scrap. From an economic point of

TABLE 20.6 New Developments on the Recovery of Metals from Electronic Waste by Hydrometallurgical Techniques

Metals recovered	Main process features	Main product	Year	References
Au	Computer chips were treated in nitric acid to dissolve base metals; the residue was leached with aqua regia; ferrous sulphate precipitation was used for gold recovery	Gold flakes	2007	[46]
Au and Ag	E-waste with size −0.5 mm was treated with combination of KI and I_2 or NaCl and bleaching powder; solvent extraction was used for gold and silver recovery	Au and Ag	2007	[47]
Ni	Leaching of nickel from waste multilayer ceramic capacitors was performed by using 1M HNO_3 at 90°C, 90-min reaction time, and 5 g/l pulp density	Ni in solution	2007	[48]
Au (98%), Pd (96%), Pt (92%), Ag (84%)	H_2SO_4 and $MgCl_2$ for dissolution of base metals; HCl and bromide ions were used to dissolve precious metals; cementation of gold by zinc powder	Au and platinum group metals powders	2006	[49]
Cu (98%)	Copper was dissolved by sulphuric acid or aqua regia; electrowinning was performed for copper recovery	Cu	2006	[50]
Cu, Ag (93%), Pd (99%), Au (95%)	Sulphuric acid leaching of copper, chloride leaching of palladium, thiourea or cyanide leaching of gold and silver, activated carbon adsorption of gold, silver, and palladium	AgCl, Cu, Pd, Au	2005	[40]
92% for Au, Ag, Pd	HCl or H_2SO_4 for dissolution of base metals, leaching of silver by HNO_3, leaching of gold and palladium by HCl and $NaClO_3$; precipitation of Au by $FeCl_2$	Gold sponge	2005	[51]
Au	E-scrap was treated with a leaching solution based on NaCl, $CuCO_3$, and HCl	Gold residue	2004	[52]
Sn, Pb	Solder was dissolved with a solution comprising Ti(IV) and an acid. Tin and lead were recovered by electrowinning	Sn and Pb	2003	[53]
Cu, Pb, and Sn	HNO_3 leaching of PCBs, electrodeposition recovery of base metals	Cu, Pb and Sn	2002	[54]
Au	Thermal treatment, HNO_3 leaching and aqua regia leaching for gold dissolution, solvent extraction of gold by diethyl malonate, ferrous sulphate solution was used for gold precipitation	Metallic gold	1997	[55]
Au	Treatment with an alkali solution in an autoclave at 80−190°C to remove aluminium, treatment with a sulphuric acid solution in another autoclave under surplus oxygen pressure for removing nonferrous metals	Enriched concentrate of precious metal	1993	[56]
Au and Ni	Leaching of base metals by sulphuric acid and oxidant (ferric sulphate) and aqua regia leaching of precious metals	Ni and Au in solution	1992	[57]

view, the recovery of precious metals from e-waste is most attractive. In the past two decades, the most active research area on recovery of metals from electronic scraps is recovering precious metals by hydrometallurgical techniques [37–42]. Compared with pyrometallurgical processing, the hydrometallurgical method is more exact, more predictable, and more easily controlled.

The main steps in hydrometallurgical processing consist of a series of acid or caustic leaches of solid material. The solutions are then subjected to separation and purification procedures such as precipitation of impurities, solvent extraction, adsorption and ion-exchange to isolate, and concentrate the metals of interest. Consequently, the solutions are treated by electrorefining process, chemical reduction, or crystallization for metal recovery [43–45].

A summary of new developments in the recovery of metals from electronic waste by hydrometallurgical techniques is listed in Table 20.6. It can be seen that most of the hydrometallurgical techniques for recovery of metals are involved in acid leaching and/or halide leaching. This is because acid leaching is the most feasible approach for the removal of base metals resulting in the exposure of the precious metals surfaces. However, to develop an environment-friendly technique for the recovery of precious metals from e-waste, more attention should be paid to evaluate the environmental impact of the various techniques.

5. ENVIRONMENTAL AND HEALTH PERSPECTIVES RELATED TO ELECTRONIC WASTE RECYCLING ACTIVITIES

Electronic waste recycling activities, such as dismantling of scrap, shredding and separation, thermal, or hydrometallurgical processes may release toxic substances. In practice, polybrominated diphenyl ethers (PBDEs) and organophosphate esters are widely used in high-impact polystyrene as flame-retardant additives. Unfortunately, such additives can leak out into the environment during the lifetime or destruction of the product, because these compounds are not chemically bound to the polymer matrix. Research work by Sjödin et al. [58] demonstrated that brominated and phosphorus-containing additives to plastic materials are emitted in indoor work environments and in areas related to recycling. Eight PBDE congeners including decabromodiphenyl ether (BDE-209); decabromobiphenyl (BB-209); 1,2-bis(2,4,6-tribromophenoxy)ethane (BTBPE); tetrabromobisphenol A (TBBPA); and five arylated and six alkylated organophosphate esters were identified and quantified in the air samples from the dismantling of a hall and a shredder room of a Swedish electronics recycling plant. In the air of the dismantling plant, the concentration of triphenyl phosphate was one to two orders of magnitude higher; hepta- to deca-BDE, BTBPE, and TBBPA were several orders of magnitude higher than those observed in any other work environment involved with the assembly of circuit boards, offices containing computers, or computer repair facilities. Therefore, the dismantling of electronic scrap may encounter a challenge and must be considered carefully as a result of the potential threat that these chemicals pose to human health.

Numerous studies [59–64] have been carried out to evaluate the exposure of toxic substances from some informal recycling activities in developing countries, such as China, India, and Nigeria. Because of the manual processes involved in the materials recovery processes, the level of toxics such as dioxins and acids released is very high in some informal recycling activities. For example, e-waste recycling in Guiyu, China, resulted in the contamination of the entire region, pervading the water, air, soil, and biota of the region [65].

6. CONCLUSIONS

Because of the ever increasing generation of e-waste and the hazardous nature of this waste stream, e-waste is an emerging issue. Many countries have drafted legislation to improve the reuse, recycling, and other forms of recovery of such waste. Characterization of e-waste showed that e-waste is significantly heterogeneous and complex in terms of the type of components and materials. However, copper and precious metals make up more than 80% of the value for most of the e-waste samples. This indicates that the recovery of precious metals and copper may remain as the major economic driver for a long time.

The hierarchy of treatment of e-waste encourages the reuse of the whole equipment first, remanufacturing, then recovery of materials by recycling techniques, and as a last resort, disposal by incineration and landfilling. Recycling of e-waste can be broadly divided into three major steps: (a) selective disassembly, targeting, and singling out hazardous or valuable components for special treatment; (b) mechanical and/or metallurgical processing to upgrade desirable materials content; (c) refining recovered materials that are retreated or purified by using chemical (metallurgical) processing so as to be acceptable for further use in their original application.

Research into the environmental effects of hazardous substances released from electronic waste recycling processes indicates an urgent need for better monitoring and control of the informal recycling activities in developing countries.

ACKNOWLEDGEMENT

The authors acknowledge the Research Council of Norway (RCN) under the program FRINAT for financial support.

References

[1] C. Bastiaan, J. Zoeteman, H.R. Krikke, J. Venselaar, Handling WEEE waste flows: on the effectiveness of producer responsibility in a globalizing world, International Journal of Advanced Manufacturing Technology 47 (2010) 415–436.

[2] J. Cui, E. Forssberg, Mechanical recycling of waste electric and electronic equipment: a review, Journal of Hazardous Materials 99 (2003) 243–263.

[3] European Parliament and the Council of the European Union, Directive 2002/96/EC of the European Parliament and of the Council of 27 January 2003 on waste electrical and electronic equipment (WEEE), Official Journal of the European Union L37 (2003) 24–38.

[4] R. Widmer, H. Oswald-Krapf, D. Sinha-Khetriwal, M. Schnellmann, H. Boni, Global perspectives on e-waste, Environmental Impact Assessment Review 25 (2005) 436–458.

[5] UK Status Report on Waste from Electrical and Electronic Equipment, Industry Council for Electronic Equipment Recycling Report UK, 2000 (2000).

[6] Draft proposal for a European Parliament and Council directive on waste electrical and electronic equipment, European Commission Report Brussels, 2000 (2000).

[7] Plastics—a material of innovation for the electrical and electronic industry—insight into consumption and recovery in Western Europe 2000, Association of Plastics Manufacturers in Europe Report Brussels, 2001, (2001), 6–7.

[8] J.E. Hoffmann, Recovering Precious Metals from Electronic Scrap, JOM 44 (1992) 43–48.

[9] M. Teller, et al., Recycling of electronic waste material, in: A. Gleich (Ed.), Sustainable Metals Management: Securing our Future - Steps Towards a Closed Loop Economy, Springer, 2006, pp. 563–576.

[10] C. Hageluken, Improving metal returns and eco-efficiency in electronics recycling - a holistic approach for interface optimisation between pre-processing and integrated metals smelting and refining, Proceedings of the 2006 IEEE International Symposium on Electronics and the Environment (2006), Institute of Electrical and Electronics Engineers (IEEE) Inc, Washington DC, 2006, pp. 218–223.

[11] J.B. Legarth, L. Alting, G.L. Baldo, Sustainability issues in circuit board recycling, Proceedings of the 1995 IEEE International Symposium on Electronics and the Environment (1995), Institute of Electrical and Electronics Engineers (IEEE) Inc, Washington DC, 1995, pp. 126–131.

[12] S. Zhang, E. Forssberg, Electronic scrap characterization for materials recycling, Journal of Waste Management and Resource Recovery 3 (1997) 157–167.

[13] J. Cui, E. Forssberg, Characterization of shredded television scrap and implications for materials recovery, Waste Management 27 (2007) 415–424.
[14] S. Ilyas, M.A. Anwar, S.B. Niazi, M. Afzal Ghauri, Bioleaching of metals from electronic scrap by moderately thermophilic acidophilic bacteria, Hydrometallurgy 88 (2007) 180–188.
[15] Draft Guidelines for Environmentally Sound Management of Electronic Waste, Central Pollution Control Board, Indian Ministry of Environment and Forests Report, Delhi, 2007.
[16] E.Y.L. Sum, The Recovery of Metals from Electronic Scrap, JOM 43 (1991) 53–61.
[17] L.D. Busselle, T.A. Moore, J.M. Shoemaker, R.E. Allred, Separation processes and economic evaluation of tertiary recycling of electronic scrap, Proceedings of the 1999 IEEE International Symposium on Electronics and the Environment (1999), Institute of Electrical and Electronics Engineers (IEEE) Inc, Washington DC, 1999, pp. 192–197.
[18] L. Theo, Integrated recycling of non-ferrous metals at Boliden Ltd. Ronnskar smelter, Proceedings of the 1998 IEEE International Symposium on Electronics and the Environment (1998), Institute of Electrical and Electronics Engineers (IEEE) Inc, Washington DC, 1998, pp. 42–47.
[19] A.K. Bhuie, O.A. .Ogunseitan, J.-D.M. Saphores, A.A. Shapiro, Environmental and economic trade-offs in consumer electronic products recycling: a case study of cell phones and computers, Proceedings of the 2004 IEEE International Symposium on Electronics and the Environment (2004), Institute of Electrical and Electronics Engineers (IEEE) Inc, Denver, CA, 2004, pp. 279–284.
[20] J. Williams, L.H. Shu, Analysis of Remanufacturer Waste Streams for Electronic Products, Institute of Electrical and Electronics Engineers Inc, Denver, CA, 2001, pp. 279–284.
[21] S. Schwarzer, A.D. Bono, G. Giuliani, S. Kluser, P. Peduzzi, E-Waste, the Hidden Side of IT Equipment's Manufacturing and Use, UNEP, Geneva, 2005.
[22] Scrap Recycling: Where Tomorrow Begins, Institute of scrap recycling industries Inc., Washington, DC, 2003.
[23] S.M. Gupta, C.R. McLean, Disassembly of products, Computers and Industrial Engineering 31 (1996) 225–228.
[24] B.A. Wills, Mineral Processing Technology, fourth ed., Pergamon Press, Oxford, 1988.
[25] J.-c. Lee, H.T. Song, J.-M. Yoo, Present status of the recycling of waste electrical and electronic equipment in Korea, Resources, Conservation and Recycling 50 (2007) 380–397.
[26] I. Dalrymple, N. Wright, R. Kellner, N. Bains, K. Geraghty, M. Goosey, et al., An integrated approach to electronic waste (WEEE) recycling, Circuit World 33 (2007) 52–58.
[27] S.A. Shuey, P. Taylor, Review of pyrometallurgical treatment of electronic scrap, Mining Engineering 57 (2005) 67–70.
[28] H. Veldbuizen, B. Sippel, Mining discarded electronics, Industry and Environment 17 (1994) 7.
[29] T. Lehner, E and HS aspects on metal recovery from electronic scrap profit from safe and clean recycling of electronics, Proceedings of the 2003 IEEE International Symposium on Electronics and the Environment, Institute of Electrical and Electronics Engineers (IEEE) Inc., Washington DC, 2003, pp. 318–322.
[30] Plastics Recovery from Waste Electrical and Electronic Equipment in Non-Ferrous Metal Processes, Association of Plastics Manufacturers in Europe Report (2000), Brussels, 2000.
[31] C. Hageluken, Recycling of electronic scrap at Umicore's integrated metals smelter and refinery, Erzmetall 59 (2006) 152–161.
[32] C. Hageluken, Recycling of e-scrap in a global environment: opportunities and challenges, in: K.V. Rajeshwari, S. Basu, R. Johri (Eds.), Tackling E-Waste Towards Efficient Management Techniques, TERI Press, New Delhi, 2007, pp. 87–104.
[33] J. Brusselaers, C. Hageluken, F. Mark, N. Mayne, L. Tange, An Eco-efficient Solution for Plastics-Metals-Mixtures from Electronic Waste: the Integrated Metals Smelter, 5th Identiplast, the Biennial Conference on the Recycling and Recovery of Plastics Identifying the Opportunities for Plastics Recovery, Brussels, 2005.
[34] J. Dunn, E. Wendell, D.D. Carda, T.A. Storbeck, Chlorination process for recovering gold values from gold alloys, US Patent, US5004500, (1991).
[35] F.G. Day, Recovery of platinum group metals, gold and silver from scrap, US Patent, US4427442 (1984).
[36] S. Aleksandrovich, E. Nicolaevich, E. Ivanovich, Method of processing of products based on ahalcogenides of base metals containing metals of platinum group and gold, Russian Patent, RU2112064 (1998).
[37] K. Gloe, P. Muhl, M. Knothe, Recovery of Precious Metals from Electronic Scrap, in Particular from Waste Products of the Thick-Layer Technique, Hydrometallurgy 25 (1990) 99–110.
[38] H. Baba, An Efficient Recovery of Gold and Other Noble-Metals from Electronic and Other Scraps, Conservation and Recycling 10 (1987) 247–252.
[39] B. Kolodziej, Z. Adamski, Ferric chloride hydrometallurgical process for recovery of silver from electronic scrap materials, Hydrometallurgy 12 (1984) 117–127.

[40] P. Quinet, J. Proost, A. Van Lierde, Recovery of precious metals from electronic scrap by hydrometallurgical processing routes, Minerals and Metallurgical Processing 22 (2005) 17–22.
[41] L.E. Macaskie, N.J. Creamer, A.M.M. Essa, N.L. Brown, A new approach for the recovery of precious metals from solution and from leachates derived from electronic scrap, Biotechnology and Bioengineering 96 (2007) 631–639.
[42] T. Ogata, Y. Nakano, Mechanisms of gold recovery from aqueous solutions using a novel tannin gel adsorbent synthesized from natural condensed tannin, Water Research 39 (2005) 4281–4286.
[43] M. Sadegh Safarzadeh, M.S. Bafghi, D. Moradkhani, M. Ojaghi Ilkhchi, A review on hydrometallurgical extraction and recovery of cadmium from various resources, Minerals Engineering 20 (2007) 211–220.
[44] G.M. Ritcey, Solvent extraction in hydrometallurgy: Present and future, Tsinghua Science and Technology 11 (2006) 137–152.
[45] L.L. Tavlarides, J.H. Bae, C.K. Lee, Solvent extraction, membranes, and ion exchange in hydrometallurgical dilute metals separation, Separation Science and Technology 22 (1985) 581–617.
[46] P.P. Sheng, T.H. Etsell, Recovery of gold from computer circuit board scrap using aqua regia, Waste Management and Research 25 (2007) 380–383.
[47] J. Shibata, S. Matsumoto, Development of Environmentally Friendly Leaching and Recovery Process of Gold and Silver from Wasted Electronic Parts, (1999). http://www.environmental-expert.com/articles/article320/article320.htm [cited 2007-10-29].
[48] E. Kim, J. Lee, B. Kim, M. Kim, J. Jeong, Leaching behavior of nickel from waste multi-layer ceramic capacitors, Hydrometallurgy 86 (2007) 89–95.
[49] V. Kogan, Process for the recovery of precious metals from electronic scrap by hydrometallurgical technique, International Patent, WO/2006/013568, (2006).
[50] H.M. Veit, A.M. Bernardes, J.Z. Ferreira, J.A.S. Tenorio, C.d. F. Malfatti, Recovery of copper from printed circuit boards scraps by mechanical processing and electrometallurgy, Journal of Hazardous Materials 137 (2006) 1704–1709.
[51] P. Zhou, Z. Zheng, J. Tie, Technological process for extracting gold, silver and palladium from electronic industry waste, Chinese Patent, CN1603432A, 2005.
[52] M. Olper, M. Maccagni, S. Cossali, Process for recovering metals, in particular precious metals, from electronic scrap, European Patent, EP1457577A1, (2004).
[53] R.W. Gibson, P.D. Goodman, L. Holt, M. Dalrymple, D.J. Fray, Process for the recovery of tin, tin alloys or lead alloys from printed circuit boards, US Patent, US6641712B1, (2003).
[54] A. Mecucci, K. Scott, Leaching and electrochemical recovery of copper, lead and tin from scrap printed circuit boards, Journal of Chemical Technology and Biotechnology 77 (2002) 449–457.
[55] A.G. Chmielewski, T.S. Urbanski, W. Migdal, Separation technologies for metals recovery from industrial wastes, Hydrometallurgy 45 (1997) 333–344.
[56] A.V. Feldman, Method for hydrometallurgical processing of friable concentration products, US Patent, US5190578, (1993).
[57] J. Zakrewski, R. Chamer, A. Koscielniak, P. Kapias, J. Ciosek, A. Bednarek, Method for recovering gold and nickel from electronic scrap and waste, Polish Patent, PL158889B, (1992).
[58] A. Sjödin, H. Carlsson, K. Thuresson, S. Sjödin, C. Å. Bergman, Östman, Flame retardants in indoor air at an electronic recycling plant and at other work environments, Environmental Science and Technology 35 (2001) 448–454.
[59] K.S. Wu, X.J. Xu, J.X. Liu, Y.Y. Guo, Y. Li, X. Huo, Polyhrominated Diphenyl Ethers in Umbilical Cord Blood and Relevant Factors in Neonates from Guiyu, China, Environmental Science and Technology 44 (2010) 813–819.
[60] L. Chen, C.N. Yu, C.F. Shen, C.K. Zhang, L. Liu, K.L. Shen, X.J. Tang, Y.X. Chen, Study on adverse impact of e-waste disassembly on surface sediment in East China by chemical analysis and bioassays, Journal of Soils and Sediments 10 (2010) 359–367.
[61] G.F. Ren, Z.Q. Yu, S.T. Ma, H.R. Li, P.G. Peng, G.Y. Sheng, et al., Determination of Dechlorane Plus in Serum from Electronics Dismantling Workers in South China, Environmental Science and Technology 43 (2009) 9453–9457.
[62] W.L. Han, J.L. Feng, Z.P. Gu, D.H. Chen, M.H. Wu, J.M. Fu, Polybrominated Diphenyl Ethers in the Atmosphere of Taizhou, a Major E-Waste Dismantling Area in China, Bulletin of Environmental Contamination and Toxicology 83 (2009) 783–788.
[63] N.N. Ha, T. Agusa, K. Ramu, N.P.C. Tu, S. Murata, K.A. Bulbule, et al., Contamination by trace elements at e-waste recycling sites in Bangalore, India, Chemosphere 76 (2009) 9–15.
[64] Y. Guo, C.J. Huang, H. Zhang, Q.X. Dong, Heavy Metal Contamination from Electronic Waste Recycling at Guiyu, Southeastern China, Journal of Environmental Quality (2009).
[65] B.H. Robinson, E-waste: An assessment of global production and environmental impacts, Science of the Total Environment 408 (2009) 183–191.

CHAPTER

21

Tyre Recycling

Valerie L. Shulman

European Tyre Recycling Association (ETRA), avenue de Tervueren 16 1040 Brussels, BELGIUM

OUTLINE

Waste Doi: 10.1016/B978-0-12-381475-3.10021-X

1. INTRODUCTION

Tyres are an essential part of the economy of every country that relies on vehicular and/or air transport to move people and goods. Although the overall rubber sector is small in comparison with others, tyres are a significant part of international and domestic trade – while frequently imported, they are often retreaded and/or recycled locally. It is evident that the total quantity of rubber used is relatively small; however, the product units are comparatively large, and the wastes are conspicuous and unattractive. Table 21.1 illustrates the relative consumption of key material and product streams in the European Union.

The annual global production of natural and synthetic rubber is estimated at approximately 23,000,000 tonnes; of which, about 75% is consumed by various sectors of the automotive industries. The preponderance, almost 60%, is used in the production of tyres for passenger cars and trucks. Smaller diverse categories are grouped as 'other', which includes tyres for off-road vehicles, such as agricultural, aeroplane, civil engineering, industrial, mining, among others, as well as bicycles and motor cycles [1]. Hundreds of non-tyre automotive products, such as, appearance items, belts, hoses, housings, mouldings, rings, seals, among others, utilise the remaining 15%.

TABLE 21.1 Examples of EU Waste Streams (in Tonnes)

Product	Use	Sub-set
Paper	±79,000,000	
Plastics	±37,000,000	
Packaging		±19,980,000
Glass	±15,000,000	
Aluminium	±8,860,000	
Automotive/construction		±5,316,000
Packaging		±1,594,800
Rubber	±5,000,000	
Tyres		±3,500,000

Since the late 1990s, global tyre production is estimated at about a billion units per annum (1,000,000,000 a^{-1}) [2]. In units sold, which are somewhat less than those produced, passenger car tyres account for slightly more than 90%, whereas truck and other categories together constitute about 10% [3]. The general rule is that for each tyre sold, whether newly manufactured, part-worn or retreaded, one tyre becomes waste. In 2009, ±3,400,000 tonnes of tyres were classified as waste in the European Union, with comparable quantities arising in North America and Asia.

2. THE TYRE: THE RAW MATERIAL FOR RECYCLING

A pneumatic tyre is often described as an engineering marvel. It is basically a large, round, hollow shell filled with compressed air that can support more than 50 times its own weight. Although the external features have not changed perceptibly since radial tyres were introduced over 50 years ago, numerous internal changes have improved durability, performance, and environmental quality. Many ingredients that improve longevity, resistance to abrasion etc., during on-road life also contribute to the effectiveness of recycled tyre materials downstream.

Material composition varies by category, i.e. passenger car, truck, and nominally, from continent to continent (Asia, Europe and the United States), although ingredient ratios remain relatively the same. The variations have negligible impact on use, wearability or recyclability.

No matter where they are produced, all tyres contain four fundamental material groups: natural and synthetic rubbers, carbon blacks/

TABLE 21.2 Composition by Weight of EU Car/Truck Tyres

Material	Car/utility (%)	Truck (%)
Rubber/Elastomers[a]	±45	±42
Carbon black and silica	±23	±24
Metals	±16	±25
Textiles	±6	±4
Zinc oxide	±1	±2
Sulphur	±1	±1
Additives	±8	±4

[a]*natural/sythetic rubber ratio: truck tyres, about 1:2; car tyres, about 4:3.*

silicas, reinforcing materials (metals/textiles) and facilitators. A generic profile of tyres produced for the EU market is given in Table 21.2.

Once the ingredients are compounded, the tyre structure is assembled and vulcanised, which cures the rubber, transforming it into a strong, elastic, rubbery hard state. The final structure is an integrated whole. Figure 21.1 illustrates the critical elements of the tyre and key material uses.

A tyre can reach the end of its on-road life at almost any point after production. Once a tyre has been permanently removed from a vehicle without the possibility of being returned to the road, it is defined as 'waste'. From this point forward, it enters into a waste management system. Tyres, within and outside the EU, are considered a 'priority waste stream' and must be handled according to international, as well as national waste management legislation.

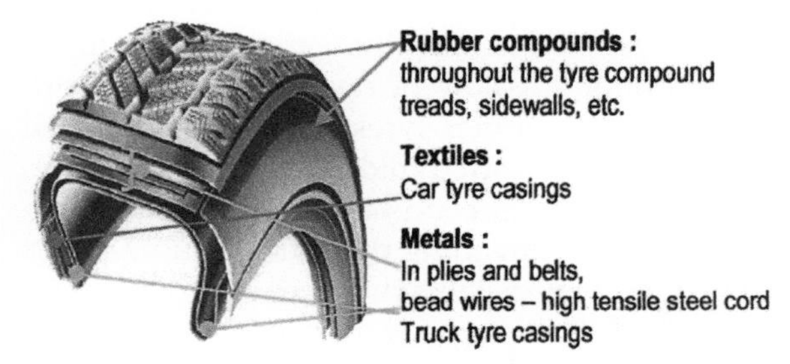

FIGURE 21.1 Examples of EU waste streams illustrate the relative consumption of key material and product streams in the European Union.

3. MANAGEMENT OF POST-CONSUMER TYRES

The management of post-consumer tyres is greatly influenced by environmental legislation at International, EU and State levels. Each has had an important impact on recycling – and recyclers – from sourcing raw materials, to product development and marketing, and the disposal of recycling residues.

Three Directives were enacted almost simultaneously in the EU to ensure that the greatest volume of tyres would be diverted from landfills – and treated in an environmentally sound and sustainable manner. The Landfill Directive (1999) [4] banned the landfilling of whole passenger car and truck tyres as from 2003 and tyre by-products from 2006. Tyres identified as *other* were excluded from the ban. The overall aim was to direct the tyres to material recycling or energy recovery facilities.

The End-of-Life Vehicle directive (2000) [5] required the removal of tyres from vehicles prior to demolition to ensure that tyre remnants would not end up in landfills. Finally, the Incineration of Waste directive (2000) [6] aimed at reducing dioxin emissions 90% by 2005 resulted in controls on the use of tyres for energy recovery and inadvertently diverted tyres towards material recycling. Cement kilns were brought into line with other incineration facilities. Box 1 briefly describes the 'four Rs' of sustainable waste management for tyres, which through their use, minimises waste and reduces reliance on natural resources.

The four Rs are reflected in the EU five-step waste management hierarchy that places material recycling at the top of the list. The revised

BOX 1

SUSTAINABLE OPTIONS

Reuse: Includes the sale of part-worn tyres for domestic on-road and other uses as well as for export to countries with less restrictive road-use requirements.

Retreading: Re-manufactures a tyre using as the core, a carefully selected, undamaged casing, which reduces production energy as well as virgin resources.

Recycling: Transforms a waste into a raw material that can be reintegrated into the economic stream as a resource to substitute the use of virgin resources.

Recovery: Transforms a waste into energy or fuel, which can be reintegrated into the economic stream as a resource to substitute the use of other energy sources.

Framework Directive on Waste (2008) [7] has also had a critical impact by directing post-consumer tyres towards recycling and away from landfilling. It also provides a definition of 'recycling' and offers the means for recycled materials to exit the waste designation:

> Recycling means any recovery operation by which waste materials are reprocessed into products, materials or substances whether for the original or other purposes. It includes the reprocessing of organic material but does not include energy recovery and the reprocessing into materials that are to be used as fuels or for backfilling operations.

Five waste streams were identified as 'specified wastes (to) cease to be a waste'. Each will require 'specific criteria' to be prepared: *aggregates, paper, glass, metal, tyres and textiles.* Thus, although tyres are not the subject of specific legislation, they are identified as among those materials and products that require special consideration.

4. MATERIAL RECYCLING: TREATMENTS AND TECHNOLOGIES

Tyre recycling is unique in comparison with other recycling sectors for several reasons:

- First, it is among the smallest waste streams;
- Second, tyres are essentially homogeneous, in other words all are made from the same materials, using virtually the same generic formulae with nominal variations – worldwide. The principal variation is size;
- Third, there is an interim treatment, retreading, which can extend the on-road life of a tyre;
- Fourth, recycling outputs are not used to produce the same or similar products as the original but have become strategic materials in over 50 different market sectors;
- Finally, under EU legislation, tyres, tyre pieces and recycling residues cannot be placed in landfills, compelling the development of new options for valorisation.

Nonetheless, the starting point for tyre recycling is the same as other industries – the sourcing of a continuous flow of raw materials. Historically, post-consumer tyres have been collected separately and sorted into two sectors: the part-worn market (second-hand) and retreading. However, as retreading dwindled, reuse/export stagnated, material recycling and energy recovery operations began to absorb the tyres. In many countries with large retreading industries, the infrastructure for collecting, sorting and transporting tyres became the means of servicing new markets, by providing raw materials for recycling and energy recovery.

Today, in most EU States, North and South America, many parts of Asia and elsewhere, tyres destined for recycling are collected under commercial contract from regular sources including garages, retail outlets and vehicle dismantlers, among others. In some regions, they are also removed from long-standing stockpiles or 'mono-fills'.

Generally, prior to delivery to a treatment facility, the tyres are sorted by category, including car, truck and others, and then by size. Road-worthy part-worn tyres are removed for domestic reuse or export under OECD regulations. Retreadable casings are diverted to appropriate facilities according to industry criteria. Non-retreadable tyres are delivered to a treatment site. As post-consumer tyres are a waste, they must be shipped in compliance with Basel Convention and OECD regulations. However, as they are non-hazardous waste destined for recovery, the documentation and information required are minimal.

Once delivered to a treatment facility, the tyres come under the jurisdiction of the national or local authority(ies) that regulate the quantity, location and circumstances under which they may be stored on the premises. Appropriate zoning and land-use permission, as well as permits and licenses must be obtained. Under certain circumstances, a waste management permit may also be required before the tyres can be processed or the materials used.

Prior to processing, the tyres are cleaned of debris such as glass, stones or miscellaneous items. Tyres acquired from stockpiles or other long-term storage often must be washed prior to use.

Tyre recycling treatments range from the simplest mechanical cutting devices to sophisticated, complex multi-phase chemical, mechano-chemical and/or thermal processes, which overcome many of the principal obstacles inherent in the recycling of thermoset rubbers. Specifically, the treatments and technologies that have evolved do not attempt to dissolve or melt the rubber into the virgin compound. Rather, they aim to exploit and enhance selected properties of the compound. It must be noted that the following discussions refer only to recycled tyre materials that will re-enter the economic flow for use in applications and products. They do not include materials that are prepared for use for energy, cement kilns or other similar purposes – although in some instances the treatments do overlap.

There are four basic levels of treatment. Each can be described in terms of its basic functions, which become increasingly complex as they progress through successive levels. Capabilities can be expanded by linking two or more technologies to operate in tandem to produce the desired result. Levels 1 and 2 treatments currently produce approximately 80% of all recycled tyre materials used today and, thus, will be the focus of this discussion.

Figure 21.2 illustrates the progressive, often inclusive nature from the 'first cut', by mechanical means, through the 'higher level' specialised treatments and technologies that add distinctive characteristics or properties to the material outputs.

4.1. Level 1 Treatments: Destruction of the Structure of the Tyre

These treatments can utilise car and truck tyres as feedstock. They are primarily simple mechanical means that destroy one or more of the physical attributes of the tyre, i.e. shape, weight-bearing capacity, rigidity, among others. They are designed to produce materials that will be directly used for simple products, in civil engineering applications, or as feedstock for further recycling. The most common methods include bead, sidewall or tread removal, compression, baling or cutting.

Equipment can be stationary or mobile, dependent on how the material will ultimately

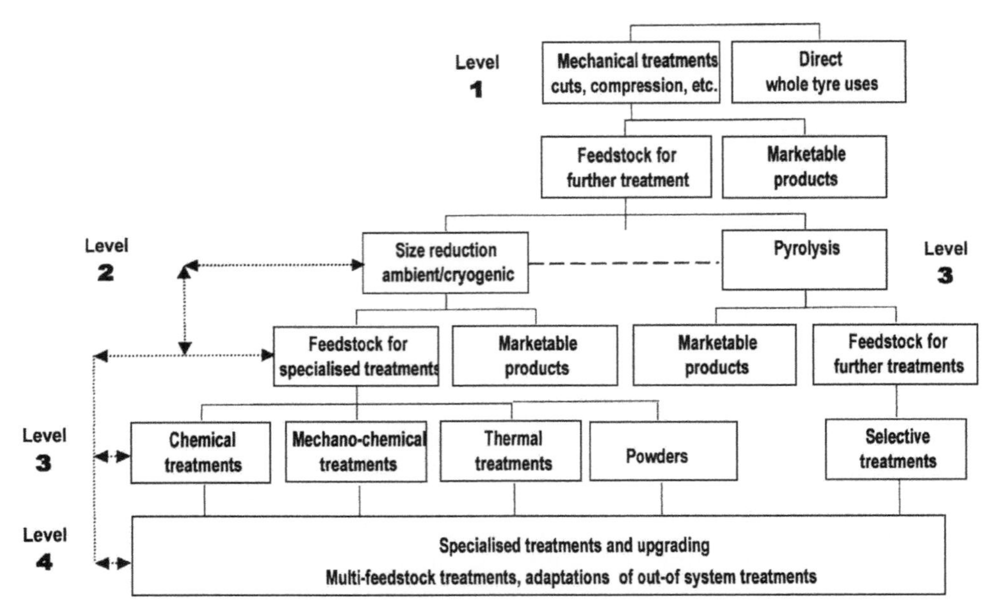

FIGURE 21.2 Schematic of the four levels of treatment, illustrates the progressive, often inclusive nature from the 'first cut', by mechanical means, through the 'higher level' specialised treatments and technologies which add distinctive characteristics or properties to the material outputs.

be used. Thus, a baler could be placed on-site during an erosion control project. Other than in debeading, neither the bead wires nor the belts are removed prior to or during processing.

4.2. Level 2 Treatments: Liberation and Separation of the Elements of the Tyre

These treatments separate out the principle components of the material, i.e. the rubber, metals and textiles. The most common technologies are ambient and cryogenic size reduction as well as some newer technologies such as microbes, water jetting, among others, to treat the tyres.

Whole tyres and Level 1 outputs are generally used as feedstock. The outputs can be used directly in applications or products or as feedstock for Level 3 processing.

Shredding and chipping are used on whole car or truck tyres. Shredding is a treatment that uses different technologies to fragment the tyre. Most often, a set of knives is used to produce material ranging between ±50 mm and ±300 mm that is irregularly shaped or equidimensional. Neither the bead wires nor the belts are removed prior to, during or after processing unless it is accomplished as the first step in size reduction processing.

Chipping is generally a second processing of shred, which results in material ranging from ±10 mm to ±50 mm that is either irregularly shaped or equidimensional. Chips are often used directly in civil engineering applications and as feedstock for ongoing Level 2 treatments.

Ambient grinding uses whole or pre-treated car or truck tyres in the form of shred, chips, sidewalls or treads. Ambient grinding is a multi-step technology. Processing takes place at or above normal room temperature. The rubbers, metals and textiles are sequentially separated out. First, the material is sheared with a system of knives. If the reinforcing and bead wires are not removed prior to processing, the metals are magnetically separated out during the granulation process. The material may continue through one or more sequential granulators to further reduce it in size. The material passes through a series of screens and sifting stations to remove the final vestiges of impurities and ensure consistency of size. During the final phase, the textile residues are removed by air separators.

Cryogenic processing generally uses pre-treated car or truck tyres as feedstock, most often in the form of chips or ambiently produced granulate. Processing takes place at very low temperature using liquid nitrogen or commercial refrigerants to embrittle the rubber. It can be a four-phased system that includes initial size reduction, cooling, separation and milling. The material enters a freezing chamber where liquid nitrogen is used to cool it below the point where rubber ceases to behave as a flexible material. The cooling process embrittles the rubber and allows it to be fractured to the desired size resulting in a smooth and regular shape. Because of its brittle state, fibres and metal are easily separated out in a hammer mill. The granulate then passes through a series of magnetic screens and sifting stations to remove the last vestiges of impurities.

Both ambient and cryogenic processing can be repeated to produce finer particles. Increasingly, with their attendant technologies, the two are combined into one continuous system to benefit from the characteristics and advantages of each and to reduce overall costs. An ambient system is generally used in the initial size reduction phases. The cryogenic system is used to further reduce the material in size and then to remove the metal and textiles. The outputs from either or both systems can be used directly or as feedstock for further processing.

4.3. Level 3 Treatments: Multi-treatment Technologies

These are treatments and technologies that further process the material to modify one or more characteristics by means of mechanical, thermal, chemical, mechano-chemical or multi-treatment procedures. The outputs of Level 2 are most often used as feedstock. Reclaim, surface activation, devulcanisation and pyrolysis are representative examples of the range of treatments used. The outputs can be used directly in applications or products or as feedstock for Level 4 processing [8].

4.4. Level 4 Treatments: Material Upgrading

Level 4 treatments refine, upgrade, modify or generate specific characteristics or properties in materials produced by Level 3 treatments, which most often provide the feedstock. Upgraded reclaim, reactivated/surface-modified/devulcanised materials, upgraded char (carbon products) and new compounds are among the most representative.

Table 21.3 shows the increasing percentage of the total weight of the tyre that is lost during processing. The principal loss is due to the removal of the metals and to a lesser extent removal of the textiles. Generally, smaller materials retain fewer impurities from metal or textile. As speciality treatments most often use smaller granulate or powders as feedstock, there is usually no further loss during processing.

TABLE 21.3 Per cent of Processed Material per Tonne

Output	±% product	±% loss
Shred and chips (unseparated)	95	5
Large granulate ±7–12 mm	70	30
Granulate or powder (truck)	60	40
Granulate or powder (car)	40	60

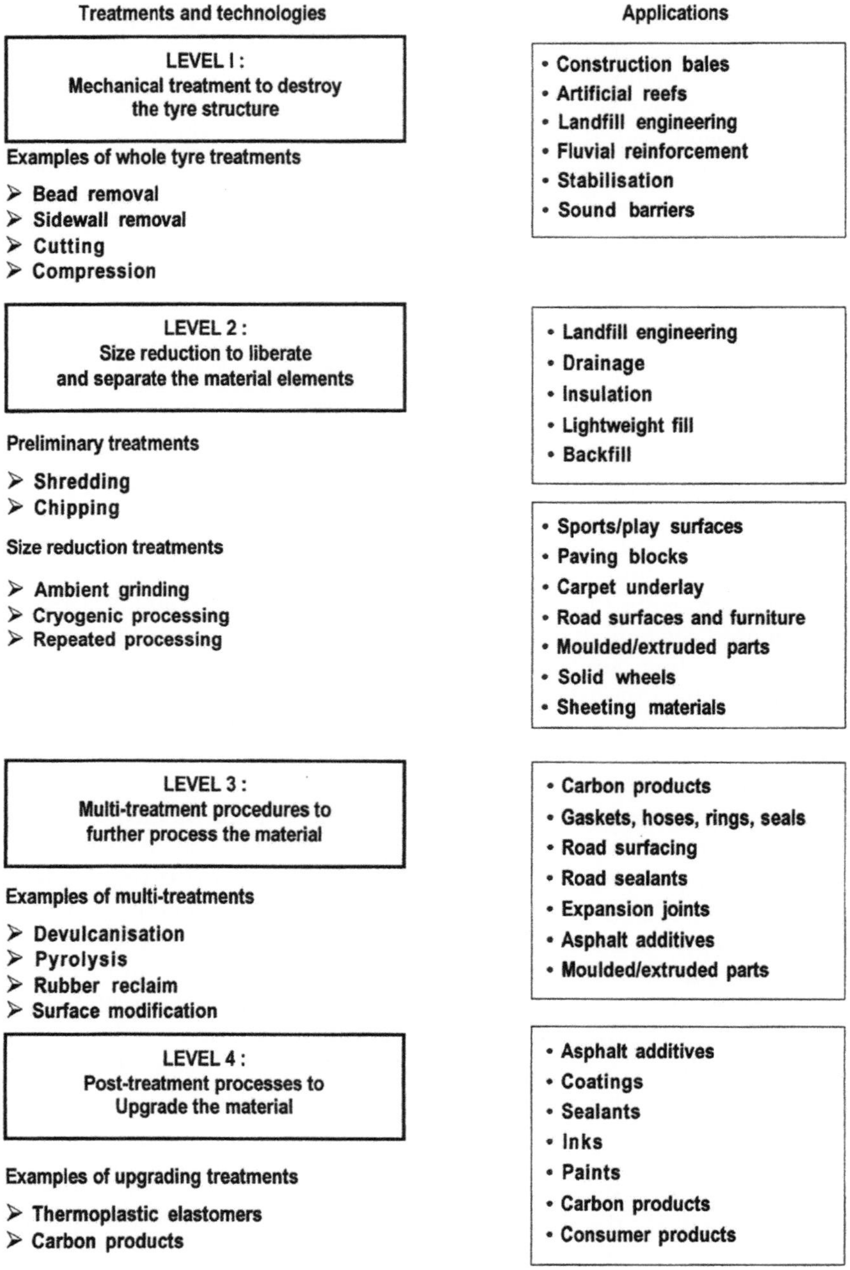

Treatments and technologies	Applications
LEVEL I: Mechanical treatment to destroy the tyre structure Examples of whole tyre treatments ➢ Bead removal ➢ Sidewall removal ➢ Cutting ➢ Compression	• Construction bales • Artificial reefs • Landfill engineering • Fluvial reinforcement • Stabilisation • Sound barriers
LEVEL 2: Size reduction to liberate and separate the material elements Preliminary treatments ➢ Shredding ➢ Chipping	• Landfill engineering • Drainage • Insulation • Lightweight fill • Backfill
Size reduction treatments ➢ Ambient grinding ➢ Cryogenic processing ➢ Repeated processing	• Sports/play surfaces • Paving blocks • Carpet underlay • Road surfaces and furniture • Moulded/extruded parts • Solid wheels • Sheeting materials
LEVEL 3: Multi-treatment procedures to further process the material Examples of multi-treatments ➢ Devulcanisation ➢ Pyrolysis ➢ Rubber reclaim ➢ Surface modification	• Carbon products • Gaskets, hoses, rings, seals • Road surfacing • Road sealants • Expansion joints • Asphalt additives • Moulded/extruded parts
LEVEL 4: Post-treatment processes to Upgrade the material Examples of upgrading treatments ➢ Thermoplastic elastomers ➢ Carbon products	• Asphalt additives • Coatings • Sealants • Inks • Paints • Carbon products • Consumer products

FIGURE 21.3 Summary of treatment levels provides a quick reference of the four treatment levels and provides examples of some of the potential outputs and applications from each of the technologies and treatments.

One of the few exceptions could be the upgrading of pyrolytic char.

Figure 21.3 provides a quick reference of the four treatment levels and provides examples of some of the potential outputs and applications from each of the technologies and treatments.

5. MATERIALS OUTPUTS

The material outputs from the four levels of treatment are broadly classified into six categories: cuts, shred, chips, granulate, powders and fine powders. Table 21.4, illustrates the continuum of materials that result from recycling treatments.

Each category is composed of one or more sub-categories with different parameters, creating a continuum from less than ±500 μm to >±300 mm. The apparent overlap between the larger granulate and the smaller chips is a function of processing specifications. As an example, granulate is characterised by multi-step processing in which metals and textiles are removed, whereas chips are characterised by processing which merely fragments the tyre leaving the metals and textiles intact.

Evaluations and standardised testing procedures are used to ensure material consistency. The methods used vary according to material size. Larger materials, such as shred or chips, are generally evaluated visually. For example, attention is paid to the size of the material and the presence of extraneous materials such as glass, free wires or stones. Other criteria may be agreed between the producer and the user.

With smaller materials, such as granulate or powder, more exacting methods are used. As can be seen, each category and sub-category is characterised as a range. However, the fragments in each sub-category are not homogeneous but constitute a range of different size particles. Thus, a sample of 1-mm powder contains particles ranging from less than 0.30 mm to more than 1 mm. The distribution in the sample should show that more than 90% of the particles are within the 0.30–1 mm range.

Particle distribution is determined with a series of screens. Dependent on the size of the material, a set number of screens is placed in descending order. The material flows through the tier within a specified time. The material that remains on each screen is weighed and expressed as a percentage of the sample. While the size and particle distribution are often critical to the selection of a material, the chemical and physical properties provide information about the material content and how it will react physically in response to certain conditions.

The four columns in Table 21.5 summarise the principal material characteristics and gives a comparison of the different material outputs in terms of:

1. *Size:* each material is described as a range of particle sizes, for example, granulate can be described as a range from 0.5 to 15 mm, whereas devulcanisates are described as less than 1 mm. Whole tyres are discussed as a unit and bales by the number of tyres required to produce a unit.
2. *Key characteristics:* describe some of the principal characteristics of the particular material that can distinguish it from other materials of the same size, for example,

TABLE 21.4 The Range of Recycled Tyre Materials

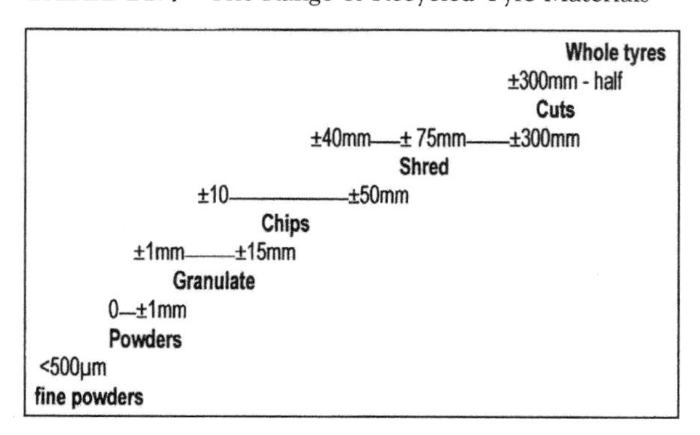

TABLE 21.5 Summary of Material Characteristics

Material	Size	Key characteristics	Traditional materials
Whole tyres	Unit	Lightweight, low-compacted density, high void ratio, good compressibility, water permeability, thermal insulation	Concrete block, clay, quarried aggregate, gravel-filled drums
Construction bale	100–125 tyres	Lightweight, low-compacted density, good thermal insulation, limited deflection, exceeds specifications for specific gravity, gravity, compressibility, deformation, creep, hydraulic conductivity	Construction block, stone riprap, gravel, packed earth, crushed rock in wire cages or other containers
Shred	±50–±300 mm	Lightweight, low-compacted density ±0.5 $t\ m^{-3}$, high void	Crushed rock or gravel, large
Chips	±10–±50 mm	Ratio with good water permeability between 10^{-1} and $10^{-3}\ m\ s^{-1}$. Good thermal insulation and compressibility, low earth pressure and high friction road characteristics	Grain sand, lightweight clay or light expanded clay aggregate, fly ash from coal burning
Size reduced		*Dependent on use*	
Ambient	±0.50–±15 mm	Cross-linked macro structure retains same characteristics as the tyre; irregular shape; some thermal degradation, temperature stressed. May exhibit nominal degree of reduced cross-linking	Product: Virgin rubbers, EPDM, CE: clay, sand, gravel, fly ash from coal burning polyurethane
Cryogenic	±0.5–±5 mm	Clean surfaces, regular particle size and shape, glossy, smooth surface, no surface decomposition or thermal stress	
Powder	<±1 mm	Dependent on technology: ambient/cyrogenic characteristics as above	Virgin rubber, EPDM, other compounding materials
Fine powder	±0–±500 μm	Most often produced by specialised treatments as below	Virgin rubber, EPDM, carbon fillers
Specialty materials		*Dependent on process*	
Reclaim	av. 0.360	Some characteristics of virgin rubber, limited ageing and deterioration or flex fatigue; resistance easily modified during processing, improves flow, mould filling, lower die-swell, can be added to virgin rubber and revulcanised	Virgin butyl rubber, EPDM, other commercial rubbers

TABLE 21.5 Summary of Material Characteristics—*cont'd*

Material	Size	Key characteristics	Traditional materials
Devulcanisate	<±1 mm	Some characteristics of virgin rubber, increased surface reactivity, good fatigue properties, improved tensile strength, reduced cross-links, permanent elongation <50%	Virgin rubber, EPDM
Surface modified	<±1 mm	Activated surface gives improved cross-link density, permanent particle modification, return of some virgin characteristics allowing bonding with matrix during revulcanisation	Virgin rubbers, EPDM, other commercial rubbers
Pyrolytic char	<±0.4–±1000 μm	Coarse particles until treated, some structural change is possible	Carbon materials, fillers, coal burning residues
Upgraded materials		*Dependent on process*	
Thermoplastic elastomer (TPE)	<±1 mm	Mechano-physical characteristics similar to TPEs, about the same shore hardness and polymer base or virgin TPEs	Virgin rubber, TPEs, virgin plastics
Carbon products	>±250 nm	Similar size, shape, surface area, structure, and dispersion as standard carbon blacks, all particles are <30 μm, 50% <1 μm can be used to replace N550, N660, and N770	Traditionally produced carbon blacks

granulate produced ambiently and cryogenically present very different characteristics.

3. *Traditional materials:* lists some of the traditional materials for which particular recycled tyre materials could be used as a substitute for virgin resources.

6. TRADITIONAL AND EVOLVING MARKETS

During the past 20 years, tyre recycling has made great strides towards meeting its goal to produce environmentally sound materials from a priority waste stream, so that they can be used in substitution of virgin resources. Significant quantities of material are produced annually with indications that capacity will continue to grow, at least within the near future. Treatments and technologies have evolved to new levels of sophistication and efficiency that allow the production of finer and cleaner materials, at more competitive prices.

Three traditional large markets coupled with smaller niche markets consistently consume more than 75% of the materials produced. Newer niche markets, many of which have the capacity to use increasingly sophisticated materials, are beginning to evolve, albeit very slowly. Until now, production and use have maintained a relatively even pace. In general, there has been some

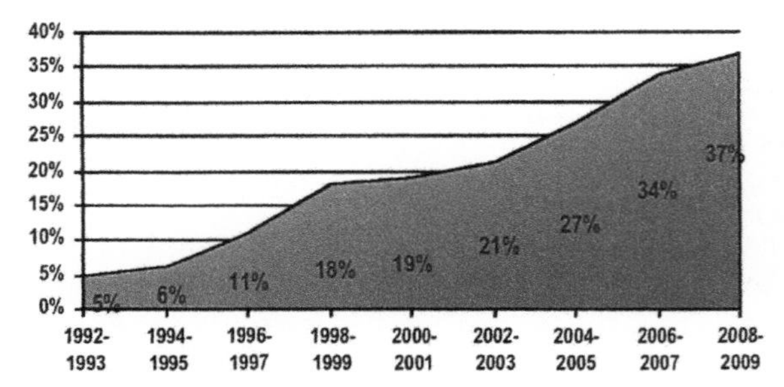

FIGURE 21.4 EU recycling capacity 1992–2009.

expansion into new realms, and there are strong indications that this pattern could continue.

In 2009, European recyclers processed slightly more than 12,200,000 tonnes of post-consumer tyres, approximately 37% of the total annual arisings in the 27 Member States and Norway. Figure 21.4 illustrates the steady expansion of material recycling in the EU during almost 20 years [9].

Figure 21.5, illustrates the percentage of the total that was produced in 2009 for each category of material. The 'miscellaneous' category includes the range of materials used in public and private product development and research activities.

Whole tyres accounted for slightly more than 1220 tonnes. The quantity used for recycling has increased nominally each of the past 5 years with a total increase for the period of about 2% [10]. It is important to note that in the EU, whole tyres, shred or chips destined for energy recovery are not included in calculations for recycling as they are most frequently delivered directly to the recovery facility and treated on site.

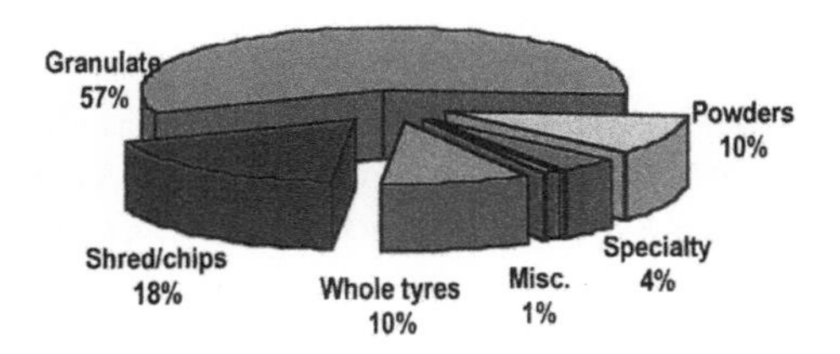

FIGURE 21.5 EU material production by category.

Approximately 219,000 tonnes of post-consumer tyres were used to produce shred and chips; of which, somewhat more than 200,000 tonnes were available for use in a variety of applications. The quantity of shred/chips produced has increased marginally during each of the past 5 years, with a total increase for the period of almost 8%.

By far, the largest quantities of tyres are used for the production of granulate. The quantity has remained relatively stable during the past 4 years, although the ratio to other materials has changed. The 695,000 tonnes of tyres input resulted in about 280,000 tonnes of clean material containing less than ±5% of impurities, textiles or metals. The granulate market has remained stable over the past 5 years although material use is now divided across a broader array of products and applications.

Powders, including specialty powders, used about 125,000 tonnes of tyres; of which, somewhat more than 50,000 tonnes of clean material became available for use. Together, both categories of powders increased just under 5% during the past 5 years. However, the data available on this category are not complete. It is assumed that at least some of the material produced is being used in very small-scale undocumented niche markets, whereas the bulk of the material is used in more traditional applications. The principal markets remain compounding for use in mainstream applications and in the preparation of innovative new materials.

The category miscellaneous utilised about 12,000 tonnes of tyres. However, very little information is available about how the materials were treated, if at all. There are very limited data on this market other than the quantities used for testing purposes and product development activities. It is assumed that this category includes diverse projects such as the 'earthship houses', sculptures or the designer clothing of several years ago.

Production and use patterns vary outside the EU. In Canada, China, Japan, Korea, the United

States, among others, the production of shred and chips far exceeds that of granulate and powders combined. Although civil engineering is more developed in these countries and consistently consumes vast quantities of tyre materials, a large portion of shred and chip production is used as tyre-derived fuel (TDF) in energy recovery facilities.

7. APPLICATIONS AND PRODUCTS

Markets for recycled tyre materials have grown exponentially over the years. They have increased on the merit of the products, primarily because they have been shown to be the most appropriate and effective for a particular application. The fact that the material is easily accessible and often the most cost-effective compared with more traditional ones has also contributed to the expansion. Recycling-for-recycling's sake is not an issue.

7.1. Whole Tyres

Whole tyres are used in a broad range of civil engineering applications because of their unique physical properties. Compared with many traditional materials, tyres are lightweight, free draining, relatively compressible, exert low lateral pressure and low thermal conductivity. They are particularly suited to installations on wet or unstable soil bases that require lightweight materials.

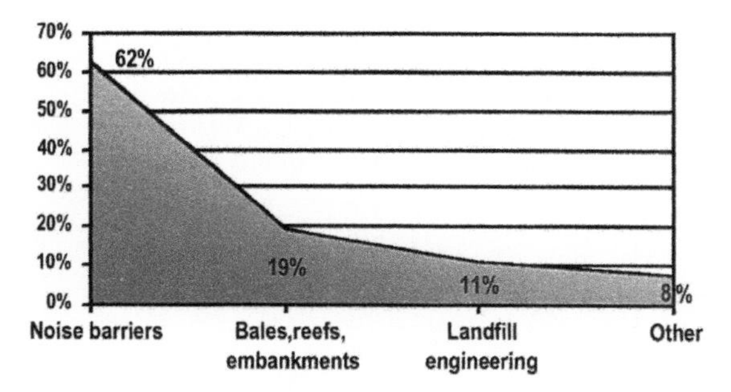

FIGURE 21.6 Whole tyre uses in the European Union illustrates the main categories of whole tyre use in the European Union during 2009.

Figure 21.6 illustrates the main categories of whole tyre use in the EU during 2009. The category 'other' includes boat fenders and silage, among others. Three examples illustrate some ways in which whole tyres contribute to attaining broad sustainable goals.

7.1.1. *Noise Barriers*

These are becoming an increasingly familiar sight along the more than 300,000 km of regional roads in the EU. With noise levels adjacent to major highways and secondary country roads estimated at between 70 and 85 dB(A), sound barriers can reduce the noise by at least 10 dB (A), resulting in levels below the threshold for nuisance or fatigue. National and local governments are demanding the installation of barriers to reduce road noise adjacent to motorways, particularly those located in rural areas or bordering residential neighbourhoods. The barriers are often constructed on soft, marshy or otherwise unstable ground that cannot sustain traditional construction designs and must rely on alternative treatments and materials. This application has been successfully used in the EU, in Scandinavia, and particularly in Finland. As the EU Directive on noise becomes effective, greater quantities of tyres may be consumed by these low-cost structures.

7.1.2. *Artificial Reefs*

Artificial reefs, marine barriers, coastal protection systems, fish hatcheries, breakwaters and sea fender systems have been successfully used since the 1960s particularly in the South Pacific and in Asia. Many installations require that the tyres first be filled with marine cement prior to construction to anchor them, so that they will not break away [10]. Because of their inert structure, tyres are easily colonised by algae, corals and shellfish. Furthermore, their open shape is an advantage when creating structures that require multiple niches where fish can hide from predators or breed. Research indicates that there is a limited amount of leaching of heavy metals and organic compounds [10]. The resilience and availability of tyres has made them popular for building artificial reefs for fishery enhancement both in fresh and sea water. As greater emphasis is placed on coastal defence and the rehabilitation of damaged coastlines, more states may attempt to use them.

7.1.3. Bales

Bales have been used in the United States since the 1980s and have gained acceptance because they are lightweight in comparison with many traditional materials and can be used effectively to rehabilitate sensitive areas such as eroded slopes, waterways and dams. They were first introduced into the United Kingdom in 2001. A bale is less than one fifth the weight of cement, one seventh the weight of gravel and one -third the weight of packed earth, making it an effective alternative material for sensitive areas. They are non-biodegradable, which has made them effective for long-term placement. Recently, they have been tested as replacement for gravel and riprap in roadway foundations and are being used for that purpose by the US Army Corps of Engineers. A number of test sites have been built in the United Kingdom and have received support from local authorities [11,12]. Bales could also be implicated in EU legislation concerning construction in sensitive areas, which could help this yet undeveloped market. Bales could have broad appeal for a variety of rehabilitative applications and could also be of particular interest to other island or peninsular regions specifically to overcome coastal erosion and defence problems.

7.2. Shred and Chips

Shred and chips are lightweight, relatively compressible and compactable, exert low lateral pressure, low thermal conductivity and provide free draining in comparison with other materials. Figure 21.7 illustrates the evolving markets for these materials in the EU.

Shred is virtually non-biodegradable even though exposed steel wires could begin to dissolve if continuously submerged below the water table in certain aqueous environments, depending on the specific pH conditions. Studies indicate that even under those conditions, the exposed steel and zinc oxide leach only trace amounts at levels that are too low to be of concern. Based on its unique characteristics, shred has been used in numerous projects across the United States, Canada, Asia (particularly Korea), Australia and South Africa. Although the EU market is still small, it is

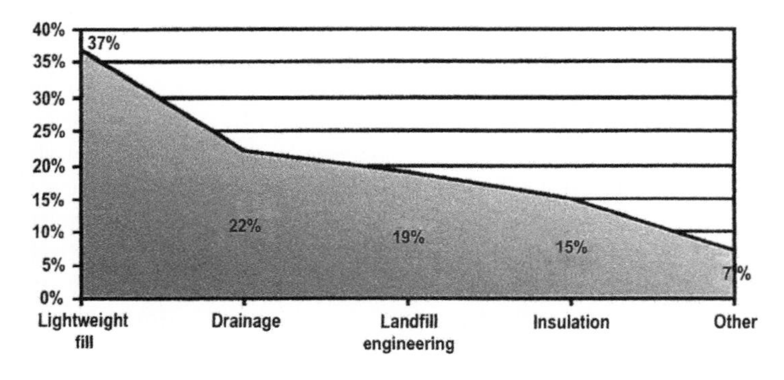

FIGURE 21.7 EU uses of shred and chips illustrate the evolving markets for these materials in the European Union.

growing solidly. Much of the ongoing research has been done at the University of Maine, at Orono [13], which was also instrumental in preparing the ASTM standard for the use of these materials [14].

7.2.1. *Lightweight Fill*

Use as lightweight fill is the largest market for shred. It is widely used in the base course for road and railway construction, as the shock absorbing layer for sports surfaces and more recently as the base for pipeline fields. Shred is used to stabilise embankments and bridge abutments, particularly on weak compressible or clayey soils. It is the primary alternative to traditional materials such as expanded clay, LECA and quarried aggregates, shale, soil or gravel when excess weight could lead to instability. Used as backfill for retaining walls and bridge abutments, shred reduces pressure on the wall reducing settlement and cracking. In some communities, loose fill is used as a covering on playgrounds to reduce injury from play or falls. Its principal benefits are the ability to reduce weight, stress, noise and vibration, which are key concerns within the European Union. These markets have expanded by almost 25% during the past 7 years with indications that they will continue to grow.

7.2.2. *Drainage*

Drainage is of strategic importance, particularly adjacent to motorways and feeder roads, at road entrances and on curves or inclines. Almost all permanent surfaces, including sports fields, playgrounds, etc., require appropriate drainage to maintain the integrity of the surface. Shred provides excellent free draining, so that water pressure does not build. Filter drains, commonly referred to as 'French drains', traditionally use graded aggregate or stone as the filtering agent. They are most often used along regional road networks to prevent the build-up of water at the side of the road and deter it from washing across the road surface, particularly during heavy downpours. When placed in layers below the road surface, the water is carried away and allowed to percolate into the soil. Traditional materials have a high risk of scatter because of vehicle overruns. Shred has been substituted for these materials with the added benefit that damage to vehicles is limited when the material is freed and becomes airborne. Other filter systems utilise a bitumen-coated layer of shred under the loose material to control water flow. These very basic systems have been used in the EU since the mid-1980s but appear to have had limited growth in recent years [12].

7.2.3. *Landfill Engineering*

Landfill Engineering provides six separate opportunities for the use of shred in strategic operations: construction of a new cell, closure of a cell, daily cover, drainage layers, leachate collection and gas drainage systems. Each of the applications relies on the unique physical and technical properties of the material. The principal requirements are non-biodegradability, lightweight in comparison with traditional materials, relative compressibility, high permeability, low density, high void ratio, low lateral pressure, low thermal conductivity and free draining. The shred allows the rapid discharge of water and gas and the reduction of the load on the landfill. The EU Landfill Directive specifies the structure to be used for the closure of a landfill, including the required water permeability. The physical properties of post-consumer tyre cuts, shred and chips are between 10^{-1} and 10^{-3} m s^{-1}, well within the minimum criteria of $>10^{-3}$ m s^{-1} [12]. In addition, shred can be used in the surrounds and to construct haulage roads. Landfill engineering has been particularly successful in Scandinavia and is being considered for use in several other States. The principal drawback is the reluctance of governments to use materials that are defined as waste in this manner.

7.2.4. *Insulation*

Insulation from frost penetration, vibration and sound are key benefits from shred in a variety of applications. The material reduces frost penetration beneath roads and under outdoor sports surface installations. The materials are readily compressible and compactable, providing good thermal insulation and preventing problems from frost and thawing, thereby reducing damage to road pavements and foundations. Shred cushions and insulates underground pipes from freezing during winter months. It was first used at test sites in Italy to reduce vibration and noise around train and tram lines. Because of its success, it is now a mainstream material that is commonly used as sub-ballast and has been introduced in other EU States. Shred was introduced in Finland for thermal and vibration insulation around the foundations of commercial structures in areas adjacent to railway lines and airports. Vibration and noise insulation are key EU themes, and it would appear that these uses could continue to increase in future.

7.3. Granulate

Granulate is used by diverse industries, each of which places different demands on the material. Figure 21.8, illustrates the primacy of granulate in EU markets in 2009, highlighting the dominance of sports surfaces.

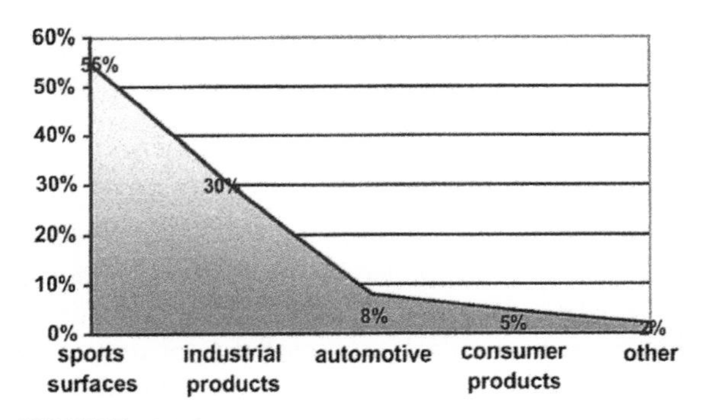

FIGURE 21.8 Granulate use in the European Union illustrates the primacy of granulate in EU markets in 2009, highlighting the dominance of sports surfaces.

7.3.1. *Sport and Play Surfaces*

Sport surfaces include, among others, hockey and soccer pitches, running tracks, tennis courts, artificial turf, turf dressings, equestrian areas and children's playgrounds. Among the key requirements for use in these structures are colour, compressibility, durability, elasticity, free drainage, impact attenuation, low moisture content, porosity, as well as size and particle distribution. The use of granulate improves safety and absorbs the energy from impact. It reduces player fatigue, the severity of injuries during play and improves game response [15].

It is estimated that more than 5,000,000 m^2 of sports fields were built last year in the European Union alone, each of which can utilise approximately 50–80 kg m^{-2} of granulate as infill and/or top cover. Further, in many states, legislation requires that primary school and municipal playgrounds be paved with shock-absorbing materials, which initiated a trend for using these surfaces at more than 150,000 primary schools in the EU. Additionally, recent UN and EU legislation have provided funds for the construction of sports fields.

Sports fields can be installed in a variety of ways i.e., with a solid surface, with a grass-like carpet, as artificial turf or with a loose unbound surface. The base can be constructed on gravel with 100% rubber infill, on a gravel base with a sand/rubber mix infill or on an elastic layer with a sand/mix infill. For solid surfaces, the materials are commonly bound with moisture curing polyurethane material or a polymer-modified bitumen.

Different size granulate is required for each part of the structure and depends on the ultimate performance criteria of the surface. At times, shred is used as lightweight under-fill. The surface materials can be wet or dry mixed, in situ or prefabricated into tiles or sheets of varying thickness and design to meet the requirements of the particular sport.

Outdoor running tracks are generally constructed with three layers including a base course and a solid surface, both made with granulate and bound with a polymer-modified bitumen. These tracks are used by schools for exercise and races. They received considerable attention when it was noted that former US President Clinton used a granulate constructed running course.

Equestrian arenas, other than practice areas that often use loose clean chips or large granulate, are generally constructed with a solid surface. As most injuries occur on entry paths, new materials have been sought. To alleviate the problem, special blocks incorporating granulate have been developed and installed in Alberta, Canada and have successfully eliminated the threat. The blocks have recently been introduced into the European Union.

7.3.2. Industrial Products and Applications

Industrial products and applications include a diverse array of products including flooring tiles, roofing materials, sound absorption and soil treatment products, animal mattresses, vibration mats and solid wheels for industrial equipment, as well as materials and products related to road construction, among many others. The markets are so diverse that there are no common material specifications. It has become apparent that the addition of granulate in all these products does improve one or more performance characteristics of the final product. However, because of space limitations, only product categories will be discussed.

Roofing and flooring tiles were among the earliest products made from granulate. They are still produced by mixing the granulate with polyurethane, placement in moulds and then heat setting. The resulting product has low tensile and tear strengths and high abrasion values. They are appropriate for use in outdoor installations and in heavily trafficked areas. In the EU they have become a choice for garden paths and patios.

Road construction including surrounds, parking areas, road furniture, traffic guidance systems and even manhole covers utilise large quantities of granulate. Many road construction projects utilise different material categories for different parts of the structure, such as, bales for foundations, lightweight fill for embankments or for sub-base drainage layers and granulate for the surface course. The well-documented benefits of granulate additives include longer life, blacker pavements, good temperature susceptibility, better adhesion, increased elasticity, improved drainage and reduced reflective cracking.

Roads surfaced with asphalt rubber also reduce noise, vibration, skidding, fog build-up and reflective glare, as well as frost penetration. Thus, rubberised asphalts could be of particular value on small winding roads, in mountainous regions, 'quiet zones' near hospitals, etc as well as surrounding historic sites where vibration is one of the principal causes of damage [16].

Rubberised asphalt can be installed in a number of different ways: gap graded, open graded, SAMI, chip seal or porous asphalt. Gap and open grading are surface course applications as are porous wearing courses are also surface applications, which increase friction while decreasing noise. SAMIs are surface maintenance treatments, used as an added layer for worn roads to prevent reflective cracking. Improved mixing procedures have reduced the cost of these applications, so that they are now cost-competitive with other materials. They have demonstrated that they are longer wearing than traditional materials.

In Australia, Canada and the United States, among other areas, road surfacing applications consume the largest amounts of granulate. It would appear that the benefits far outweigh the disadvantages of using these materials. They add up to a smoother ride for drivers and lower maintenance costs for municipalities. However, rubberised asphalt has remained a very limited market in the EU although during

2008–2009, nine new test projects were initiated in seven different States.

7.4. Powders and Specialty Powders

Powders and specialty powders include reclaim, surface-activated materials, thermoplastic elastomers, devulcanisates, pyrolytic materials and hybrid materials that are labelled other. Overall, this category of material is the most uncertain in terms of actual quantities produced or used in different products and/or applications. Figure 21.9 provides estimates of the quantities based on reports from projects and pilot applications.

Unlike the materials in other categories that are selected for use in specific products or applications, a majority of the powders and specialty powders are not used directly. They are ingredients used in compounds, which are then mixed or blended with virgin materials for use in an array of different types of products.

7.4.1. Powders

Powders produced ambiently or cryogenically retain the characteristics of the technology that produced them. Thus, ambient powders retain their irregular shape as well as the cross-linked macrostructure of the tyre from which it was produced. Because of increased processing to attain the finer size, some thermal degradation may occur that could produce a nominal reduction in cross-linking. Cryogenic powders retain their smooth, glossy, even shape and show no evidence of surface decomposition or thermal stress.

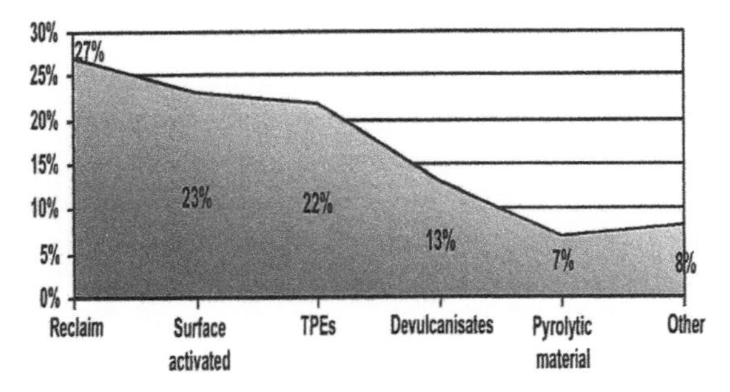

FIGURE 21.9 Uses of powders and specialty powders provide estimates of the quantities based on reports from projects and pilot applications.

These powders have been successfully used in coating materials for roads as well as for sprayed-on coatings and sealants. A unique process has been developed in which the powder is added to a solvent to form a rubber suspension. The material is used for waterproofing and can be painted, sprayed or dipped.

7.4.2. Reclaim

Reclaimed tyres is one of the oldest materials capable of restoring some of the characteristics of virgin rubber to recycled material. The addition of reclaim improves the flow of the new compound to facilitate mould filling, after which it lowers die swell. The new compound can be revulcanised. The largest market for reclaim is new tyre production. It is used as an ingredient in the inner liner and can also be incorporated into the under-tread compound. Reclaimed tyres can be added to virgin rubber compounds and then revulcanised. It is often used in compounds for a range of moulded products. Significant quantities of up to 75% of reclaim are used in the soles and other elements of footwear. The hardness, tensile strength and elongation at break contribute to the ease of moulding and are said to add to the durability and comfort of wear.

7.4.3. Surface-Modified or -Activated Materials

Surface-modified or activated materials have a highly structured surface, so that a high cross-link density can be developed, allowing it to be used in new rubber products. Key characteristics of the material include high density, high resistance to abrasion that can increase the overall resistance of the vulcanised material, high tensile and tear strength and good elongation at break. It can be used as an active filler to substitute for virgin rubbers because the high cross-link density bonds with the surrounding matrix during vulcanisation. Surface-modified powders are a active ingredients in shoe soles. They are increasingly beings selected for use in automotive parts. Steps-plates for vans have recently been added as a product with considerable potential.

7.4.4. *Thermoplastic elastomers*

Thermoplastic elastomers represent a new group of compounds that combine the material qualities of powder with the processing behaviour of thermoplastics. The key characteristics include tear and tensile strength, elongation at break, hardness, heat deflection temperature, impact resistance, rebound elasticity, bulk density and dynamic modulus. The most important among its special quality parameters are permanent elongation distinctly below 50%, with approximately the same shore hardness and polymer base as traditional thermoplastic elastomers. Both the odour and the colour can be altered to comply with customer specifications.

7.4.5. *Devulcanisates*

Devulcanisates are materials that have been reactivated, reducing the cross-links to restore some of the characteristics of virgin rubber – they are not totally devulcanised. Post-consumer tyre materials can be partially devulcanised using several different methods, each of which will leave its mark on the final material, for example, chemical activation agents can change some of the physical and/or chemical properties of the resulting material. Key characteristics include hardness, low density, decreased surface reactivity, good fatigue properties, improved tensile strength and permanent elongation of <50%.

Table 21.6 provides an estimate of the quantities of recycled tyre materials required to

TABLE 21.6 Examples of Recycled Tyre Material Applications

Application	±Quantity	±Unit of application	Format
Sea embankment	3,000 car tyres	500 m × 1.5 m high	Whole
Sound barriers	20,000 truck tyres	1 km × 3 m high	Whole/cut
Artificial reef	30,000 car tyres	1 km × 1 m high	Whole/bale
Drainage culvert bed	50,000 tyres	1 km long	Whole/cut/bale/shred*
Coastal stabilisation	2,000 bales	1.3 m high × 1 km	Bales
Embankment	1,000,000 car tyres	330 m × 3 m high	Shred (compacted)*
Backfill	80–100 car tyres	1 m^3	Shred (compacted)*
Bridge abutment fill	100,000 tyres	1 m wide × 200 mm	Shred (compacted)*
Lightweight fill	2,700–3600 km m^2	Layer thickness ± 1–6 m	Shred (compacted)*
Tram rail beds	50,000 tyres	1 km	Shred (compacted)*
Equestrian track	15 tyres	10 M2 with 15 mm thick	Shred
Thermal insulation	300,000 tyres	0.3 m thick × 10 m wide	chip/shred*
Drainage layer	300,000 tyres	0.3 m thick × 10 m wide	chip/shred*
Road surface	70,000 tyres	1 km 1 lane road	Granulate

(*Continued*)

TABLE 21.6 Examples of Recycled Tyre Material Applications—*cont'd*

Application	±Quantity	±Unit of application	Format
Play surface (25 mil)	1,400 tyres	~500 m^2	Granulate
Asphalt rubber	3,500 tyres	1 km × 12 m × 0.05 m	Granulate
Sound barriers	20,000 tyres	1 km × 3 m high	Granulate
Running tracks	2,700 tyres	400 × 7 m	Granulate
Infill for artificial turf	12,200 tyres	Normal field	Granulate
Safety tiles	4 tyres	1 m × 1 m × 0.04 m	Granulate
Elastic layers	3 tyres	1 m × 1 m × 0.03 m	Granulate
Mats or sheets	1 tyres	1 m × 1 m × 0.01 m	Granulate
Solid wheels (carts)	±1 tonne of tyres	±900 units	Granulate/Powder
Antistatic shoe soles	1 tyre	6 shoe soles (adult)	Powder
Pigments	112 tyres	30–50 pigments	Powder

* *Quantity of tyres depends on producer's formula and specifications of BRRC, BioSafe, ETRA, D. Humphrey, and La Sapienza.*

produce a broad sample of products and applications. The few examples above reflect the basic profiles of more than 500 products and applications in mainstream markets today that utilise recycled tyre materials.

8. ENERGY RECOVERY

Although energy recovery is not recycling, it is a crucial part of attaining sustainable development goals. Thus, no report on tyre recycling would be complete without mentioning energy recovery.

As material recycling, energy recovery is inextricably linked to the prevention and minimisation of waste. It is one of the two principal means of valorising the waste that does occur and, thus, reducing its environmental and economic impacts. Material recycling and energy recovery together offer alternative and complementary means of gaining the greatest sustainable benefit from natural resources and their wastes, thereby reducing the consumption of virgin resources.

Almost 40% of post-consumer tyre in the European Union and about 45% in the United States are used as a supplementary non-fossil fuel in some form of energy recovery process in cement kilns or incinerators to generate electricity, steam, etc., replacing other energy sources. Every country involved in tyre recycling has developed at least one technology for incineration with energy recovery. Recent developments that permit the use of mixed feedstock, such as tyres mixed with other post-consumer or production materials, ensure continued interest.

TDF is a broad rubric that includes whole tyres, tyre wastes, cuts, shred and chips. It is used extensively in developed as well as developing regions around the world. In Japan and the United States, among others, energy recovery is the foremost means of valorisation of post-consumer tyres and industrial wastes.

9. THE FUTURE

It is apparent that the tyre recycling industry has, at times, two mutually exclusive missions.

First, under existing International and EU legislation, it is a waste management industry. As such, it must respond to numerous, and at times onerous regulations that cover every aspect, including permitting, transport, treatment and use of its raw materials. However, second, it is a commercial industry that, to prove its worth and success, must be profitable. It must provide materials and products that meet commercial demands and performance specifications, cost-effectively.

From virtually every perspective, the tyre recycling industry has made great strides during the past 20 years. By all indications, it is successfully meeting the demands of international and EU legislation and mandates in terms of treating waste to produce viable materials and then using them in applications and products. Furthermore, today, those materials are being produced in a an environmentally sound manner and are effectively used to substitute virgin materials for increasingly diverse markets – without substituting quality. In many instances, the characteristics and properties of these materials contribute to improved performance, often more cost-effectively. Many have taken their place in the mainstream – even though they are still considered waste. Even recycling residues are treated and used in innovative ways.

During the last decade – 1999 through 2009 – tyre arisings in the 27 EU Member States (and Norway) totalled somewhat more than 30,000,000 tonnes. Approximately 8,500,000 tonnes of those arisings were materially recycled in an array of applications and products as discussed in this chapter.

Efficient, effective material recycling produces vast savings in terms of material and energy. Energy in different forms is a crucial part of each phase of a product life cycle – from the production of raw materials, to manufacturing, distribution and use, through collection during the waste phase, and, finally, for recycling and delivery of sustainable products to market.

In addition to providing cost-effective, clean, environmentally sound and durable materials for 50 industry sectors, the recycling of post-consumer tyres contributes to the EU goal of a recycling society, with reduced reliance on energy inputs. Table 21.7 illustrates the energy savings gained from material recycling.

TABLE 21.7 Comparison of Energy Use

Quantity	Action	Energy (tonne)
1 tonne	Manufacture of tyre rubber	121,000,000 BTUs
1 tonne	Recycle of post-consumer tyres	2,200,000 BTUs
1 tonne	Recovery of energy	±28,600,000 BTUs

Note: 1 BTU = 1055.06 J.

The difference in energy use between manufacturing 1 tonne of new tyre rubber and recycling 1 tonne of tyre rubber is equivalent to about 3300 L (21 barrels) of petroleum.

In terms of ***energy*** alone, the savings that accrue as a result of the material recycling of 8,500,000 tonnes of tyres, rather then manufacturing the same amount of new tyre rubber, or using tyres for 'energy', is equivalent to: 27,700,000,000 L (174,300,000 barrels) of petroleum. That amount is somewhat more than 1 full year of daily petroleum imports for 6 EU Member States (Bulgaria, Cyprus, Hungary, Luxembourg, Slovakia, Slovenia), or 425 days of petroleum imports for Greece, Or numerous trips circumnavigating the globe (each round trip uses 5000 L or 31.45 barrels). Tyre recycling makes sense from both an environmental and a commercial perspective.

There is an Appendix of relevant literature below, some of which have not been referenced in the text.

APPENDIX FOR FURTHER READING

1. Bontoux, L., Leone, F., Nicolai, M., Papameletiou, D. (1996), The Recycling Industry in the European Union – Impediments and Prospects, European Joint

Research Centre, Institute for Prospective Technological Studies, Seville, Spain.

2. Abbot, G. M. P. (2001), The use of crumb rubber from end of life tyres in sport and play, Proceedings of the International Symposium on Recycling and Reuse of Used Tyres, Thomas Telford, London, pp. 203–211.
3. Amirkhanian, S. (2003), Utilization of tyres in civil engineering applications, Proceedings of the International Symposium on Recycling and Reuse of Waste Materials, Dundee, pp. 543–552.
4. Basel Convention (1999) Technical guidelines on hazardous and other wastes: identification and management of used tyres, Basel convention series/SBC no: 99/008, Geneva.
5. Collins, K. J., Jensen, A. C., Mallinson, J. J., Mudge, S. M., Russel, A. and Smith, I. P. (2001) Scrap tyres for marine construction: environmental impact. In Dhir, R. K., Limbachiyya, M. C. and Paine, K. A. (eds.) Recycling and Reuse of Used Tyres. Thomas Telford, London, pp.149–162.
6. Deutsch Bank: European Rubber Report, 2000.
7. EC (1999), Council directive 1999/31/EC of 26 April 1999 on the landfill of waste. Official Journal of the European Communities Vol. L 182, p. 19.
8. EC (2000), Directive 2000/53/EC of the European Parliament and of the Council of 18 September 2000 on end-of life vehicles. Official Journal of the European Communities Vol. L269, p. 9.
9. EC (2000), Directive 2000/76/EC of the European Parliament and of the Council of 4 December 2000 on the incineration of waste. Official Journal of the European Communities Vol. L332, p. 22.
10. EC (2006), Directive 2006/12/EC of the European Parliament and Council of 5 April 2006 on waste, Official Journal of the European Union, Vol. L 114, pp. 9–21.
11. ERC (2006), The 'Recycling Coalition's' reaction to the Commission proposal for a directive on Waste (COM (2005)667 final): the need for a clear recycling definition in the Waste Directive. EurActiv, 27 April 2006, Published on Internet site: http://www.euractiv.com.
12. European Commission Directorate General XII (1998), The Multiple Pathway Method – A Guide to the Application of the Methodology Developed Through the Research Project: A Combined Methodology to Evaluate Recycling Processes Based on Life Cycle Assessment (LCA) and Economic Valuation Analysis (EVA). Final Report.
13. European Topic Centre on Waste and Material Flows (2004), Summary Report of Main Findings, Conclusions and Recommendations for 2003, DG Environment, Denmark.
14. Hird, A. B., Griffiths, P. J., Smith, R. A. (2002), Tyre Waste and Resource Management: A Mass Balance approach. Viridis Report VR2, Transport Research Laboratory, Crowthorne, United Kingdom.
15. HR Wallingford (2004), Sustainable Re-use PF Tyres in Port, Coastal and River Engineering: Guidance for Planning, Implementation and Maintenance, HR Wallingford, United Kingdom.
16. Humphrey, D. N., Nickels, W. L. Jr. (1994), Tire shreds as subgrade insulation and lightweight fill, 18th Annual Meeting of the Asphalt Recycling and Reclaiming Association, Asphalt Recycling and Reclaiming Association, Annapolis, MD, pp. 83–105.
17. Humphrey, D. N., Sandford, T. C., Cribbs, M. M., Gharegrat, H., Manion, W. P. (1993), Shear Strength and Compressibility of Tire Shreds for Use as Retaining Wall Backfill, Transportation Research Record No. 1422, Transportation Research Board, pp. 29–35.
18. Humphrey, D. N., Whetten, N., Weaver, J., Recker, K., Cosgrove, T. A. (1998), Tire

shreds as lightweight fill for embankments and retaining walls, Proceedings of the Conference on Recycled Materials in Geotechnical Applications, ASCE, p. 15.

19. Hylands, K. N., Shulman, V. (2003), Civil Engineering Applications of Tyres, Viridis Report VR5, Transport Research Laboratory, Crowthorne, United Kingdom.
20. Bressi, G., Ed. (1995), Recovery of Materials and Energy from Waste Tyres – Present situation and Future Trends, ISWA, Denmark.
21. Kerr, S. (1992) Artificial reefs in Australia: their construction, location and function. Bureau of Rural Resources Working Paper No. WP/8/92. Bureau of Rural Rescues, Canberra, p. 34.
22. Paine, K. A., Moroney, R. C., Dhir, R. K. (2003), Performance of concrete comprising shredded rubber tyres, Proceedings of the International Symposium on Recycling and Reuse of Waste Materials, Dundee, pp. 719–729.
23. Pilakoutas, K., Neocleous, K., Tlemat, H. (2004), Reuse of steel fibres as concrete reinforcement. Engineering Sustainability, Vol. 157, Issue ES3, pp. 131–138.
24. Shulman, V. L. (2002), The Status of Post-Consumer Tyres in the European Union. Viridis Report VR3, ISSN 1478-0143, Transport Research Laboratory, Crowthorne, United Kingdom.
25. Shulman, V. L. (2004), Tyre Recycling. RAPRA Review Report, Vol. 15, No 7, Report 175, ISBN 978-1-85957-489-8.
26. Shulman, V. L. (2010), Introduction to Tyre Recycling: 2010. European Tyre Recycling Association, Paris. Published on Internet site: http://www.etra.eu.org (January 2010).
27. Shulman, V. (2000), Recycling: the future, Tire Technology International, pp. 50–55.
28. Shulman, V. (2001), Shulman – breaking the mould, Tire Technology International, pp. 12–13.
29. Shulman, V. (2000), Tyre Recycling After 2000: Status and Options, The European Tyre Recycling Association (ETRA), Paris, France.
31. Talola, M. (2004), Tyre shred civil engineering applications in Finland, Proceedings of the 11th ETRA Conference Brussels, Belgium.
32. Tlemat, H. (2004), Steel Fibres From Waste Tyres to Concrete: Testing, Modelling and Design, PhD thesis, Department of Civil and Structural Engineering, The University of Sheffield, Sheffield, United Kingdom.
33. Tlemat, H. Pilakoutas, K., Neocleous, K. (2006), Modelling of SFRC using inverse finite element analysis. Materials and Structures (RILEM) Journal, Vol. 39, Issue 2, pp. 197–207.
34. Tlemat, H. Pilakoutas, K., Neocleous, K. (2006), Stress-strain characteristics of SFRC using recycled fibres. Materials and Structures (RILEM) Journal, Vol. 39, Issue 3, pp. 333–345.
35. Towards Sustainability, A European Community Programme of Policy and Action in relation to the Environment and Sustainable Development, COM(92) 23 final – Vol. 11, Brussels, 27 March 1992.
36. UNCTAD (1996), Draft Statistical Review of International Trade in Tyre and Tyre-Related Rubber Waste With Particular Emphasis on Trade Between OECD and in Non-OECD Countries and Trade Among Non-OECD Countries 1990–1994, Commodities Division Environmental Issues Section.
37. UTWG (2004), Sixth report of the used tyre working group. Published on Internet site of Used Tyre Working Group: http://www.tyredisposal.co.uk (2004).

References

[1] International Rubber Study Group (IRSG) Rubber Industry Report, vol. 8, No. 10–12, April–June 2009.

[2] Annual Tyre Report, European Rubber Journal, Telford, London, (October 2008).

[3] Deutch Bank, Deutsch Bank: European Rubber Report, 2000.
[4] EC, Council directive 1999/31/EC of 26 April 1999 on the landfill of waste, Official Journal of the European Communities vol. L 182 (1999) 19.
[5] EC, Directive 2000/53/EC of the European Parliament and of the Council of 18 September 2000 on end-of life vehicles, Official Journal of the European Communities vol. L269 (2000) 9.
[6] EC, Directive 2000/76/EC of the European Parliament and of the Council of 4 December 2000 on the incineration of waste, Official Journal of the European Communities vol. L332 (2000) 22.
[7] EC Directive 2008/98/EC Revised Framework Directive on Waste (2008).
[8] V.L. Shulman, Tyre Recycling, RAPRA Review Report vol. 15 (No. 7) (2004). Report 175, ISBN 978-1-85957-489-8.
[9] V.L. Shulman, Introduction to Tyre Recycling: 2010, European Tyre Recycling Association, Paris, 2010. Published on Internet site: http://www.etra.eu.org (January 2010).
[10] K.J. Collins, A.C. Jensen, J.J. Mallinson, S.M. Mudge, A. Russel, I.P. Smith, Scrap tyres for marine construction: environmental impact, in: R.K. Dhir, M.C. Limbachiyya, K.A. Paine (Eds.), Recycling and Reuse of Used Tyres, Thomas Telford, London, 2001, pp. 149–162.
[11] H.R. Wallingford, Sustainable Re-use PF Tyres in Port, Coastal and River Engineering: Guidance for Planning, Implementation and Maintenance, HR Wallingford, United Kingdom, 2004.
[12] K.N. Hylands, V. Shulman, Civil Engineering Applications of Tyres. Viridis Report VR5, Transport Research Laboratory, Crowthorne, United Kingdom, 2003.
[13] D.N. Humphrey, W.P. Manion, ASCE vol. 2 (1992) 1344–1355.
[14] ASTM D-6270-98 – Standard Practice for Use of Scrap Tires in Civil Engineering Applications, 1998.
[15] G.M.P. Abbot, The use of crumb rubber from end of life tyres in sport and play, Proceedings of the International Symposium on Recycling and Reuse of Used Tyres, Thomas Telford, London, 2001, pp. 203–211.
[16] Amirkhanian, S., Utilization of tyres in civil engineering applications, Proceedings of the International Symposium on Recycling and Reuse of Waste Materials, Dundee (2003) 543–552.

CHAPTER

22

Battery Waste

Ash Genaidy, Reynold Sequeira

WorldTek, Inc., 1776 Mentor Avenue, Suite 423, Cincinnati, OH 45212, USA

OUTLINE

1. INTRODUCTION

The past five decades have witnessed the evolution of environmental technologies and management practices from end of pipe treatment (e.g., reactive, driven by regulations, no regard for resource consumption, limited accountability), to pollution prevention (e.g., reduce, reuse, recycle), to design for environment (e.g., proactive, beyond compliance, lifecycle analyses, ISO 14000), to most recently sustainable development (e.g., triple bottom line, multifaceted accountability for both public and private sectors) [1, 2]. In today's global economy, the practice of sustainable development is of particular importance to the handling of toxic materials such as lead to encourage industrial organizations and its stakeholders to take an active role in maximizing human and environmental health.

Currently, there is limited information on recycling rates for lead-acid batteries (LAB) in the published literature. The Battery Council International (BCI) [3] determined that the recycling rates for recovering lead from spent LAB is approaching 99.2% for the 1999 to 2003 time period in the United States. Although very encouraging, such rates seem to be significantly higher than those reported for other countries. For example, Bied-Charreton [4] documented that the recycling rate for LAB is 85% in Western Europe. Therefore, the current state of recycling efforts in the United States has been unclear so as to determine the strategies for maximum lead recovery and recycling in the face of

Waste Doi: 10.1016/B978-0-12-381475-3.10022-1

significant demands for LAB particularly in the auto industry.

2. HISTORICAL USE OF LEAD

According to the US Geological Survey Mineral Yearbook [5], before 1900s lead was primarily used for ammunition, brass, burial vault liners, ceramic glazes, leaded glass and crystal, paints or other protective coatings, pewter, and water lines and pipes. In the early 1900s, lead uses expanded to include bearing metals, cable covering, caulking lead, solders, and type metal. By the mid 1900s, the growth in lead use was derived from the production of public and private motorized vehicles and associated the use of starting–lighting–ignition (SLI) lead-acid storage batteries and terne (an alloy of lead) metal for gas tanks. In addition, the use of lead was drawn from radiation shielding in medical analysis, video display equipment, and gasoline additive.

By the late 1900s, the use of lead was significantly reduced or eliminated in non-battery products in most of the developed and developing countries, including gasoline, paints, solders, and water systems. As the use of lead in non-lead battery products has continued to decline, the demand of lead has continued to grow in SLI and non-SLI LAB applications (e.g., motive sources of power for industrial forklifts, mining equipment, airport ground equipment, uninterruptible power systems in telecommunication networks). In the early 2000s, the total demand for lead in all types of lead-acid storage batteries represented around 88% of apparent lead consumption.

3. LEAD-ACID BATTERY LIFE CYCLE

Figure 22.1 presents an overview of the product lifecycle for the manufacture of LAB. The components of this process include lead mining, lead smelting, battery manufacturing, battery distribution, battery consumption, lead disposal, and lead recycling [6–8]. The case study conducted by Genaidy et al. [9] to determine the actual recycling rates for the year 2006 in the United States demonstrated that, primary lead smelting industry (Box 1, Fig. 22.1) produces 512,000 tonnes (512×10^3 t) of lead of which 78% is used in the manufacture of LAB [5]. On the other hand, secondary lead smelting (2) results in a total production of 1.025×10^6 t of lead [5] of which 93% is consumed in the manufacture of LAB [10]. As such, the ratio of lead from secondary smelting to that from primary smelting is 2.38; to put it differently, lead from secondary smelting accounts for almost 70% of the lead used in the manufacture of LAB in the United States (3).

Both US and non-US (Box 4, Fig. 22.1) battery manufacturing contribute to original equipment manufacturers (5) and battery distributors (6) which in turn satisfy the needs of US consumers (7). Consumers do not return all spent batteries to distributors. A major portion of these batteries are left with consumers in which some are disposed to landfills (8) and others to automobile junkyards (9). In some occasions, spent batteries are fed to landfills from battery distributors. Battery recycling facilities (12) receive a sizable share from both battery distributors and automobile junkyards. Automobile junkyards feed on auto wreckers (12). The feedback loop depicts that lead from spent batteries are redirected to secondary smelters from a number of sources including domestic LAB manufacturers, distributors, landfills, automobile junkyards, and recycling facilities.

A close look at the macroscopic view portrayed in Fig. 22.1 suggests three opportunities for lead recovery and recycling. These include (a) lead in spent batteries with consumers, (b) mishandled batteries sent to auto wreckers, and (c) lead in spent batteries in municipal waste. Additional opportunities are shown in

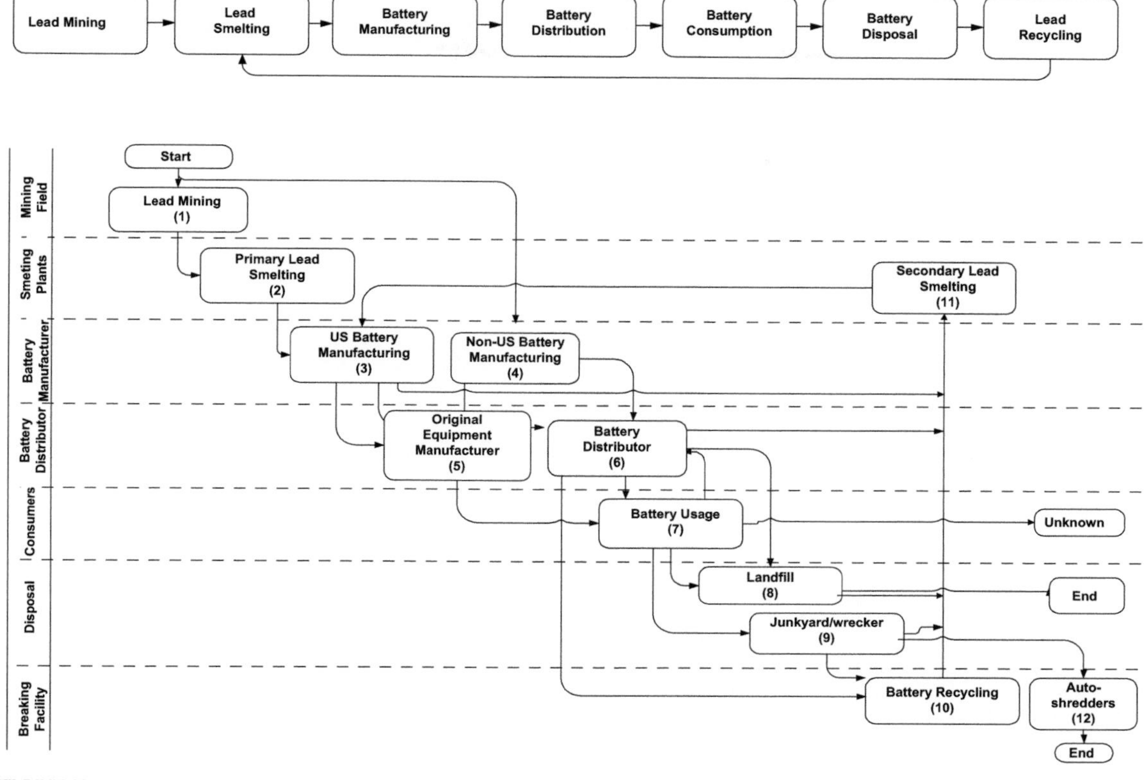

FIGURE 22.1 Overview of lead-acid battery product lifecycle.

Fig. 22.2, displaying a more detailed picture. These include lead lost in air and water emissions during the process of secondary lead smelting and LAB manufacture, which were not quantified because they represent a small fraction of total recycling loop volume. In addition, one may observe the lead lost in slag coming out of the secondary smelting operations that depend on the efficiency of recycling and recovery technologies.

It is well documented that lead is a toxic substance with significant adverse effects on human and environmental health [11–14]. Yet, the use of lead in some industrial products such as LAB is steadily growing to satisfy consumer demand. In light of increased lead prices, it makes economic and environmental sense to maximize lead recovery and recycling by establishing a strong ecologic interface between various stakeholders (e.g., smelters, distributors) and consumers.

4. LAB RECYCLING RATE

Current recycling rates in the Unite States are obtained from statistics derived from a single source, that is, the BCI. The BCI reported that the recycling rates for lead content in the LAB lifecycle for the 1995 to 1999, the 1997 to 2001, and the 1999 to 2003 time periods were 93.3%, 97.1%, and 99.2%, respectively. Although these numbers are quite remarkable and demonstrate a strong recovery and recycling of lead from

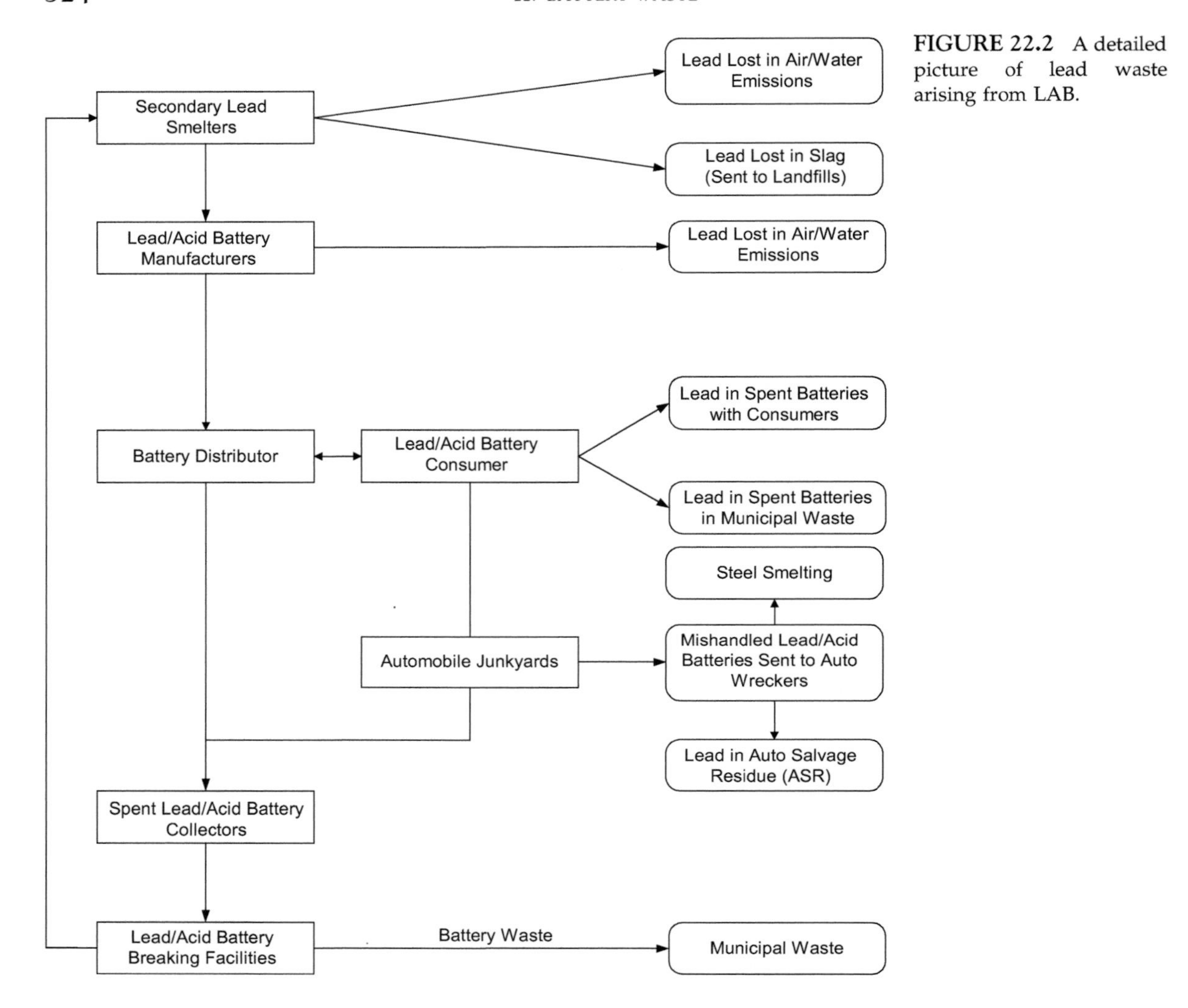

FIGURE 22.2 A detailed picture of lead waste arising from LAB.

LAB, the values are quite high compared with figures reported in other countries. For example, Bied-Charreton [4] reported an 85% recycling rate of LAB in Western Europe. The study identified marked differences in four peer-reviewed articles [4, 15–17] and three reports [3, 10, 18]. An evaluation of these articles and reports suggests that none of the reviewed models took into account lead in spent batteries left with consumers, lead lost in slag, and lead in discarded spent batteries in municipal waste. Lead found in batteries in auto junkyards and subsequently mishandled LAB sent to auto wreckers was partly considered Bied-Charreton's model [4].

Figure 22.3 provides an overview of the integrated model formulated by Genaidy et al [9] to compute lead recycling rates in the LAB product lifecycle. It accounts for all elements identified in Fig. 22.1 and adds new components with the goal to identify the maximum opportunities for lead recovery and recycling. In addition,

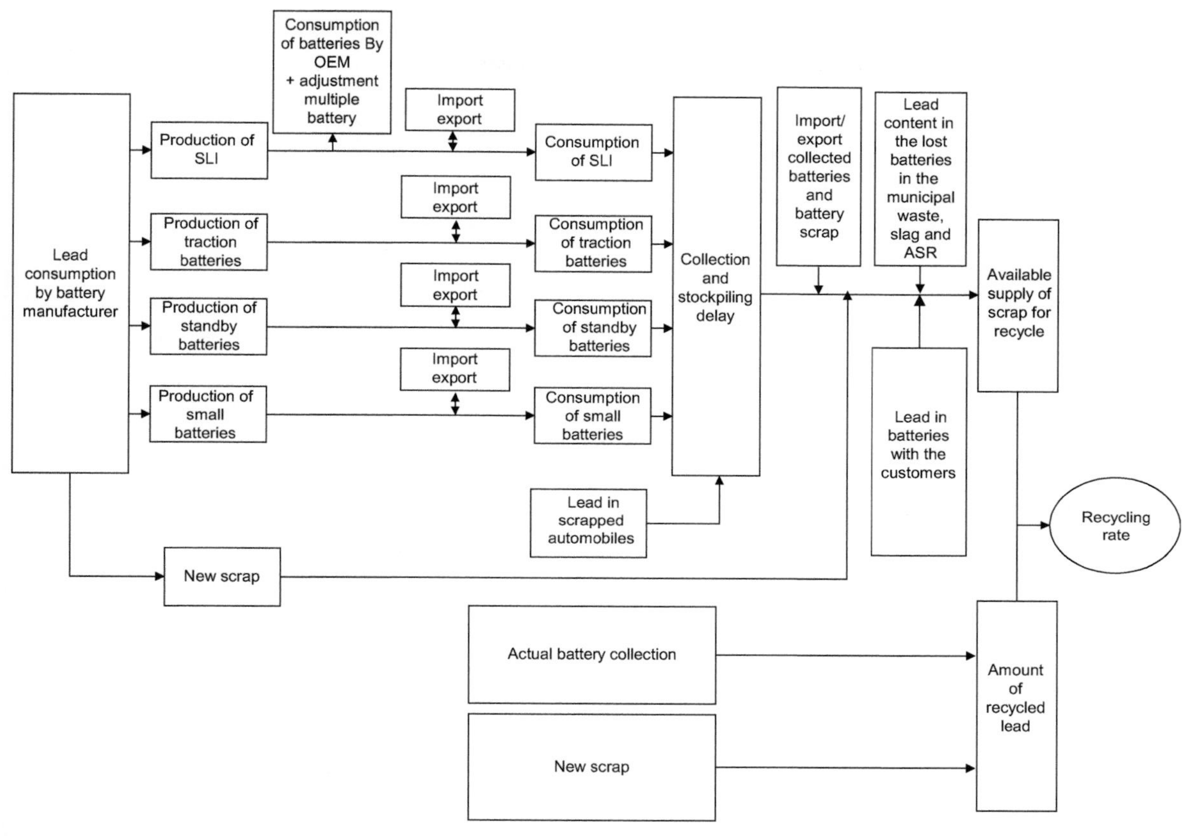

FIGURE 22.3 Integrated LAB recycling rate model [9].

new scrap is generated as a result of these activities and is defined as scrap produced during the manufacture of LAB, components for intermediate use such as drosses and castings, as well as components for final use in the manufacture of LAB such as grids and posts. In other words, new scrap is considered defective finished or intermediate articles that need rework [17].

The decline in recycling rates for the 1999 to 2006 time period was attributed to a number of reasons, including: (1) amount of lead recovered was stagnant, (2) consumption of lead was growing at a 2.25% annual rate, and (3) amount of lead in imported batteries increase at a higher rate than that for exported batteries. In light of the aforementioned, the results confirmed the need to explore opportunities for lead recovery and recycling along the path of the Lead-acid battery lifecycle. A possible explanation for the stagnation of lead recovery from spent batteries in the United States between 1999 and 2006 is that the number of secondary smelters dropped from 42 companies in 1995 [19] to 15 companies in 2006 [20] due to strict environmental regulations requiring a number of effective and sometimes costly environmental safeguards in place to achieve compliance [21]. Fewer facilities may have the effect of reducing the number of recycling channels

TABLE 22.1 Comparison of All Models Using 2006 Data for Recycling Rate Calculations

Description	Bied-Charreton	Sibley	Smith	BCI	Mao	Proposed model
Battery lead recovered (X_1)	9.61×10^5	9.61×10^5	9.61×10^5	9.61×10^5	9.61×10^5	9.61×10^5
New lead based scrap generated (X_2)			1.22×10^4			13,500
Numerator X	$X = X_1$	$X = X_1$	$X = X_1 + X_2$	$X = X_1$	$X = X_1$	$X = X_1 + X_2$
Total X	9.61×10^5	9.61×10^5	9.74×10^5	9.61×10^5	9.61×10^5	9.74×10^5
Lead consumption by US battery manufacturers (Y_1)		1.08×10^6	1.08×10^6			
Lead in domestic battery consumption in 2002 (Y_2)	1.04×10^6			1.04×10^6	1.04×10^6	1.04×10^6
Import battery lead in 2002 (Y_3)	4.76×10^5	4.76×10^5	4.76×10^5	4.76×10^5	4.76×10^5	4.76×10^5
Export battery lead in 2002 (Y_4)	1.48×10^5	1.48×10^5	1.48×10^5	1.48×10^5	1.48×10^5	1.48×10^5
Import lead scrap in 2006 (Y_5)	6.33×10^4	6.33×10^4	6.33×10^4	6.33×10^4		6.33×10^4
Export lead scrap in 2006 (Y_6)	1.12×10^5	1.12×10^5	1.12×10^5	1.12×10^5		1.12×10^5
Lead content in scrapped automobile batteries (Y_7)	1.14×10^5					
New lead based scrap generated (Y_8)			1.22×10^4		1.22×10^4	1.22×10^4
Lead found in municipal waste (Y_9)						136
lead content in auto salvage residue (Y_{10})						7.7×10^3
Lead content in slag from smelters (Y_{11})						7.78×10^3
Stock changes (Y_{12})			4.93×10^4			
Lead in spent batteries with consumers (Y_{13})						1.27×10^5
Denominator Y	$Y = Y_2 + Y_3 - Y_4 + Y_5 - Y_6 + Y_7$	$Y = Y_1 + Y_3 - Y_4 + Y_5 - Y_6$	$Y = Y_1 + Y_3 - Y_4 + Y_5 - Y_6 + Y_8 + Y_{12}$	$Y = Y_2 + Y_3 - Y_4 + Y_5 - Y_6$	$Y = Y_2 + Y_3 - Y_4 + Y_8$	$Y = Y_2 + Y_3 - Y_4 + Y_5 - Y_6 + Y_8 + Y_9 + Y_{10} + Y_{11} + Y_{13}$
Total Y	1.44×10^6	1.36×10^6	1.42×10^6	1.32×10^6	1.38×10^6	1.48×10^6
Recycling rate	67.02	70.77	68.56	72.82	69.61	66.00

All units are in metric tonnes per annum.

available to capture lead scrap. Conversely, consolidating lead production may also have the effect of making production more efficient via economies of scale and more focused regulatory scrutiny from regulatory agencies.

5. OPPORTUNITIES FOR LEAD RECOVERY

Genaidy's findings (see Table 22.1) suggest that the opportunities are centered on the following [9]: (1) lead content in spent batteries with consumers 127×10^3 t a^{-1}, (2) mishandled LAB sent to auto salvage residue 7.7×10^3 t a^{-1}, (3) slag generated due to the inefficiencies of lead recycling technologies 7.78×10^3 t a^{-1}, and (4) lead lost in municipal waste 136 t a^{-1}. Without accounting for these opportunities, the recycling rate using the proposed model approaches 87.59%, a value that is very close to the 87.41% recycling rate calculated for the BCI model for the 1999 to 2006 time period. However, accounting for these opportunities, the recycling rate for the proposed model is estimated at 77.4% which is about 10% less than that for the BCI model. The question becomes where does the 13% of unaccounted for lead go to in the product lifecycle, that is the difference between 100% recycling and the 87% recycling rate that discounts the identified opportunities? A possible answer is that the amount of lead estimated in the proposed model is significantly underestimated. For example, consider the amount of lead lost in municipal waste: it was estimated conservatively at 0.1% of the discarded LAB reported in the EPA Report [22]. In light of the recycling rate comparison above, perhaps this estimate should be higher. Nevertheless, it is evident that there are significant amounts of lead unaccounted for in the product lifecycle.

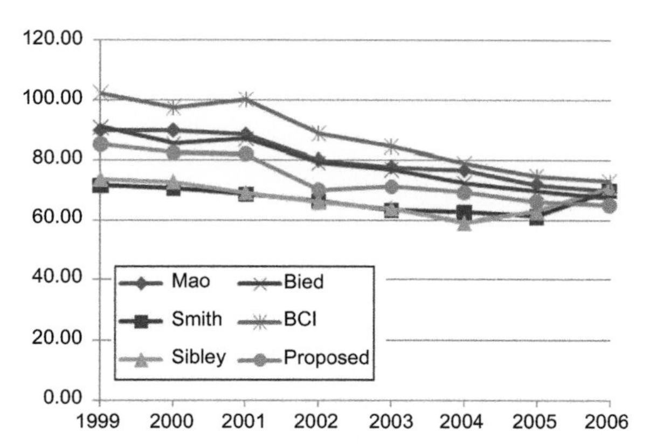

FIGURE 22.4 Revised rates for integrated model in comparison with other models. The *y*-axis refers to percentage.

6. CONCLUSIONS

Figure 22.4 shows the revised recycling rates according to the proposed model. It is shown that the recycling rates derived from the proposed model are around or above 80% from 1999 to 2001 in comparison with the +97.5% rates calculated by the BCI model. Thereafter, recycling rates were dropped by more than 10%. The opportunities for maximizing lead recovery and recycling are centered on the spent batteries left with consumers, mishandled LAB sent to auto wreckers, slag resulting from recycling technology process inefficiencies, and lead lost in municipal waste. High lead prices should provide increased incentives to reduce the lead lost to auto wreckers and improve smelter efficiencies. Spent batteries with consumers and lead lost to landfills may offer the most fruitful opportunities because consumers are less cognizant of lead prices, that is, less likely to incorporate into their decision making, at the end of the useful life of a product.

As such, one can deduce that these opportunities should be closely re-examined with different stakeholders to confirm the amounts of lead available for recovery and recycling along the product lifecycle.

References

[1] J.R. Mihelcic, J.C. Crittenden, M.J. Small, R. David, R.R. Shonnard, D.R. Hokanson, et al., Sustainable Science and Engineering: the emergence of a new metadiscipline. Environmental Science and Technology 37 (2003) 5314–5324.

[2] S.J.T. Pollard, A. Brookes, N. Earl, J. Lowe, T. Kearney, C.P. Nathanail, Science of the Total Environment 325 (2004) 15–28.

[3] BCI (Battery Council International), National Recycling Rate Study, Illinois, Chicago, 2005.

[4] B. Bied-Charreton, Journal of Power Sources 42 (1993) 331–334.

[5] P.N. Gabby, Lead: U.S. Geological Survey Minerals Yearbook, United States Geological Survey, 2005, pp. 43.1–43.20. US Department of Interior, Washington.

[6] R. Jolly, C. Rhin, Resources, Conservation and Recycling 10 (1994) 137–143.

[7] USEPA (United States Environmental Protection Agency), Locating and Estimating Air Emissions From Sources of Lead and Lead Compounds, Office of Air Quality Planning and Standards, Research Triangle Park, NC, 1998. 27711.

[8] I.K. Wernick, N.J. Themelis, Annual Reviews - Energy and Environment 23 (1998) 465–497.

[9] A.M. Genaidy, R. Sequeira, T. Tolaymat, J. Kohler, M. Rinder, Science of Total Environment 407 (2008) 7–22.

[10] BCI (Battery Council International), National Recycling Rate Study, Illinois, Chicago, 2003.

[11] Y. Finkelstein, M.E. Markowitz, J.F. Rosen, Brain Research Reviews 27 (1998) 168–176.

[12] R.A. Goyer, Annual Review of Nutrition 17 (1997) 37–50.

[13] R.A. Goyer, Environmental Health Perspectives 100 (1993) 177–187.

[14] A.C. Todd, D.R. Chettle, Environmental Health Perspectives 102 (1994) 172–178.

[15] J. Mao, Z. Lu, Z. Yang, Journal of Industrial Ecology 10 (2006) 185–197.

[16] S.F. Sibley, W.C. Butterman, Resources, Conservation and Recycling 15 (1995) 259–267.

[17] G.R. Smith, Lead Recycling in the United States in 1998, Flow studies for recycling metal commodities in the United States, United States Geological Survey, 1999.

[18] BCI (Battery Council International), National Recycling Rate Study, Illinois, Chicago, 2001.

[19] US Environmental Protection Agency, Compilation of Air Pollutant Emission Factors, Volume 1: Stationary point and area sources. AP 42, fifth ed., US Environmental Protection Agency, 1995.

[20] P.N. Gabby, Lead: US. Geological Survey Minerals Yearbook, United States Geological Survey, 2004, pp. 43.1–43.19.

[21] G. Hilson, Barriers to implementing cleaner technologies and cleaner production (CP) practices in the mining industry: A case study of the Americas, Minerals Engineering 13 (2000) 699–717.

[22] USEPA (United States Environmental Protection Agency). Materials Generated in the Municipal Waste Stream, 1960 to 2006. <http://www.epa.gov/epaoswer/non-hw/muncpl/pubs/06data.pdf>. Accessed 15 January 2008.

CHAPTER

23

Medical Waste

[†]Andrew L. Shannon, [‡]Anne Woolridge

[†]Michigan Department of Natural Resources and Environment, 525 West Allegan Street, Lansing, Michigan 48909, USA,

[‡]Independent Safety Services Limited, Dabell Avenue, Bulwell, Nottinghamshire, NG6 8WA, UK

OUTLINE

1. INTRODUCTION

This chapter provides a general overview of regulations and issues relating to the generation, storage, treatment, and disposal of regulated medical waste (RMW), generally referred to as healthcare waste (HCW) in the United States, the United Kingdom, in Europe, and elsewhere. The goal is to provide a comprehensive definition of what is generally considered to be RMW/HCW, to provide a brief history of the catalysts that prompted an international regulatory response to the potential hazards associated with this waste stream, and to discuss the development of regulations and rules relating to RMW/HCW.

In reference to regulations promulgated in the United States, beginning with increased public awareness of reported adverse pollution events in the late 1980s and improved awareness regarding infection control and safety practices in the handling of this waste stream, the US Environmental Protection Agency (EPA) instituted the Medical Waste Tracking Act of 1988 (MWTA) [1]. The US Occupational Safety and Health Administration (OSHA) subsequently developed occupational health and safety standards addressing the hazards associated with bloodborne infectious diseases (pathogens) that may be of concern in the promotion and maintenance of the health and safety of employees in the workplace [2].

Information regarding the handling, segregation, treatment, and disposal of RMW/HCW in other countries worldwide will also be discussed. The information provided will focus on current practices and problems associated with the proper management and disposal of this waste stream, as indicated by researched sources. Recurring themes are indicated in the consulted literature, which include (1) the lack of appropriate and available financial and technical resources, (2) inadequate or nonexistent regulatory frameworks, (3) educational concerns in the private and public sectors, and (4) improper disposal of RMW/HCW in poor rural areas [3–6].

2. REGULATIONS

2.1. Regulation in the United States

The MWTA of 1988 was enacted by US Congress in response to incidents of medical waste washing up on east coast beaches (Atlantic Ocean coastline) and the Great Lakes (Michigan, Illinois, Ohio, Wisconsin). These incidents led to increased concern regarding potential health hazards that may be caused by untreated and improperly disposed RMW/HCW, both generated in the public and private sectors. The MWTA amended the existing Solid Waste Disposal Act already under the purview of regulations enacted by the EPA, and specifically focused on the following noteworthy improvements and initiatives [1]:

- Defined medical waste and established which medical wastes would be subject to program regulations,
- Established a cradle-to-grave tracking system using a generator-initiated tracking form,
- Required management standards for segregation, packaging, labeling and marking, and storage of the medical waste,
- Established record keeping requirements and penalties that could be imposed for mismanagement.

Following promulgation of the MWTA regulations in 1989, regulations were fully enforced and administered in 1992. Studies relating to RMC/HCW management and treatment were conducted between 1989 and 1991 in several states throughout the eastern United States (New York, New Jersey, Connecticut, and Rhode Island) in addition to Puerto Rico. During this period, the EPA evaluated a wide array of medical waste treatment methods for RMW/HCW generated in these regions, focusing primarily on the overall effectiveness of the methods used in the inactivation of pathogenic organisms potentially present in untreated RMW/HCW before ultimate disposal. The studies conducted by the EPA included treatment methods that included incineration, autoclaving, microwave treatment, and several other chemical and mechanical treatment technologies. Following the 2-year administration, enforcement, and follow-up study conducted by the EPA, guidance documents were developed at the federal level for future use by individual US states in developing their own regulatory programs [1].

Responsibility for promulgation of regulations and enforcement jurisdiction was ultimately relegated to all individual states within the United States on a voluntary basis, and subsequently, assuming individual state government representatives chose to promulgate legislation, the majority of US states, and territories developed enforceable standards and staff provisions to administer and enforce regulations pertaining to the ultimate disposition of RMW/HCW. As a result of these studies, the EPA determined that the risk of exposure to infectious agents from medical waste was highest at the point of generation (hospitals, medical offices, etc.) and tapered off significantly after the subject waste left the generating facilities [1]. As of 2009, the majority of US states have enacted regulations regarding the handling and disposal of RMW. Most US state regulations have been in place for 10 years or more.

2.2. Regulation in the United Kingdom

All waste generated in any healthcare establishment is defined as HCW. However, not all this waste is treated as hazardous waste. Waste should be segregated at the point of generation based on the hazardous characteristics at the point of production [3]. There will be a large proportion of the waste that has no hazardous properties and will be disposed of as domestic-type waste. The World Health Organisation (WHO) estimates that 10% of this waste (HCW) has the potential to cause infection, injury, or other health impact [5]. Hospitals, Primary Care Trusts (PCTs), and dental practices within the National Health Service (NHS) are not the only source of this waste, others include private hospitals and dentists, military hospitals, funeral parlours, tattooists, veterinary surgeries, and some private clinics, for example, podiatrists. The generation and disposal of HCW are covered by a range of legislation including

- Environmental Protection Act 1990
- Environmental Protection (Duty of Care) Regulations 1991
- The Controlled Waste Regulations 1992
- Radioactive Substances Act 1993
- The Management of Health and Safety at Work Act 1999
- Ionising Radiation Regulations 1999
- Directive 2000/76/EC on the incineration of waste
- Schedule 3 of the Control of Substances Hazardous to Health 2002
- The Hazardous Waste (England and Wales) Regulations 2005
- The Carriage of Dangerous Goods and Use of Transportable Pressure Equipment Regulations 4, as amended in 2009
- Landfill (England and Wales) Regulations 2005
- Waste Electrical and Electronic Equipment Regulations (2002/96/EC)
- DH, 2005

The complexity of the legislation has meant that systems for the management of HCW have evolved steadily over time to meet the regulatory requirements, but so far there has been little emphasis on waste minimisation and implementation of best practice.

2.3. Regulations in China

According to one source [6], regulations relating to RMW/HCW have been either historically substandard or nonexistent in certain regions of China until 2003, when regulations were adopted to enforce proper management of this waste stream. In recent years, regulations have been developed to be increasingly comprehensive in scale, and government-sourced allocations for development of effective treatment devices and facilities have been appropriated. Studies are currently underway nationwide to survey and identify pollution sources (including RMW/RCW), and the Chinese State Environmental Protection Administration (SEPA) is expected to publish a national report on the effectiveness of current programs and recommendations for improvements for the tracking and disposition of RMW/HCW.

2.4. Regulations in Africa

The history regarding RMW/HCW disposal issues in Africa indicates mounting problems that have developed over decades due to lack of proper controls and legislation. One source estimates that approximately 45% of RMW/HCW is not being accounted. This may indicate that large volumes of RMW/HCW may be improperly treated, buried, burned, or dumped in areas that could affect the health of the people and the environment [7].

Many hospitals and clinics currently use inadequate or improperly maintained incinerators to burn RMW/HCW, resulting in the release of hazardous chemicals such as dioxins and furans created by the incineration of plastics used in the manufacture of sharps (needles and other sharp instruments), containers to mercury contamination released from the incineration of thermometers, and sphygmomanometers that contain mercury, which in many cases are still in use. In addition, in some poor rural areas, individuals are exposed to untreated medical waste when picking through dumpsites for usable resources [7,8].

2.5. Worldwide Concerns

Although current regulations and controls over this waste stream vary around the world in complexity and efficacy, several key concerns remain ubiquitous in the consulted literature, including, but not limited to, the following:

- Treating large volumes of waste as RMW/HCW, even though they may be safely disposed of as general waste. These include items such as used latex gloves, gowns, cardboard, and other items that may be generated in a health facility that do not pose a risk to public health, and in fact may pose more of a risk to public health and the environment during treatment processes through the release of emissions (dioxins, furans, etc.),
- Failure to reduce all waste types by creating and instituting improved policies and procedures regarding reuse and recycling of items when possible, and educating healthcare workers on appropriate practices,
- Education of the public and private sectors regarding potential hazards of exposure to contaminated waste,
- Lack of government and organizationally sponsored funding for public sharps disposal programs. Failure to institute these programs leads to higher exposure rates to the public, sanitation workers, and landfill workers to hazards associated with needle stick injuries.

3. DEFINITIONS OF RMW AND HCW

3.1. Definition of RMW in the United States

Although variations of what is considered to be RMW do exist among states within the United States, the basic designation of waste types considered to be part of this waste stream were developed using guidance established by the MWTA of 1988. A listing of standard categories of medical wastes, as reflected in the regulations promulgated in Michigan's Medical Waste Regulatory Act, Part 138 of the Michigan Public Health Code, 1978 PA 368, as amended (MWRA) [9], closely mirror those regulated by individual states, and include the following categories:

- "Cultures and stocks of infectious agents and associated biologicals, including laboratory waste, biological production wastes, discarded live and attenuated vaccines, culture dishes, and related devices,"
- "Liquid human and animal waste, including blood and blood products and body fluids,

but not including urine or materials stained with blood or body fluids,"

- "Pathological waste ... includes human organs, tissues, body parts other than teeth, products of conception, and fluids removed by trauma or during surgery or autopsy or other medical procedure, and not fixed in formaldehyde,"
- "Sharps, including needles, syringes, scalpels, and intravenous tubing with needles attached,"
- "Contaminated waste from animals subjected to pathogens infectious to humans, primarily originating from laboratory research testing."

3.2. Definition of HCW in Europe

In Europe, there is a single classification system for all wastes that depends on the properties of the waste; this is known as the European Waste Catalogue (EWC). HCW falls within Chapter 18 of the EWC. There are 14 hazard groups in the Hazardous Waste Regulations (2005), these groups include properties such as being toxic, infectious, carcinogenic, toxic for reproduction ecotoxic, and so on. Wastes that are not domestic-type wastes fall broadly into two categories, waste that has the potential to cause infection and those that are medicinal. These are the wastes that need to be treated to render them safe. Everyone who generates waste has a "duty of care" to ensure that it is managed correctly from the point of generation, this means that those producing the waste have to dispose of it in an appropriate container that will be sent for correct disposal. There are a number of routes that HCW can take, depending on its hazardous properties.

Hazardous wastes include infectious waste, pharmaceutical waste, fluorescent tubes, specified chemicals, oils, batteries, waste electronic and electrical equipment, asbestos, paints, and solvents. Nonhazardous wastes include domestic-type waste, food waste, offensive/hygiene waste, packaging wastes, recyclates, and furniture [3].

4. EXAMPLE OF VOLUME OF REGULATED FACILITIES IN THE US STATE OF MICHIGAN

Although Table 23.1 represents a very limited picture of regulations worldwide, it provides an example of the tracking of different types of facilities that generate RMW/HCW. This information has been collected from the regulation and registration of facilities in one US state (Michigan) by a program that has been in place for almost 20 years. It can be estimated from this data that most of the regulated facilities in the United Sates are private practices in the medical, veterinary, and dental professions (~78%). However, this data does not represent percentages by volume RMW/HCW generated, because hospitals, nursing homes, and other large healthcare facilities produce a majority of RMW/HCW but represent a small percentage of total facilities that generate RMW/HCW.

5. REGULATED FACILITIES AND TYPES OF MEDICAL WASTE GENERATED

The types of facilities in which medical waste fall under regulatory purview may include, but are not limited to, the following (see Table 23.2):

6. APPROVED METHODS OF TREATMENT FOR MEDICAL WASTE (UNITED STATES AND EUROPE)

6.1. In the United States

In September of 1997, the US EPA adopted new source performance standards (NSPS) and

TABLE 23.1 Tabulation of Medical Waste Regulatory Program Active Registrations in Michigan—September 30, 2005–September 30, 2007

Producing facility category	September 30, 2005	September 30, 2006	September 30, 2007	Average (%)
Medical (Private Physician's Offices, MDs, DOs, Podiatry, etc.)	5,821	6,702	6,793	50
Dental offices	3,363	3,710	3,633	28
Veterinary offices and hospitals	858	917	904	7
Funeral homes/mortuaries	508	534	529	4
Nursing homes	412	418	410	>3
Clinical and analytical laboratories	163	166	162	>1
Hospitals	165	165	165	>1
Mental health facilities	112	124	124	<1
Ambulance/paramedic/fire departments	83	89	93	<1
Pharmacies	31	34	33	<1
Other: dialysis/blood collection/medical education/body art	201	211	232	>1
Total active registrations	11,717	13,070	13,078	100

Source: 2007 Medical Waste Annual Report, Michigan Department of Environmental Quality [10].

emissions guidelines (EG) for hospital/medical/infectious waste incinerators (HMIWI), established under Sections 111 and 129 of the Clean Air Act [11]. An HMIWI is generally defined by the EPA as a device used to treat wastes generated in hospitals or medical/infectious waste, a typical unit being a small dual-chambered incineration device that burns approximately 350 kg h^{-1} of medical waste. Smaller devices may burn as little as 7 kg h^{-1} and larger units up to about 1650 kg h^{-1}. Before the adoption of these NSPS and EG in 1997, the majority of RMW treated in the United States was treated by incineration. Because of strict performance standards and EG, most of the medical waste incinerators in the United States have shut down since that time and other methods have been used (autoclaving, microwaving, etc.). The EPA reported that in 1997 approximately 2300 HWIMI existed in the United States; in 2009 only 57 HMIWI were known to exist [11].

Any remnants of RMW/HCW that remain following the incineration process (ash) are typically disposed of in sanitary landfills. Air pollution remains the most significant problem relating to the incineration of medical waste, as multiple pollutants can be generated from this waste stream either unchanged or as a result of the incineration process. The EPA identifies the following pollutants as being generated in the treatment of medical waste by incineration:

- Particulate matter
- Heavy metals, including lead (Pb)
- Cadmium (Cd)
- Mercury (Hg)

TABLE 23.2 Commonly Regulated Medical Waste Producing Facilities and Wastes Generated

Facility type	Types of medical/healthcare generated waste
Hospitals	I, II, III, IV
Physician offices	I, II, III, IV
Dental offices	I, II
Mortuaries	I, II, III
Veterinary facilities	I, II, IV, V
Nursing homes, hospices	I, II
Ambulatory, EMS services	I, II, III
Tattoo/body art	I, II
Transfer stations	I, II, III, IV, V
Treatment facilities	I, II, II, IV, V
Mental health facilities	I
Outpatient surgical facilities	I, II, III
County and University Health Departments and Clinics	I, II
Dialysis facilities	I, II
Clinical and analytical laboratories	I, II, III, IV, V
Blood/plasma collection facilities	I, II
Industrial clinics	I, II

Key: Type I: sharps (syringes, IV tubing with needles attached, lancets, scalpels, etc.); Type II: blood, body fluids, and items saturated with blood, body fluids; Type III: pathological waste (human tissues, body parts, products of conception, tissues not placed in a fixative agent, organs removed during trauma, surgery, autopsy, or other medical procedure); Type IV: cultures and stocks of infectious agents (laboratory waste, biological production wastes, discarded live and attenuated vaccines, culture dishes, and related devices); Type V: contaminated waste from animals subjected to pathogens infectious to humans, primarily originating from laboratory research testing.

- Chlorinated dibenzo-*p*-dioxins/dibenzofurans
- Carbon monoxide (CO)
- Nitrogen oxides (NO_x)
- Hydrogen chloride (HCl) gas
- Sulfur dioxide (SO_2) gas

In October 2009, the Federal Register was updated, and new NSPS and EG were adopted for HMIWI with more stringent requirements, reducing permissible pollutant levels for these types of incinerators [11]. As mentioned earlier, the adoption of these requirements may lead to closure of additional HMIWI in the future.

The Healthcare Environmental Resource Center recommends that the following pollution prevention methods be performed to alleviate the pollution risks associated with incineration of medical waste, which include the following [12]:

- Ensure that wastes are appropriately segregated at the source of generation to ensure that only wastes that are required to be incinerated by each governmental agency are actually sent out for treatment by incineration. This most often includes pathological waste and medical waste contaminated with trace amounts of chemotherapy drugs. Sharps may also be required to be incinerated by some regulatory agencies.
- Minimize the volumes of PVC plastics that are incorporated into this waste stream, including plastic packaging used in the healthcare field. This type of waste should be recycled whenever possible and eliminated from waste that is treated by incineration, because of the potential formations of dioxins during the process.
- Eliminate any mercury-containing wastes from this waste stream.
- Do not allow any wastes containing chlorofluorocarbons to be incinerated.
- Use alternative treatment methods whenever allowable under government agency regulations.

A significant issue that will arise with the elimination of HMIWI in the United Sates is a lack of devices capable of meeting current state regulations for treatment. Many states require that pathological waste must be treated by incineration only, along with trace chemotherapy waste. Although most of the medical waste generated in hospitals and other regulated facilities can be treated using alternative methods, legislation will need to allow for the future treatment of these types of wastes that is not incineration. According to "healthcare without harm" [13], approximately 15% of all waste generated in hospitals consists of infectious/RMW, whereas pathological waste represents approximately 2% of that waste.

Many US states require that any RMW/HCW treatment technology (that is not incineration or autoclaving) receive a review and approval from the state agency responsible for regulating medical waste before its installation and use in commercial treatment. Standards were developed for states following the MWTA of 1988 to use in the evaluation of these technologies by the EPA and other independent professional associations that are still used today in the approval process.

6.1.1. Alternatives to Incineration: Other RMW/HCW Treatment Technologies

AUTOCLAVES

Autoclaves are devices used to treat RMW using superheated steam under pressure. At increased temperatures, induced using high pressure, pathogens in medical waste are destroyed, leaving the waste safe for routine disposal as solid waste. To make the treated waste suitable for disposal in a sanitary landfill, many state regulations require that waste that has been autoclaved be rendered unrecognizable by shredding or by some other method.

MICROWAVES

Microwave treatment is similar in nature to autoclave treatment in that the waste is subjected to steam heated using microwave technology. The overall process is very similar in that the waste is often shredded to achieve a higher penetration level and infiltration of dense waste materials.

CHEMICAL TREATMENT DEVICES

Chemical treatment decontaminates the medical waste to compounds that are safe for disposal, either via a sanitary sewer system or disposal in a sanitary landfill. Tissue digestion technologies for laboratory waste and animal carcasses typically uses a strong alkali compound (NaOH or KOH) to slowly decompose the waste material over several hours, and the ending pH of the sanitary sewer discharge material is brought within acceptable discharge limits via neutralization that occurs during digestion of the materials. Oxidizing compounds such as chlorine dioxide and other chemicals may also be used to treat medical waste, followed by grinding and disposal of any solid remaining materials, and finally disposal in a sanitary landfill.

OTHER TREATMENT METHODS

Other treatment methods, including pyrolysis (extremely high-temperature treatment) and dry heat technologies, among others, have also been evaluated and approved for use by some regulatory agencies.

THE HISTORY AND FURTHER EMERGING PROBLEMS OF PHARMACEUTICAL AND PERSONAL CARE PRODUCTS

"Pharmaceutical waste" [(aka Pharmaceutical and Personal Care Product (PPCPs)], which includes used and unused expired prescription pharmaceuticals, home-use personal care products, and over the counter medications, have emerged since the development of standard medical waste regulations as being a new major public and environmental health concern. Significant research has been performed worldwide and data have been collected regarding the complex issues associated with this waste stream. In the United States, the EPA has performed multiple studies and research relating to the existence and effects of PPCPs in the environment. Links to important worldwide research, data, and literature references can be found on the EPA website [14].

6.2. Example of Pharmaceutical Waste Management: England

The WHO estimates that around 3% of waste from Primary Care Treatment Centers comprises pharmaceutical products and chemicals [15]. This amounts to significant quantities and value when taken across a single county in England and when scaled up to a national level it is clearly a burden on NHS resources in England. Waste pharmaceutical and medicinal products arise in two different ways; there are those products that are dispensed to the public and returned either part used or completely unused, and there are wastes that are produced by the pharmacy themselves—these are generally medicines that have been ordered and held in stock but never dispensed [16]. This could be due to the products being ordered for a specific patient, but no longer required due to changes in their medication regime or their condition.

Disposal of unwanted medicines and pharmaceutical products from the general public falls within the community pharmacy contract in England. In November 2005, The UK Waste Regulations (SI 2005/2900) stated that householders have a "duty of care" to ensure that they transfer the pharmaceutical waste that they generate from their home to an authorized person for disposal or transport. The financial penalty for inappropriate disposing of household waste is £5000 [17]. Since 2005, community pharmacies need to have in place a system to take in unwanted pharmaceutical and medical products [16].

The PCTs, the bodies responsible for the general health and well being of the general population, have the responsibility of ensuring that the waste is collected in a timely manner and that the contractor is authorized to carry that waste stream. The "public" in this context involves individuals and those in residential care homes, but not care homes with nursing [16]. Under the Environmental Protection Act of 1990, household waste is defined as "*Waste from domestic property, that is to say a building or self-contained part of a building which is used solely for the purpose of living accommodation.*" Nonclinical waste arising as the result of primary care, for example, waste medicines could also be included in this definition. However, for safety, patients are encouraged to return unwanted medicines to their local pharmacy [18]. The reason that waste from nursing homes does not fall into this category is that exemption from Waste Management Regulations 1994 only covers waste medicines from households and individuals and does not cover industrial wastes, or those generated as the result of commercial enterprise [18]. This also means that nonclinical pharmaceutical wastes generated by nurse prescribers and district nurses cannot be returned to the community pharmacy [19].

The waste legislation changed in July 2005, and as a result, producers of waste have had to address the issue of how their waste is handled and what needs to be done to comply with the new legislation. For community pharmacies, the Pharmaceutical Services Negotiating Committee produced some guidance on the legal framework and what is required of them to meet the new legislation [20]. In 2007, the Department of Health published guidance on the disposal of wastes from community pharmacies as a part of the series of Environment and Sustainability: Health and Technical Memoranda [21].

7. CONCLUSIONS AND FUTURE CONSIDERATIONS

Currently established frameworks and standards relating to the characterization, packaging, treatment, and disposal of RMW/HCW from an international perspective demonstrate a range of disparities according to the cited literature, and they are generally dependent on several factors:

- Existing socioeconomic issues and government provisions (or lack thereof) available for the appropriate segregation and treatment of this potentially hazardous waste stream,
- The level and quality of education and resources provided to the public,
- The provision and use of appropriate treatment technologies capable of affording the most viable and effective disposition of this waste stream,
- Regulatory initiatives and enforcement provisions available to curtail the inappropriate handling, treatment, and disposal of items within this waste stream.

Many industrialized nations have developed regulations that adequately and effectively provide safe and environmentally viable means for the treatment and disposal of RMW/HCW, but financial and regulatory burdens remain significant obstacles for nations lacking necessary resources.

Other emerging environmental concerns originating from the evolution, production, and provision of new and existing drug treatments (pharmaceuticals) in the healthcare industry may also prove to require urgent and immediate attention and response by environmental regulatory agencies. Although substantial progress has been made by some countries regarding the proper segregation and disposal of pharmaceutical compounds, the disposal of unused, used, and expired pharmaceuticals and other personal care products is still largely unregulated in other countries.

References

[1] US Environmental Protection Agency, Medical Waste Tracking Act of 1988 (2010), <http://www.epa.gov/waste/nonhaz/industrial/medical/tracking.htm>. Accessed on 11/04/2010.

[2] US Department of Labor, Occupational Safety and Health Administration, Standards 29 Code of Federal Regulations, Standard 1910.1030, Bloodborne Pathogens (2010), <http://www.osha.gov/pls/oshaweb/owadisp.show_document?p_table=STANDARDS&p_id=10051>. Accessed on 11/04/2010.

[3] Department of Health, Environment and Sustainability Health Technical Memorandum 07-01: Safe Management of Healthcare Waste, Crown Copyright (2006).

[4] Department of Health, Safe Management of Healthcare Waste: A Public Consultation, Crown Copyright (2005).

[5] P. Rushbrook, R. Zghondi, Better Healthcare Waste Management: An Integral Component of Health Investment, World Health Organisation, Amman, 2005.

[6] Samantha Jones, China Environment Forum's partnership with Western Kentucky University on the US AID-Supported China Environmental Health Project, A China Environmental Health Project Fact Sheet: Medical Waste and Health Challenges in China (2007).

[7] Leonard, L. Health care waste in Southern Africa: a civil society perspective, Proceedings of the International Health Care Waste Management Conference and Exhibition, Johannesburg, South Africa (2004).

[8] Patience Aseweh Abor, Medical Waste Management Practices in a Southern African Hospital, Journal of Applied Sciences and Environmental Management 11 (2007) 91–96.

[9] Medical Waste Regulatory Act, Part 138 of the Michigan Public Health Code, 1978 PA 368, as amended, Definitions (1990).

[10] Michigan Department of Environmental Quality, 2007 Medical Waste Regulatory Program Annual Report (2008).

[11] Standards of Performance for New Stationary Sources and Emissions Guidelines for Existing Sources, Hospital/Medical/Infectious Waste Incinerators, Environmental Protection Agency, 40 CFR Part 60, Federal Register Vol. 74 (No. 192, Tuesday) (October 6, 2009). Rules and Regulations.

[12] Health Care Environmental Resource Website (2010), <http://www.hercenter.org/facilitiesandgrounds/incinerators.cfm>. Accessed on 11/04/2010.
[13] Medical Waste Treatment Technologies, Evaluating Non-Incineration Alternatives—A Tool for Health Care Staff and Concerned Community Members, Health Care Without Harm (May 2000).
[14] U.S. Environmental Protection Agency, Pharmaceuticals and Personal Care Products (PPCPs) (2010), <http://www.epa.gov/ppcp/>. Accessed on 11/04/2010.
[15] World Health Organisation, Management of Solid Health-Care Waste at Primary Health-Care Centres: A Decision-Making Guide, WHO Document Production Services, WHO Document Production Services, Geneva, Switzerland, 2005.
[16] C. Bellingham, How to dispose of unwanted medicines, The Pharmaceutical Journal 273 (686) (2004).
[17] Queen's Printer of Acts of Parliament, The Waste (Household Waste Duty of Care) (England and Wales) Regulations 2005, Statutory Instrument 2005 No 2900, HMSO, London, 2005.
[18] R. Griffith, C. Tengnah, Legal regulation of clinical waste in the community. British Journal of Community Nursing 11 (2006) 33–37.
[19] R. Griffith, Legal regulations for management of health care waste, Nurse Prescribing 5 (2007) 409–412.
[20] PSNC, Pharmacy contractor briefing on waste, Pharmaceutical Society Negotiating Committee, Aylesbury, UK, 2005.
[21] Department of Health, Environment and Sustainability Health Technical Memorandum 07-06: Disposal of pharmaceutical waste in community pharmacies, The Stationery Office, Leeds, 2007.

CHAPTER

24

Agricultural Waste and Pollution

R. Nagendran

Centre for Environmental Studies, Anna University, Chennai 600025, India

OUTLINE

1. INTRODUCTION

Agriculture refers to the production of food material and related goods through farming. It is perhaps the oldest contribution of man for the survival and welfare of human race. From a humble beginning marked by 'gathering' of food, man has developed agriculture into a massive, technology driven industry. With ever increasing global human population, the scope and application of knowledge gained from fields such as chemistry, engineering, and even mathematics and law to agriculture are on the rise. In the last few decades, industry-like dimension of agriculture has lead to environmental and ecological backlashes in both developed and developing countries. To counter such insurgencies, scientists and administrators are leaning on management sciences. Thanks to sustained research interest in agro science, enormous literature on various aspects of agriculture-environment nexus is available. International organizations such as Food and Agriculture Organization (FAO) of United Nations generate

Waste Doi: 10.1016/B978-0-12-381475-3.10024-5

state-of-the-art reports at frequent intervals to assist planners and decision makers to understand and address issues related to agriculture.

Doubtless, the modern agricultural methods have played a pivotal role in increasing food production all over the world. At the same time, like other facets of developmental activities, agriculture too has been a major source of environmental pollution and waste generation. These two inter-related outcomes of agriculture result from a multitude of activities, and materials were used to enhance efficiency and increase global agro production. Conversion of vast stretches of waste lands into arable lands, development of ground water resources and consequent over extraction of water, excessive use of inorganic fertilizers, unscientific deployment of pesticides and adoption of incompatible agro methods are some of the issues that have resulted in often irreparable and irreversible changes in the environment. In this context, this chapter gives a glimpse of the complexities of agricultural wastes, their impacts and possible options to manage them.

2. AGRICULTURAL WASTE

United Nations [1] defines agricultural waste as:

> waste produced as a result of various agricultural operations. It includes manure and other wastes from farms, poultry houses and slaughterhouses; harvest waste; fertilizer run- off from fields; pesticides that enter into water, air or soils; and salt and silt drained from fields

This definition drives home the fact that the quantity and composition of agricultural waste are dictated by the geographical and cultural aspects of a country or a region and also the extent of land used for agriculture. Table 24.1 presents data on agricultural land area and waste generation in selected countries and Table 24.2 lists the composition of general agricultural waste. Types of agricultural wastes are indicated in Fig. 24.1. A review of sources and types of non-natural agricultural wastes is available in Ref. [2].

TABLE 24.1 Total Agricultural Land and Total Waste Generation

Country	Total Agricultural Area* (Million Hectares)	Total Agricultural Waste Generation (Million Tonnes Per Annum)
Australia	425.44	RNA
China	552.83	56.2†
Egypt	3.53	27‡
Germany	16.9	RNA
India	179.9	RNA
Indonesia	48.5	8.65†
Nigeria	34	RNA
UK	17.6	RNA
USA	411.16	RNA

* *Ongley [3].*
† *Agamuthu [4].*
‡ *GTZ [5].*
RNA – Reference Not Available.

TABLE 24.2 Composition of Agricultural Waste

S.No	Source	Composition
1	Rice*	Husk, Bran
2	Wheat†	Bran, straw
3	Oat†	Straw
4	Maize*	Stover, Husk, Skins trimming, Cobs
5	Millet*	Stovers
6	Pineapple*	Outer peel, crown, bud ends, fruit trimming
7	Sugarcane*	Sugarcane tops, bagasse, molasses

* *Phonbumrung and Khemsawas [6].*
† *Arvanitoyannis and Tserkezou [7].*

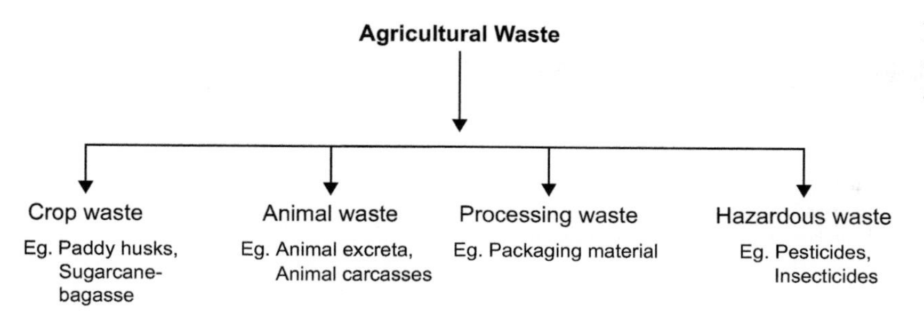

FIGURE 24.1 Types of agricultural waste. *Source: Agamuthu [4].*

3. AGRICULTURAL POLLUTION

Air pollution resulting from agriculture is comparatively minimal, being accounted by emissions from agriculture machinery and open burning of agro wastes – a common practice in many developing countries. Agriculture is a major source of pollution of water and land resources. In view of the enormity of agriculture driven water pollution, special emphasis is given to this aspect here.

It is an acknowledged fact that agriculture is the largest user of freshwater resources, using a global average of 70% of all surface water supplies [3]. As a result, agriculture is a major cause of degradation of surface and groundwater resources. The source of pollutants could be 'point or non point'. Erosion and post-precipitation run-off of chemicals used in fertilizer and pesticide formulations are the major processes responsible for degrading the quality of receiving water bodies. A number of comprehensive reviews and research papers on these and related aspects are readily available [3, 8–10]. Agriculture-driven water quality problems are complex in nature and require multidisciplinary input and approach for proper understanding of the same. One such representation of complexity developed by Rickert [11] is reproduced as Fig. 24.2, and the impacts of

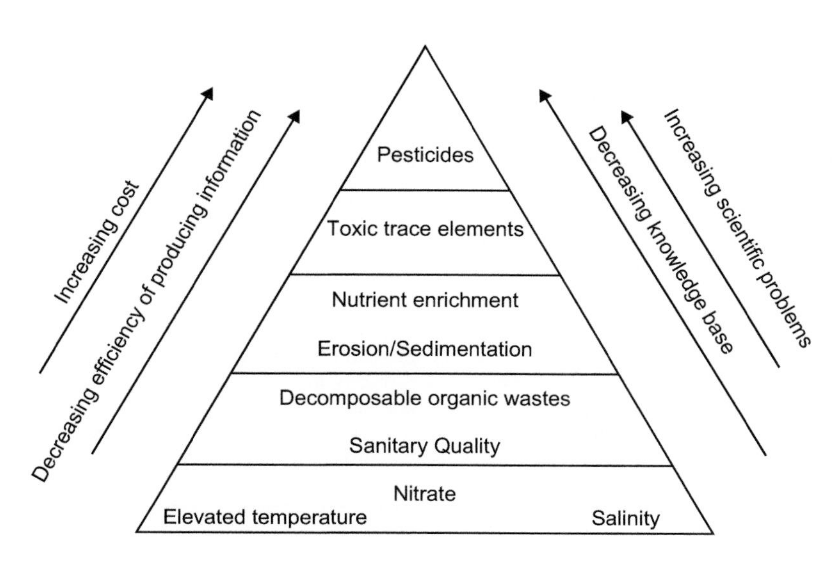

FIGURE 24.2 Hierarchial complexity of agriculturally related water quality problems. *Source: Rickert [11].*

TABLE 24.3 Agricultural Impacts on Water Quality

	Impacts	
Agricultural Activity	**Surface Water**	**Groundwater**
Tillage/ploughing	Sediment/turbidity: sediments carry phosphorus and pesticides adsorbed to sediment particles; siltation of river beds and loss of habitat, spawning ground, etc.	
Fertilizing	Runoff of nutrients, especially phosphorus, leading to eutrophication causing taste and odour in public water supply, excess algae growth leading to deoxygenation of water and fish kills.	Leaching of nitrate to groundwater; excessive levels are a threat to public health.
Manure spreading	Carried out as a fertilizer activity; spreading on frozen ground results in high levels of contamination of receiving waters by pathogens, metals, phosphorus and nitrogen, leading to eutrophication and potential contamination.	Contamination of ground-water, especially by nitrogen.
Pesticides	Runoff of pesticides leads to contamination of surface water and biota; dysfunction of ecological system in surface waters by loss of top predators due to growth inhibition and reproductive failure; public health impacts from eating contaminated fish. Pesticides are carried as dust by wind over very long distances and contaminate aquatic systems thousands of miles away (e.g. tropical/subtropical pesticides found in Arctic mammals).	Some pesticides may leach into groundwater causing human health problems from contaminated wells.
Feedlots/animal corrals	Contamination of surface water with many pathogens (bacteria, viruses, etc.) leading to chronic public health problems. Also contamination by metals contained in urine and faeces.	Potential leaching of nitrogen, metals, etc. to groundwater.
Irrigation	Runoff of salts leading to salinization of surface waters; runoff of fertilizers and pesticides to surface waters with ecological damage, bioaccumulation in edible fish species, etc. High levels of trace elements such as selenium can occur with serious ecological damage and potential human health impacts.	Enrichment of groundwater with salts, nutrients (especially nitrate).
Clear cutting	Erosion of land, leading to high levels of turbidity in rivers, siltation of bottom habitat, etc. Disruption and change of hydrologic regime, often with loss of perennial streams; causes public health problems due to loss of potable water.	Erosion of land, leading to high levels of turbidity in rivers, siltation of bottom habitat, etc. Disruption and change of hydrologic regime, often with loss of perennial streams; causes public health problems due to loss of potable water.

Source: Ongley [3].

agricultural waste on water quality are summarized in Table 24.3.

Agricultural run-off may also cause human health problems. For example, pathogen load may significantly increase in receiving flowing and stagnant water bodies. When these sources are used for meeting drinking needs, water-borne infections may surface, especially in rural populations.

4. AGRICULTURE AND IRRIGATION

Irrigation agriculture is projected as a major global strategy to meet agricultural production targets, especially in developing economies. As early as in 1996, the FAO had projected that an estimated 90% of agricultural land will be under irrigation by 2000. As presented in Fig. 24.3, irrigation is a top priority component of water managers in developing countries including India and China and this trend is likely to continue owing mainly to increasing population levels. In countries like India the same water source, be it surface or underground, is used for irrigation as well as other needs. This imposes economic as well as environmental pressures while executing developmental plans.

The problems associated with irrigation agriculture include erosion of top soil, increase in salinity, waterlogging and desertification. Whenever hydro-geological conditions favor, ground water contamination due to toxic leachates may take place. According to the US-EPA [12], agriculture is the most important land use causing wetland degradation in the United States. Given the low awareness levels and economic constraints, the situation will not be different in developing countries. Galbraith et al. [9] provided an in-depth review of the literature on the effects of agricultural irrigation on wetland ecosystems.

In some developing countries, untreated or partially treated sewage is used for irrigation. This is known to cause microbial contamination of food crops often leading to outbreak of diseases such as typhoid, cholera, amoebiasis and so on [13].

5. AGRICULTURE AND SALINIZATION

Salinization is the most striking effect of agriculture in all parts of the world. It results from natural and human activities, the latter including land clearing and excessive irrigation. Salinity due to such activities is called

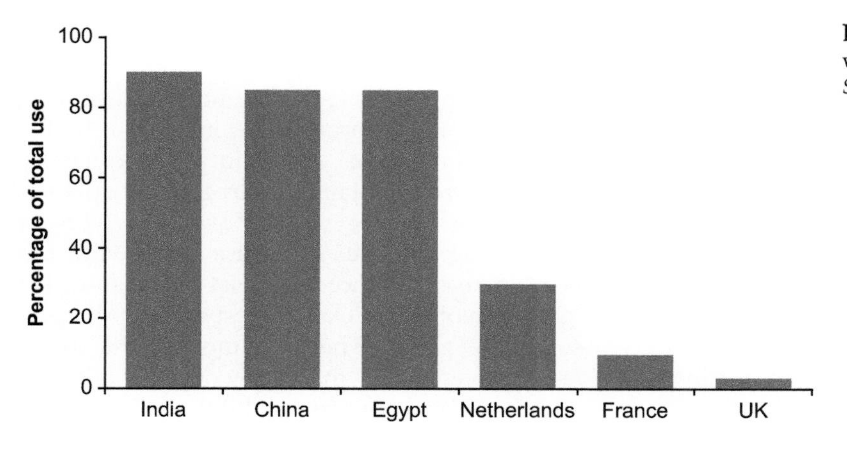

FIGURE 24.3 Percentage of total water used for irrigation. *Source: Saeijs and Van Berkel [14].*

TABLE 24.4 Global Estimate of Secondary Salinization in the World's Irrigated Lands

Country	Cropped Area/(Mha)	Irrigated Area/(Mha)	Share of Irrigated to Cropped Area (%)	Salt-Affected Land in Irrigated Area (Mha)	Share of Salt-Affected to Irrigated Land (%)
China	96.97	44.88	46.2	6.70	15.0
India	168.99	42.10	24.9	7.00	16.6
Commonwealth of independent states	232.57	20.46	8.8	3.70	18.1
United States	189.91	16.10	9.5	4.16	23.0
Pakistan	20.76	16.08	77.5	4.22	26.2
Iran	14.83	5.74	38.7	1.72	30.0
Thailand	20.05	4.00	19.9	0.40	10.0
Egypt	2.69	2.69	100.0	0.88	33.0
Australia	47.11	1.83	3.9	0.16	8.7
Argentina	35.75	1.72	4.8	0.58	33.7
South Africa	13.17	1.13	8.6	0.10	8.9
Subtotal	842.80	158.70	18.8	29.62	20.0
World	1473.70	227.11	15.4	45.4	20.0

Source: Ghassemi et al [16].

secondary salinity. The rising ground water during these activities dissolves salts that get deposited at or near the surface. Table 24.4 presents the global estimate of secondary salinization in the world's irrigated lands. Vast literature on this phenomenon and its effects is readily available, a notable one being the exhaustive review by Omami [15]. Saline soils not only bring down productivity but also render the soil useless for other purposes as well (e.g., construction). It is a documented fact that high salt concentration in the soil would restrict the uptake of water and nutrients by plants. Other notable effects include damage to farm machinery, water and gas pipelines, railways, and buildings. The social dimension of soil salinity is reflected by decreasing family income and higher health expenses. Physical, biological, and engineering solutions have been developed for managing dry land and irrigation salinity conditions, and these are detailed in the study by Ongley [3].

6. AGRICULTURE AND FERTILIZERS

According to the FAO, chemical fertilizers are the single most important contributor to the increase in world agricultural productivity [17]. Fertilizers containing nitrogen, phosphorus and potassium are viewed as the drivers of modern agriculture. Their use worldwide has been increasing since the onset of the so-called 'green revolution'. Data in respect of India presented in Table 24.5 bear testimony to this trend, and it is very likely that many other developing and developed countries exhibit similar trends.

TABLE 24.5 All-India Consumption of Fertilizers in Terms of Nutrients ('000 Tonnes)

Year	N	P	K	Total
1950–1951	158.7	6.9	–	165.6
1955–1956	107.5	13.0	10.3	130.8
1960–1961	210.0	53.1	29.0	292.1
1965–1966	574.8	132.5	77.3	784.6
1970–1971	1,487.0	462.0	228.0	2177.0
1975–1976	2,148.6	466.8	278.3	2893.7
1980–1981	3,678.1	1,213.6	623.9	5515.6
1985–1986	5,660.8	2,005.2	808.1	8474.1
1986–1987	5,716.0	2,078.9	850.0	8644.9
1987–1988	5,716.8	2,187.0	880.5	8784.3
1988–1989	7,251.0	2,720.7	1,068.3	11,040.0
1989–1990	7,386.0	3,014.2	1,168.0	11,568.2
1990–1991	7,997.2	3,221.0	1,328.0	12,546.2
1991–1992	8,046.3	3,321.2	1,360.5	12,728.0
1992–1993	8,426.8	2,843.8	883.9	12,154.5
1993–1994	8,788.3	2,669.3	908.4	12,366.0
1994–1995	9,507.1	2,931.7	1,124.7	13,563.5
1995–1996	9,822.8	2,897.5	1,155.8	13,876.1
1996–1997	10,301.8	2,976.8	1,029.6	14,308.1
1997–1998	10,901.8	3,913.6	1,372.5	16,187.9
1998–1999	11,353.8	4,112.2	1,331.5	16,797.5
1999–2000	11,592.7	4,798.3	1,678.7	18,069.7
2000–2001	10,920.2	4,214.6	1,567.5	16,702.3
2001–2002	11,310.2	4,382.4	1,667.1	17,359.7
2002–2003	10,474.1	4,018.8	1,601.2	16,094.1
2003–2004	11,077.0	4,124.3	1,597.9	16,799.1
2004–2005	11,713.9	4,623.8	2,060.6	18,398.3
2005–2006	12,723.3	5,203.7	2,413.3	20,340.3
2006–2007	13,772.9	5,543.3	2,334.8	21,651.0
2007–2008	14,419.1	5,514.7	2,636.3	22,570.1
2008–2009	15,090.5	6,506.2	3,312.6	24,909.3

Source: Agricultural Statistics at a Glance, 2009, Government of India [18].

Unscientific and overzealous use and application of nitrogenous and phosphorus fertilizers in agriculture have led to the well-known eutrophication of all types of water bodies. Agricultural use of N and P very often lead to problems of 'non-point' pollution. Figure 24.4 captures the phases and pathways involved in the process. Data on N and P discharges to surface waters in the United States are available from Ref. [19], and the negative effects of eutrophication on aquatic systems and humans are in Ref. [20].

In some parts of the world (e.g., Asia), organic fertilizers derived from animal and human excreta are used to supplement inorganic ones. Their eco-friendly dimension notwithstanding, they are associated with some human and animal health problems. There have been reports, especially from developing countries, of water quality deterioration caused by their discharge into water bodies. Infiltration of nitrates and to a very little extent, phosphates, may contaminate groundwater as well. Volatilizing ammonia may lead to acidification of water and land. Heavy metal fractions occurring in excreta may add to the mess.

From a futuristic view point, there is an urgent need to reverse the trend in N and P flows into and through the agro systems before they reach their ultimate reservoirs. Nitrogen input–output analyses coupled with modeling studies may help agro managers to achieve this (see also Refs. [21, 22]). Other efforts should include urban runoff management, proper understanding and modeling of atmospheric deposition of N, biocapturing of excess N and P in the system, and optimal dosing of nutrients.

7. AGRICULTURE AND BIOCIDES

In agriculture, the term biocide includes rodenticides, insecticides, nematocides, fungicides and herbicides. Among these, herbicides and pesticides are often considered as agro

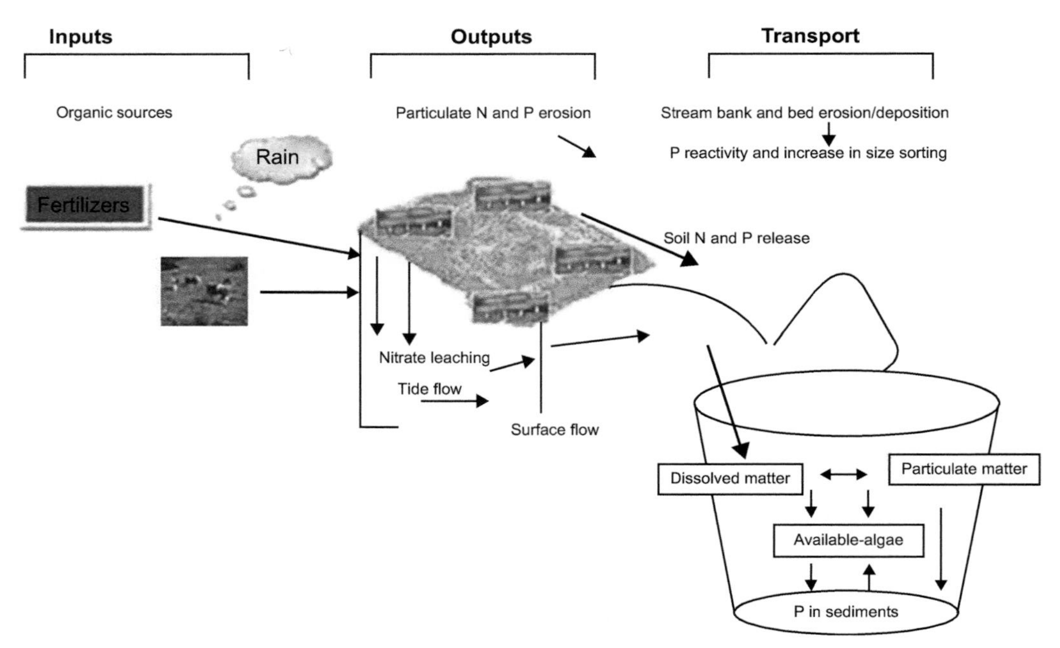

FIGURE 24.4 Inputs, outputs and transport of P and N from agricultural land modified after Carpenter et al. [20].

saviors. A majority of these are chemical formulations that are prey–predator specific. Their use during the last five or six decades has thrown much light on their anticipated benefits as well as inevitable negative impacts. Exhaustive information on pesticides and agricultural chemicals is available in Ref. [23]. Unlike fertilizers, the impacts of pesticides get compounded by independent or co-action of other ingredients such as wetting agents, emulsifiers, solvents and so on, which are commonly used. In some cases, substances produced during the degradation of the main active ingredient join hands and increase the magnitude of impacts. In irrigated agriculture, use of multiple biocides at different phases of production results in the discharge and retention of a multitude of toxic chemicals and their derivatives. These are known to have caused unwanted ecological consequences. The criteria for determining such ecological impacts and their health effects are well documented [3]. Persistent toxic substances (PTS) are attracting the attention of researchers all over the world. Significant contributions on these include those of Ding et al. [24] on PAH, Mai et al. [25] on PCB, Cao et al. [26] on DDT and Hu et al. [27] on HCHs, DDT, PCBs and PAHs. These studies highlight the significant relationships found between different PTS.

Selected classes of pesticides have been shown to cause genetic and hormonal imbalances and disruptions in animals and humans. This aspect needs urgent attention in the context of developing countries as agricultural contamination in food and water may have serious implications on human health and reproduction.

8. AGRICULTURAL WASTE MANAGEMENT

Agriculture management is a complex process involving individual attention to address the issues related to all functional components such as water, fertilizers, biocides and so on, as outlined in preceding sections. In view of this specific objective of the book, this section specifically addresses the management of agricultural wastes.

A significant constraint encountered in managing agriculture waste is lack of data pertaining to different geographical regions. Therefore, the urgent need is to formulate a tangible mechanism to create an international database on the quantity, composition and characteristics of agro wastes. Preparation of a compendium flagging the success stories and challenges in respect of developed and developing economies would serve well for officials in governance and on the field.

Scattered information on the current and futuristic options for managing agro waste is available. Forced perhaps by the economic considerations and conducive environmental conditions, developing countries such as Vietnam are taking lead in managing agro wastes following the principle of 3Rs [28]) and other local/indigenous methods. Table 24.6 outlines similar efforts taken to utilize common

TABLE 24.6 Agricultural Waste Utilization

Waste	Utilization
Rice husk ash and charcoal	Additive in cement mixesWater glass manufactureActive carbon
Rice husk	Electricity production
Banana peel and sugarcane fibers	Paper making pulp
Husk, bagasse	Mushroom cultivation
Bagasse, banana fruit reject	Ethanol production animal feed
Animal waste	Compost fertilizer
Husk, straw, cow dung	Biogas production
Sugarcane ethanol	green polythene

Sources: Agamuthu [4] and Braskem Ltd. [30].

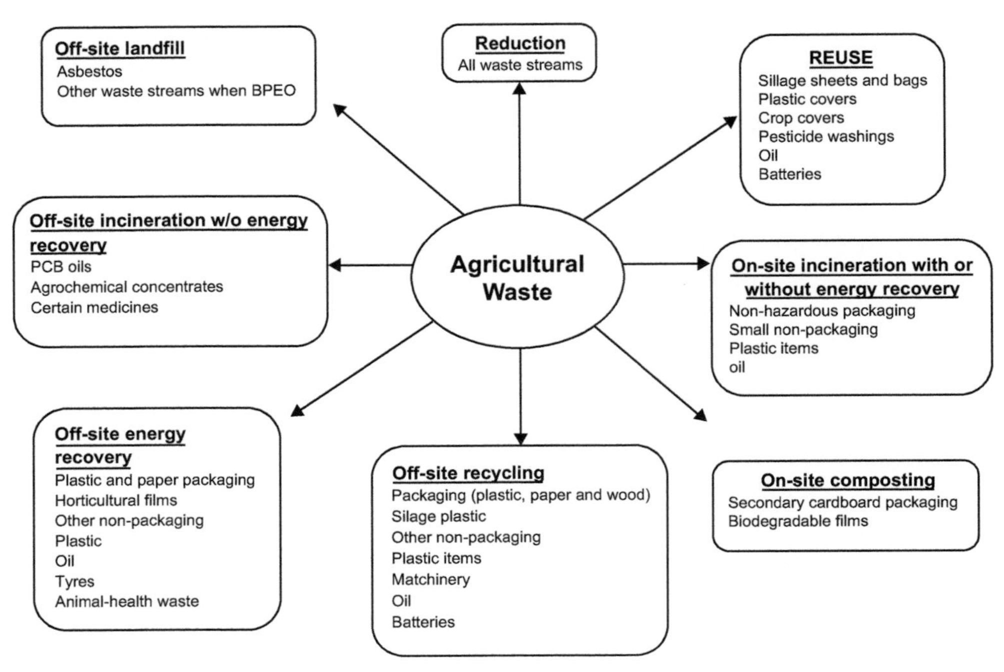

FIGURE 24.5 Overview of best options for agricultural waste management modified after [2].

agricultural wastes in Malaysia and Brazil. A success story of creating fields of gold using agricultural waste in rural India is available in Ref. [29]. Some of the ecological and engineering options available for effective management of agricultural wastes are captured in Fig. 24.5.

9. AGRICULTURE AND CLIMATE CHANGE

The strongest and loudest warning bell ever rung by *Nature* to call the attention of humans is probably 'Climate Change'. In fact, every category of anthropogenic activity has been shown to contribute to global warming and climate change. Being a multiactivity practice, agriculture has multiple sources for green house gas (GHG) emissions and understandably contributes significantly to global warming. As detailed in Ref. [31], globally, agricultural CH_4 and N_2O emissions have increased by nearly 17% from 1990 to 2005, agriculture accounting for 5.1 to 6.1 $\times$ 10^9 tonnes of CO_2 or equivalent per year (approximately 10–12% of total global anthropogenic emissions of GHG), in 2005 alone. The case of CH_4 is more alarming. It is reported that CH_4 emissions cause nearly half of the Earth's human-induced warming and its number 1 source is animal agriculture [32].

Although energy intensive activities involving machinery contribute to emissions in developed economies, vast cultivated areas and biomass do the same in developing countries. Major sources of 'direct' GHG emissions in agriculture are indicated in Fig. 24.6. Biomass combustion, often in open fields, is a common practice across the world. Reasons and social compulsions (e.g., domestic fuel) notwithstanding the resulting quantum of GHG emissions would be significant as can be inferred from the data in Table 24.7.

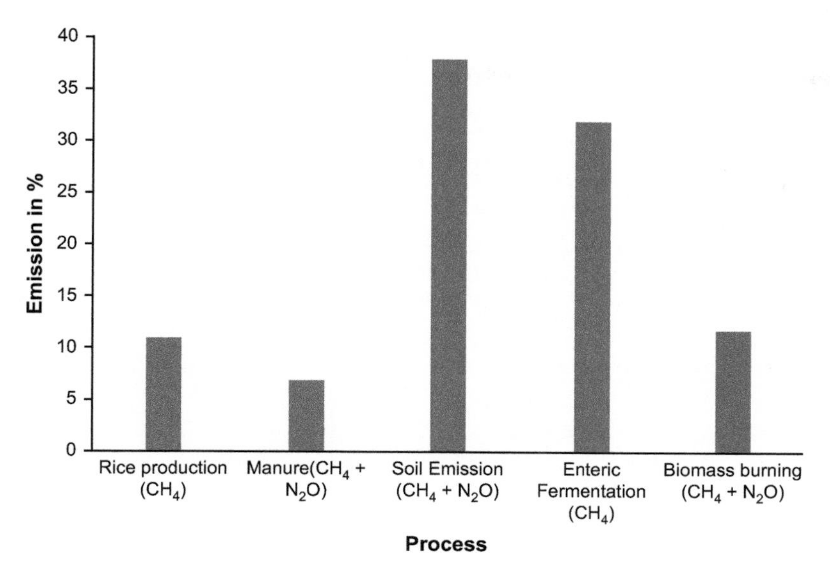

FIGURE 24.6 Major sources of 'direct' GHG emissions in the agricultural sector in 2005. *Source: Smith et al. [31].*

Research and development activities and international cooperation to counter global warming from agro sector have opened up newer technical avenues to increase the mitigation potential. The estimated potential from agriculture, excluding fossil fuel offsets from biomass, is an encouraging 5.5–6.0 $\times$ 10^9 tonnes of CO_2 or equivalent per year [31]. A number of technomanagerial options exist to reduce, if not eliminate, agricultural GHG emissions. The major ones include effective land and crop management through scientific nutrient and residue use, restoration of soil quality through organic amendments, better irrigation and water management practices, promotion of ecological forestry and application of traditional agro technologies that are often site and country specific. Of course, the successful implementation of the option(s), especially in developing countries, is dictated by political will and economic feasibility.

'Soil carbon sequestration' is emerging as the most sought after option to prevent and reduce GHG emissions. Despite the fact that the process is controlled by a multitude of variables such as existing land use, soil type and chemistry, local climatic and weather conditions, crop pattern, input of carbon containing supplements and so on, a number of attempts are being made in different geographical locations to assess its efficacy. Table 24.8 presents the highlights of a few case studies compiled in Luske's study [34].

TABLE 24.7 Agro Biomass Combustion in Selected Countries

Country	Amount of Dry Matter Burned in Field (Mt)
Africa	49
Asia (excluding India and China)	274
India	81
China	6
Latin America	85
Brazil	42
USA	36
Australia	7

Source: Yevich and Logan [33].

TABLE 24.8 Summary of a Few Case Studies on Soil Carbon Sequestration

Case study	Country	System	Organic Practices Used	Mitigation Benefits (Tonnes of Carbon per Hectare per Annum)
Farming systems trial	USA	Arable crops	Organic fertilization Crop rotation Cover crops	2.3
Pure Graze	The Netherlands	Pasture-based diary system	Use of natural reproductive cycles Locally produced fodder and feed Reduced use of concentrate feed	0.4 10% (per kg produce) and 40% (per hectare) less GHG emissions than conventional farming
Composting in Egypt's desert	Egypt	Vegetable crops	Composting	0.85
DOK trial	Switzerland	Arable crops	Organic fertilization Crop rotation including grass clover Cover crops	0.25
Agroforestry	Indonesia	Cocoa/forest	Intercropping with trees and shrubs	11

Source: Luske [34].

Organic agriculture (OA) is a viable alternative with great promise to mitigate climate change. Being ecological in origin, it relies on natural cyclic material processes and unidirectional flow of energy through diverse biosystems that are adapted to local conditions. With a broad canvas to innovate, experiment and use endemic knowledge, it can work even at a very small spatial scale. Adoption of OA has been reported to have the potential to sequester up to the equivalent of more than 30% of anthropogenic GHG emissions [34]. Usefulness of OA in building robust soils to counter the effects of extreme conditions and its possible role in mitigating climate change are detailed in Ref. [35].

10. AGRICULTURE AND BIOTECHNOLOGY

A 'never dead and ever alive' debate in the realm of agriculture has been the good and bad impacts of the 'invasion' of biotechnology into the sector. Both science and non-science seem to lend their support to both sides, often in near equal measures! Neo-agriculturists, if such a term can be used, in the developed western countries, introduced gene manipulation/modification techniques developed by genetic engineers to produce the (in)famous genetically modified (GM) foods. The projected features of GM plants included resistance to diseases and pests and tolerance to cold climate, herbicides and salinity (see Ref. [36]). They were also projected as potential candidates to produce vitamin-rich grains [37] and vaccines [38] and to decontaminate metal laden soils [39].

Although projected as a boon to developing countries, United States continues to increase the area for cultivating GM crops whereas China and India, the two most populous developing countries, lag far behind as illustrated in Fig. 24.7. Contradictory public and political views, lack of demonstrated benefits under local conditions and very high investment needs are the likely reasons for the low momentum.

Worldwide experience in GM food production during the last two decades and extensive

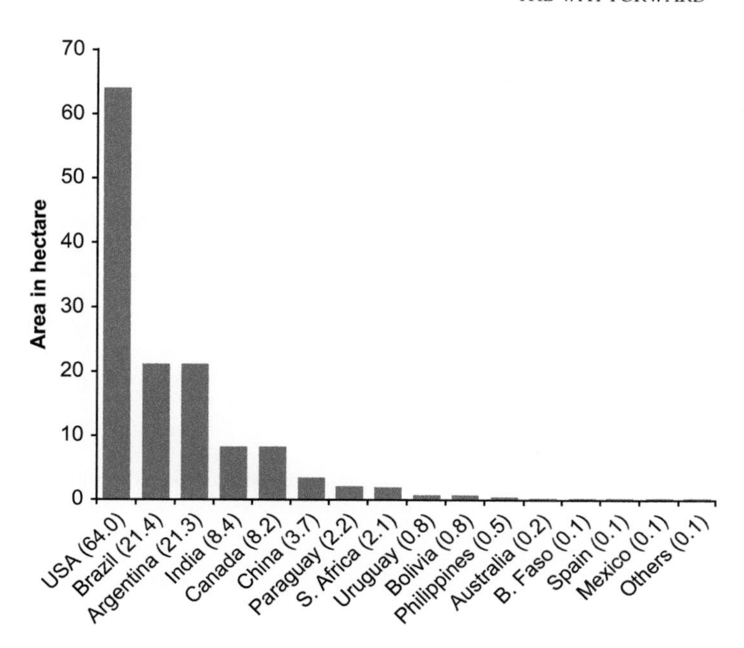

FIGURE 24.7 Global area of biotech crops in 2009 by country (million hectares). *Source: James [40].*

laboratory and field research under varying geographical conditions have brought to light several negative aspects of the technology. For instance, the findings of a 10-year field research on ecological impacts of GM crops are available in Ref. [41]. GM crops have been shown to interfere with 'soil–plant' interactions involving microbial processes [42,43].

The major environmental impacts as documented in Refs. [41,44–46] include: (i) spread and persistence of transgenic plants outside the cultivated area due to development of resistance; (ii) changes in ecosystem dynamics caused by the effects on non-target organisms and uptake of transgenic products through feeding; (iii) contamination of receiving water bodies caused by persisting transgenics and their degradation products; (iv) undirected and uncontrolled flow of genes beyond cultivation domain leading to the formation and spread of unanticipated and undesirable hybrids; and (v) invasion of GM crops outside cultivated areas.

Questions relating to the moral and ethical aspects of GM agriculture are being raised in many quarters and the exuberant promoters of agro biotechnology are obliged to respond. The current global understanding suggests that despite being a very productive tool, large-scale acceptance and application of biotechnology in agro sector may have to wait for a while.

11. THE WAY FORWARD

From the foregoing, it emerges that the oldest profession of man seems to be posing newer challenges in his quest to satisfy the food needs of the hungry world. The onus is fully on the so-called *man the wise* or *Homosapiens* to maintain the ecological balance and environmental integrity of the planet as we continue to exploit the resources to augment and enhance agricultural productivity. The road to success seems to lie on sound science, agro efficiency aided by engineering interventions and guided by wise management principles to create a seamless and hunger-free society.

ACKNOWLEDGEMENTS

I thank the authorities of Anna University, Chennai, India, for permitting me to accept this assignment. I thank my research students Ms. S. Meenu Krithika and Mr. C. Madhan for their whole-hearted help during the preparation of the manuscript.

References

[1] United Nations, Glossary of Environment Statistics, Studies in Methods, Series F, 67, Department for Economic and Social Information and Policy Analysis, Statistics Division, New York, USA, 1997. 96.

[2] EA-UK R&D Technical Report P1-399/1 (2001), Towards Sustainable Agricultural Waste Management, R&D Technical Report P1-399/1, Environment Agency, Bristol, UK, 2001. 106.

[3] E.D. Ongley, Food and Agriculture Organization of the United Nations, Rome, 1996. 101.

[4] P. Agamuthu, An Asian Perspective Meeting of First Regional 3R Forum in Asia, 11-12 Nov 2009, Tokyo, Japan (2009).

[5] GTZ International Services, Technical/Policy Note on Agricultural Waste Management in Egypt, Preliminary Report (2006).

[6] T. Phonbumrung, C. Khemsawas, Proceedings of Sixth Meeting of Regional Working Group on Grazing and Feed Resources for Southeast Asia (October 5-9, 1998) 183–187.

[7] I.S. Arvanitoyannis, P. Tserkezou, Wheat, barley and oat waste: a comparative and critical presentation of methods and potential uses of treated waste, Int J Food Sci Tech 43 (2008) 694–725.

[8] D.G. Abler, J.S. Shortle, Technology as an Agricultural Pollution Control Policy, Am. J. Agric. Econ. 77 (1995) 20–32.

[9] H. Galbraith, P. Amerasinghe, A. Huber-Lee, CA Discussion paper 1 Colombo, Srilanka: Comprehensive Assessment Secretariat Available at: <http://www.iwmi.cgiar.org/assessment/files_new/publications/Discussion%20Paper/CADiscussionPaper1.pdf>, (2005). Accessed on 2nd June 2010.

[10] M.S. Gaballah, K. Khalaf, A. Beck, J. Lopez, J. Appl. Sci. Res. 1 (2005) 9–17.

[11] D. Rickert, Proceedings of the FAO Expert Consultation, Santiago, Chile, 20-23 October, 1992, Water Report 1. FAO, Rome, 1993. 171–194.

[12] Us-Epa, National Water Quality Inventory: 1992, Report to Congress. EPA-841-R-94-001, Office of Water, Washington, DC, 1994.

[13] F.M. Reiff, Proceedings of the Interregional Seminar on Non-conventional Water Resources Use in Developing Countries, 22–28 April 1985, Series No. 22, United Nations, New York, 1987. 245–259.

[14] H.F.L. Saeijs, M.J. Van Berkel, Corporate Water Policies, 5 26–40. [Global Water Crisis, the Major Issue of the 21st Century, European Water Pollution and Control, 5(1995) 26–40]. (1995)

[15] E.N. Omami, Doctoral Thesis, University of Pretoria, South Africa, 2005.

[16] F. Ghassemi, A.J. Jakeman, H.A. Nix, Salinization of Land and Water Resources: Human Causes, Extent, Management, and Case Studies, CAB International, Wallingford, Oxon, UK, 1995.

[17] J. Williams, P.R.E.M.I.U.M. 2005 Research Experience for Undergraduates Sponsored by the National Science Foundation and Michigan State University (2005).

[18] Agricultural Statistics at a Glance 2009, Directorate of Economics and Statistics, Department of Agriculture and Cooperation, Government of India, 2009. 282.

[19] K.E. Havens, A.D. Steinmen, Aquatic Systems, in: J.E. Rechgl (Ed.), CRC Press, Lewis, Boca Raton, Florida, 1995, pp. 121–151.

[20] S.R. Carpenter, N.F. Caraco, D.L. Correll, R.W. Howarth, A.N. Sharpley, V.H. Smith, Non point pollution of surface waters with Phosphorus and Nitrogen, Ecological Application 8 (1998) 559–568.

[21] G. McIsaac, Surface Water Pollution by Nitrogen Fertilizers, Encyclopedia of Water Science (2003) 950–955.

[22] R.W. Howarth, G. Billen, D. Swaney, A. Townsend, N. Jaworski, K. Lajtha, et al., Regional Nitrogen Budgets and Riverine N&P Fluxes for the Drainages to the North Atlantic Ocean: Natural and Human Influences, Biogeochemistry 35 (1996) 75–139.

[23] S.A. Green, R.P. Pohanish, Sittig's Handbook of Pesticides and Agricultural Chemicals, William Andrew Inc., Norwich (2005) 1213.

[24] A.F. Ding, G.X. Pan, X.H. Zhang, Contents and Origin Analysis of PAHs in Paddy Soils of Wujiang County, Journal of Agro-Environmental Science 24 (2005) 1166–1170.

[25] B.X. Mai, E.Y. Zeng, X.J. Luo, Q.S. Yang, G. Zhang, X.D. Li, et al., Abundance, depositional fluxes, and homologue patterns of polychlorinated biphenyls in dated sediment cores from the Pearl River Delta, China, Environmental Science and Technology 39 (2005) 49–56.

[26] H. Cao, T. Liang, S. Tao, C. Zhang, Simulating the temporal changes of OCP pollution in Hangzhou, China, Chemosphere 67 (2007) 1335–1345.

[27] G.J. Hu, S.L. Chen, Y.G. Zhao, C. Sun, J. Li, H. Wang, Persistent Toxic Substances in Agricultural Soils of Lishui County, Jiangsu Province, China, Bulletin of Environmental Contamination and Toxicology 82 (2009) 48–54.

[28] H.T. Hai, Inaugural Workshop of "Asia Resource Circulation Policy Research", Institute for Global Environmental Strategies (IGES), Japan, 9–10 November 2009.

[29] Bid Network <http://www.bidnetwork.org/page/12736/en>, (2004). Accessed on 5th June 2010.

[30] Braskem Ltd, 'First certified green Polyethylene in the World', a press release dated 21 (June 2007). www.betalabservices.com/PDF/Braskem%20Press%20Release.pdf. Accessed on 5th June 2010.

[31] P. Smith, D. Martino, Z. Cai, D. Gwary, H. Janzen, P. Kumar, et al., in: B. Metz, O.R. Davidson, P.R. Bosch, R. Dave, L.A. Meyer (Eds.), The Fourth Assessment Report of the Intergovernmental Panel on Climate Change, Cambridge University Press, 2007. NY.

[32] D. Kruger, Workshop on U.S. and Japanese climate policy in Washington, D.C. 12 February 2004.

[33] R. Yevich, J.A. Logan, An assessment of biofuel use and burning of agricultural waste in the developing world, Global Biogeochem Cycles 17 (2003) 1095.

[34] B. Luske, International Federation of Organic Agriculture Movements - Case Studies <http://www.ifoam.org/growing_organic/1_arguments_for_oa/environmental_benefits/pdfs/IFOAM-CC-Mitigation-Web.pdf>, (2009). Accessed on 2nd June 2010.

[35] U. Niggli, A. Fliessbach, P. Hepperly, N. Scialabba. Low Greenhouse Gas Agriculture: Mitigation and Adaptation Potential of Sustainable Farming Systems. FAO, April 2009, Rev. 2, (2009). <http://www.ifoam.org/growing_organic/1_arguments_for_oa/environmental_benefits/pdfs/IFOAM-CC-Mitigation-Web.pdf>, Accessed June 09, 2010.

[36] F. Jamal, Q.S. Haque, T. Qidwai, A.K. Paliwal, Genetically modified (GM) foods: a brief perspective, Int. J. Biotech. Biochem. 6 (2010) 13–24.

[37] D. Normile, Rockefeller to End Network After 15 Years of Success, Science 286 (1999) 1468–1469.

[38] H. Daniell, S.J. Streatfield, K. Wyckoff, Medical molecular farming: production of antibodies, biopharmaceuticals and edible vaccines in plants, Trend Plant Sci 6 (2001) 219–226.

[39] S.P. Bizily, C.L. Richard, B. Meagher, Phytodetoxification of hazardous organomercurials by genetically engineered plants, Nat. Biotechnol. 18 (2000) 213–217.

[40] C. James, Global Status of Commercialized Biotech/GM Crops: 2009, ISAAA Briefs No. 41. ISAAA: Ithaca, NY (2009). <http://www.isaaa.org/resources/publications/briefs/41/executivesummary/default.asp>. Accessed on 11th June 2010.

[41] O. Sanvido, J. Romeis, F. Bigler, Ecological impacts of genetically modified crops: Ten years of field research and commercial cultivation, Adv Biochem Eng. Biotechnol. 107 (2007) 235–278.

[42] P.P. Motavalli, R.J. Kremer, M. Fang, N.E. Means, Impact of genetically-modified crops and their management on soil microbially-mediated plant nutrient transformations, J. Environ. Qual. 33 (2004) 816–824.

[43] K.E. Dunfield, J.J. Germida, Impact of genetically modified crops on soil and plant associated microbial communities, J. Environ. Qual. 33 (2004) 806–815.

[44] L.L. Wolfenbarger, P. Phifer, The Ecological Risks and Benefits of Genetically Engineered Plants, Science 290 (2000) 2088–2093.

[45] P.J. Dale, B. Clarke, E.M.G. Fontes, Potential for the environmental impact of transgenic crops, Nat. Biotechnol. 20 (2002) 567–574.

[46] A.A. Snow, D.A. Andow, P. Gepts, E.M. Hallerman, A. Power, J.M. Tiedje, et al., Genetically engineered organisms and the environment: Current status and recommendations, Ecol. Appln. 15 (2005) 377–404.

CHAPTER

25

Military Solid and Hazardous Wastes—Assessment of Issues at Military Facilities and Base Camps

Victor F. Medina, Scott A. Waisner

U.S. Army Corps of Engineers, Engineers Research and Development Center, Vicksburg MS, USA

OUTLINE

Waste Doi: 10.1016/B978-0-12-381475-3.10025-7

1. INTRODUCTION

This Chapter explores military waste issues. For the purpose of narrowing the scope, this will focus on the issues associated with the United States Army from existing military facilities and from base camp operations.

2. FACILITIES

The Army operates a wide range of facilities. The Army has facilities to house and train troops, store munitions, support operational commands, conduct research and development, and engage in the production of munitions. Of course, many facilities combine several aspects of these activities.

Within these facilities, a wide range of activities can take place, each of which may result in the generation of various waste streams. Many of these activities are similar to those found in the civilian sector. These include typical household wastes from housing areas, office wastes from various administrative and command activities, landscaping wastes, and construction and demolition (C&D) wastes. In fact, large bases are in many ways such as small cities or towns.

On the other hand, these facilities can also have some very unique aspects. First, they may experience drastic changes in population over time. Facilities that train personnel may experience sharp spikes in population during active training periods, and facilities that serve as bases for Army units will experience sharp declines in populations during deployments. Second, these facilities can have unique waste streams, particularly munitions, munitions constituents, and residuals of munitions.

Much of the data available is given as Federal Fiscal Years (FY). The Federal FY runs from 1 October to 30 September. In some cases, data may be given as calendar year (CY)—from 1 January to 31 December. In the United States, disposal terms are commonly given in lbs or tons. Soil disposal is typically given in tons or cubic yards (yd^3) (note: 1 lb = 454 g; 1 ton = 0.907 t and 1 yd^3 = 0.765 m^3).

2.1. Solid Waste Management at Military Facilities

In 1994, Griggs and Kemme [1] conducted a study evaluating alternative waste disposal options for the United States Military Academy at West Point. At that time, local landfills were nearing their planned lifetimes. Options included incinerating the solid waste either off-site or onsite, creation of an onsite landfill, and finding other landfill space. Although other landfill space would result in increased costs, particularly from increased transportation costs, this option was deemed the best. A similar study was conducted by Baker in 1973 [2], who evaluated solid waste disposal options for the Atlanta Army Depot. Although the solid waste generated by Atlanta Army Deport was determined to be of excellent quality for incineration, the study concluded that landfilling was the most cost-effective option.

In 2008, Boone et al. [3] studied waste management associated with the U.S. Department of Defense Base Realignment and Closure (BRAC) program. They found that diversion of C&D wastes aids in landfill sustainability and that high diversion rates are an achievable goal.

In 1999, Stankoff and White [4] studied disposal of sewage sludge from a domestic wastewater treatment system at Eielson Air Force Base located in Alaska. Options included incineration, composting, landfilling, and land application by spreading. Although the cold weather limited the effective treatment season for land application, this option was deemed most cost effective.

Energy is becoming a critical driving force in decision making for the U.S. Army [5]. In 2008, Holcomb et al. [6] detailed several approaches for the implementation of waste to energy at military facilities. The report was inconclusive in terms of economics, but indicated several waste to energy opportunities exist at most facilities.

2.2. Hazardous Wastes

In 1999, Ray et al. [7] evaluated the potential for waste minimization and recycling at five military facilities. They found a range of activities that generated hazardous wastes, including motor pools, industrial shops, small arms shops, aviation maintenance shops, paint shops, photography facilities, hospitals, clinics, and laboratories. Waste streams included used motor oil, solvents, paint thinners, antifreeze, and batteries. Source reduction, waste minimization, and recycling opportunities were identified and life cycle analyses indicated potential for cost savings.

In 1984, Speakman and McCauley [8] studied siting requirements for hazardous waste management facilities from a DoD (Naval, Marine Corps, and Army) perspective. Long-term liability issues precluded the DoD from maintaining hazardous waste landfills. However, the need does exist to establish and maintain treatment processes and storage facilities.

2.3. Training Ranges and Munitions Wastes

The Army maintains hundreds of training facilities throughout the United States [9]. These training ranges have virtually no civilian counter part, and they represent vastly different waste generation and contaminant characteristics compared with civilian waste streams. Solid wastes and hazardous wastes generated in the United States must be managed by the Resource Conservation and Reclamation Act (RCRA). Training ranges represent a special challenge for RCRA because when a munition was fired, it was not clear whether it was at that time it rendered a waste material. If they became wastes immediately after firing, then they would be subject to RCRA management requirements. The Military Munitions Rule (MMR) was promulgated in 1997 to address this issue [10]. The rule created a conditional exemption for waste munitions provided they are used for the intended purpose and left in place or destroyed or treated on range. However, used or fired military munitions can be classified as solid wastes if they are: (1) transported off-range for reclamation, treatment, disposal, or storage, (2) if they are recovered, collected, or disposed of by burial, landfilling, or land treatment, or (3) by RCRA requirements if a munition lands off-range and is not promptly rendered safe and/or retrieved [11].

In 2004, Mills [12,13] discussed management of debris from Air Force training ranges including wood, metal canisters, fiberglass, drag parachutes, and spent targets. These wastes were separated into those with and without recycling possibilities. Those without recycling capabilities were landfilled. Those that could be recycled were taken to a landfill and stored. A special exemption was obtained to the daily cover rule, so that the materials would not have to be excavated as recycling opportunities were obtained.

Over time, munitions will age and become out of specifications. When this occurs, it is often safest to destroy them. This is frequently conducted in Open Burning/Open Detonation (OB/OD) sites [14]. Open burning (OB) and OD refer to methods to destroy propellants, explosives, ordinance, and munitions. OB involves combustion of the contaminants and is frequently used for propellants. The contaminants are ignited, usually using a supplementary ignition source until the materials reach a self-sustaining burn. In OD, which is used for explosive materials, the materials are detonated to achieve destruction. The purpose of these operations is twofold: to remove the threat of explosive interactions of the materials and to destroy potential environmental contaminants found in the munitions. Another method for dealing with off-specification munitions includes using high-pressure water to mine out the explosives from opened munitions. This method has fallen out of favor because of concerns the explosives may become mobile and contaminate soil and groundwater. Munitions can also be destroyed using specialized incineration equipment. This approach reduces environmental issues associated with OB/OD, but the equipment is much more expensive to both purchase and operate. Furthermore, using incineration equipment generally results in more handling of the munitions, which can create additional safety issues.

2.4. Remediation

Soil and groundwater remediation projects have a large impact on the amount of hazardous waste generated by a facility. Increasingly, facilities are investigating methods to minimize wastes from remediation projects. In 2010, Wrobel and Gross [15] listed several approaches, including *comprehensive sampling*, to allow for more refined separation and classification of excavated soil and other material as hazardous or nonhazardous, *surgical excavation* to allow for less hazardous waste generation, the use of removed *trees as fuel* instead of landfilling, *segregation of metallic materials* for recycling, and *beneficial reuse of materials* on post whenever possible, such as reusing excavated soil for on-site grading and backfill. The application of these approaches to a large soil remediation project at the Aberdeen Proving Grounds resulted in a reduction of 1700 yd^3 (1300 m^3) of total waste generated, a 700 yd^3 (536 m^3) reduction of hazardous waste produced, a saving of 75 trees from removal, heating energy equivalent to eight homes for 1 year produced from the use of wood from excavated trees for fuel, 15,000 lbs (6.8 t where t refers to metric tonne) of aluminum recycled, and the recycling of 14×10^3 tons (12×10^3 t) of excavated soil as clean fill.

The following case studies of Joint Base Lewis-McCord and the Pictanny Arsenal illustrate the solid and hazardous waste generation at two different types of military facilities.

2.5. Example Facility 1. Joint Base Lewis-McCord

2.5.1. Base Description

Joint Base Lewis-McCord (JBLM) is a U.S. military facility located in Pierce and Thurston Counties in the State of Washington. JBLM represents the merging of two adjacent facilities: the Army facility Fort Lewis and the Air Force Base of McCord Field. The integration of the two facilities was started on 1 February 2010. Full integration is expected by 1 October 2010. Fort Lewis is the larger of the two facilities and it is the home to the I Corps, the second, third, and fourth Brigades, and the second Infantry Division. It also houses several other smaller commands. McCord Field is the home of the sixty-second and four hundred forty sixth Airlift Wings and other units and commands.

JBLM has extensive lands reserved for training, including small arms training, artillery firing, and areas for maneuver exercises. The cantonment area includes areas for housing, shopping, recreation, physical training, and extensive office facilities for both military command operations and administration and operation of the base itself.

Waste data were obtained from the Department of Public Works.

2.5.2. Solid Waste Generation

2.5.2.1. BASE POPULATION

Like many military facilities, the population of JBLM fluctuates substantially over time due to deployments, training requirements, and other military needs (Table 25.1). From FY2003 to FY2009, the base population (resident and nonresident) has ranged from 31,495 to 51,132—a difference close to 20,000. The civilian

TABLE 25.1 Population at Fort Lewis/Joint Base Lewis McCord Since FY2003

	FY03			FY04			FY05			FY06			FY07			FY08		
	Military	Civilian	Total	Military	Civilian	Total	Military	Civilian	Total	Military	Civilian	Total	Military	Civilian	Total	Military	Civilian	Total
Resident	8,990	9,192	18,182	10,101	9,594	19,695	21,010	5,148	26,158	23,915	5,815	29,730	18,964	6,446	25,410	20,916	8,459	29,375
Nonresident	8,401	4,912	13,313	19,393	8,357	27,750	4,484	4,810	9,294	3,422	5,523	8,945	12,642	4,477	17,119	15,538	6,219	21,757
Total	17,391	14,104	31,495	29,494	17,951	47,445	25,494	9,958	35,452	27,337	11,338	38,675	31,606	10,923	42,529	36,454	14,678	51,132

Notes: Population data are a single point in time, and population totals fluctuate on a regular basis. Population data do not include military units that come on post for training only.

population ranged from 9,958 to 17,951, whereas the military population varied from 17,391 to 38,454. In a similar manner, there have been large variations in the resident population over time.

2.5.2.2. SOLID WASTE AND RECYCLING

Table 25.2 is a comprehensive solid waste report for JBLM from FY2002 to FY2009. Total solid waste generation varied from 23 $\times$ 10^3 tons (21 $\times$ 10^3 t) (FY2003) to nearly 120 $\times$ 10^3 tons (109 $\times$ 10^3 t) (FY2008). Overall, it seems that the total solid waste generation at JBLM has been increasing over time. A key factor is that the base population has increased over time (Table 25.1). There are many waste generating activities at JBLM. There are also numerous efforts for waste minimization. The goal is zero net waste by 2025.

The data indicate that JBLM has been successful at recycling much of their solid waste production. Thirty six to 77% of potential solid waste produced at JBLM has been diverted to recycling. This diversion has steadily increased over time.

Table 25.2 also indicates that C&D wastes have varied significantly over time, from 1.1 $\times$ 10^3 tons (1 $\times$ 10^3 t) to over 64 $\times$ 10^3 tons (58 $\times$ 10^3 t). Much of the C&D wastes are diverted for other uses, close to 100% for most years. This high rate of diversion is consistent with the findings of Boone et al. [3], who concluded that C&D waste material can be beneficially diverted for other uses.

JBLM has instituted recycling programs and maintains a recycle center on post. Currently, a wide range of materials are recycled (Table 25.3). In addition, JBLM has an earthworks/environmental education training center that conducts composting activities, wood shredding for beautification projects, soil amendments, crushed concrete, and asphalt. Many projects on the post use these materials from the earthworks/environmental center.

2.5.2.3. HAZARDOUS WASTE GENERATION AND RECYCLING

Table 25.2 indicates that hazardous waste generation has also steadily increased from FY2003 to 2009—from just over 1 $\times$ 10^6 lbs (454 t) in FY2003 to close to 3 $\times$ 10^6 lbs (1362 t) in FY2009. However, recycling of hazardous wastes has also increased. The diversion rate increased from 24% in FY2003 to 69% in FY2009. Several hazardous waste reuse projects have been instituted. One example is a hazardous material control center where excess hazardous materials are collected (as opposed to being categorized as waste) and with the goal of providing them for use for other organizations that can use the materials beneficially. This program has substantially reduced the disposal of excess hazardous material and has reduced procurement costs as well. In 2009, the estimated total cost savings/avoidance at JBLM was $1,683,801. This savings was from the reuse of 293,286 pounds (133 t) of HM material, and the reduced purchase of 2300 55-gallon

TABLE 25.2 Comprehensive Solid Waste Reports, FY 2002 to 2009, for JBLM

SWAR Comprehensive Report Data (tons)	FY02	FY03	FY04	FY05	FY06	FY07	FY08	FY09
Section 8. Management Summary								
C&D generation	3,503.880	1,186.470	9,262.324	4,712.928	36,970.506	15,622.580	64,197.480	10,749.507
C&D disposal	0.000	0.000	2.410	0.000	0.000	0.000	21,438.000	18.680
C&D diversion	3,503.880	1,186.470	9,259.914	4,712.928	36,970.506	15,622.580	42,759.610	10,760.827
C&D diversion rate (%)	100.00	100.00	99.97	100.00	100.00	100.00	66.61	99.83
Non-C&D generation	30,752.758	22,527.508	26,844.090	26,454.663	24,531.367	25,359.000	53,108.810	48,431.493
Non-C&D disposal	21,845.079	14,384.974	13,955.718	13,364.771	17,028.905	12,832.010	21,438.170	13,312.557
Non-C&D diversion	8,907.679	8,142.534	12,888.371	13,089.891	7,502.461	12,526.990	31,670.640	35,118.937
Non-C&D diversion rate (%)	28.97	36.14	48.01	49.48	30.58	49.40	59.63	72.51
Without C&D and composting generation (MSW)		22,527.508	26,834.470	24,957.022	24,490.997	25,359.000	53,108.810	48,431.493
Without C&D and composting disposal (MSW)		14,384.974	13,955.718	13,364.771	17,832.010	12,832.010	21,438.170	13,312.557
Without C&D and composting diversion (MSW)		98,142.534	12,878.751	11,592.251	7,462.092	12,526.990	31,670.640	35,118.937
Without C&D and composting rate (MSW) (%)		36.14	47.99	46.45	30.47	49.40	59.63	72.51
Total SW generated (sum of C&D and non-C&D generation)	34,256.638	23,713.978	36,106.414	31,167.591	61,501.873	40,981.580	117,306.290	59,181.000

Total SW generated (sum of C&D and non-C&D disposal)	21,845.079	14,384.974	13,958.128	13,364.771	17,028.905	12,832.010	42,876.170	13,331.237
Total SW generated (sum of C&D and non-C&D Diversion)	12,411.559	9,329.004	22,148.285	17,802.819	44,472.967	28,149.570	74,430.250	45,879.763
Total SW diversion rate (percentage rate) (%)	36.23	39.34	61.34	57.12	72.31	68.69	63.45	77.47
MSW (without C&D and composting)								
These fields auto-calculate from totals above								
MSW Generated		22,527.51	26,834.47	24,957.02	24,491.00	25,359.00	53,108.81	48,431.49
MSW Recycled		14,384.97	13,955.72	13,364.77	17.028.91	12,832.01	21,438.17	13,312.56
MSW Disposal		98,142.53	12,878.75	11,592.25	7,462.09	12,526.99	31,670.64	35,118.94
Diversion rate (%)		36.14	47.99	46.45	30.47	49.40	59.63	72.51
Hazardous waste totals (lbs)								
All HW generated (D,E,X)		1,002,015	946,751	1,206,916	1,942,020	1,888,177	2,691,733	2,990,698
total recycle and diversion		244,541	361,300	753,568	745,689	1,030,957	1,213,823	2,055,182
Diversion rate (%)		24	38	62	38	55	45	69
HM Usage (313 Chemicals)	266,896	392,893	392,974	365,615	311,153			

The units are in U.S. tons (1 ton = 0.907 t) and lbs (1 lb = 0.454 kg).

TABLE 25.3 Items Recycled at JBLM

Item	Comments
Antifreeze	
Fuel	Diesel, JP-8, Kerosene, etc.
Mixed Paper	
Commingled Products	Aluminum cans, tin cans, glass bottles and jars (no window glass, ceramic plates or cups) Milk and juice cartons Plastic bottles #1 to #7
Cardboard	
Wood	
Plastic film/bags	
Tennis shoes	No metal eyelets or imbedded lights
Metals	Silverware, pots and pans, curling irons, blow dryers, electric fans, all kitchen gadgets with cords, telephones, telephone cords, Christmas lights, CD players, DVD players, microwaves, clocks, radios, ceiling fan motors, small plug in transformers, extension cords, vacuum cleaners without cloth bags, and all other household metal items
Carpet	
Plastic/metal materials	Baby strollers, baby swings, car seats, large playground toys, big wheels, kitchen play toys, kids basketball hoops, plastic pools, Rubbermaid storage items, laundry baskets, plastic hangers, plastic silverware trays, and dish drains
Computer plastics	Loose computer parts, circuit boards, mice, keyboards, fax machines, printers, cell phones, PDA's and computer cables
Clothes	Clothing, blankets, sheets, curtains, bags, tents, duffle bags, and any other items made of fabric

drums that would have been used to dispose of that material.

Table 25.4 details disposal costs for hazardous wastes at JBLM from CY1998 to CY2009. Disposal costs have ranged from $177,000 in CY2003 to more than $800,000 in CY2007. There has been substantial variation from year to year. In CY2001, a tank-remediation project contributed to an increase in costs compared with CY2000. CY2003 had an increase in process-generated hazardous waste. During this year, Fort Lewis was very active preparing units to support Operations Enduring Freedom and Iraqi Freedom. A large portion of these process wastes resulted from the preparation of vehicles for overseas shipment. The process waste generation decreased in CY2004, as the base had less use because many units were deployed at that time. Beginning CY2006, there has been an increase in hazardous waste generation and costs because of troop increases and activities to support these troops. In CY2007, disposal costs were particularly high because of a project involving the disposal of refrigerant.

Table 25.5 details types of hazardous wastes and those materials diverted for recycling from CY2003 to CY2009. Recycling of antifreeze, oil, fuel, solvent, and dry sweep materials has increased dramatically during this time period. Although some of this increase is due to increased base activity, the magnitude of the increases are, in fact, much greater than what would be expected just from the base population data, suggesting that the bulk of the increase was due to improvement in identifying and applying recycling opportunities.

2.6. Example Facility 2. Picatinny Arsenal

2.6.1. Facility Description

The Picatinny Arsenal is located on over 6,000 acres (2430 ha) in northern New Jersey. The facility is home to the Army Armaments Research, Development and Engineering Command, which has the mission of developing new munitions for the Army, improving existing munitions, and developing improved manufacturing processes.

TABLE 25.4 Hazardous Waste Generation and Costs at JBLM

	Process generated HW	Remediation, one-time, and intermittent HW	FTL HW generated	DRMS disposal cost	HW received by DRMO	HW shipped off-site	UW generated	PCB waste	Cost billed for all waste disposal, supplies, contracts, etc.
CY09	568,693	85,977	654,670	$578,318.00	0	654,670	41242	11,667	
CY08			569,452	$382,241.60	0	569,452	42855	113,966	
CY07			415,300	$830,516.00	0	415,300	42871	32,997	$925,509.24
CY06	281,333	281,203	694,490	$353,133.49	0	694,490	80743	55,379	$815,108.00
CY05	163,782	43,302	254,594	$205,092.66	0	254,594	40367	359	$540,842.67
CY04	268,875	7,396	276,271	$202,560.58	0	276,271	44688	2148	$696,204.00
CY03	510,207	3,029	513,056	$177,706.00	41,083	554,139	26758	902	$714,657.00
CY02	127,620	529,301	656,921	$210,076.00	33,728	690,649	42341	963	$600,448.00
CY01	164,189	452,917	617,106	$227,313.00	48,429	665,535	66678	29454	$422,895.21
CY00	233,591	558,571	792,162	$187,650.18	115,998	908,160	39620	12227	$632,657.00
CY99	272,313	585,670	583,226	$273,303.00	162,866	746,092	68471	13424	$451,028.00
CY98	655,229	940,523	1,593,981	Unknown	123,113	1,717,094	49415	31643	Unknown

TABLE 25.5 Detailed Breakdown of Hazardous Wastes and Materials Diverted for Recycling

Hazardous waste totals	FY03	FY04	FY05	FY06	FY07	FY08	FY09
All generated (D, E, X, UW, RR Clin)	789,977	632,324	497,216	1,388,839	1,027,878	1,548,868	1,151,655
Waste total (D, E, and X)	757,474	585,451	453,348	1,196,331	857,220	1,477,910	935,516
HW (D and E)	513,056	276,271	254,594	694,490	415,300	536,452	654,652
Non-Haz (X)	244,418	309,180	198,754	501,841	441,920	941,458	280,864
UW	26,758	44,688	40,367	80,743	42,871	41,218	59,996
Recycled (RR Clin)	5,745	2,185	3,501	111,765	127,787	29,740	156,139
Total recycled	32,503	46,873	43,868	192,508	170,658	70,958	216,139
Recycled percent (of haz waste)	6	17	17	28	41	13	33
PCB Waste	902	2,148	359	55,379	32,997	113,966	11,667
Recycling data from P2							
Antifreeze	100,101	106,333	146,479	80,517	111,040	190,478	320,181
Oil	6,226	6,226	386,688	283,721	316,315	455,800	597,100
Fuel	102,468	171,424	150,867	167,779	422,674	480,666	880,000
Solvent	3,242	2,523	3,107	1,804	770	1,284	2,242
Dry Sweep	0	27,920	22,559	19,360	9,500	14,637	39,520
Totals	212,038	314,427	709,700	553,181	860,299	1,142,865	1,839,043
All generated waste (D, E, X)	1,002,015	946,751	1,206,916	1,942,020	1,888,177	2,691,733	2,990,698
Total recycled	244,541	361,300	753,568	745,689	1,030,957	1,213,823	2,055,182
Recycled percent	24.40	38.16	62.44	38.40	54.60	45.09	68.72

The units are in lbs (1000 lbs = 454 kg).

The base includes a cantonment area with offices and research laboratories. The cantonment area also has indoor small arms ranges for testing purpose. Picatinny Arsenal also has firing ranges for artillery and projectiles. There is no housing area. Waste generation was provided by Picatinny Arsenal's Department of Public Works.

2.6.2. Solid Waste Generation

Figure 25.1 summarizes solid waste generation and recycling at the Picatinny Arsenal. This figure indicates that solid waste generation has increased at Picatinny since 2005. This has been mitigated to a large degree by an increase in recycling, which has increased

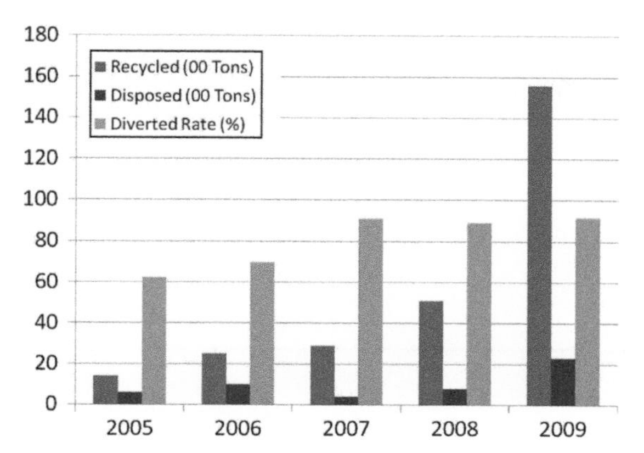

FIGURE 25.1 Solid waste disposal and recycling at the Picatinny Arsenal from 2005 to 2009. The units are U.S. tons (1 ton = 0.907 t).

nearly 10-fold [from about 1600 tons (1450 t) in FY2005 to nearly 16,000 tons (1.45×10^3 t) in FY2009].

Several factors have resulted in increased solid waste generation at Picatinny Arsenal. First, research at the arsenal has supported wars in Iraq and Afghanistan and research associated with munitions used in these wars has resulted in more activity and more waste generation over time. In particular, this research has greatly increased the use of both indoor and outdoor testing ranges at the Arsenal, which increased the need for cleanups and remediation. Further, in 2005, the BRAC program directed the Picatinny Arsenal to become the receiving installation for consolidating research in guns and ammunition. As a result, several other organizations and programs, particularly naval programs, have relocated to the Arsenal. This necessitated new construction and refurbishing of older buildings resulting in an increase in construction and demolition debris. Both these increases have resulted in a dramatic increase in the generation of metallic process streams (i.e., brass, steel, etc.). Fortunately, these materials are amenable to recycling and doing so has resulted in a big fiscal return for the Arsenal.

2.6.2.1. HAZARDOUS WASTES

Picatinny Arsenal performs research and development work for conventional ammunition and numerous laboratories and indoor and outdoor test facilities are maintained. The types of chemicals used number in the thousands. Picatinny Arsenal is a permitted treatment and storage facility for hazardous waste and has more than 100 areas on base which generate thousands of pounds of hazardous waste. In FY2009, the arsenal permitted an explosive hazardous waste incinerator and maintains OB/OD operations for the destruction of explosive wastes in two different areas. Picatinny Arsenal also has numerous Environmental Protection Agency (EPA) Superfund sites around the base.

Table 25.6 summarizes hazardous waste production at the Picatinny Arsenal from FY 2005 to FY2009. The wastes are divided into three categories: remediation wastes, hazardous waste (other than remediation) shipped off-site, and hazardous wastes treated onsite. The amount of onsite-treated wastes was relatively stable, ranging from 21,000 to 29,000 lbs (from 9.5 to 13 t) annually. Remediation wastes and off-site shipped quantities, on the other hand, varied drastically from year to year. Remediation waste ranged from 0 to more than 1.5 million lbs per annum (680 t a^{-1}) whereas offsite shipped wastes varied from 21,000 to just under 1 million lbs (from 9.5 to 454 t).

These fluctuations were the result of the variation of the continually changing research programs, as well as simply the total amount of research being conducted. This variation is greater for a research facility such as Picatinny Arsenal compared to a production facility. Furthermore, remediation projects can greatly increase the amount of hazardous waste in a given year. For example, in 2009, remediation waste mostly came from soils removed during a cleanup operation under CERCLA performed

TABLE 25.6 Hazardous Waste (HW) Generation (lbs) at the Picatinny Arsenal from FY2005 to FY 2009

Year	Remediation waste (lbs)	HW shipped off-site (lbs)	HW treated On-site (lbs)
2005	5000	31,000	22,000
2006	0	944,000	29,000
2007	43,000	135,000	22,000
2008	2000	21,000	21,000
2009	1,514,000	144,000	20,000

The units are in U.S. lbs (1000 lbs = 454 kg).

at the arsenal. Another factor contributing to fluctuations in hazardous waste production are routine maintenance programs that have schedules beyond an annual frequency. For example, in FY 2009, Picatinny conducted routine cleanout of indoor small arms range sand that was contaminated with lead. This activity resulted in an increase in the hazardous waste shipped off-site. Small arms range cleanout occurs every few years, depending on the frequency of use of the various indoor testing facilities on site.

2.7. Discussion and Conclusions

The literature review revealed that bases typically landfill their wastes using offsite facilities. In many ways, wastes generated at military facilities are similar to civilian residential communities and commercial facilities. However, training areas in particular provide substantially different waste issues than commonly found in civilian environments. Another area of substantial difference is the management of explosive munitions, particularly the decommissioning of out-of-specification materials.

Several things stand out in studying the two example facilities (JBLM and the Picatinny Arsenal). First, there seems to be a general increase in solid and hazardous waste over time as these facilities have continued to support operations in Iraq and Afghanistan. Base activity is likely to vary significantly over time. In the case of JBLM, this is due to training and preparation activities associated with the various units it supports. Deployments particularly generate large amounts of wastes, largely due to preparation of motor vehicles. Waste generation has also increased at the Picatinny Arsenal due to expansion in the number and types of research programs. The Picatinny Arsenal also had a substantial increase in C&D waste due to construction activities to support its new role supporting naval research.

Remediation cleanup projects play a large role in the total amount of waste generation in a given year. Most of these are one-time cleanups, but in some cases, such as the cleanout of the indoor small arms ranges at Picatinny Arsenal, the cleanup is recurring. In some cases, the disposal costs of a remediation or cleanup greatly increase the waste disposal costs of a given year, such as in CY2007 when the disposal of refrigerant nearly tripled typical disposal costs at JBLM.

Both facilities have clearly instituted programs that have resulted in a large increase in potential wastes, both solid and hazardous, diverted for recycling—both in terms of total quantity and percentages. Improvements in this area indicate that the U.S. Army is committed to responsible management of its waste and is looking for ways to reduce waste generation while continuing in its defense mission.

3. WASTE MANAGEMENT FOR BASE CAMPS DURING EXPEDITIONARY, CONTINGENCY, AND FULL SPECTRUM OPERATIONS

During operations, the Army typically maintains base camps that support various field operations. An expedition, which by definition is conducted in a foreign country, requires guidance and analysis from the senior commander to determine the applicable policy and legal requirements. Many foreign nations have no stated environmental regulations to follow. A contingency operation, such as a response to a natural disaster conducted in the continental United States will require that the military follow U.S. environmental laws and regulations with limited exceptions. Full Spectrum Operations is a recent Army term to encompass missions that may include all or some of the following: combat, reconstruction (nation building), humanitarian and disaster assistance and training assistance. It covers the "full spectrum" of what soldiers may be called on to do during any given deployment. The Army is increasingly becoming more sensitive to its impact on the local environment and this is resulting in increased focus on waste management during operations. The U.S. Army Environmental Command publishes Army Environmental Requirements and Technology Assessments (AERTA) [16]. AERTA ER-10-07-01—Avoidance of Environmental Risk during Contingency Operations, supports responsible waste management at base camps. Army Field Manual FM-3-34.5 stresses the importance of environmental considerations in military operations and includes a section on Base Support Operations [17].

3.1. Nonhazardous Solid Wastes

3.1.1. Waste Characterization

The types and quantities of waste generated by the military units deployed to long-term base camps are similar to those generated in garrison and those of a small municipality. These include nonhazardous solid wastes, wastewater, hazardous wastes, and medical wastes. However, types and quantities are different for units on the move, in temporary camps established for only a few days, or under low-security conditions. The most significant differences between large mature base camps and other units are likely to be found in the methods of handling and disposal of wastes.

From 2003 to 2004, several surveys were conducted of nonhazardous solid waste generated by mature Army base camps in the Balkan states of Bosnia, Kosovo, and Bulgaria. The surveys are documented in a U.S. Army Corps of Engineers report [18], and they provide snapshots of the waste generated at brigade and battalion size populations during this time period. The results of these surveys are taken from the report and are provided in Table 25.7. Another report prepared for U.S. Army RD&E Command Natick Soldier Center [19] studied the quantity and types of the waste produced by soldiers in the Force Provider Training Module in Fort Polk, Louisiana, over a 5-day period. The Force Provider Module (FPM) consisted of a total population of 605 (550 soldiers and 55 Force Provider Staff). The results of this are summarized in Table 25.8 and include a calculated annual waste generation rate for a battalion-size base camp for comparison with the values reported in Table 25.7.

The FPM study provides significantly lower estimates of solid waste production for several reasons. The FPM is a containerized system of prefabricated parts that does not require the use of additional construction materials. The study was conducted for a very short duration at a stateside training area and it is assumed

TABLE 25.7 Base Camp Solid Waste Production [18]

	Soldier (lbs/yr)	Company 170 capita (t a^{-1})	Battalion 750 capita (t a^{-1})	Brigade 3,000 capita (t a^{-1})	Fraction of total (%)
Plastic bottles[1]	295	25	111	443	5.1
Polystyrene	9.3	0.79	3.5	14	0.2
Other plastics	143	12	54	215	2.5
Aluminum	10	0.89	3.9	16	0.2
Other metals	11	0.89	4.0	16	0.2
Corrugated paper	349	30	131	523	6.0
Other paper	179	15	67	268	3.1
Scrap wood	4,151	353	1,557	6,227	72.0
Kitchen food waste	328	29	123	491	5.7
Post-consumer food waste	51	4.3	19	76	0.9
WWTP sludge (dry weight)[2]	70	6.0	26	105	1.2
Sawdust	47	4.0	18	71	0.8
Grass clippings	39	3.3	15	58	0.7
Glass	40	3.4	15	59	0.7
Textiles	25	2.1	9.5	38	0.4
Medical waste	13	1.1	5.0	20	0.2
Rubber	3.9	0.33	1.5	5.9	0.1
Miscellaneous	5.3	0.45	2.0	7.9	0.1
Total	5,769.5	491.55	2165.4	8,653.8	100.1

[1]*Reflects 100% drinking water distribution via disposable bottled water.*
[2]*WWTP sludge weight expressed as 100% solids – multiply by 5 for a cake and by 50 for a liquid*
Note: 1 U.S. ton = 0.907 t. Survey includes all discarded solid waste except hazardous waste, recycled scrap metal, and salvaged construction material and equipment. Above values do not reflect additional loadings due to TOA rotations (estimated to increase annual waste production by approximately 1 month for bi-annual TOAs). Above values are based on relatively short-term studies and reflect a population "snapshot." It is not known whether this table accurately includes the fraction of solid wastes generated by host-nation contract employees and transient combatants.

TABLE 25.8 Waste Weight and Volume Rates from Force Provider Module [19]

Waste	Soldier (lbs/day)	Soldier (ft^3/day)	Battalion 750 capita (t a^{-1})
Trash and kitchen waste	3.2	1.12	438
Slop food	0.7	0.02	96
Cooking oil	0.2	0.004	27
Total	4.1	1.14	561

Note: 1 lb = 0.454 kg; 1 ft^3 = 0.0283 m^3: 1 ton = 0.907 t.

that border protection was not constructed around the camp during the study. These two factors alone eliminate most if not all of the scrap lumber and sawdust production, which accounts for 72% of the waste in the study of bases in the Balkans. Other factors that do not seem to be accounted for in the FPM study are grass clippings and wastewater treatment plant (WWTP) sludge. Because of the short duration of the deployment of troops for stateside training, textiles in the solid waste due to uniform replacement or repair were also likely to be negligible. These items together accounted for 75% of the waste generated during the study of bases in the Balkans. When these waste sources are eliminated, the total annual SW production from the Balkans study was 541 ton per annum (491 t a^{-1}), which is very close to the 561 ton per annum estimated from the FPM study.

The U.S. Army Engineer School has recently published a handbook entitled "Waste Management for Deployed Forces" [20]. Table 25.9 is a reproduction from this document of nonhazardous solid waste generation rates used for planning purposes by smaller units to report to higher commands and do not account for recycling and reuse of any materials. Total generation rates for units on the move in this table are very close to those provided in Table 25.8, which were measured for Force Provider modules. However, the rates proposed for base camp planning purposes are below those measured at the large mature base camps in the Balkans. This is likely due to items not included in Table 25.9, which are found at large mature camps such as: WWTP sludge, grass clippings, and medical waste. These items would be accounted for by planners at a higher command.

It is unclear whether other solid waste items generated during combat were included in this list, because they are not strictly base camp wastes. One such possibility is spent shell casings. Although small-arms shell casings are likely to be discarded, larger shell casings such as those from artillery or tanks are likely to be recovered and transported for appropriate disposal or recycling.

TABLE 25.9 Nonhazardous SW Generation Rates Used for Planning [20]

Component	Soldier (lbs per capita per day)	Battalion 750 capita (t a^{-1})
Generation rates on the move		
General refuse	1.5	205
Food waste	2.5	342
Total nonhazardous SW	4.0	547
Generation rates in base camps		
Plastic bottles	0.54	74
Other plastic	1.38	189
Aluminum	0.13	18
Cardboard	1.45	198
Paper	2.67	365
Food waste	1.67	229
Textiles	0.26	36
Glass	0.10	14
Scrap wood	2.95	404
Miscellaneous	2.30	315
Total solid waste	13.45[1]	1842

[1]*This value is listed as 18.2 in the reference; however, the total of the values listed is 13.45.*
Note: 1 lb = 0.454 kg; 1 ton = 0.907 t.

3.1.2. *Waste Disposal*

The method of solid waste disposal by deployed military forces is very dependent on three factors: the duration of a camp site, the current security situation of a particular site, and the ability and willingness of local host nation infrastructure to handle these wastes. Under transient conditions of less than a week or when other conditions make it necessary, field-expedient methods of disposal are used.

Units on the move for 3 days or less typically bag and haul solid waste until appropriate disposal facilities are available.

Field expedient disposal methods can take several forms. Garbage burial pits or trenches can be used where soil types, ground water levels, and environmental considerations allow. These burial sites are shallow, typically 4-feet (1.2 m) deep, landfills without a liner. When closed they are marked with a sign indicating the type of burial pit/trench, date of closure, and responsible unit if the situation allows. The location of each burial pit/trench is recorded and reported to higher headquarters. When waste burial is not possible, solid waste is openly burned in shallow pits, barrel incinerators, or inclined-plane incinerators.

For long-term encampments and permanent bases, the use of local facilities and contracted support is the preferred method of waste disposal. The determination to use local or nonlocal contractors is highly dependent on the security situation due to necessary access to areas inside the military base by trucks and personnel performing the waste removal functions. If available, local landfills or other disposal facilities must be of sufficient capacity and effectiveness to handle the additional load from the military bases. Local public opinion concerning handling of the base's waste must also be considered.

In the event that local services are not capable of handling the solid waste, the military must construct a disposal facility for the waste. This task is typically contracted to outside organizations whenever possible. All standard disposal methods are evaluated for potential use including: reuse, recycling, burial, composting, and incineration.

3.1.3. *Waste Reduction/Reuse*

Because of the change of nature of military operations in recent decades toward long-term involvement in host nations and changes in the world opinion on environmental matters, the military has invested significant efforts to reduce, reuse, and recycle solid wastes at base camps. These practices directly reduce logistical support required for base camps and improve relations with the local population. Many systems are being developed and tested to shred, compact, compost, and incinerate and reclaim energy from base camp wastes. Many of the systems are briefly described in *Waste Management for Deployed Forces*.

One notable system that underwent a recent field trial in Iraq is the Tactical Garbage to Energy Refinery (TGER), which was developed and fabricated at Purdue University under a Small Business Technology Transfer Research grant [21]. This system was evaluated at a base camp in Iraq in by the Army's Rapid Equipping Force during 2008. This system uses a combination of a fermenter and downdraft gasifier to convert solid waster to usable energy. Solids are shred, pelletized, and fed into the gasifier to produce a syngas. Liquids are fed into the fermenter to produce ethanol. The unit is coupled with a 60 kW generator to produce electrical power using the derived fuel. The test results showed that the TGER unit as tested produced 54 kW of net usable power while reducing 1440 lbs (654 kg) of solids and 312 lbs (142 kg) of liquids to a small amount of ash. Use of the system directly saved the consumption of 86 gallons (326 l) of diesel fuel per day, which does not account for fuel savings associated with reduced waste transportation. Although the system did have some issues, the demonstration illustrates the feasibility of this type of technology at operating bases.

3.2. Hazardous and Special Waste Generated at Base Camps

While in garrison, the handling and disposal of hazardous and special wastes by the military is treated by federal and state regulations in the same manner as for any other small or large quantity generator. Special wastes or hazardous materials created during training operations on military ranges are typically covered under the "MMR" [10].

Deployed military forces must also store and dispose hazardous and special wastes under federal guidelines with the additional complication of host-nation regulations. The types of hazardous wastes generated by the military are similar to those generated by municipalities; however, the rates at which these wastes are generated will vary greatly depending on the nature, pace, and geography of the operations being conducted. Two waste categories that may greatly increase are fluids and materials used for vehicle and generator maintenance (e.g. motor oil, antifreeze, tires, and track pads), and biological wastes from hospital units. Typically units must also collect any soils contaminated due to military operations, such as spilled fuel.

In general, deployed forces handle these wastes and materials by standard waste separation and collection at the source and segregated storage until the wastes are transported to larger units for disposal. Additional precautions are typically necessary for the storage of these wastes due to the increased security risks associated with military operations near hostile forces. These increased risks can also make the safe transportation of these materials more difficult. All hazardous wastes must be removed from military bases before closure or transfer of the bases.

Final treatment or disposal of these materials can be very problematic and costly for the military depending on the capabilities and support of the host nation. Often these wastes are transported out of the country for satisfactory treatment or disposal. Appropriate incinerators can typically be purchased and operated at or near deployed military hospitals to handle biologically contaminated wastes. Some special wastes such as used motor oils can also be processed and used for heating or blended in small percentages into diesel fuel for reuse.

3.3. Wastewater

For the military, wastewater generation and handling while in garrison is essentially the same as any municipality. Smaller bases near urban areas frequently discharge wastewater directly to local publicly or privately owned municipal WWTPs. Large bases or bases in remote areas typically establish their own WWTPs. These plants may be owned and operated by DoD civilians or the operation and maintenance may be contracted to a private-sector business. In recent years, the increased demand on local water resources has instigated water conservation and wastewater reuse programs on many bases.

For deployed military forces, wastewater treatment, disposal, and reuse have also gained renewed interest. This has been caused by both increased scrutiny of environmental impacts and military operations in extremely arid environments such as those in Iraq and Afghanistan. For enduring base camps, local wastewater treatment infrastructure will be used if available. However, large enduring base camps frequently contain populations and wastewater flows that greatly exceed the design flows of local municipal WWTPs. Therefore, WWTPs must be constructed for most of these bases by the military or contractors. The type of wastewater treatment processes is chosen based on the same parameters used for municipalities with additional security concerns taken into account.

For deployed forces in transit or in small or temporary bases, field expedient methods are often used. Field expedient methods used are dependent on the length of stay, size of the unit, and type of wastewater. Discussion of these methods is divided into separate sections based on the type of wastewater, gray water or black water. Based on past experience, the military has generated estimated wastewater generation rates for deployed forces. Table 25.10 provides estimated gray water generation rates used by the military to aid in the planning for

TABLE 25.10 Gray Water Generation Rates Within Base Camps Used for Planning [20]

Source	Gray water generation rate
Enduring base camps (total gray water rule of thumb)	42 gpd/person
Showers	5 to 10 gpd/shower
Kitchens	1 to 5 gpd/person
Quartermaster laundry company	64,000 gpd
Hospitals	200 gpd/bed
Force provider (500-person module)	Total 30,000 gpd
Containerized latrine[1]	3,465 gpd
Containerized batch laundry	5,200 gpd
Containerized shower (one 10-min shower/day)	12,100 gpd
Food services	1,375 gpd

[1] *Latrine wastewater is actually considered black water.*
The units of gpd refer to U.S. gallons per day and 1 U.S. gallon = 3.79 L.

base camp wastewater disposal. These estimated rates do not seem to account for water usage by vehicle wash racks.

3.4. Gray Water

Gray water includes the wastewater from laundry, showers, hand-washing, vehicle wash-racks, and dining facilities. Plans for reduction, recycling, and reuse are developed to both reduce the level of wastewater for treatment and disposal and the demand for fresh water. One example is the collection and treatment of gray water from laundries and showers by reverse osmosis. This water can be used for most potable water purposes that do not include ingestion of the water. These uses include laundry, vehicle washing, fire fighting, and dust suppression. Although Reverse Osmosis (RO) treatment of this water does require energy, these energy demands are often offset by the energy required to transport or treat potable water from other sources.

Two field-expedient methods of gray water disposal are soakage pits or trenches and evaporation beds. The use of these methods is dependent on local environmental variables such as the permeability of the soil, water table level, temperatures, and environmental concerns. If the water table is shallow and a receiving water body is near, wetlands may be constructed for gray water treatment. Effective use of all these treatment methods requires the use of grease traps for effluent from dining facilities and oil-water separators from vehicle wash-racks.

If local treatment and disposal methods are not an option, gray water must be temporarily collected in containers, such as expandable pillows or drums, and transferred to a WWTP. This option is the least desirable due to additional logistics support required on a continual basis, and it is usually only an interim solution until a more permanent solution is established.

3.5. Black Water

Black water refers to water containing human or animal waste such as urine and feces. When fixed facilities are not available, units must use field-expedient methods. When on the move, short-term field-expedient methods such as commercially available human waste bags (such as Brief Relief® bags) or cat holes or straddle-trench latrines may be used. A cat-hole latrine is a shallow (approximately 1 ft or 0.3 m) deep hole dug in the earth, which is buried with the excavated earth after use. The straddle-trench latrine is dug approximately 1-ft (0.3 m) wide, 2-ft (0.76 m) deep, and 4-ft (1.2 m) long. The excavated earth is piled at the end of the trench and is used to bury the waste after each use. Generally, these short-term methods become unsuitable where units are stationary for more than 72 hours.

When initially establishing a base camp, the use of chemical latrines is the preferred option. When chemical latrines are not available, deep-pit, bored-hole, or mound latrines are used. All these methods rely on soakage of liquids into the surrounding soil. Closure of a latrine typically involves simple burial. The mound latrine can be used in areas where the water table is shallow or it is not possible to dig. Burn-out latrines capture the waste in a container, which are removed on a regular basis. A fuel mixture is added to the containers to completely burn the waste. Any remaining ash in the container is buried, and the container is reused. Pail latrines are similar to burn-out latrines, except the waste is typically transported to another area for disposal. Urine disposal facilities for males are typically located near latrines to minimize the use of latrines. Several different types of field-expedient urinals are used, but all rely on some type of soakage pit to absorb the urine into the earth.

As small base camps mature, treatment systems for wastewater may be constructed. Sewage lagoons or oxidation ponds and septic tanks with drain fields may be used. Commercially available package units are also being developed. One such system is the Deployable Aerobic Aqueous Bioreactor, which is shipped and operates in two 20-ft (6.1 m) long ISO shipping containers. The units treat more than 20,000 gpd (76 m^3 d^{-1}) of municipal wastewater and can be fully operational within 48 hours of placement on site. System effluent meets EPA standards for municipal wastewater and for non-potable reuse in most states in United States. Six of these units were recently purchased by the U.S. Army for use at small operating bases in Afghanistan. The units are semiautonomous and have on-board power generation. This system was developed by the Texas Research Institute for Environmental Studies with grants from the Air Force Institute for Operational Health and the U.S. Army Engineer Research and Development Center. The military continues to search for new solutions to the age-old problem of waste disposal for deployed forces.

During base camp closure or transfer, all latrines, soakage pits, and septic systems are closed and marked. Closure methods may simply involve burial, but depending on host nation agreements more detailed methods of closure and monitoring may be required.

ACKNOWLEDGEMENTS

Ms. Tracie Minnard provided data and background information for Joint Base Lewis McCord. Mr. Joseph Clark provided data and background information for the Picatinny Arsenal. Al Vargesko of the Army Directorate of Environmental Integration at Fort Leonard Wood provided references and expertise on base camp waste management.

References

[1] K.L. Griggs, M.R. Kemme. Solid Waste Disposal Alternatives for the U.S. Military Academy. U.S. Army Construction Engineering Research Laboratory Technical Report. FE-94/11, 1994.

[2] R.L. Baker. Comprehensive Solid Waste Evaluation Study for the Atlanta Army Depot. U.S. Army Construction Engineering Research Laboratory Letter Report, E-18, 1973.

[3] B. Boone, M. Shami, H. Weinick, Waste meangement strategies and global sustainability of deconstruction, International Journal of Environmental Technology and Management 8 (2008) 229–260.
[4] R. Stankoff, D.M. White, Sewage Sludge Management at Eielson AFB, Alaska, Proceedings of the International Conference on Cold Regions Engineering (1999) 317–328.
[5] T.J. Hartranft, Energy security and independence for military installations: Candidate mission-focused vision and policy measures, ASME 2007 Energy Sustainability Conference (ES2007), Long Beach, California, USA., July 27–30, 2007.
[6] F.H. Holcomb, R.S. Parker, T.J. Hartranft, K. Preston, H.R. Sanborn, P.J. Darcy, Proceedings of the 1st Army Installation Waste to Energy Workshop, US Army Engineer Research & Development Center. ERDC/CERL TR-08-11, Champaign, IL, 2008.
[7] C. Ray, R. Jain, B.A. Donahue, E.D. Smith, Hazardous waste minimization through life cycle costs analysis at federal facilities, Journal of the Air and Waste Association 49 (1999) 17–27.
[8] J.N. Speakman, J.L. McCauley. Siting Hazardous Management Facilities at Military Installations—Case Histories. In: Young, R.A. Proceedings of the Second Annual Hazardous Materials Management Conference. Philadelphia, PA, 1982.
[9] T.F. Jenkins, S. Thiboutot, G. Ampleman, A.D. Hewitt, M.E. Walsh, T.A. Ranney, et al., Identity and Distribution of Residues of Energetic Compounds at Military Live-Fire Training Ranges, U.S. Army Engineer Research and Development Center, 2005. TR-05–10.
[10] United States Environmental Protection Agency (USEPA). Military Munitions Rule: Hazardous Waste Identification and Management; Explosives Emergencies; Manifest Exemption for Transport of Hazardous Waste on Right-of-Ways on Contiguous Properties; Final Rule. 40 CFR Parts 260, 261, 262, 263, 264, 265, 266, and 270. Federal Register: February 12, 1997 (Volume 62, Number 29, Page 6621-6657). Available at: <http://www.epa.gov/fedrgstr/EPA-WASTE/1997/February/Day-12/f3218.htm>. Accessed July 2010.
[11] R. Cox, Regulatory encroachment on range sustainment; another pressure? Presentation Slides from the Proceedings of the Sustainable Range Management Conference. New Orleans, LA., 2004.
[12] D.W. Mills, Environmental Site Characterization on Active Bombing Range: Challenges and Strategies. Presentation Slides from the Sustainable Range Management Conference. New Orleans, LA. January 5-8, 2004.
[13] D.W. Mills, Environmental site investigation on active ranges: challenges and strategies, in: B.C. Alleman, S.A. Downes (Eds.), Sustainable Range Management, Battelle, Columbus, OH, 2004. 237–260.
[14] A. Wood, D. Wallace, J. Nocera, L. Blake, Progressive approach to closure of subpart X EOD burn sites, in: B.C. Alleman, S.A. Downes (Eds.), Sustainable Range Management, Battelle, Columbus, OH, 2004. 938–958.
[15] J. Wrobel, J.P. Gross, Sustainable Superfund remediation, The Military Engineer 102 (2010) 59–60.
[16] United States Army Environmental Command (USAEC). Army Environmental Requirements and Technology Assessments. Aberdeen Proving Grounds, MD, 2009.
[17] United States Department of the Army (USDOA). Environmental Considerations in Military Operations. U.S. Army Field Manual. FM-3–34.5. 2010.
[18] G.L. Gerdes, A.L. Jantzer, Base Camp Solid Waste Characterization Study. U.S. Army Engineer Research and Development Center, ERDC/CERL TR-06–24 (ADB323774), 2006.
[19] W.H. Ruppert, T.A. Bush, D.P. Verdonik, J.A. Geiman, and M.A. Harrison, Force Provider Solid Waste Characterization Study. U.S. Army Research, Development and Engineering Command – Natick Soldier Center. Natick/TR-04/017 (ADA427565), 2004.
[20] United States Army Engineer School (USAES), Waste Management for Deployed Forces – Commander's Handbook, U.S. Army Engineer School Fort, Leonard Wood, Missouri, 2010.
[21] J.J. Valdes, J. Warner. Tactical Garbage to Energy Refinery. Edgewood Chemical Biological Center, Aberdeen Proving Ground, Maryland. ECBC-TR-713 (ADA510944), 2009.

CHAPTER

26

Space Waste

Gene Stansbery

NASA Orbital Debris Program Office, NASA/Johnson Space Center/KX2, USA

OUTLINE

1. INTRODUCTION

On February 10, 2009, at an altitude of 790 km over Siberia, two satellites collided at a relative velocity of 11 km s^{-1} [1], creating almost 2000 fragments larger than ~10 cm and many thousands of smaller fragments. Iridium 33, launched in 1997, was an operational communications satellite that was part of a constellation of 66 satellites that provide voice and data services worldwide. The other satellite was Cosmos 2251, a decommissioned communication satellite launched by the Commonwealth of Independent States (Russia) in 1993. This was a landmark event in the history of orbital debris as it was the first accidental hypervelocity collision of two large, intact satellites.

Earth-orbiting spacecraft have become an integral part of our everyday lives. We depend on them for communications, weather forecasts, scientific research, and national security. A real and growing concern for the safety and reliability of these satellites is the threat from collision with other orbiting objects, including space debris. Even small particles can damage, degrade, or destroy spacecraft due to the very

Waste Doi: 10.1016/B978-0-12-381475-3.10026-9

high velocities involved in a collision, on average about 11 km s^{-1} in low-Earth orbit (LEO).

For example, during the Space Shuttle flight of STS-7 in 1983, a paint fleck only 0.2 mm in size impacted the Shuttle window and created a pit of 0.4 mm depth, which exceeded the allowable damage criteria for reuse of the window outer pane during subsequent launches (see Fig. 26.1) [2]. Although this was the first documented example of damage to the Space Shuttle from an orbital debris impact, the orbital debris environment in this size regime has grown considerably. Currently, the Space Shuttle program replaces, on average, one to two windows per shuttle flight due to hypervelocity impacts from space debris, including micrometeoroids.

In another example, the STS-115 flight in 2006 returned with damage to the starboard payload bay radiator number 4 from a hypervelocity debris impact. The debris penetrated both walls of the honeycomb structure and the shock wave from the penetration created a crack in the rear surface of the radiator 6.8 mm long (see Fig. 26.2) [3]. Scanning electron microscopy and energy dispersive X-ray detection analyses of residual material around the hole and in the interior of the radiator show that the impactor was a small fragment of circuit board material.

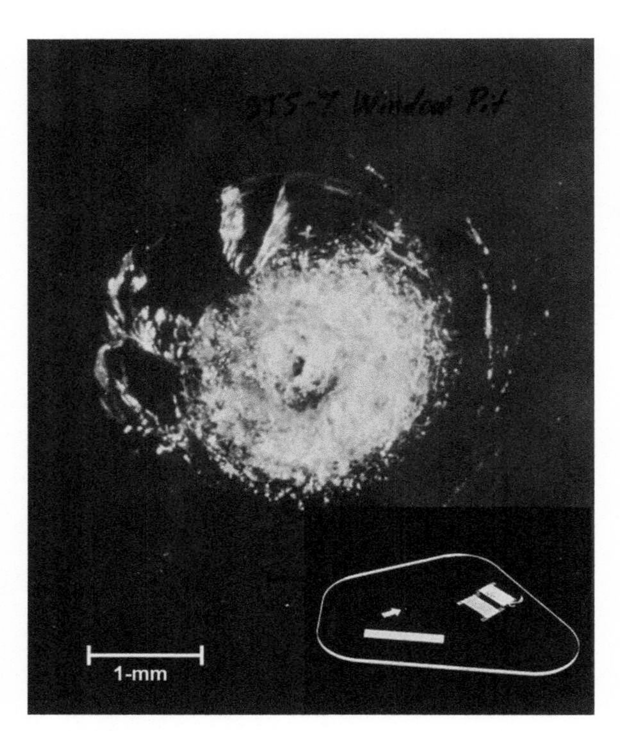

FIGURE 26.1 STS-7 window strike. Damage caused by a paint fleck only 0.2 mm in size.

2. THE CURRENT ORBITAL DEBRIS ENVIRONMENT

2.1. Number of Debris Objects

The most reliable historical source of information on the growth and composition of the orbital debris environment is provided by the catalog of artificial orbiting objects maintained by the U.S. Strategic Command's (USSTRATCOM) Space Surveillance Network (SSN), a worldwide network of radar, optical telescope, and passive radio frequency sensors. This catalog and ephemeris of orbital objects and debris is limited to sizes as small as 10 cm diameter in LEO and about 1 m diameter at geosynchronous (GEO) altitudes (the altitude where the satellite's orbit matches the Earth's rotation).

Note: The Russian Federation also maintains a network of sensors and a catalog. However, most of the sensors in the Russian network are located within the borders of the former Soviet Union and are more heavily weighted toward optical sensors than radars [4].

For sizes smaller than 10 cm, statistical sampling of the environment is done using special remote sensing radars, optical telescopes, and returned spacecraft surfaces.

As of January 1, 2010, the SSN was actively tracking ~21,000 artificial objects within 40,000 km of the Earth's surface. This number includes ~6000 objects that have not yet been entered into the USSTRATCOM official catalog. Statistical

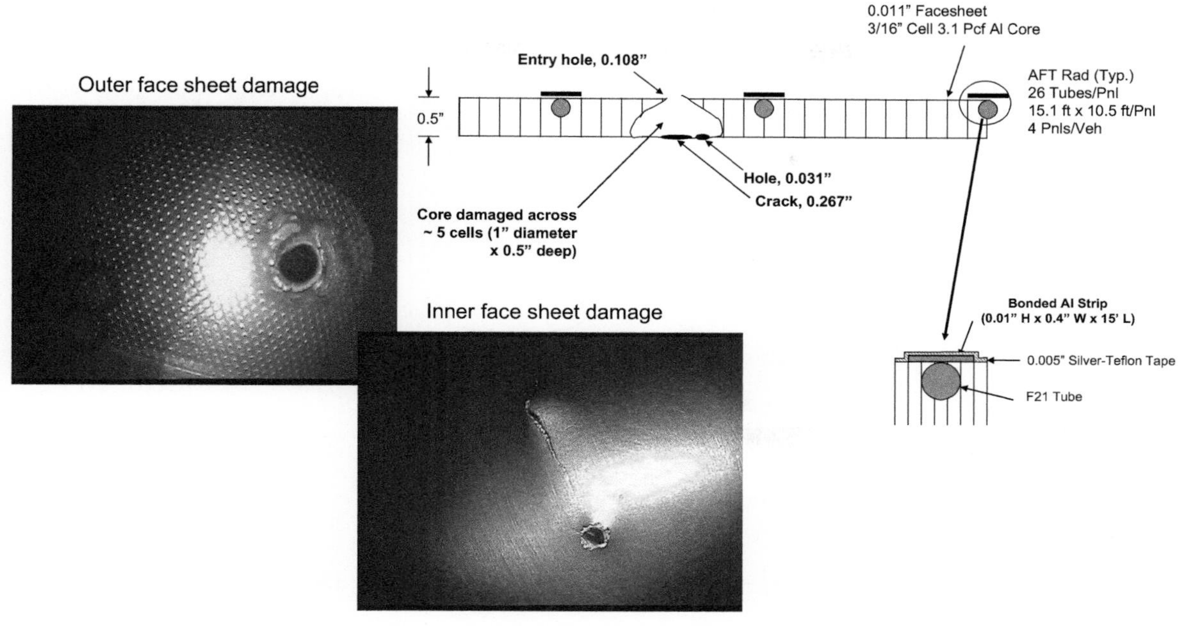

FIGURE 26.2 Radiator damage on STS-115. This damage was caused by a small piece of circuit board material.

measurements indicate that there are an additional 500,000 objects between 1 and 10 cm diameter and millions of objects smaller than 1 cm. Tracked objects include operational (functional) spacecraft, non-operational (inactive or retired) spacecraft, and rocket bodies, as well as debris from a variety of sources. Any object that is not performing its intended function is considered to be orbital debris, including upper stages that have already delivered their payloads and spacecraft that are no longer functional. Of the 21,000 orbiting objects mentioned above, less than 1000 are functioning spacecraft. Figure 26.3 shows the growth in the number of objects in Earth orbit since the beginning of the Space Age, based on the USSTRATCOM catalog [5].

Note: In addition to the Iridium/Cosmos collision in 2009, the other significant event shown in Fig. 26.3 is the intentional destruction of the Fengyun-1C weather satellite in January 2007 when it collided with a direct ascent anti-satellite (ASAT) weapon launched by the Chinese. The Fengyun-1C was in a near circular orbit at an altitude of ~850 km. This stands as the single largest fragmentation event in space with more than 2700 cataloged fragments to date. Furthermore, because both this event and the Iridium/Cosmos collision occurred at relatively high altitudes, much of the debris from both events will have orbital lifetimes of many decades [6,7].

2.2. Distribution of Orbital Debris

Figure 26.4 shows the spatial density of orbiting objects as a function of altitude. The region below 2000 km altitude, LEO, is a natural divide since the spatial density drops significantly above this altitude. Altitudes near 20,000 km are home to the NAVSTAR GPS and GLONASS satellite constellations used for navigation. The spike in spatial density near 36,000 km shows the location of GEO altitude, where the satellite orbit period approximately equals

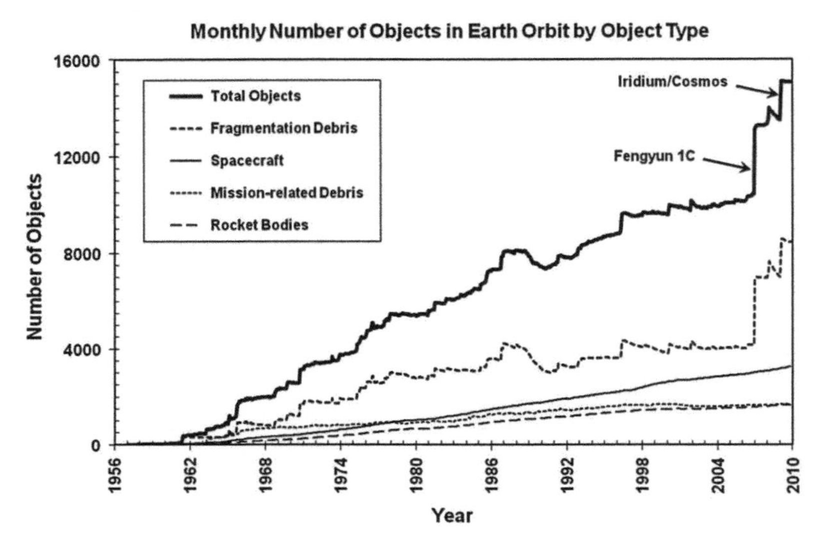

FIGURE 26.3 Growth of the number of objects in Earth orbit from the USSTRATCOM catalog. These totals represent the number of objects 10 cm and larger in LEO and 1 m and larger at GEO altitudes.

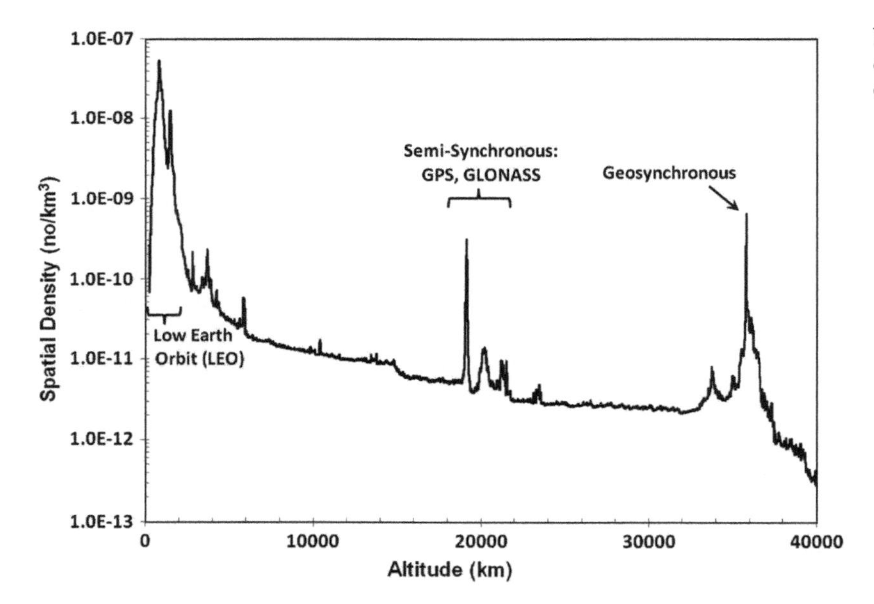

FIGURE 26.4 Spatial density of objects in the USSTRATCOM catalog as a function of altitude.

1 day, keeping the satellite over the same region of Earth.

Figure 26.5 shows the distribution of the inclination of objects in the USSTRATCOM catalog with perigees below 2000 and eccentricities less than 0.3. Inclinations are usually chosen depending on the mission of the satellite, but sometimes, the inclination is related to the launch site. Launching due east from the launch site takes advantage of the eastward velocity of the ground itself as the Earth rotates. This minimum energy launch would place the satellite in an orbit with the inclination equal to the launch site's latitude.

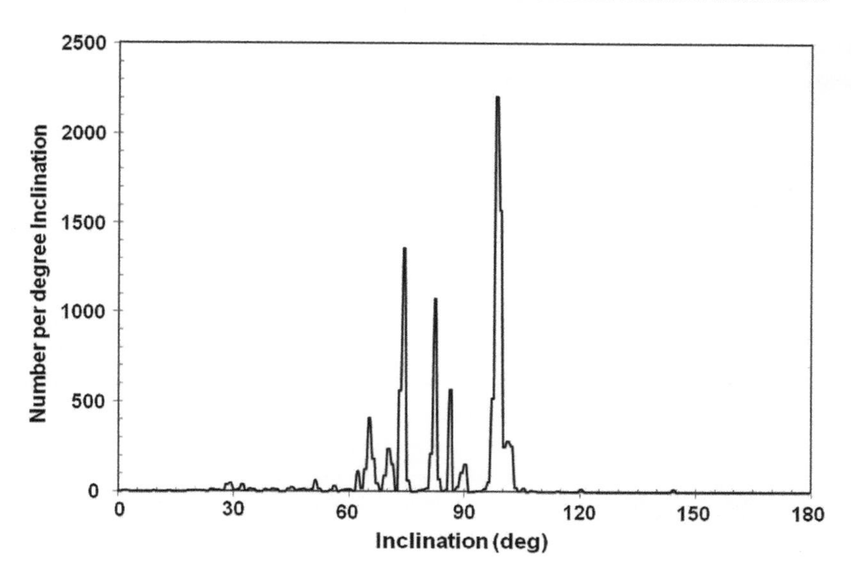

FIGURE 26.5 Distribution of objects in or passing through LEO (LEO defined at altitudes below 2000 km).

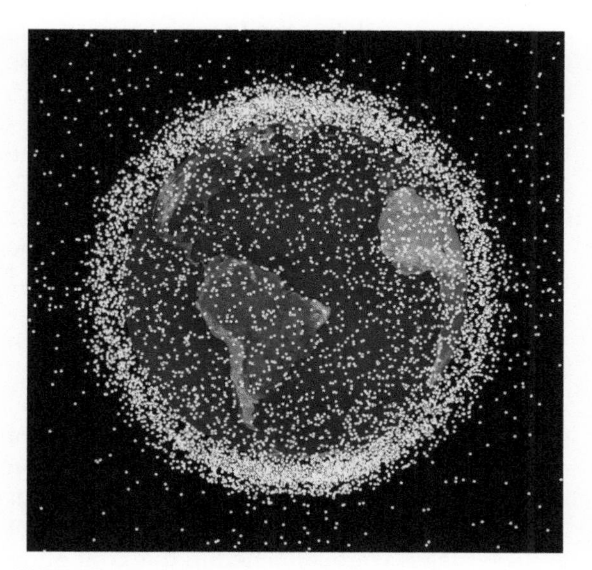

FIGURE 26.6 Representation of the location and distribution of the LEO debris population at a snapshot in time (midnight, January 1, 2010).

Although launched into specific inclinations, visualizations of the locations or motions of satellites in LEO exhibit a random appearance (see Fig. 26.6). In general, it can be assumed that the right ascension of ascending node (RAAN), the location where a satellite crosses the equator traveling south-to-north, is randomized. When a fragmentation event occurs, a delta velocity is imparted to the fragments. This delta velocity will change the orbit of the fragment relative to the pre-event parent body orbit. The oblateness of the Earth causes orbits to precess. Because the precession rate is different for objects in different orbits, this spreads out the RAAN of objects over time, producing the random orbit appearance.

The behavior of objects is somewhat different in GEO. Typically, operational satellites are placed into GEO at 0° inclination at a specific longitude. At the correct altitude, the orbit period is matched to the rotation rate of the Earth, and with very small maneuvers, the satellite is kept stationary over the equator at that longitude. However, once maneuvering fuel runs out (if the satellite has not been moved to a disposal orbit), the object will be perturbed out of its position. The object's inclination will evolve over time and oscillate around a $7^1/_2°$ stable plane caused by the Earth–Moon–Sun configuration (see Fig. 26.7). Objects released

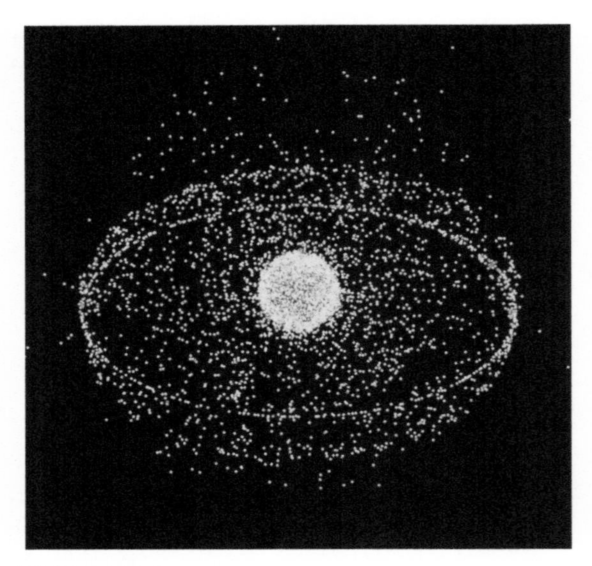

FIGURE 26.7 Representation of the location and distribution of the debris population, including GEO altitudes, at a snapshot in time (midnight, January 1, 2010).

from 0° inclination will increase in inclination to a maximum of ~15° and then return to 0° again over a period of ~52 years [8]. The objects will also drift in longitude, bringing them near to other objects in the GEO region.

2.3. Sources of Orbital Debris

2.3.1. Fragmentation Debris

On-orbit fragmentations can result from accidental or intentional explosions, accidental or intentional collisions, or aerodynamic forces as an object nears atmospheric reentry. More than half of all the USSTRATCOM cataloged objects are fragmentation debris. Since the first recognized fragmentation in June 1961, more than 200 satellites have experienced on-orbit breakups (excluding aerodynamic breakups). Most of the fragmentations have been the result of explosive release of stored energy: unused propellant, pressurized gasses, or batteries. Aerodynamic breakups, while relatively common, produce very few long-lived debris and are a nonfactor in the long-term orbital debris problem. Collisions, both deliberate and accidental, have been rare during the space age, but have created some of the larger debris clouds, including the largest debris producing event, the Fengyun-1C ASAT test. Table 26.1 shows the 10 largest debris producing events as defined by the number of objects cataloged as of January 1, 2010.

Note: The International Designator comprised the year of launch, the sequential launch number for that year, and the sequential piece cataloged associated to the launch. For example, the Fengyun-1C has an international designator of 1999-025A. The Fengyun-1C was the 25th launch in 1999 and A represents that it was the first object associated with the launch.

Most of the ~200 satellite fragmentations have occurred in LEO or in highly elliptical orbits that pass through LEO, which accounts for the higher spatial density in this region. Only two known breakups have occurred in GEO: a Titan transtage rocket body (International Designator 1968-081E) that delivered the OV2 satellite and the EKRAN 2 satellite (1977-092A). However, some surveys of GEO indicate that there may be additional breakups in the region [9]. No known breakups have occurred in the circular semisynchronous orbits.

In the early and mid-1970s, Delta launch vehicle second stages were left in orbit with residual fuel on board. Over time, thermal stresses caused by the passage in and out of the Earth's shadow would rupture the common bulkhead between the propellant and oxidizer tanks, resulting in an on-orbit explosion. Eight of these stages were launched in this configuration before recognition and correction of the problem. The last of these exploded in 1991. Correction of the problem by venting any residual fuel, once on-orbit, was one of the early successes of orbital debris mitigation efforts [10].

However, the problem with residual fuels was not recognized by all launch operators.

TABLE 26.1 Top 10 Orbital Debris-Producing Events Based On the Number of Cataloged Debris as of January 1, 2010

Name	International Designator	Launch Date	Event Date	Debris Cataloged	Remaining On-Orbit	Apogee (km)	Perigee (km)	Inclination (deg)	Assessed Cause
Fengyun-1C	1999-025A	10 May 1999	11 January 2007	2691	2623	865	845	98.6	Deliberate collision
Cosmos 2251	1993-036A	16 June 1993	10 February 2009	1142	1102	800	775	74.0	Accidental Collision
Iridium 33	1997-051C	14 September 1997	10 February 2009	490	473	780	775	86.4	
STEP II R/B (Pegasus HAPS)	1994-029B	19 May 1994	3 June 1996	712	62	820	585	82	Propulsion
Cosmos 2421	2006-026A	25 June 2006	14 March 2008	508	27	420	400	65	Unknown explosion
SPOT1/VIKING R/B (Ariane 1)	1986-019C	22 February 1986	13 November 1986	492	33	835	805	98.7	Propulsion
OV2-1/LCS-2 R/B (Titan 3C Transtage)	1965-082DM	15 October 1965	15 October 1965	472	36	790	710	32.2	Propulsion
NIMBUS 4 R/B (Agena D)	1970-025C	8 April 1970	17 October 1970	373	247	1085	1065	99.9	Unknown explosion
TES R/B (PSLV)	2001-49D	22 October 2001	19 December 2001	368	116	675	550	97.9	Propulsion
CBERS-1/SACI-1 R/B (CZ-4)	1999-057C	14 October 1999	11 March 2000	341	187	745	725	98.5	Propulsion
Cosmos 1275	1981-053A	4 June 1981	24 July 1981	310	256	1015	960	83	Battery

In 1986, the SPOT 1 and Viking satellites were launched into sun-synchronous orbits by an Ariane 1 rocket. Nine months later, the Ariane upper stage exploded, likely due to residual fuels left on-board [11]. This event led to improved cooperation and information exchange on orbital debris among various national space agencies, leading to the eventual creation of the Inter-Agency Space Debris Coordination Committee (IADC).

2.3.2. *Mission-Related Debris*

Mission-related debris are objects intentionally discarded during satellite delivery or satellite operations, including lens caps, separation and packing devices, spin-up mechanisms, empty propellant tanks, or a few objects thrown away or dropped during human activities. Most missions have very few pieces of this type of debris, and with the growing awareness of debris environmental issues and relatively recent national and international guidelines for reducing the growth of orbital debris, there has been very little growth of this class of debris during the last decade.

2.3.3. *Anomalous/Deterioration*

As spacecraft spend many years in the space environment, many deteriorate to the point where pieces or parts separate from the original spacecraft, resulting in what is referred to as "anomalous debris." An anomalous event is the unplanned separation, usually at low velocity, of one or more detectable objects from a satellite that remains relatively intact [12].

The disintegration of spacecraft exterior paint is thought to create many sub-millimeter pieces of debris. The cause of this paint flaking is believed to be due to radiation effects and/or to atomic oxygen erosion of the organic binder in the paint and this erosion is another major source of small debris in LEO. Stage and spacecraft separation processes that occur in orbit also frequently release small debris.

2.3.4. *Slag/NaK/Other*

Many small orbital debris particles are created by solid rocket motor (SRM) firings, which produce aluminum oxide (Al_2O_3) particles called slag. These particles are formed during SRM tail-off, or termination of burn, by the rapid expansion, dissemination, and solidification of the molten Al_2O_3 slag accumulated during the burn. Since the transfer orbits are elliptical orbits, most of the particles reenter quickly because of the effects of atmospheric drag and other forces at the orbit perigee. However, the propensity of SRMs to generate particles of 100 μm and larger has caused concern regarding their contribution to the debris environment. Particle sizes as large as 1 cm have been witnessed in ground tests, and comparable sizes have been estimated via ground-based and in-situ observations of suborbital tail-off events [13]. Because of the large number of particles ejected by each motor, these Al_2O_3 particles can represent a significant surface erosion and contamination threat to spacecraft.

Another class of debris is the sodium–potassium (NaK) droplets that were released from Radar Ocean Reconnaissance SATellite (RORSAT) spacecraft launched between 1971 and 1988 by the former Soviet Union. On a number of RORSATs, the nuclear reactor core was ejected from the spacecraft, which apparently caused a breach in the NaK cooling loop. The liquid metal NaK spewed out and formed spherical droplets. Very few of the drops were large enough to be detected by ground sensors until statistical measurements of the environment with the Haystack and Goldstone radars in the 1990s. There may be more than 50,000 NaK droplets that are 5 mm or larger in orbit today [14].

3. COUNTER MEASURES

Despite the rapid increase in the number of debris in orbit since 2007, the absolute risks to operational spacecraft from collisions with man-made debris remain low, as evidenced by

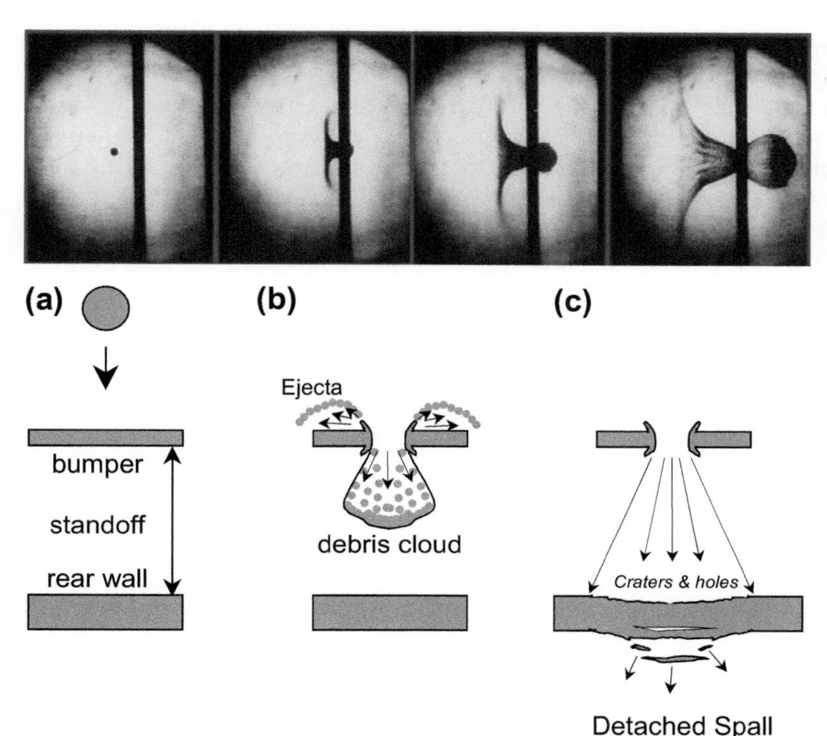

FIGURE 26.8 Simple Whipple Shield.
(a) Whipple shields consist of a BUMPER, standoff (gap or spacing), and rear wall.
(b) Hypervelocity impacts will generate a cloud of BUMPER and projectile debris that can contain solid fragments, liquid, and vapor particles.
(c) The rear wall must survive the fragments and debris cloud impulsive loading. It could fail by perforation from solid fragments, spall, or tear and petal from the impulsive loading.

only one known permanent loss of a functioning spacecraft (Iridium 33) to date. However, the threat to the approximately 1000 operational satellites now in orbit around the Earth increases with the growth of the orbital debris population. The three principal counter measures to this threat that are being used are shielding, collision avoidance, and curtailment of the creation of new debris through design and operational practices.

3.1. Shielding Against Orbital Debris

Small, untracked debris, due to their very large numbers, pose the greatest threat to operational spacecraft. The inherent structure of satellites provide some protection from the smallest orbital debris, that is, less than 1 mm, but most satellites are vulnerable to impacts from particles in the millimeter-class size regime and larger.

The Whipple Shield was the first spacecraft shield ever implemented. It was suggested by Fred Whipple as early as the 1947 [15] and is still in use today. A simple Whipple shield consists of a thin "sacrificial" wall mounted at a distance from a rear wall (see Fig. 26.8). The function of the first sheet or "bumper" is to break up the projectile into a cloud of material containing both projectile and bumper debris. This cloud expands while moving across the standoff, resulting in the impactor momentum being distributed over a wide area of the rear wall. The back sheet must be thick enough to withstand the blast loading from the debris cloud and any solid fragments that remain. For most conditions, a Whipple shield results in a significant weight reduction over a single plate, which must contend with deposition of the projectile kinetic energy in a localized area.

Shielding technologies have advanced considerably beyond the simple Whipple shield. For example, the Stuffed Whipple shield has layers of Kevlar and Nextel sandwiched between the outer and inner walls. This increases the effectiveness of the shield and decreases the standoff distance needed between the two walls.

The International Space Station (ISS) is the most heavily shielded spacecraft in history. Because the orbital debris environment presents a very asymmetric flux to the ISS, not all critical surfaces need the same level of protection. Therefore, the ISS uses hundreds of different, custom-made shields to protect critical components of the extensive structure, including habitable compartments and vital external units and lines, and to minimize the volume and mass of the shields.

The latest concepts for improved orbital debris shields involve replacing typical Whipple stuffing materials and aluminum honeycomb with new metallic foams. These new designs offer greater protection from the meteoroid and orbital debris environments with less overall structural mass [16].

However, shielding techniques are practical only for particles on the order of 1 cm or less. Moreover, many robotic spacecraft cannot afford the mass penalty associated with orbital debris shields, and payload elements frequently cannot host shields for operational reasons.

3.2. Collision Avoidance

As a result of the collision of Iridium 33 and Cosmos 2251 in early 2009, the Joint Space Operations Center of the U.S. Strategic Command now conducts conjunction assessments for all operational spacecraft in Earth orbit, regardless of ownership nationality. Any prediction of a close approach, typically within 1 km, will be shared with the spacecraft owner/operator freely and immediately [17].

Because of inherent uncertainties in space surveillance measurements, the dynamic state of the atmosphere and, in many cases, the instability of at least one of the conjuncting objects predicting the collision of two satellites remains a probabilistic endeavor. Typical maneuver probability thresholds for collision avoidance are 1 in 10,000 for human space flight and 1 in 1000 (or more) for robotic satellites.

Contrary to common belief, collision avoidance maneuvers are normally not highly disruptive to satellite operations. Moreover, collision avoidance maneuvers are typically very small, that is, involve changes in velocity of less than 1 m s^{-1}, and in most cases, it can be conducted in a manner that does not waste propellant resources. For example, collision avoidance maneuvers performed by the ISS almost always result in a small increase in orbital altitude and thus, simply constitute an unscheduled anti-drag maneuver. Similar procedures are used for robotic satellites.

In addition to the obvious due diligence aspect of protecting operational spacecraft, one long-term benefit of collision avoidance is the prevention of collisions between two large objects, which in turn could further degrade near-Earth space with large numbers of new debris, as was the case with the collision of Iridium 33 and Cosmos 2251. On the other hand, over 99% of the risk to operational spacecraft from collisions with orbital debris comes from objects too small to track on a routine basis, that is, smaller than 10 cm. Hence, only an improvement in the orbital debris environment itself can dramatically reduce the risks to operational spacecraft.

3.3. Orbital Debris Mitigation Policies and Practices

The purpose of mitigation policies is to outline spacecraft and launch vehicle design and operational procedures that would reduce the amount of unnecessary orbital debris being generated accidentally or intentionally, thus promoting the safe and reliable operation of spacecraft.

One of the first comprehensive sets of orbital debris mitigation recommendations was issued by the American Institute of Aeronautics and Astronautics in 1981. Orbital debris mitigation was first mentioned in President Reagan's National Space Policy in 1988 and has been in each subsequent Space Policy. Following the first U.S. government interagency report on orbital debris in 1989, NASA issued the first formal orbital debris mitigation guidelines for a U.S. government agency in 1995. These guidelines served as the basis for the U.S. Orbital Debris Mitigation Standard Practices, which were adopted in 2001 after a multi-year coordination with the U.S. aerospace industry [18].

In 2002 the IADC, which comprised the space agencies of 10 nations plus the European Space Agency, adopted space debris mitigation guidelines, which were then submitted for consideration to the Scientific and Technical Subcommittee of the United Nations' Committee on the Peaceful Uses of Outer Space. After several years of debate and negotiations, space debris mitigation guidelines were accepted by the United Nations in 2007.

The IADC Space Debris Mitigation Guidelines emphasize four major areas that require the attention of the aerospace community [19]:

(1) limitation of debris released during normal operations;
(2) minimization of the potential for on-orbit break-ups during and after space operations;
(3) post-mission disposal recommendations for vehicles in LEO, GEO, and other orbital regimes; and
(4) prevention of on-orbit collisions.

The first area concentrates on avoiding the creation of unnecessary orbital debris during the deployment and operational phases of a space mission. For example, the release of payload attachment devices, sensor covers, explosive bolt fragments, and straps and wires should be avoided whenever possible. During the past 25 years, approximately 15% of all cataloged objects in Earth orbit have fallen into this category; although, since the turn of the century, the absolute number of mission-related debris in Earth orbit has declined due to increasing adoption of design modifications.

The second area discusses the potential for on-orbit break-ups during and after space operations. Before the collisions of 2007 and 2009, the greatest source of debris in orbit was from violent explosions of spacecraft and rocket bodies. Faulty propulsion systems are by far the most prevalent source of explosion debris. In fact, the vast majority of these propellant- or pressurant-induced break-ups occur after a spacecraft or launch vehicle orbital stage has successfully completed its mission.

Fortunately, simple end-of-mission passivation, especially the release or burning of residual propellants and pressurants, of a spacecraft or orbital stage can eliminate the potential for future explosions. Such post-mission action is now routine for virtually all space launch vehicles of the world and, consequently, the rate of creation of long-lived debris has fallen. Spacecraft operators, particularly those responsible for GEO spacecraft, are also following suit.

The third area of interest is the disposal of spacecraft and orbital stages after their useful lives have ended. The ultimate objective whether in LEO or in GEO, is to limit the probability of derelict objects colliding with other space objects (operational or not), which can result in the creation of large numbers of additional debris. Spacecraft and orbital stages operating in LEO should be left in orbits that will reenter the atmosphere within 25 years of their end of mission. Vehicles operating in or near GEO should be placed in disposal orbits above GEO, which prevent their drifting back within 200 km above GEO, typically for at least 100 years.

The fourth topic of space debris mitigation is active collision avoidance already discussed in Section 3.2. Some space launch operators also rely on conjunction assessments to avoid colliding with a resident space object during the initial few hours of orbital insertion and deployment of new missions.

Although not rising to the status of an international treaty, the UN space debris mitigation guidelines are recommended for implementation via national procedures. For example, the UN guidelines are compatible with the U.S. Orbital Debris Mitigation Standard Practices [20], which are implemented for government-sponsored space missions through directives from NASA and the Department of Defense and for commercial space operations through the regulations of the Department of Transportation, the Federal Communications Commission, and the Department of Commerce. Several other nations have invoked similar, nonvoluntary orbital debris mitigation requirements.

4. FUTURE ORBITAL DEBRIS POPULATION AND ACTIVE DEBRIS REMOVAL

The mitigation measures adopted by the international community to date will slow the growth of the orbital debris environment, but will not stop it.

NASA performed a study in 2005 that projected the debris environment 200 years into the future, assuming no new launches after 2004 [21]. In addition, no future disposal maneuvers were allowed for existing spacecraft, few of which currently have such a capability. Satellite explosions were permitted at their current historical rates. The study used NASA's orbital debris evolutionary model, LEGEND (LEO-to-GEO Environment Debris model), which is a three-dimensional physical model that is capable of simulating the historical satellite environment, as well as the evolution of future debris populations. The historical record, including launches, reentries, and all known fragmentation events, was used to recreate the known cataloged environment at the end of 2004. Collision probabilities among objects are estimated with a fast, pair-wise comparison algorithm in the projection component. Only objects 10 cm and larger are considered for potential collisions. This size threshold is the detection limit of the SSN sensors, and more than 95% of the debris population mass is in objects 10 cm and larger. A total of 50 200-year future projection Monte Carlo simulations were executed and evaluated.

The simulated LEO debris populations (10 cm and larger objects) between 1957 and the end of a 200-year future projection period are shown in Fig. 26.9. The results indicate that collision fragments replace other decaying debris (because of atmospheric drag and solar radiation pressure) through 2055, keeping the total LEO population approximately constant. Beyond 2055, however, the creation of new collision fragments exceeds the number of decaying debris, forcing the total satellite population to increase. An average of 18.2 collisions (10.8 catastrophic, 7.4 noncatastrophic) would be expected in the next 200 years. Of these, 10.8 collisions would be catastrophic, involving total fragmentation of the satellites. Detailed analysis of the results indicates that the predicted catastrophic collisions and the resulting population increase are non-uniform throughout LEO. About 60% of all catastrophic collisions would occur between 900 and 1000 km altitudes, leading to a major population increase in the same region.

The results from the 2005 study show a best-case scenario. The study was done before the Fengyun-1C ASAT test in January 2007 and the Iridium/Cosmos collision of February 2009. Further, the no new launch scenario is unrealistic. Because the best case scenario still shows continued growth in the population, the only way to stabilize the number of debris objects is to actively remove objects already in orbit.

Initially in 2008, and then updated in 2009, a series of simulations was conducted that

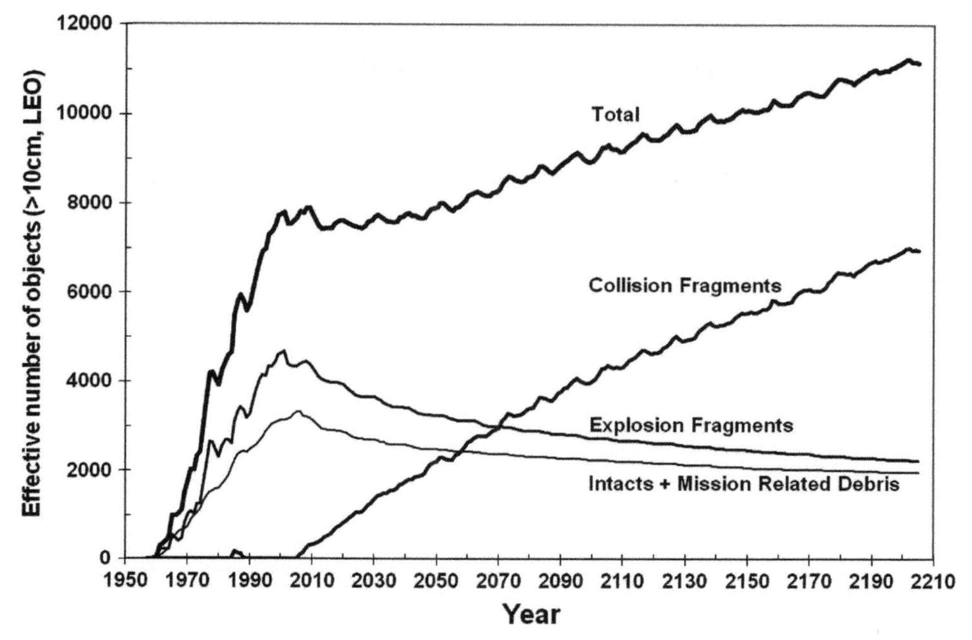

FIGURE 26.9 The simulated LEO debris population. The historical component is based on recorded launches and on-orbit fragmentations through the end of 2004. The future projection is the average of 50 LEGEND Monte Carlo simulations based on the "no new launches" assumption.

looked at active debris removal (ADR) options [22]. For this series of studies, future launch traffic was simulated by repeating the 1999 to 2006 launch cycle. Postmission disposal (PMD) mitigation measures were allowed and simulated. Rocket bodies, after launch, were moved to 25-year decay orbits or to LEO storage orbits (above 2000 km altitude), depending on which option required the lowest change in velocity for the maneuvers. In most cases, the 25-year decay orbit was the preferred choice for vehicles passing through LEO. The mission lifetimes of future payloads were set to 8 years. At the end of the mission lifetime, each payload was moved either to the 25-year decay orbit or to an LEO storage orbit. The PMD success rate was set to 90%. No explosions or deliberate breakups were allowed for future rocket bodies and payloads.

To maximize the effectiveness of ADR, objects with the greatest potential of contributing to future collision activities and generating the largest amount of debris were removed first – these were objects with the highest mass and collision probability products. Operating payloads (assuming a nominal lifetime of 8 years) and breakup fragments were excluded from removal consideration. The study concluded that the LEO population could be stabilized by removing as few as five selected objects per year (see Fig. 26.10).

To reduce the near term risk to existing spacecraft, removal of the numerous small fragments is needed. To keep the environment from continuing unchecked, removal of a few large, massive, intact satellites or rocket bodies in high-risk orbits will stabilize the environment. The first International Conference on Orbital Debris Removal, cosponsored by the Defense Advanced Research Projects Agency (DARPA) and NASA, was held in December 2009. The conference looked at legal

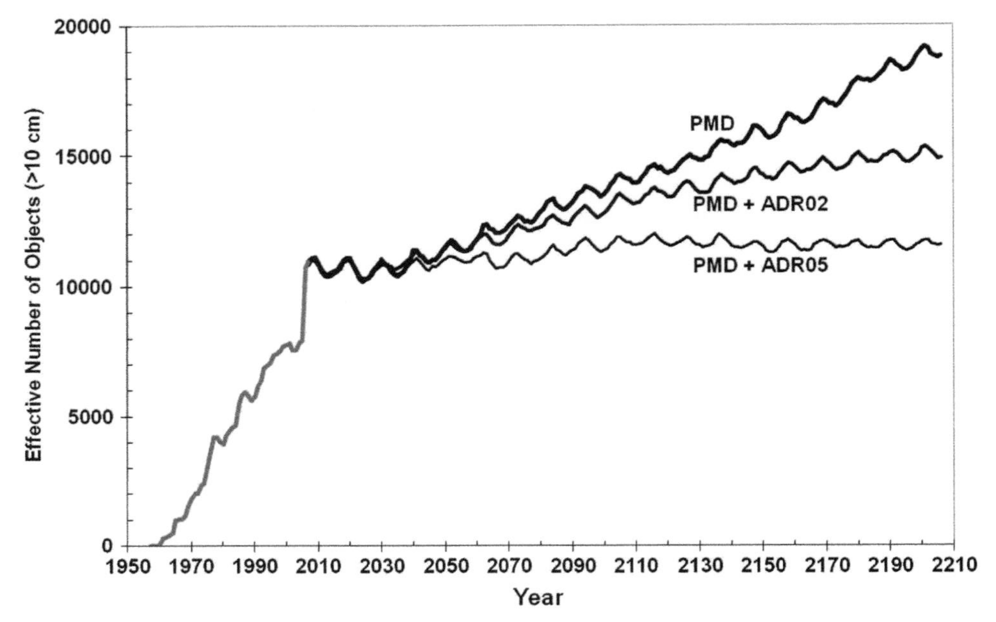

FIGURE 26.10 Comparison of three different scenarios. From top to bottom: Postmission disposal (PMD) only, PMD and Active Debris Removal (ADR) of two objects per year, and PMD and ADR of five objects per year, respectively.

and economic issues as well as examined technological concepts for removing large and small orbital debris. However, removing existing debris or satellites from orbit is a very difficult task. No method presented at the conference was deemed both economically sound and technologically mature. For the time being, not creating long-lived debris (i.e., following existing debris mitigation guidelines and requirements) remains the best method for slowing the growth of the orbital debris environment.

5. CONCLUSIONS

Humankind is a long way from the cynical view of the distant future depicted in such movies as *WALL-E*, where it is almost impossible to launch through the cloud of Earth-orbiting satellites without collision. The probability of a satellite accidentally colliding with another orbiting object is still very small. However, environment models show that collisions will inevitably occur, producing more objects than are removed from the environment by atmospheric drag. Therefore, the number of objects will slowly increase over a period of many hundreds of years unless objects already on-orbit are removed.

But, because the probability of collision for an active satellite is so low, and the cost of debris removal is so high, it is unlikely that a serious effort will be made in the near future to clean up space. It can only be hoped that technological advances in the future change the cost/benefit equation in a positive manner.

References

[1] Anom, Satellite Collision Leaves Significant Debris Clouds, NASA Orbital Debris Quarterly News, 13 (2) 1–2. Available at: <http://orbitaldebris.jsc.nasa.gov/newsletter/pdfs/ODQNv13i2.pdf>, April 2009, accessed on 1 January 2010.

[2] D.S.F. Portree, J.P. Loftus Jr., Orbital Debris: A Chronology, NASA/TP-1999-208856 (January 1999) 42.
[3] J. Hyde, E. Christiansen, D. Lear, et al., Investigation of MMOD impact on STS-115 Shuttle Payload Bay Door Radiator, NASA Orbital Debris Quarterly News 11 (3) 2–5, 2007 July. Available at: <http://orbitaldebris.jsc.nasa.gov/newsletter/pdfs/ODQNv11i3.pdf>, July 2007, accessed on 1 January 2010..
[4] G. Batyr, V. Veniaminov, S. Dicky, et al., The current state of the Russian Space Surveillance System and its capability in surveying space debris, Proceedings of the First European Conference on Space Debris, ESA SD-01, pp. 43–47, July 1993.
[5] Anom, Monthly Number of Objects in Earth Orbit by Object Type, NASA Orbital Debris Quarterly News, 14 (1) 12. Available at: <http://orbitaldebris.jsc.nasa.gov/newsletter/pdfs/ODQNv14i1.pdf>, January 2010, accessed on 15 January 2010.
[6] Anom, Chinese Anti-satellite Test Creates Most Severe Orbital Debris Cloud in History, *NASA Orbital Debris Quarterly News,* 11(2) 2–3. Available at: <http://orbitaldebris.jsc.nasa.gov/newsletter/pdfsODQNv11i2.pdf>, April 2007, accessed on 1 January 2010.
[7] N.L. Johnson, E. Stansbery, J.-C. Liou, M. Horstman, C. Stokely, D. Whitlock, The Characteristics and Consequences of the Break-up of the Fengyun-1C Spacecraft, IAC-07–A6.3.01, 58th International Astronautical Congress, Hyderabad, India, September 2007.
[8] L. Friesen, A. A Jackson, H.A. Zook, D.J. Kessler, Results in Orbital Evolution of Objects in the Geosynchronous Region, AIAA 90–1362, AIAA/NASA/DoD Orbital Debris Conference, Technical Issues and Future Directions, Baltimore, MD, 1990.
[9] R. Jehn, S. Ariafar, T. Schildknecht, R. Musci, M. Oswald, Acta Astronautica 59 (2006) 84–90.
[10] A.A. Gray Jr., Current Operational Practice for the Delta Second Stage, in: P. Joseph Loftus Jr. (Ed.), Debris from Upper Stage Breakup, AIAA, Washington, DC, 1989, pp. 201–203.
[11] R. Hergott, E. Perez, Ariane Third Stage, Current Operational Practice, Modifications Expected, in: P. Joseph Loftus Jr. (Ed.), Debris from Upper Stage Breakup, AIAA, Washington, DC, 1989, pp. 203–211.
[12] N.L. Johnson, J.R. Gabbard, G.T. DeVere, E.E. Johnson, History of On-Orbit Satellite Fragmentations, in: 14, NASA/TM-2008-214779, NASA JSC, Texas, Houston, June 2008.
[13] M.F. Horstman, M. Mulrooney, An Analysis of the Orbital Distribution of Solid Rocket Motor Slag, 58th International Astronautical Congress, IAC-07 – 6.2.03, Hyderabad, India, September 2007.
[14] J.L. Foster, Paula H. Krisko, Mark J. Matney, NaK Droplet Source Modeling, Space Debris and Space Traffic Management Symposium 2003, American Astronautical Society. IAA 03–5.2.02, Bremen Germany, 29 September–3 October 2003.
[15] D.S.F. Portree, J.P. Loftus Jr., Orbital Debris: A Chronology, NASA/TP-1999-208856 (January 1999) 16.
[16] S. Ryan, E.L. Christiansen, D.M. Lear, Shielding Against Micrometeoroid and Orbital Debris Impact with Metallic Foams, NASA Orbital Debris Quarterly News, 14 (1) 4–7. Available at: <http://orbitaldebris.jsc.nasa.gov/newsletter/pdfs/ODQNv14i1.pdf>, accessed on 1 January 2010.
[17] T. Payne, R. Morris, The Space Surveillance Network (SSN) and Orbital Debris, 33rd Annual AAS Guidance and Control Conference, American Astronautical Society, Breckenridge, Colorado, February 2010. AAS 10–012.
[18] N.L. Johnson, Orbital Debris: The Growing Threat to Space Operations, , In: 33rd Annual AAS Guidance and Control Conference, American Astronautical Society, Breckenridge, Colorado, February 2010. AAS 10–012.
[19] Anom, IADC Space Debris Mitigation Guidelines, Rev. 1, IADC-02–01, Available at: <http://www.iadc-online.org/index.cgi?item=docs_pub>, September 2007.
[20] Anom, U.S. Orbital Debris Mitigation Standard Practices, 2001. Available at: <http://orbitaldebris.jsc.nasa.gov/library/USG_OD_Standard_Practices.pdf>, accessed on 1 January 2010.
[21] J.-C. Liou, N.L. Johnson, Risks in space from orbiting debris, Science 311 (2006) 340–341.
[22] J.-C. Liou, N.L. Johnson, N.M. Hill, Controlling the growth of future LEO debris populations with active debris removal, 59th International Astronautical Congress, Glasgow, Scotland, September–October 2008. IAC-08-A6.2.9.

CHAPTER

27

Hazardous Wastes

Daniel A. Vallero

Pratt School of Engineering, Duke University, Durham, North Carolina, USA

OUTLINE

1. INTRODUCTION

Any waste stream mentioned in this book can contain "hazardous" waste in varying amounts. A specific chemical constituent of a waste stream can cause particular kinds of harm, as codified in rules and regulations. For the most part, hazardous waste has been considered as a subset of solid waste and has been distinguished from municipal wastes and nonhazardous industrial wastes. For example, in the United States, the Resource Conservation and Recovery Act (RCRA) [1] requires proper handling of wastes from their production to

Waste Doi: 10.1016/B978-0-12-381475-3.10027-0

their transport to their ultimate disposal, that is the so-called "cradle to grave" manifest system. RCRA divides wastes into two basic types, that is nonhazardous solid waste and hazardous waste. Subtitle D of the Act covers solid waste and Subtitle C regulates hazardous waste. Therefore, even though it is a subset of "solid" waste, a hazardous waste may be of any physical phase; that is solid, liquid, gas, or mixtures.

The hazardous attributes of the waste are usually based on its inherent physicochemical properties, including its likelihood to ignite, explode, and react with water (see Table 27.1). The hazardous inherent properties of a waste can also be biological, such as the infectious nature of medical wastes, or a chemical compound that has been shown to elicit acute effects, for example skin irritations and chronic effects, such as cancer; harm to the endocrine, immune, or nervous system; interference with tissue development and reproduction; or mutations, birth defects, or other toxic endpoints.

2. MANAGING HAZARDOUS WASTES

Hazardous waste management involves reducing the amount of hazardous substances produced, treating hazardous wastes to reduce their toxicity, and applying sound engineering controls to reduce or eliminate exposures to these wastes. From a chemical engineering perspective, any wastes, but especially hazardous wastes, are produced unintentionally. Facilities that generate the chemicals are attempting to produce substances demanded by the marketplace, but in the process and under the wrong circumstances these substances or their components become hazards. If the same substances were produced intentionally, it would be a *hazardous material*, but not a *hazardous waste.*

Engineers design and operate facilities to prevent, contain, decontaminate, and otherwise treat these wastes before they are released. However, hazardous waste engineering is most often associated with interventions after the wastes have been released. Engineers are frequently called on to address these wastes once they are released to the environment.

A hazard is expressed as the potential of an unacceptable outcome (Table 27.1). For chemicals, the most important hazard is the potential for disease or death (morbidity and mortality, respectively). The hazards to human health are collectively referred to as toxicity in the medical and environmental sciences. Human toxicology is the study of these health outcomes and their potential causes. Likewise, the study of hazards to ecosystems is known as ecological toxicology or simply ecotoxicology, which is further subdivided by discipline, for example, aquatic toxicology and mammalian toxicology.

2.1. Evaluating Hazards

In the United States, as in many countries, the term hazardous waste is a legal term defined by the regulatory agencies. Under the RCRA hazardous waste regulations, the U.S. Environmental Protection Agency (EPA) is principally responsible for the permitting of hazardous waste treatment, storage, and disposal facilities. However, the EPA has authorized most states to operate portions or all their hazardous waste programs.[1] Arguably, the most important definition is found in Section 1004(5) of the RCRA, which describes a hazardous waste as a solid

[1]This differs from Subtitle D, nonhazardous wastes. In fact, the U.S. Congress intended that permitting and monitoring of municipal and nonhazardous waste landfills shall be a state responsibility. Information on the permitting process and on individual landfills must be obtained by contacting the state agencies (and in some states the local health departments) and the local municipality. See: *Managing Non-Hazardous Municipal and Solid Waste (RCRA)*; http://yosemite.epa.gov/r10/owcm.nsf/RCRA/nonhaz_waste. Accessed on June 12, 2010.

TABLE 27.1 Properties of Hazardous Wastes

H1	"Explosive": substances and preparations that may explode under the effect of flame or that are more sensitive to shocks or friction than dinitrobenzene
H2	"Oxidising": substances and preparations that exhibit highly exothermic reactions when in contact with other substances, particularly flammable substances
H3A	"Highly flammable" - liquid substances and preparations having a flashpoint of below 21 °C (including extremely flammable liquids), or - substances and preparations that may readily catch fire after brief contact with a source of ignition, or - gaseous substances and preparations that are flammable in air at normal pressure, or - substances and preparations that are at normal pressure when in contact with water or damp air, or - substances and preparations that, in contact with water or damp air, evolve highly flammable gases in dangerous quantities
H3B	"Flammable": liquid substances and preparations having a flashpoint equal to or greater than 21 °C and less than or equal to 55 °C
H4	"Irritant": noncorrosive substances and preparations that, through immediate, prolonged, or repeated contact with the skin or mucous membrane, can cause inflammation
H5	"Harmful": substances and preparations that, if inhaled or ingested or if penetrated through skin, may involve limited health risks
H6	"Toxic": substances and preparations (including very toxic substances and preparations) that, if inhaled or ingested or if penetrated through skin, may involve serious, acute, or chronic health risks and even death
H7	"Carcinogenic": substances and preparations that, if inhaled or ingested or if penetrated through skin, may induce cancer or increase its incidence
H8	"Corrosive": substances and preparations that may destroy living tissue on contact
H9	"Infectious": substances containing viable micro-organisms or their toxins that are known or reliably believed to cause disease in man or other living organisms
H10*	"Toxic for reproduction": substances and preparations that, if inhaled or ingested or if penetrated through skin, may produce or increase the incidence of nonheritable adverse effects in the progeny and/or of male or female reproductive functions or capacity
H11	"Mutagenic": substances and preparations that, if inhaled or ingested or if penetrated through skin, may induce hereditary genetic defects or increase their incidence
H12	Substances and preparations that release toxic or very toxic gases in contact with water, air, or an acid
H13	Substances and preparations capable by any means, after disposal, of yielding another substance, e.g. a leachate, which possesses any of the characteristics listed above
H14	"Ecotoxic": substances and preparations that present or may present immediate or delayed risks for one or more sectors of the environment

** The Revised European Waste Catalogue (http://eur-lex.europa.eu/LexUriServ/LexUriServ.do?uri=CONSLEG:2000D0532:20020101:EN:PDF; accessed on December 20, 2010) states that "Toxic for reproduction" is considered to be in line with the hazardous property H10 "Teratogenic" in the Hazardous Waste Directive (Council Directive of 12th December 1991 on hazardous waste (91/689/EEC) as amended by the Council Directive 94/31/EEC of 27th June 1994), which sets the framework within Member States of the European Community for controlling hazardous wastes.*

Source: United Kingdom Environment Agency (2003). Interpretation of the Definition and Classification of Hazardous Waste. Technical Guidance WM2, Hazardous Waste Directive Annex III.

waste that may pose a substantial present or potential threat to human health and the environment when improperly treated, stored, transported, or otherwise managed.

The EPA has developed standard approaches and set criteria to determine whether substances exhibit hazardous characteristics [2]. Because RCRA defines a hazardous waste as a waste that presents a threat to human health and the environment when it is improperly managed, the government identified a set of assumptions that would allow for a waste to be disposed if it is not subject to the controls mandated by Subtitle C of RCRA. This mismanagement scenario was designed to simulate a plausible worst case.

Under a worst-case scenario, a potentially hazardous waste is assumed to be disposed along with municipal solid waste in a landfill with actively decomposing substances overlying an aquifer. Hopefully, few wastes would be managed in this manner but a dependable set of assumptions are needed to ensure that the hazardous waste definition was implemented. The Hazardous and Solid Waste Amendments of 1984 established the Toxicity Characteristic Leaching Procedure (TCLP) to provide replicable results for contaminants commonly found in hazardous waste sites (Table 27.2). The specific medium used in the test is dictated by the alkalinity of the waste. The liquid extracted from the waste is analyzed for the 39 listed toxic constituents and the concentration of each contaminant is compared with the TCLP standards specific to each contaminant.

Hazards other than toxicity are also important and regulated accordingly. These include threats to public and personal safety, which may result from a substance's potential to ignite, its corrosiveness, flammability, or explosiveness. A substance may be hazardous to public welfare if it damages property values or physical materials, for example, expressed as its corrosiveness or acidity. The hazard may also relate to environmental quality, such as an ecosystem stress indicated by diminished productivity and biodiversity, loss of important habitats, and decreases in the size of the population of sensitive species. The hazard may be inherent to the substance. But more than likely, the hazard depends on the situation and conditions in which the exposure may occur. The substance is most hazardous when a number of conditions exist simultaneously, such as the hazard to firefighters using water in the presence of oxidizers.

Since 1980, specific methods for testing of hazards are delineated in the EPA publication SW-846, *Test Methods for Evaluating Solid Chemical Methods/Chemical Methods*, which is a compendium of analytical and sampling methods that comply with the RCRA regulations. The SW-846 provides guidance, although not necessarily a legal requirement, when conducting RCRA-related sampling and analysis requirements. This guidance is updated and revised as new information becomes available. It is currently in its third edition as a fully integrated 3500 page manual [3].

2.2. Waste Management Hierarchy

A typical sequence for managing hazardous wastes involves source reduction, recycling, treatment, and disposal [4]. The waste generator is encouraged to begin at the top of the hierarchy and proceed downward, if necessary (obviously, if the waste is eliminated, there is no further need to follow the sequence). In the United States, the term pollution prevention is often preferred to source reduction, and waste minimization is used as an umbrella term for strategies and technologies that use source reduction and recycling techniques (Fig. 27.1).

2.2.1. *Source Reduction*

Reducing the amount or severity of a hazardous waste can be achieved via modifications in process, substitutions in feedstock, improvements in feedstock purity, changes in housekeeping and management practice, increases in the efficiency of equipment, and recycling within

TABLE 27.2 Toxicity Characteristic Chemical Constituent Regulatory Levels for 39 Hazardous Chemicals

Contaminant	Regulatory level (mg L^{-1})	Contaminant	Regulatory level (mg L^{-1})
Arsenic	5.0	*m*-Cresol	200.0
Barium	100.0	*p*-Cresol	200.0
Cadmium	1.0	Cresol	200.0
Chromium	5.0	1,4-Dichlorobenzene	7.5
Lead	5.0	1,2-Dichloroethane	0.5
Mercury	0.2	1,1-Dichloroethylene	0.7
Selenium	1.0	2,4-Dinitrotoluene	0.13
Silver	5.0	Heptachlor (and its hydroxide)	0.008
Endrin	0.02	Hexachloroethane	3.0
Lindane	0.4	Hexachlorobutadiene	0.5
Methoxychlor	10.0	Hexachloroethane	3.0
Toxaphene	0.5	Methyl ethyl ketone	200.0
2,4-D	10.0	Nitrobenzene	2.0
2,4,5 TP (Silvex)	1.0	Pentachlorophenol	100.0
Benzene	0.5	Pyridine	5.0
Carbon tetrachloride	0.5	Tetrachloroethylene	0.7
Chlordane	0.03	Trichloroethylene	0.5
Chlorobenzene	100.0	2,4,5-Trichlorophenol	400.0
Chloroform	6.0	2,4,6-Trichlorophenol	2.0
o-Cresol	200.0	Vinyl chloride	0.2

Source: Resource Conservation and Recovery Act of 1976 regulations. Code of Federal Regulations, Part 40: Protection of the Environment, Sections 261 and 262.11.

a process [5]. In the United States, there is a requirement in the federal environmental regulations that all large hazardous waste generators must have a program at their facility to encourage source reduction and recycling. Other countries, such as Austria, Germany, and Denmark, have initiated direct subsidies to encourage preferable waste-management options.

2.2.2. Recycling

The use or reuse of hazardous waste as an effective substitute, for a commercial product, or as an ingredient or feedstock in an industrial process can be advantageous not only from a public health and environmental standpoint but also from a financial perspective. Recycling includes the reclamation of useful constituent fractions within a waste material or the removal of contaminants from a waste to allow it to be reused.

2.2.3. Disposal and Treatment

Hazardous waste disposal consists of the discharge, deposition, injection, dumping, spilling, leaking, or placing of hazardous waste

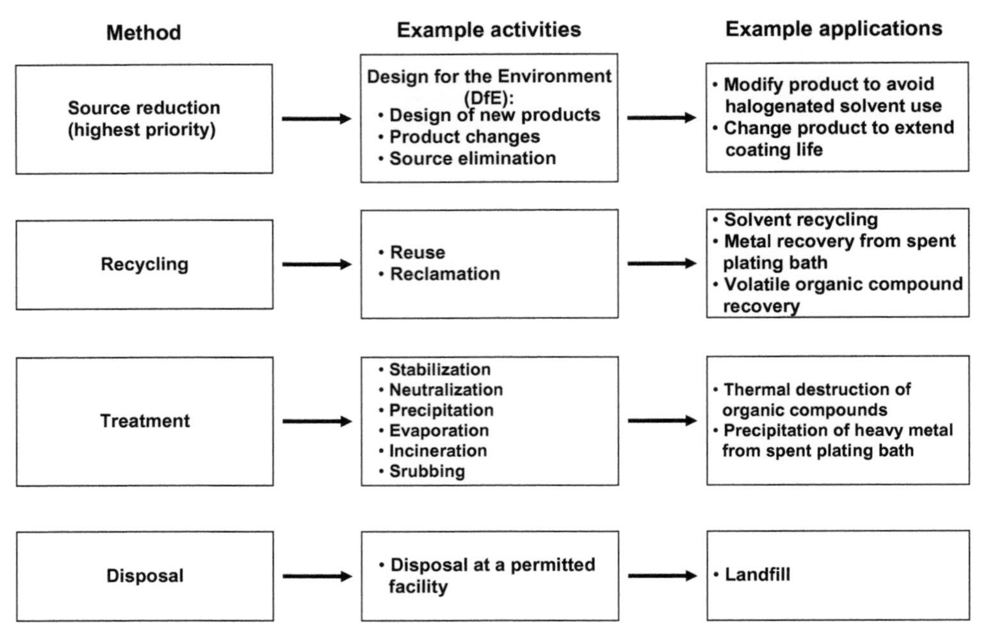

FIGURE 27.1 Flow chart for management of waste: reduction from manufacture, use, treatment, and disposal. *Source: D.A. Vallero (2009). Hazardous wastes. In: McGraw-Hill Encyclopedia of Science and Technology. McGraw-Hill: New York, NY. Adapted from: U.S Environmental Protection Agency (1992). Facility Pollution Prevention Guide, EPA/600/R-92/088, 1992.*

into or on any land or body of water, so that the waste or any constituents may enter the air or be discharged into any waters, including ground water.

The portion of the hazardous waste that cannot be eliminated or recycled must undergo a method, technique, or process that changes the physical, chemical, or biological character of the substance to neutralize it; to recover energy or material resources from the waste; or to render the waste nonhazardous, less hazardous, safer to manage, amenable for recovery, amenable for storage, or reduced in volume.

2.3. Hazard Incident Response

This section stresses the environmental hazards, but other paradigms also exist. For example, in the United States, the Occupational Health and Safety Administration (OSHA) is interested in protecting workers from exposure to hazards at waste sites. This calls for an incident–response model that consists of five components [6]:

1. Recognition
2. Evaluation
3. Information
4. Safety
5. Control

Before entering a potentially dangerous situation at a waste site, one needs to recognize potential hazards, including the physical, chemical, and biological hazards mentioned previously. However, this is not enough for safe entry. An evaluation of the site depends on sound information. This can be drawn from current and previous site investigations; interviews with suppliers, customers, and employees; and scientifically sound sampling and monitoring before, during, and after any incident response. Controls include containment, suppression, and treatment, and depending on the severity of a hazard, for example, immediate evacuation from a potentially explosive situation, or shutoff of a contaminated drinking water supply.

This process may involve either physical or health hazards. Physical hazards include reactivity, fire, explosion, corrosivity, and radioactivity. Health hazards include various types of effects, including acute effects such as poisoning to chronic effects such as cancer or neurotoxicity.

Anyone involved in any emergency or incident response must be properly trained to recognize these hazards. Thus, the EPA and OSHA require completion of an intensive course in hazardous waste operations and emergency response. Employers must provide this training and must ensure that workers be recertified annually.

3. TREATMENT TECHNOLOGIES

Waste-treatment technologies include physical treatment, chemical treatment, biological treatment, incineration, and solidification or stabilization treatment. These processes recycle and reuse waste materials, reduce the volume and toxicity of a waste stream, or produce a final residual material that is suitable for disposal.

The selection of the most effective technology depends on the characteristics of the wastes being treated [7]. Similarly, the characteristics of the media in need of treatment determine the performance of any contaminant treatment or control. For example, if the waste is found in sediment, sludge, slurries, and soil, these characteristics of the media will influence the efficacy of treatment technologies and include particle size, solids content, and high contaminant concentration (Table 27.3).

As is common in most environmental situations, particle size is an important limiting characteristic for applying treatment technologies to porous media. Most treatment technologies work well on sandy soils and sediments. The presence of fine-grained material reduces the effectiveness of treatment system emission controls because it increases particulate generation during thermal drying; it is more difficult to de-water, and it has a greater attraction for the contaminants (especially fine-grained clays). Clayey sediments that are cohesive also present materials handling problems in most processing systems.

Solids content generally range from high (30–60% solids by weight) to low (10–30% solids by weight). Treatment of slurries is better at lower solids contents, but this can be achieved for high solids contents by adding water at the time of processing. It is more difficult to change from a lower to a higher solids content, but evaporative and de-watering approaches, such as those used for municipal sludges, may be used. Also, thermal and dehalogenation processes are decreasingly efficient as the solids content is reduced. More water means increased chemical costs and an increased need for wastewater treatment.

Chemical composition of the wastes must also be considered in the treatment of hazardous wastes. Elevated levels of organic compounds or heavy metals in high concentrations may require special safeguards. Organic matter can be a bonus, given that higher total organic carbon (TOC) content favors incineration and oxidation processes. The TOC can be the contaminant of concern or any organic, because they are combustibles with caloric value. Conversely, higher metal concentrations may make a technology less favorable by increasing contaminant mobility of certain metal species after processing.

Numerous other factors affect the selection of a hazardous waste treatment technology (Table 27.4). For example, vitrification and supercritical water oxidation have only been used for relatively small projects and have not been proven for use in full-scale sediment projects. Regulatory compliance and community perception

TABLE 27.3 Effect of Particle Size, Solids Content, and Extent of Contamination on Decontamination Efficiencies

	Predominant particle size			Solids content		High contaminant concentration	
Treatment technology	**Sand**	**Silt**	**Clay**	**High (slurry)**	**Low (in situ)**	**Organic compounds**	**Metals**
Conventional incineration	N	X	X	F	X	F	X
Innovative incineration	N	X	X	F	X	F	F
Pyrolysis	N	N	N	F	X	F	F
Vitrification	F	X	X	F	X	F	F
Supercritical water oxidation	X	F	F	X	F	F	X
Wet air oxidation	X	F	F	X	F	F	X
Thermal desorption	F	X	X	F	X	F	N
Immobilization	F	X	X	F	X	X	N
Solvent extraction	F	F	X	F	X	X	N
Soil washing	F	F	X	N	F	N	N
Dechlorination	U	U	U	F	X	X	N
Oxidation	F	X	X	N	F	X	X
Bioslurry process	N	F	N	N	F	X	X
Composting	F	N	X	F	X	F	X
Contained treatment facility	F	N	X	F	X	X	X

Note: F, sediment characteristic favorable to the effectiveness of the process; N, sediment characteristic has no significant effect on process performance; U, effect of sediment characteristic on process is unknown; X, sediment characteristic may impede process performance or increase cost.
Source: U.S. Environmental Protection Agency (2003). Remediation Guidance Document, EPA-905-B94-003 Chapter 7.

are always a part of decisions regarding an incineration system. Land use considerations, such as the area needed, are commonly confronted in solidification and solid-phase bioremediation projects, as well as in sludge farming and land application. Disposing of ash and other residues after treatment must be part of any process. Treating water effluent and air emissions must be a part of the decontamination decision-making process.

3.1. Physical Treatment

Processes are available to separate components of a waste stream or change the physical form of the waste without altering the chemical structure of the constituent materials. Physical treatment techniques are often used to separate the materials within the waste stream so that they can be reused or detoxified by chemical or biological treatment or destroyed by high-temperature incineration. These processes are very useful for separating hazardous materials from an otherwise nonhazardous waste stream so that they may be treated in a more concentrated form, separating various hazardous components for different treatment processes, and preparing a waste stream for ultimate destruction in a biological or thermal treatment process.

TABLE 27.4 Factors for Selecting Decontamination and Treatment Approaches

Treatment technology	Implementability at full scale	Regulatory compliance	Community acceptance	Land requirements	Residuals disposal	Wastewater treatment	Air emissions control
Conventional incineration		✓	✓				✓
Innovative incineration		✓	✓				✓
Pyrolysis		✓					✓
Vitrification	✓	✓					✓
Supercritical water oxidation	✓						
Wet air oxidation							
Thermal desorption					✓	✓	✓
Immobilization				✓			✓
Solvent extraction					✓	✓	
Soil washing					✓	✓	
Dechlorination							✓
Oxidation	✓						
Bioslurry process	✓						✓
Composting				✓			✓
Contained treatment facility				✓		✓	✓

Note: ✓, the factor is critical in the evaluation of the technology.
Source: U.S. Environmental Protection Agency (2003). Remediation Guidance Document, EPA-905-B94-003 Chapter 7.

Most integrated waste-treatment systems use physical treatment regardless of the nature of the waste materials or the ultimate technologies used for treatment or destruction. The processes commonly include screening, sedimentation, flotation, filtration, centrifugation, dialysis, membrane separations, ultrafiltration, distillation, solvent extraction, evaporation, and adsorption.

3.1.1. *Physical Processes*

3.1.1.1. HEATING

Thermal technologies are discussed in detail in Chapter 16. Heat changes the characteristics of the waste. Thus, if done properly, the waste will decrease in volume, become less toxic, and have a decreased likelihood of migrating offsite.

3.1.1.2. SOLIDIFICATION AND STABILIZATION

Physical treatment systems can be designed to improve handling and the physical characteristic of the waste, decrease the surface area across which transfer or loss of contained pollutants can occur, and limit the solubility of, or detoxify, any hazardous constituents contained in the wastes. In solidification, these results are obtained primarily, but not exclusively, by producing a monolithic block of treated waste with high structural integrity. Stabilization techniques limit the solubility or detoxify waste contaminants even though the physical characteristics of the waste may not be changed. Stabilization usually involves the addition of materials that ensure that the hazardous constituents are maintained in their least soluble or least toxic form.

3.1.1.3. DISPOSAL

After treatment is completed, an inorganic valueless residue remains that must be disposed of safely. There are five options for disposing hazardous waste. (1) Underground injection wells are steel- and concrete-encased shafts placed deep below the surface of the earth into which hazardous wastes are deposited by force and under pressure. Some liquid waste streams are commonly disposed in underground injection wells. (2) Surface impoundment involves natural or engineered depressions or diked areas that can be used to treat, store, or dispose hazardous waste. Surface impoundments are often referred as pits, ponds, lagoons, and basins. (3) Landfills are disposal facilities where hazardous waste is placed in or on land. Properly designed and operated landfills are lined to prevent leakage and contain systems to collect potentially contaminated surface water runoff. Most landfills isolate wastes in discrete cells or trenches, thereby preventing potential contact of incompatible wastes. (4) Land treatment is a disposal process in which hazardous waste is applied onto or incorporated into the soil surface. Natural microbes in the soil break down or immobilize the hazardous constituents. Land treatment facilities are also known as land-application or land-farming facilities. (5) Waste piles are noncontainerized accumulations of solid, nonflowing hazardous waste. Although some are used for final disposal, many waste piles are used for temporary storage until the waste is transferred to its final disposal site.

Of the hazardous waste disposed on land, nearly 60% is disposed in underground injection wells, approximately 35% in surface impoundments, 5% in landfills, and less than 1% in waste piles or by land application.

3.1.1.4. COMPREHENSIVE TREATMENT FACILITIES

In the United States, Canada, and many western European countries, there are large facilities capable of treating and disposing many types of hazardous wastes. These facilities are able to realize economies of scale by incorporating many treatment processes that might not be economical for individual generators. Comprehensive facilities are also able to exploit the synergistic opportunities made possible by having many different types of waste present at a single site. These include using waste acids and alkalies to neutralize one another; waste oxidants to treat cyanides; organic contaminants in water, salts, and acids to salt out organic compounds from wastewater; onsite incinerators to dispose of organic vapors generated by other onsite processes; ash and calcium and magnesium oxides to aid in stabilization processes; and combustible solids and liquids to produce blended liquid fuels.

3.2. Chemical Treatment

The chemical structure of the constituents of the waste can be altered to produce either an innocuous or a less-hazardous material by taking advantage of chemical reactions. This can be said of many biological processes as well, but so-called chemical treatment processes are understood to be abiotic (that is, chemical reactions anywhere outside a living organism). Four categories of chemical reactions are available to treat hazardous wastes: synthesis or combination, decomposition, single replacement, and double replacement.

3.2.1. *Synthesis or Combination*

In combination reactions, two or more substances react to form a single substance:

$$A + B \rightarrow AB \quad (27.1)$$

Two types of combination reactions are important in environmental systems: formation and hydration. Formation reactions are those in which elements combine to form a compound. An example is the formation reaction of ferric oxide:

$$4Fe(s) + 3O_2(g) \rightarrow 2Fe_2O_3(s) \quad (27.2)$$

Hydration reactions involve the addition of water to synthesize a new compound, for example, in reaction when calcium oxide is hydrated to form calcium hydroxide:

$$CaO(s) + H_2O(l) \rightarrow Ca(OH)_2(s) \quad (27.3)$$

Hydration reactions also occur when phosphate is hydrated to form phosphoric acid:

$$P_2O_5(s) + 3H_2O(l) \rightarrow 2H_3PO_4(aq) \quad (27.4)$$

3.2.2. *Decomposition*

Often referred to as degradation when discussing organic compounds in toxicology, environmental sciences, and engineering, in decomposition reactions:

$$AB \rightarrow A + B \quad (27.5)$$

where one substance breaks down into two or more new substances. For example, in decomposition reaction, calcium carbonate breaks down into calcium oxide and carbon dioxide:

$$CaCO_3(s) \rightarrow CaO(s) + CO_2(g) \quad (27.6)$$

3.2.3. *Single Replacement (Single Displacement)*

This reaction commonly occurs when one reactant replaces another reactant:

$$A + BC \rightarrow AC + B \quad (27.7)$$

For example, a less-toxic ion can be used to replace a more toxic ion, or if the cationic replacement makes collection, disposal, or additional treatment easier. A simple replacement occurs when trivalent chromium replaces monovalent silver:

$$3AgNO_3(aq) + Cr(s) \rightarrow Cr(NO_3)_3(aq) + 3Ag(s) \quad (27.8)$$

3.2.4. *Double Replacement (Metathesis or Double Displacement)*

In metathesis reactions, molecules exchange their cations and anions:

$$AB + CD \rightarrow AD + CB \quad (27.9)$$

These reactions are commonly used to precipitate metals out of solution, such as when lead is precipitated, as in the reaction of a lead salt with an acid such as potassium chloride:

$$Pb(ClO_3)_2(aq) + 2KCl(aq) \rightarrow PbCl_2(s) + 2KClO_3(aq) \quad (27.10)$$

Abiotic chemical processes are attractive because they usually produce minimal air emissions, they can often be carried out on the site of the waste generator, and some processes can be designed and constructed as mobile units. A few commonly applied chemical treatment processes are neutralization, precipitation (e.g., waste steams containing heavy metals), ion exchange, dehalogenation (especially dechlorination and debromination), and oxidation–reduction for detoxifying toxic wastes. Some processes can be categorized as either physical or chemical, such as precipitation and co-precipitation to remove heavy metals from water. Often, such physicochemical processes involvement enhancement of other mechanisms. For example, pH and electronegativity can be enhnanced by adding alum or otherwise changing physical (e.g. temperature) and chemical (e.g. ionic strength) conditions.

3.3. Biodegradation of Hazardous Waste

Environmental biotechnology is the use of microorganisms to decompose organic wastes into water, carbon dioxide, and simple inorganic substances, or into simpler organic substances such as aldehydes and acids. Contaminants, if completely organic in structure are, in theory, fully destructible using microorganisms with the engineering inputs and outputs summarized as:

$$\text{Hydrocarbons} + O_2 + \text{microorganisms} + \text{energy} \rightarrow CO_2 + H_2O + \text{microorganisms} \quad (27.11)$$

Hazardous wastes are mixed with oxygen and aerobic microorganisms, sometimes in the presence of an external energy source in the form of added nutrition for the microorganisms. In time, the byproducts of gaseous carbon dioxide and water are produced, which exit the top of the reaction vessel, while a solid mass of microorganisms is produced to exit the bottom of the reaction vessel.

If the waste contains other chemical constituents, in particular chlorine and/or heavy metals, and if the microorganisms are able to withstand and flourish in such an environment, the simple input and output relationship is modified to reaction (27.12).

$$\begin{aligned}&\text{Hydrocarbons} + O_2 + \text{microorganisms}\\&\quad + \text{energy} + \text{Cl or heavy metal(s)} + H_2O\\&\quad + \text{inorganic salts} + \text{nitrogen compounds}\\&\quad + \text{sulfur compounds} + \text{phosphorus compounds}\\&\quad \rightarrow CO_2 + H_2O + \text{chlorinated hydrocarbons or}\\&\qquad \text{heavy metal(s) inorganic salts}\\&\quad + \text{nitrogen compounds} + \text{sulfur compounds}\\&\quad + \text{phosphorus compounds}\end{aligned} \quad (27.12)$$

If the microorganisms do survive in this complicated environment, the potential exists for the transformation to a potentially more toxic molecule that contains chlorinated hydrocarbons, higher heavy-metal concentrations, and more mobile or more toxic chemical species of heavy metals.

All bioreactor systems have some similar attributes. All rely on a population of microorganisms to metabolize organic contaminants, ideally into the harmless byproducts of CO_2 + H_2O. Reactors also generate heat and other energy byproducts. In all the systems, the microorganisms must be either initially cultured in the laboratory to be able to metabolize a specific organic waste or must be given sufficient time to evolve to be able to digest the contaminant. The engineer must plan for and undertake extensive and continual monitoring and fine-tuning of each microbiological processing system during its operation.

Bioremediation success depends on the growth and survival of microbial populations and the ability of these organisms to come into contact with the substances that need to be degraded into less-toxic compounds. Sufficient numbers of microorganisms must be present and active to make bioremediation successful and the preferred cleanup choice. Also, the microbial environment must be habitable for the microbes to thrive. Concentrations of compounds can be so high that the environment is toxic to microbial populations. Therefore, the design must either offer a method other than bioremediation or modify the environment (e.g., dilution, change in pH, pumped oxygen, adding organic matter) to make it habitable. An important modification is the removal of nonaqueous phase liquids (NAPLs) since the microbes' biofilm and other mechanisms usually work best when the microbe is attached to a particle. Thus, most of the NAPLs need to be removed by vapor extraction.

Physical, chemical, and biological factors drive the selection of bioremediation strategy. For example, low permeability soils, such as clays, are difficult to treat, since liquids (water, solutes, and nutrients) are difficult to pump through these systems. Usually, bioremediation works best in soils that are relatively sandy, allowing mobility and greater likelihood of contact between the microbe and the contaminant. Thus, an understanding of the environmental conditions sets the stage for problem formulation (i.e., identification of the factors at work and the resulting threats to health and environmental quality) and risk management (i.e., what various options are available to address these factors and how difficult it will be to overcome obstacles or to enhance those factors that make remediation successful).

Bioremediation is a process of optimization by selecting options among a number of biological, chemical, and physical factors. These include correctly matching the degrading microbes to conditions, understanding, and controlling the movement of the contaminant (microbial food) so as to come into contact with the microbes, and characterizing the abiotic conditions controlling both these factors. Optimization can vary among options, such as artificially adding microbial populations known to break down the compounds of concern. Compound degradation can be quite specific to microbial genera. Only a few species can break down certain organic compounds. The number of microbe genera that can degrade a molecule decreases with the complexity of the molecule, including the presence of certain functional groups (e.g., Cl and Br substitution).

Two major limiting factors of any biodegradation process are toxicity to the microbial population and inherent biodegradability of the compound. As the biology of microbial biodegradation has become better understood (e.g., mechanisms and modes of action), efficacy has improved. For example, the methanogenic granular sludge structure has been found to play a key role in high-rate anaerobic processes [8]. Granular sludge is an excellent example of a system, because it is an aggregation of several metabolic groups of bacteria living in synergism. The factors that make granular sludge environments hospitable for these bacteria may be replicated in other environments, both in situ and ex situ.

The benefits of biological treatment systems include: (1) the potential for energy recovery; (2) volume reduction of the hazardous waste; (3) detoxification as selected molecules are reformulated; (4) the basic scientific principles, engineering designs, and technologies are well understood from a wide range of other applications including municipal wastewater treatment at facilities; (5) application to most organic contaminants that, as a group, compose a large percentage of the total hazardous waste generated nationwide; and (6) the possibility to scale the technologies to handle a single liter or kilogram or millions of liters or tons of waste per day.

The drawbacks of the biotreatment systems include: (1) the operation of the equipment requires very skilled operators and is more costly as input contaminant characteristics change over time and correctional controls become necessary; and (2) ultimate disposal of the waste microorganisms is necessary and particularly troublesome and costly if heavy metals and/or chlorinated compounds are found during the monitoring activities. With some exceptions, hazardous waste containing heavy metals should not be bioprocessed.

Variations on three different types of bioprocessors are available to treat hazardous wastes: trickling filter, activated sludge, and aeration lagoons.

3.3.1. *Mixed Hazardous Waste Reactors*

In nature, the degradation varies in space and time according to available oxygen, moisture, nutrients, and biota. Thus, natural systems are usually a mixture of aerobic and anaerobic systems. Engineers have applied this to mixed reactor systems, most notably the trickling filter, which is a bed of rocks (~10 cm diameter, but highly variable in size and shape) enclosed in a rectangular or cylindrical structure, through which wastes are carried (Fig. 27.2). Biofilms are selected from laboratory studies and encouraged to grow on the rocks. As the liquid waste moves with gravity downward through the bed, the microorganisms come in contact with the organic contaminant/food source and ideally metabolize the waste into $CO_2 + H_2O$ + microorganisms + energy. Oxygen is supplied by blowers from the bottom of the reactor and passed upward through the bed. The treated waste moves downward through the bed, and subsequently enters a quiescent tank where the microorganisms that are sloughed off the rocks are settled, collected, and disposed. Trickling filters are actually considered mixed-treatment systems because aerobic bacteria grow in the upper, high-oxygen layers of the media, whereas anaerobes grow in the lower regions of the system.

Another mixed-reactor system used to treat hazardous waste is the landfill. In North America, landfill regulations call for a system that minimizes liquid infiltration into the solid waste mass by controlling the amount of moisture allowed into these landfills. A hazardous waste landfill is defined as a disposal facility or part of a facility where hazardous waste is placed in or on land and which is not a pile, a land treatment facility, a surface impoundment, an underground injection well, a salt dome formation, a salt bed formation, an underground mine, a cave, or a corrective action management unit [9]. A hazardous waste landfill is a type of treatment, storage, and disposal facility (TSDF) and as such must be appropriately permitted, and the permit will specify all design and operating practices necessary to ensure compliance. All hazardous waste landfills must include

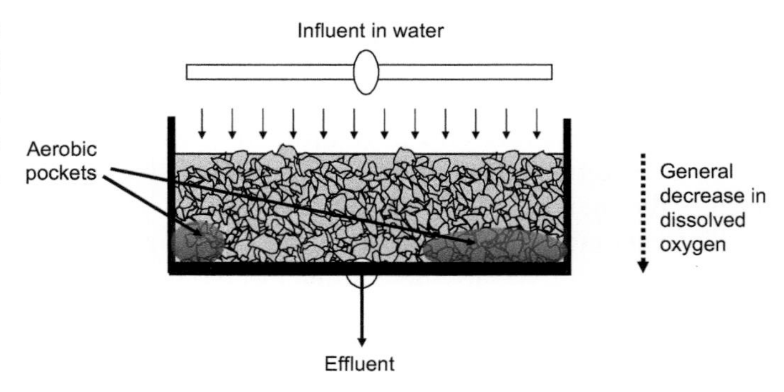

FIGURE 27.2 Trickling filter treatment system, which includes zones of varying concentrations of dissolved oxygen, that is it is a mixed reactor for treating hazardous wastes. *Adapted from: D. Vallero (2003). Engineering the Risks of Hazardous Wastes. Butterworth-Heinemann. Boston, MA.*

a run-on control system, a run-off management system, and a control of the wind dispersal of particulate matter. The capacity of the run-on control system must prevent flow onto the active portion of the landfill during peak discharge of a 25-year storm event. The run-off management system must be able to collect and control water from a 24-h or a 25-year storm and the chemical constituents of the run-off must be tested. The landfill design must be based on this information. In addition to the design, the landfill must be operated to prevent migration of contaminants off-site. For example, holding tanks and basins designed to retain run-on and run-off control systems must be emptied expeditiously after storm events.

The design and operation of landfills must also be properly documented. Each hazardous waste landfill needs a map with the exact topological and dimensional information, for example, depth of each cell with respect to permanently surveyed benchmarks. The contents of each cell and the approximate location of each hazardous waste type within the cell must be recorded.

The contents of a hazardous waste landfull are strictly enforced. For example, bulk or noncontainerized liquid waste or waste containing free liquids are not permitted. The exemption is containers holding free liquids if they meet one of the following standards:

- All free-standing liquids have been removed by decanting, or other methods, mixed with sorbent or solidified so that free-standing liquid is no longer observed, or otherwise eliminated;
- The container is very small, such as an ampule (i.e., hermetically sealed vial);
- The container is designed to hold free liquids for use other than storage (e.g., a battery or capacitor);
- The container is a lab pack.

To dispose the sorbents used to treat free liquids in an hazardous waste landfill, the sorbent must be nonbiodegradable. Approved sorbents include the following:

- Inorganic minerals, other inorganic materials, and elemental carbon;
- High-molecular-weight synthetic polymers, except for polymers derived from biological material or polymers specifically designed to be degradable;
- Mixtures of nonbiodegradable materials.

Unless they are very small, containers must be at least 90% full and crushed, shredded, or similarly reduced in volume to the maximum practical extent.

3.3.1.1. LANDFILL DESIGN CONSIDERATIONS

So-called "dry tomb" landfill designs produce strata within the bioreactor system with low moisture content. Although this decreases the amount of leachate, it also severely limits biodegradation, because moisture is a limiting factor for biofilm production and microbial metabolism. Thus, the likelihood of undegraded contaminants in the treated hazardous waste increases, as does their associated risk since the exposure integration time is protracted. These exposures include emissions from fugitive dust, combustion from the flare, microbial releases, and the migration of contaminants through soil, ground, and surface waters. This is the rationale for long-term monitoring around these sites; for example, current regulations require leachate and gas emissions to be monitored for at least 30 years after closure of a municipal landfill site or even longer if there is a reason to believe that the risks continue after that time, which is likely the case for hazardous waste landfills, because the volume and toxicity of the wastes are likely of more concern than those of a typical municipal solid waste landfill.

The rate of microbial degradation in landfills can be enhanced, especially by increasing the moisture content of the waste. This can be performed by increasing the amount of water in the strata and by leachate recirculation [10]

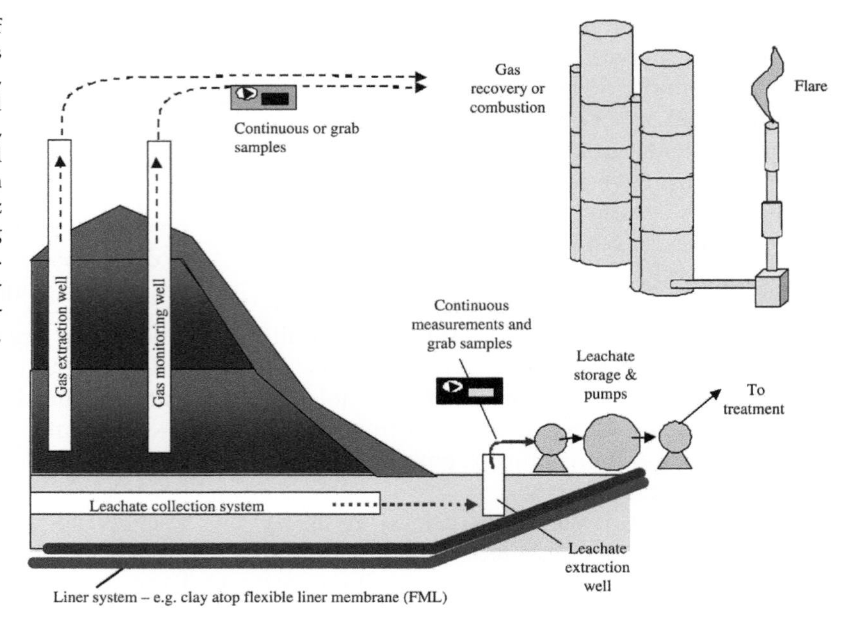

FIGURE 27.3 Schematic of a landfill system showing gas collection and monitoring wells, leachate collection system, and treatment processes. In addition, a hazardous waste landfill will require redundancies and extra protection to prevent any haz ardous substances from migrating off-site, such as barriers and berms. *Source: D.A. Vallero, 2010,* Environmental Biotechnology: A Biosystems Approach. *Academic Press. Amsterdam, NV.*

(the leachate is now collected and treated; see Fig. 27.3).

Waste arrives and is stored in a landfill on an ongoing basis so the age of the waste between and within (e.g., cells within layers) is quite variable. The different landfill stabilization phases often overlap and can be viewed systematically. The initial phase results in aerobic decomposition followed by four stages of anaerobic degradation. Thus, the majority of landfill decomposition by volume occurs under anaerobic conditions. Generally, biodegradation follows three basic stages [11]:

1. The organic material in solid phase (represented by chemical oxygen demand, i.e., COD_S) decays rapidly as larger organic molecules degrade into smaller molecules.
2. These smaller organic molecules in the solid phase undergo dissolution and move to the liquid phase (COD_L), with subsequent hydrolysis of these organic molecules.
3. The smaller molecules are transformed and volatilize as CO_2 and CH_4, with remaining biomass in solid and liquid phases.

During the first two phases, little material volume reaches the leachate. However, the biodegradable organic matter of the waste undergoes rapid reduction. Meanwhile, the leachate COD accumulates as a result of excesses of more recalcitrant compounds compared with the more reactive compounds in the leachate.

These three steps can be further grouped into five phases by which degradation occurs in a landfill bioreactor system, as shown in Fig. 27.4. Successful conversion and stabilization of the waste depends on how well microbial populations function in *syntrophy* (i.e., an interaction of different populations that supply each other's nutritional needs).

PHASE I—INITIAL ADJUSTMENT

In environmental microbiology, this phase is referred to as the lag phase. As the waste is placed in the landfill, the void spaces contain high volumes of molecular oxygen (O_2). As additional wastes are added and compacted, the O_2 content of the landfill bioreactor strata gradually decreases. With increasing moisture,

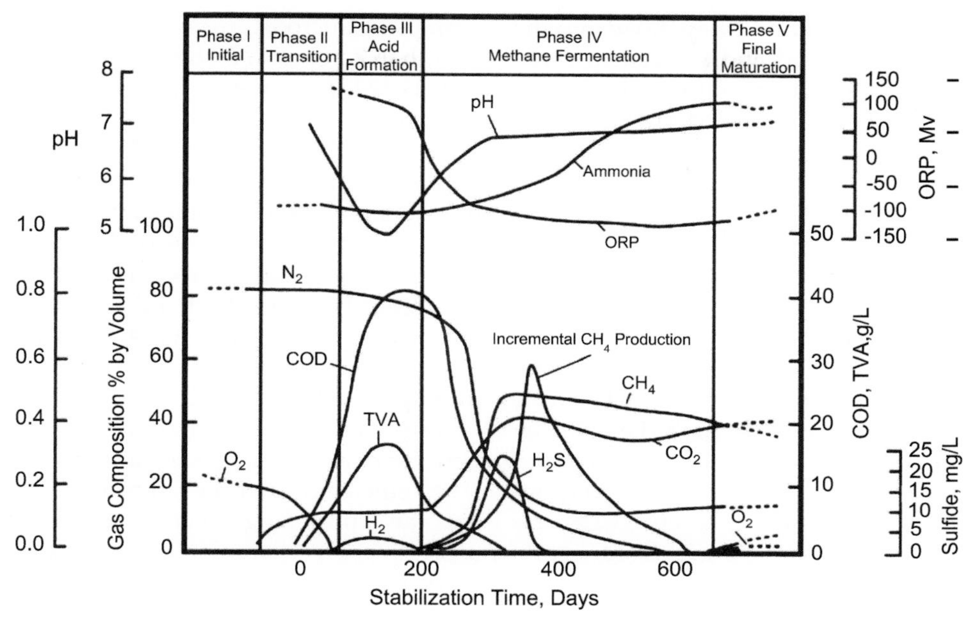

FIGURE 27.4 Phases of waste decomposition in a landfill, showing changes in released compounds and landfill conditions. Note: COD = chemical oxygen demand; TVA = total volatile acids; ORP = oxidation–reduction potential. *Source: U.S. Environmental Protection Agency (2007). National Risk Management Research Laboratory. Landfill Bioreactor Performance: Second Interim Report: Outer Loop Recycling & Disposal Facility - Louisville, Kentucky. Report No. EPA/600/R-07/060. Cincinnati, Ohio.*

the microbial population density increases, initiating aerobic biodegradation, that is, the primary electron acceptor is O_2.

PHASE II—TRANSITION

This phase is short lived as the O_2 is rapidly degraded by the existing microbial populations. The decreasing O_2 results in a transition from aerobic to anaerobic conditions in the stratum. The primary electron acceptors during transition are nitrates and sulfates, since O_2 is rapidly displaced by CO_2 in the effluent gas.

PHASE III—ACID FORMATION

Hydrolysis of the biodegradable fraction of the solid waste begins in the acid formation phase, which leads to rapid accumulation of volatile fatty acids (VFAs) in the leachate. The increased organic acid content decreases the leachate pH from approximately 7.5 to 5.6 [12]. During this phase, the decomposition intermediate compounds such as the VFAs contribute much to the COD. Long-chain volatile organic acids (VOAs) are converted to acetic acid ($C_2H_4O_2$), CO_2, and hydrogen gas (H_2). High concentrations of VFAs increase both the biochemical oxygen demand and VOA concentrations, which initiates H_2 production by fermentative bacteria, which stimulates the growth of H_2-oxidizing bacteria. The H_2 generation phase is relatively short because it is complete by the end of the acid-formation phase.

As seen in Fig. 27.4, this phase is also accompanied by an increase in the biomass of acidogenic bacteria and rapid degradation of substrates and consumption of nutrients. Because the aqueous solubility of metals generally increases with decreasing pH, metallic compounds may become more mobile during this phase. This can be either beneficial or problematic. For example, the more mobile compounds may move into the leachate and be collected and segregated more easily. Or, the more mobile compounds may be highly toxic and, if not collected and treated properly, migrate off-site, contaminating soil, aquifers, and surface waters.

PHASE IV—METHANE FERMENTATION

The acid-formation phase intermediary products (e.g., acetic, propionic, and butyric acids) are converted to CH_4 and CO_2 by *methanogenic* microorganisms. As VFAs are metabolized by the methanogens, the landfill water pH returns to neutrality. The organic strength (i.e., oxygen demand) of the leachate decreases at a rapid rate as gas production increases in correspondence with increases in CH_4 and CO_2 gas production. This is the longest-lived waste decomposition phase.

PHASE V—FINAL MATURATION AND STABILIZATION

The rate of microbiological activity slows in the last phase of waste decomposition in conjunction with nutrient limits, for example, bioavailable phosphorus. CH_4 production almost completely disappears, with O_2 and oxidized species gradually reappearing in the gas wells as O_2 permeates from the troposphere. This changes the oxidation—reduction potential of leachate, making it more oxidative. The residual organic materials may incrementally be converted to the gas phase and composting of organic matter to humic-like compounds, although this has not yet been scientifically documented.

On-site degradation is carried out by different genera of microbes, making the kinetics difficult to predict for any facility. Such microbial populations occur naturally in soil (e.g., natural soil bacteria). However, varying temporal and spatial site-specific conditions to achieve optimal performance is often a heuristic process. The kinetics of microorganisms in landfill bioreactors have not been widely investigated, likely because these groups of microorganisms are much more difficult than aerobes to culture. Advances in molecular-based non-culture techniques will likely advance the quality of this information. To date, landfill bioreactor performance has been evaluated and controlled by indirect evidence, for example, waste stabilization is characterized by monitoring the outcome of the decomposition process. For example, recirculation of leachate has been shown to shorten the initial lag phase, even though the microbial species responsible for this accelerated decomposition have not been precisely identified [13].

Improvements to land-based bioreactors require attention to the sensitivity of variables. The efficiency can be improved depending on the optimizing microbiological metabolic processes. For example, the amount of leachate to be recirculated affects the quantity of organic acids. If VFA concentrations are high, methanogenesis can be inhibited. This is not a direct inhibition of microbial metabolism, but a response of the microbial population to the lower pH induced by the VFAs. Thus, the volume of recirculated leachate must be adjusted to minimize the accumulation of VFAs (see Fig. 27.5).

Regulatory agencies often require two or three pairs of systems, for example, for leachate collection and treatment, as well as other design redundancies to protect the integrity of a hazardous waste storage or treatment facility. For example, the primary leachate collection and treatment system must be designed like the bottom of the landfill bathtub. This leachate collection system must be graded to promote the flow of liquid within the landfill from all points in the landfill to a central collection point where the liquid can be pumped to the surface for subsequent monitoring and treatment. Crushed stone and perforated pipes are used to channel the liquid along the top layer of this compacted clay liner to the pumping locations.

Thus, effective degradation depends on matching the abiotic and biotic conditions in a bioreactor with the needs of the microbial

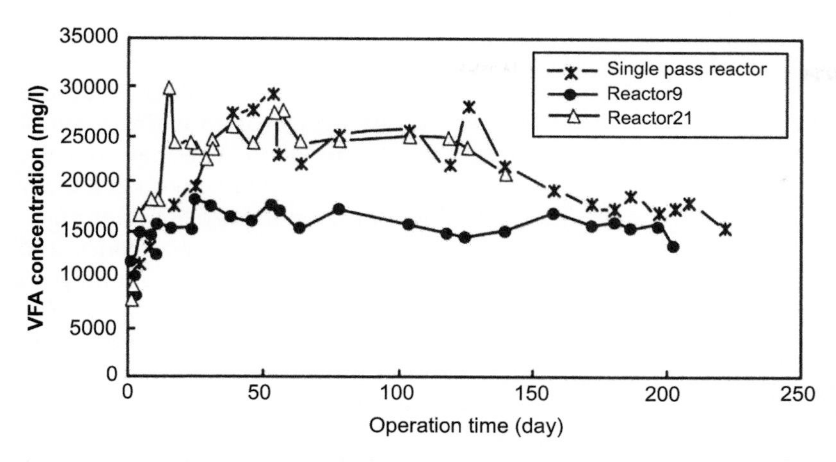

FIGURE 27.5 Effects of leachate recirculation on volatile fatty acid (VFA) accumulation compared with VFA generation under conventional landfill management. The leachate in "Reactor 9" was recirculated at 9 L d^{-1} (or 2.4 gal d^{-1}) = 13% of the reactor volume. The leachate in "Reactor 21" was recirculated at 21 L d^{-1} = 30% of the reactor volume. The VFA buildup in the reactor with a higher leachate recirculation rate of 21 L d^{-1} was nearly as high as the VFAs generated in the single-pass (i.e., conventional) reactor. The 21 L d^{-1} bioreactor experienced a spike of 30,000 mg L^{-1} of VFA within 30 days, which can be detrimental to methanogenic bacteria. *Sources: U.S. Environmental Protection Agency (2007). National Risk Management Research Laboratory. Landfill Bioreactor Performance: Second Interim Report: Outer Loop Recycling & Disposal Facility - Louisville, Kentucky. Report No. EPA/600/R-07/060. Cincinnati, Ohio; and D. Sponza and O. Agdag (2004). Impact of leachate re-circulation and re-circulation volume on stabilization of municipal solid wastes in simulated anaerobic bioreactors," Process Biochemistry, 39(12): 2157-2165.*

population. Biotreatment of hazardous waste must optimize these variables, which calls for very careful and accurate monitoring, operation, and maintenance of land-based reactors.

3.3.2. *Aerobic Hazardous Waste Reactors*

The presence and quantity of oxygen are always key factors in designing hazardous waste treatment systems. Numerous hazardous wastes are best degraded under aerobic, that is, excess oxygen conditions. In an activated sludge system, microorganisms are recycled within the system (Fig. 27.6). This reuse enables the microorganisms to evolve over time and adapt to the changing characteristics of the hazardous waste.

In an activated sludge system, a tank of liquid hazardous waste is injected with a mass of microorganisms. Oxygen is supplied through pipes in the aeration basin as the microorganisms come in contact, become adsorbed, and metabolize the waste. For complete and ideal conditions, this would convert all organic matter into CO_2 + H_2O + microorganisms + energy. The heavy (well-fed) microorganisms then flow into a quiescent tank in which they settle by gravity and are collected and disposed. Depending on the operating conditions of the facility, some or many of the settled (now hungry) and active microorganisms are returned to the aeration basin where they feed again. Liquid effluent from the activated-sludge system may require additional microbiological and/or chemical processing before release into a receiving stream or sewer system.

Aeration lagoons or ponds treat liquid and dissolved contaminants for long terms, from months to years (Fig. 27.7A). Persistent organic molecules, those not readily degraded in trickling filter or activated sludge systems, are potentially broken down by certain microbes into CO_2 + H_2O + microorganisms + energy, if given enough time. The ponds are open to the weather and ideally oxygen is supplied directly to the microorganisms from the atmosphere. Their dimensions vary but some general configurations based on laboratory experiments and pilot studies include:

- Design: Pond size: 1 to 8 ha (0.45 to 20 acres);
- Design: Pond depth: 0.3 to 9 m (1 ft to 30 ft);
- Design: Detention time = days to months to possibly years;

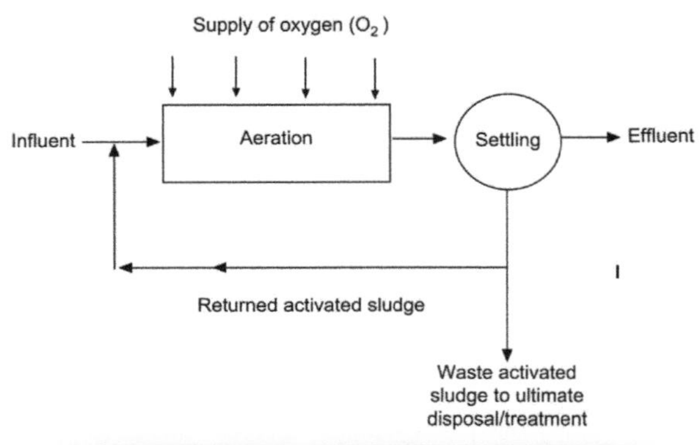

FIGURE 27.6 Aerobic treatment approach to degrade hazardous waste. In this activated sludge treatment system, the waste is combined with recycled biomass and aerated to maintain a target dissolved oxygen (DO) content. Organisms use the organic components, expressed as biochemical oxygen demand (BOD) of waste as food, decreasing the organic levels in the wastewater. Oxygen concentrations must be controlled to maintain optimal treatment efficiencies. One means of achieving optimal DO content is tapered aeration. The tapered system provides high concentrations of oxygen near the influent to accommodate the large oxygen demand from microbes as waste is introduced to the aeration tank (photo). *Adapted from: D. Vallero (2003). Engineering the Risks of Hazardous Wastes. Butterworth-Heinemann, Boston, MA; photo courtesy of D.J. Vallero.*

- Operation: In series with other treatment systems, other ponds, or not;
- Operation: The flow to the pond is either continuous or intermittent;
- Operation: The supply of additional oxygen to the system through blowers and diffusers may be required (i.e., active systems).

A similar aerobic system is slurry-phase lagoon activation, where air and contaminated soil contact one another to promote biodegradation. Such a system can treat an entire batch of sludge in a single operation. These are usually less than 0.8 ha (2 acres) in area, with geometry and depth depending on the type of liner material, as well as the sludge characteristics and thickness. Larger systems require sectioning off into smaller lagoon compartments (Fig. 27.7B). Total solids content ranges from 5% to 20% [14].

Slurry systems are also used ex situ in tanks in which higher degradation rates and greater engineering controls are needed (e.g., to treat soils and other media contaminated with very toxic and/or particularly recalcitrant organic contaminants). These stirred-tank bioreactors can receive sequenced batches or continuously fed sludge.

Another pond-like biological treatment system (Fig. 27.7C) is the engineered and constructed wetland, which operates as a biofilter. Wetlands naturally contain diverse and abundant microbial populations. These features are combined with growth, nutrient extraction, photosynthesis, and ion-exchange processes of plants (Fig. 27.8). For example, gasoline-contaminated groundwater has been bioremediated using a radial-flow constructed wetland system in Casper, Wyoming. In this system, subsurface beds of crushed concrete reclaimed from a closed refinery were insulated with 15-cm layers of mulch. Above these layers, bulrushes, switch grass, and cordgrass were planted in separate sections of the wetland system. About 2650 m^3 (700,000 U.S. gallons) of water are passively treated each day, with large reductions in benzene and other hydrocarbons. The rock/mulch insulated layers allow biological activity to continue throughout the year, even in the cold Wyoming winters. The passive system is also cost effective (estimated construction costs were about 20% of a pump and treat system that can meet similar criteria, e.g., air stripping and catalytic oxidation) [15].

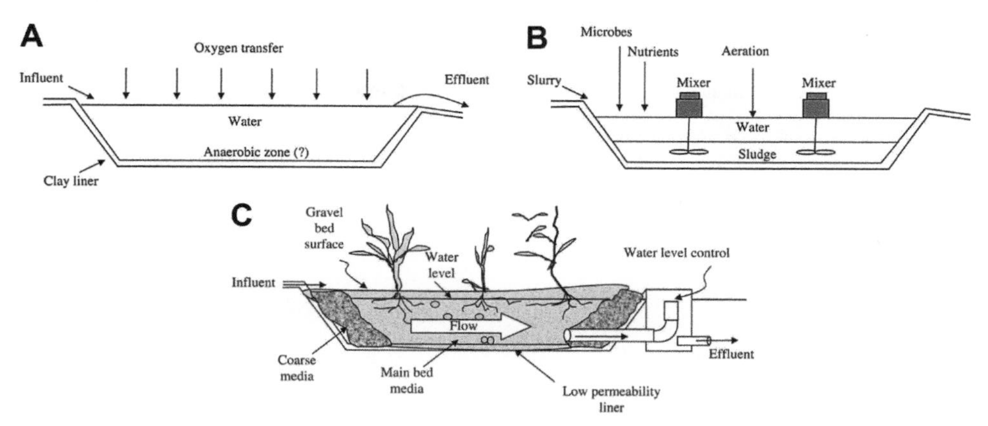

FIGURE 27.7 (A) Passive system: aeration pond. (B) Active system: in situ slurry-phase lagoon. (C) Combined active and passive system: engineered and constructed wetland. All three systems make use of microbes, nutrients, and oxygen, but the active system increases the contact between the microbes and organic matter by mechanically mixing the sludge layer (which contains the organic contaminants to be degraded) with the water in the lagoon. The engineered wetland also incorporates plant processes in the degradation process. *Sources: D.A. Vallero, 2010,* Environmental Biotechnology: A Biosystems Approach. *Academic Press. Amsterdam, NV. U.S. Environmental Protection Agency (1993). Pilot-scale demonstration of slurry-phase biological reactor for creosote-contaminated soil: Applications analysis report. Report No. EPA/540/A5-91/009. Cincinnati, Ohio; and S. Wallace (2004). Engineered wetlands lead the way. Land and Water. 48(5);http://www.landandwater.com/features/vol48no5/vol48no5_1.html. Accessed on September 9, 2009.*

FIGURE 27.8 Engineered and constructed wetland system in San Diego, California, utilizing macrophytes, that is water hyacinths (*Eichhornia spp.*). The system takes advantage of both microbial and larger plant species (bioremediation and phytoremediation, respectively). *Source: C.P Gerba, I.L. Pepper (2009). Wastewater treatment and biosolids reuse. In: R.M. Maier, I.L. Pepper, C.P. Gerba (2009). Environmental Microbiology. 2nd Edition. Elsevier Academic Press. Burlington, Massachusetts.*

The crucial engineering concerns in the design and operation of ponds and other environmental biotechnologies are the identification and maintenance of microbial populations that metabolize the specific contaminant of concern. Once selected, the conditions most favorable to the microbial populations can be determined and controlled.

The fluid dynamics and biological principles of the slurry-phase lagoon (Fig. 27.7B) and the engineered wetland (Figs. 27.7C and 27.8) can be put to use to treat air pollutants in a simple biofilter (Fig. 27.9). However, polluted air rather than water is pumped into the system and mixed with water and pushed into the bottom of a 1-m deep trench covered with unconsolidated material (soil or compost).[2] The saturated air and water containing the organic contaminants

[2]Note that highly hazardous wastes often will not be allowed to be treated in these open systems, without very strong containment systems surrounding them in all directions, that is, including the need for covers and collectors for any volatilized or aerosolized material emitted to the air. In addition to scientific and engineering constraints, legal restrictions and public perception may prevent in situ treatment.

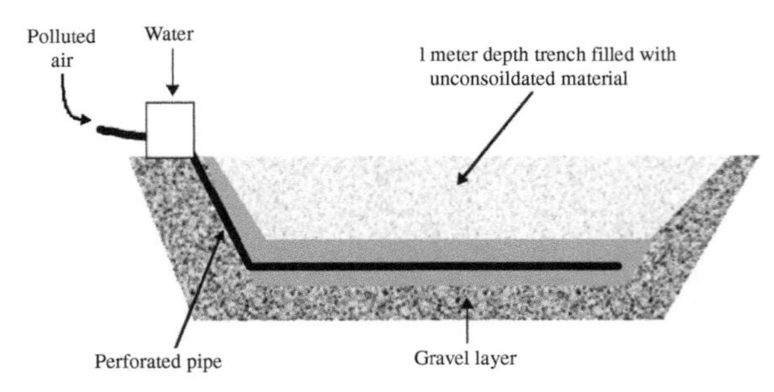

FIGURE 27.9 Biofilter system used to treat air contaminated with organic pollutants. *Adapted from: A. Scragg (2004). Environmental Biotechnology. 2nd Edition. Oxford University Press, Oxford, UK.*

move into the gravel and come in contact with microbial biofilm. The vapor pressure allows the more volatile compounds to move first through the media, but with time, less-volatile compound vapors will percolate through the unconsolidated material. During percolation, the air contacts the biofilm of microbes sorbed to the particles, which degrade the organic contaminants. A more intricate version is the packed bed biological control system to treat volatile compounds anaerobically (discussed in the next section).

3.3.3. *Anaerobic Hazardous Waste Reactors*

Anaerobic and aerobic processes frequently occur in different parts of the same system, as mentioned in the trickling filter discussions. On the negative side, one of the challenges of an aerobic system is to keep aerobes in contact with molecular oxygen. If the oxygen level of the reactor drops or if the tank is not completely mixed, pockets of anoxic and reduced conditions can lead to localized anaerobic conditions within the bioreactor. This is unacceptable if the bioreactions depend solely on oxygen as the electron acceptor. In fact, the foul smells from wastewater treatment facilities can usually be attributed to some part of the plant or the receiving water "going anaerobic," meaning that sulfur compounds are being reduced to odiferous forms, for example, hydrogen sulfide and mercaptans. The benefit of anaerobic conditions is that they can be used to degrade otherwise recalcitrant organic compounds (Fig. 27.10).

Anaerobic degradation consists of a series of steps whereby polysaccharides, proteins, fats and other complex polymeric materials are hydrolyzed by the microbes. These microbial reactions, catalyzed by enzymes, generate products with greater aqueous solubility. The new compounds are secreted by the microorganisms into the biofilm where they can be transported, predominantly by advection, diffusion, and receptor binding,[3] across cellular membranes.

Ultimately, if the anaerobic biodegradation is successful, the production and ensuing escape of methane indicates that the organic material has been stabilized and degraded. The design of anaerobic systems needs to ensure that the retention time of the solids is sufficient for contact and reaction by the microbes with the substrate [16]. For example, if the input of relatively easily degradable organics is too rapid, this can lead to acidic and toxic conditions, with the buildup of organic acids, which can foul the reactor by inhibiting methanogenesis (i.e., instead of reaching the desired methane–water products, the system is stuck in the acid production steps).

[3]Advective transport includes the transport with the flow of the mass, that is, in this instance the water flowing into and through the cell membrane. Diffusion is the molecular transport, that is, a chemical compound in the waste reaches the biofilm from which its molecular concentration is higher than that of the cell membrane and enters the cell. However, the actual degradation of the chemical is by receptor binding. In this case, the chemical (known as the ligand) is usually aided by a transport protein (enzyme–biological catalyst) that is specific to a particular class of chemicals. The enzyme–ligand complex binds with a receptor on the cell membrane and is transported into the cell. This process can be likened to a key (ligand) that fits into a lock (receptor), which allows the chemical to enter the cell.

FIGURE 27.10 Anaerobic treatment system. The lighter substances have migrated to the surface. These may be fats that have been separated physically during the treatment process, or bubbles from gases, such as methane, that are produced when the anaerobes degrade the wastes. Note that although this is an anoxic chamber, a thin film layer at the surface will be aerobic because it is in contact with the atmosphere. Photo courtesy of D.J. Vallero.

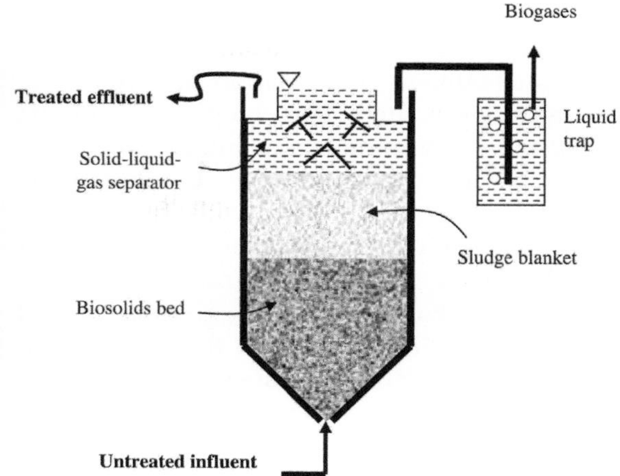

FIGURE 27.11 Schematic of upflow anaerobic sludge blanket system. *Adapted from D.R. Christensen, J.A. Gerick, J.E. Eblen (1984). Design and operation of an upflow anaerobic sludge blanket reactor. Journal of the Water Pollution Control Federation, 56 (9): 1059–1062.*

Acclimatization of the microorganisms to a substrate can take several weeks. The biochemodynamics of the bioreactor affect anaerobic digestion rates, including temperature, pH, and concentration of toxic substances. The conventional anaerobic treatment process consists of a reactor containing waste and biosolids (sludge containing large microbial populations). As in the aerobic, activated sludge systems, these biosolids can be added continuously or in semi-batches, whereupon they are mixed in the bioreactor. In theory, the anaerobic digester is a once-through, completely mixed, reactor. If so, the hydraulic retention time (HRT) equals the solids retention time (SRT). This means that efficiency depends directly on the contact with the biosolids, that is, the SRT. However, HRT and SRTcan be decoupled for increased bioremediation efficiency. For example, an anaerobic upflow filter can substantially improve anaerobic degradation volumes and rates, because the filter can catch and sustain high concentrations of biosolids. By holding the solids, attenuated SRTs allow for much larger throughput which is needed to degrade low-strength organic wastes under feasible, environmental conditions (e.g., ambient temperatures and barometric pressures).

A bed upflow reactor is another anaerobic system, but depends on the sorption of biomass on the surfaces of media. This is done by passing liquid solutions of the organic compounds to be treated upward through a bed of sand-sized particles at a velocity necessary to fluidize and partially expand the sand bed. More recently, the upflow anaerobic sludge blanket process (UASB) takes advantage of the inherent properties of flocculation and settling of anaerobic sludge, allowing for much higher HRT loadings and partitioning of gases (e.g., H_2, CH_4) from the sludge solids. The UASB is based on two biochemodynamic processes (see Fig. 27.11): separation of solids and gases from the liquid; and degradation of biodegradable organic matter. If heavy mechanical agitation is avoided, unlike the other aerobic digesters, a separate settler with biosolids return pump is not needed, so reactor volume can flow consistently through the systems. Unlike the fluidized bed reactor, high-rate effluent recirculation and pumping are also eliminated. The biogas production enhances continuous contact between substrate and anaerobes. The UASB reactor, under optimal conditions, can be assumed to be a completely mixed reactor. The contact of the biosolids with the incoming organic liquids is enhanced by the agitation caused by the release of gases from the biodegradation, along with an inlet that evenly distributes incoming materials in the lower level of the reactor.

Because biological treatment systems do not alter or destroy inorganic substances, and high concentrations of such materials can severely inhibit decomposition activity, chemical or physical treatment may be required to extract inorganic materials from a waste stream before biological treatment.

4. ABANDONED DISPOSAL SITES

Many countries have sites where hazardous waste has been disposed improperly in the past and where cleanup operations are needed to restore the sites to their original state. In the United States, about 20,000 of these sites have been identified and approximately 2000 sites require immediate action. In the United States, these sites are treated differently from those regulated under RCRA. This is because there is often no clear means of removing wastes or remediating problems (e.g., the ownership is unknown, the former owner is bankrupt, or there are numerous potentially responsible parties). The Comprehensive Environmental Response, Compensation and Liability Act of 1980 (better known as Superfund) was enacted in response to the discovery of toxic waste dumps in the 1970s [17]. The law allows the EPA to clean up abandoned sites and to compel responsible parties to perform cleanups or reimburse the government for cleanups [18]. Many European sites have also been addressed. Denmark, the Netherlands, and Sweden have listed sites requiring immediate attention.

Environmental regulations lay out explicit steps for cleaning up contamination (Fig. 27.12). The first step of a contaminant cleanup is a preliminary assessment (PA). During the PA of a site, readily available information about a site and its surrounding area are collected to determine the threat to human health and the environment. Any possible emergency response actions may also be identified. If information beyond the preliminary data in the PA is needed, a site inspection is performed.

Cleaning up abandoned disposal sites involves isolating and containing contaminated material, removing and redepositing contaminated sediments, and in situ and direct treatment of the hazardous wastes involved. The same physical, chemical, and biological treatment technologies discussed previously are applied to remediating abandoned sites.

4.1. Ex Situ and In Situ Treatment

Contaminated soil and sediment must often be removed and treated off-site (ex situ). Contaminated soil may be excavated and transported to kilns or other high-temperature operations, where the contaminated soil is mixed with combustible material. High-sand-content soils may be used as part of an asphalt mix. Contaminated soil may also be distributed onto an impermeable surface, allowing the more volatile compounds to evaporate. Microbial biodegradation can be accelerated and enhanced by adding nutrients and moisture to the soil. A faster process, thermal desorption, entails heating the soil to evaporate the contaminants and capture the compounds, and then burn them in a vapor-treatment device.

Generally, groundwater is treated by drilling recovery wells to pump contaminated water to the surface. Commonly used groundwater

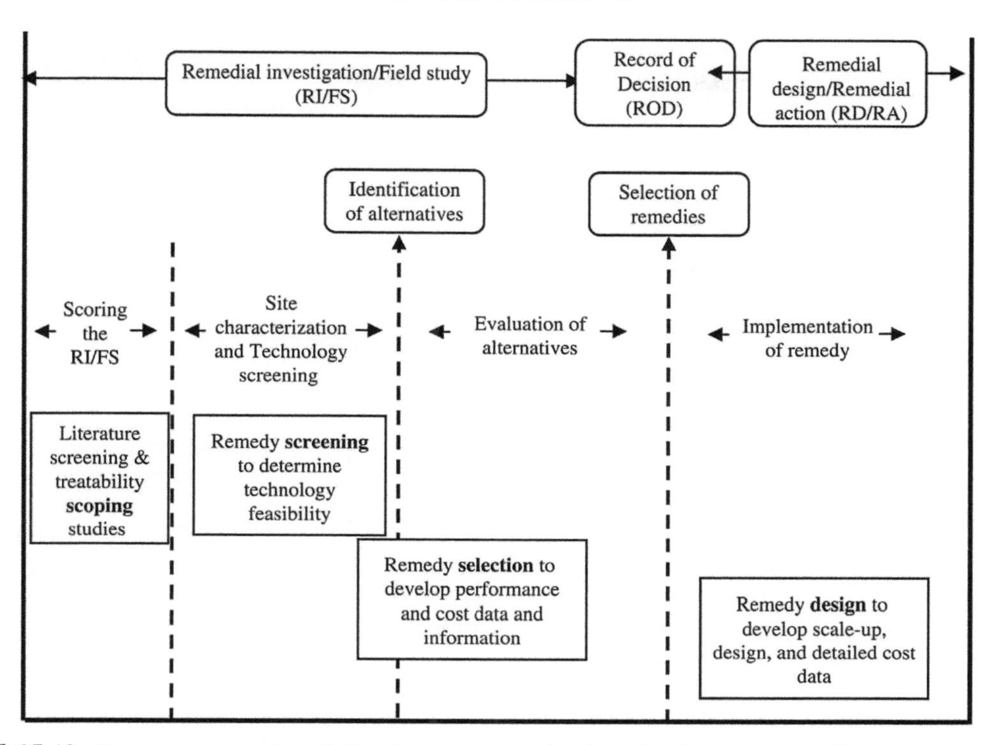

FIGURE 27.12 Steps in a contaminated site cleanup, as mandated by the Comprehensive Environmental Response, Compensation and Liability Act of 1980 (Superfund). *Source: U.S. Environmental Protection Agency, 1992, Guide for Conducting Treatability Studies under CERCLA: Thermal Desorption, EPA/540/R-92/074 B.*

treatment approaches include air stripping, filtering with granulated activated carbon (GAC), and air sparging. Air stripping transfers volatile compounds from water to air. Groundwater is allowed to drip downward in a tower filled with a permeable material through which a stream of air flows upward. Another method bubbles pressurized air through contaminated water in a tank. Filtering groundwater with GAC entails pumping the water through the GAC to trap the contaminants. In air sparging, air is pumped into the groundwater to aerate the water. Most often, a soil venting system is combined with an air sparging system for vapor extraction.

Hazardous wastes treated where they are found without first removing them is known as in situ remediation. Bioremediation makes use of living microorganisms to break down toxic chemicals or to render them less hazardous. This is often done by using bacteria, and in some instances algae and fungi, which are already living in the soil, sediment, or water. These microbes are exposed to incrementally increasing amounts of the chemical, so that they adapt to using the chemical as an energy (food) source. This process is known as acclimation. The acclimated microbes can then be taken from the laboratory and applied to the waste in the field. The most passive form of bioremediation is natural attenuation, where no direct engineering intervention is used, and the contaminants are allowed to be degraded by resident microbes over time. The only role for the engineer is to monitor the soil and groundwater to measure the rate at which the chemicals are degrading. Natural attenuation can work well for compounds that are found in the laboratory to break down under the conditions found at the site. For example, if a compound is degraded under reduced, low pH conditions in the laboratory, it may also degrade readily in soils with these same conditions (e.g., in deeper soil layers where bacteria have adapted to these conditions naturally). Natural attenuation can sometimes be enhanced, for example, when soil moisture levels are increased.

In situ treatment can be a combination of numerous technologies. For example, the leachate collection systems of a hazardous waste landfill need to be treated provide a way to collect wastes that can then be treated. This is often via "pump and treat" systems, which can produce air pollutants (Fig. 27.3). Leachate and ground water can be treated by drilling recovery wells to pump contaminated groundwater to the surface. Commonly used groundwater treatment approaches include air stripping, filtering with GAC, and air sparging. Air stripping transfers volatile compounds from water to air (Fig. 27.13).

Ground water is allowed to drip downward in a tower filled with a permeable material through which a stream of air flows upward. Another method bubbles pressurized air through contaminated water in a tank. The air leaving the tank (i.e., the off-gas) is treated by removing vapor-phase pollutants. Filtering groundwater with GAC entails pumping the water through the GAC to trap the contaminants. In air sparging, air is pumped into the ground water to aerate the water. Usually, a soil-venting system is combined with an air-sparging system for vapor extraction, with the gaseous pollutants treated, as in air stripping.

Certainly, microbial activity can even enhance the treatment of vapor phase compounds, for example, the volatile organic compounds (VOCs) such as benzene, vinyl chloride, and acrolein (propenal), in wastes. In any biological system, biofilms of microorganisms

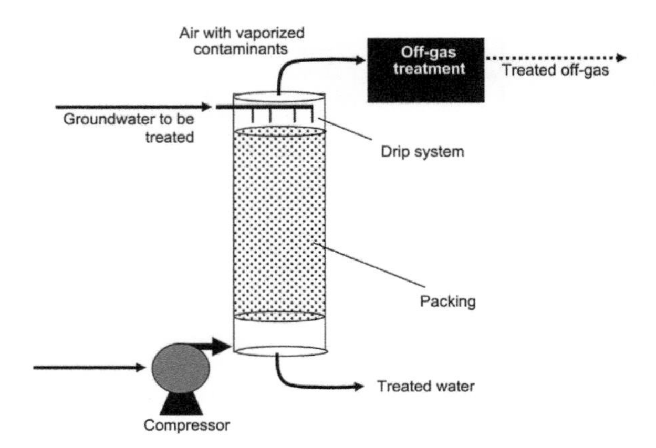

FIGURE 27.13 Schematic diagram of air-stripping system to treat volatile compounds in water.

(bacteria and fungi) are grown on porous media. The air or other gas containing the VOCs is passed through the biologically active media, where the microbes break down the compounds into simpler compounds, eventually to carbon dioxide (if aerobic), methane (if anaerobic), and water. The major difference between biofiltration and trickling systems is how the liquid interfaces with the microbes. The liquid phase is stationary in a biofilter (see Fig. 27.14), but liquids move through the porous media of the system (i.e., the liquid "trickles" as it does in Fig. 27.2). A particularly novel biotechnological method in biofiltration (see Fig. 27.15) uses compost as the porous media. Compost contains numerous species of beneficial microbes that are already acclimated to organic wastes. Industrial compost biofilters have achieved removal rates at the 99% level. Biofilters are also the most common method for removing VOCs and odorous compounds from air streams.

Success is highly dependent on the degradability of the compounds present in the air stream, their fugacity, and solubility needed to enter the biofilm and pollutant-loading rates. Care must be taken in monitoring the porous media for incomplete biodegradation, the presence of substances that may be toxic to the microbes, excessive concentrations of organic

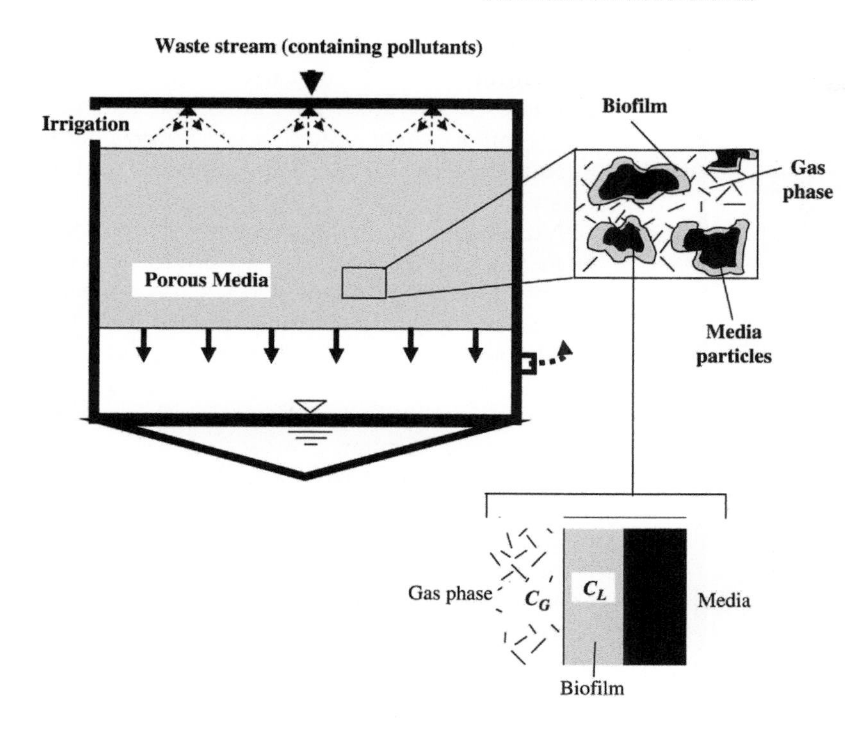

FIGURE 27.14 Schematic representation of packed bed biological control system to treat volatile compounds. Air containing gas-phase pollutants (C_G) traverse porous media. The soluble fraction of the volatilized compounds in the air stream partition into the biofilm (C_L) according to Henry's Law: $C_L = \frac{C_G}{H}$; where H is the Henry's Law constant. *Source: D.A. Vallero. Fundamentals of Air Pollution, 4th Edition, Academic Press, Burlington, MA, 2007; adapted from S.J. Ergas, S and K.A. Kinney. "Biological Control Systems" in: Air and Waste Management Association, Air Pollution Control Manual, 2nd Edition, W. T. Davis (ed.), John Wiley & Sons, Inc., New York, pp. 55–65, 2000.*

acids and alcohols, and pH. The system should also be checked for shock and the presence of dust, grease, or other substances that may clog the pore spaces of the media [19].

4.2. Phytoremediation

Plants are also used in hazardous waste cleanup. Phytoremediation is bioremediation by way of plant life. It is usually in situ and is almost always dependent of available air (i.e., it is almost exclusively an aerobic process).

Phytoremediation takes advantage of plants' absorption of CO_2 for photosynthesis, the process whereby plants convert solar energy into biomass and release O_2 as a byproduct. Thus, the essential oxygen is actually the waste product of photosynthesis and is derived from carbon-based compounds. Respiration generates carbon dioxide as a waste product of oxidation that takes place in organisms, so there is a balance between green plants' uptake of CO_2 and release of O_2 in photosynthesis and the uptake of O_2 and release of CO_2 in respiration by animals, microbes, and other organisms.

Phytoremediation [21] can be used for a wide range of contaminants and soil types. It is frequently used to remediate metal-contaminated sites (e.g., nickel and its compounds). Phytoextraction (or phytoaccumulation) involves the uptake of contaminants from soil by roots into the transfer, that is, translocation, to above-ground portions of the plants. Various species of plants, depending on the contaminant, soil, climate, and local conditions are planted and grown. Some are harvested similar to crops (e.g., grasses) and others maintained for removal and sequestration of the contaminants for longer time periods. The harvested materials can be treated and recycled by various methods, including composting and other bioremediation approaches and thermal processes, depending

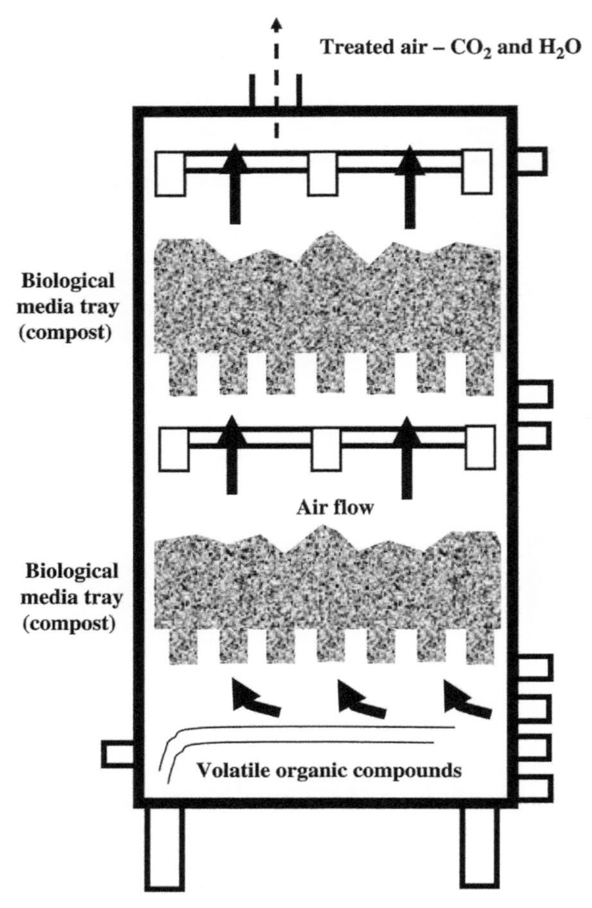

FIGURE 27.15 Biofiltration without a liquid phase used to treat vapor-phase pollutants. Air carrying the volatilized contaminants upward through porous media (e.g., compost) containing microbes acclimated to break down the particular contaminants. The wastes at the bottom of the system can be heated to increase the partitioning to the gas phase. Microbes in the biofilm surrounding each individual compost particle metabolize the contaminants into simpler compounds, eventually converting them into carbon dioxide and water vapor. *Source: S. J. Ergas and K. A. Kinney (2000). Biological control systems. In: W. T. Davis (Ed), Air and Waste Management Association. Air Pollution Control Manual (2nd Edition). John Wiley & Sons, Inc., New York, NY.*

on the sequestered pollutants (e.g., content and form of halogenated and metallic compounds) or recycling of the metals. This procedure may be repeated as necessary to decrease soil contaminant levels down to target cleanup concentrations. If thermally treated, for example, incinerated, emissions and ash residues must meet regulatory requirements, including disposal as a possible hazardous waste.

Nickel, zinc, and copper are metals that have been particularly amenable to phytoextraction since hundreds of plant species have mechanisms for their uptake and absorption. This would seem to indicate that numerous other metals would also be candidates because the micronutrient cycling and translocation mechanisms may be similar.

Rhizofiltration (rhizo = root) is the adsorption or precipitation of dissolved compounds onto plant roots or absorption into the roots. Rhizofiltration usually addresses contaminated ground water, while phytoextraction targets contaminated soil. Thus, the plants are usually grown hypoponically in greenhouses and exposed to water from the contaminated site once the plants grow mature root systems. This acclimation step allows the plant's endogenous processes to become acclimated to the chemicals in the water. After acclimation, the plants are transplanted in soil above the contaminated aquifer or perched water table, and when their uptake is considered to have reached a threshold, they are harvested and the biomass treated similarly to those from phytoextraction.

Phytostabilisation immobilizes contaminants in the soil and ground water by sorption and accumulation by roots, by adsorption onto roots, or by precipitation within the root zone of plants (rhizosphere). The biochemodynamics hinder transport, for example, to the ground water or air, as well as decreasing bioavailability of the contaminants.

In addition to these physical translocation and filtering processes, plants can degrade pollutants, using mechanisms similar to microbial degradation. Phytodegradation or phytotransformation is the process of degrading compounds after uptake by the plant's

TABLE 27.5 Examples of Successful Projects

Location	Application	Contaminants	Medium	Plant(s)
Edgewood, MD	Phytovolatilization Rhizofiltration Hydraulic control	Chlorinated solvents	Ground water	Hybrid poplar
Fort Worth, TX	Phytodegradation Phytovolatilization Rhizodegradation	Chlorinated solvents	Ground water	Eastern cottonwood
New Gretna, NJ	Phytodegradation Hydraulic control	Chlorinated solvents	Ground water	Hybrid poplar
Ogden, UT	Phytovolatilization Rhizofiltration	Petroleum hydrocarbons	Soil, ground water	Alfalfa, poplar juniper, fescue
Portsmouth, VA	Phytovolatilization Rhizofiltration	Petroleum	Soil	Grasses, clover
Portland, OR	Phytovolatilization	PCP, PAHs	Soil	Ryegrass
Trenton, NJ	Phytovolatilization	Heavy metals, radionuclides	Soil	Indian mustard
Anderson, SC	Phytostabilization	Heavy metals	Soil	Hybird poplar, grasses
Chernobyl, Ukraine	Rhizofiltration	Radionuclides	Ground water	Sunflowers
Ashtabula OH	Rhizofiltration	Radionuclides	Ground water	Sunflowers
Upton, NY	Phytoextraction	Radionuclides	Soil	Indian mustard cabbage
Milan, TN	Phytodegradation	Explosives wastes	Ground water	Duckweed parrot feather
Beaverton, OR	Vegetative cover	Metals, nitrates	Not applicable	Cottonwood
Texas City, TX	Vegetative cover Rhizodegradation	PAHs	Soil	Mulberry
Amana, IA	Riparian corridor Phytodegradation	Nitrates	Ground water	Hybrid poplar

Source: U.S. Environmental Protection Agency (1998). A Citizen's Guide to Phyto remediation. Report No. EPA 542-F-98-011. Washington, DC.

metabolism. It may also involve the breakdown of compounds externally by exogenous secretions, such as enzymes. Complex organic molecules follow similar degradation pathways to those of microbes. That is, the complex molecules are broken down into simpler compounds to gain energy and to build plant biomass, similar microbial metabolic mechanisms.

Rhizodegradation goes by a number of names, including enhanced rhizosphere biodegradation, phytostimulation, and plant-assisted bioremediation/degradation, is actually the same microbial process as those described earlier in this chapter, that is the degradation results from microbes using the contaminants as their energy and carbon sources. Microbes degrade soilborne contaminants microbially, but these mechanisms are enhanced by the biochemodynamics of the root zone (the rhizosphere). This process is usually slower than phytodegradation. This process can enhance cometabolism because plant roots release natural substances, for example, sugars, alcohols, and acid, which are food for soil microorganisms. The fixation and release of nutrients are also a natural type of biostimulation for the microbes. In addition, the roots provide conduits and physical loosening of soil, improving microbial contact with oxygen and nutrients—both a transport and transformation process. A plant takes up a compound from the soil or aquifer and transpires it to the troposphere. This can be solely the parent compound, if it resists degradation (i.e., metabolism and growth products), or it may be the metabolic degradation products, or both. For example, a poplar may volatilize as much as 90% of the VOC taken up by the tree.

Phytoremediation has been successfully used on a variety of compounds in numerous locations (see Table 27.5). The local conditions and properties of the contaminants determine the degree of degradation.

Thus, directly treating hazardous wastes physically and chemically, as with thermal systems and indirectly controlling air pollutants as and when gases are released from pump and treat systems, requires a comprehensive approach. Otherwise, we are merely moving the pollutants to different locations or even making matters worse by either rendering some contaminants more toxic or exposing receptors to dangerous substances.

Plant life may also be used to reduce the amount of hazardous wastes. In phytoremediation, contaminated areas are seeded and plant roots extract the chemicals from the soil. The harvested plants are either treated onsite or transferred to a treatment facility. In reality, both microbial and macrophytic processes occur simultaneously. Poplar trees can help to treat areas contaminated with agricultural chemicals. Plants, such as grasses and field crops, have even been used to treat the very persistent polychlorinated biphenyls, wood preservatives, and petroleum. Plants have also been used to extract heavy metals and radioactive substances from contaminated soil. Bioremediation has been used successfully to treat numerous other organic and inorganic compounds.

5. CONCLUSIONS

Hazardous wastes exist in various forms. A wide range of methods are available for treating substances after they have been released into the environment. The engineer must take great care in selecting, designing, operating, and maintaining systems to collect, transport, store, and dispose of hazardous wastes. However, eliminating the wastes before they are released is the best means of reducing hazards and the associated risks, especially by applying the principles of green chemistry and sustainable design.

References

[1] Resource Conservation and Recovery Act of 1976. Public Law 94-580, 42; United States Code, Sections 901 *et seq*.

[2] U.S. Environmental Protection Agency, Test Methods for Evaluating Solid Waste, Volumes I and II(SW-846), third ed., U.S. Environmental Protection Agency, 1995. November 1986.

[3] U.S. Environmental Protection Agency, Wastes-Hazardous Waste-Test Methods: SW-846; Test Methods for Evaluating Solid Waste, Physical/Chemical Methods; <http://www.epa.gov/wastes/hazard/testmethods/sw846/index.htm.>. Accessed on June 5, 2010 (2010)

[4] U.S Environmental Protection Agency, Facility Pollution Prevention Guide. EPA/600/R-92/088 (1992)

[5] H.M. Freeman, Hazardous Waste, AccessScience@McGraw-Hill, 2002. <http://accessscience.com>. Accessed on July 3, 2009.

[6] Occupational Safety and Health Administration, Code of Federal Regulations 1910.120. <http://www.osha.gov/pls/oshaweb/owadisp.show_document?p_table=STANDARDS&p_id=9765.>, Accessed on June 28, 2010 (2010).

[7] U.S. Environmental Protection Agency, Remediation Guidance Document, EPA-905-B94-003 (2003). Chapter 7.

[8] N.P. Cheremisinoff, Biotechnology for Waste and Wastewater Treatment, Noyes Publications, Westwood, NJ, 1996.

[9] Code of Federal Regulations - Part 40: Protection of the Environment. Section 260. Hazardous Waste Management System. Amended November 7, 1986.

[10] F.G. Pohland, Sanitary Landfill Stabilization with Leachate Recycle and Residual Treatment. EPA-600/2-75-043, U.S. Environmental Protection Agency, 1975.

[11] Y. Long, Y.-Y. Long, H.-C. Liu, D.-S. Shen, Degradation of Refuse in Hybrid Bioreactor Landfill, Biomedical and Environmental Sciences 22 (2009) 303–310.

[12] F. Pohland, W. Cross, J. Gloud, D. Reinhart, Behavior and Assimilation of Organic and Inorganic Priority Pollutants Co-disposed With Municipal Refuse. Report No. EPA/600/R-93/137a, Risk Reduction Engineering Laboratory. Office of Research and Development, Cincinnati, Ohio, 1993.

[13] R. Amman, W. Ludwig, K.-H. Schleifer, Microbiological Reviews 59 (1995) 143–169;
P. Hugenholtz, B. Goebel, N. Pace, Journal of Bacteriology 180 (1998) 4765–4774;
P. Jjemba, Environmental Microbiology: Principles and Applications, Science Publishers, Enfield, New Hampshire, 2004.

[14] U.S. Environmental Protection Agency, Pilot-Scale Demonstration of Slurry-Phase Biological Reactor for Creosote Contaminated Soil: Applications Analysis Report. Report No. EPA/540/A5-91/009, Cincinnati, OH (1993).

[15] S. Wallace. Engineered wetlands lead the way. *Land and Water*, 48 (5); <http://www.landandwater.com/features/vol48no5/vol48no5_1.html;>. Accessed September 9, 2009 (2004).

[16] A.S. Bal, N.N. Dhagat, Indian Journal of Environmental Health 43 (2001) 1–82.

[17] Public Law 96-510, enacted December 11, 1980; enlarged and reauthorized by the Superfund Amendments and Reauthorization Act of 1986 (SARA, P.L. 99-499). United States Code: 42 U.S.C. 9601-9675.

[18] U.S. Environmental Protection Agency, Superfund: Basic Information; <http://www.epa.gov/superfund/about.htm>. Accessed on June 12, 2010 (2010).

[19] D. Vallero, C. Brasier, Sustainable Design: The Science of Sustainability and Green Engineering, fourth ed., John Wiley & Sons, Hoboken, NJ, 2008.

[20] D.A. Vallero, C. Brasier, Fundamentals of Air Pollution, fourth Edition., Academic Press, Burlington, MA, 2007.

[21] U.S. Environmental Protection Agency, A Citizen's Guide to Phyto-remediation. Report No. EPA 542-F-98-011. Washington, DC. (1998).

CHAPTER

28

Thermal Pollution

Daniel A. Vallero

Pratt School of Engineering, Duke University, Durham, N. Carolina, USA

OUTLINE

1. INTRODUCTION

From a waste management perspective, heat is both a friend and foe. When properly designed and operated, incinerators and other thermal technologies reduce municipal and industrial wastes in volume and change their physical and chemical properties to make these wastes less toxic and more easily manageable. However, if not operated properly, these same technologies can form very toxic chemical compounds. In addition, most industrial processes make ample use of chemical reactions at high temperatures, which not only release excess heat to the environment but also often generate chemical pollutants.

Organisms and ecosystems survive within a finite range of environmental conditions. One of the key factors in these conditions is the temperature range. Anthropogenic activities can alter the heat balances within the environment. Thus, heat can be considered a pollutant in situations where its release into an environmental system adversely affects the optimal temperature ranges or indirectly changes other conditions that harm organisms, including humans. Such input can occur at any environmental scale, from cellular (e.g., small-scale changes to a portion of an aquatic system that interferes with microbial metabolism) to planetary (e.g., large-scale changes in global heat balances that increase seasonal ambient

Waste Doi: 10.1016/B978-0-12-381475-3.10028-2

temperatures, leading to increased incidences of heat stress or indirect climate changes that adversely impact ecosystems).

Thermal pollutants can affect the environment in every phase and environmental media. Heat may be a water pollutant if its addition directly or indirectly harms the biota living in surface wards. The raised temperatures in water, for example, can alter the biodiversity of an ecosystem in two ways. Increased temperature may not be tolerable for aquatic biota and/or the increased temperature increases microbial growth, which in turn decreases dissolved oxygen (DO), makes metals more bioavailable, or in other ways increases the harm from nutrients and toxins.

2. CUMULATIVE EFFECTS OF THERMAL POLLUTION

Direct heating of substrates, for example, water, air, and soil, can cause environmental harm. Microclimatological changes can occur due to the release of heated plumes from combustion facilities and vehicles. Even soil that receives added heat can be polluted if it changes the habitat (e.g., changes to freeze-thaw cycles, seasonal variations, and selectivity) of certain soil biota (e.g., bacteria, plant root systems, and fauna-like earthworms and burrowing and nesting animals). Such pollutants need to be treated by cooling towers before heated water reaches surface water and other sensitive habitat.

Heat can initiate cumulative environmental impacts, such as the heat exchange and changing conditions of receiving water bodies [1]. In fact, the value of a fishable stream can be directly related to water temperature as it defines the DO content, which is a limiting factor of the type of fish communities that can be supported by a water body (see Tables 28.1 and 28.2). A trout stream is a highly valued

TABLE 28.1 Relationship Between Water Temperature and Maximum Dissolved Oxygen Concentration in Water (At 1 atm)

Temperature (°C)	Dissolved oxygen (mg L^{-1})	Temperature (°C)	Dissolved oxygen (mg L^{-1})
0	14.60	23	8.56
1	14.19	24	8.40
2	13.81	25	8.24
3	13.44	26	8.09
4	13.09	27	7.95
5	12.75	28	7.81
6	12.43	29	7.67
7	12.12	30	7.54
8	11.83	31	7.41
9	11.55	32	7.28
10	11.27	33	7.16
11	11.01	34	7.16
12	10.76	35	6.93
13	10.52	36	6.82
14	10.29	37	6.71
15	10.07	38	6.61
16	9.85	39	6.51
17	9.65	40	6.41
18	9.45	41	6.41
19	9.26	42	6.22
20	9.07	43	6.13
21	8.90	44	6.04
22	8.72	45	5.95

Sources: D.A. Vallero (2010). Environmental Biotechnology: A Systems Approach Elsevier Academic Press. Amsterdam, NV; and U.S. Environmental Protection Agency (1997)). Volunteer Stream Monitoring Methods Manual. Report No. EPA 841-B-97-003. Chapter 5. Monitoring and assessing water quality: 5.2. Dissolved oxygen and biochemical oxygen demand.

TABLE 28.2 Normal Temperature Tolerances of Aquatic Organisms

Organism	Taxonomy	Range in temperature tolerance (°C)	Minimum dissolved oxygen (mg L^{-1})
Trout	*Salma, Oncorhynchus and Salvelinus spp.*	5–20	6.5
Smallmouth bass	*Micopterus dolomieu*	5–28	6.5
Caddisfly larvae	*Brachycentrus spp.*	10–25	4.0
Mayfly larvae	*Ephemerella invaria*		4.0
Stonefly larvae	*Pteronarcys spp.*	10–25	4.0
Catfish	*Order Siluriformes*	20–25	2.5
Carp	*Cyprinus spp.*	10–25	2.0
Water boatmen	*Notonecta spp.*	10–25	2.0
Mosquito larvae	*Family Culicidae*	10–25	1.0

Source: D.A. Vallero (2010). Environmental Biotechnology: A Systems Approach. Elsevier Academic Press. Amsterdam, NV; and Vernier Corporation (2009). Computer 19: Dissolved oxygen in water. http://www2.vernier.com/sample_labs/BWV-19-COMP-dissolved_oxygen.pdf. Accessed on October 19, 2009.

resource that is adversely impacted if mean temperatures increase. Rougher, less-valued fish (e.g., carp and catfish) can live at much lower ambient water body temperatures than can salmon, trout, and other cold-water fish populations. Thus, net increase in heat may directly stress the game fish population. That is, fish species vary in their ability to tolerate higher temperatures, meaning that the less-tolerant, higher value fish will be inordinately threatened.

The threat may not be completely explained as heat stress is also related to an increase in temperature [2]. Much can be explained by the concomitant decrease in the stream's DO concentrations (see Figs. 28.1 and 28.2), which deems the water body hostile to the fish. Even if the adult fish can survive at the reduced DO levels, their reproductive capacities decrease. Or, the reproduction is not adversely affected, but the survival of juvenile fish can be reduced.

The increased temperature can also increase the solubility of substances toxic to organisms, which increases the exposure. For example, greater concentrations of mercury and other toxic metals will occur with elevated temperatures. The lower DO concentrations will lead to a reduced environment where the metals and compounds could form sulfides and other compounds that can be toxic to the fish. Thus, the change in temperature, the resulting decrease in DO and increase in metal concentrations, and the synergistic impact of combining the hypoxic water and reduced metal compounds is a cascade of harm to the stream's ecosystems. (see Fig. 28.3).

Biota also plays a role in the heat-initiated effect. Combined abiotic and biotic responses occur. Notably, the growth and metabolism of the bacteria result in even more rapidly decreasing DO levels. Algae both consume DO for metabolism and produce DO by photosynthesis.

The first-order abiotic effect (i.e., increased temperature) results in an increased microbial population. However, the growth and metabolism of the bacteria results in decreasing the DO

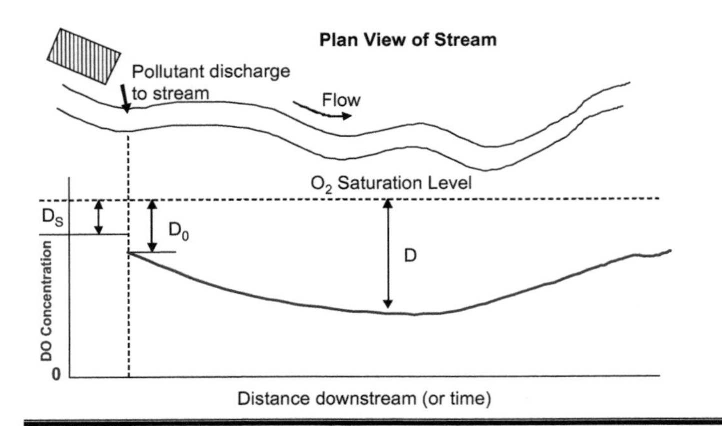

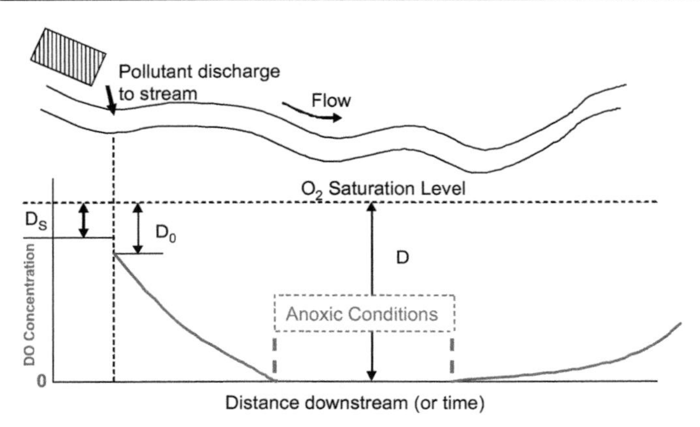

FIGURE 28.1 Dissolved oxygen (DO) deficit downstream from a heated effluent. The increased temperature can result in an increase in microbial kinetics as well as more rapid abiotic chemical reactions, both consuming DO. The concentration of DO in the top curve remains above 0, so although the DO decreases, the overall system DO recovers. The bottom sags where DO falls to 0, and anoxic conditions result and continue until the DO concentrations begin to increase. D_S is the background oxygen deficit before the pollutants enter the stream. D_0 is the oxygen deficit after the pollutant is mixed. D is the deficit for contaminant A that may be measured at any point downstream. The deficit is overcome more slowly in the lower curve (smaller slope) because the reoxygenation is dampened by the higher temperatures and changes to the microbial system, which means the system has become more vulnerable to another insult; for example, another downstream source could cause the system to return to anoxic conditions.

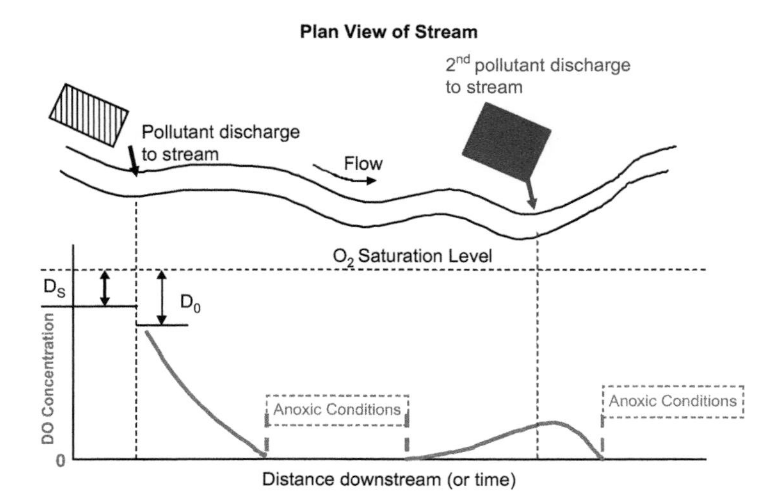

FIGURE 28.2 Cumulative effect of a second heat source, causing an overall system to become more vulnerable. The rate of reoxygenation is suppressed, with a return to anoxic conditions.

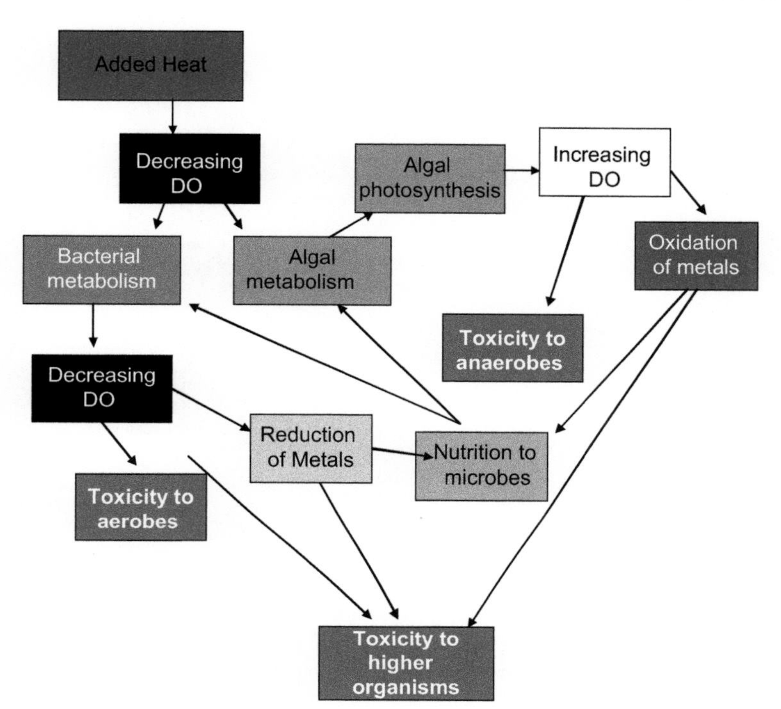

FIGURE 28.3 Adverse effects in the real world usually result from a combination of conditions. In this example, the added heat results in an abiotic response (i.e., decreased dissolved oxygen [DO] concentrations in the water). *Source: D.A. Vallero (2010). Environmental Biotechnology: A Biosystems Approach. Elsevier Academic Press. Amsterdam, NV.*

levels, but the growth of the algae both consume DO for metabolism and produce DO by photosynthesis. Meanwhile, a combined abiotic and biotic response occurs with the metals. The increase in temperature increases their aqueous solubility and the decrease in DO is accompanied by redox changes, for example, formation of reduced metal species, such as metal sulfides. This is also being mediated by the bacteria, some of which will begin reducing the metals as the oxygen levels drop (reduced conditions in the water and sediment). However, the opposite is true in the more oxidized regions, that is, the metals are forming oxides. The increase in the metal compounds combined with the reduced DO, combined with the increased temperatures can act synergistically to make the conditions toxic for higher animals, for example, a fish kill [3]. Predicting the likelihood of a fish kill can be quite complicated, with many factors that either mitigate or exacerbate the outcome (see Figs. 28.4 and 28.5).

The first law of thermodynamics requires, for example, that allowing heated water to be released in any amount, even the permitted level, would increase the overall temperature of the receiving stream. Up to the 1970s, every power plant along the major rivers of the United States was releasing heated water to a stream (see Fig. 28.6). This meant that the incremental effect of all the permitted releases led to a cumulative increase in temperature. In the late 1970s, once-through cooling, that is, letting water pass through turbines and then discharged to adjacent streams, was no longer allowed in the U.S. waters (Fig. 28.6A). Other cooling systems, for example, cooling towers and cooling lakes, had to be installed and operated, which meant power-plant water systems became more closed, both from a fluid dynamics and thermodynamic perspective (Fig. 28.6B).

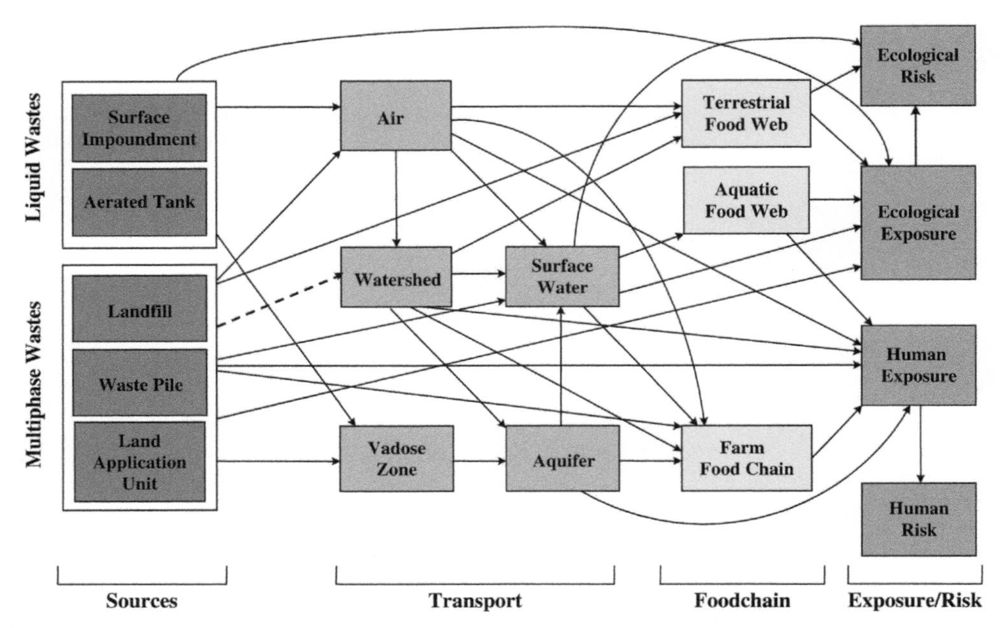

FIGURE 28.4 Environmental transport pathways can be affected by net heat gain. Compounds (nutrients, contaminants), microbes, and energy (e.g., heat) follow the path through the environment indicated by arrows. The residence time within any of the boxes is affected by conditions, including temperature. *Adapted from D.A. Vallero, K.H. Reckhow, A.D. Gronewold (2007). Application of multimedia models for human and ecological exposure analysis. International Conference on Environmental Epidemiology and Exposure. October 17, 2007. Durham, NC. Graphic adapted from U.S. Environmental Protection Agency.*

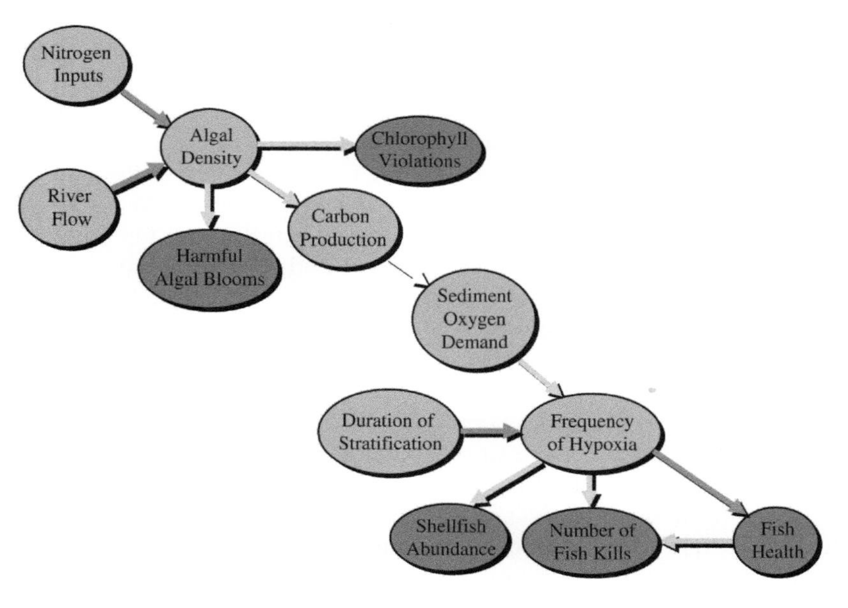

FIGURE 28.5 Flow of events and conditions leading to fish kills, indicating some of the points where added heat can exacerbate the likelihood of a fish kill or other adverse environmental event. *Source: D.A. Vallero, K.H. Reckhow, A.D. Gronewold (2007). Application of multimedia models for human and ecological exposure analysis. International Conference on Environmental Epidemiology and Exposure. October 17, 2007. Durham, NC.*

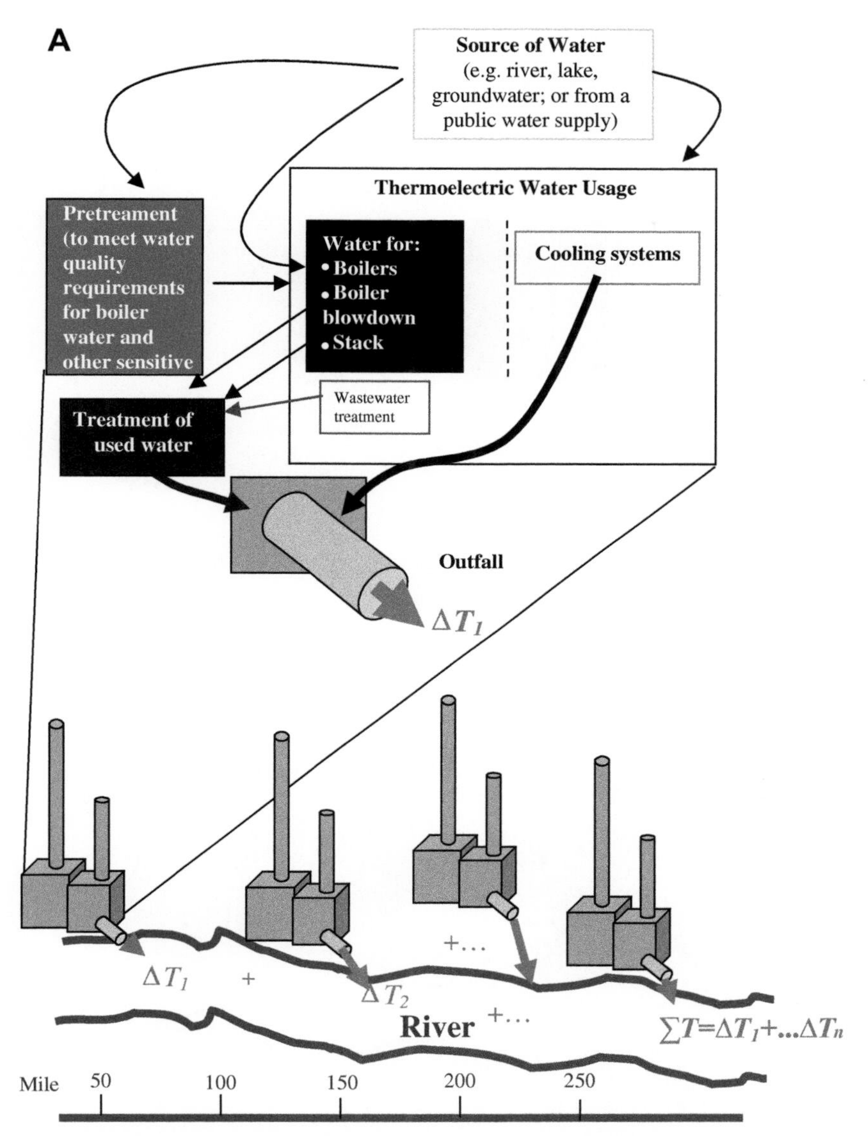

FIGURE 28.6 Difference in cumulative heat contribution to a river from electric generating plants using a once-through cooling system (A) versus the same plants using a cooling water return system (B). The cumulatively added heat is greatly reduced with the closed water return systems compared to the once through cooling systems (hypothetical scenario). *Source: D.A. Vallero (2010). Environmental Biotechnology: A Biosystems Approach. Elsevier Academic Press. Amsterdam, NV.*

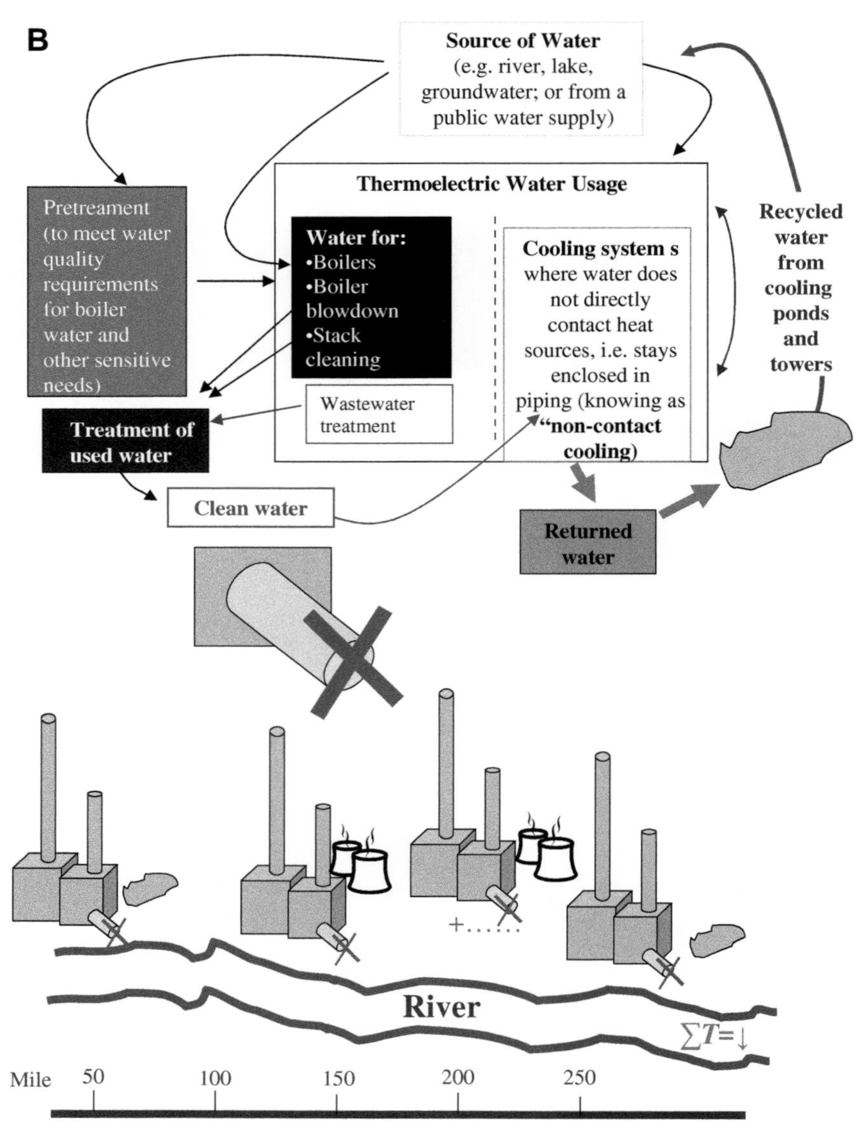

FIGURE 28.6 —(*continued*).

3. THERMOCHEMICAL POLLUTION

Heating processes have generated almost every class of pollutants discussed in Section 2. Indirect thermal pollution usually involves changing the physical, chemical, and biological integrity of a system. From a thermodynamic perspective, the environment is subdivided into a network of control volumes. One important type of thermal pollution is indirect thermochemical pollution that results from changes in the heat budgets within these control volumes. For example, the secondary gases generated in thermal technologies discussed in Chapter 16 can cause or exacerbate the generation of chemical air pollutants, which in turn may be released to the atmosphere, and in turn cause harm when deposited in the ecosystem or in human populations who breathe the air. These effects can be mediated by control technologies, such as secondary combustion chamber, for more complete combustion of incinerator gases or wet scrubbers for their removal, as well as electrostatic precipitators and fabric filters to remove particulate matter (see Chapter 17). Better still, by designing systems that require less heat or wherein heat is managed properly, the thermal pollutants can be prevented from being generated in the first place.

3.1. Thermochemical Formation of Carbon Compounds

Most greenhouse gases are formed in thermal reactions. In fact, complete combustion means that the most prominent greenhouse gas, carbon dioxide (CO_2), will be released. However, this is just one of the carbon compounds that are cycled continuously biogeochemically through the environment [4,5] (see Fig. 28.7).

After the release of CO_2 and other carbon compounds, the carbon may remain sequestered in the soil, roots, sediment, and other compartments; be released to the atmosphere; or otherwise be cycled throughout the environment. Therefore, even a small net addition of greenhouse gases can profoundly increase the atmosphere's greenhouse potential.

CO_2 is an inorganic compound because its carbon atom does not contain a covalent bond

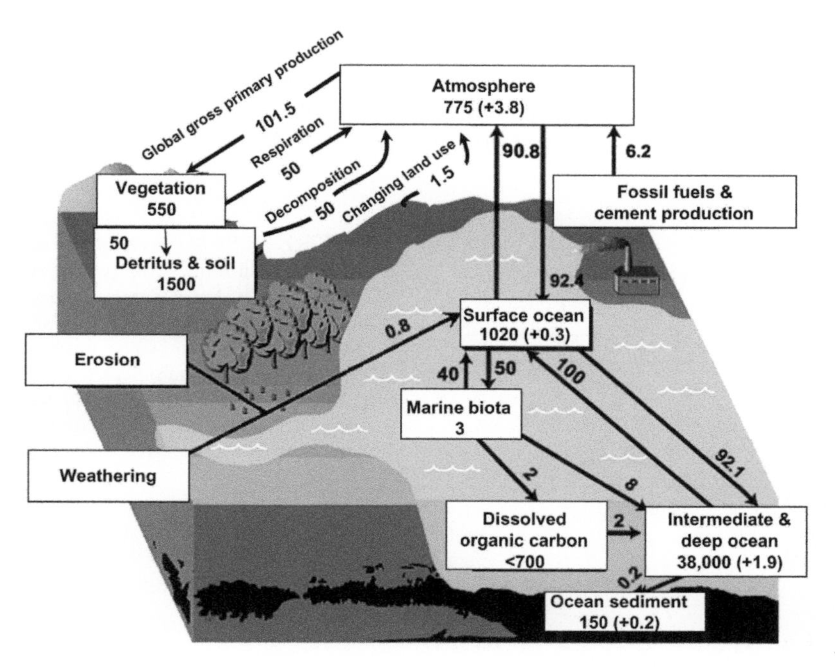

FIGURE 28.7 Global carbon cycle from 1992 to 1997. Carbon pool are boxes, expressed in gigatonnes (Gt) of carbon (Note: Gt = 10^{15} g). Annual increments are expressed in Gt per annum of carbon (shown in parentheses). All fluxes indicated by the arrows are expressed in Gt a^{-1} of carbon. The inferred net terrestrial uptake of 0.7 Gt a^{-1} of carbon considers gross primary production (~101.5), plant respiration (~50), decomposition (~50), and additional removal from the atmosphere directly or indirectly, through vegetation and soil and eventual flow to the ocean through the terrestrial processes of weathering, erosion, and runoff (~0.8). Net ocean uptake (~1.6) considers air/sea exchange (~92.4 gross uptake, −90.8 gross release). As the rate of fossil fuel burning increases the CO_2 released to the atmosphere, it is expected that the fraction of this C remaining in the atmosphere will increase resulting in a doubling or tripling of the atmospheric amount in the coming century. *Source: M. Post. Oak Ridge National Laboratory; http://cdiac.ornl.gov/pns/graphics/c_cycle.htm. Accessed on January 29, 2009.*

with other carbon or hydrogen atoms. Other important inorganic carbon compounds include the pesticides sodium cyanide (NaCN) and potassium cyanide (KCN), and the toxic gas carbon monoxide (CO). Inorganic compounds also include inorganic acids, such as carbonic acid (H_2CO_3) and cyanic acid (HCNO), and compounds derived from reactions with the anions carbonate (CO_3^{2-}) and bicarbonate (HCO_3^-). However, most carbon compounds are organic.

Figure 28.8 demonstrates the equilibrium among carbonates, bicarbonates, organic compounds, carbonic acid, and CO_2. On a global scale, uncontaminated rain's mean pH is about 5.6, owing to the CO_2 that has been dissolved in the rain. As the water droplets fall through the air, the CO_2 in the atmosphere becomes dissolved in the water, setting up an equilibrium condition:

$$CO_2(\text{gas in air}) \leftrightarrow CO_2(\text{dissolved in the water}) \quad (28.1)$$

The CO_2 in the water reacts to produce hydrogen ions as:

$$CO_2 + H_2O \leftrightarrow H_2CO_3 \leftrightarrow H^+ + HCO_3^- \quad (28.2)$$

$$HCO_3^- \leftrightarrow 2H^+ + CO_3^{2-} \quad (28.3)$$

Assuming the mean partial pressure CO_2 in the air to be 3.0×10^{-4} atm, it is possible to calculate the pH of water in equilibrium. Heat is not only a factor in the formation of CO_2 but is also a secondary factor; that is, if the atmosphere is heated, it will speed up most reactions. For example, if we assume an ambient temperature of 25 °C and an adjusted mean concentration of CO_2 in the troposphere of 350 ppm, we would have to adjust the temperature assumption if this concentration is rising as some estimates at a rate of 1 ppm per year.

Henry's Law states that the concentration of a dissolved gas is directly proportional to the partial pressure of that gas above the solution:

$$p_a = K_H[c] \quad (28.4)$$

where K_H is the Henry's Law constant, p_a is the partial pressure of the gas, and $[c]$ is the molar concentration of the gas.

The brackets indicate the chemical concentration in mol L^{-1}. More commonly in

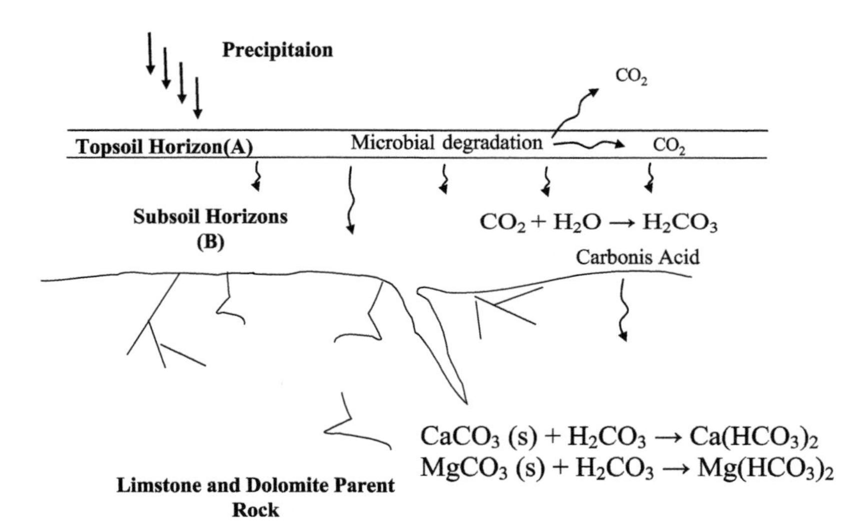

FIGURE 28.8 Biogeochemistry of carbon equilibrium. The processes that release carbonates are responsible for much of the buffering capacity of natural soils against the effects of acid rain. *Source: D.A. Vallero (2010). Environmental Biotechnology: A Biosystems Approach. Elsevier Academic Press. Amsterdam, NV.*

environmental publications, concentration is expressed in units of mass per volume, such as:

$$p_a = K_H C_W \quad (28.5)$$

where C_W is the concentration of the nanogram of the gas per liter of water.

The Henry's Law coefficient is therefore a function of a substance's solubility in water and its vapor pressure and expresses the proportionality between the concentration of a dissolved contaminant and its partial pressure in the open atmosphere at equilibrium. That is, the Henry's Law constant is an example of an *equilibrium constant*, which is the ratio of concentrations when chemical equilibrium is reached in a reversible reaction. This equilibrium state is attained when the rate of the forward reaction is the same as the rate of the reverse reaction.

The CO_2 concentration of the water droplet at equilibrium with air is obtained from the partial pressure of Henry's Law constant:

$$p_{CO_2} = K_H[CO_2]_{aq} \quad (28.6)$$

The change from CO_2 in the atmosphere to carbonate ions in water droplets follows a sequence of equilibrium reactions:

$$CO_2(g) \overset{K_H}{\longleftrightarrow} CO_2(aq) \overset{K_r}{\longleftrightarrow} H_2CO_3(aq)$$

$$\overset{K_{a1}}{\longleftrightarrow} HCO_3^-(aq) \overset{K_{a2}}{\longleftrightarrow} CO_3^{2-}(aq) \quad (28.7)$$

The processes that release carbonates increase the buffering capacity of natural soils against the effects of acidic water ($pH < 5$). Thus, carbonate-rich soils like those of central North America are able to withstand even elevated acid deposition compared to the thin soil areas, such as those in the Canadian Shield, the New York Finger Lakes region, and much of Scandinavia.

The concentration of CO_2 is constant, as the CO_2 in solution is in equilibrium with the air that has a constant partial pressure of CO_2. The two reactions and ionization constants for carbonic acid are as follows:

$$H_2CO_3 + H_2O \leftrightarrow HCO_3^- + H_3O^+ \rightarrow K_{a1} = 4.3 \times 10^{-7} \quad (28.8)$$

$$HCO_3^- + H_2O \leftrightarrow CO_3^{2-} + H_3O^+ \rightarrow K_{a2} = 4.7 \times 10^{-11} \quad (28.9)$$

K_{a1} is four orders of magnitude greater than K_{a2}, so the second reaction can be ignored for environmental acid rain considerations. The solubility of gases in liquids can be described quantitatively by Henry's Law, so for CO_2 in the atmosphere at 25 °C, we can apply the Henry's Law constant and the partial pressure to find the equilibrium. The K_H for $CO_2 = 3.4 \times 10^{-2}$ mol L^{-1} atm^{-1}. The partial pressure of CO_2 is the fraction of CO_2 of all gases in the atmosphere. If CO_2 in the earth's troposphere is 350 ppm by volume in the atmosphere, the fraction of CO_2 must be 350 divided by 1,000,000 or 0.000350 atm.

Thus, the CO_2 and carbonic acid molar concentration can now be found by:

$$\begin{aligned}[CO_2] &= [H_2CO_3] \\ &= 3.4 \times 10^{-2} \text{ mol L}^{-1} \text{ atm}^{-1} \\ &\quad \times 0.000350 \text{ atm} \\ &= 1.2 \times 10^{-5} \text{ M}\end{aligned}$$

The equilibrium is $[H_3O^+] = [HCO^-]$. Taking this and our CO_2 molar concentration gives us the following:

$$\begin{aligned}K_{a1} &= 4.3 \times 10^{-7} = \frac{[HCO_3^-][H_3O^+]}{CO_2} \\ &= \frac{[H_3O^+]^2}{1.2 \times 10^{-5}}\end{aligned}$$

$$[H_3O^+]^2 = 5.2 \times 10^{-12}$$

$$[H_3O^+] = 2.6 \times 10^{-6} \text{M}$$

Or, the droplet pH is about 5.6.

Thus, changes in CO_2 concentrations will result in changes to the mean acidity of precipitation. For example, many models expect a rather constant increase in tropospheric CO_2 concentrations. For example, the increase from the present 350 ppm to 400 ppm tropospheric CO_2 concentrations would be accompanied by a proportional decrease in precipitation pH. The molar concentration can be adjusted using the previous equations:

$$3.4 \times 10^{-2}\ \text{mol L}^{-1}\text{atm}^{-1} \times 0.000400\ \text{atm}$$
$$= 1.4 \times 10^{-5}\text{M, so } 4.3 \times 10^{-7}$$
$$= \frac{[H_3O^+]^{\ 2}}{1.4 \times 10^{-5}} \text{ and } [H_3O^+]^2$$
$$= 6.0 \times 10^{-12} \text{ and } [H_3O^+] = 3.0 \times 10^{-6}\ \text{M}$$

With a 50 ppm increase in CO_2, average water droplet pH would decrease to about 5.5. This means that an increase in atmospheric CO_2 will result in an increase in the acidity of natural rainfall. The precipitation rates themselves would also be affected if greenhouse gas concentrations continue to increase, so any changes in atmospheric precipitation rates would also, on average, be expected to be more acidic. The forcing factors for these interrelationships are shown in Fig. 28.9. This is an interesting example of the earth acting as a very large reactor. Changing one variable can profoundly change the entire system; in this instance, the release of one gas changes numerous physical (e.g., temperature) and chemical (e.g., precipitation pH) factors, which in turn evoke a biological response (biome and ecosystem diversity). This also leads to second-order changes, such as that between two prominent greenhouse gases, CO_2 and methane (CH_4). The increased amounts of CO_2 will likely affect global temperature, which affects biomes and the kinetics within individual ecosystems. Ecological structure, such as tree associations, canopies, and forest floors, as well as wetland structures may change so that the reduced conditions may increase, with an attendant increase in anaerobic microbial decomposition, meaning greater releases of CH_4, which would mean increasing global temperatures, all other factors being held constant. However, if greater biological activity and increased photosynthesis is triggered by the increase in CO_2 and wetland depth is decreased, CH_4 global concentrations would fall, leading to less global temperature rise. Conversely, if this increased biological activity and photosynthesis leads to a decreased forest floor detritus mass, then less anaerobic activity may also lead to lower releases of CH_4. In actuality, there will be increases and decreases at various scales, so the net effects on a complex, planetary system is highly uncertain.

Global warming is actually a thermal effect of chemical pollution. That is, as chemicals are added to the atmosphere, the conversion of incoming short-wave radiation to longer wave infrared radiation is increased, as is the absorption of the heat. Thus, the increased concentrations of greenhouse chemicals lead to a thermal effect, that is, surface and atmospheric heating. The uncertainties in predicting the amount of heating include the actual radiative capacity of these chemicals; the cooling affect of other chemicals, for example, sulfates; and the accuracy in giving weights to these and other factors to a very complex atmosphere.

3.2. Thermochemical Formation of Sulfur and Nitrogen Compounds

Much of the concern for acid rain has been rightly concerned about compounds other than CO_2 and CH_4, notably oxides of sulfur and nitrogen. These also emcompass a number of greenhouse gases (e.g., sulfates may actually be cooling agents, by reflecting the sun's rays).

Many compounds contain both nitrogen and sulfur along with the typical organic elements

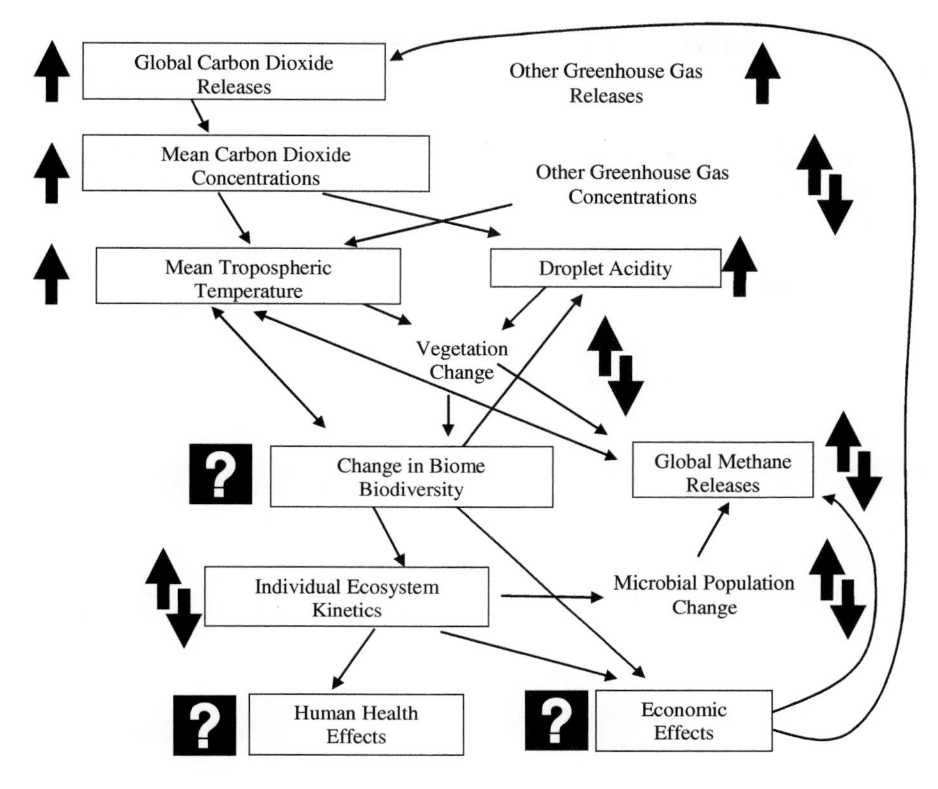

FIGURE 28.9 Systematic view of changes in tropospheric carbon dioxide. Thick arrows indicate whether this factor will increase (up arrow), decrease (down arrow), or will vary depending on the specifics (e.g., some greenhouse gas releases have decreased such as the chlorofluorocarbons and some gases can cool the atmosphere such as sulfate aerosols). Question mark indicates that the type and/or direction of change are unknown or mixed. Thin arrows connect the factors as drivers toward downstream effects. *Source: D.A. Vallero (2010). Environmental Biotechnology: A Biosystems Approach. Elsevier Academic Press. Amsterdan, NV.*

(carbon, hydrogen, and oxygen). The reaction for the combustion of sulfur and nitrogen compounds in general form is:

$$C_aH_bO_cN_dS_e + (4a + b - 2c) \xrightarrow{\Delta} aCO_2 + \left(\frac{b}{2}\right)H_2O + \left(\frac{d}{2}\right)N_2 + eS \quad (28.10)$$

Thermal processes often generate oxides [e.g., sulfur dioxide (SO_2) and nitrogen dioxide (NO_2)]. These, in turn, form acids as they react with water that ultimately cause environmental problems (i.e., acid deposition). Adding such compounds to the environment has both chemical and physical impacts. For example, some soils easily buffer large amounts of acidic rain, whereas others have very limited buffering capacities. Some organisms have large pH tolerance ranges, whereas others may not be able to withstand even a slight change in pH. Whether a compound will remain in the physical and chemical state in which it was deposited is also dependent on the environmental conditions. In the real world, the environment is filled with mixtures of elements, compounds, substrates, and matrices, so pure reactions are almost never the case.

Comparing Reaction (28.10) to that of complete combustion of a simple hydrocarbon ($C_aH_b + O_2 \xrightarrow{\Delta} CO_2 + H_2O$) reactions can occur in sequence, parallel, or both. For example, a feedstock to a municipal incinerator usually contains myriad types of wastes, from garbage to household chemicals to commercial wastes, and even small (and sometimes large) industrial wastes that may be illegally dumped. This variability of feedstock can, however, be improved with proper collection, sorting, segregation, and handling, but it never can be completely eliminated.

Even when the feedstock is a single waste stream, there can be considerable variability. For example, the nitrogen content of typical cow manure is about 5 kg per metric tonne (about 0.5%). If the fuel used to burn the waste also contains sulfur along with the organic matter, then the five elements will react according to the stoichiometry of Reaction (28.10).

Certainly, combustion specifically and oxidation generally are very important processes that generate nitrogen and sulfur pollutants. But they are certainly not the only ones. In fact, we need to explain what oxidation really means. In the environment, oxidation *and* reduction occur. The formation of two SO_2 and nitric oxide by acidifying molecular sulfur is a redox reaction:

$$S(s) + NO_3^-(aq) \rightarrow SO_2(g) + NO(g) \quad (28.11)$$

The designations in parentheses give the physical phase of each reactant and product: "s" for solid, "aq" for aqueous, and "g" for gas.

The oxidation half-reactions for this reaction are as follows:

$$S \rightarrow SO_2$$

$$S + 2H_2O \rightarrow SO_2 + 4H^+ + 4e^-$$

The reduction half-reactions for this reaction are as follows:

$$NO_3^- \rightarrow NO$$

$$NO_3^- + 4H^+ + 3e^- \rightarrow NO + 2H_2O$$

Therefore, the balanced oxidation–reduction reactions are as follows:

$$4NO_3^- + 3S + 16H^+ + 6H_2O \rightarrow 3SO_2 + 16H^+ + 4NO + 8H_2O$$

$$4NO_3^- + 3S + 4H^+ \rightarrow 3SO_2 + 4NO + 2H_2$$

A reduced form of sulfur that is highly toxic and an important pollutant is hydrogen sulfide (H_2S). Certain steps in thermal processes can be anoxic (i.e., lacking O_2) where nitrogen and sulfur is reduced. Thus, the release of highly toxic compounds, like H_2S, would be released without proper treatment. In this case, redox principles can be used to treat H_2S contamination. Strong oxidizers, like molecular oxygen and hydrogen peroxide, most effectively oxidize the reduced forms of S, N, or any reduced compound.

3.3. Toxic Byproducts

Thermal processes are used to break down harmful molecules into smaller, less-harmful molecules (see Chapter 16). Conversely, these same processes can also lead to the formation of toxic compounds that are even more harmful than the parent compounds in the original waste. Thermal reactions can generate some very toxic byproducts, such as carbon monoxide (CO), polycyclic aromatic hydrocarbons (PAHs) and other products of incomplete combustion (PICs). The chlorinated dioxins are a particularly toxic class of PICs, consisting of 75 different dioxin forms and 135 different chlorinated furans, delineated by the number and arrangement of chlorine (Cl) atoms on the molecules. The compounds can be separated into groups that have the same number of Cl atoms attached to the furan or dioxin ring. There are also numerous other dioxins with varying combinations of halogens, for example, brominated dioxins or bromochlorodioxins. Each form varies in its chemical, physical, and toxicological characteristics (see Fig. 28.10).

Dioxins are highly toxic compounds that are created unintentionally during combustion processes. The most toxic form is the 2,3,7,8-tetrachlorodibenzo-*p*-dioxin (TCDD) isomer. Other isomers with the 2,3,7,8 configuration are also considered to have higher toxicity than the dioxins and furans with different Cl atom arrangements. The chemical and physical mechanisms that lead to the production of dioxin involve Cl. Incinerators of chlorinated wastes are the most common environmental sources of dioxins, accounting for about 95% of the volume.

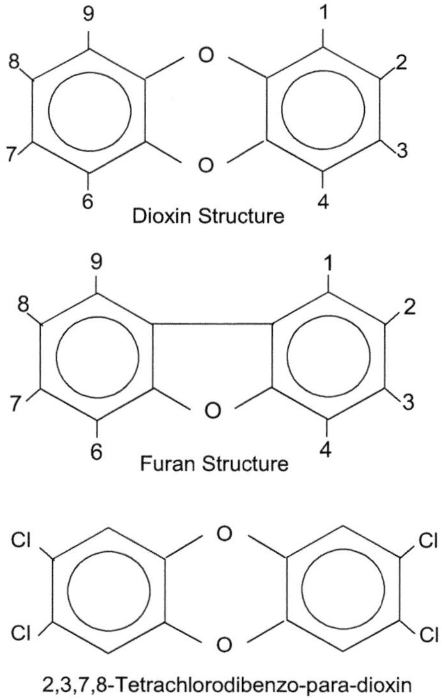

Dioxin Structure

Furan Structure

2,3,7,8-Tetrachlorodibenzo-para-dioxin

FIGURE 28.10 Molecular structures of dioxins and furans. Bottom structure is of the most toxic dioxin congener, tetrachlorodibenzo-*para*-dioxin (TCDD), formed by the substitution of chlorine for hydrogen atoms at positions 2, 3, 7, and 8 on the molecule.

The emission of dioxins and furans from thermal processes may follow three general physicochemical pathways. The first pathway is actually not a thermal process, but one where the thermal treatment is ineffective. That is, the feed material entering the combustion unit contains dioxins or furans, and a fraction of these compounds survives thermal breakdown mechanisms, allowing these compounds to pass through, and ultimately emitted from vents or stacks. This is not considered to account for a large volume of dioxin released to the environment, but it may account for the release of some dioxin-like, coplanar polychlorinated biphenyls (PCBs) from incinerators.

Dioxins and furans can also be formed during the thermal breakdown and molecular rearrangement of precursor compounds, such as the chlorinated benzenes, chlorinated phenols (e.g., pentachlorophenol), and PCBs, which are chlorinated aromatic compounds with chemical structures similar to those of the chlorinated dioxin and furan molecules. Dioxins may well form after the precursor has condensed and is adsorbed onto the surface of particles, such as fly ash. This heterogeneous process occurs on the active sorption sites on the particles, allowing for the chemical reactions that are catalyzed by the presence of inorganic chloride compounds and ions sorbed to the particle surface. The process occurs within the range of 250 to 450 °C, so most of the dioxin formation under the precursor mechanism takes place away from the high temperature zone in the incinerator. This is where the gases and smoke derived from combustion of the organic materials have cooled during conduction through flue ducts, heat exchanger and boiler tubes, air pollution control equipment, or the vents and the stack.

The third means of synthesizing dioxins ensues as a de novo process within the so-called "cool zone" of the incinerator. Here, the dioxins and furans form from moieties different from those of the molecular structure of dioxins, furans, or precursor compounds. Generally, these can include a wide range of both halogenated compounds like polyvinylchloride (PVC) and nonhalogenated organic compounds like petroleum products, nonchlorinated plastics (polystyrene), cellulose, lignin, coke, coal, and inorganic compounds like particulate carbon and hydrogen chloride gas. No matter which de novo compounds are involved, a Cl atom must be donated to the reaction, which leads to the formation and chlorination of a chemical intermediate that is a precursor. The reaction steps, after this precursor is formed, can be identical to the precursor mechanism discussed in the previous paragraph.

De novo formation of dioxins and furans may involve even more fundamental substances than those moieties mentioned above. For example, dioxins may be generated [6] by heating of carbon particles absorbed with mixtures of magnesium–aluminum silicate complexes when the catalyst copper chloride ($CuCl_2$) is present (see Table 28.3 and Fig. 28.11). The de novo formation of chlorinated dioxins and furans from the oxidation of carbonaceous particles seems to occur at around 300 °C. Other chlorinated benzenes, chlorinated biphenyls, and chlorinated naphthalene compounds are also generated by this type of mechanism.

Other processes generate dioxin pollution. A source that has been greatly reduced in the

TABLE 28.3 De Novo Formation of Chlorinated Dioxins and Furans After Heating Mg-Al Silicate, 4% Charcoal, 7% Cl, 1% $CuCl_2$ H_2O at 300 °C

	Concentrations (ng g^{-1})				
	Reaction time (h)				
Compound	**0.25**	**0.5**	**1**	**2**	**4**
Tetrachlorodioxin	2	4	14	30	100
Pentachlorodioxin	110	120	250	490	820
Hexachlorodioxin	730	780	1600	2200	3800
Heptachlorodioxin	1700	1840	3500	4100	6300
Octachlorodioxin	800	1000	2000	2250	6000
Total chlorinated dioxins	3342	3744	7364	9070	17,020
Tetrachlorofuran	240	280	670	1170	1960
Pentachlorofuran	1360	1670	3720	5550	8300
Hexachlorofuran	2500	3350	6240	8900	14,000
Heptachlorofuran	3000	3600	5500	6700	9800
Octachlorofuran	1260	1450	1840	1840	4330
Total chlorinated furans	8360	10,350	17,970	24,160	38,390

Source: L. Stieglitz, G. Zwick, J. Beck, H. Bautz, and W. Roth, 1989, Chemosphere 19:283.

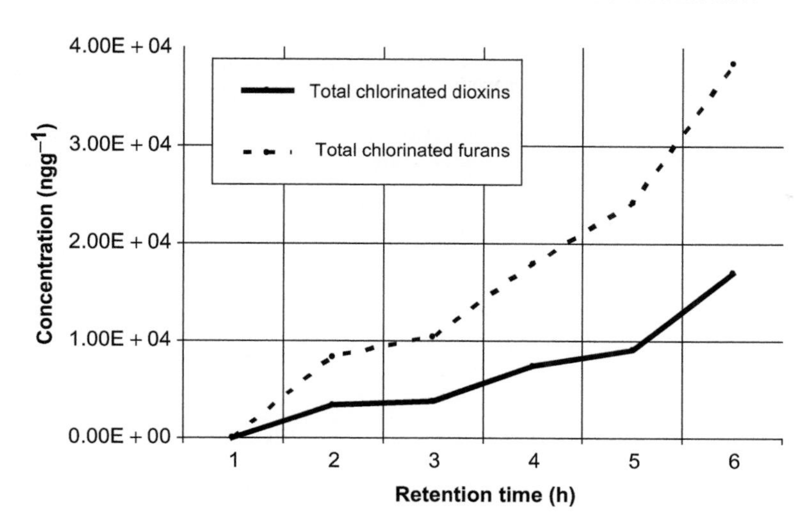

FIGURE 28.11 De novo formation of chlorinated dioxins and furans after heating Mg-Al Silicate, 4% Charcoal, 7% Cl, 1% $CuCl_2 \cdot H_2O$ at 300 °C. *Source: L. Stieglitz, G. Zwick, J. Beck, H. Bautz, and W. Roth, 1989, Chemosphere 19:283.*

TABLE 28.4 Conservative Estimates of Heavy Metals and Metalloids Partitioning to Flue Gas as a Function of Solids Temperature (t) and Chlorine (Cl) Content*

Metal or metalloid	*t* (871 °C)		*t* (1093 °C)	
	Cl = 0%	Cl = 1%	Cl = 0%	Cl = 1%
Antimony (%)	100	100	100	100
Arsenic (%)	100	100	100	100
Barium (%)	50	30	100	100
Beryllium (%)	5	5	5	5
Cadmium (%)	100	100	100	100
Chromium (%)	5	5	5	5
Lead (%)	100	100	100	100
Mercury (%)	100	100	100	100
Silver (%)	8	100	100	100
Thallium (%)	100	100	100	100

* *The remaining percentage of metal is contained in the bottom ash. Partitioning for liquids is estimated at 100% for all metals. The combustion gas temperature is expected to be 40–500 °C higher than the solids temperature.*

Source: U.S. Environmental Protection Agency, 1989, EPA 1989, Guidance on Setting Permit Conditions and Reporting Trial Burn Results: Volume II, Hazardous Waste Incineration Guidance Series, EPA/625/6-89/019. Washington, DC: EPA.

last decade is the paper production process, which formerly used Cl bleaching. This process has been dramatically changed, so that most paper mills no longer use the Cl bleaching process. Dioxin is also produced in the making of PVC plastics, which may follow chemical and physical mechanisms similar to the second and third processes discussed above.

TABLE 28.5 Metal and Metalloid Volatilization Temperatures

	Without chlorine		With 10% chlorine	
Metal or metalloid	**Volatility temperature (°C)**	**Principal species**	**Volatility temperature (°C)**	**Principal species**
Chromium	1613	CrO_2/CrO_3	1611	CrO_2/CrO_3
Nickel	1210	$Ni(OH)_2$	693	$NiCl_2$
Beryllium	1054	$Be(OH)_2$	1054	$Be(OH)_2$
Silver	904	Ag	627	AgCl
Barium	841	$Ba(OH)_2$	904	$BaCl_2$
Thallium	721	Tl_2O_3	138	TIOH
Antimony	660	Sb_2O_3	660	Sb_2O_3
Lead	627	Pb	−15	$PbCl_4$
Selenium	318	SeO_2	318	SeO_2
Cadmium	214	Cd	214	Cd
Arsenic	32	As_2O_3	32	As_2O_3
Mercury	14	Hg	14	Hg

Source: Agency for Toxic Substances and Disease Registry, 2002, B. Willis, M. Howie, and R. Williams, Public Health Reviews of Hazardous Waste Thermal Treatment Technologies: A Guidance Manual for Public Health Assessors.

As all dioxin and dioxin-like compounds are lipophilic and persistent, they accumulate in soils, sediments, and organic matter and can persist in solid and hazardous waste disposal sites [7,8]. These compounds are semivolatile, so they may migrate away from these sites and be transported in the atmosphere either as aerosols (solid and liquid phase) or as gases (the portion of the compound that volatilizes). Therefore, the engineer must take great care in removal and remediation efforts not to unwittingly cause releases from soil and sediments via volatilization or via perturbations, such as landfill and dredging operations.

Incineration is frequently used to decontaminate solid matrices, for example, soils, with elevated concentrations of organic hazardous constituents. Many of these substances can be completely combusted at much lower temperatures than larger, more recalcitrant compounds like the dioxins. For example, high-temperature incineration may not be needed to treat wastes contaminated with many volatile organic compounds (VOCs). In addition, high-temperature incineration of wastes with heavy metals will likely increase the volatilization of some of these metals into the combustion flue gas (see Tables 28.4 and 28.5). High concentrations of volatile trace metal compounds in the flue gas poses increased challenges to air pollution control. Thus, other thermal processes, that is, thermal desorption and pyrolysis, can provide an effective alternative to incineration.

When successful in decontaminating wastes to the necessary treatment levels, thermally desorbing contaminants from the solid matrix has the additional benefits of lower fuel consumption, no formation of slag, less volatilization of metal compounds, and less-complicated air pollution control demands. So, beyond monetary costs and ease of operation, a less energy (heat)-intensive system can be more advantageous in terms of actual pollutant removal efficiency.

4. CONCLUSIONS

No environmental system can be understood without ample knowledge of temperature. Every chemical reaction is affected by temperature. Complete destruction of a toxic compound versus the release of even more toxic compounds in an incinerator can be determined by a relatively small temperature range.

Adding heat to a stream or other ecosystem can completely alter its biological integrity. Food chains and the health of human populations are affected by direct heating or by the indirect effects of added heat. Thus, waste management must always devote attention to these and other possible impacts from heat in the design and operation of systems.

References

[1] D.A. Vallero, Environmental Biotechnology: A Systems Approach. Elsevier Academic Press. Amsterdam, NV; and U.S. Environmental Protection Agency (1997), 2010.
[2] Volunteer Stream Monitoring Methods Manual. Report No. EPA 841-B-97-003. Chapter 5. Monitoring and assessing water quality: 5.2.
[3] D.A. Vallero, K.H. Reckhow, A.D. Gronewold, Application of multimedia models for human and ecological exposure analysis. International Conference on Environmental Epidemiology and Exposure. Durham, NC, USA, 2007.
[4] M. Battle, M.L. Bender, P.P. Tans, J.W.C. White, T. Conway, R.J. Francey, Global carbon sinks and their variability inferred from atmospheric O_2 and d13C, Science 287 (2000) 2467–2470.
[5] N. Gruber, J.L. Sarmiento, Large-scale biogeochemical-physical interactions in elemental cycles, in: A.R. Robinson, J. McCarthy, B.J. Rothschild (Eds.), The Sea, Vol. 12, John Wiley & Sons, New York, NY, 2002, pp. 337–399.
[6] L. Stieglitz, G. Zwick, J. Beck, H. Bautz, W. Roth, On the de-novo synthesis of PCDD/PCDF on fly ash of municipal waste incinerators, Chemosphere 19 (1989) 283.
[7] C. Koester, R. Hites, Wet and dry deposition of chlorinated dioxins and furans, Environmental Science and Technology 26 (1992) 1375–1382.
[8] R. Hites, Atmospheric transport and deposition of polychlorinated dibenzo-p-dioxins and dibenzofurans. U.S. Environmental Protection Agency. Report No. EPA/600/3-91/002, 1991.

CHAPTER

29

Land Pollution

Daniel J. Vallero †, *Daniel A. Vallero* ‡

† Public Works Department, Engineering Section, City of Durham, N. Carolina, USA,
‡ Pratt School of Engineering, Duke University, Durham, N. Carolina, USA

OUTLINE

1. INTRODUCTION

Land is susceptible to pollution in two basic ways. It can be contaminated by chemical pollutants, or it can be altered physically in a manner that renders it less useful or sustainable. Previous chapters have discussed how soil and water become contaminated by the chemical constituents of wastes, such as leachate from landfills, hazardous waste storage and treatment facilities, and even from atmospheric pollutants, for example, sulfur and ntirogen compounds that lead to acid rain. This chapter focuses on how land can be harmed by human activities, such as construction, agriculture, and transportation. These and other human activities lead to the release of chemical contaminants, but they also result in landscape damage, such as soil erosion, habitat destruction, and loss of resources, such as wetlands and coastal ecosystems.

Waste Doi: 10.1016/B978-0-12-381475-3.10029-4

2. THE LAND ETHIC

A full accounting of the value of a land must include cultural and social factors. In fact, as mentioned in chapter 3, environmental impact statements (EISs) under the National Environmental Policy Act (NEPA) require that these factors be part of the decision as to whether a federal project would have a significant effect on the environment. These have included historic preservation, economics, psychology (e.g., open space, green areas, and crowding), aesthetics, urban renewal, and the so-called "land ethic." Aldo Leopold in his famous essays, posthumously published as *A Sand County Almanac*, argued for a holistic approach toward land:

> A thing is right when it tends to preserve the integrity, stability and beauty of the biotic community. It is wrong when it tends otherwise [1].

NEPA did not really begin as a technical law. It was signed into law in 1970 after contentious hearings in the US Congress. NEPA was not really a technical law. It did two main things. In addition to requiring EISs, it established the Council on Environmental Quality (CEQ) in the Office of the President. Of the two, the EIS represented a sea of change in how the federal government was to conduct business. Agencies were required to prepare EISs on any major action that they were considering that could "significantly" affect the quality of the environment. From the outset, the agencies had to reconcile often-competing values, that is, their mission and the protection of the environment.

The CEQ was charged with developing guidance for all federal agencies on NEPA compliance, especially when and how to prepare an EIS. The EIS process combines scientific assessment with public review. The process is similar for most federal agencies. Agencies often strive to receive a so-called finding of no significant impact, so that they may proceed unencumbered on a mission-oriented project.[1]

The agencies may well perceive their projects as improvements to the status quo and be tempted to view land as a "blank slate" or that buildings are simply three-dimensional structures ready to be built, changed, or demolished as means to engineering ends. Indeed, developers and real estate professionals often refer such structures as "improvements" (see Discussion Box: Failure in Land Development: The Case of Pruitt–Igoe).

3. THE COMPLAINT PARADIGM

In many places, environmental response is often precipitated first by a complaint. The underlying assumption of fairness is that not only everyone has a voice in the process but also that voice is loud enough to be heard. However, if a certain group of people has had little or no voice in the past, they are likely to feel and be disenfranchised. For example, for centuries African-American communities were not successful in voicing concerns about environmentally unacceptable conditions in their neighborhoods. Presently, Hispanic-Americans may have even less voice in environmental and public health matters, because the government may not be seen as a viable place to lodge a complaint and the perceived risks of complaining (e.g., questions about citizenship and possible deportation) may outweigh the expected benefit (e.g., cleaner environment).

Aldo Leopold's "land ethic" is a reminder that the use of land is dependent on the values placed on it. One means of determining land's value is

[1] This is understandable if the agency is in the business of something not directly related to environmental work, but even the natural resources and environmental agencies have asserted that there is no significant impact to their projects. It causes the cynic to ask, then, why are they engaged in any project that has no significant impact? The answer is that the term "significant impact" is really understood to mean "significant adverse impact" to the human environment.

DISCUSSION BOX: FAILURES IN LAND DEVELOPMENT: THE CASE OF PRUITT–IGOE

Regrettably, it is not difficult to find ample examples of improper development of land no matter where one resides. Land waste can result from subdivisions with improper erosion and sediment control systems, clear-cutting trees, and complete disregard for and elimination of sensitive ecosystems, as discussed later in this chapter. Another type of improper land development has to do with social values. The Pruitt–Igoe public housing project provides an example of failing to include land as an application of the social sciences in planning, design, construction, and maintenance. This housing development in St. Louis, Missouri, was supposed to be emblematic of advances in fair housing and progress in the war on poverty. Instead, Pruitt–Igoe has become an icon of failure of imagination, especially imagination that properly accounts for the human condition. Although we think of public housing projects in terms of housing, they often represent elements of environmental justice. Contemporary understanding of environmental quality is often associated with physical, chemical, and biological contaminants, but in the formative years of the environmental movement, aesthetics and other "quality of life" considerations were essential parts of environmental quality.

The project was designed before EISs, and the land ethic was yet to be articulated prior to the construction of the Pruitt–Igoe project, so the designers did not benefit from the insights of Leopold and his contemporaries. However, the problems that led to the premature demolition of this costly housing experiment may have been anticipated intuitively if the designers had taken the time to understand what people expected.

In such catastrophic failure, there is usually plenty of culpability to go around. Some blame the inability of the modern architectural style to create livable environments for people living in poverty, largely because they "are not the nuanced and sophisticated 'readers' of architectural space the educated architects were" [2]. This is a telling observation and an important lesson for anyone developing land. We need to make sure that the use and operation of whatever is designed is sufficiently understood by those living with it, long after the designers have left.

Design incompatibility was almost inevitable for high-rise buildings and low-income families with children. It appears that socioeconomics play a large role in the success of housing developments. Most large cities have large populations of families with children living in such environments. Indeed, St. Louis had successful luxury townhomes not too far from Pruitt–Igoe, with many of the same design features. Part of the problem can be traced to racism in the era when the project was conceived. In fact, when originally inhabited, the Pruitt section was restricted to African-Americans and Igoe to Whites.

Costs always become a factor in land waste. The building contractors' bids in the Pruitt–Igoe project were increased to a level when the project construction costs in St. Louis exceeded the national average by 60%. The response to the local housing authority's refusal to raise unit cost ceilings to accommodate the elevated bids was to reduce room sizes, eliminate amenities, and raise densities [3]. As originally designed, the buildings were to become "vertical neighborhoods" with nearby playgrounds, open air hallways, porches, laundries, and storage areas. The compromises eliminated these features. Furthermore, some of the removal of "amenities" led to dangerous situations. Elevators were undersized and stopped only every third floor, and lighting was inadequate in the stairwells. Therefore, another lesson must be to know the difference

(Continued)

DISCUSSION BOX: FAILURES IN LAND DEVELOPMENT: THE CASE OF PRUITT–IGOE—*Cont'd*

between desirable and essential design elements. No self-respecting structural engineer involved in the building design would have shortcut the factors of safety built into load bearing. Conversely, human elements essential to a vibrant community were eliminated without much, if any, accommodation [4].

Finally, the project was mismatched to the people who would live there. Many came from single family residences. They were moved to a very large, imposing project with 2800 units and almost 11,000 people living there. This quadrupled the size of the next largest project of the time.

When the failure of the project became overwhelmingly clear, the only reasonable decision was to demolish it, and this spectacular implosion became a lesson in failure for planners, architects, and engineers. In the designers' own words:

> I never thought people were that destructive. As an architect, I doubt if I would think about it now. I suppose we should have quit the job. It's a job I wish I hadn't done [5].

a scientific tabulation of the land's resources and threats to those resources. Since NEPA's passage, land use decisions in the United States must follow from an environmental assessment. Most environmental impact assessment handbooks prior to the late 1990s contained little information and guidelines related to fairness issues in terms of housing and development. They were usually concerned about open space; wetland and farmland preservation; housing density; ratios of single versus multiple family residences; owner-occupied housing versus rental housing; building height, signage, and other restrictions; designated land for public facilities such as landfills and treatment works; and institutional land uses for religious, health care, police, and fire protection.

When land uses evolve (usually to become more urbanized), the environmental impacts may be direct or indirect. Examples of direct land use effects include *eminent domain,* which allows land to be taken with just compensation for the public good. Easements are another direct form of land use impacts, such as a 100-m right-of-way for a highway project that converts any existing land use (e.g., farming, housing, or commercial enterprises) to a transportation use. Land use change may also come about indirectly, such as so-called secondary effects of a project that extend, in time and space, the influence of a project. For example, a wastewater treatment plant and its connected sewer lines will create accessibility that spawns suburban growth [6]. People living in very expensive homes may not even realize that their building lots were once farmland or open space and, had it not been for some expenditure of public funds and the use of public powers such as eminent domain, there would be no subdivision.

3.1. Competing Values: Affordable Housing versus Conservation

Environmentalists are generally concerned about increased population densities, but housing advocates may be concerned that once the land use has been changed, environmental and zoning regulations may work against affordable housing. Even worse, environmental protection can be used as an excuse for some elitist and exclusionary decisions. In the name of environmental protection, certain classes of people are economically restricted from living in certain areas. This problem first appeared in the United States in the 1960s and 1970s in a search for ways to preserve open spaces and green areas. One measure was the minimum lot size. The idea was that rather than having the public sector securing land through easements or outright purchases (i.e., fee simple) to preserve open spaces, developers could either set aside open areas or require large lots to have their subdivisions approved. Thus, green areas would exist without the requisite costs and O&M entailed by public parks and recreational areas. Such areas have numerous environmental benefits such as wetland protection, flood management, and aesthetic appeal. However, minimum lot size translates into higher costs for residences. The local rules for large lots that result in less affordable housing is called *exclusionary zoning*. One value (open space and green areas) is pitted against another (affordable housing). In some cases, it could be argued that preserving open spaces is simply a tool for excluding people of lesser means or even people of minority races [7].

Land use plans must somehow balance the need to preserve economic value of homes and to enhance the quality of our neighborhoods with fair and affordable housing. Thus, engineers and land use planners must take great care that their ends (environmental protection) are not used as a rationale for unjust means (unfair development practices). Environmental laws and policies, such as zoning ordinances and subdivision regulations, should not be used as a means to keep lower socioeconomic groups out of privileged neighborhoods.

Environmental quality continues to be used, knowingly or innocently, to compete with fairness. This may be most dramatically demonstrated when people who benefit little bear too large a burden in terms of pollution and waste. Often, exposures to the hazards brought about by land use decisions are set in place well before the actual land use decisions are made. In fact, because everything that engineers do may have an impact on health, safety, and welfare, inclusiveness and affordability should be factors in all standard operating procedures for all designs that potentially affect the public. Engineers and planners must find ways to ensure plans are fair and just, often looking beyond property value "bottom lines." Sustainable development, therefore, is the result of applications of both physical and social sciences.

Environmental injustice can result from factors other than the profit motive and its driving corporate decisions to site environmentally hazardous facilities where people are less likely to complain. Public decisions have also brought lower socioeconomic communities into environmental harm's way. Although public agencies, such as housing authorities and public works administrations do not have a profit motive, per se, they do need to address budgetary and policy considerations. If open space is cheaper and certain neighborhoods are less likely to complain (or by extension, vote against elected officials), the "default" for unpopular facilities such as landfills and hazardous waste sites may be to locate them in low-income neighborhoods.

It may be easier to site other types of unpopular projects, such as public housing projects, in areas where complaints are less likely to be put forth or where land is cheaper. In 1954, for example, the Dallas, Texas, Housing Authority built a large public housing project on land immediately adjacent to a lead smelter. The project had 3500 living units and became a predominantly African-American community. During the 1960s, the lead smelter stacks emitted more than 180 tons (20 US tons) of lead annually into the air. Recycling companies had owned and operated the smelter to recover lead from as many as 10,000 car batteries per day. The lead emissions were associated with blood lead levels in the housing project's children, and these were 35% higher than in children from comparable areas.

Lead is a particularly insidious pollutant, because it can result in developmental damage.

Study after study showed that the children at this project were in danger of higher lead levels, but nothing was done for more than 20 years. Finally, in the early 1980s, the city brought suit against the lead smelter, and the smelter immediately initiated control measures that reduced its emissions to allowable standards. The smelter also agreed to clean up the contaminated soil around the smelter and to pay compensation to people who had been harmed.

This case illustrates two issues of environmental racism and injustice. First, the housing units should never have been built next to a lead smelter. The reason for locating the units there would have been justified on the basis of economics. The land was inexpensive and, thus, this saved the government money. The second issue was delays by the city in insisting that the smelter clean up the emissions. Once the case had been made, within 2 years the plant was in compliance. By 2003, blood lead levels in West Dallas were below the national average. Thus, time is often of the essence in land use decisions.

4. ADDRESSING LAND POLLUTION

Local governments address land pollution using a number of tools, including land use plans, zoning ordinances, and development rules, such as subdivision regulations, that must undergo significant public review. An illustrative example is that of Durham, NC. The City of Durham Public Works mission statement is based on the City of Durham's *Customer Bill of Rights*, which is a citizen-focused initiative "to provide excellent service at all times by determining the needs of our citizens and customers and satisfying those needs beyond their expectations [8]." Human health and environmental quality are top priorities in fulfilling the city's mission. The chief instrument in the environmentally responsible approach is achieved by the city's growth adhering to the *Unified Development Ordinance*, which is designed to promote the health, safety, and general welfare of the residents of Durham City and County. It does so by:

A. Protecting existing neighborhoods, preventing their decline, and promoting their livability;
B. Addressing future needs, growth, and change in the jurisdiction;
C. Conserving land and water resources;
D. Preserving groundwater quality and supply;
E. Recognizing geologic features, soil, and topography;
F. Improving air quality;
G. Minimizing congestion in the streets and reducing reliance on automobiles by providing options for walking, bicycling, and transit use;
H. Securing safety from fire and other dangers;
I. Providing adequate light and air;
J. Preventing overcrowding of land and undue concentrations of population;
K. Providing adequate transportation, water supplies, sewer service, schools, parks, open space, and public facilities;
L. Conserving the value of buildings;
M. Examining the most appropriate use of the land;
N. Regulating the location of business and industry;
O. Regulating the height and bulk of buildings;
P. Protecting the capacity of floodways and non-encroachment areas to prevent loss or damage to homes or property;
Q. Regulating the area of yards and open spaces for buildings;
R. Providing for the needs of agriculture;
S. Protecting historic sites and areas;
T. Encouraging an aesthetically attractive community; and
U. Preventing secondary effects from land uses that could negatively impact nearby land uses, consistent with prior ordinances restricting such uses and evidence supporting such restrictions.

This list of expectations provides a yardstick for land use. If an activity adheres to all these factors, it is likely to be acceptable environmentally. The ordinance provides for the orderly, efficient, and economic development of the City and County by coordinating streets, highways, and other public facilities within proposed subdivisions with existing or planned streets and highways or other public facilities. It also provides for the dedication and reservation of rights-of-way, easements or sites for streets, utilities, open space, recreation areas, and other public facilities; the protection of historic resources and the natural environment; and the distribution of population and traffic to avoid congestion and overcrowding and which shall create conditions essential to public health, safety, and the general welfare.

5. CHARACTERIZING LAND POLLUTION

Land and water are intricately connected. In addition to the damage caused to the land when topsoil is lost after a precipitation event, the stormwater transports both suspended material (e.g., soil and debris) and soluble matter (e.g., organic and organic chemicals). These materials either directly enter surface waters or reach a storm sewer system. The material reaching the water is harmful in two ways. First, it harms the integrity of the receiving stream, increasing turbidity, suspended and dissolved solids content, and increases the organic load. Second, the material can be hazardous.

The material in the runoff is collectively referred to as "sediment," which can substantially decrease the ability of light to penetrate the water column, that is, increased turbidity. Sediment loading diminishes the growth capacity of aquatic plants and disrupts normal processes, such as seasonal turnover and nutrient cycling in the surface waters.

Household hazardous wastes, such as insecticides, pesticides, paint, solvents, used motor oil, and other auto fluids are also frequently part of the sediment load. These are often highly toxic to aquatic life and human populations. These wastes are discussed in detail in Chapter 13.

Pathogenic microbes are suspended or sorbed to suspended particles. Notably, *Escherichia coli* can cause major health problems and indicate the presence of enteric bacteria, that is, those that grow in the intestines of animals, so they are also known as fecal coliform bacteria. This can be an indication of agricultural and human waste sources, for example, septic tanks or even poorly treated effluent from municipal waste water treatment systems.

Sediment loading can change the species composition and diversity of lakes and other surface waters. For example, the increase input of nitrogen, phosphorus, and other nutrients encourages the growth of algae, leading to algal blooms. When these algae die, they fall to the lake bottom. As they continue to decompose, more oxygen is demanded by the bacteria and abiotic processes, so the concentrations of dissolved oxygen (DO) decline. As a result, fish and other organisms at higher trophic levels in the food chain are stressed. This can result in fish kills and complete alterations to aquatic habitats.

5.1. Land-Based Stress to Water Bodies: Biochemical Oxygen Demand

The biochemical oxygen demand (BOD) is the amount of oxygen that bacteria will consume in the process of decomposing organic matter under aerobic conditions. The BOD is measured by incubating a sealed sample of water for 5 days and measuring the loss of oxygen by comparing the O_2 concentration of the sample at time $= 0$ (just before the sample is sealed) to the concentration at time $= 5$ days (known specifically as BOD_5). Samples are commonly diluted before incubation to prevent the bacteria from depleting all the oxygen in the sample before the test is complete [9]. BOD_5 is merely the measured DO at the beginning time, that is, the initial DO (D_1), measured immediately after it is taken from the source) minus the DO of the same water measured exactly 5 days after D_1, that is, D_5:

$$\text{BOD} = \frac{D_1 - D_5}{P} \tag{29.1}$$

where P = decimal volumetric fraction of water utilized. D units are in mg L^{-1}. If the dilution water is seeded, the calculation becomes:

$$\text{BOD} = \frac{(D_1 - D_5) - (B_1 - B_5)f}{P} \tag{29.2}$$

where B_1 = initial DO of seed control; B_5 = final DO of seed control; and f = the ratio of seed in sample to seed in control $= \frac{\%\text{ seed in } D_1}{\%\text{ seed in } B_1}$. B units are in mg L^{-1}.

For example, to find the BOD_5 value for a 10-mL water sample added to 300 mL of dilution water with a measured DO of 7 mg L^{-1} and a measured DO of 4 mg L^{-1} 5 days later:

$$P = \frac{10}{300} = 0.03$$

$$\text{BOD}_5 = \frac{7-4}{0.03} = 100 \text{ mg L}^{-1}$$

Thus, the microbial population in this water is demanding 100 mg L^{-1} DO over the 5-day period. Therefore, if a conventional municipal wastewater treatment system is achieving 95% treatment efficiency, the effluent discharged from this plant would be 5 mg L^{-1}.

Chemical oxygen demand (COD) does not differentiate between biologically available and inert organic matter, and it is a measure of the total quantity of oxygen required to oxidize all organic material completely to carbon dioxide and water. COD values always exceed BOD values for the same sample. COD (mg L^{-1}) is measured by oxidation using potassium dichromate ($K_2Cr_2O_7$) in the presence of sulfuric acid (H_2SO_4) and silver. By convention, 1 g of carbohydrate or 1 g of protein accounts for about 1 g of COD. On average, the ratio BOD:COD is 0.5. If the ratio is <0.3, the water sample likely contains increased concentrations of *recalcitrant* organic compounds, that is, compounds that resist biodegradation [10], that is, there are numerous carbon-based compounds in the sample, but the microbial populations are not efficiently using them for carbon and energy sources. This is the advantage of having both BOD and COD measurements. Sometimes, however, COD measurements are conducted simply because they require only a few hours compared with the 5 days for BOD.

Because available carbon is a limiting factor, the carbonaceous BOD reaches a plateau, that is, *the ultimate carbonaceous BOD* (see Fig. 29.1). However, carbonaceous compounds are the only substances demanding oxygen. Microbial populations will continue to demand O_2 from the water to degrade other compounds, especially nitrogenous compounds, which account for the bump in the BOD curve. Thus, in addition to serving as an indication of the amount

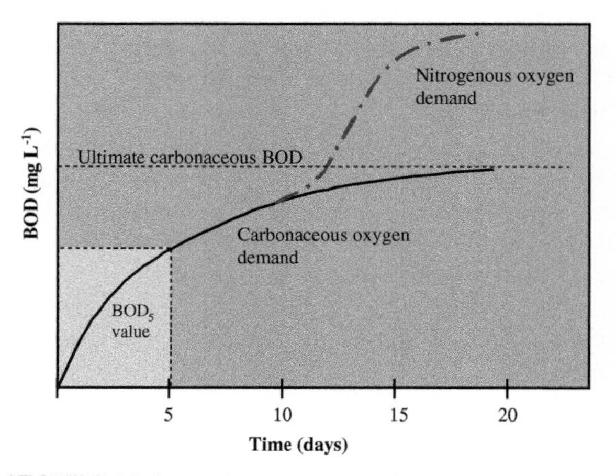

FIGURE 29.1 Biochemical oxygen demand (BOD) curve, showing ultimate carbonaceous BOD and nitrogenous BOD. *Adapted from Ref. [10].*

of molecular oxygen needed for biological treatment of the organic matter, BOD also provides a guide to sizing a treatment process, assigning its efficiency, and giving operators and regulators information about whether the facility is meeting its design criteria and is complying with pollution control permits.

If effluent with high BOD concentrations reaches surface waters, it may diminish DO to levels that are lethal to some fish and many aquatic insects. As the water body re-aerates as a result of mixing with the atmosphere and by algal photosynthesis, O_2 is added to the water, the oxygen levels will slowly increase downstream. The drop and rise in DO concentrations downstream from a source of BOD is known as the DO sag curve, because the concentration of DO "sags" as the microbes deplete it. Therefore, the O_2 concentrations decrease with both time and distance from the point where the high BOD substances enter the water.

5.2. Predicting Sediment-Induced Stress

The stress from decreasing DO is usually indicated by the BOD. Like most environmental systems, the water bodies that receive sediment loads are complex in their response to increased input of materials. The DO will respond both positively and negatively to increased nutrient levels, because the biota have unique optimal ranges of growth and metabolism that vary among species (e.g., algae will add some O_2 with photosynthesis but use some O_2 for metabolism, whereas the bacteria will generally be net consumers of molecular oxygen).

The most widely applied environmental indices are those that follow the framework of an *index of biological integrity*. In biological systems, *integrity* is the capacity of a system to sustain a balanced and healthy community. This means the community of organisms in that system meets certain criteria for species composition, diversity, and adaptability, often compared with a *reference site* that is a benchmark for integrity. As such, biological integrity indices are designed to integrate the relationships of chemical and physical parameters with each other and across various levels of biological organization. They are now used to evaluate the integrity of environmental systems using a range of metrics to describe the system.

Thus, environmental indices combine attributes to determine a system's condition (e.g., diversity and productivity) and to estimate stresses. The original index of biotic integrity developed by Karr [11] was based on fish fauna attributes and has provided predictions of how well a system will respond to a combination of stresses. In fact, the index is completely biological, with no direct chemical measurements. However, the metrics are indirect indicators of physicochemical factors (e.g., the abundance of game fish is directly related to DO concentrations). The metrics provide descriptions of a system's structure and function.

An example of the data that is gathered to characterize a system is provided in Table 29.1. The information that is gleaned from these data is tailored to the physical, chemical, and biological conditions of an area (in Table 29.2), for example, for large spatial regions. The information from a biologically based index can be used to evaluate a system, as shown in Fig. 29.2.

Systems involve scale and complexities in both biology and chemistry. In land habitats, scale and complexity must be part of any analysis of threats and risks; with considerations of both abiotic and biotic factors.

For example, a fish's direct aqueous exposure (AE in $\mu g\ d^{-1}$) is the product of the organism's ventilation volume, that is, the flow Q (in mL d^{-1}), and the compound's aqueous concentration, C_w ($\mu g\ mL^{-1}$). The fish's exposure by its diet (DE, in $\mu g\ d^{-1}$) is the product of its feeding rate, F_w ($g\ d^{-1}$, wet weight), and the compound's concentration in the fish's prey, C_p ($\mu g\ g^{-1}$, wet weight). If the fish's food consist of single type of prey that is at equilibrium with the water, fish's aqueous and dietary exposures

TABLE 29.1 Biological Metrics Used in the Original Index of Biological Integrity (IBI)

Integrity aspect	Biological metric
Species richness and composition	Total number of fish species (total taxa)
	Number of *Catostomidae* species (suckers)
	Number of darter species
	Number of sunfish species
	Number of intolerant or sensitive species
Indicator species metrics	Percentage of individuals that are *Lepomis cyanellus* (Centrarchidae)
	Percentage of individuals that are omnivores
Trophic function metrics	Percentage of individuals that are insectivorous, Cyprinidae
	Percentage of individuals that are top carnivores or piscivores
	Percentage of individuals that are hybrids
Reproductive function metrics	Abundance or catch per effort of fish
Abundance and condition metrics	Percentage of individuals that are diseased, deformed, or that have deformities, eroded fins, lesions and tumors (DELTs)

Source: Ref. [11].

and the bioconcentration factor (BC) can be calculated when they are equal:

$$\mathrm{AE} = \mathrm{DE};\ QC_{\mathrm{w}} = F_{\mathrm{w}}C_{\mathrm{p}};\mathrm{BCF} = \frac{Q}{F_{\mathrm{w}}} \quad (29.3)$$

The ventilation-to-feeding ratio for a 1-kg trout has been found [12] to be on the order of $10^{4.3}$ mL g^{-1}. Assuming the quantitative structure activity relationship for the trout's prey is BCF $= 0.048$ times the octanol–water coefficient (K_{ow}); it appears that the trout's predominant route of exposure for any chemical with a $K_{\mathrm{ow}} > 10^{5.6}$. Exposure must also account for the organism's assimilation of compounds in food, which for very lipophilic compounds, will probably account for the majority of exposure compared with that from the water. Even though chemical exchange occurs from both food and water via passive diffusion (Fick's Law relationships), the uptake from food, unlike direct uptake from water, does not necessarily relax the diffusion gradient into the fish. The difference between digestion and assimilation of food can result in higher contaminant concentrations in the fish's gut.

Predicting expected uptake where the principal route of exchange is dietary can be further complicated by the fact that most fish species exhibit well-defined size-dependent, taxonomic, and temporal trends regarding their prey. Thus, a single bioaccumulation factor (BAF) may not be universally useful for risk assessments for all fish species. Indeed, the BAF may not even apply to different sizes of the same species.

The systematic biological exchange of materials between the organism, in this case various species of fishes, is known as uptake, which can be expressed by the following three differential equations for each age class or cohort of fish [13]:

$$\frac{dB_{\mathrm{f}}}{dt} = J_{\mathrm{g}} + J_{\mathrm{i}} + J_{\mathrm{bt}} \quad (29.4)$$

where $B_{\mathrm{f}} =$ compound's body burden (microgram per fish), J_{g} represents the net chemical exchange (microgram per day) across the fish's gills from the water; J_{i} represents the net chemical exchange (microgram per day) across the fish's intestine from food; and J_{bt} represents the compound's biotransformation rate (microgram per day). The amount of contaminant uptake depends on the organism, the habit and the complexity of processes, as evidenced by change in biomass:

$$\frac{dW_{\mathrm{d}}}{dt} = F_{\mathrm{d}} - E_{\mathrm{d}} - R - \mathrm{EX} - \mathrm{SDA} \quad (29.5)$$

where W_{d} = dry body weight (gram dry weight per fish) of the average individual

TABLE 29.2 Biological Metrics That Apply to Various Regions of North America[a]

Alternative IBI metrics	Midwestern United States	Central Appalachians	Sacramento-San Joaquin	Colorado front range	Western Oregon Ohio	Ohio headwater sites	Northeastern United States	Ontario	Central corn belt plain	Wisconsin-warm water	Wisconsin-cold water	Maryland coastal plain	Maryland Nontidal
1. Total number of species	X	X	X	X			X		X			X	X
Number of native fish species					X	X	X	X		X			
Number of salmonid age classes[b]				X	X								
2. Number of darter species	X	X		X		X			X	X			
Number of sculpin species					X								
Number of benthic insectivore species								X					
Number of darter and sculpin spices							X						
Number of darter, sculpin, and madtom species										X			
Number of salmonid juveniles (individuals)[b]			X		X			X					
Percentage of round-bodied suckers						X[c]							
Number of sculpins (individuals)			X										
Number of benthic species												X	X
3. Number of sunfish species	X			X		X			X	X			
Number of cyprimid species					X								
Number of water column species								X					
Number of sunfish and trout species								X					
Number of salmonid species			X						X				
Number of headwater species							X						

(*Continued*)

TABLE 29.2 Biological Metrics That Apply to Various Regions of North America[a]—*cont'd*

Alternative IBI metrics	Midwestern United States	Central Appalachians	Sacramento-San Joaquin	Colorado front range	Western Oregon Ohio	Ohio headwater sites	Northeastern United States	Ontario	Central corn belt plain	Wisconsin-warm water	Wisconsin-cold water	Maryland coastal plain	Maryland Nontidal
Percentage of headwater species							X		X				
4. Number of sucker species	X				X	X		X	X	X			
Number of adult trout species[b]			X		X								
Number of minnow species				X			X		X				
Number of sucker and catfish species								X					
5. Number of intolerant species	X			X	X	X		X		X	X	X	X
Number of sensitive species							X		X				
Number of amphibian species			X										
Presence of brook trout								X					
Percentage of stenothermal cool and cold water species											X		
Percentage of salmonid ind. as brook trout											X		
6. Percentage of green sunfish	X												
Percentage of common carp					X								
Percentage of white sucker				X				X					
Percentage of tolerant species						X	X		X	X	X	X	X
Percentage of creek chub		X											
Percentage of dace species								X					
Percentage of eastern mudminnow												X	
7. Percentage of omnivores	X			X		X	X	X X	X	X			

Percentage of generalist feeders		X												
Percentage of generalist, omnivores, and invertivores														X
8. Percentage of insectivorous cyprinids	X												X	
Percentage of insectivores					X			X		X	X		X	X
Percentage of specialized insectivores		X		X										
Number of juvenile trout			X											
Percentage of insectivorous species						X	X							
9. Percentage of top carnivores	X					X		X	X	X	X	X		
Percentage of catchable salmonids					X									
Percentage of catchable trout			X											
Percentage of pioneering species							X			X			X	
Density catchable wild trout			X											
10. Number of individuals (or catch per effort)	X	X	X	X	X	X[d]	X[e]		X	X	X			
Density of individuals								X						X
Percentage of abundance of dominant species													X	X
Biomass (per m^2)														X[f]
11. Percentage of hybrids	X							X						
Percentage of introduced species				X	X									
Percentage of simple lithophills						X				X	X			X
Percentage of simple lithophills species							X							
Percentage of native species			X											

(Continued)

TABLE 29.2 Biological Metrics That Apply to Various Regions of North America[a]—*cont'd*

Alternative IBI metrics	Midwestern United States	Central Appalachians	Sacramento-San Joaquin	Colorado front range	Western Oregon Ohio	Ohio headwater sites	Northeastern United States	Ontario	Central corn belt plain	Wisconsin-warm water	Wisconsin-cold water	Maryland coastal plain	Maryland Nontidal
Percentage of native wild individuals			X										
Percentage of silt-intolerant spawners												X	
12. Percentage of diseased individuals (deformities, eroded fins, lesions, and tumors)	X	X		X	X	X	X	X X	X	X		X	X

[a]*Taken from Karr et al. (1986), Leonard and Orth (1986), Moyle et al. (1986), Fausch and Schrader (1987), Hughes and Gammon (1987), Ohio EPA (1987), Miler et al. (1988), Steedman (1988), Simon (1991), Lyons (1992a), Barbour et al. (1995), Simon and Lyons (1995), Hall et al. (1988), Lyons et al. (1996), Roth et al. (1997).*
[b]*Metric suggested by Moyle et al. (1986) or Hughes and Gammon (1987) as a provisional replacement metric in small western salmonid streams.*
[c]*Boat sampling methods only (i.e., larger streams/rivers).*
[d]*Excluding individuals of tolerant species.*
[e]*Noncoastal plain streams only.*
[f]*Coastal plain streams only.*
Note: X = metric used in region. Many of these variations are applicable elsewhere.
Source: M. T. Barbour, J. Gerritsen, B. D. Snyder, and J. B. Stribling (1999). Rapid Bioassessment Protocols for Use in Streams and Wadeable Rivers: Periphyton, Benthic Macroinvertebrates and Fish, 2nd ed, Report No. EPA 841-B-99-002, U.S. Environmental Protection Agency, Office of Water, Washington, DC.

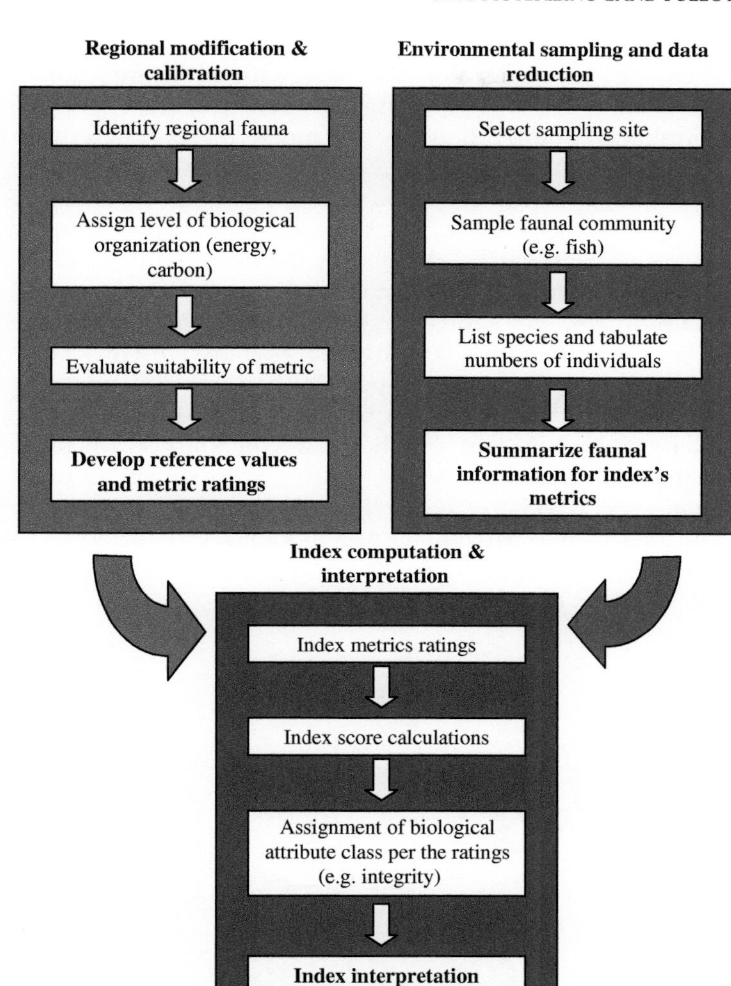

FIGURE 29.2 Sequence of activities involved in calculating and interpreting an Index of Biotic Integrity (IBI) [14,15].

within the cohort; F_d = the fish's feeding; E_d = egestion (i.e., expulsion of undigested material); R = routine respiration; EX = excretion; and SDA = specific dynamic action (i.e., the respiratory expenditure in excess of R required to assimilate food). All these parameters have units of gram per day, dry weight. The population of a particular species is also affected by predator-prey complexities:

$$\frac{dN}{dt} = \text{EM} - \text{NM} - \text{PM} \tag{29.6}$$

where N is the cohort's population density (fish per hectare), EM is the emigration/dispersal (fish per hectare per day), NM is non-predatory mortality (fish per hectare per day).

Physiologically based models for fish growth are often formulated in terms of energy content and flow (e.g., kilograms per fish and kilojoules per day), Equation 29.3 is basically the same as such bioenergetic models because energy densities of fish depend on their dry weight [16]. Obviously, feeding depends on the availability of suitable prey, so the mortality of the fish is a function of the individual feeding levels and population densities of its predators. Thus, the fish's dietary exposure is directly related to the organism's feeding rate and the concentration chemicals in its prey.

5.3. Point and Nonpoint Sources of Sediment

The amount of any substance, for example, nutrient, pollutant, or microbes, that is discharged into a system is known as the *load.* Nutrient loading, pollutant loading, and microorganism loading can be quantified, not only by the gross mass of the substance entering a system, for example, a water body, but also by the response of that system to the load. Thus, the relationship between a substance's entry and its impact on the receiving river, lake, wetland, estuary, ocean, or aquifer. However, these receptors can also be compartmentalized into their various media, for example, soil, sediment, water, and biota.

The first major division of loads is between *point sources* and *nonpoint sources.* This is actually a distinction of convenience. What some may call a point source, others may consider to be a nonpoint source. For example, if a 1-ha field is near a small creek, this may be a relatively large and dispersed source of a pesticide being released to the creek. However, a 1-ha trickling filter system near a large river may be classified as a point source, because the majority of the discharge to the river comes through a conduit (i.e., the *outfall structure*) from the system. However, it is highly likely that pollutants are being released from the treatment facility from sources other than the outfall structure.

Thus, nonpoint sources are generally associated with *runoff,* that is, multiple sources in a given two-dimensional space that leave these sources and are transported toward a receptor system. These receiving bodies can be found above the ground (e.g., streams and lakes) or below the ground (aquifers). They are usually cascades of systems, such as a contaminant moving atop the soil, with some infiltration. The infiltrated fraction may find its way to the aquifer and moves much more slowly than the runoff found above the ground. However, the contaminated ground water may recharge a stream flow. The same stream may have already been contaminated by the overland flow, so the stream is receiving a continuous flow (ground water source to stream) and a plug or episodic flow (overland runoff), as shown in Fig. 29.3.

For biota, especially bacteria, the nonpoint and point contribution is even more complex than that of chemical compounds. Microbes may be directly loaded into a stream or other receiving body (Fig. 29.3A). In addition, they may accumulate as nonpoint source loads from a

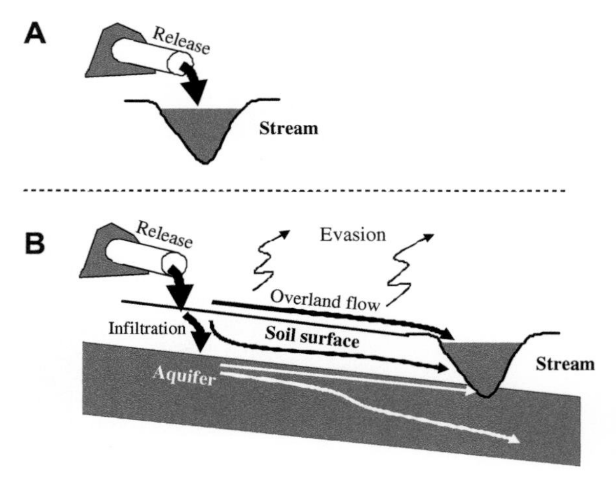

FIGURE 29.3 (A) Point source—material is released directly to the receiving body. (B) Nonpoint source—material runs off, infiltrates through soil column, follows ground water flow lines, and eventually reaches the receiving body. However, along the way, a fraction of the material may be chemically transformed and physically held (e.g., sorbed to soil particles). For volatile compounds, a fraction will also be evaded to the atmosphere. Thus, the net that reaches the water body is a function of numerous factors, including vegetative cover, soil porosity and permeability, ground water flow rates, soil texture and Henry's Law, and other equilibrium coefficients for the released material. *Source: D. A. Vallero (2010). Environmental Biotechnology: A Biosystems Approach, Elsevier Academic Press, Amsterdam, NV.*

combination of sources. As they are transported over land and under the soil surface, the microbes meet hostile and accommodating conditions, which will decrease and increase their numbers, respectively. However, the hostile conditions may lead to the formation of cysts and spores, which are more likely than the bacteria themselves to be transported by advection in the air and water or carried by fauna (see Fig. 29.4). When the conditions change to become more accommodating, the spores will generate increased microbial populations. Thus, the actual loading to the body consists of a complicated set of transport and fate processes.

Loading calculations must take into account background sources (sometimes referred to as "natural" sources). Atmospheric loading is similar, but usually the distinctions are made between *mobile sources* and *stationary sources*. Two systems, hydrologic and atmospheric, join at the interfaces between earth and water bodies and the atmospheric, that is, terrestrial and tropospheric fluxes. For example, atmospheric deposition is an important source of nutrient loading (e.g., N, P, S, and K) to ecosystems. Ecosystems, for example, wetlands, load nutrients back into the atmosphere, but often as different chemical species, that is, the ecosystems are acting as control volumes in which reactions are taking place. Many of these reactions are biotic and are mediated by microbes and plants.

The linkages between biota and their environments are influenced by the availability and forms of nutrients. When rain containing nitrates falls to the wetland water and land surfaces, for instance, plant roots take it up and metabolize it into

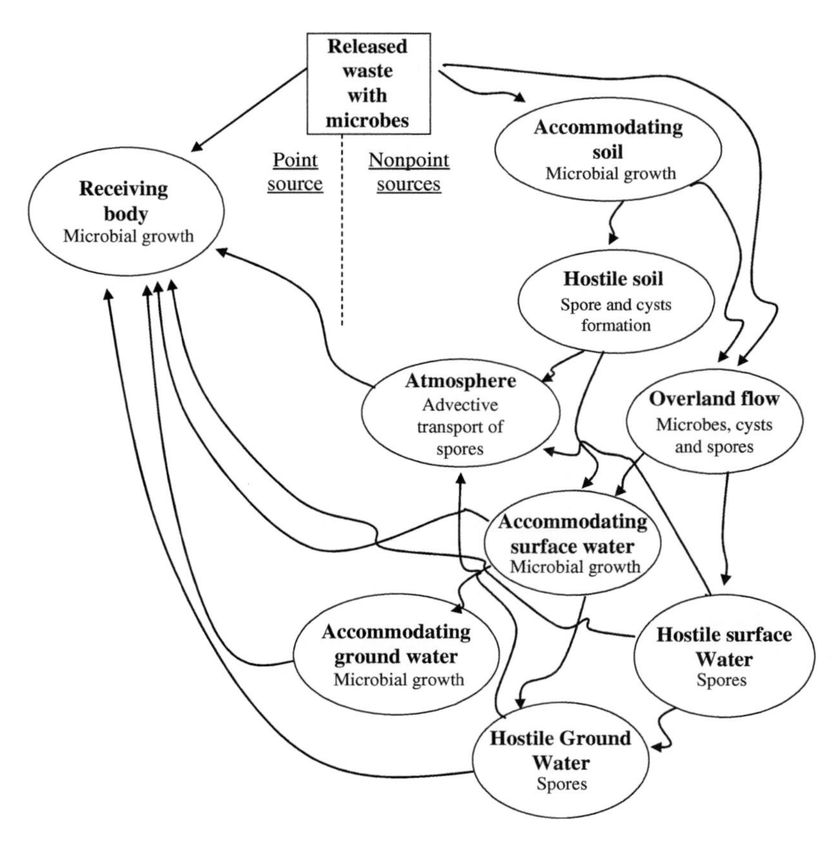

FIGURE 29.4 Difference between point source and nonpoint source releases of microbes. In a point source, the growth, metabolism, and formation of cysts and spores depend on the conditions of the receiving water body, which varies in time but not in space. In a nonpoint source scenario, the growth, metabolism, and formation of spores and cysts are a cascade in time and space. In accommodating environments, the microbes will undergo metabolism and growth, which may allow for larger microbial populations delivered to other compartments and eventually to the receiving body. However, various compartments in the flow may be hostile (insufficient nutrients, water, pH, temperature, etc.). These may induce the formation of spores that can be advectively transported in the atmosphere or with the flow of ground and surface waters. *Source: D. A. Vallero (2010). Environmental Biotechnology: A Biosystems Approach, Elsevier Academic Press, Amsterdam, NV.*

organic forms of nitrogen. Meanwhile, bacteria in sediments reduce it to ammonia. In the opposite direction, the reduced forms are oxidized by other bacteria (e.g., *Nitrosomonas* converts ammonium ions [NH_4^+] to nitrite [NO_2^-], *Nitrobacter* converts the NO_2^- to nitrate [NO_3^-]). Thus, two opposite reactions are at play constantly in loading scenarios. In this instance, these are *ammonification* (or *deamination*) and *nitification:*

Ammonification:

$$NH_4^+ + OH^- \Leftrightarrow NH_3 + H_2O \quad (29.7)$$

Nitrification:

$$NH_4^+ + 2O_2 \rightarrow NO_3^- + 2H^+ + H_2O \quad (29.8)$$

Note that ammonification is written as an equilibrium reaction, and nitrification is an oxidation reaction. However, the loading depends on numerous environmental conditions, including pH. Thus, under acidic conditions, most of the ammonia nitrogen will ionize to ammonium, and under basic conditions, the non-ionized ammonia concentrations will increase in proportion to the ammonium. This is important because nonionized ammonia is very toxic to aquatic biota, whereas ionized species are nutrients needed for plants and algae. Add to this, the decomposition of organic matter, and the loading becomes a complicated mix of reactions:

$$\text{Organic nitrogen} + O_2 \rightarrow NH_3 \text{ nitrogen} + O_2$$

$$\rightarrow NO_2 \text{ nitrogen} + O_2 \rightarrow NO_3 \text{ nitrogen} + O_2 \quad (29.9)$$

Even this is a gross oversimplification, because the oxidation and reduction depend on the types of bacteria present. For example, the O_2 required for nitrification is theoretically 4.56 mg of O_2 per square milligram of NH_4^+, but this is an *autotrophic* reaction. Thus, O_2 is being produced by the nitrifying bacteria, decreasing the amount needed. However, the growth rate of nitrifying microbes is far less than that of *heterotrophic* microbes decomposing organic wastes. Therefore, when large amounts of organic matter are being degraded, the nitrifiers' growth rate will be sharply limited by the heterotrophs, which means the rate of nitrification will commensurately be decreased [17]. Furthermore, these same responses to conditions exist for all other nutrients in the system, for example, organic P will be oxidized (and oxidized P species will be reduced). Thus, the kinetics of nutrient loading are quite complex.

5.3.1. *Total Maximum Daily Loading*

In the United States, the Clean Water Act restricts the total maximum daily load (TMDL) of pollutants released to water bodies that have been deemed to be *impaired* by the state in which technology-based and other engineering controls are not adequate to achieve water quality standards. Thus, the TMDL reflects the amount of a pollutant that can be discharged from point, nonpoint, and natural background sources, including a margin of safety (MOS) for any water quality-limited water body. The TMDL process consists of five steps:

1. Selection of the pollutant in need of consideration;
2. Estimation of the water body assimilative capacity (i.e., loading capacity);
3. Estimation of the pollutant loading from all sources to the water body;
4. Analysis of current pollutant load and determination of needed reductions to meet assimilative capacity; and
5. Allocation, including a MOS, of the allowable pollutant load among the different pollutant sources in a manner that water quality standards are achieved

A beneficial use that has been impaired is a function of a change in the chemical, physical, or biological integrity of surface waters, such as the following:

- Restrictions on fish and wildlife consumption
- Tainting of fish and wildlife flavor

- Degradation of fish wildlife populations
- Fish tumors or other deformities
- Bird or animal deformities or reproduction problems
- Degradation of benthos
- Restrictions on dredging activities
- Eutrophication or undesirable algae
- Restrictions on drinking water consumption or taste and odor problems
- Beach closings
- Degradation of aesthetics
- Added costs to agriculture or industry
- Degradation of phytoplankton and zooplankton populations
- Loss of fish and wildlife habitat

The general components needed to develop a TMDL are shown in Fig. 29.5. Meeting water quality standards is a way to protect beneficial uses, including recreational activities, aquatic life, and water supply. For example, recreational beneficial use can be based on protection against fecal coliform, including *E. coli* bacteriological criteria. For example, if a lake is designated for primary contact recreation water, potential exposures to bacteria and a baseline level of disinfection will be required. This can be more straightforward for point sources, for example, building wastewater treatment systems, than for nonpoint sources. The latter often requires management approaches, such as measures to prevent soil erosion, because soil may carry coliform bacteria.

6. HABITAT LOSS AND DESTRUCTION

Obviously, terrestrial species depend on both the quantity and quality of land for their survival. In addition, all creatures, to some extent, are affected by land habitats; some because a life stage is terrestrial, for example, the amphibians; and some because the land is the source of stress on the water.

6.1. Protecting and Enhancing Land Habitats

Plans must include measures to complement the natural and cultural environment and to avoid destruction or adverse alteration of the land. This means that good practice should always consider approaches to minimize impacts of construction, development, and facility operation on the surrounding watershed, by preventing water pollution and actions that disrupt natural drainage patterns.

In addition, land use planning must directly address habitat destruction by recommending ways that any development conserve land that serves as habitat for native flora and fauna, both on- and off-site, and directly and indirectly. An example of indirect destruction or secondary impacts from infrastructure development is location of new sewer lines or roads in previously undeveloped or low-density land. This invites new residential and commercial development, which in turn calls for more roads, structures, and impervious surfaces. These changes then lead to less soil infiltration and increase the risks of flooding on-site and downstream, as the holding capacity for water during rain events decreases.

Another special consideration for land development is the presence of threatened or endangered species and the potential that their habitats may be affected by developing a site.

7. WASTE SITES

Love Canal was the key event that led to the passage of groundbreaking environmental legislation. As mentioned in Chapter 27, the Resource Conservation and Recovery Act of 1976 was passed in the US Congress to ensure that wastes are tracked from "cradle to grave" for active hazardous waste facilities, that is, those that are still operational and where an owner/operator is identified [18]. The Comprehensive Environmental Response,

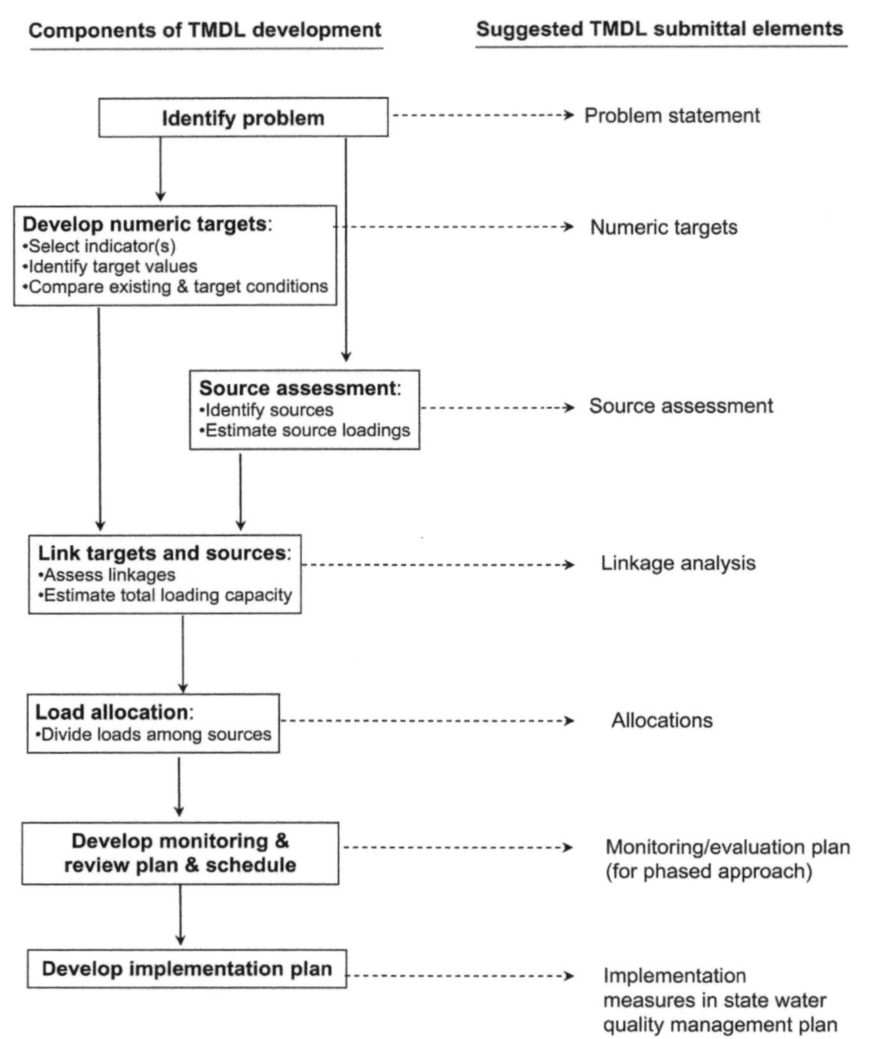

FIGURE 29.5 Steps in developing a total maximum daily load. *Source: US Environmental Protection Agency (2001). Protocol for developing pathogen TMDLs. Report No. EPA 841-R-00-002.*

Compensation and Liability Act of 1980, better known as "Superfund," was passed to address and clean up abandoned waste sites [19]. These laws authorized the thousands of pages of regulations and guidance that declared that when pollution thresholds are crossed, the design has failed to provide adequate protection.

Numerous other waste sites were discovered in the late 1970s, including dioxin contaminated Times Beach, Missouri and abandoned chemicals in the Valley of the Drums, Kentucky. These discoveries prompted regulators to approach pollutants from the perspective of risk.

People living near hazardous waste sites, for good reason, want to be assured that they will be "safe" but safety is a relative term. Calling something "safe" integrates a value judgment that is invariably accompanied by uncertainties. The safety of a product or process can be described in objective and quantitative terms.

Factors of safety are a part of every waste management decision (see Chapter 32).

7.1. Restoration of Contaminated Land

A key question when discovering a contaminated site is the extent to which it would be possible to locate and rehabilitate or restore native habitats. For example, the U.S. Environmental Protection Agency's (EPA) Land Revitalization Program has been designed to restore land and other natural resources into sustainable community assets that maximize beneficial economic, ecological, and social uses and ensure protection of human health and the environment [20].

Projects have included sustainable cleanup and redevelopment; green building; sustainable development; alternatives for affordable housing, incorporating the reuse of construction and demolition of debris into redevelopment projects; and comprehensive "hands on" training to local government officials on assessing, remediating, and revitalizing contaminated properties in a sustainable manner. When a previously contaminated site is remediated to the point that it has regained a commercial or other beneficial use, this is known as a brownfield. Several sites now exist throughout the United States.

8. BEST PRACTICES

A question crucial to any waste management decision is how well does the plan fit with land development practices, codes, and design criteria? Land use planning is actually an "applied social science." Proper land development calls for a set of criteria that draws from the ample reservoir of safety criteria, especially the following four rules [21]:

1. The design must comply with applicable laws.
2. The design must adhere to "acceptable engineering practice."
3. Alternative designs must be sought to see whether better practices are available.
4. Possible misuse of the design must be foreseen.

Land use decisions in the United States are predominantly made at the local level. Durham's, NC, development plan [22], for example, incorporates several land use considerations, including the following:

1. Intensity/density of the proposed development, that is, any human-caused change to improved or unimproved real estate that requires a permit or approval from any agency of the City or County of Durham, including but not limited to, constructing or changing buildings or other structures, mining, dredging, filling, grading, paving, excavation or drilling operations, and storage of equipment or materials;
2. Sensitive areas and related protection;
3. Any limitations on number, type, or range of uses, that is, the purpose for which a building, structure, or area of land may be arranged or occupied or the activity conducted or proposed in a building, structure, or on an area of land;
4. Dedications, that is, the transfer without payment of ownership or other interest in real property from a private entity to a public agency or reservations;
5. Design elements if required or otherwise provided; and
6. Development phasing if required or otherwise provided.

Each of these considerations requires that land use decision be based on thorough and scientifically sound information and that the planners, engineers, and others involved in implementing the plan use sound judgment and follow best practices.

9. CONCLUSIONS

As real estate professionals are fond of saying about land, "they aren't making any more of it."

With the small exception of the Zuiderzee and a few other sedimentation projects in the Persian Gulf and elsewhere, this is generally true. Therefore, wasting land by losing it through erosion, diminishing its beneficial uses, and contaminating it is wholly unacceptable. Protecting land relies on sound application of not only the physical and natural sciences but also the social sciences. Where people choose to live is based on environmental quality and sociological and psychological factors. Land use planners, engineers, and waste managers each have vital roles in preventing and addressing land pollution.

References

[1] A. Leopold, A Sand County Almanac, Oxford University Press (1987), New York, NY, 1949.

[2] E. Birmingham, Position Paper: "Reframing the Ruins: Pruitt-Igoe, Structural Racism, and African American Rhetoric as a Space for Cultural Critique," Brandenburgische Technische Universität, Cottbus, Germany (1998). See also: C. Jencks, The Language of Post-Modern Architecture, fifth ed., Rizzoli, New York, NY (1987).

[3] A. von Hoffman, Why They Built Pruitt-Igoe, Taubman Centre Publications, A. Alfred Taubman Centre for State and Local Government, Harvard University, Cambridge, MA, 2002.

[4] J. Bailey, A case history of failure, Architectural Forum 122 (9) (1965).

[5] Ibid.

[6] B.B. Marriott, Environmental Impact Assessment: A Practical Guide, Chapter 5, 'Land Use and Development,', McGraw-Hill, New York, NY, 1997.

[7] M. Ritzdorf, Locked Out of Paradise: Contemporary Exclusionary Zoning, the Supreme Court, and African Americans, 1970 to the Present, in: J.M. Thomas, M. Ritzdorf (Eds.), Urban Planning and the African American Community: In the Shadows, Sage Publications, Thousand Oaks, CA, 1997.

[8] M. Woolfolk, Stormwater 201—Water Quality Program, Joint City-County Planning Committee, Durham, NC, 2008.

[9] State of Georgia, Watershed Protection Plan Development Guidebook, Georgia Department of Natural Resources, Environmental Protection Division, Atlanta, GA, 2003.

[10] C. P Gerba, I.L. Pepper, Wastewater treatment and biosolids reuse, in: R.M. Maier, I.L. Pepper, C.P. Gerba (Eds.), Environmental Microbiology, 2nd ed., Elsevier Academic Press, Burlington, MA, 2009.

[11] J.R. Karr, Assessment of biotic integrity using fish communities, Fisheries 6 (1981) 21–27.

[12] R.L. Erickson, J.M. McKim, A model for exchange of organic chemicals at fish gills: flow and diffusion limitations, Aquatic Toxicology 18 (1990) 175–198; D.J. Stewart, D. Weininger, D.V. Rottiers, T.A. Edsall, An energetics model for lake trout *Salvelinus namaycush*: Application to the Lake Michigan population, Canadian Journal of Fisheries and Aquatic Sciences 40 (1983) 681–698.

[13] M.C. Barber, Bioaccumulation and Aquatic System Simulator (BASS). User's Manual, Version 2.2, U.S. Environmental Protection Agency, Athens, GA, 2008. Report No. EPA 600/R-01/035, update 2.2, March 2008.

[14] M.T. Barbour, J. Gerritsen, B.D. Snyder, J.B. Stribling, Rapid Bioassessment Protocols for Use in Streams and Wadeable Rivers: Periphyton, Benthic Macroinvertebrates and Fish, 2nd ed., U.S. Environmental Protection Agency, Office of Water, Washington, DC, 1999. Report No. EPA 841-B-99-002.

[15] J.R. Karr, Biological monitoring and environmental assessment: A conceptual framework, Environmental Management 11 (1987) 249–256.

[16] J.A. Kushlan, S.A. Voorhees, W.F. Loftus, P.C. Frohring, Length, mass, and calorific relationships of Everglades animals, Florida Scientist 49 (1986) 65–79; K.J. Hartman, S.B. Brandt, Estimating energy density of fish, Transactions of the American Fisheries Society 124 (1995) 347–355; K. Schreckenbach, R. Knösche, K. Ebert, Nutrient and energy content of freshwater fishes, Journal of Applied Ichthyology 17 (2001) 142–144.

[17] V. Novotny, Water Quality, second ed., John Wiley & Sons, Inc., Hoboken, NJ, 2003.

[18] Resource Conservation and Recovery Act of 1976 (42 U.S.C. s/s 321 et seq.).

[19] Comprehensive Environmental Response, Compensation and Liability Act of 1980 (42 U.S.C. 9601-9675). December 11, 1980. In 1986, CERCLA was updated and improved under the Superfund Amendments and Reauthorization Act (42 U.S.C. 9601 et seq), October 17, 1986.

[20] U.S. Environmental Protection Agency, Land revitalization: Restoring land for America's communities; <http://www.epa.gov/landrevitalization/download/lrbrochure08.pdf>, accessed on July 20, 2010 (2010).

[21] C.B. Fleddermann, *Engineering Ethics*, Chapter 5, 'Safety and Risk', Prentice-Hall, Upper Saddle River, NJ, 1999.

[22] City of Durham, NC, Unified Development Ordinance. Section 3.5.6; <http://www.durhamnc.gov/udo/mainPage.asp>, accessed on July 20, 2010 (2010).

PART III

BEST PRACTICE AND MANAGEMENT

From the information gleaned from the many streams we have followed, we can now investigate recent innovations that will help us in our journey over the next 20 years.

A waste management professional provides best practice if the service is optimal to the client; in this case the client is the public. Best practice means that the recommended controls and management approaches are appropriate, accepted, and widely used according to expert consensus and they embody an integrated, comprehensive and continuously improving approach. Thus, this last section of the handbook provides a wide swath of tools needed to manage waste. Obviously, this is just the tip of the iceberg. Each waste scenario is unique, but the information in this section should help point to ways to assess the risks posed by wastes, steps needed to manage those risks, and metrics for deciding whether the selected approaches are proper and functioning as designed.

CHAPTER

30

Landfills – Yesterday, Today and Tomorrow

Geoffrey Blight

School of Civil and Environmental Engineering, University of the Witwatersrand, Johannesburg, South Africa

OUTLINE

1. DUMPS, TIPS, LANDFILLS AND SANITARY LANDFILLS

The people of the ever-expanding population of the world seek an improved or improving standard of life that, in turn, demands the acquisition of new goods and the discarding of old. As most of the world's population is urbanized, an ever-increasing volume of municipal solid waste (MSW) is produced, most of which has to be disposed of in what have become known as landfills.

The need for the collection and sanitary disposal of MSW was not recognized until recently. The science and engineering of MSW management is also relatively new. Fifty years ago, throughout the world, most municipal waste was disposed of in open dumps or tips, even by authorities serving very large communities, for example, New York. It was only in the early 1930s that the need for improved methods of MSW disposal started to raise concern in Europe [1]. In 1959, the first recognized definition of sanitary landfilling,

Waste Doi: 10.1016/B978-0-12-381475-3.10030-0

more or less as we know it now, was published by the American Society of Civil Engineers (ASCE) [2]. The ASCE defined 'sanitary landfilling' as a controlled operation in which MSW is deposited in defined layers, each layer being compacted and covered with soil before depositing the next layer. Although very strict regulations relating to sanitary landfilling have been formulated and are now being applied in various parts of the developed world [3,4], it is also apparent that these new rules have and are being applied mainly to new landfills and that conditions, at many pre-existing landfills, fall short of present ideals. It is also inevitable that current 'ideal' regulations and practices will become outdated and be regarded as rudimentary and inadequate within the space of a few years. Hence, even in developed countries, the professions concerned with waste disposal are in a constant state of trying to catch up with and introduce the latest trends and regulations. In most developing countries, open dumping remains the most widespread method for disposing of MSW. To the environmentally conscious municipal engineers working in developing countries, the problems of attempting to bring the municipal waste problem under control often appear almost insuperable, yet work must continue in the direction of more efficient collection and transportation and safer and more environmentally acceptable disposal.

What distinguishes a *developing* country from a *developed* one? A formal and succinct definition [5] is that a *developing country is one where the per capita gross domestic product is lower than the average for the world.* Thus, a developing country is one where the people are poor, on average. However, there are many countries for which this definition may be inadequate, because industrialized urban areas in a country may be "developed", while country areas are still "developing". It is rare to find wealth evenly distributed within large cities or between town and country. These countries are usually referred to as having *mixed economies* (e.g. China and India).

In an open dump or tip, MSW is deposited in a conveniently situated pre-existing excavation (e.g. an abandoned quarry pit), on a convenient piece of unused land or on a convenient hillside. The term 'landfilling' arises from the operation of filling a hole, that is, filling to build a 'new' land surface, whereas the operation of dumping on a level or sloping surface was originally termed landbuilding. Now, however, the word 'landfilling' is used for both types of operation.

The basic difference between a dump or tip and a landfill is that in a dump there is no attempt to separate the waste from the underlying soil or rock strata and where the hole extends to below the groundwater level, waste is dumped directly into the groundwater. As a result, pollution of the groundwater by substances leached from the waste is almost inevitable, unless (by chance) the underlying soil or rock is highly impervious. There is also no attempt to cover the waste to prevent odours and exclude breeding flies and scavenging birds (e.g. crows) and animals (e.g. rats) or to seal the surface of the dump against the ingress of rain. In contrast, a sanitary landfill is constructed on a sealed impervious base that is covered with a drainage system designed to collect leachate that usually (but not always) emanates from the waste. Each day's deposition is formed into a layer about 2 m thick and is compacted to reduce its volume (and therefore maximize the utilization of space). Each layer is covered at the end of the day to exclude rain, insects and scavenging birds and animals. The collected leachate is treated to purify it before discharging it into the surface water system. Figure 30.1 shows sections through a hillside dump and a sanitary landfill to show the differences in their construction.

In summary, the simple but inadequate waste dumps of yesterday are still with us in most of the developing world, whereas the developed world is using the vastly improved sanitary

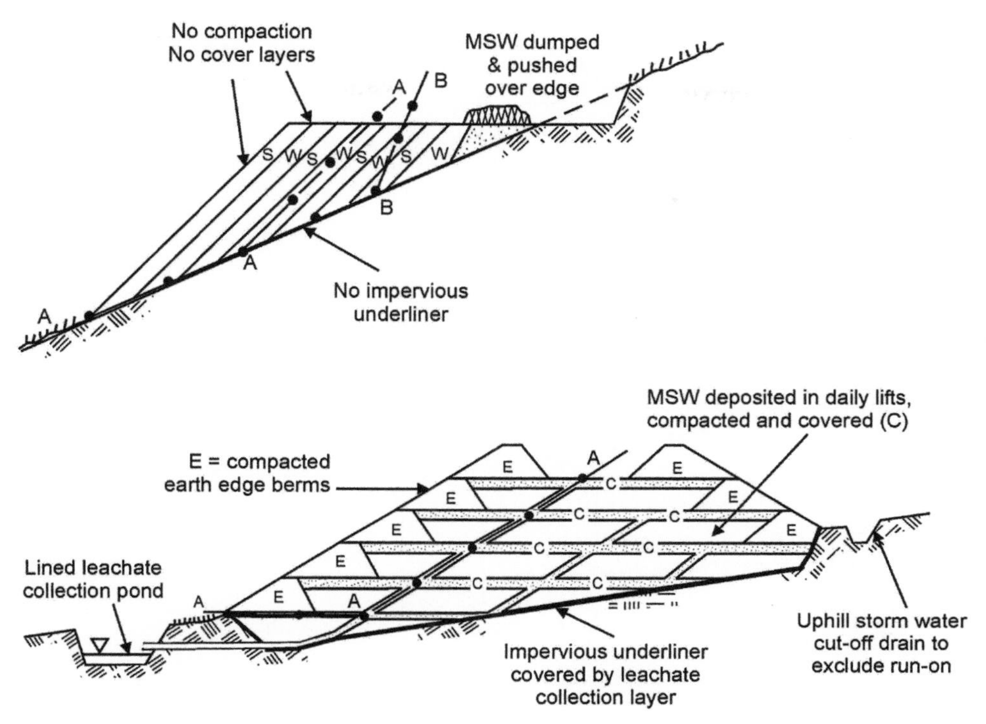

FIGURE 30.1 (a) Construction of a MSW dump placed on a hillside. The layers W, S, W,... indicate differences in the composition of the waste between winter and summer. (b) Construction of a sanitary landfill, with MSW compacted and covered to exclude rain, insects and scavenging birds and mammals. (Lines AAA and BBA will be explained later.)

landfill of today while planning and experimenting with and tentatively applying the still more improved landfill technology of tomorrow.

Figure 30.2 shows a typical layout of the leachate drainage collection system of a sanitary landfill, leading to a leachate and contaminated water collection sump, while Fig. 30.3 shows the construction of a typical impervious underliner, with its leachate collection layers. The leachate collection layers consist of a large-size gravel or crushed rock of 38 to 50 mm size to reduce the risk of the voids between the rock particles becoming blocked by chemical gels of iron hydroxide or by bio-slimes. The flexible membrane liner (FML) or geomembrane usually consists of a 2-mm thick synthetic rubber (neoprene) or plastic [e.g. flexible polypropylene (fPP)] membrane protected from accidental damage by a sand layer above and laid in direct contact with a compacted clay layer. The geomembrane should be completely impervious to seepage of the leachate, but because it has to be constructed by welding a series of sheets together, it is always possible that a seam may leak, in which case the compacted clay liner will prevent the leakage from escaping to the groundwater. There is also a possibility that the clay liner may crack, and for this reason, it rests on a leakage detection and collection layer. This is constructed in defined zones so as to enable the position of any leak to be established and if possible repaired, or if not possible isolated, and the leakage separately collected.

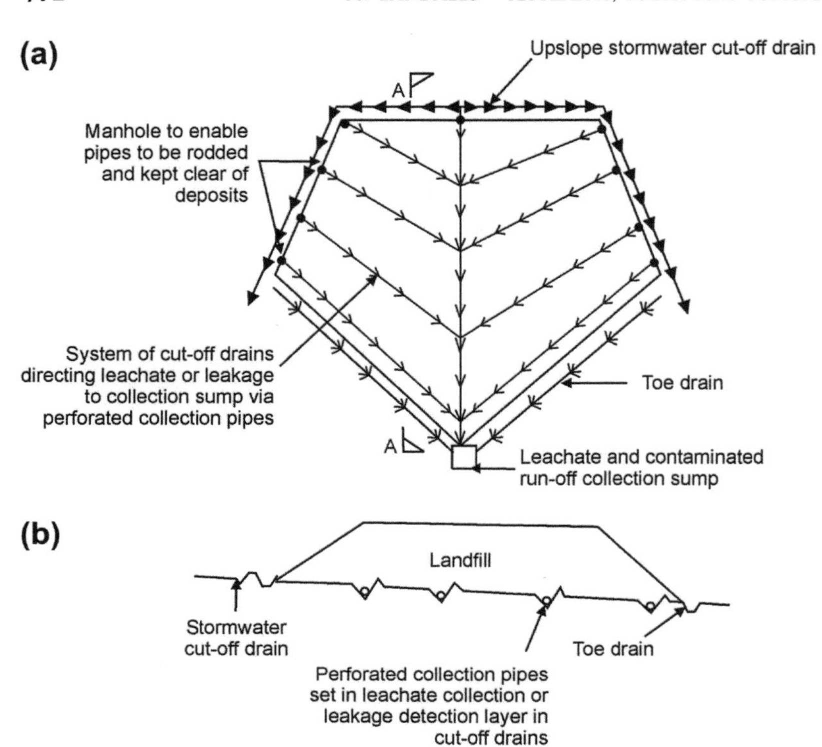

FIGURE 30.2 (a) Typical layout in plan of a leachate collection system for a sanitary landfill. (b) Layout in section along line AA.

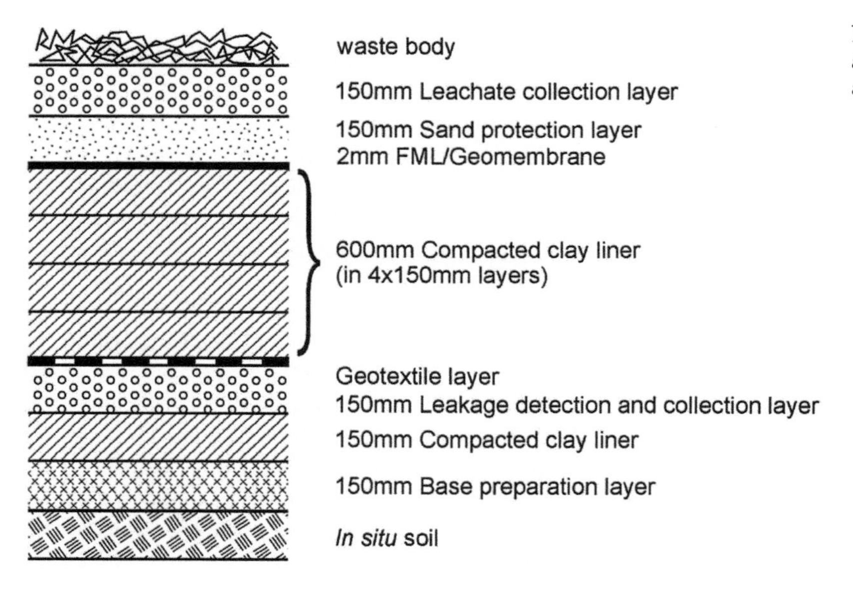

FIGURE 30.3 Construction of a typical impervious lining for a sanitary landfill.

2. GENERATION AND CHARACTERISTICS OF MSW

The type and quantity of refuse generated by a community depends on its culture and the per capita income. Wealthy communities form *throw away* societies, whereas poor communities have less to throw away and are more ingenious in reusing, recycling and refurbishing articles that a wealthier community would discard. The amount of waste produced per head of population varies from as low as 0.05 kilogram per day per capita (0.05 kg d^{-1} ca^{-1}) for the very poor to as much as 2 kg d^{-1} ca^{-1} for the rich (note that here and elsewhere in this chapter, d is the symbol for day and ca is the symbol for capita) [6,7]. The relationship between wealth and waste is illustrated by Fig. 30.4. The urbanized, very poor invariably live in overcrowded conditions, whereas the rich have a generous living space. The net effect is that very poor communities generate about 45 tonne per hectare per annum (45 t ha^{-1} a^{-1}) of MSW, whereas rich communities generate only 10 t ha^{-1} a^{-1} [7]. This is also indicated in Fig. 30.4. A very poor urban area in a developing country measuring, say, 10 km $\times$ 10 km may generate 450,000 t a^{-1} of MSW (note that here and elsewhere in this chapter, a is the symbol for annum, that is, year). As the size of cities, especially in the developing world, is rapidly increasing, the disposal of MSW has become a problem of insuperable proportions in many cities of the developing world, for example, Manila (Philippines) and Jakarta (Indonesia).

Figure 30.4 enables a rough estimate to be made of the present total annual generation of household waste (i.e. waste generated in either town or country). If the world's population is taken as 7×10^9 and if most of these people are poor and on average generate only 0.2 kg d^{-1} ca^{-1} the annual global generation of household waste would be $7 \times 10^9 \times 0.2 \times 365$ kg a^{-1} = 511×10^9 kg a^{-1}, or 511×10^6 t a^{-1}, or say 450×10^6 to 550×10^6 t a^{-1}. This is quite small

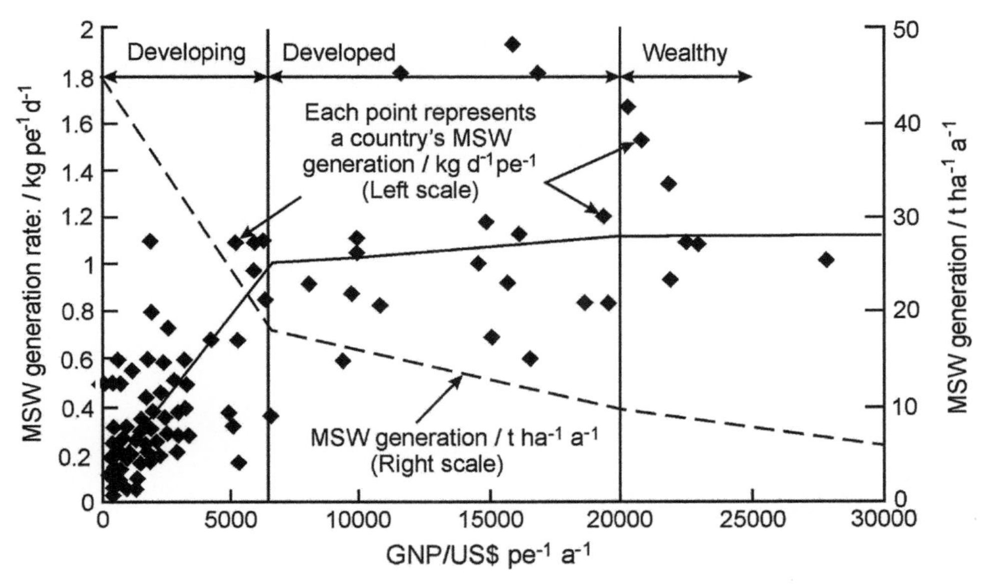

FIGURE 30.4 Relationship between rate of generation of municipal solid waste (MSW) as a personal generation rate (kg d^{-1} ca^{-1}) and gross national product ($ a^{-1} ca^{-1}) as well as an areal generation rate (t ha^{-1} a^{-1}). (After [5] and [6]). Here $ refers to U.S. dollar, ca to capita, d to day, ha to hectare and a to annum.

(15%) in comparison with the estimated annual quantity of mine waste (see Chapter 3) of up to 35×10^9 t a^{-1}, but if uncontrolled or poorly controlled, MSW can adversely affect the health, quality of life and safety of the population to an even greater extent than does mine waste.

Table 30.1 (based on [7] and other sources) illustrates some of the differences and similarities that may exist in refuse composition between developed and developing countries and mixed economies. The most notable differences between refuse compositions, for developed and developing countries, lie in the percentages of paper and plastics (mainly packaging materials, newspapers, etc.). The content of unidentifiable material, ashes, dust and so on is also characteristically larger in waste from developing communities.

There are also differences between the wastes of developing communities in different parts of the world. The differences in refuse composition between developing countries arise from such factors as culture, climate, differences in fuel, variations in diet and so on. For example, in Delhi, many people are vegetarians and dung is a common cooking fuel. Relatively little fuel is required for heating. In Wattville (a poor township, by South African standards, south of Benoni in South Africa), coal is commonly used for cooking and heating, and everything that is combustible tends to get burned as fuel. This largely accounts for differences such as those in Table 30.1 in the entries for vegetable matter of 47% for Delhi and only 9% for Wattville. Also, note the differences in refuse generation rates between, say, the United States, Delhi, Wattville and Benoni (a diverse-populated mining town on the East Rand of South Africa). An affluent community generates 3 to 5 times as much waste per capita as does a poor community. There are usually

TABLE 30.1 Examples of the Composition (Mass %) and Generation of Municipal Solid Waste in Developed and Developing Countries and Mixed Economies

					Mixed economy	
Composition in % by mass	**Developed countries**		**Developing countries**		**South Africa**	
	US	**UK**	**Delhi, India**	**Wuhan, China**	**Benoni**	**Wattville**
Vegetable	22	25	47	16	20	9
Paper	34	29	6	2	36	9
Metals	13	8	1	0.5	10	3
Glass	9	10	0.6	0.6	11	12
Textiles	4	3	–	0.6	1	1
Plastic	10	7	0.9	0.5	7	3
Wood	4			1.8		
Dust, ash and other unidentified material	4	18	44.5	78	15	63
Refuse density/kg m^{-3} (uncompacted)	100+	150	420	600	170	400
Refuse generation rate/t a^{-1} ca^{-1}	0.65	0.65	0.14	0.20	0.58–0.73	0.15

a refers to annum.

also differences in waste composition and volume between summer and winter, with less organic waste arising during winter in temperate climates.

It must also be emphasized that the composition of MSW is not constant with time, but is ever changing as lifestyles evolve. This is illustrated by Fig. 30.5, which shows (a) how the composition of MSW in the United Kingdom evolved in the seven decades from the 1930s to the 1990s [8]. Comparing Fig. 30.5a with Table 30.1 shows that MSW in the United Kingdom during the 1930s to late 1940s was fairly similar in composition to present-day MSW in Wattville. Figure 30.5b [8] shows how the uncompacted (as collected) density of the United Kingdom's MSW reduced from nearly 300 kg m^{-3} in the 1930s and 1940s to about 150 kg m^{-3} today, mainly as a result of a decreasing ash content and increasing quantities of plastic film, cardboard and foam packaging.

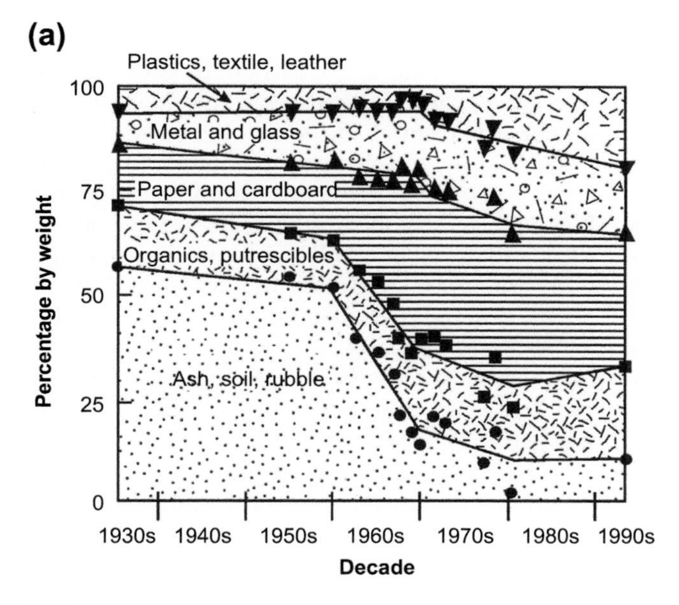

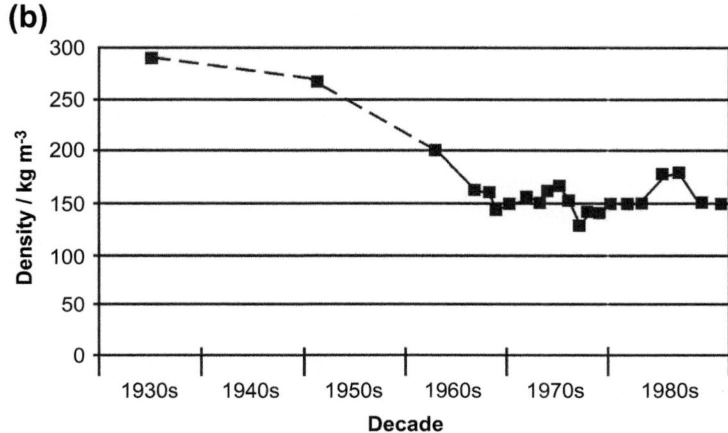

FIGURE 30.5 Changes in the composition and collected density of MSW in the United Kingdom from 1935 to 1990. (a) Composition, mass %. (b) Loose (as collected) density.

TABLE 30.2 Design Guide for Refuse Generation in a South African Urban Region of Mixed Economy

Income of community	Population density (ca ha^{-1})	Waste density (uncompacted) (kg m^{-3})	Waste generation rate (t a^{-1} ca^{-1})	Waste generation (t ha^{-1} a^{-1})
High	18	170	0.58	10
Medium to high	25	200	0.51	13
Low to medium	50	300	0.36	18
Low	200	400	0.22	44
Very low (informal settlements)	300	400	0.15	45

a refers to annum and t refers to metric tonne, 10^3 kg.

The generation rate of MSW is an important consideration in the selection of a suitably large disposal site and in the design of landfills, whether the economy is developed, developing or mixed. Table 30.2, for example, is a design guide for the selection of new landfill sites in South Africa, which takes account of the effects of income and population density on MSW generation.

A landfill being designed to serve a total area comprising suburbs of varying income must take into account variations such as those shown in Table 30.2. The effect of population growth must also be taken into account by escalating the annual tonnages requiring disposal, by the estimated population growth rate, applied as compound escalation.

3. THE GENERATION OF GAS IN LANDFILLED MSW

Once MSW has been deposited on a dump or in a sanitary landfill, it continues to decompose. The products of decomposition are mainly gas (known as landfill gas or LFG) and leachate. In the period shortly after the waste has been landfilled, the pore space is filled mainly with nitrogen and oxygen having much the same proportions as the ambient air. As decomposition of the organic and putrescible components proceeds, the oxygen is very quickly converted by aerobic bacteria to carbon dioxide (CO_2). The gas in the waste passes from an aerobic to an anaerobic state, and anaerobic bacteria start to proliferate and produce methane (CH_4). The effects of these processes are illustrated by Fig. 30.6, which shows the relative volumes of CO_2 and CH_4 in LFG and how they change with time after landfilling in a 50:50 mixture of MSW generated in Benoni and Wattville (see Table 30.1). Note the rapidity with which the CO_2 content rises initially, while no CH_4 is produced until the oxygen in the pores of the waste has been converted to CO_2 and the anaerobic bacteria are able to proliferate. The main diagram in Fig. 30.6 shows the composition of the gas. The inset diagram shows (schematically) the duration and quantity of LFG generation, which usually reaches a maximum within the first decade and then declines to a negligible amount after 30 years. The two main gas components, CO_2 and CH_4, do not add up to 100%. The rest of the gas component consists of nitrogen.

The CH_4 in LFG is flammable, and this can be a danger if, for example, the waste is ignited by spontaneous combustion. CH_4 can also explode. A case occurred in South Africa where CH_4 seeped out of an MSW dump and passed along the line of a badly backfilled trench into the basement of a nearby building where it accumulated. An electrical spark caused by switching on the light in the basement set off an explosion that wrecked the building. The LFG can be

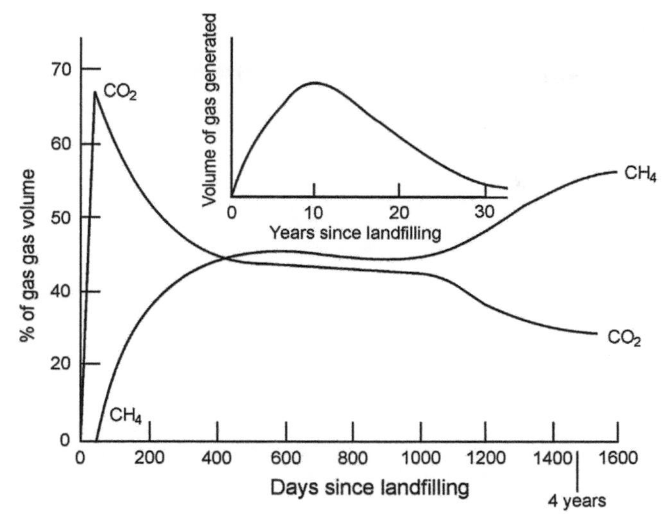

FIGURE 30.6 Generation of carbon dioxide, CO_2, and methane, CH_4, in landfilled waste. Variation of relative gas composition by volume with time after landfilling. Inset: Variation of gas quantity generated with time.

extracted from the waste by means of vertical gas collector wells or horizontal collector drains and then be used to power internal combustion gas engines to generate electricity. It is, however, more common to burn off the collected CH_4 in a flare, both as safety and environmental protection measures. Both CO_2 and CH_4 are greenhouse gases and the increase of their concentration in the atmosphere is currently being blamed for global warming. If the global warming potential (GWP) of CO_2 is taken as 1, that of CH_4 has been determined as 25 [9]. When CH_4 is burned, it combines with oxygen to give CO_2 and water ($CH_4 + 2O_2 \rightarrow CO_2 + 2H_2O$), thus reducing its GWP from 25 to 1. This is one of the reasons for flaring off the gas.

4. THE GENERATION AND POLLUTION POTENTIAL OF LEACHATE

Leachate arises from squeezing out of the water content of the incoming waste ('squeezate') and from rain that percolates through the cover layers and accumulates in the waste. Landfills in wet climates and those that store large quantities of high water content waste therefore generate more leachate than those in dry climates.

In very dry climates, no leachate may be generated. The covered surfaces of sanitary landfills should be, and usually are, sloped to encourage rainfall to run off, thus reducing the quantity that seeps into the waste. Water is also lost by evaporation from the landfill surface. Hence, the quantity of leachate, expressed as litres per m^2 (or mm of water), is always less than the rainfall. In climates where evaporation from a free water surface exceeds rainfall, no or very little leachate may be generated. In South Africa, sanitary landfill sites are classified by climate into those where landfills are likely to produce leachate and those that are not. Landfills in areas of high-potential evaporation and low rainfall are not required to be equipped with an impervious underlining. Figure 30.7 shows the method of classification used in South Africa [10]. Values for the average annual rainfall and the average annual A-pan evaporation are available for all of the meteorological stations in the

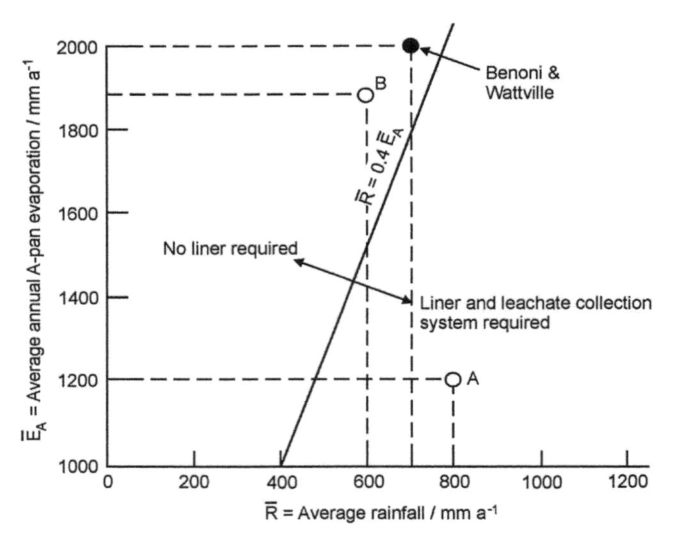

FIGURE 30.7 Chart used in South Africa to classify landfill sites by climate. A site plotting at point A would require an underliner and facilities for leachate treatment to be provided. One plotting at B would require no liner, as leachate would not be generated.

country. (An A-pan consists of a shallow galvanized steel pan, 1.8 m in diameter by 0.3 m deep. The pan is filled with water and the water level is measured daily to determine the evaporation, E_A, compensating for any rain, R, that fell during the past day.) The constant of proportionality 0.4 in the equation $R = 0.4\bar{E}_A$ in Fig. 30.7 would vary from climatic area to climatic area on the Earth. As indicated in Fig. 30.7, if a point such as that for Site A falls below and to the right of the dividing line, leachate is expected to be generated and an underliner is required. If it falls above and to the left, as for Site B, no leachate would be likely and no underliner would be required.

Figure 30.8 shows examples of leachate generation compared with rainfall in a series of field experiments carried out at Benoni and Wattville, and it also shows the potentially polluting nature of leachate if it should escape from the closed system of landfill – leachate collection layers – leachate treatment and disposal.

The construction of the experimental cells will be described in detail to show the possible effect of size and construction details on the results obtained from field experiments. The experiment comprised two large and two small experimental landfill cells, one of each being filled with waste from Wattville and the other two with waste from Benoni.

The large cells were constructed as part of the prototype landfill. The cells were lined with high-density polyethylene (HDPE) geomembranes covered by a gravel drainage layer draining to an outlet that was closed by a gate valve. This allowed the leachate to be drained off and measured at regular intervals. The cells measured an average of 12 m × 12 m in plan and were 2.5 m deep. The cover layers consisted of a 300 mm layer of the same compacted clayey soil used on the landfill, with a water infiltration rate of 30 mm d^{-1}. The surface of each cell was domed to shed rainfall and further limit infiltration. The small cells were 3.12 m^2 in plan and 1.06 m deep, constructed with vertical brick walls around the cells and with floors sloping to one side, lined with an HDPE geomembrane and covered with a gravel drainage layer.

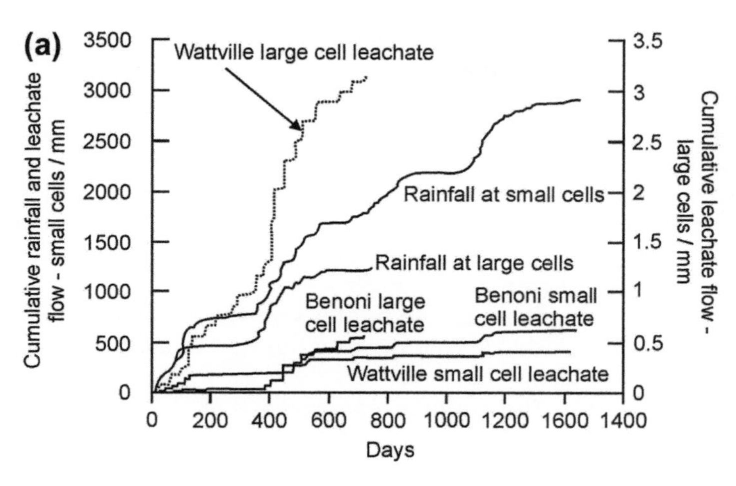

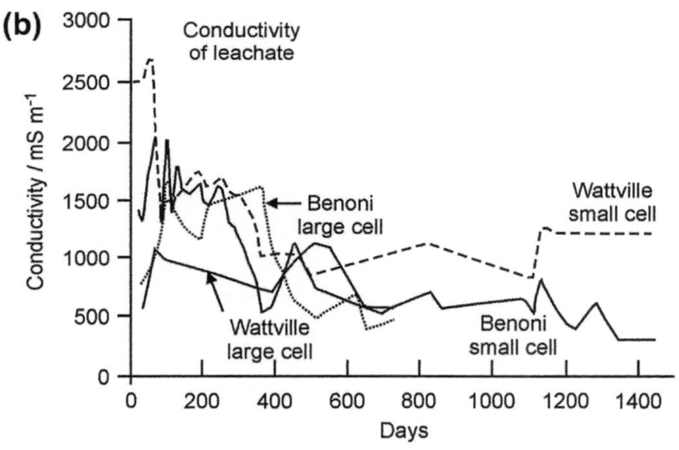

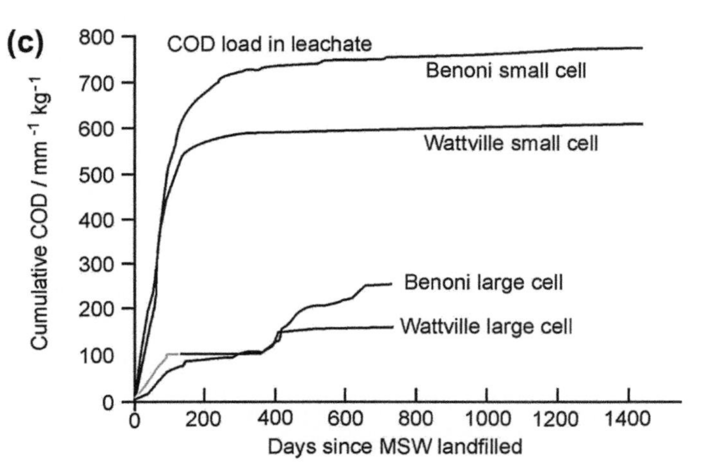

FIGURE 30.8 (a) Relationships between rainfall, leachate generated and time since landfilling for small and large experimental landfill cells. (b) Relationship between conductivity of leachate and time. (c) Relationship between cumulative chemical oxygen demand of leachate and time.

Leachate could be tapped off from each cell via a gate valve. The cover layer on the small cells had an infiltration rate in excess of 200 mm d^{-1} and run-off was prevented by the surrounding brick walls.

Figure 30.8 shows a selection of results from these tests. At the time, the tests on the small cells had been in progress for 1425 days (3.9 years) and those on the large cells for 740 days (2.0 years).

Figure 30.8a compares rainfall and leachate production (leachate on the right hand scale). The cumulative rainfall at the small cells from 725 days onwards (the start of the measurements on the large cells) amounted to 1200 mm, whereas for the large cells it was 1300 mm, hence the two sites had comparable rainfalls. The leachate flow is always much less than the rainfall, as it represents the difference between the rain infiltration and the water that is absorbed by the waste and subsequently partly or completely re-evaporated.

Figure 30.8b shows the variation of the leachate conductivity with time. The conductivity correlates with the total dissolved substances (TDS) in the leachate and therefore gives an indication of the potential for escaping leachate to pollute groundwater. The diagram shows that the maximum pollution potential exists early in the life of a cell of landfilled waste but declines to a much lower value after about 4 years. This is not to say that after 4 years the leachate is innocuous.

Figure 30.8c shows how the total quantities of pollutants in the leachate such as chemical oxygen demand (COD, shown here), chloride, ammonia and nitrates build up rapidly in the first year after landfilling and thereafter grow more slowly. This figure corroborates the data shown in Fig. 30.8b.

Table 30.3 compares some of the results of the study at 725 days after the start of each series of tests.

The comparison shows that the large cells produced very much less leachate than the small cells. This was to be expected because of the differences between the large and small cells (lower cover permeability and larger depth of waste to absorb water). The small cells each produced about the same leachate as a proportion of rainfall, but the large cell of Wattville waste produced much more leachate than the corresponding cell of Benoni waste, probably because of its lesser capacity to absorb and store water.

Judging by the large cells, the Wattville waste, with its lower organic and higher inorganic contents (Table 30.1), produced leachate with a lower conductivity and COD than did the Benoni waste. The two wastes produced almost the same loads of ammonia and chloride and overall produced leachates with much the same pollution potential. Hence, this experiment seems

TABLE 30.3 Comparison of Results of Small and Large Cell Tests on MSW From Benoni and Wattville at 725 Days After the Start of Each Series

Source of MSW	Cell size	Cumulative rain (mm)	Cumulative leachate (mm)	Leachate/rain (%)	Leachate conductivity (mS m^{-1})	Leachate conductivity Wattville/Benoni	COD load in leachate (mg mm^{-1} kg^{-1})	COD Wattville/Benoni
Wattville	Small	1800	300	17	1000	2	590	0.79
	Large	1200	3.1	0.3	300	0.6	130	0.5
Benoni	Small	1800	400	22	500	–	750	–
	Large	1200	0.6	0.05	500	–	260	–

Note that the units for COD refer to milligrams per millimetre height of water with a surface area of 1 m^2/kg of dry waste.

to show that the pollution potential of landfilled MSW is relatively little affected by quite large differences in waste composition.

The fact that the leachate to rainfall ratios in the large and small tests was so different, although the rainfall was similar, points to the obvious conclusion that field experiments must represent a prototype situation very closely if they are to give results representative of field conditions.

The average annual A-pan evaporation and rainfall for Benoni and Wattville plot is as shown on the climatic classification diagram in Fig. 30.7, and one would therefore expect little or no leachate to be produced from a landfill in this area. The maximum leachate flow of 3 mm in 2 years is only 10% of the expected leakage rate from a geomembrane with welded seams [11] and is therefore acceptable without the need for an underliner.

Finally, the question arises of to what distances downstream and for how long an unlined landfill emitting leachate into groundwater will continue to cause pollution. If groundwater pollution occurs, is it a short-term transient problem or does it persist? Figure 30.9 provides an illustrative example.

Three clay pits, used for brick making, were closed in 1906 because of the ingress of groundwater. In 1928, the water-filled pits were filled by tipping MSW into them, and the levelled site was then used for an above surface MSW dump. In the late 1960s, the dump was converted to a sanitary landfill which was finally closed in 1978. Figure 30.9a shows the areas eventually covered by the landfill, as well as the positions of the original clay pits. In 1976, Ball [12] started a project to establish the pollution status of the landfill, taking groundwater samples by hand augering 1.5 m deep holes at intervals along the banks of the stream that originates upstream of the landfill, passes under the 1960s landfill in a concrete culvert, and flows down the valley downstream of the landfill. The 30 sampling positions are marked on Fig. 30.9a. In 1976/1977, a 1.2 m diameter hole was augered through the landfill. Samples taken from the hole showed that the post 1960s sanitary landfill had been separated from the underlying waste dump by a 2.5 m thick layer of compacted soil. The landfill is located in a climatic area for which the A-pan evaporation and rainfall are very similar to those of Benoni–Wattville and therefore does not generate leachate. Samples from the underlying soil confirmed this by showing it to be quite unpolluted. Hence, if any pollution was to be found, it must have originated from the waste deposited prior to the 1960s. The groundwater samples from close to the toe of the landfill showed levels of pollution, relative to the then current South African standards for surface and groundwater quality that slightly exceeded allowable values. The groundwater sampling was repeated in 1993/1994 [13]. The two sets of measurements, taken 17 years apart, gave almost identical results, as can be seen from Fig. 30.9b in which most of the pairs of measurements coincide, except for a few where two points appear at the same horizontal distance from Point E (e.g. at 560 and 1300 m). The more detailed inset in Fig. 30.9b showed that unacceptable values of ammonia persisted to a distance along the stream of 50 to 80 m from the toe of the landfill and much the same applied to conductivity and COD.

This study and other work [14] show that pollution plumes from landfills may not progress very far from the source, but they may persist for many years, in this case for more than 30 years.

5. THE SAFETY AND STABILITY OF DUMPS AND LANDFILLS

Mention was made in Section 3 of some of the dangers posed by LFG, namely, fires and explosions. There is also a danger of a person, usually a scavenger, in a developing country drowning in CH_4 or CO_2 if they fall asleep in a hollow,

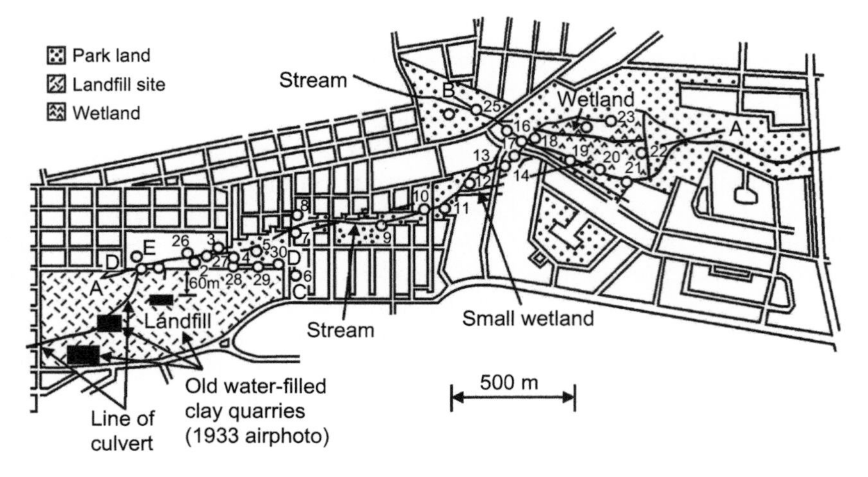

Plan showing landfill, stream and groundwater sampling holes (numbered 1 to 30)

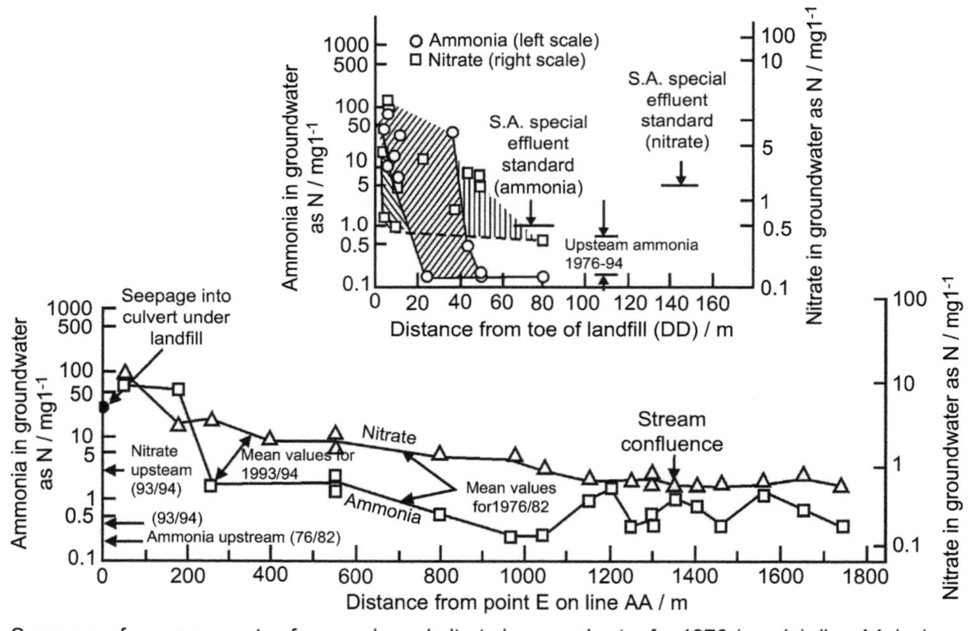

Summary of measurements of ammonia and nitrate in groundwater for 1976 (see (a), line AA / m)

FIGURE 30.9 Extent and duration of groundwater pollution by leachate from old landfill.

possibly kept clear of gas by the wind in the day, that then fills with gas when the wind drops at night. This also happens on coal waste dumps, and a number of people die in this way each year.

Dumps and landfills are also dangerous because of items such as broken glass and sharp metal objects, for example, pieces of razor wire and so on, that may be, and often are, mixed up with the MSW. Landfill compactors, trucks dumping waste and dozers levelling waste are also extremely dangerous to persons working or scavenging on the dump or landfill. Scavengers and workers on landfills are also prone to fungal skin and lung diseases picked up by handling the waste with bare hands and from spores in dust.

Stability of the dump or landfill as a whole also represents a danger. As mentioned in Chapter 3, MSW dumps and landfills can be every bit as dangerous as tailings storages. Dumps, especially hillside dumps, may fail by sliding of a wedge of dumped waste such as those defined in Fig. 30.1a and 1b by surfaces AAA or BBAA. Between 1977 and 1997, seven major failures of MSW dumps and landfills occurred worldwide [15]. Of these, four slides occurred in uncompacted waste dumps in developing countries and three took place in supposedly carefully engineered and operated sanitary landfills, one each in Colombia, South Africa and the United States of America [16]. The last was reputed (before the failure) to be the biggest and best operated sanitary landfill on the Earth. No lives were lost in the failures of the three sanitary landfills and one of the dumps. In failures of dumps in Turkey, Philippines and Indonesia between 1993 and 2007, at least 39,278 confirmed and 350 missing and at least 147 deaths occurred, respectively, a minimum total of 464.

The following is a brief description of the disaster with the highest death toll, the slide at Quezon City, Manila, Phillipines in 2000 [15], in which between 278 and 628 died.

By the year 2000, the dump covered an area of 12.7 ha and was 18 to 30 m high. Incoming waste was dumped on the top of the existing MSW fill and, after being picked over by informal scavengers, was bulldozed over the tipping edge to form a face at the angle of repose. The surface of the ground on which the dump had been formed, sloped gently at about 2° away from the dump. In late 1999, the contractor operating the dump began to bulldoze waste from the centre of the top of the dump towards and over the edges with the result that a basin was created on the top and the slopes around the edges were covered with freshly deposited MSW at its angle of repose.

In the 10 days preceding the failure, exceptionally heavy rain, amounting to 750 mm, fell in the Payatas area and reportedly created a lake of storm water in the basin on top of the dump. The contractor cut a deep trench to drain the lake and allow the water to run down the steep outer dump slopes. At some time earlier, a 2–3 m deep open trench had been cut along the toe of the face to intercept the streams of leachate running from it. The waste was also burning in places. What made the situation particularly dangerous was that large numbers of informal scavengers, working on the dump, had established their makeshift shack homes right at the toe of the waste slope. The failure began at 04.30 h on 10 July 2000 when loud cracking noises were reportedly heard coming from the waste body. The slide occurred 3 h later when an estimated 9 to 10 $\times$ 10^3 m^3 of waste slid forward, burying or destroying all the shack homes in its path. The waste does not, however, appear to have liquefied as it only moved forward a relatively short distance and the failure has the appearance of a translational wedge failure of the freshly tipped MSW, similar to that outlined in Fig. 30.1a.

Figure 30.10 is an aerial view of the dump after the failure. At the toe of the dump, to the right of the failure, a large area of shack homes can be seen. This area extended further to the left, and houses in the extension were deeply buried by the wedge of sliding waste. The burying of these houses was the cause of the enormous death toll. Fifty-eight people were rescued, and after weeks of searching and digging, 278 bodies had been recovered, leaving between 80 and 350 missing, believed dead. Thus, the death toll may have exceeded 600. The smouldering waste, exposed to oxygen by the slide, started to burn, adding to the difficulties of would-be rescue workers.

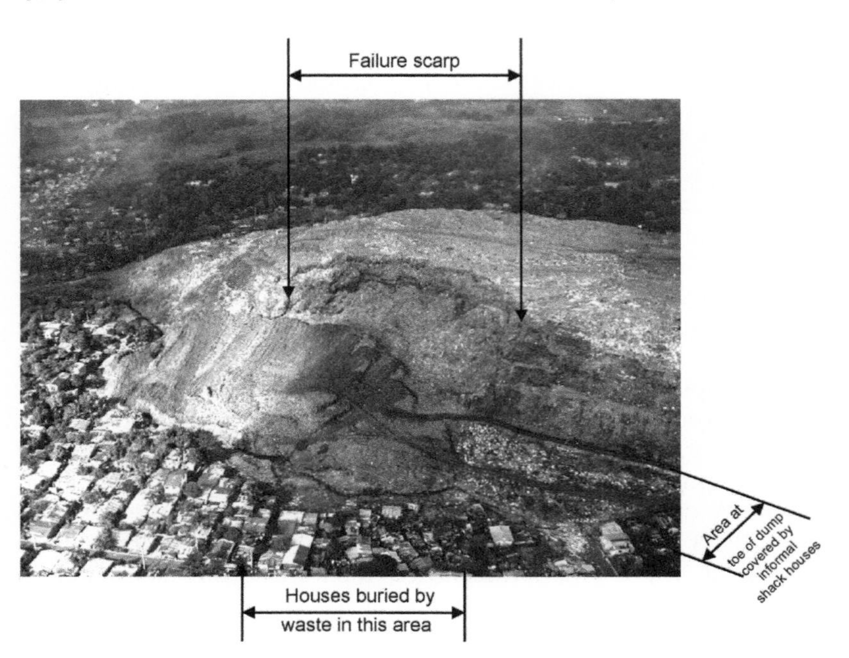

FIGURE 30.10 View of failure of Payatas waste dump at Quezon City, Manila, Phillipines, in 2000. A strip of shack homes at the toe of the dump was buried by the failure. (Photo provided by Professor Manassero).

6. CONCLUSION

Uncompacted, uncovered dumps of MSW should be the landfills of yesterday. Unfortunately, they are still with us today, and in developing countries they are all too often the only form of MSW disposal in use. Even though it is 50 years since the concept of the compacted, covered sanitary landfill was introduced and formalized, it remains the landfill of the distant future over much of the Earth. In Europe and North America, tomorrow's forms of sanitary landfill are being introduced. In the bio-reactor landfill, for example, leachate is recirculated through the waste to accelerate the rate of decomposition. Increasingly, separation of recyclable materials at source is reducing the quantities of waste being landfilled and changing the composition of the waste. [Unfortunately, one of the major landfill failures (mentioned above) in Colombia was being operated as a bio-reactor landfill when it failed, to some extent because the operators did not appreciate that increasing the liquid content of the waste would drastically decrease the stability of the landfill.]

In Europe and China, more and more MSW is being used as refuse-derived fuel (RDF) for the thermal generation of electricity. Another option is to compost the organic content of MSW and recycle the paper, plastic and glass contents. This requires separation at source by the householder. This is very feasible, as schemes of this sort are in successful operation in North and South America, Europe, Australia, New Zealand and South Africa. Another option is to convert the organic fraction of MSW to liquid bio-diesel that can be used to power heavy transport or agricultural tractors. Thus, the sanitary landfill and the MSW of the future should have much reduced organic contents, and landfills should therefore generate less LFG and (possibly) a less polluting leachate. Unfortunately, some residues will always remain, (e.g. the ash from RDF fuelled power stations) and hence the landfill will probably always be with us. It is our challenge to make the landfills of tomorrow much better, safer and less polluting than those of today.

References

[1] B.B. Jones, F. Owen. Notes Published by City of Mancheser, Manchester, UK, 1934.

[2] American Society of Civil Engineers (A.S.C.E.), Sanitary Landfill Manual of Practice, ASCE, New York, 1959.

[3] United States Environmental Protection Agency (USEPA), Solid Waste Disposal Criteria, USEPA Report 40 CFR, USEPA, Washington, DC, 1992.

[4] European Union, Proposal for a Council Directive on the Landfill of Waste – Common Position. Official Journal, European Community, European union. No 6692/97 ENV 82 (1997).

[5] D.V.J. Campbell, Proceedings of the 4th International Waste Management and Landfill Symposium, Cagliari, Italy, 1993, V.2, 1851–1866.

[6] L. Medina, Personal communication, 1997.

[7] G.E. Blight, C. Mbande, Journal of Solid Waste Technology and Management 23 (1) (1996) 19–27.

[8] K.S. Watts, J.A. Charles, Proceedings, Institution of Civil Engineers, UK, Geotechincal Engineering Geotech 137 (1999) 225–234.

[9] Intergovernmental Panel on Climate Change (IPCC), 4th Assessment Report (2007), Cambridge University Press, Cambridge, UK, 2007.

[10] G.E. Blight, Waste Management & Research 24 (2006) 482–490.

[11] R.M. Koerner, Civil Engineering (ASCE), 2 (2001), 96.

[12] J.M. Ball, Degradation of Ground and Surface Water Quality in Relation to a Sanitary Landfill, MSc(Eng) Thesis, University of Witwatersrand, Johannesburg, 1984.

[13] G.E. Blight, Proceedings of the 5th International Landfill Symposium, Cagliari, Italy, 1995, V.3, 593–599.

[14] T.H. Christensen, R. Cossu, R. Stegmann (Eds.), Landfilling of Waste: Leachate, Elsevier Applied Science, London, UK, 1992.

[15] G.E. Blight, Waste Management & Research 26 (2008) 448–463.

[16] B.O. Schmucker, D.M. Hendron, Proceedings of the 12th GRI Conference, Folsom, PA, USA, 1998, 269–295.

CHAPTER

31

Pollution Management and Responsible Care

Nicholas P. Cheremisinoff

N&P Ltd., Willow Spring Road, Charles Town, West Virginia, USA

OUTLINE

1. INTRODUCTION

All manufacturing operations generate pollution because every process devised by man produces waste. The waste may be in the form of air emissions, solids and semi-solids, process liquid effluents, and thermal energy losses. Some of this waste is toxic and hazardous and is regulated by environmental agencies and ministries in industrialized countries. Not all wastes that are harmful to the environment are regulated, and indeed many wastes that are regulated continue to be released in significant quantities that are harmful to communities and the environment.

There are two broad areas of pollution that pose ongoing negative impacts to the public. These are legacy pollution and ongoing releases. Legacy pollution refers to toxic chemical releases into the environment during a time pre-dating strict environmental enforcement practices, which in the United States is 1972 with the promulgation of the U.S. Environmental Protection Agency (U.S. EPA). The introduction of environmental regulations along with enforcement tools during the early 1970s made many voluntary industry standards aimed at responsible waste and pollution management obligatory. Environmental statutes also introduced for the first time the concepts of

Waste Doi: 10.1016/B978-0-12-381475-3.10031-2

joint and several liability, and retroactive responsibility for legacy pollution. For companies to be good corporate citizens they must take responsibility for legacy pollution and be proactive in the management of the negative impacts of ongoing operations. This chapter provides an overview of responsible care as it relates to environmental stewardship.

2. RESPONSIBLE CARE

"Standard of Care" is a term of art that is generally defined as the degree of prudence and caution required of an individual who is under a duty of care. Generally, it is thought to concern the degree of caution that a reasonable person should exercise in a given situation so as to avoid causing injury. It is the watchfulness, attention, caution, and prudence that a reasonable person under the same or similar circumstances would exercise.

If a person's actions do not meet this standard of care, then the acts are considered negligent. Consider a hypothetical reasonable person or entity that provides an objective by which the conduct of others is judged. A reasonable person is not an average person or a typical person but a composite of the community's judgment as to how the typical community member should behave in situations that might pose a threat of harm to the public, workers, or individuals. Even though the majority of people or companies in a community may behave in a certain way, that alone does not establish the standard of conduct of the reasonable person or entity. For example, a majority of people in a community may jaywalk, but jaywalking may still fall below the community's standards of safe conduct.

The conduct of any one company in managing the environmental aspects of its business should not be compared with how other companies conduct their affairs, except for benchmarking purposes. We may observe for example that before the regulatory enforcement era (post 1972), many companies did not invest into pollution controls, often because it was believed that there were few if any financial incentives to do so and regulatory enforcement was poor. General belief 30 or so years ago was that pollution controls and source reduction based on process changes constitute sunk costs that could not be recovered; hence, some companies were not willing to make such investments or improvements into older operating facilities. In contrast, industry trade associations such as the American Petroleum Institute and the Chemical Manufacturers Association, among many others, promoted sound environmental management practices based on the technologies of the day, which shows that industry has always understood its responsibility to protect the public and environment from harmful pollution despite the absence of statutory obligations. Decisions not to make investments into pollution controls and best management practices (BMPs) were irresponsible.

Strict environmental enforcement has made the single most difference in overall improved environmental performance by industry. The fear of fines, penalties, and even imprisonment of owners and operators who violate statutory obligations under pollution control permits have driven industry on the whole to adopt BMPs and to make investments into controls.

Despite the gains in a greater degree of responsible care, corporate profitability and greed can still take precedence even today; and where there is a will there is a way to circumvent the cost of doing business in a responsible manner. Here are a few examples:

- On Monday, December 22, 2008, a dike containing the Tennessee Valley Authority (TVA) Kingston Fossil Plant (KFP) coal ash dredge cells failed, releasing approximately 4.1×10^6 m^3 (5.4 million cubic yards) of fly ash and bottom ash into adjacent waterways and over land. The KFP is located near the confluence of the Emory and Clinch Rivers on Watts Bar Reservoir near Kingston, Tennessee. The ash flow covered approximately 120 ha (300 acres) to varying thicknesses, including the Swan Pond Embayment on the north side of the KFP property. Fly ash also entered the channel and overbank areas of the riverine section of the Emory River. The initial response focused on providing temporary housing for affected residents and protection of public health, restoring essential services, stabilization of released ash, and environmental monitoring of ash, air, surface water, municipal drinking water supplies, and ground water. Figures 31.1–31.3 show photographs of the disaster site. N&P Limited analyzed samples of the ash from the disaster site finding it to contain arsenic, cadmium, various low-level concentrations of toxic metals, and radionuclides. Hardly an act of God. TVA saved itself money by not inspecting the walls of the cells and not implementing recommended structural improvements. And to add further to its irresponsible actions, it was in the process of hauling off 2.7×10^6 metric tonnes (2.7×10^6 t) (3 million U.S. tons) of ash to the Uniontown Landfill located in Alabama, which is located within 0.8 km (mile) of a residential community. The community is being exposed to constant airborne fugitive dust emissions from the disposal operation.

- The Tonawanda Coke Chemical (TCC) Plant is located in New York State. This is a community of about 80,000 residents nestled along the Niagara River about 1.6 km (1 mile) north of the city of Buffalo. Despite the clean appearances of homes and parks, the air is filled with the stench of benzene, naphthalene, ammonia, and polyaromatic hydrocarbons (PAHs). After decades of citizen complaints, the U.S. EPA finally got around to raid the TCC's offices in late 2009. A review of records through the Freedom of Information Act (FOIA) shows that this facility has a history stretching back into the 1970s of fines, penalties, notices of violations,

FIGURE 31.1 The power plant in the background and the collapsed 40 acre ash pond in the foreground. Thousands of front-end loaders were dispatched within a month to begin excavating buried homes.

FIGURE 31.2 Photograph taken along the bank of lake area in which the ash pond collapsed over.

FIGURE 31.3 Aerial reconnaissance flight taken by N&P Limited showing a small section of the release. Note: home partially buried in ash in southeast corner.

and records of decisions against it for violations of air pollution control permits. A review of the facility's self-reported emissions on the Toxic Release Inventory shows that the facility never reported any industrial solid or liquid waste releases or offsite disposal from its facility over the entire history of the TRI program. Either TCC eats its industrial waste or it is simply not accurately reporting its emissions. The TRI self-reported emissions from the facility also show typically less than 4.5 kg a^{-1} (10 pounds per year) of fugitive emissions from its manufacturing operations, yet it handles millions of tonnes of coal and generates billions of cubic meters of coke oven gas—none of which TCC admits ever gets into the community. It is certainly no leap in logic to conclude that if one smells the chemicals that are produced as by-products from a manufacturing operation, that air emissions are being released into the community and people are at risk from exposure to chemical toxins. But one may visit the Web site of TCC and see the words "TCC is committed to the preservation of a clean and healthy environment. By utilizing the resources available TCC resolves to implement applicable environmental regulations with innovation, determination and effectiveness. To this end TCC has employed ammonia stills... to remove ammonia from the wastewater that is discharged to the POTW (Publically Owned Treatment Works)." TCC releases nearly 0.45×10^6 kg a^{-1} (1 million pounds per year) of ammonia into the community's atmosphere in addition to benzene, PAHs, and particulate matter. It uses ammonia stills to control this chemical for its sewer discharges because its wastewater permit will not allow it to destroy the municipal treatment plant works. TCC has simply gotten around the law by transforming a water pollution problem into an air pollution problem.

- Santa Maria, a coastal city in Santa Barbara County, sits on top of an oil field. Since 1930, the oil field has been operated by many different oil companies and produced 206 million barrels of oil (2007 Preliminary Report on California Oil and Gas Production Statistics, California Department of Conservation, Division of oil, gas and geothermal resources, -January 2008). In the 1950s, large oil well sumps were built to collect byproducts of drilling including water, drilling mud, and oil. Each oil well had at least 1 sump, varying in size from the size of a house to the size of a football field (Doane-Allmon, Remediation Technologies Symposium 2005, 19–21 October 2005, Fairmont Banff Springs. 2005). After its peak in oil production in the 1950s, parts of the oilfield began being decommissioned and the city of Santa Maria began to grow on top of it. Over the next couple decades, 1707 oil wells were abandoned. As wells were decommissioned, the responsible oil company removed the oil and covered the sumps with 0.3 to 1.2 m (1 to 4 ft) of clean soil. Without first being decontaminated, the land was taken over by houses, agriculture, and industry. As a result, many residents in Santa Maria live on top of the oil sumps and were exposed to petroleum waste chemicals. It was not until the turn of this century that clean-up of the sumps was instigated. Crude oil is made up of several hundred compounds, collectively known as total petroleum hydrocarbon (TPH). The specific composition is based on the geology of the region where the oil was originally excavated; however, it generally contains benzene, xylene, jet fuel, toluene, and hexane. These compounds can enter the body through inhalation, ingestion, or dermal contact and have been associated with negatively impacting the blood, immune system, lungs, skin, nervous system, and fetal development [The Agency for Toxic Substance and Disease Registry (ATSDR), 1999, 4770 Burford Hwy NE, Atlanta GA, 30341]. Benzene is a known carcinogen, specifically causing acute myeloid leukemia with long-term exposure. Hexane causes peripheral neuropathy, a disorder of the nervous system characterized by numbness or paralysis (US DHHS, 1999). Xylene exposure can affect the kidneys and liver (US DHHS, 1999). Toluene can cause respiratory, liver, and kidney damage (US DHHS, 1999). These are just a few examples of negative health effects that specific TPH compounds can have. Because there can be hundreds of chemicals in crude oil and the human toxicity of the majority of the TPH compounds is not available, the actual health effects of crude oil is not known (US DHHS, 1999).

- Roy O. Martin has owned and operated the Colfax Wood Treating Plant since 1948 in Pineville, Louisiana. The plant makes utility poles using pentachlorophenol (PCP) and coal tar creosote. The PCP contains dioxins that are super toxins. The product has been banned in nearly a dozen countries. Coal tar creosote contains more than 300 toxic chemicals, of which the most potent toxins are the PAHs. International Agency for Research on Cancer (IARC) has determined that coal tar creosote is a probable human carcinogen. The U.S. EPA has also determined that coal tar creosote is a probable human carcinogen, and that coal tar pitch is a confirmed human carcinogen. The EPA classified coal tar creosote as a carcinogen in the 1992 Toxics Release Inventory (TRI). National Institute of Occupational Safety and Health (NIOSH) reported "from the epidemiologic and toxicological evidence on coal tar, coal tar pitch, and creosote, NIOSH has concluded that they are carcinogenic and can increase the risk of lung and skin cancer in workers."[1] Beginning in 1984, the facility was listed on the National Priority List as a Superfund Site. Superfund sites are the worst of the worst. The facility was delisted and EPA has published on its Web site for more than a decade that legacy pollution problems associated with poor housekeeping practices have been remediated and that there are no pathways of exposure to the community. Through FOIA requests, N&P discovered that Roy O. Martin has claimed that there is no contaminated groundwater off site. It based this reporting to state and federal regulatory agencies by installing its monitoring wells only within its own property lines and reporting contamination to mysteriously stop at its fence line. Other documents revealed that it used PCP and PAH contaminated soils, and it was ordered to excavate from solid waste management units to backfill areas around its drip pad. Soil borings from various locations on its property report through present times contamination of dioxins and PAHs in the league of parts per hundred; and despite the property being in the middle of a 100 year flood plain, Roy O. Martin has argued that there are no pathways for toxins to enter into the community. This facility also operated an illegal landfill on its site until 2001, at which time it removed the waste and reported an amazing 315 kg (700 pounds) of dioxins removed as a part of the waste taken off site. It later retracted this statement claiming the amount to be 4.7 kg. This lower waste estimate represents more dioxins than reported by all U.S. industry for 2001 under the National Emissions Inventory. In 2008, when Hurricane Gustav struck the town, astronomical levels of dioxins were found in soil and backed up sewer water in the surrounding community; however Roy O. Martin claims that these toxins are not from its plant operations. Through present times, this facility has operated with no air pollution controls. Figure 31.4 shows a photograph of the emissions from its treating cylinders. These emissions are not measured, monitored, or controlled.

[1]Criteria for Recommended Standard: Occupational Exposure to Coal Tar Products; National Institute for Occupational Safety & Health, Washington, D.C., September 1977.

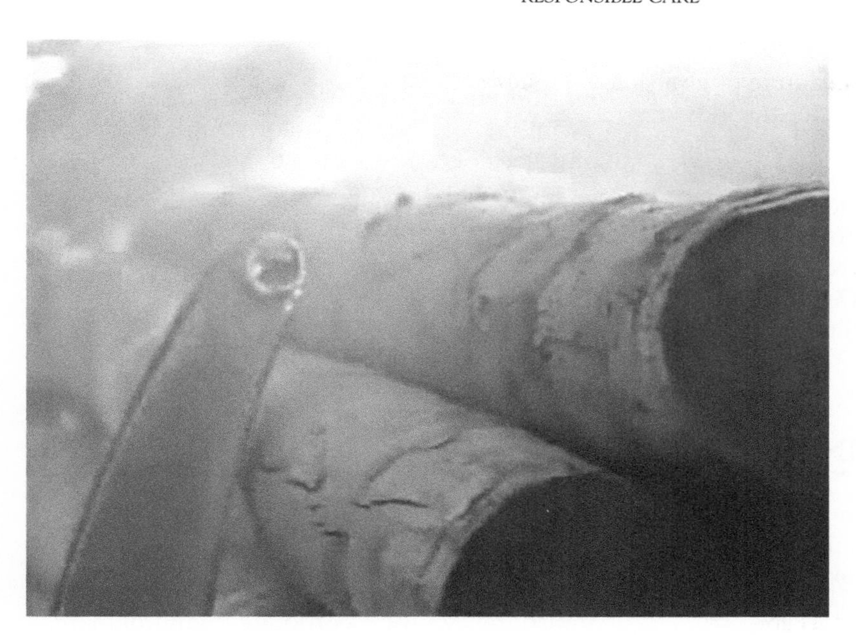

FIGURE 31.4 Close-up photo showing uncontrolled air emissions from one of Roy. O Martin's wood treating cylinders.

FIGURE 31.5 Photograph showing treating cylinder operations in an uncovered area where emissions are allowed to freely discharge into the atmosphere without any controls or monitoring.

Figure 31.5 shows that Colfax operates its treating cylinders in an unenclosed area and allows its emissions to freely discharge into the community. There are more than a dozen large point and area sources of emissions at this facility—none of which are monitored. In the twenty-first century and in a country which often time claims to have the strictest environmental laws among industrialized countries, Roy O. Martin's Colfax Treating

Plant stands out. The company does note on its Web site that it sets an example of good environmental management for the industry and further touts a 2004 award from the governor's office for leading the state in pollution prevention practices. The conditions and actions of this company, as well as the lack of aggressive enforcement on the part of the Louisiana Department of Environmental Quality fall far short of being responsible.

3. TOXIC RELEASE INVENTORY

The *Bhopal disaster* was an industrial catastrophe of epic proportions. It took place at a pesticide plant owned and operated by Union Carbide (UCIL) in Bhopal, Madhya Pradesh, India, on December 3, 1984. Around 12 AM, the plant released methyl isocyanate gas and other toxins, resulting in the exposure of more than 500,000 people. Estimates on the death toll range from 2,259 to 15,000. Some have estimated that 8,000 to 10,000 died within 72 h and 25,000 have since died from gas-related diseases [1]. The effects of this disaster are still being reported today since full remediation of soil and groundwater was never completed.

Shortly after the Bhopal disaster, there was a serious chemical release at a sister plant in West Virginia. These incidents underscored demands by industrial workers and communities in several states for information on hazardous materials. Public interest and environmental organizations around the country accelerated demands for information on toxic chemicals being released "beyond the fence line"—that is, outside of the facility. Against this background, the Emergency Planning and Community Right-to-Know Act (EPCRA) was enacted in 1986.

The primary purpose of EPCRA is to inform communities and citizens of chemical hazards in their areas. Sections 311 and 312 of EPCRA require businesses to report the locations and quantities of chemicals stored on-site to state and local governments to help communities prepare to respond to chemical spills and similar emergencies. Section 313 of EPCRA requires EPA and the States to annually collect data on releases and transfers of certain toxic chemicals from industrial facilities, and make the data available to the public in the TRI. In 1990, Congress passed the Pollution Prevention Act, which required that additional data on waste management and source reduction activities be reported under TRI. The intended goal of TRI is to empower citizens, through information, to hold companies and local governments accountable in terms of how toxic chemicals are managed. The TRI is intended to provide transparency of corporate environmental stewardship.

The EPA compiles the TRI data each year and makes it available through several data access tools, including the TRI Explorer and Envirofacts. Over the years, the TRI program has expanded. U.S. EPA has issued rules to roughly double the number of chemicals included in the TRI to approximately 650. Seven new industry sectors have been added to expand coverage significantly beyond the original covered industries, that is, manufacturing industries. Most recently, EPA reduced the reporting thresholds for certain persistent, bioaccumulative, and toxic chemicals to be able to provide additional information to the public on these chemicals.

Unfortunately, the TRI falls far short of its intended goal of keeping communities informed and bringing transparency to corporate environmental management practices. The following are some of its drawbacks:

1. Many companies inconsistently report their releases. An example of a coke chemical company was given in the last section. Many hundreds of examples of inconsistent and even illogical reporting of releases and

emissions can be found among historical and current TRI-reported releases.

2. Air emissions are calculated. The accuracy of calculated emissions is dependent on the assumptions applied, for which there is wide flexibility. There are no measured emissions reported on the TRI and verification based on measurements is not a requirement.
3. TRI-reported releases do not account for episodic releases. All manufacturing operations experience transient periods of operations in which there are episodic releases, yet these situations are not accounted for and their mass emissions are not included in reported yearly average releases.
4. TRI is self-reported and not independently verified. Some state regulatory agencies review and comment on the underpinning calculations applied by companies, but it is done so in an inconsistent and infrequent manner.
5. Data are manipulated. Some companies have reported substantial reductions in emissions claiming that these are the result of pollution prevention and waste reduction programs, when in reality the reductions were largely achieved through lower production during soft market periods and by reclassifying some waste streams as process streams to avoid reporting certain chemical releases.

Affording industry the freedom to self-report toxic chemical releases without measurement and independent monitoring is not different from having the fox guard the chicken coup.

4. EMISSION FACTORS

U.S. EPA publishes *Compilation of Air Pollutant Emission Factors* best known as AP-42. The first official version was published in 1972. Supplements to AP-42 have been routinely published to add new emission source categories and to update existing emission factors.

According to the EPA:

> an emission factor is a representative value that attempts to relate the quantity of a pollutant released to the atmosphere with an activity associated with the release of that pollutant. Emission factors usually are expressed as the weight of pollutant divided by the unit weight, volume, distance, or duration of the activity that emits the pollutant.

The EPA further reports that the emission factors presented in AP-42 may be appropriate to use in a number of situations, including making source-specific emission estimates for area-wide inventories for dispersion modeling, developing control strategies, screening sources for compliance purposes, establishing operating permit fees, and making permit applicability determinations.

It is reasonable to expect that any methodology and procedure used for the purpose of quantifying air emissions should be reproducible and provide representative emissions. It is however, impossible to provide precise air emissions in any industry sector based on the current limitations of available knowledge. The reason for this is that emissions are calculated based on emission factors that are arithmetic averages of releases reported from different plants. Although technologies within any one industry sector can be the same, operational practices can vary widely, thereby resulting in emission factors that significantly vary from one facility to another. A good example is the refining industry.

U.S. EPA has reported in industry sector books that no two refineries are the same. Since no two refineries are the same, then the average emission factors reported in the refining sector AP-42 supplement cannot be viewed as precise representations of any one facility. The same may be said for wood treating facilities, pesticide manufacturing plants, cement making plants, or just about any other industry sector one may think of.

The argument that has been adopted in support of AP-42 methodology for emissions quantification is that application of recommended emission factors on the average represents the mass emissions from any one facility. But again, if no two plants are the same, especially in terms of the age and operational efficiencies of equipment, this assumption cannot be reasonable.

In developing the collection of emission factors, U.S. EPA relied on industry reported emissions but screened out unusable test reports, documents, and information from which emission factors could not be developed. It applied the following criteria:

1. Emission data must be from a primary reference:
   a. Source testing must be from a referenced study that does not reiterate information from previous studies.
   b. The document must constitute the original source of test data. For example, a technical paper was not included if the original study was contained in the previous document. If the exact source of the data could not be determined, the document was eliminated.
2. The referenced study should contain test results based on more than one test run. If results from only one run are presented, the emission factors must be down rated.
3. The report must contain sufficient data to evaluate the testing procedures and source operating conditions (e.g., one-page reports were generally rejected).

AP-42 notes that the use of these criteria is somewhat subjective and depends to an extent on the individual reviewer. AP-42 fails to highlight or scrutinize the fact that all the reported emission factors are based on industry-collected data. Again, we see the fox left in charge of the chicken coup.

5. PREPARING EMISSIONS INVENTORIES

The term "emissions inventory" refers to the mass rate accounting of priority pollutants from the different sources within a manufacturing process. Both fugitive and point sources of emissions are required to be accounted. These are not by any means total emissions, but only those emissions that are required to be reported. The reader should note that our discussions are restricted only to air emissions, but there are other regulated waste forms such as liquid, wastewater, and solid wastes that industry is required to report on.

Emissions inventories are prepared largely by means of applying emission factors to volume or mass production rates. In other words, the vast majority of reporting of air pollutants in the United States is by means of calculation and not actual monitoring using field instrumentation.

The U.S. EPA Protocol, dated November 1995, entitled *1995 Protocol for Equipment Leak Emission Estimates* (EPA-453/R-95-017, "the 1995 EPA Protocol") presents four different methods for estimating equipment leak emissions. The methods, in order of increasing refinement, are: Method 1: Average Emission Factor Method; Method 2: Screening Value Range Method; Method 3: Correlation Equation Method; and Method 4: The Unit-Specific Correlation Equation Method.

In general, a more refined method requires more data and provides more reliable fugitive emission estimates. It is also more costly to implement and hence is not relied on by many facilities. In the Average Emission Factor Method and the Screening Value Range Method, emission factors are combined with equipment counts to estimate emissions. This is the least cost methodology. To estimate emissions with the Correlation Equation Method, OVA-measured concentrations (screening values) for all equipment components are individually entered into correlation equations or counted as either default zeros or pegged components. The reader can find detailed discussion of the four methods in the book by Cheremisinoff and Rosenfeld (*Best Practices in the Petroleum Industry, Handbook of Pollution Prevention and Cleaner Production*, Volume 1, Elsevier Publishers, UK, 2009).

The basis for the application of any industry-published emission factor is the assumption that all facilities on the average will generate about the same amount of pollution per unit of production if the same technologies and controls are used. Further to this assumption, most facilities argue that since published emission factors are averages, calculated emissions should be viewed as yearly, or rather long-term averages. In other words, although there may be excursions in releases, the argument is that if one were to take measurements of actual emissions over a sufficiently long period of time, the average of the measurements would be in agreement with calculated mass emissions.

Published emission factors for the most part are based on industry-reported averages. We must assume that a sufficiently large enough population of facilities has been sampled to obtain representative emission factors such that variations among equipment performance are included in the published factor. To our knowledge, this has never been substantiated. Although AP-42 provides a semi-qualitative rating of the accuracy of published emission factors, it neither publishes a range nor a standard deviation for emission factors. It only publishes a single value and advises the user that there is a higher level of confidence in some values versus others. But again, we emphasize that published emission factors are based on values obtained and reported by the industry itself.

The U.S. EPA has acknowledged in discussions on its Web site and in AP-42 itself that emission factors can be site specific. Not only are there variations among the same technologies used, especially controls, but the age and condition of equipment as well as site-specific practices can also dramatically impact the value of any emission factor.

The most serious shortcoming in relying on any published emission factor is that published values are based on measurements obtained during steady-state, continuous operations. Published emission factors do not capture upsets, breakdowns, excursions, or other operational conditions that can lead to episodic releases. Older equipment and operations are prone to these events and certainly more frequently than newer controls and process equipment that come with warranties.

The age and condition of equipment, especially air pollution controls, unquestionably have an impact on the accuracy of emission factors that are used for calculating air pollution levels for a facility. The World Bank Organization (WBO), as an example, reports ranges of emissions factors and makes the distinction that the higher values reported are more typical of older facilities with aging infrastructure (see Table 31.1 as an example of emission factors for oil refineries). Relying on default values that are reported by state regulatory agencies or AP-42 without any independent verification as to whether those values truly characterize the average conditions for a facility is not a reasonable approach. In examining the WBO's published values, we see that there are several orders of magnitude differences between the high and low values for some reported emissions. Relying on an average value as in the case of the AP-42 procedure does not support a general industry argument that calculated emissions are conservative and overstate the pollution from a facility.

A memorandum issued by Shine [3] of the U.S. EPA documents numerous incidents throughout the oil industry where omissions and misrepresentations of fugitive emissions are ongoing. The following summarizes the various incidents that are noted as typical omissions in reporting on the TRI:

- Exclusion of upsets, malfunctions, startups, and shutdowns from emissions inventories;

TABLE 31.1 World Bank Organization Reported Emission Factors for Oil Refineries [2][2]

Pollutant	Average	Low	Upper
Particulate matter/kg m^{-3}	4.464 (0.2778)	0.5581 (3.473×10^{-2})	16.74 (1.042)
Sulfur oxides/kg m^{-3}	7.256 (0.4515)	1.116 (6.946×10^{-2})	33.48 (2.084)
Sulfur oxides with sulfur recovery/kg m^{-3}	0.5581 (3.473×10^{-2})		
Nitrogen oxides/kg m^{-3}	1.674 (0.1042)	0.3348 (2.084×10^{-2})	2.790 (0.1736)
VOCs/kg m^{-3}	5.581 (0.3473)	2.790 (0.1736)	33.48 (2.084)
BTX/kg m^{-3}	1.395×10^{-2} (8.682×10^{-4})	4.186×10^{-2} (2.605×10^{-4})	3.348×10^{-2} (2.084×10^{-3})

[2]*The units are kilogram per cubic meter crude and in parenthesis (lbs per bbl crude).*

- Omission of sources that are unexpected or not measured, such as leaks in heat exchanger systems or emissions from process sewers;
- Exclusion of emission events such as tank roof landings;
- Improper characterization of input parameters for emission models such as not using actual tank or material properties in the AP-42 tank emission estimation methodologies.

Shine's memorandum points out that the current U.S. National Emissions Inventory does not identify upsets, startups, or shutdowns as emission events, nor is the data specifically requested from the reporters (the states). To understand the order of magnitude of these omissions and upsets in relation to routine operations, the EPA reviewed the emission inventory data from the Texas Commission on Environmental Quality (TCEQ) for the 2004 reporting year. This data set contains emissions data for 30 of the approximately 150 U.S. refineries and accounts for more than 25% of the U.S. refining capacity. Additionally, the TCEQ inventory identifies emissions from routine events separately from upsets, startups, and shutdowns, so a comparison of reported emissions is possible.

Shine reported that in general, the quantity of emissions reported as nonroutine is smaller than the routine emissions. For VOC-unclassified contaminant, emissions of upsets and various malfunctions, startups, and shutdowns were 5% of the emissions reported from routine events [578 tons per annum (524 t a^{-1}) vs. 10006 t a^{-1}]. However, for some compounds, such as 1,3-butadiene, emissions from these incidents accounted for as much as 20% of the routine emissions (18.0 t a^{-1} vs. 82.5 t a^{-1}). The investigator goes on to note that for certain types of emission points, emissions from startups, shutdowns, and malfunctions comprise the majority of the emissions.

Shine has further noted that the comparison was done between *reported* upsets and the unstable or transient events and *reported* routine emissions. This comparison does not consider events such as upsets and shutdowns/startups/malfunction events that are not properly characterized and reported to begin with.

The U.S. EPA memorandum notes that there are emission events that are not measured and further that there are many events that are not even characterized or reported in inventories. For example, monitoring of cooling tower water return for VOC is required at some refineries because of state permitting rules, but these are not required for refineries at the Federal level.

Shine has noted that in one release report submitted to the National Response Center in 2006, a facility initially reported potential emissions of 315 kg per day (700 lbs per day) each of benzene, toluene, and xylene from a reformer unit cooling tower, based on sampling of their cooling water return and the expected composition of the process streams that were being cooled. On further analysis and speciation of the cooling water, however, the facility submitted a final report indicating that the exchanger had leaked 360 kg per day (800 lbs per day) of propane and isobutane for approximately 8 days. The U.S. EPA's memorandum noted that the subject facility monitored the tower and this is the reason why the leak was identified and reported. However, many refineries do not conduct routine cooling tower water monitoring.

In a sampling of the refining industry to be used to supplement EPA's emissions inventory for the purpose of risk modeling, the EPA surveyed 22 refineries and requested emissions of benzene. Of the 22 facilities surveyed, only three indicated that they have sampled their cooling towers for leaks. The remaining facilities that did report emissions used AP-42 VOC emission factors for cooling towers and an assumed speciation for benzene. Five facilities simply reported no emissions at all.

Additional omissions were identified by Shine regarding the wastewater treatment emission estimates provided by U.S. refineries. A recent study evaluated collection system emissions for five Bay Area refineries. Using extensive sampling, flow measurements, and detailed TOXCHEM+ modeling, the study showed that four of the five refineries underestimated the VOC emissions from their wastewater collection system. Two refinery estimates were within a factor of two of the regulatory agency estimate (one higher and one lower), but one refinery had underestimated its emissions by a factor of 40 and another refinery underestimated its emissions by a factor of 1400. In reviewing the emission estimates reported by the residual risk survey respondents for wastewater collection and treatment systems, Shine also noted low estimates for several refineries.

One of the more disturbing observations reported by Shine is that in an Alberta refinery measurement study, emissions of VOC were 30 times higher and emissions of benzene were 100 times higher than the emissions calculated using AP-42 equations. The reason for this underreporting is that the AP-42 equations require a number of inputs about the tank and material characteristics and storage conditions. Mischaracterization of these inputs leads to erroneous results. API points out that when actual measurements indicate unexpectedly high emissions, environmental conditions may be outside the scope of the method. But then, how is the refinery to know when and when not to apply the calculation method properly? These concerns are sources of uncertainty that can explain differences in the order of two or three, but they do not explain differences that are in the order of 30 to 100. Given the magnitude of the difference, either emissions are zero most of the time (when events are not on the high side) or the annual estimates of the emissions are grossly understated.

Further to this, Shine has noted that there are numerous examples of tank maintenance issues that, if not characterized properly, would lead to erroneous results. One example cited is on March 11, 2003, the South Coast Air Quality Management District (SCAQMD) filed suit against BP West Coast Products, LLC. Most of the allegations accuse the company of failing to properly inspect and maintain 26 storage tanks equipped with floating roofs, as required under SCAQMD Rule 463. SCAQMD inspections revealed that more than 80% of the tanks had numerous leaks, gaps, torn seals, and other defects that caused excess emissions [4].

6. RESPONSIBLE STEWARDSHIP

Good corporate governance includes environmental stewardship as a foundation. From the most general standpoint, Responsible Care means acting in a responsible manner toward public and worker safety, protecting the environment, and respecting the properties and quality of life of others. In short, responsible care means that all companies have an obligation to eliminate, or most certainly reduce, the negative impacts of the environmental aspects of their businesses operations. Failure to follow reasonable care is a breach of duty and may be viewed in the legal system as negligence.

The U.S. EPA manages more than 1240 active Superfund Sites across the country [5]. Most of these sites were created by the practices followed by industry before the creation of strict environmental enforcement policies. There are no accurate figures published on either the total costs for cleanup to the public or the numbers of responsible parties. A review of the Superfund Sites listed on the U.S. EPA Web site reveals that a number of sites are so badly contaminated that remediation costs are incalculable. There are also no reliable published estimates on the costs of litigations that range from disputes over cleanup liabilities, to third-party damages, to toxic torts and class action suits involving medical monitoring of entire communities.

There are a variety of excuses made by companies for past poor environmental management practices that are universally understood by the public today to be environmentally unfriendly. Among the most frequent are:

- Little information existed on the long-term and acute health risks of industrial chemicals at the time environmental damages took place.
- The environment was believed to have infinite capacity to dilute and break down harmful chemicals.
- Few industry technical standards and pollution control technologies existed until the 1970s.
- Highly toxic chemicals such as dioxins were not known or could not be detected until the late 1970s.
- Principles of pollution prevention (P2) were not invented and their benefits not understood until the early 1990s.
- Environmental Management Systems and the principles of ISO 14001 were not introduced until the mid 1990s and hence a formalized approach to managing environmental obligations did not exist.
- No information on leachate formation, dense non-aqueous phase liquids, and the fate and transport of chemicals in the environment existed.

Although some of the arguments are justified, many are simply excuses, because industry overall has had a keen understanding of toxicology and fate and transport concepts beginning in the 1930s. Organizations such as the NIOSH, the IARC, and the American Conference of Governmental Hygienists (ACGIH) have been active for decades long before the introduction of strict environmental enforcement policies. ACGIH® has been considered a well-respected organization by individuals in the industrial hygiene and occupational and environmental health and safety industry since the late 1930s. The best known activities of ACGIH®, the Threshold Limit Values for Chemical Substances (TLV®-CS) Committee were established in 1941. This group was charged with investigating, recommending, and annually reviewing exposure limits for chemical substances. It became a standing committee in 1944. Two years later, the organization adopted its first list of 148 exposure limits, then referred to as Maximum Allowable Concentrations. The term "Threshold Limit Values (TLVs®)" was introduced in 1956. The first *Documentation of the Threshold Limit Values* was published in 1962 and is now in its seventh edition.

Stewardship of environmental aspects should not be based on simply following governmental rules and regulations. Companies that focus only on meeting their statutory compliance obligations are not being responsible, because it is the same argument used by responsible parties who created Superfund sites. These parties have argued that they did everything the government expected of them when it came to managing and controlling pollution. But that unto itself is unreasonable. And it is especially unreasonable in this century where the present and future generations are left to wrestle with legacy pollution.

In this chapter, we have sited what hopefully are some extreme examples of poor environmental stewardship. To be fair, industry faces significant problems in meeting statutory obligations across multimedia: air emissions, wastewater control and releases, and solid waste management. Across all the media, there are issues concerning omissions in reporting, applying questionable assumptions to estimate off-site transfers and releases of regulated wastes; but even with companies that do take their environmental statutory obligations seriously, the regulations themselves are complex and require almost endless resources at times.

Within the last 15 years, improved management tools have evolved which responsible companies are relying more and more on to improve not just compliance records, but overall environmental performance. In this regard, the best managed companies recognize that environmental performance is not simply meeting statutory reporting obligations—it means applying principles of Total Quality Environmental Management to the business. In a practical sense, this means responsible companies need to exceed their minimum statutory requirements—which in fact is common sense, since after all, environmental statutes offer a minimum level of protection for the public. Management systems and principles such as those embodied under the ISO 14000 Environmental Management Standards and the application of IT-based Environmental Management Information Systems are becoming standard tools that enable responsible companies to more rapidly identify, prioritize, and systematically eliminate the negative impacts of their manufacturing operations. Extending the application of these tools toward greening a company's supply chain further helps to reduce the negative impacts of industrial activities. These tools enable companies to integrate environmental performance into the bottom line financial performance of a business. This has been achieved by a number of companies that apply performance-based metrics that track environmental performance in ways that offer greater transparency to business leaders, environmental managers, investors, regulators and communities. Being a good corporate citizen requires examining and carefully evaluating the environmental aspects associated with the manufacturing and business activities. Proper evaluations should identify present and even future impacts to neighboring communities and the environment. Companies need to devise, implement, and continually revise their environmental action plans to reduce the impacts of their business operations to better protect and preserve the environment and to reduce impacts on communities. Although the implementation of Environmental Management Systems such as ISO14001 establishes a protocol for compliance and managing the permitting requirements of companies, it does not ensure that facilities can demonstrate good environmental performance. Companies that truly make good environmental performance a part of their core values and strategic business plans exceed statutory obligations. They manage their companies by making investments into green technologies, reducing pollution at the sources, and making a direct connection between bottom-line financial performance and overall environmental performance of the business.

References

[1] Industrial Disaster Still Haunts India—South and Central Asia—msnbc.com.< http://www.msnbc.msn.com/id/34247132/ns/world_news-south_and_central_asia/page/2/>. Retrieved December 3, 2009; *No Takers for Bhopal Toxic Waste*. BBC. 2008-09-30. <http://news.bbc.co.uk/2/hi/south_asia/7569891.stm>.

[2] Pollution Prevention and Abatement Handbook, World Bank Group, Washington, DC, July 1998.

[3] B. Shine. Potential Low Bias of Reported VOC Emissions from the Petroleum Refining. Technical Memorandum, 27 July 2007, EPA Docket, No. 2PA-HQ-OAR-2003-0146

[4] C. Whetzel, South Coast Air District Seeks $319 For Violations at Los Angeles Area Refinery, The Bureau of National Affairs, Inc, Washington D.C, 2003.

[5] USEPA Superfund Web site. <http://www.epa.gov/superfund/sites/npl/npl.htm;>. Accessed 2 June 2010.

CHAPTER

32

Risk Assessment, Management, and Accountability

Daniel A. Vallero

Pratt School of Engineering, Duke University, Durham, N. Carolina, USA

OUTLINE

1. INTRODUCTION

Risk is a common metric for waste management. One means of determining whether a waste is handled properly is if it has exceeded some level of acceptable risk. This measure of risk, then, is an expression of operational success or failure. Too much risk means that the waste management has failed society. Societal expectations of acceptable risk are mandated by the standards and specifications of certifying authorities. For wastes, this includes health codes and regulations, zoning and building codes and regulations, design principles, canons of professional engineering and medical practice, national and international promulgating bodies (e.g., ISO, or the International Standards Organization). In the United

Waste Doi: 10.1016/B978-0-12-381475-3.10032-4

States, for example, standards can come from a federal agency, such as hazardous waste landfill guidelines of the US Environmental Protection Agency or material specifications for equipment, such as those of the National Institute of Standards and Testing (NIST). Standards are also articulated by private groups and associations, such as those from ASTM (formerly the American Society for Testing and Materials).

2. RISK ESTIMATION

Risk is generally understood to be the likelihood that an unwelcome event will occur. Risk is also investigated scientifically within well-established frameworks. *Risk analysis* addresses the factors that lead to a risk. The reduction of this risk (e.g., building containment structures around a landfill or enacting local rules to limit exposure to leachate) is an example of *risk management*. Risk management includes all the ways to address a particular risk, e.g. legal, political, and financial. Risk management is often differentiated from *risk assessment*, which is comprised of the scientific consideration hazards and exposures that make up a risk.

2.1. Hazard

Articulating the hazard (the agent of harm) is matched against the ways that people or other receptors may come into contact with that hazard, that is, exposure. Thus, risk management includes the policies, laws, and other societal endeavors that limit these two components of risk. From the hazard perspectives, regulatory agencies may decide that a product is too hazardous when it is manufactured, used, or when it becomes a waste product. Thus, the hazard may occur before a waste is generated, such as a component of a manufacturing process. For example, if 1,1,1-trichloroethane (TCE) is used as solvent in a chemical processing plant, it may be hazardous to the workers because it is carcinogenic. It may also be hazardous if it finds its way to a landfill (in drums or contaminated sawdust after a cleanup).

2.2. Exposure

Articulating the exposure is the other aspect of risk assessment. In the previous example, TCE, people can come into contact with the substance in occupational and environmental settings. Thus, the exposure to TCE varies by activities (high for workers who use it, less for workers who may not work with TCE but are nearby and breathe the vapors, and even less for other workers). Usually, worker exposure is based on a 5-workday exposure (e.g., 8 or 10 h), whereas environmental exposures, especially for chronic diseases like cancer, are based on lifetime, 24-h per day exposures. Thus, environmental regulations are often more stringent than occupational regulations when aimed at reducing exposure to a substance.

2.3. Waste Risk Characterization

To ascertain possible risks from wastes, the first step is to identify a general hazard (a potential threat) and then to develop a scenario of events that could take place to unleash the potential threat and lead to an effect. This means that not only there is a need for a sound characterization of what is in a waste but also a characterization of how these waste constituents change in time and space and how they may act synergistically or antagonistically within the waste. To assess the importance of a given scenario, the severity of the effect and the likelihood that it will occur in that scenario are calculated. This combination of the hazard particular with the scenario constitutes the risk.

The relationship between the severity and probability of a risk follows a general equation (32.1) [1]:

$$R = f(S, P) \tag{32.1}$$

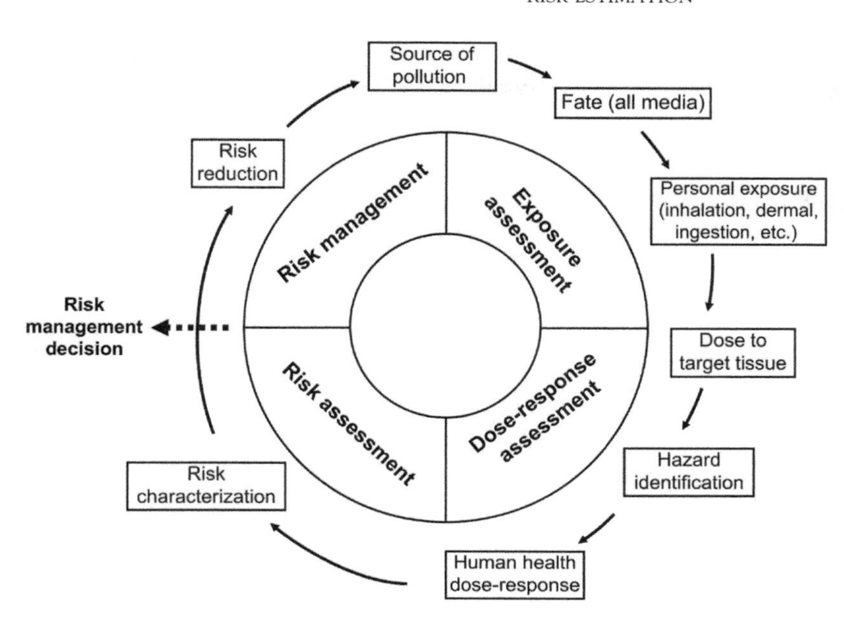

FIGURE 32.1 Risk assessment and management paradigm as employed by environmental agencies in the United States. The inner circle includes the steps recommended by the National Research Council [2]. The outer circle indicates the activities (research and assessment) that are currently used by regulatory agencies to meet these required steps.

where risk (R) is a function (f) of the severity (S) and the probability (P) of harm. The risk equation can be simplified to be a product of severity and probability:

$$R = S \times P \tag{32.2}$$

The traditional chemical risk assessment paradigm (see Figure 32.1) is generally a stepwise process. It begins with the identification of a hazard, which is comprised of a summary of an agent's physicochemical properties and routes and patterns of exposure and a review of toxic effects [2]. The tools for hazard identification take into account the chemical structures that are associated with toxicity, metabolic and pharmacokinetic properties, short-term animal and cell tests, long-term animal (in vivo) testing, and human studies (e.g., epidemiology, such as longitudinal and case-control studies). These comprise the core components of hazard identification; however, additional hazard identification methods have been emerging that increasingly provide improved reliability of characterization and prediction.

One complicating factor for most wastes is that they are mixtures. There is no unanimity in defining "mixtures." The chemical definition distinguishes mixtures from compounds. Until relatively recently, toxicologists studied mixtures in a step-wise manner, adding one substance at a time to ascertain the response of an organism with each interaction. A number of recent toxicological studies have begun to look at multicomponent mixtures. From an exposure perspective, a mixture is actually a co-exposure. People and ecosystems are exposed to an array of compounds simultaneously [3]. A key question is how do the individual constituent's physical and chemical properties affect those of other waste constituents and vice versa? Are there additive, synergistic, or antagonistic effects when they are exposed to different substances simultaneously.

Thus, characterizing the inherent properties of an individual constituent of a waste is but the first step in waste risk assessment. A number of tools have emerged to assist in this characterization. Risk assessors now can apply biomarkers of genetic damage (i.e., *toxicogenomics*) for more immediate assessments, as well as improved *structure–activity relationships* (SAR), which have incrementally been quantified in terms of stereochemistry and other chemical descriptions, that is, using *quantitative structure–activity relationships* (QSAR) and computational chemistry. For the most part, however, health-effects research has focused on early indicators of outcome, making it possible to shorten the time between exposure and observation of an effect [4].

Wastes often possess nonchemical hazards. Notably biological and infectious wastes present hazards from biological agents that differ from those posed by chemical-laden wastes. Of course, biological agents range from beneficial to extremely hazardous. The Safety in Biotechnology Working Party of the European Federation of Biotechnology [4] has identified four risk classes for microorganisms[1]:

1. Risk class 1. No adverse effect or very unlikely to produce an adverse effect. Organisms in this class are considered to be safe.
2. Risk class 2. Adverse effects are possible but are unlikely to represent a serious hazard with respect to the value to be protected. Local adverse effects are possible, which can either revert spontaneously (e.g., owing to environmental elasticity and resilience) or be controlled by available treatment or preventive measures. Spread beyond the application area is highly unlikely.
3. Risk class 3. Serious adverse local effects are likely with respect to the value to be protected, but spread beyond the area of application is unlikely. Treatment and/or preventive measures are available.
4. Risk class 4. Serious adverse effects are to be expected with respect to the value to be protected, both locally and outside the area of application. No treatment or preventive measures are available.

These classes indicate that even the safest microbes carry some risk and with more uncertainty about an organism, one cannot assume it to be safe. That said, most microbes are not considered to be pathogenic, but certain human subpopulations are susceptible to hazards that may not exist for a large proportion of the population. Also, the risks such as a change induced by the release of organism into an environment where there are no natural predators may not be direct. Thus, risk scenarios include not only the effects resulting from the intended purpose of the environmental application but also the downstream effects and side effects that are not part of the desired purpose.

Like the hazard-identification process for chemicals, the microbe is classified according to inherent properties. It is in the next stage that environmental conditions are taken into account: characterizing different responses to dose in different populations. Both the hazard identification and dose-response information are based on research that is used in the risk analysis. For microbes, the highest score for any one determines the overall risk class for environmental application.

Managing exposures to biological wastes (and any waste for that matter) must consider protecting the most vulnerable members of society, especially pregnant women and their yet-to-be-born infants, neonates, and immunocompromised subpopulations. Also, the exposure protections vary by threat. For example, adolescents may be particularly vulnerable to hormonally active agents, including many pesticides.

Risk can be extrapolated from available knowledge of similar chemical or biological

[1]The Working Party applied these groupings to genetically modified organisms, but they are also helpful for other microbial hazards.

agents with similar characteristics or of yet untested, but similar environmental conditions (e.g., a field study's results in one type of field extrapolated to a different agricultural or environmental remediation setting). In chemical hazard identification, this is accomplished by *structural activity relationships*.

In the United States, ecological risk assessment paradigms have differed from human health risk assessment paradigms. The ecological risk assessment framework (see Figure 32.2) is based mainly on characterizing exposure and ecological effects. Both exposure and effects are considered during problem formulation [5].

Interestingly, the ecological risk framework is driving current thinking in human risk assessment. The process shown in the inner circle of Figure 32.1 does not target the technical analysis of risk so much as it provides coherence and connections between risk assessment and risk management. In the early 1980s, there was confusion and mixing of the two. For example, a share of the criticism of federal response to environmental disasters, such as those in Love Canal, New York and Times Beach, Missouri

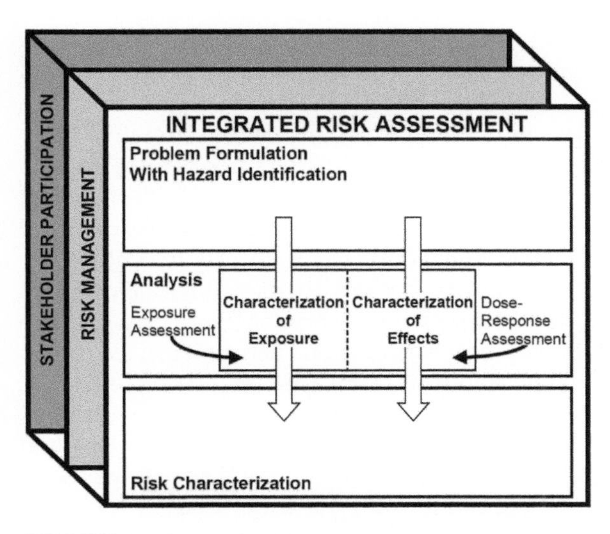

FIGURE 32.2 Framework for integrated human health and ecological risk assessment. [6].

related to the mixing of scientifically sound studies (risk assessment) and decisions on whether to pursue certain actions (risk management). When carried out simultaneously, decision making could be influenced by the need for immediacy, convenience, or other political and financial motivations, as opposed to a rational and scientifically credible assessment that would underpin management decisions.

At some point, risk assessment and waste management must merge. That is, the final step of the risk assessment process is referred to as "characterization" to mean that "both quantitative and qualitative elements of risk analysis, and of the scientific uncertainties in it, should be fully captured by the risk manager" [7]. The problem-formulation step in the ecological framework has the advantage of providing an analytic—deliberative process early on, because it combines sound science with input from various stakeholders inside and outside of the scientific community. This can be helpful not only in siting waste facilities but also in nonstructural solutions to waste problems (e.g., waste minimization, changes in use scenarios, product substitution, and life cycle perspectives).

The ecological risk framework calls for the characterization of ecological effects instead of hazard identification used in human health risk assessments. This is because the term "hazard" has been used in chemical risk assessments to connote either intrinsic effects of a stressor or a margin of safety by comparing a health effect with an estimate of exposure concentration. Thus, the term becomes ambiguous when applied to nonchemical hazards, such as those encountered in biological systems.

Specific investigations are needed in the laboratory and field when adverse outcomes may be substantial and small changes may lead to very different functions and behaviors from unknown and insufficiently known chemicals or microbes. For example, a chemical compound may have only been used in highly controlled experiments with little or no information about how it would behave inside an organism. Often, the proponents of a product would have done substantial research on the benefits and operational aspects of the chemical constituents, but the regulatory agencies and the public may call for more and better information about unintended and yet-to-be-understood consequences and side effects [1].

Even when much is known, there often remain large knowledge gaps when trying to estimate environmental impacts. The bacterium *Bacillus thuringiensis,* for instance, has been applied for several decades as a biological alternative to some chemical pesticides. It has been quite effective when sprayed onto cornfields to eliminate the European corn borer. The current state of knowledge indicates that this bacterium is not specific in the organisms that it targets. What if in the process, *B. thuringiensis* also kills honeybees? Obviously, this would be a side effect that would not be tolerable from either an ecological or agricultural perspective (the same corn crop being protected from the borer needs the pollinators). Furthermore, physical, chemical, and biological factors can influence these effects, for example, type of application of Bt can influence the amount of drift toward nontarget species. Downstream effects can be even more difficult to predict than side effects, because they not only occur within variable space but also in variable time regimes. For example, risk can arise from both the application method and from the build up of toxic materials and gene flow following the pesticide drift.

3. SUCCESS IN WASTE MANAGEMENT

Waste managers must not only control the crucial factors after the wastes have been generated but should also inform decisions at all points in the product life cycle. The public expects waste managers to "give results, not excuses" [8], and risk and reliability are accountability measures of their success. Engineers design systems to reduce risk and look for ways to enhance the reliability of these systems. Thus, every waste manager deals directly or indirectly with risk and reliability. Risk means different things to different people.

Although risk has some very precise definitions within the scientific community, the various scientific disciplines have divergent concepts of risk (see Table 32.1).

3.1. Modeling to Support Accountability

Risk management depends on models to estimate exposures. Such models range from "screening level" to "high tiered." Screening models included generally overpredicted exposures because they are based on conservative default values and assumptions. They provide a first approximation that screens out exposures not likely to be of concern [9]. Conversely, higher-tiered models typically include algorithms that provide specific site characteristics, time-activity patterns, and are based on relatively realistic values and assumptions. Such models require data of higher resolution and quality than the screening models and, in return, provide more refined exposure estimates [9].

Risk involves a stressor, a receptor, and an outcome. The stressor can be physical, chemical, or biological. If a water body's temperature increases beyond some threshold value, the fish may die or fail to reproduce. This is an example of physical stressor (heat) on a population of receptors (various fish genera), leading to a deleterious outcome (fish kill). Sometimes, the outcome is not specific to any particular stressor. For example, the fish kill could have resulted from a chemical stressor, such as the release of organic matter into the water body, which was used as food by the aquatic microorganisms, using the available oxygen. In this case, the

TABLE 32.1 Comparison of Definitions of Risk in Technical Publications versus Social Vernacular

Technical definitions of risk compiled by the Society of Risk Analysts [10].	1. Possibility of loss, injury, disadvantage, or destruction; to expose to hazard or danger; to incur risk of danger
	2. An expression of possible loss over a specific period of time or number of operational cycles
	3. Consequence per unit time = frequency (events per unit time) × magnitude (consequences per event)
	4. Measure of the probability and severity of adverse effects
	5. Conditional probability of an adverse effect (given that the necessary causative events have occurred)
	6. Potential for unwanted negative consequences of an event or activity
	7. Probability that a substance will produce harm under specified conditions
	8. Probability of loss or injury to people and property
	9. Potential for realization of unwanted, negative consequences to human life, health, or the environment
	10. Product of the probability of an adverse event times the consequences of that event were it to occur
	11. Function of two major factors: (a) probability that an event or series of events of various magnitudes, will occur and (b) the consequences of the event(s)
	12. Probability distribution over all possible consequences of a specific cause which can have an adverse effect on human health, property, or the environment
	13. Measure of the occurrence and severity of an adverse effect to health, property, or the environment
Social definitions of risk (Compiled in: S.M. Macgill, Y.L. Siu (2005). A new paradigm for risk analysis. *Futures*. 37: 1105–1131).	1. Probability of an adverse event amplified or attenuated by degrees of trust, acceptance of liability, and/or share of benefit
	2. Opportunity tinged with danger
	3. A code word that alerts society that a change in the expected order of things is being precipitated
	4. Something to worry about/have hope about
	5. An arena for contending discourses over institutional relationships, socio-cultural issues, political, and economic power distributions
	6. A threat to sustainability/current lifestyles
	7. Uncertainty
	8. Part of a structure of meaning based in the security of those institutional settings in which people find themselves

(*Continued*)

TABLE 32.1 Comparison of Definitions of Risk in Technical Publications versus Social Vernacular—*cont'd*

	9. The general means through which society envisages its future
	10. Someone's judgment on expected consequences and their likelihood
	11. What people define it to be is something different to different people
	12. Financial loss associated with a product, system, or plant
	13. The converse of safety

stressor is of first and second order. The first-order stress was the release of organic material; the second-order stress was the decreased dissolved oxygen (DO). The fish kill could also be the result of a biological stressor, such as a dinoflagellate, *Karenia brevis* (red tide) or *Pfiesteria piscicida*. The outcome in all three scenarios is the same, that is, a fish kill, but the path to this outcome is different.

Environmental stressors can be modeled in a unidirectional and one-dimensional fashion. A conceptual framework can link exposure to environmental outcomes across levels of biological organization (Figure 32.3). Thus, environmental exposure and risk assessment consider coupled networks that span multiple levels of biological organization that can describe the interrelationships within the biological system. Mechanisms can be derived by characterizing and perturbing these networks (e.g., behavioral and environmental factors) [11]. This can apply to a food chain or food web model (see Figure 32.4) or a kinetic model (see Figure 32.5) or numerous other modeling platforms.

To assess the risks associated with a waste management decision, three questions [12] must be asked:

1. What are the specific environmental concerns or harm that will or can occur?
2. What is the probability that the concerns will be realized or harm will occur?
3. What are the adverse outcomes (e.g., to health and the environment) when the harm occurs, including how widespread in time and space?

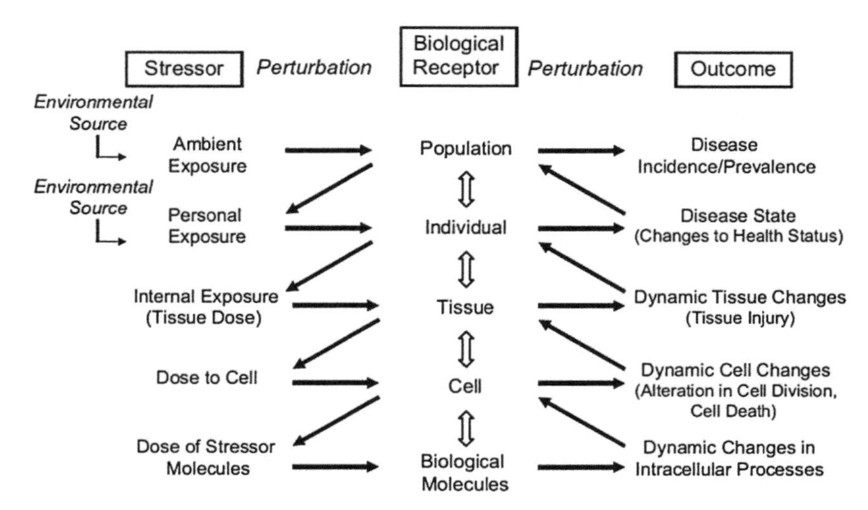

FIGURE 32.3 Systems' cascade of exposure–response processes. In this instance, scale and levels of biological organization are used to integrate exposure information with biological outcomes. The stressor (chemical or biological agent) moves both within and among the levels of biological organization, reaching various receptors, thereby influencing and inducing outcomes. The outcome can be explained by physical, chemical, and biological processes (e.g., toxicogenomic mode-of-action information). *Source: Ref. [11].*

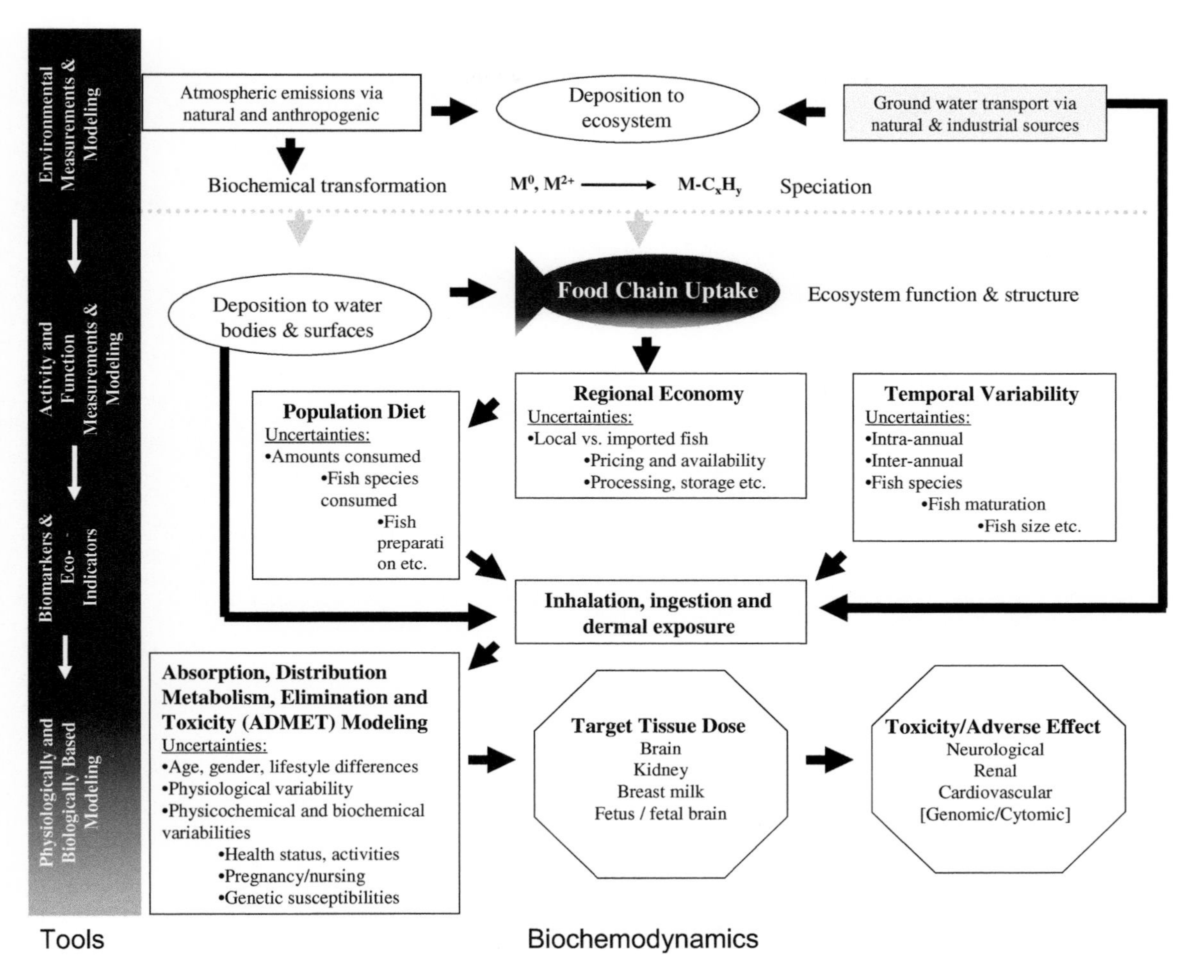

FIGURE 32.4 Biochemodynamic pathways for a substance (in this case a single chemical compound). The fate is mammalian tissue. Various modeling tools are available to characterize the movement, transformation, uptake, and fate of the compound. Similar biochemodynamic paradigms can be constructed for multiple chemicals (e.g., mixtures) and microorganisms. *Source: Ref. [18].*

These are deceptively simple questions. They are only easy to answer when an action is clearly wrong with no benefits whatsoever. For example, if a scientist simply wants to design a bandage that stays sterile during its entire use, few would oppose it. However, if the scientist used a neurotoxic metal like mercury to achieve this sterility, many would oppose such research on the basis of its likelihood to cause other problems (e.g., to the central nervous system). However, the scientist may argue that he really needs such a bandage for a flesh-eating virus that seems to resist every other substance available for the bandage. Whether we agree or not, this is actually a question of short-term risks versus long-term advancement in medical

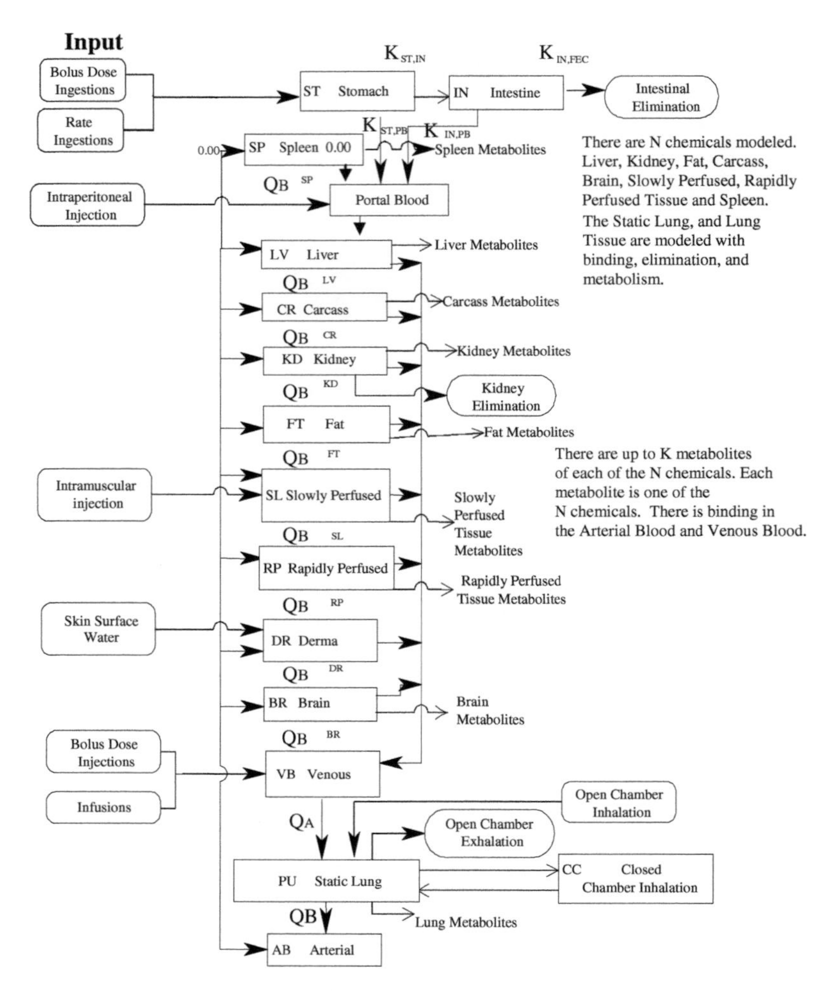

FIGURE 32.5 Toxicokinetic model used to estimate dose as part of an environmental exposure. This diagram represents the static lung, with each of the compartments (brain, carcass, fat, kidney, liver, lung tissue, rapidly and slowly perfused tissues, spleen, and the static lung) having two forms of elimination, an equilibrium binding process, and numerous metabolites. Notes: K refers to kinetic rate; Q to mass flow; and Q_B to blood flow. A breathing lung model would consist of alveoli, lower dead space, lung tissue, pulmonary capillaries, and upper dead space compartments. Gastro-intestinal (GI) models allow for multiple circulating compounds with multiple metabolites entering and leaving each compartment, that is, the GI model consists of the wall and lumen for the stomach, duodenum, lower small intestine, and colon, with lymph pool and portal blood compartments included. Bile flow is treated as an output from the liver to the duodenum lumen. All uptaken substances are treated as circulating. Nonspecific ligand binding, for example, plasma protein binding, is represented in arterial blood, pulmonary capillaries, portal blood, and venous blood. *[Source: ref. [13], adapted from [14].*

and other societal knowledge. This is exactly what has happened in introductions of species, such as kudzu and multiflora roses, which were only later found to have dramatically negative ecological impacts. If we could go back in time and have answers to the three questions above from our full knowledge of the implications observed in the field, these answers would have in all likelihood told the scientists and agricultural extension agents that the risks outweighed the benefits and the decision should be a "no go."

4. MAKING DECISIONS

Having all the data needed for a *completely* informed risk decision is impossible. Scientific objectivity and humility dictate that risk assessors are upfront about uncertainties. The three questions can only be answered by looking for patterns and analogies from events that are similar to the potential threat being considered. From there, scenarios can be developed to follow various paths to good, bad, and indifferent outcomes. This is known as a decision tree, which we will discuss later in this chapter. Risk assessment is a way to estimate the importance of each scenario and select the one with the most acceptable risk. This is not the same, necessarily, as the one with the most benefits compared to risks, that is, a benefit to risk ratio or relationship, or benefit to cost ratio or relationship. However, this is indeed one of the more widely used approaches. The challenge is how to quantify many of the benefits and risks, because risk is a function of likelihood and severity of a particular adverse outcome (Equation (32.2)). As mentioned, environmental risk is often considered to be the product of the hazard (H) and the exposure (E) to that hazard:

$$R = H \times E \tag{32.3}$$

Waste risk characterization follows four basic steps: hazard identification, dose-response estimation, exposure assessment, and effects assessment.

4.1. Hazard Decisions

Anything with the potential to cause harm is a hazard. Some things are inherently hazardous, whereas others are hazardous in one scenario but essential or desirable in another. Liquid water is a drowning hazard. Ice is a slipping hazard. Sharps, like syringe needles, are infection hazards. Pesticides are health hazards. At least a portion of the hazard is an intrinsic property of a substance, product, or process, that is, a concept of potential harm. For example, a biochemical hazard is an absolute expression of a substance's properties, because all substances have unique physical and chemical properties. These properties can render the substance hazardous. Conversely, Equation (32.3) shows risk can only occur with exposure. So, if one walks on a street in the summer, little likelihood of slipping on ice is near zero. One's total slipping risk is not necessarily zero (e.g., a person could step on an oily surface or someone could throw ice in one's path). If not in a medical facility, one's infection risk from sharps may be near zero, but a person's total infection risk is not zero (e.g., people may be exposed to the same infection from a person sneezing in their office). If a person does not use pesticides, the pesticide health risk is also lower. However, because certain pesticides are persistent and can remain in the food chain, the person's exposure is not zero. Also, even if a person's pesticide exposure is near zero, that person's cancer risk is not zero, because he or she may be exposed to other cancerous hazards.

Of all the environmental hazards, the most attention has been devoted to toxicity. Other important environmental hazards are shown in Table 32.2. Hazards can be expressed according to the physical and chemical characteristics, as in Table 32.2, as well as in the ways they may affect living things. For example, Table 32.3 summarizes some of the expressions of biologically based criteria of hazards. Other hazards, such as flammability, are also important to waste management. However, the chief hazard in most environmental situations has been toxicity.

TABLE 32.2 Hazards Defined by the Resource Conservation and Recovery Act

Hazard type	Criteria	Physical/chemical classes in definition
Corrosivity	A substance with an ability to destroy tissue by chemical reactions	Acids, bases, and salts of strong acids and strong bases. The waste dissolves metals, other materials, or burns the skin. Examples include rust removers, waste acid, alkaline cleaning fluids, and waste battery fluids. Corrosive wastes have a pH of <2.0 or >12.5. The U.S. EPA waste code for corrosive wastes is "D002."
Ignitability	A substance that readily oxidizes by burning	Any substance that spontaneously combusts at 54.3 °C in air or at any temperature in water, or any strong oxidizer. Examples are paint and coating wastes, some degreasers, and other solvents. The U.S. EPA waste code for ignitable wastes is "D001."
Reactivity	A substance that can react, detonate, or decompose explosively at environmental temperatures and pressures	A reaction usually requires a strong initiator (e.g., an explosive like TNT, trinitrotoluene), confined heat (e.g., salt peter in gunpowder), or explosive reactions with water (e.g., Na). A reactive waste is unstable and can rapidly or violently react with water or other substances. Examples include wastes from cyanide-based plating operations, bleaches, waste oxidizers, and waste explosives. The U.S. EPA waste code for reactive wastes is "D003."
Toxicity	A substance that causes harm to organisms. Acutely toxic substances elicit harm soon after exposure (e.g., highly toxic pesticides causing neurological damage within hours after exposure). Chronically toxic substances elicit harm after a long period of time of exposure (e.g., carcinogens, immunosuppressants, endocrine disruptors, and chronic neurotoxins).	Toxic chemicals include pesticides, heavy metals, and mobile or volatile compounds that migrate readily, as determined by the Toxicity Characteristic Leaching Procedure (TCLP), or a "TC waste." TC wastes are designated with waste codes "D004" through "D043."

4.1.1. Dose-Response Curves

The first means of determining exposure is to identify *dose*, the amount (e.g., mass) of a contaminant that comes into contact with an organism. Dose can be the amount administered to an organism (so-called "applied dose"), the amount of the contaminant that enters the organism ("internal dose"), the amount of the contaminant that is absorbed by an organism over a certain time interval ("absorbed dose"), or the amount of the contaminants or its metabolites that reach a particular "target" organ ("biologically effective dose" or "bioeffective dose"), such as the amount of a hepatotoxin (a chemical that harms the liver) that finds its way to liver cells or a neurotoxin (a chemical that harms the nervous system) that reaches the nerve or other nervous system cells. Theoretically, the higher the concentration of a hazardous substance or microbe that comes into contact with an organism, the greater the expected adverse outcome. The pharmacological and toxicological gradient is the so-called "dose-response" curve (Figure 32.6). Generally,

TABLE 32.3 Biologically Based Classification Criteria for Chemical Substances [15]

Criterion	Description
Bioconcentration	The process by which living organisms concentrate a chemical contaminant to levels exceeding the surrounding environmental media (e.g., water, air, soil, or sediment)
Lethal dose (LD)	A dose of a contaminant calculated to expect a certain percentage of a population of an organism (e.g., minnow) exposed through a route other than respiration (dose units are mg [contaminant] kg^{-1} body weight). The most common metric from a bioassay is the lethal dose 50 (LD_{50}), wherein 50% of a population exposed to a contaminant is killed
Lethal concentration (LC)	A calculated concentration of a contaminant in the air that, when respired for 4 h (i.e., exposure duration = 4 h) by a population of an organism (e.g., rat) will kill a certain percentage of that population. The most common metric from a bioassay is the lethal concentration 50 (LC_{50}), wherein 50% of a population exposed to a contaminant is killed (Air concentration units are mg [contaminant] L^{-1} air)

increasing the amount of the dose means a greater incidence of the adverse outcome.

Dose-response assessment generally follows a sequence of five steps [16,17]:

1. Fitting the experimental dose-response data from animal and human studies with a mathematical model that fits the data reasonably well;
2. Expressing the upper confidence limit (e.g., 95%) line equation for the selected mathematical model;
3. Extrapolating the confidence limit line to a response point just below the lowest measured response in the experimental point (known as the "point of departure"), that is, the beginning of the extrapolation to lower doses from actual measurements;
4. Assuming the response is a linear function of dose from the point of departure to zero response at zero dose; and,
5. Calculating the dose on the line that is estimated to produce the response.

The risk assessor can use published physical and chemical hazard characteristics for all of the chemicals used in the life cycle of a biotechnology. The process is useful but may not be completely applicable to all wastes. For example, if a microbe in a waste is harmful to

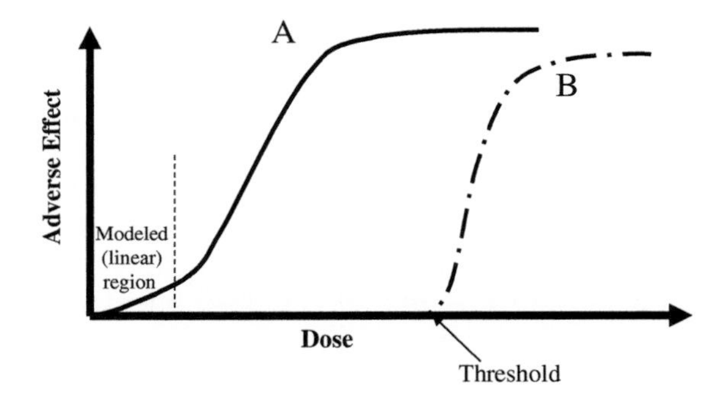

FIGURE 32.6 Prototypical dose-response curves. Curve A represents the "no-threshold" curve, which predicts a response (e.g., cancer) even if exposed to a single molecule ("one-hit" model). As shown, the low end of the curve, below which experimental data are available, is linear. Thus, Curve A represents a linearized multistage model. Curve B represents toxicity above a certain threshold (no observable adverse effect level (NOAEL) is the level below which no response is expected). Another threshold is the no observable effect concentration (NOEC), which is the highest concentration where no effect on survival is observed ($NOEC_{survival}$) or where no effect on growth or reproduction is observed ($NOEC_{growth}$). Note that both curves are sigmoidal in shape because of the saturation effect at high dose (i.e., less response with increasing dose). [Source: *Adapted from ref. [18]*].

a particular type of cell (e.g., a nerve), it may follow the steps just as a neurotoxic chemical. However, if the microbial modifications change microbial populations in an organism or in an ecosystem, the dose response may become much more complex than a single, abiotic chemical hazard.

The curves in Figure 32.6 represent those generally found for toxic chemicals [19]. Once a substance is suspected of being toxic, the extent and quantification of that hazard are assessed.[2] This step is frequently referred to as a dose-response evaluation because this is when researchers study the relationship between the mass or concentration (i.e., dose) and the damage caused (i.e., response). Many dose-response studies are ascertained from animal studies (in vivo toxicological studies), but they may also be inferred from studies of human populations (epidemiology). To some degree, "Petri dish" (i.e., in vitro) studies, such as mutagenicity studies like the Ames test [20] of bacteria complement dose-response assessments, are mainly used for screening and qualitative or, at best, semi-quantitative analysis of responses to substances. The actual name of the test is the "Ames *Salmonella*/microsome mutagenicity assay" which shows the short-term reverse mutation in histidine dependent *Salmonella* strains of bacteria. Its main use is to screen for a broad range of chemicals that induce genetic aberrations leading to genetic mutations. The process works by using a culture that allows only those bacteria whose genes revert to histidine interdependence to form colonies. As a mutagenic chemical is added to the culture, a biological gradient can usually be determined. That is, the greater the amount of the chemical that is added, the greater the number of microbes and the larger the size of colonies on the plate. The test is widely used to screen for mutagenicity of new or modified chemicals and mixtures. It is also a "red flag" for carcinogenicity, because cancer is a genetic disease and a manifestation of mutations.

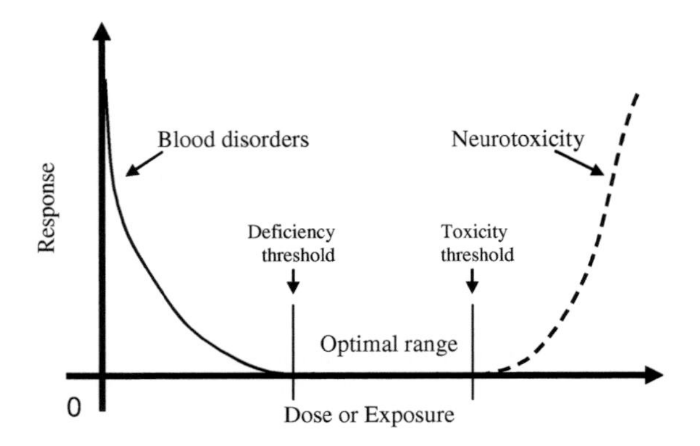

FIGURE 32.7 Hypothetical dose-response curve showing deficiency at low doses and toxicity at high doses.

The toxicity criteria include both acute and chronic effects, and include both human and ecosystem effects. These criteria can be quantitative. For example, a manufacturer of a new chemical may have to show that there are no toxic effects in fish exposed to concentrations below 10 mg L^{-1}. If fish show effects at 9 mg L^{-1}, the new chemical would be considered to be toxic.

A contaminant is acutely toxic if it can cause damage with only a few doses. Chronic toxicity occurs when a person or ecosystem is exposed to a contaminant over a protracted period of time, with repeated exposures. The essential indication of toxicity is the dose-response curve. The curves in Figure 32.6 are sigmoidal because toxicity is often concentration dependent. As the doses increase the response cannot mathematically stay linear (e.g., the toxic effect

[2]Actually, another curve could be shown for essential compounds like vitamins and certain metallic compounds. In such a curve (Figure 32.7), the left-hand side (low dose or low exposure) of the curve would represent deficiency and the right-hand side (high dose or exposure) would represent toxicity, with an optimal, healthy range between these two adverse responses.

cannot double with each doubling of the dose). So, the toxic effect continues to increase but at a decreasing rate (i.e., decreasing slope). Curve A is the classic cancer dose-response curve, that is, any amount of exposure to a cancer-causing agent may result in an expression of cancer at the cellular level (i.e., no safe level of exposure). Thus, the curve intercepts the x-axis at 0.

Curve B is the classic noncancer dose-response curve. The steepness of the three curves represents the potency or severity of the toxicity. For example, Curve B is steeper than Curve A, so the adverse outcome (disease) caused by chemical in Curve B is more potent than that of the chemical in Curve A. Obviously, potency is only one factor in the risk. For example, a chemical may be very potent in its ability to elicit a rather innocuous effect, like a headache, and another chemical may have a rather gentle slope (lower potency) for a dreaded disease like cancer.

With increasing potency, the range of response decreases. In other words, as shown in Figure 32.8, a severe response represented by a steep curve will be manifested in greater mortality or morbidity over a smaller range of dose. For example, an acutely toxic contaminant's dose that kills 50% of test animals (i.e., the LD_{50}) is closer to the dose that kills only 5% (LD_5) and 95% (LD_{95}) of the animals. The dose difference of a less acutely toxic contaminant will cover a broader range, with the differences between the LD_{50} and LD_5, and LD_{95} being more extended than that of the more acutely toxic substance.

4.1.1.1. UNCERTAINTY AND FACTORS OF SAFETY

The major distinction of toxicity is between carcinogenic and noncancer outcomes. The term "noncancer" is commonly used to distinguish cancer effects (e.g., bladder cancer, leukemia, or adenocarcinoma of the lung) from other maladies, such as neurotoxic disorders, immune system disorders, and endocrine disruption. The policies of many regulatory agencies and international organizations treat cancer differently than noncancer effects, particularly in how the dose–response curves are drawn. As discussed in the dose-response curves, there is no safe dose for carcinogens. Cancer dose-response curve is almost always a nonthreshold curve, that is, no safe dose is expected while, theoretically at least, noncancer effects can have a dose below which the adverse outcomes do not present themselves. So, for all other diseases, safe doses of compounds can be established. These are known as reference doses (RfD), usually based on the oral exposure route. If the substance is an air pollutant, the safe dose is known as the reference concentration (RfC), which is calculated in the same manner as the RfD, using units that apply to air (e.g., $\mu g\ m^{-3}$). These references are calculated from thresholds below which no adverse effect is observed in animal and human studies. If the models and data were perfect, the safe level would be the threshold, known as the NOAEL.

The term "noncancer" has a completely different meaning from the terms "anticancer" and "anticarcinogens." Anticancer procedures include radiation and drugs that are used to attack tumor cells. Anticarcinogens are chemical substances that work against the processes that lead to cancer, such as antioxidants and essential substances that help the body's immune, hormonal, and other systems to prevent carcinogenesis.

Any hazard identification or dose-response research is never perfect and so the data derived from these investigations are often beset with various forms of uncertainty. Chief reasons for this uncertainty include variability among the animals and people being tested, as well as differences in response to the compound by different species (e.g., one species may have decreased adrenal gland activity, whereas another may have thyroid effects). Although

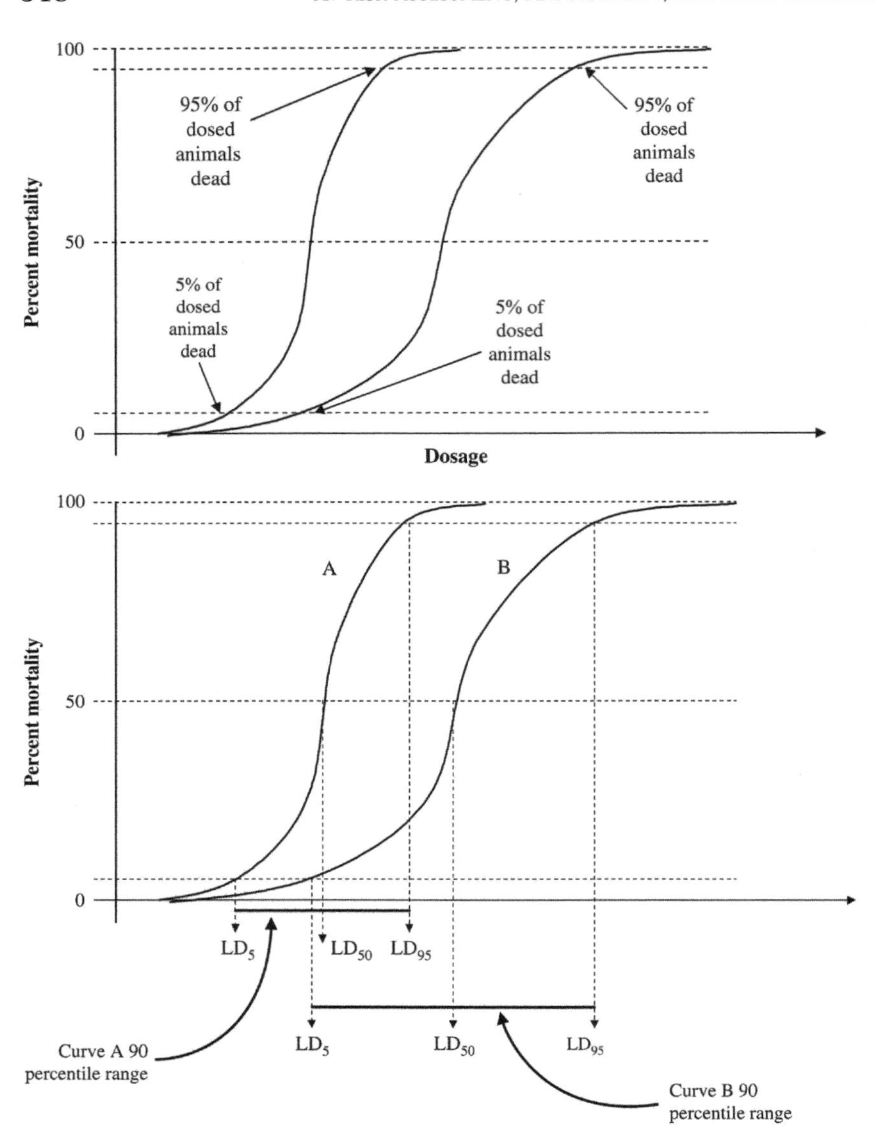

FIGURE 32.8 The greater the potency or severity of response (i.e., steepness of the slope) of dose-response curve the smaller the range of toxic response (90 percentile range shown in bottom graph). Also, note that both curves have thresholds and that Curve B is less acutely toxic based upon all three reported lethal doses (LD_5, LD_{50}, and LD_{95}). In fact, the LD_5 for Curve A is nearly the same as the LD_{50} for Curve B, meaning that about the same dose, contaminant A kills nearly half the test animals, but contaminant B has only killed 5%. Thus, contaminant A is much more acutely toxic. *[Source: D.A. Vallero, 2004,* Environmental Contaminants: Assessment and Control, *Elsevier Academic Press, Burlington, MA].*

this is usually associated with chemical risk, these uncertainties can also be part of microbial data sets. For example, certain immunocompromised subpopulations may respond adversely to microbial exposures that are below thresholds for the general population.

Sometimes, studies only indicate the lowest concentration of a contaminant that causes the effect, that is, the lowest observed adverse effect level (LOAEL), but the NOAEL is unknown. If the LOAEL is used, one is less certain about how close this is to a safe level where no effect is expected. Often, there is temporal incongruence, such as most of the studies taking place in a shorter timeframe than in the real world. Thus, in lieu of long-term human studies, hazards and risks may have to be extrapolated from acute or subchronic studies of the same or similar agents. Likewise, routes and pathways of exposure used to administer the agent to subjects may differ from the likely real-world exposures. For example, if the dose of substance in a research study is administered orally, but the pollutant is more likely to be inhaled by humans, this route-to-route extrapolation adds uncertainty. This is particularly problematic for microbial exposures; for example, inhalational anthrax (*Bacillus anthracis*) is more virulent in human populations than is ingestional anthrax.

Finally, the hazard and exposure data themselves may be weak because the studies from which they have been gathered lack sufficient quality or the precision, accuracy, completeness, and representativeness, or they may not be directly relevant to the risk assessment at hand.

The factors underlying the uncertainties are quantified as specific uncertainty factors (UFs). The uncertainties in the RfD are largely due to the differences between results found in animal testing and expected outcomes in human population. As in other engineering operations, a factor of safety must be added to calculations to account for UFs. So, for environmental risk analyses and assessments, the safe level is expressed in the RfD, or in air, the RfC. This is the dose or concentration below which regulatory agencies do not expect a specific unacceptable outcome. Thus, all the uncertainty factors adjust the actual measured levels of no effect (i.e., the threshold values, e.g., NOAELs and LOAELs) in the direction of a zero concentration. This is calculated as

$$\text{RfD} = \frac{\text{NOAEL}}{\text{UF}_{\text{inter}} \times \text{UF}_{\text{intra}} \times \text{UF}_{\text{other}}} \tag{32.4}$$

The first of the three types of uncertainty is the one resulting from the difference between the species tested and *Homosapiens* (UF_{inter}). Humans may be more or less sensitive than the tested species to a particular compound. The second uncertainty factor is associated with the fact that certain human subpopulations are more sensitive to the effects of a compound than the general human population. These are known as intraspecies uncertainty factors (UF_{intra}). The third type of uncertainties (UF_{other}) results when the available data and science are lacking, such as when a LOAEL is used rather than a NOAEL. That is, data show a dose at which an effect is observed, but the "no effect" threshold has to be extrapolated. Because the UFs are in the denominator, the greater the uncertainties, the closer the safe level (i.e., the RfD) is to zero, that is, the threshold is divided by these factors. The UFs are usually multiples of 10, although the UF_{other} can range from 2 to 10.

A particularly sensitive subpopulation is that of children, because they are growing and tissue development is much more prolific in them than in adults. To address these sensitivities, the Food Quality Protection Act (FQPA) now includes what is known as the "10X" rule. This rule requires that the RfD for products regulated under FQPA, for example, pesticides, must include an additional factor of 10 of protection of infants, children, and females between the ages of 13 and 50 years. This factor is included in the RfD denominator along with the other three UF values. The RfD that includes the UFs and the 10X protection is known as the population-adjusted dose (PAD). A risk estimate that is less than 100% of the acute or chronic PAD does not exceed the US Environmental Protection Agency's risk concern.

An example of the use of an RfD as a factor of safety can be demonstrated by the US Environmental Protection Agency's decision making regarding the re-registration of the organophosphate pesticide, chlorpyrifos. The acute dietary scenario had a NOAEL of 0.5 mg kg^{-1} d^{-1} (where d refers to day), and the three UF values equaled 100. Thus, the acute RfD = 5×10^{-3} mg kg^{-1}d^{-1}, but the more protective acute PAD is 5×10^{-4} mg kg^{-1} d^{-1}. The chronic dietary scenario is even more protective, because the exposure is for long term. The chronic NOAEL was found to be 0.03 mg kg^{-1} d^{-1}. Thus, the chronic RfD for chlorpyrifos is 3×10^{-4} mg kg^{-1} d^{-1}, and the more protective acute PAD is 5×10^{-5} mg kg^{-1} d^{-1}. Therefore, had the NOAEL threshold been used alone without the safety adjustment of the RfD, the allowable exposure would have been three orders of magnitude higher [21].

Uncertainty can also come from error. Two errors can occur when information is interpreted in the absence of sound science. The first is the *false negative*, or reporting that there is no problem when one in fact exists. The need to address this problem is often at the core of the positions taken by environmental and public health agencies and advocacy groups. They ask questions like

- What if a leachate monitor's level of detection is above that needed to show that a contained chemical is actually being released?
- What if the leak detector registers zero, but in fact toxic substances are being released from the tank?
- Is there a way to determine the amount of waste in space quantitatively and with known levels of confidence?
- What if this substance really does cause cancer, but the tests are unreliable?
- What if people are being exposed to a contaminant from a waste product but via a pathway other than the ones being studied?
- What if there is a relationship that is different from the laboratory when this substance is released into the "real world," such as the difference between how a chemical behaves in the human body by itself as opposed to when other chemicals are present (i.e., the problem of "complex mixtures")?

The other concern is, conversely, the *false positive*. This can be a major challenge for public health agencies with the mandate to protect people from exposures to environmental contaminants. For example, what if previous evidence shows that an agency had listed a compound as a potential endocrine disruptor, only to find that a wealth of new information is now showing that it has no such effect? This can happen if the conclusions were based upon faulty models, or models that only work well for lower organisms, but subsequently developed models have taken into consideration the physical, chemical, and biological complexities of high-level organisms, including humans. False positives may force public health officials to devote inordinate amount of time and resources to deal with so-called "nonproblems." False positives also erroneously scare people about potentially useful products. False positives, especially when occur frequently, create credibility gaps between engineers and scientists and the decision makers. In turn the public, those whom we have been charged to protect, lose confidence in us as professionals.

Environmental risk assessment calls for high quality, scientifically based information. Put in engineering language, the risk assessment process is a "critical path" in which any unacceptable error or uncertainty along the way will decrease the quality of the risk assessment and, quite likely, will lead to a bad environmental decision.

5. EXPOSURE ESTIMATION

An exposure is any contact with an agent [22]. For chemical and biological agents this contact can come about from a number of exposure pathways, that is, routes taken by a substance, beginning with its source to its end point (i.e., a target organ, like the liver, or a location short of that, such as in fat tissues).

Exposure results from sequential and parallel processes in the environment, from release to

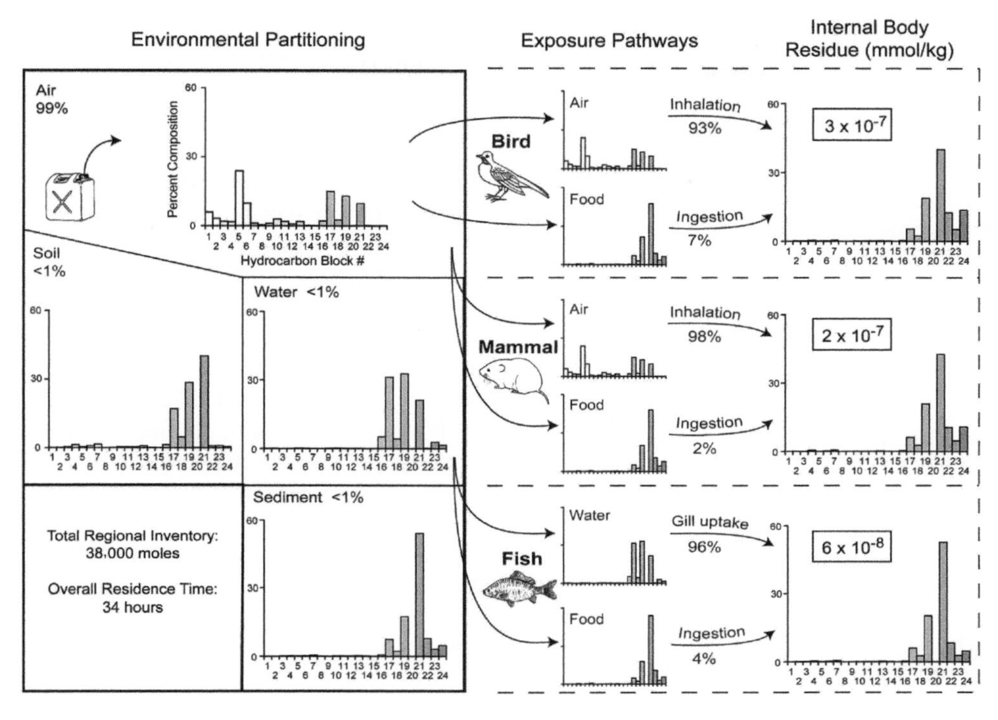

FIGURE 32.9 Processes leading to organismal uptake and fate of chemical and biological agents after release into the environment. In this instance, the predominant sources are air emissions, and predominant pathway of exposure is through inhalation. However, due to deposition on surface waters and the agent's affinity for sediment, the ingestion pathways are also important. Dermal pathways, in this case, do not constitute a large fraction of potential exposure. [Source: *T. McKone, R. Maddalena, W. Riley, R. Rosenbaum and D. Vallero (2006). Significance of partitioning, kinetics, and uptake at biological exchange surfaces. International Conference on Environmental Epidemiology & Exposure Analysis. Paris, France].*

environmental partitioning to movement through pathways to uptake and fate in the organism (see Figure 32.9). The substances often change to other chemical species as a result of the body's metabolic and detoxification processes. Certainly, genetic modifications can affect such processes. New substances, known as degradation products or metabolites, are produced as cells use the parent compounds as food and energy sources. These metabolic processes, such as hydrolysis and oxidation, are the mechanisms by which chemicals are broken down.

Physical agents, such as electromagnetic radiation, ultraviolet (UV) light, and noise, do not follow this pathway exactly. The contact with these sources of energy can elicit a physiological response that may generate endogenous chemical changes that behave somewhat like the metabolites. For example, UV light may infiltrate and damage skin cells. The UV light helps to promote skin-tumor by activating the transcription factor complex activator protein-1 (AP-1) and enhancing the expression of the gene that produces the enzyme cyclooxygenase-2 (*COX2*). Noise, that is, acoustical energy, can also elicit physiological responses that affect an organism's chemical messaging systems, that is, endocrine, immune, and neural. It is possible that genetically modified organisms will respond differently to these physical agents than their unmodified counterparts.

The exposure pathway also includes the ways that humans and other organisms can come into contact with the hazard. The pathway has five parts:

1. The source of contamination (e.g., fugitive dust or leachate from a landfill);
2. An environmental medium and transport mechanism (e.g., soil with water moving through it);
3. A point of exposure (such as a well used for drinking water);
4. A route of exposure (e.g., inhalation, dietary ingestion, nondietary ingestion, dermal contact, and nasal);
5. A receptor population (those who are actually exposed or who are where there is a potential for exposure).

If all five parts are present, the exposure pathway is known as a completed exposure pathway. In addition, the exposure may be short term, intermediate, or long term. Short-term contact is known as an acute exposure, that is, occurring as a single event or for only a short period of time (up to 14 days). An intermediate exposure is one that lasts from 14 days to less than one year. Long-term or chronic exposures are for more than one year in duration.

Determining the exposure for a neighborhood can be complicated. For example, even if we do a good job identifying all the contaminants of concern and its possible source (no small task), we may have little idea of the extent to which the receptor population has come into contact with these contaminants (steps 2 through 4). Thus, assessing exposure involves not only the physical sciences but also the social sciences, for example, psychological and behavioral sciences. People's activities greatly affect the amount and type of exposures. That is why exposure scientists use a number of techniques to establish activity patterns, such as asking potentially exposed individuals to keep diaries, videotaping, and using telemetry to monitor vital information, for example, heart and ventilation rates.

General ambient measurements, such as air pollution monitoring equipment located throughout cities, are often not good indicators of actual population exposures. For example, metals and their compounds comprise the greatest mass of toxic substances *released* into the U.S. environment. This is largely due to the large volume and surface areas involved in metal extraction and refining operations. However, this does not necessarily mean that more people will be exposed at higher concentrations or more frequently to these compounds than to others. A substance that is released or even resides in the ambient environment is not tantamount to its coming in contact with a *receptor*. Conversely, even a small amount of a substance under the right circumstances can lead to very high levels of exposure (e.g., handling raw materials and residues at a waste site).

A recent study by the Lawrence Berkeley Laboratory demonstrates the importance of not simply assuming that the released or even background concentrations are a good indicator of actual exposure [23]. The researchers were interested in how sorption may affect microenvironments, so they set up a chamber constructed of typical building materials and furnished with actual furniture like that found in most residential settings. A number of air pollutants were released into the room and monitored (see Figure 32.10). With the chamber initially sealed, the observed decay of xylene, a volatile organic compound, in vapor phase concentrations results from adsorption onto surfaces (walls, furniture, etc.). The adsorption continues for hours, with xylene concentrations reaching a quasi-steady state. At this point the chamber was flushed with clean air to free the vapor phase xylene. The xylene concentrations shortly after the flush began to rise again until reaching a new steady state. This rise must be the result of desorption of the previously sorbed xylene, as the initial source is gone. Sorption is one of the biochemodynamic processes that must be considered to account

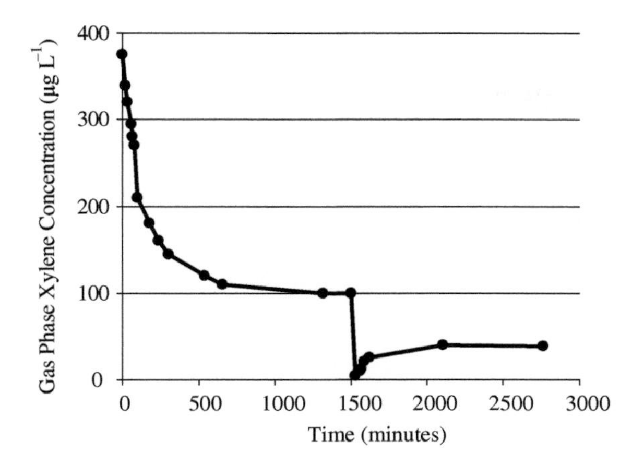

FIGURE 32.10 Concentrations of xylene measured in its vapor phase in a chamber sealed during adsorption and desorption periods. [Source: *Adapted from B. Singer (2003). "A tool to predict exposure to hazardous air pollutants,"* Environmental Energy Technologies Division News. *4(4): 5].*

for differences in the temporal pattern of microenvironmental (e.g., occupational) and ambient concentrations.

The simplest quantitative expression of exposure is:

$$E = D/t \tag{32.5}$$

where E is the human exposure during the time period t (units of concentration, mg kg^{-1}d^{-1}); D is the mass of pollutant per body mass (mg kg^{-1}) and t is time (d).

D, the chemical concentration of a pollutant, is usually measured near the interface of the person and the environment, during a specified time period. This measurement is sometimes referred to as the potential dose (i.e., the chemical has not yet crossed the boundary into the body but is present where it may enter the person, such as on the skin, at the mouth, or at the nose).

Expressed quantitatively, exposure is a function of the concentration of the agent and time. It is an expression of the magnitude and duration of the contact. That is, exposure to a contaminant is the concentration of that contact in a medium integrated over the time of contact:

$$E = \int_{t=t_1}^{t=t_2} C(t)\, dt \tag{32.6}$$

where, E is the exposure during the time period from t_1 to t_2 and $C(t)$ is the concentration at the interface between the organism and the environment, at time t.

The concentration at the interface is the potential dose (i.e., the agent has not yet crossed the boundary into the body but is present at a point from where it may enter the receptor). Because the amount of a chemical agent that penetrates from the ambient atmosphere into a control volume affects the concentration term of the exposure equation, a complete mass balance of the contaminant must be understood and accounted for; otherwise exposure estimates will be incorrect. Recall that the mass balance consists of all inputs and outputs as well as chemical changes to the contaminant:

$$\begin{aligned} &\text{Accumulation or loss of contaminant } A \\ &\quad = \text{Mass of } A \text{ transported in} \\ &\qquad - \text{Mass of } A \text{ transported out} \pm \text{Reactions} \end{aligned} \tag{32.7}$$

The reactions may be either those that generate substance A (i.e., *sources*) or those that destroy substance A (i.e., *sinks*). Thus, the amount of mass transported in is the inflow to the system that includes pollutant discharges, transfer from other control volumes and other media (e.g., if the control volume is soil, the water and air may contribute to mass of chemical A), and formation of chemical A by abiotic chemistry and biological transformation. Conversely, the outflow is the mass transported out of the control volume, which includes uptake, by biota, transfer to other compartments (e.g., volatilization to the atmosphere) and abiotic and biological degradation of chemical A. This means that the rate of change of mass in a control volume is equal to the rate of chemical A transported in less the rate of chemical A transported out, plus the rate of production from sources, and minus the rate of elimination by sinks. Stated as a differential equation, the rate of change contaminant A is:

$$\frac{d[A]}{dt} = -v \cdot \frac{d[A]}{dx} + \frac{d}{dx}\left(\Gamma \cdot \frac{d[A]}{dx}\right) + r \quad (32.8)$$

where v is the fluid velocity; Γ is a rate constant specific to the environmental medium; $d[A]/dx$ is the concentration gradient of chemical A; and r refers to the internal sinks and sources within the control volume.

Reactive compounds can be particularly difficult to measure. For example, many volatile organic compounds in the air can be measured first by collecting in stainless steel canisters and analyzed by chromatography in the lab. However, some of these compounds, like the carbonyls (notably aldehydes like formaldehyde and acetaldehyde) are prone to react inside the canister, meaning that by the time the sample is analyzed, a portion of the carbonyls is degraded (under-reported). Therefore, other methods, such as trapping the compounds with dinitrophenyl hydrazine (DNPH)-treated silica gel tubes that are frozen until being extracted for chromatographic analysis. The purpose of the measurement is to see what is in the air, water, soil, sediment, or biota at the time of sampling, so any reactions before the analysis give measurement error.

If the released chemical is reactive, some or all of it may have changed into another form (i.e., *speciated*) by the time it is measured. Even relatively nonreactive compounds may speciate between the time when the sample is collected (e.g., in a water sample, an air canister, a soil core, or a bag) and when the sample is analyzed. In fact, each contaminant has unique characteristics which vary according to the type of media in which it exists, and extrinsic conditions like temperature and pressure (see Table 32.4).

TABLE 32.4 Preservation and Holding Times for Anion Sampling and Analysis

Analyte	Preservation	Holding time
PART A: Common anions		
Bromide	None required	28 d
Chloride	None required	28 d
Fluoride	None required	28 d
Nitrate-N	Cool to 4 °C	48 h
Nitrite-N	Cool to 4 °C	48 h
Ortho-Phosphate-P	Cool to 4 °C	48 h
Sulfate	Cool to 4 °C	28 d
PART B: Inorganic disinfection by-products		
Bromate	50 mg L^{-1} EDA	28 d
Bromide	None required	28 d
Chlorate	50 mg L^{-1} EDA	28 d
Chlorite	50 mg L^{-1} EDA, cool to 4 °C	14 d

[Source: *U.S. Environmental Protection Agency, 1997, EPA Method 300.1: Determination of Inorganic Anions in Drinking Water by Ion Chromatography, Revision 1.0*]

The general exposure Equation (32.7) is rewritten to address each route of exposure, accounting for chemical concentration and the activities that affect the time of contact. The exposure calculated from these equations is actually the chemical intake (I) in units of concentration (mass per volume or mass per mass) per time, mg kg^{-1} d^{-1}:

$$I = \frac{C \times CR \times EF \times ED \times AF}{BW \times AT} \quad (32.9)$$

where C is the chemical concentration of contaminant (mass per volume); CR is the contact rate (mass per time); EF is the exposure frequency (number of events, dimensionless); and ED is the exposure duration (time).

These factors are further specified for each route of exposure, such as the lifetime average daily dose (LADD) as shown in Table 32.5. The

TABLE 32.5 Equations for Calculating Lifetime Average Daily Dose (LADD) for Various Routes of Exposure

Route of exposure	Equation LADD (in mg kg^{-1} d^{-1})	Definitions
Inhaling aerosols (particulate matter)	$\frac{C \times PC \times IR \times RF \times EL \times AF \times ED \times 10^{-6}}{BW \times TL}$	C = concentration of the contaminant on the aerosol/particle (mg kg^{-1}); PC = particle concentration in air (gm m^{-3}); IR = inhalation rate (m^{-3} h^{-1}); RF = respirable fraction of total particulates (dimensionless, usually determined by aerodynamic diameters, e.g., 2.5 μm); EL = exposure length (h d^{-1}); ED = duration of exposure (d); AF = absorption factor (dimensionless); BW = body weight (kg); TL = typical lifetime (d); 10^{-6} is a conversion factor (kg to mg)
Inhaling vapor phase contaminants	$\frac{C \times IR \times EL \times AF \times ED}{BW \times TL}$	C= concentration of the contaminant in the gas phase (mg m^{-3}); other variables are the same as above
Drinking water	$\frac{C \times CR \times ED \times AF}{BW \times TL}$	C = concentration of the contaminant in the drinking water (mg L^{-1}); CR= rate of water consumption (L d^{-1}); ED = duration of exposure (d); AF = portion (fraction) of the ingested contaminant that is physiologically absorbed (dimensionless); other variables are the same as above
Contact with soil-borne contaminants	$\frac{C \times SA \times BF \times FC \times \times ED \times 10^{-6}}{BW \times TL}$	C = concentration of the contaminant in the soil (mg kg^{-1}); SA = skin surface area exposed (cm^{-2}); BF = bioavailability (percent of contaminant absorbed per day); FC = fraction of total soil from contaminated source (dimensionless); SDF = soil deposition, the mass of soil deposited per unit area of skin surface (mg cm^{-1} d^{-1}); other variables are the same as above

[Source: *M. Derelanko, 1999,* CRC Handbook of Toxicology, *"Risk Assessment," M. J. Derelanko and M.A. Hollinger, editors, CRC Press, Boca Raton, FL.*]

LADD is obviously based on a chronic, long-term exposure.

Acute and subchronic exposures require different equations, since the exposure duration (ED) is much shorter. For example, instead of LADD, acute exposures to noncarcinogens may use maximum daily dose (MDD) to calculate exposure (see Discussion Box). However, even these exposures follow the general model given in Equation (32.9).

Once the hazard and exposure calculations are done, risks can be characterized quantitatively.

DISCUSSION BOX

EXPOSURE CALCULATION

In the process of synthesizing pesticides over an 18-year period, a polymer manufacturer has contaminated the soil on its property with vinyl chloride. The plant closed two years ago, but vinyl chloride vapors continue to reach the neighborhood surrounding the plant at an average concentration of 1 mg m^{-3}. Assume that people are breathing at a ventilation rate of 0.5 m^3 h^{-1} (about the average of adult males and females over 18 years of age [24]). The legal settlement allows neighboring residents to

(Continued)

DISCUSSION BOX—*Cont'd*

evacuate and sell their homes to the company. However, they may also stay. The neighbors have asked for advice on whether to stay or leave, because they have already been exposed for 20 years.

Vinyl chloride is highly volatile, so its phase distribution will be mainly in the gas phase rather than the aerosol phase. Although some of the vinyl chloride may be sorbed to particles, we will use only vapor phase LADD equation, because the particle phase is likely to be relatively small. Also, we will assume that outdoor concentrations are the exposure concentrations. However, this is unlikely, because people spend very little time outdoors compared to indoors, so this may provide an additional factor of safety. To determine how much vinyl chloride penetrates living quarters, indoor air studies would have to be conducted. For a scientist to compare exposures, indoor air measurements should be taken.

Find the appropriate equation in Table 32.5 and insert values for each variable. Absorption rates are published by the EPA and the Oak Ridge National Laboratory's Risk Assessment Information System http://rais.ornl.gov/tox/profiles/vinyl.html#t21 (accessed on November 19, 2010). Vinyl chloride is well absorbed, so in a worst case we can assume that AF = 1. We will also assume that the person staying in the neighborhood is exposed at the average concentration of 24 h a day (EL = 24) and a person remains exposed the rest of his typical lifetime at the measured concentration.

TABLE 32.6 Commonly Used Human Exposure Factors [25]

Exposure Factor	Adult Male	Adult Female	Child (3–12 years of age) [26]
Body weight (kg)	70	60	15–40
Total fluids ingested ($L\ d^{-1}$)	2	1.4	1.0
Surface area of skin, without clothing (m^2)	1.8	1.6	0.9
Surface area of skin, wearing clothes (m^2)	0.1–0.3	0.1–0.3	0.05–0.15
Respiration/ventilation rate ($L\ min^{-1}$)—Resting	7.5	6.0	5.0
Respiration/ventilation rate ($L\ min^{-1}$)—Light activity	20	19	13
Volume of air breathed ($m^3\ d^{-1}$)	23	21	15
Typical lifetime (years)	70	70	NA
National upper-bound time (90th percentile) at one residence (years)	30	30	NA
National median time (50th percentile) at one residence (years)	9	9	NA

[*Sources: U.S. Environmental Protection Agency, 2003,* Exposure Factor Handbook; *and Agency for Toxic Substances and Disease Registry, 2003, ATSDR Public Health Assessment Guidance Manual.*]

DISCUSSION BOX—*Cont'd*

Although the ambient concentrations of vinyl chloride may have been higher when the plant was operating, the only measurements we have are those taken recently. Thus, this is an area of uncertainty that must be discussed with the clients. The common default value for a lifetime is 70 years, so we can assume that the longest exposure would be 70 years (25,550 days). Table 32.6 gives some of the commonly used default values in exposure assessments. If the person is now 20 years old and has already been exposed for that time, and lives the remaining 50 years exposed at 1 mg m^{-3}, then

$$\begin{aligned} \text{LADD} &= \frac{\text{C} \times \text{IR} \times \text{EL} \times \text{AF} \times \text{ED}}{\text{BW} \times \text{TL}} \\ &= \frac{1 \times 0.5 \times 24 \times 1 \times 25550}{70 \times 25550} \\ &= 0.2 \text{ mg kg}^{-1} \text{ day}^{-1} \end{aligned}$$

If the 20-year-old leaves today, the exposure duration would be for the 20 years that the person lived in the neighborhood. Thus, only the ED term would change, that is, from 25,550 days to 7300 days (i.e., 20 years).

Thus, the LADD falls to 2/7 of its value:

$$\text{LADD} = 0.05 \text{ mg kg}^{-1} \text{ day}^{-1}$$

There are two general ways by which risk characterizations are used in environmental problem solving, that is, direct risk assessments and risk-based cleanup standards.

6. DIRECT RISK CALCULATIONS

Although in its simplest form, risk is the product of the hazard and exposure, the assumptions can greatly affect risk estimates. For example, cancer risk can be defined as the theoretical probability of contracting cancer when continually exposed for a lifetime (e.g., 70 years) to a given concentration of a substance (carcinogen). The probability is usually calculated as an upper confidence limit. The maximum estimated risk may be presented as the number of chances in a million of contracting cancer.

Two measures of risk are commonly reported. One is the individual risk, that is, the probability of a person developing an adverse effect (e.g., cancer) due to the exposure. This is often reported as a "residual" or increased probability above background. For example, if we want to characterize the contribution of all the power plants in the United States to increased cancer incidence, the risk above background would be reported. The second way that risk is reported is population risk, that is, the annual excess number of cancers in an exposed population. The maximum individual risk might be calculated from exposure estimates based upon a "maximum exposed individual" or MEI. The hypothetical MEI lives an entire lifetime outdoors at the point where pollutant concentrations are highest. Assumptions about exposure will greatly affect the risk estimates. For example, the cancer risk from power plants in the United States has been estimated to be 100- to 1000-fold lower for an average exposed individual than that for the MEI.

For cancer risk assessments, the hazard is generally assumed to be the slope factor and the long-term exposure is the lifetime average daily dose:

$$\text{Cancer risk} = \text{SF} \times \text{LADD} \qquad (32.10)$$

Therefore, cancer risk can be calculated if the exposure (LADD) and potency (slope factor) are known (see Discussion Box).

DISCUSSION BOX

CANCER RISK CALCULATION

Using the lifetime average daily dose value from the vinyl chloride exposure calculation in the previous section, estimate the direct risk to the people living near the abandoned polymer plant. What information needs to be communicated?

Substitute the calculated LADD values and the vinyl chloride inhalation slope factor. In fact, Two slope factors are available for this chemical, one that accounts for exposure occurring during early life and one that accounts for exposure occurring later in life. Since this is a lifetime exposure, let us assume the steeper slope for continuous lifetime exposure from birth, i.e. 1.5 (http://www.epa.gov/iris/subst/1001.htm; accessed on November 19, 2010). The calculated cancer risk to the neighborhood exposed for an entire lifetime (exposure duration = 70 years) is 0.2 mg kg^{-1} $d^{-1} \times$ 1.5 (mg kg^{-1} $d^{-1})^{-1}$ = 0.3. This is an incredibly high risk! The threshold for concern is often one in a million (0.000001); this is a probability of 30%.

Even at the shorter duration of period (20 years of exposure instead of 70 years), the risk is calculated as $0.05 \times 1.5 = 0.075$ or a 7.5% risk. The combination of a very steep slope factor and very high lifetime exposures leads to a very high risk. Vinyl chloride is a liver carcinogen, so unless corrective actions significantly lower the ambient concentrations of vinyl chloride the prudent course of action is that the neighbors accept the buyout and leave the area.

Incidentally, vinyl chloride has relatively high water solubility and can be absorbed into soil particles, so ingestion of drinking water (e.g., people on private wells drawing water from groundwater that has been contaminated) and dermal exposures (e.g., children playing in the soil) are also conceivable. The total risk from a single contaminant like vinyl chloride is equal to the sum of risks from all pathways (e.g., vinyl chloride in the air, water, and soil):

$$\text{Total risk} = \sum \text{risks from all exposure pathways} \tag{32.11}$$

Requirements and measures of success are seldom if ever as straightforward as the vinyl chloride example. In fact, the waste manager would be ethically remiss if the only advice given is whether or not the local community should accept the buyout. Of course, one of the engineering canons is to serve a "faithful agent" to the clientele. However, the first engineering canon is to hold paramount the health and safety of the public. Thus, the engineer must balance the client's desire for anonymity to be protected with the need to protect public health. In this case, the engineer must inform the client and prime contractors, for example, that the regulatory agencies need to know that even if the current neighbors are moving, the threat to others, including future populations, are threatened.

Waste cleanup must consider both space and time. The contaminant may remain in untreated or poorly treated areas, which may later be released (e.g., future excavation, long-term transportation from groundwater to surface or from groundwater to drinking water sources). Thus, postclosure monitoring should be designed and operated based on worst-case scenarios. The risk of adverse outcome other than cancer (so-called "noncancer risk") is generally called the "hazard quotient" (HQ). It is calculated by dividing the MDD by the acceptable daily intake (ADI):

$$\text{Noncancer risk} = \text{HQ} = \frac{\text{MDD}}{\text{ADI}} = \frac{\text{Exposure}}{\text{RfD}} \tag{32.12}$$

DISCUSSION BOX—*Cont'd*

Note that this is an index, not a probability, so it is really an indication of relative risk. If the noncancer risk is greater than 1, then the potential risk may be significant, and if the noncancer risk is less than 1, then the noncancer risk may be considered to be insignificant (see Discussion Box). Thus, the reference dose, RfD, is one type of ADI.

DISCUSSION BOX

NONCANCER RISK CALCULATION

Assume that the dermal chronic RfD for an acid mist is 6.00×10^{-3} mg kg^{-1} day^{-1}. If the actual dermal exposure of people living near a metal-processing plant is calculated (e.g., by intake or LADD) to be 4.00×10^{-3} mg kg^{-1} day^{-1}, calculate the hazard quotient for the noncancer risk of the chromic acid mist to the neighborhood near the plant and interpret the meanings.

From Equation (32.12),

$$\frac{\text{Exposure}}{\text{RfD}} = \frac{4.00 \times 10^{-3}}{6.00 \times 10^{-3}} = 0.67$$

Because this is less than 1, one would not expect people chronically exposed at this level to show adverse effects from skin contact. However, toxicity varies by route. The oral RfD is 3.00 × 10^{-}3 (Integrated Risk Information System; http://www.epa.gov/IRIS/subst/0144.htm; accessed on November 19, 2010). So, at this same chronic exposure, that is, 4.00×10^{-3} mg kg^{-1} d^{-1}, to the acid mists via oral route, the HQ = 4/3 or 1.3. The value is greater than 1, so we cannot rule out adverse noncancer effects.

If a population is exposed to more than one contaminant, the hazard index (HI) can be used to express the level of cumulative noncancer risk from pollutants 1 through n:

$$HI = \sum_{1}^{n} HQ \tag{32.13}$$

The HI is useful in comparing risks at various locations, for example, benzene risks in St. Louis, Cleveland, and Los Angeles. It can also give the cumulative (additive risk) in a single population exposed to more than one contaminant. For example, if the HQ for benzene is 0.2 (not significant), toluene is 0.5 (not significant), and tetrachloromethane is 0.4 (not significant), the cumulative risk of the three contaminants is 1.1 (potentially significant).

Realistic estimates of the hazard and the exposures are desirable in such calculations. However, precaution is the watchword for risk. Estimations of both hazard (toxicity) and exposure are often worst-case scenarios, because the risk calculations can have large uncertainties. Models usually assume effects to occur even at very low doses. Human data are usually gathered from epidemiological studies that, no matter how well are designed, are fraught with error and variability (science must be balanced with the rights and respect of subjects, population change, activities may be missed, and confounding variables are ever present).

Uncertainties exist in every phase of risk assessment, from the quality of data, to limitations and assumptions in models, to natural variability in environments and populations.

7. RISK-BASED CLEANUP STANDARDS

For most of the second half of the twentieth century, environmental protection was based on two types of controls, that is, technology and quality. Technology-based controls are set according to what is "achievable" from the current state of the science and engineering. These are feasibility-based standards. The Clean Air Act has called for "best achievable control technologies" (BACT) and more recently for "maximally achievable control technologies" (MACT). Both standards reflect the reality that even though from an air quality standpoint it would be best to have extremely low levels of pollutants, technologies are not available or are not sufficiently reliable to reach these levels. Requiring unproven or unreliable technologies can even exacerbate the pollution, such as in the early days of wet scrubbers on coal-fired power plants. Theoretically, the removal of sulfur dioxide could be accomplished by venting the power plant flue through a slurry of carbonate, but the technology at the time unproven and unreliable, allowing all-too-frequent releases of untreated emissions when the slurry systems were being repaired. Selecting a new technology over older proven techniques is unwise if the tradeoff of the benefit of improved treatment over older methods is outweighed by the numerous failures (i.e., no treatment).

Technology-based standards are a part of most environmental programs. Wastewater treatment, groundwater remediation, soil cleaning, sediment reclamation, drinking water supply, air emission controls, and hazardous waste site cleanup are in part determined by availability and feasibility of control technologies.

Quality-based controls are those that are required to ensure that an environmental resource is in good enough condition to support a particular use. For example, a stream may need to be improved so that people can swim in it and it can be a source of water supply. Certain streams may need higher levels of protection than others, such as the so-called "wild and scenic rivers." The parameters will vary but usually include minimum levels of dissolved oxygen and maximum levels of contaminants. The same goes for air quality, where ambient air quality must be achieved, so that concentrations of contaminants listed as National Ambient Air Quality Standards, as well as certain toxic pollutants, are below the levels established to protect health and welfare.

Risk-based approaches to environmental protection, especially contaminant target concentrations, are designed for engineering controls and preventive measures to ensure that risks are not exceeded. The risk-based approach actually embodies elements of both technology-based and quality-based standards. The technology assessment helps us to determine how realistic it will be to meet certain contaminant concentrations, whereas the quality of the environment sets the goals and means to achieve cleanup. Waste managers are often asked, "How clean is clean?" When do we know that we have done a sufficient job of cleaning up a spill or hazardous waste site? It is often not possible to have nondetectable concentrations of a pollutant. Commonly, the threshold for cancer risk to a population is one in a million excess cancers. However, one may find that the contaminant is so difficult to remove that we almost give up on dealing with the contamination and put in measures to prevent exposures, that is, fencing the area in and prohibiting no access. This is often done as a first step in remediation but is unsatisfying and controversial (and usually politically and legally unacceptable). Thus, even if costs are high and technology unreliable, the engineer must find suitable and creative ways to cleanup the mess and meet risk-based standards.

Risk-based target concentrations can be calculated by solving for the target contaminant concentration in the exposure and risk equations. Since risk is the hazard (e.g., slope factor) times the exposure (e.g., LADD), a cancer risk-based cleanup standard can be found by enumerating the exposure equations within the risk equation (in this instance, the drinking water equation from Table 32.5) gives:

$$\text{Risk} = \frac{C \times CR \times EF \times ED \times AF \times SF}{BW \times AT} \quad (32.14)$$

and solving for C:

$$C = \frac{\text{Risk} \times BW \times AT}{CR \times EF \times ED \times AF \times SF} \quad (32.15)$$

This is the target concentration for each contaminant needed to protect the population from the specified risk, for example, 10^{-6}. In other words, this is the concentration that must not be exceeded to protect a population having an average body weight and over a specified averaging time from an exposure of certain duration and frequency that leads to a risk of one in a million. Although one-in-a-million added risk is the commonly used benchmark, clean up may not always be required to achieve this level. For example, if a site is considered to be a "removal" action, the principal objective is to get rid of a sufficient amount of contaminated soil to reduce possible exposures. This means that the risk reduction target may be as high as one additional cancer per 10,000 (i.e., 10^{-4}). In fact, removal actions may be based on other thresholds besides cancer, such as the Drinking Water Equivalent Level (DWEL, which is calculated by multiplying the oral RfD by 70 kilograms (standard adult body weight) and dividing by the average volume of water (2 liters for adults) consumed per day. Other removal action cleanup levels are based on drinking water standards and health advisories.

The decision regarding the actual cleanup level, including whether to approach a contaminated site or facility as a removal or a remedial action is not risk assessment but falls within the province of risk management. The waste manager will have input to the decision but will not be the only party in that decision.

DISCUSSION BOX

RISK-BASED CONTAMINANT CLEANUP

A well is the principal water supply for the town of Apple Chill. A study has found that the well contains 80 mg L^{-1} of a halogenated organic compound ("CX_4"). Assuming that the average adult in the town drinks 2 L d^{-1} of water from the well and lives in the town for an entire lifetime, what is the lifetime cancer risk to the population if no treatment is added? What concentration is needed to ensure that the population's cancer risk is below 10^{-6}?

The lifetime cancer risk added to Apple Chill's population can be estimated using the LADD and slope factor for CX_4. In addition to the assumptions given, we will use default values from Table 32.6. We will also assume that people live in the town for their entire lifetimes, and their exposure duration is equal to their typical lifetime.

Thus, ED and TL terms cancel, leaving the abbreviated:

$$LADD = \frac{C \times CR \times AF}{BW}$$

As we have not specified male or female adults, we will use the average body weight,

(Continued)

DISCUSSION BOX—*Cont'd*

assuming that there are about the same number of males as females. We look up the absorption factor for CX_4 and find that it is 0.85, so the adult lifetime exposure is:

$$\text{LADD} = \frac{80 \times 2 \times 0.85}{65} = 4.2 \text{ mg kg}^{-1}\text{ d}^{-1}$$

Using the midpoint value between the default values ((15 + 40)/2=27.5 kg) for body weight and default CR values (1 L d^{-1}) the children lifetime exposure is:

$$\text{LADD} = \frac{80 \times 1 \times 0.85}{27.5} = 2.5 \text{ mg kg}^{-1}\text{ d}^{-1}$$

for the first 13 years, and the adult exposure of 4.2 mg kg^{-1} d^{-1} thereafter.

The oral SF for Cx_4 is 1.30×10^{-1} kg d^{-1}, so the added adult lifetime risk from drinking the water is $4.2 \times (1.30 \times 10^{-1}) = 5.5 \times 10^{-1}$. And, the added risk to children is $2.5 \times (1.30 \times 10^{-1}) = 3.3 \times 10^{-1}$.

Some subpopulations are more vulnerable to exposures than others. For example, for children, environmental and public health agencies recommend an additional factor of safety beyond what would be used to calculate risks for adults. This is known as the "10X" rule, that is, children need to be protected 10 times more than adults because they have longer life expectancies (so latency periods for cancer need to be accounted for), and their tissue is developing prolifically and changing. So, in this case, with the added risk, our reported "risk" would be 3.3. Although this is statistically impossible (i.e., one cannot have a probability greater than one because it would mean that the outcome is more than 100% likely, which of course is impossible!), it is actually an adjustment to the cleanup concentration. Because the combination of a very high slope of the dose response curve and a very high LADD increases the risk, children need a measure of protection beyond the general population. This is accomplished either by removing the contaminants from the water or the provision of a new water supply. In any event, the city public works and/or health department should mandate another source of drinking water (e.g., bottled water) immediately.

The cleanup of the water supply to achieve risks below one in a million can be calculated from the same information and reordering the risk equation to solve for *C*:

$$\text{Risk} = \text{LADD} \times \text{SF}$$

$$\text{Risk} = \frac{C \times \text{CR} \times \text{AF} \times \text{SF}}{\text{BW}}$$

$$C = \frac{\text{BW} \times \text{Risk}}{\text{CR} \times \text{AF} \times \text{SF}}$$

Based on adult LADD, the well water must be treated so that the tetrachloromethane concentrations are below:

$$C = \frac{65 \times 10^{-6}}{2 \times 0.85 \times 0.13} = 2.9 \times 10^{-4} \text{ mg L}^{-1}$$
$$= 290 \text{ ng L}^{-1}$$

Based on children's LADD, and the additional "10X," the well water must be treated so that the tetrachloromethane concentrations are below:

$$C = \frac{27.5 \times 10^{-7}}{1 \times 0.85 \times 0.13} = 2.5 \times 10^{-5} \text{ mg L}^{-1}$$
$$= 25 \text{ ng L}^{-1}$$

The town must remove the contaminant so that the concentration of CX_4 in the finished water will be at a level six orders of magnitude less than the untreated well water, that is, lowered from 80 mg L^{-1} to 25 ng L^{-1}. Cleanup standards are part of the arsenal needed to manage risks. However, other considerations

DISCUSSION BOX—*Cont'd*

needed to be given to a contaminated site, such as how to monitor the progress in lowering levels and how to ensure that the community stays engaged and participates in the cleanup actions, where appropriate. Even when the engineering solutions are working well, the engineer must allot sufficient time and effort to these other activities, otherwise skepticism and distrust can arise. Note that these calculations are for illustrative purposes only. In fact, there are numerous ways to calculate exposure besides the LADD. Also, the toxicity values also vary by source and time, e.g. RfDs and cancer slope factors are updated with new data and models. In addition, cleanup measures must be tailored to the needs of each site.

The risk assessment data and information will comprise much of the scientific underpinning of the decision, but legal, economic, and other societal drivers will also be considered to arrive at cleanup levels. It is not unusual, for example, for a legal document, such as a consent decree, to prescribe cleanup levels that are more protective than typical removal or remedial levels.

Table 32.7 describes a number of the general principles that have been almost universally adopted by public health regulatory agencies. The principles particularly apply to cancer risks from environmental exposures.

As mentioned, many risks are not directly tied to a single agent. The risks to public health, such as cancer and noncancer endpoints, are only one class of outcomes of concern. Others include ecosystem changes from gene flow and other ecological endpoints and opportunity risks. The latter risks include those posed by not allowing biological pesticides that may be safer than abiotically derived pesticides.

A waste-related decision must be comprehensive. Even when one aspect seems to be favorable for a particular site, the overall risk of one waste operation may be greater than that of another (e.g., less reliance of toxic raw materials, decreased likelihood of release of chemical toxins). Thus, a complete life cycle must be constructed and the composite and individual risks compared among various alternatives.

One of most difficult risks to quantify is the opportunity risk. For example, how can the loss of the opportunity that a technology affords be compared to the risk it poses? For example, if regulatory agencies are overly cautious about approving a cancer drug that may entail the introduction of carcinogens from possible environmental releases, the risk assessment may indicate that the overall cancer risk is actually improved if the cancers that are prevented and treated with the new drug are part of the risk calculation. Usually, however, it is not that straightforward. For example, what if the decision involves ecological values versus public health, such as not using a pesticide that would protect humans from a mosquito-borne disease because the use of the pesticide may endanger sensitive species? The risk decision and management process must always consider these tradeoffs.

7.1. Causality

Just from the arithmetic, it should be noted that zero risk can be calculated when either the hazard (e.g., toxicity) does not exist or when the exposure to that hazard is zero. A substance found to be associated with cancers based upon animal testing or observations of human

TABLE 32.7 General Principles Applied to Health and Environmental Risk Assessments Conducted by Regulatory Agencies in the United States

Principle	Explanation
Human data are preferable to animal data	For purposes of hazard identification and dose-response evaluation, epidemiological and other human data better predict health effects than animal models
Animal data can be used in lieu of sufficient, meaningful human data	Although epidemiological data are preferred, agencies are allowed to extrapolate hazards and generate dose-response curves from animal models
Animal studies can be used as basis for risk assessment	Risk assessments can be based on data from the most highly sensitive animal studies
Route of exposure in animal study should be analogous to human routes	Animal studies are best if from the same route of exposure as those in humans, for example, inhalation, dermal, or ingestion routes. For example, if an air pollutant is being studied in rats, inhalation is a better indicator of effect than if the rats are dosed on the skin or if the exposure is dietary
Threshold is assumed for noncarcinogens	For noncancer effects, for example, neurotoxicity, endocrine dysfunction, and immunosuppression, there is assumed to be a safe level under which no effect would occur (e.g., "no observed adverse effect level", NOAEL, which is preferred, but also "lowest observed adverse effect level," LOAEL)
Threshold is calculated as a reference dose or reference concentration (air)	Reference dose (RfD) or concentration (RfC) is the quotient of the threshold (NOAEL) divided by factors of safety (uncertainty factors and modifying factors; each usually multiples of 10): $$\text{RfD} = \frac{\text{NOAEL}}{\text{UF} \times \text{MF}}$$
Sources of uncertainty must be identified	Uncertainty factors (UFs) address: • Interindividual variability in testing • Interspecies extrapolation • LOAEL-to-NOAEL extrapolation • Subchronic-to-chronic extrapolation • Route-to-route extrapolation • Data quality (precision, accuracy, completeness, and representativeness) and modifying factors (MFs) address uncertainties that are less explicit than the UFs
Factors of safety can be generalized	The uncertainty and modifying factors should follow certain protocols, for example, 10 = extrapolation from a sensitive individual to a population; 10 = rat-to-human extrapolation, 10 = subchronic-to-chronic data extrapolation) and 10 = LOAEL used instead of NOAEL
No threshold is assumed for carcinogens	There is no safe level of exposure assumed for cancer-causing agents
Precautionary principle applied to cancer model	A linear, no-threshold dose-response model is used to estimate cancer effects at low doses, that is, to draw the unknown part of the dose-response curve from the region of observation (where data are available) to the region of extrapolation
Precautionary principle applied to cancer exposure assessment	The most highly exposed individual is generally used in the risk assessment (upper-bound exposure assumptions). Agencies are reconsidering this worst-case policy and considering more realistic exposure scenarios

[Source: U.S. Environmental Protection Agency, 2001, General Principles for Performing Aggregate Exposure and Risk Assessment, Office of Pesticides Programs, Washington, DC.]

populations can be further characterized. Association of two factors, such as the level of exposure to a compound and the occurrence of a disease, does not necessarily mean that one necessarily "causes" the other. Often, after study, a third variable explains the relationship. However, it is important for science to do what it can to link causes with effects. Otherwise, corrective and preventive actions cannot be identified. So, strength of association is a beginning step toward cause and effect. A major consideration in strength of association is the application of sound technical judgment of the weight of evidence. For example, characterizing the weight of' evidence for carcinogenicity in humans consists of three major steps [27]:

1. Characterization of the evidence from human studies and animal studies individually;
2. Combination of the characterizations of these two types of data to show the overall weight of evidence for human carcinogenicity; and,
3. Evaluation of all supporting information to determine whether the overall weight of evidence should be changed.

Note that none of these steps is absolutely certain.

8. COMMUNICATION

Risk information must be presented in a meaningful way without violating overextending the interpretation of the data (see Discussion Box: Waste Communications). Assigning causality when none really exists must be avoided but, if all we can say is that the variables are associated, the public is going to want to know more about what may be contributing an adverse affect (e.g., learning disabilities and blood lead levels). This was particularly problematic in early cancer research. Possible causes of cancer were being explored and major

DISCUSSION BOX

WASTE COMMUNICATIONS

Waste managers are called upon to produce scientifically sound products and systems in a way that involve diverse perspectives, especially of those most directly or indirectly affected by our decisions. Thus, managers must be able to communicate effectively to arrive at adequate designs, ensure that these technically sound designs are accepted by clients and stakeholder, and convey sufficient information to users so that the designs are operated and maintained satisfactorily.

Technical communication can be seen as a critical path, where the manager sends a message and the audience receives it (see Figure 32.11). The means of communication can be either perceptual or interpretive [29]. Perceptual communications are directed toward the senses. Human perceptual communications are similar to that of other animals; that is, we react to sensory information (e.g., reading body language or assigning meaning to gestures, such as a hand held up with palms out, meaning "stop" or smile conveying approval).

Interpretive communications encode messages that require intellectual effort by the receiver to understand the sender's meanings. This type of communication can either be verbal or symbolic. Scientists and engineers draw heavily on symbolic information when communicating amongst themselves. If you have ever

(Continued)

DISCUSSION BOX—*Cont'd*

mistakenly walked into the seminar where experts are discussing an area of science not memorable to you, using unrecognizable symbols and vernacular, this is an example of symbolic miscommunication. In fact, the experts may be using words and symbols that are used in your area of expertise but with very different meanings. For example, psychologists speak of "conditioning" with a very different meaning from that of an engineer.

Thus, the manager must be aware the venue of communication to apply designs properly.

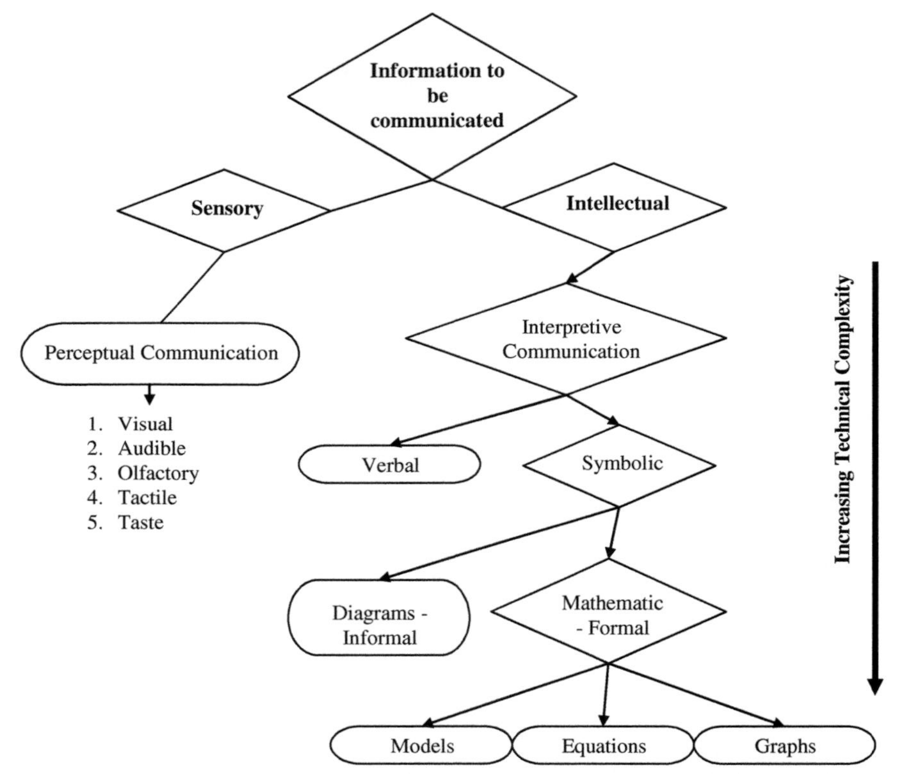

FIGURE 32.11 Risk communication techniques. All humans use perceptual communication, such as observing body language of a colleague or smelling an animal feeding operation. The right side of the figure is the domain of technical communication. Thus, the public may be overwhelmed by perceptive cues or may not understand the symbolic, interpretive language being used by a waste manager and others in the risk-communication process. Thus, the type of communication in a scientific briefing is quite different from that of a public meeting or a briefing for a neighborhood group potentially affected by a risk management decision. *Adapted from: M. Myers and A. Kaposi (2004)* The First Systems Book: Technology and Management. *2nd Edition, Imperial College Press. London, UK; and T.R.G. Green (1989). "Cognitive Dimensions of Notations," in: A. Sutcliffe and L. Macaulay (Eds.)* People and Computers V. *Cambridge University Press, Cambridge, UK.*

TABLE 32.8 Factors to be Considered in Determining whether a Waste Constituent Elicits an Effect Based on Hill's Criteria for Causality

Criterion	Description
Strength of association	For exposure to cause an effect, the exposure must be associated with that affect. Strong associations provide more certain evidence of causality than is provided by weak associations. Common epidemiological metrics used in association include risk ratio, odds ratio, and standardized mortality ratio
Consistency	If the chemical exposure is associated with an effect consistently under different studies using diverse methods of study of assorted populations under varying circumstances by different investigators, the link to causality is stronger. For example, the carcinogenic effects of chemical X is found in mutagenicity studies, mouse and Rhesus monkey experiments, and human epidemiological studies, there is greater consistency between chemical X and cancer than if only one of these studies showed the effect
Specificity	The specificity criterion holds that the cause should lead to only one disease and the disease should result from only this single cause. This criterion appears to be based in the germ theory of microbiology, where a specific strain of bacteria and viruses elicits a specific disease. This is rarely the case while studying most chronic diseases, since a chemical can be associated with cancers in numerous organs, and the same chemical may elicit cancer, hormonal, immunological, and neural dysfunctions
Temporality	Timing of exposure is critical to causality. This criterion requires that exposure to the chemical must precede the effect. For example, in a retrospective study, the researcher must be certain that the manifestation of a disease was not already present before the exposure to the chemical. If the disease were present prior to the exposure, it may not mean that the chemical in question is not a cause, but it does mean that it is not the sole cause of the disease (see "Specificity" above)
Biologic gradient	This is another essential criterion for chemical risks. In fact, this is known as the "dose-response" step in risk assessment. If the level, intensity, duration, or total level of chemical exposure is increased a concomitant, progressive increase should occur in the toxic effect
Biological plausibility	Generally, an association needs to follow a well-defined explanation based on known biological system. However, "paradigm shifts" in the understanding of key scientific concepts do change. A noteworthy example is the change in the latter part of the twentieth century of the understanding of how the endocrine, immune, and neural systems function, from the view that these are exclusive systems to today's perspective that in many ways they constitute an integrated chemical and electrical set of signals in an organism. For example, Candace Pert, a pioneer in endorphin research, has espoused the concept of mind/body, with all the systems interconnected, rather than separate and independent systems
Coherence	The criterion of coherence suggests that all available evidence concerning the natural history and biology of the disease should "stick together" (cohere) to form a cohesive whole. By that, the proposed causal relationship should not conflict or contradict information from experimental, laboratory, epidemiologic, theory, or other knowledge sources
Experimentation	Experimental evidence in support of a causal hypothesis may come in the form of community and clinical trials, in vitro laboratory experiments, animal models, and natural experiments
Analogy	The term analogy implies a similarity in some respects among things that are otherwise different. It is thus considered one of the weaker forms of evidence

[Source: *A.B. Hill, 1965, "The environment and disease: Association or causation?" Proceedings of the Royal Society of Medicine,* Occupational Medicine, *58, 7295.]*

research efforts were being directed at myriad physical, chemical, and biological agents. So, there needed to be some manner of sorting through findings to see what might be causal and what is more likely to be spurious results. Sir Austin Bradford Hill is credited with articulating key criteria (see Table 32.8) that need to be satisfied to attribute cause and effect in medical research [28].

In assessing risks, some of Hill's criteria are more important than others. Risk assessments rely heavily on strength of association, for example, to establish dose-response relationships. Coherence is also very important. When data from epidemiology, animal studies (*in vivo*) and petri dishes (*in vitro* studies) disagree, this must be explained (e.g. selective responses from different species or differences within sensitive individuals in the same species). Biological gradient is crucial, because this is the basis for dose response (the more the dose, the greater the biological response).

Temporality is crucial to all scientific research, that is, the cause must precede the effect. However, this is sometimes difficult to see in some instances, such as when the exposures to suspected agents have been continuous for decades and the health data are only recently available.

The key is that sound engineering and scientific judgment, based on the best available and most reliable data, be used to estimate risks. Linking cause and effect is often difficult in environmental matters. It is best for information to be transparent and coherent, including the uncertainties.

Environmental risk by nature addresses unwanted outcomes. Risk characterization is the stage where the waste manager pulls together the necessary assumptions, describes the scientific uncertainties, and determines the strengths and limitations of the analyses. The risks are articulated by integrating the analytical results, interpreting adverse outcomes, and describing the uncertainties and weights of evidence.

As mentioned earlier, risk assessment is a process distinct from risk management, where actions are taken to address and reduce the risks. However, the two are deeply interrelated and require continuous feedback from each other, with waste managers having to navigate both processes.

Waste managers need tools to support the risk analysis, but none or all of these tools provide a surefire answer to environmental and health risks. They are too complicated and complex for simple risk calculations. Certainly, waste components can indeed be evaluated for their roles in human and ecosystem risks. For example, many compounds have been studied sufficiently so that they have well-defined dose-response curves and other hazard metrics.

There are numerous ways to evaluate the performance of waste management actions. Does an action "work" (effectiveness)? Is it the best way to reach the end for which we strive (efficiency)? If it works and if it is the best means of providing the outcome, what is the probability of benefit. For example, the efficacy of a waste treatment technology is often assessed from pilot studies. However, such studies may be highly controlled, so that an approach may be more or less effective, depending on site conditions and operator performance [30].

Waste management requires a few more steps. Consideration must be given to whether the technology will likely continue to "work" (reliability) and, further, we must consider the hazards that can come about when the new technology is used. Risk is a function of likelihood that the hazard will in fact encounter, so we must also try to predict the adverse implications that society might face (risk). Thus, the "risk" associated with an action refers to the possibility and likelihood of undesirable and possibly harmful effects. Errors in risk prediction can range from not foreseeing outcomes that are merely annoying (e.g., scrubby vegetation thriving on a landfill cap, instead of the expected grass) to those that are devastating (e.g., the release of carcinogens into the environment) [31].

Unanticipated outcomes may not appear for years or even decades. For example, the decision to build a needed facility, such as a park or school on a site where toxic wastes have been buried. Arguably, Love Canal, New York is the classic, worst case of such a decision. The school board and other local groups were probably delighted to be given ample land on which to build a school. In due time, however, they rued that decision in light of the toxic substances that emerged. A classic example of such a failure, albeit not a biotechnological one, was the decision making related to Ford Pinto, a subcompact care produced by Ford Motor Company between 1971 and 1980. The car's fuel tank was poorly placed in such a way that it increased the probability of a fire from fuel spillage by a rear collision. A confluence of events, including the poor design, the likelihood of rear-end crashes, and lack of public knowledge of the risk, resulted in injuries and fatalities. The ensuing adverse implications manifested in the series of injuries that resulted from the defect that could have been predicted at some level of accuracy. Often, the problem is not a question of right versus wrong but is of one right versus another right.

Risk managers must often rely in part on semiquantitative or qualitative risk assessments. Retrospective failure analyses can be more quantitative, because there are forensic techniques available to tease out and assign weights to the factors that led to the outcome. There are even methods to calculate outcomes had other steps been taken.

9. CONCLUSIONS

Waste management success is a function of the amount of risk that has been reduced or avoided. Thus, the success of all waste operations is, to some extent, a reflection of the operation's hazards and potential exposures to these hazards. Decisions must be based on scientifically credible information. This information is part of the risk assessment that informs waste management decisions.

References

[1] The Safety in Biotechnology Working Party of the European Federation of Biotechnology and O. Doblhoff-Dier, Safe biotechnology 9: values in risk assessment for the environmental application of microorganisms, Trends in Biotechnology 17 (8) (1999) 307–311.

[2] NRC (1983). Risk Assessment in the Federal Government. National Academy of Sciences. Washington, DC

[3] A. Kortenkamp, T. Backhaus, M. Faust, State of the Art Report on Mixture Toxicity. Final Report, Part 1. European Commission. Study contract no. 070307/2007/485103/ETU/D.1 (2009).

[4] National Academy of Sciences, Biosolids Applied to Land: Advancing Standards and Practices, National Academy Press, Washington, DC, 2002.

[5] U.S. Environmental Protection Agency, Framework for ecological risk assessment. Report no. EPA/630/R-92/001. Risk Assessment Forum. Washington, DC (1992).

[6] U.S. Environmental Protection Agency. 1998. Guidelines for ecological risk assessment. EPA/630/R-95/002F, Washington, DC, 1998.

[7] National Research Council, Science and Decisions: Advancing Risk Assessment, National Academies Press, Washington, DC, 2009.

[8] C. Mitcham, R.S. Duval, Engineering Ethics, Chapter 8, Responsibility in Engineering, Prentice-Hall, Upper Saddle River, NJ, 2000.

[9] U.S. Environmental Production Agency, Guidelines for exposure assessment, u.s. environmental production agency, Washington, DC, 1992. Report no. EPA/600/Z-92/001. Risk Assessment Forum.

[10] S.M. Macgill, Y.L. Siu, A new paradigm for risk analysis, Futures 37 (2005) 1105–1131.

[11] E.A. Cohen Hubal, A.M. Richard, S. Imran, J. Gallagher, R. Kavlock, J. Blancato, et al., Exposure science and the US EPA National Center for Computational Toxicology, Journal of Exposure Science and Environmental Epidemiology (2008). doi:10.1038/jes.2008.70 [Online: November 5, 2008].

[12] S.H. Morris, EU biotech crop regulations and environmental risk: a case of the emperor's new clothes, Trends in Biotechnology 24 (1) (2006) 2–6;

A.J. Conner, T.R. Glare, J.P. Nap, The release of genetically modified crops into the environment, Plant Journal 33 (2003) 19–46.

[13] C.C. Dary, P.J. Georgopoulos, D.A. Vallero, D.A., R. Tornero-Velez. M. Morgan, M. Okino, M. Dellarco, F.W. Power and J.N. Blancato (2007). Characterizing chemical exposure from biomonitoring data using the exposure related dose estimating model (ERDEM), 17th Annual Conference of the International Society of Exposure Analysis. Durham, NC. October 17, 2007.

[14] J.N. Blancato, F.W. Power, R.N. Brown and C.C. Dary (2006). Exposure Related Dose Estimating Model (ERDEM): A Physiologically-Based Pharmacokinetic and Pharmacodynamic (PBPK/PD) Model for Assessing Human Exposure and Risk. Report No. EPA/600/R-06/061. U.S. Environmental Protection Agency. Las Vegas, Nevada.

[15] P. Aarne Vesilind, J. Jeffrey Peirce, Ruth F. Weiner, Environmental Engineering, third ed., Butterworth-Heinemann, Boston, MA, 1993.

[16] U.S. Environmental Protection Agency, Guidelines for carcinogen risk assessment. Report No. EPA/630/R-00/004, Federal Register 51 (185) (1986) 33992–34003. Washington, DC.

[17] R.I. Larsen, An Air Quality Data Analysis System for Interrelating Effects, Standards, and Needed Source Reductions: Part 13 – Applying the EPA Proposed Guidelines for Carcinogen Risk Assessment to a Set of Asbestos Lung Cancer Mortality Data, Journal of the Air & Waste Management Association 53 (2003) 1326–1339.

[18] D.A. Vallero, Environmental Contaminants: Assessment and Control, Elsevier Academic Press, Burlington, MA (2004).

[19] J. Duffus, H. Worth, Training program: The Science of Chemical Safety: Essential Toxicology—4; Hazard and Risk IUPAC Educators' Resource Material, International Union of Pure and Applied Chemistry, 2001.

[20] For an excellent summary of the theory and practical applications of the Ames test, see, K. Mortelmans, E. Zeiger, The Ames *Salmonella*/Microsome Mutagenicity Assay, Mutation Research 455 (2000) 29–60.

[21] U.S. Environmental Protection Agency, Interim reregistration eligibility decision for chlorpyrifos. Report no. EPA 738-R-01-007. Washington, DC (2002).

[22] The principal source for this section is, D.A. Vallero, Environmental Biotechnology: A Biosystems Approach, Elsevier Academic Press, Amsterdam, NV, 2010.

[23] B. Singer, A tool to predict exposure to hazardous air pollutants, Environmental Energy Technologies Division News 4 (4) (2003) 5.

[24] U.S. Environmental Protection Agency, Exposure factors handbook. Report no. EPA/600/P-95/002Fa, Washington, DC (1997).

[25] These factors are updated periodically by the U.S. EPA in the Exposure Factor Handbook at www.epa.gov/ncea/exposfac.htm.

[26] The definition of child is highly variable in risk assessment. The Exposure Factors Handbook uses these values for children between the ages of 3 and 12 years.

[27] U.S. Environmental Protection Agency, Guidelines for carcinogen risk assessment. Report no. EPA/630/R-00/004, Federal Register 51(185):33992-34003, Washington, DC (1986).

[28] A. Bradford Hill, The environment and disease: association or causation? Proceedings of the Royal Society of Medicine, Occupational Medicine, 58, p. 295. A. Bradford-Hill, 1965, The environment and disease: association or causation? President's Address: Proceedings of the Royal Society of Medicine, 9, (1965) 295–300.

[29] The principal sources of this discussion are M. Myers, A. Kaposi, The First Systems Book: Technology and Management, second ed., Imperial College Press, London, UK, 2004;
T.R.G. Green, Cognitive dimensions of notations, in: A. Sutcliffe, L. Macaulay (Eds.), People and Computers V, Cambridge University Press, Cambridge, UK, 1989.

[30] See, for example, France Biotech, FROM GMP TO GBP: Fostering bioethics practices (GBP) among the European biotechnology Industry: <http://cordis.europa.eu/fetch?CALLER=FP6_PROJ&ACION=D&DOC=16&CAT=PROJ&QUERY=1172750150639&RCN=80077>; Accessed on July 19, 2009 (2009).

[31] H. Petroski, To Engineer is Human: The Role of Failure in Successful Design, St. Martin's Press, New York, NY, 1985.

Epilogue

Our journey from exploitation to sustainability has taken us along many paths and for the moment it is over. We see that over the past 20 years many new areas of expertise have been developed and much energy has been expended to contain old and the new waste forms. Here are some reflections on the questions asked in the Prologue:

1. *How much is known about each waste stream? What is the state-of-the-science?*

   The general consensus is that for the moment most of our waste streams are in reasonably good shape. For a few areas, this is not the case, such as the looming threat of poorly managed nuclear wastes and the ominous scenarios that may arise from space wastes. In addition, new problems are breaking the horizon which we have not dealt with in our book. For example, nanotechnology is just beginning to make a major splash in our industrial world, and so far we have not considered the waste issues surrounding its implementation. We also found that technological advances may be a two-edged sword. The food supply, for example, may be enhanced but new chemical and biological stresses may be introduced as a result. Emerging applications of other technologies, such as genetically modified organisms, may present threats and even, once again, redefine wastes (e.g., waste organisms?).
2. *What levels of certainty are needed to take actions?*

   The challenge for the waste professional is to balance hubris and boldness. We must boldly confront these emerging challenges but with the humility that solutions have often presented problems that were even worse than the previous problems. Notable among such "wrong things for the right reasons" were PCB incineration leading to the formation and release of dioxins and furans; chemical additives like MTBE intending to reduce carbon monoxide emissions polluting ground water; and tall stacks designed to eliminate localized effects from sulfur dioxide playing a large part in the formation of acid rain.

   These mistakes are evidence that waste management is chaotic in its every connotation. All decisions *can* lead to unanticipated consequences. One could argue that every decision in fact *does* lead to unanticipated outcomes. For example, moving from the circumscribed and controlled conditions of the laboratory is always different than moving to the prototype stage, which is always different from moving to actual conditions in the environment. We are never completely certain about the extent of how things will change once they reach the environment. This is chaos. Initial conditions may be relatively well defined, but almost immediately after, new factors appear which may significantly alter the outcome in an unacceptable ways.

Some argue that any major environmental decision proceed carefully by following a "precautionary principle" which states that if the consequences of an action, such as the application of a new technology, are unknown but the possible scenario is sufficiently devastating, then it is prudent to avoid the action. However, the precautionary approach must also be balanced against opportunity risks. In other words, by our extreme caution are we missing opportunities that would better serve the public and future generations? The key to balancing these connections can be a complete and accurate characterization of risks and benefits. Unfortunately, waste decisions are often not fully understood until after the fact (and viewed through the prism of lawsuits and media coverage).

3. *How do the various professional and scientific disciplines vary in their approaches?*

   The interdependence of science, engineering, and technology is important in waste management, as in almost any technical field. Interestingly, thermodynamic terms are frequently used, and sometimes abused by the lay public and scientific community. The system and its boundary, surroundings, constraints, and drivers must be known to engage in biotechnological design. Waste management also calls up other connotations of the word "system," especially that used by the lay public for a mental construct that influences the way one thinks. Waste problems require that we keep in mind how best to measure success "systematically."

4. *How should waste decisions be evaluated?*

   Chaos is defined as a dynamical system that is extremely sensitive to its initial conditions. Small differences early in a process can profoundly and extensively change the final state of the system even over small timescales. Such unintended change can be considered an *artifact* of the system. Waste management is vulnerable to artifacts. That is, even well-conceived designs may lead to impacts down the road. Waste risk assessment should not be limited to engineers. For example, waste solutions can be harmed by failure in any engineering discipline, as well as by other professions.

5. *What constitutes best practices in waste management?*

   Most engineers' codes of ethic require them be competent within an area of expertise but is not an excuse not to consider possible flaws that result down the road. In fact, just the opposite is true. That is, part of the competence requires consideration of how a system may be intentionally abused (e.g., terrorism) or unintentionally misused, so-called "offsetting behavior." This means that all the solutions proposed in this book, under certain circumstances, could lead to artifacts and unintended consequences. The US Food and Drug Administration and others offer some insights about how to avoid offsetting behaviors, which can be paraphrased to apply to waste management decisions:

   1. Beware of overconfidence—Engineers must overcome the temptation to assume that problems can be resolved using off-the-shelf approaches and a particular waste is inherently similar to another (small differences can be amplified through a product's life cycle).
   2. Do not confuse reliability with safety—A solution (e.g., software predicting chemical transport or a reactor designed to break down a pollutant) can work thousands of times without error, but even these probabilities are unacceptable when consequences are potentially widespread and for long term.
   3. Include a defensive design—Redundancies as well as self-checks, trouble-shooting, and error detection and correction systems are vital.

A worst-case design scenario should be identified.

4. **Failure to address root causes**—Causes may be misidentified and the corrections may solve problems other than the ones leading to the real failures.
5. **Avoid complacency**—In many engineering failures the contingency that led to the artifact was part of a routine process. Continued success does not mean due diligence can be a shortcut.
6. **Conduct realistic risk assessments**—The failure analysis should not assume independence and must be attentive to key factors. Seemingly mundane or benign technologies can have adverse effects [1].

Waste management is an inexact science that attempts to be as exact as possible in addressing ill-posed problems. That is part of the reason that this book is more about guidance than about "plug and chug" tools. Best practices must be tailored to the specific conditions of each waste. Managing wastes is decision-making with large uncertainties about numerous variables. Thus, waste management is at once exciting and frightening. The famous engineer, Norm Augustine, has put this dichotomy in perspective:

> The bottom line is that the things engineers do have consequences, both positive and negative, sometimes unintended, often widespread, and occasionally irreversible.... Engineers who make bad decisions often don't know they are confronting ethical issues [2].

Thus, waste management is not only a technically challenging and exciting endeavor, but it is also a moral imperative.

References

[1] N. Leveson, "Medical Devices: The Therac-25", in: Safeware: System Safety and Computers, Addison-Wesley Professional Publications, New York, NY, 1995.

[2] N. Augustine, Ethics and the second law of thermodynamics, The Bridge 32 (2002) 4–7.

Index

I

M

N

O

S

Color Plates

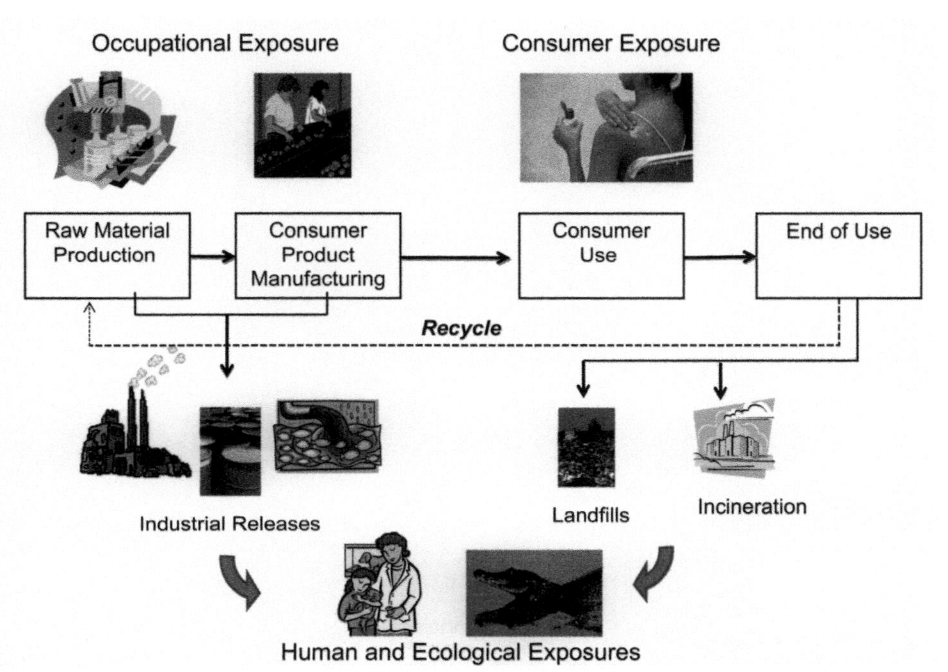

FIGURE 2.4 Life cycle of product, showing points of exposure and potential risk, including after the useful life of the product [6].

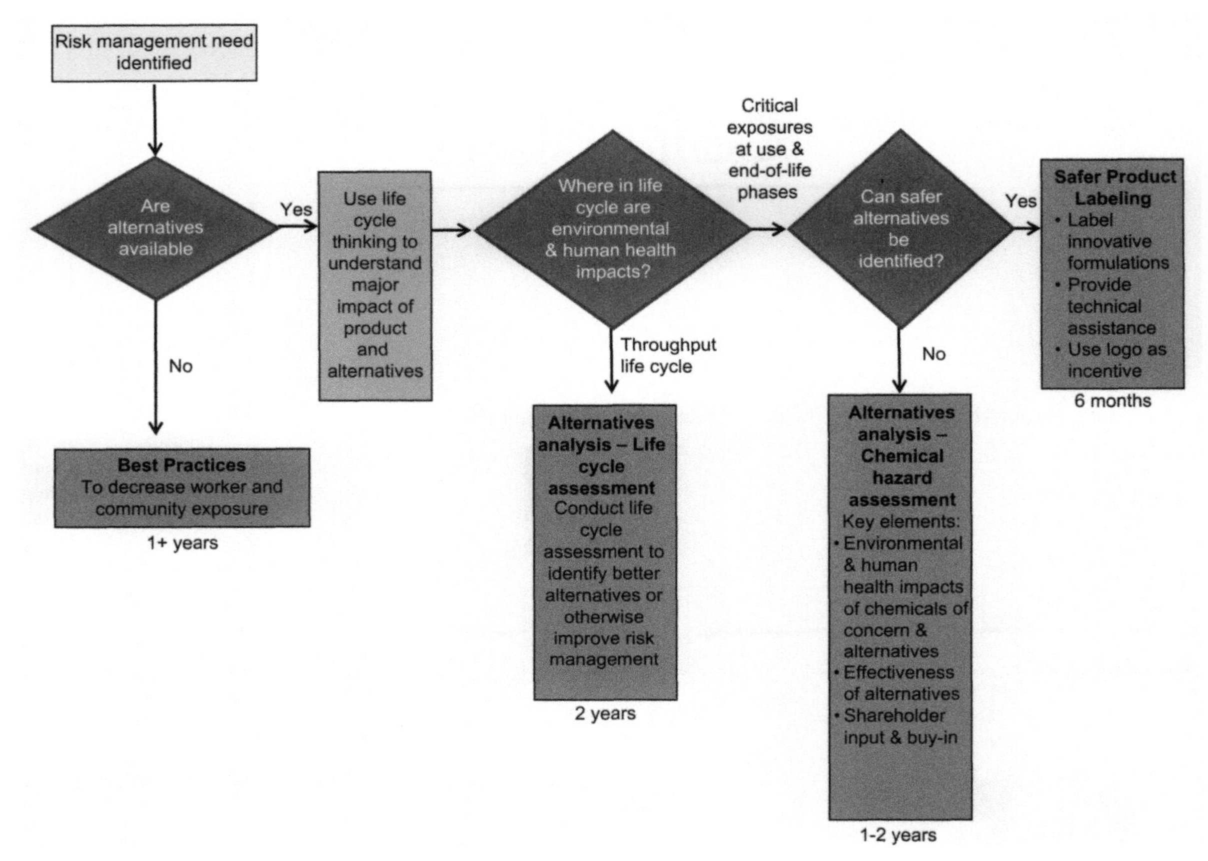

FIGURE 2.5 Decision logic for DfE approaches [7].

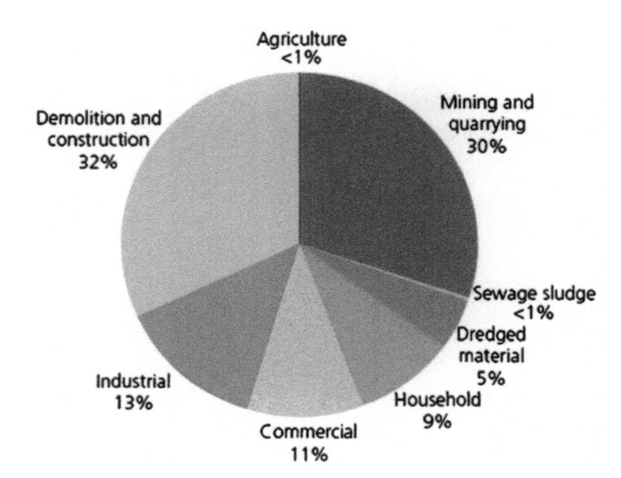

FIGURE 4.1 Waste collected in England by sector by weight. *Source*: *Ref.* [1].

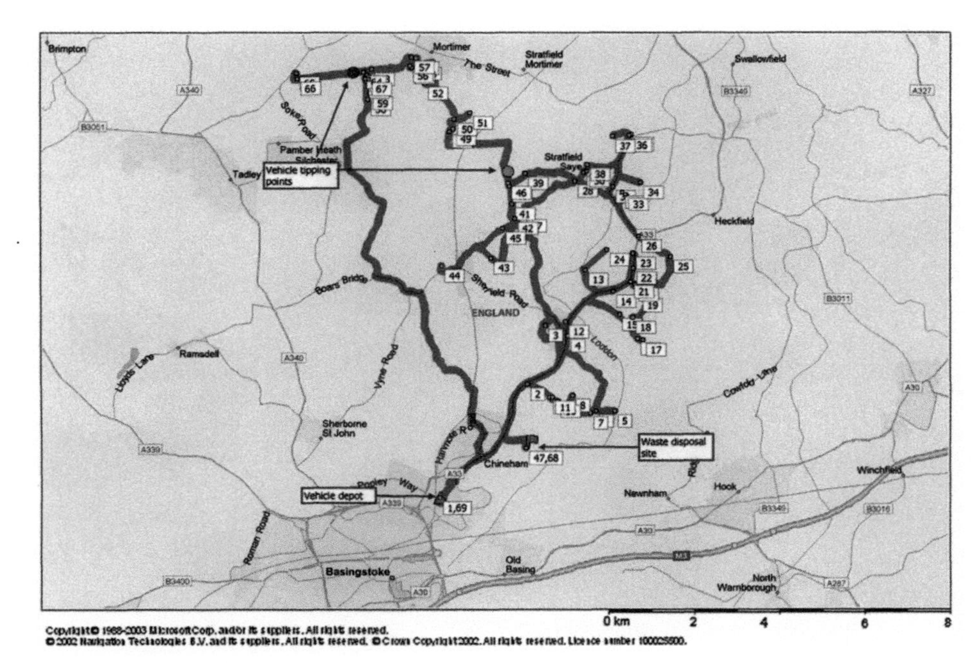

FIGURE 4.4A Original modelled route with poorly placed tipping points.

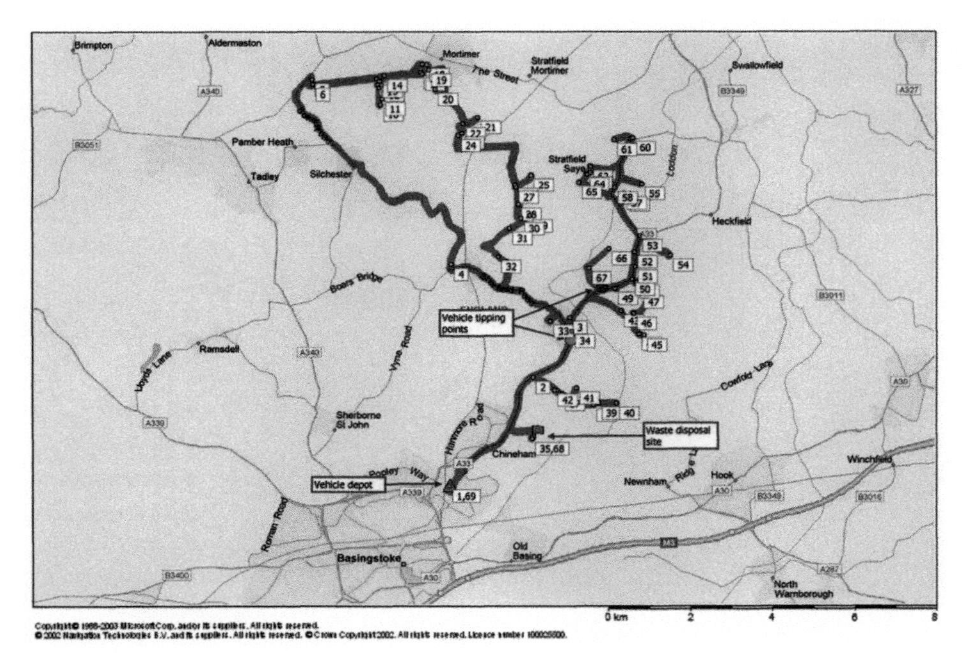

FIGURE 4.4B Redesigned route with tipping points closer to the waste disposal site.

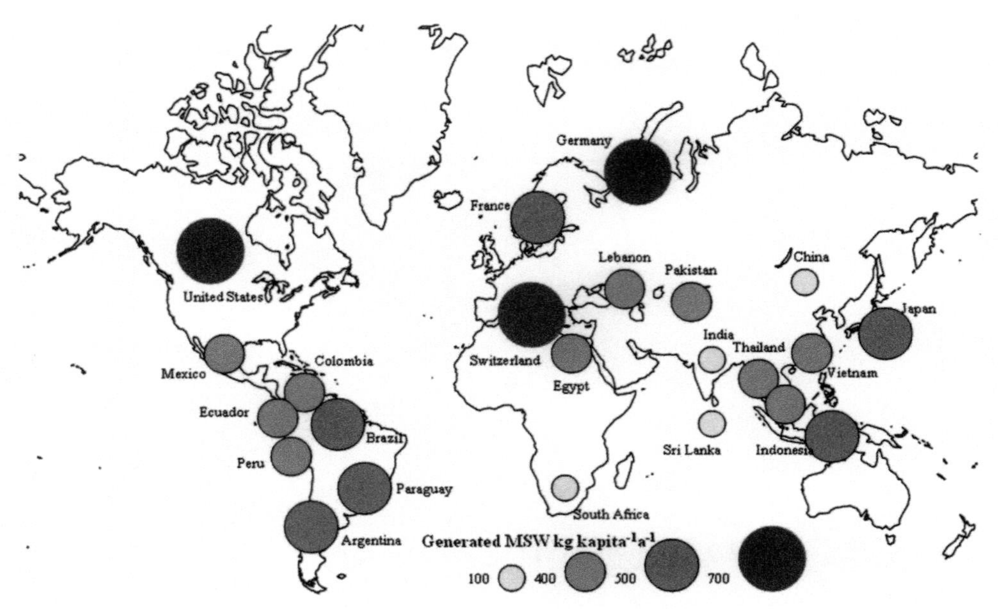

FIGURE 8.1 Municipal solid waste generation (kg capita $^{-1}$ a $^{-1}$) in 25 countries grouped according to their gross national income (GNI). *Source: eawag: Swiss Federal Institute of Aquatic Science and Technology.*

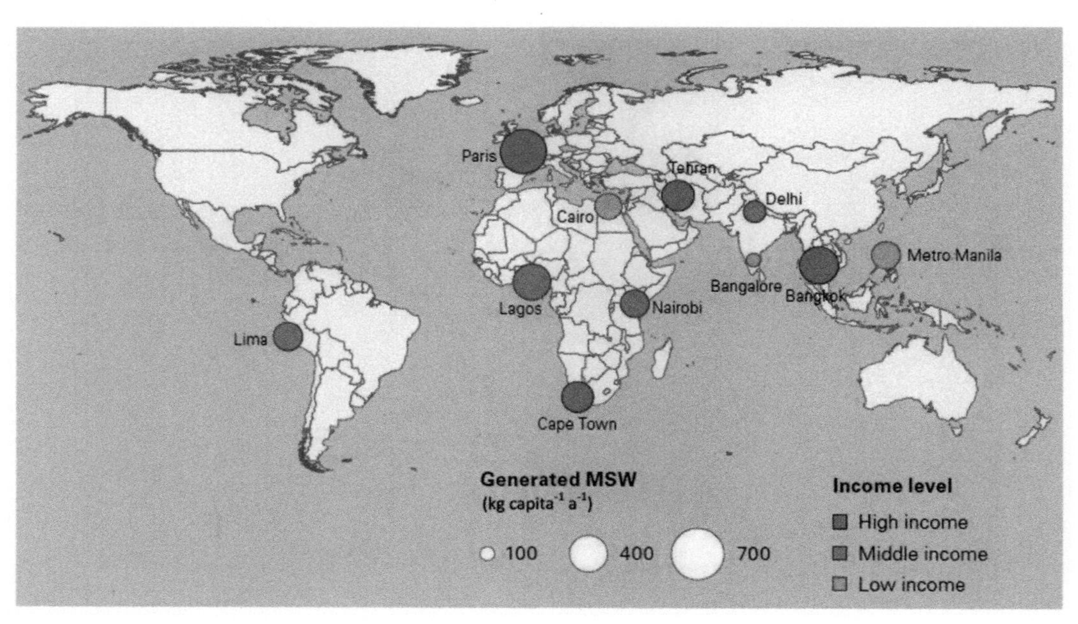

FIGURE 8.2 Generation of MSW (kg capita $^{-1}$ a $^{-1}$) in 11 cities and their gross domestic product in 2005 (in US$, using *purchasing power parity* exchange rates) per capita according to the World Bank's income classification of 2006. Source: eawag: Swiss Federal Institute of Aquatic Science and Technology.

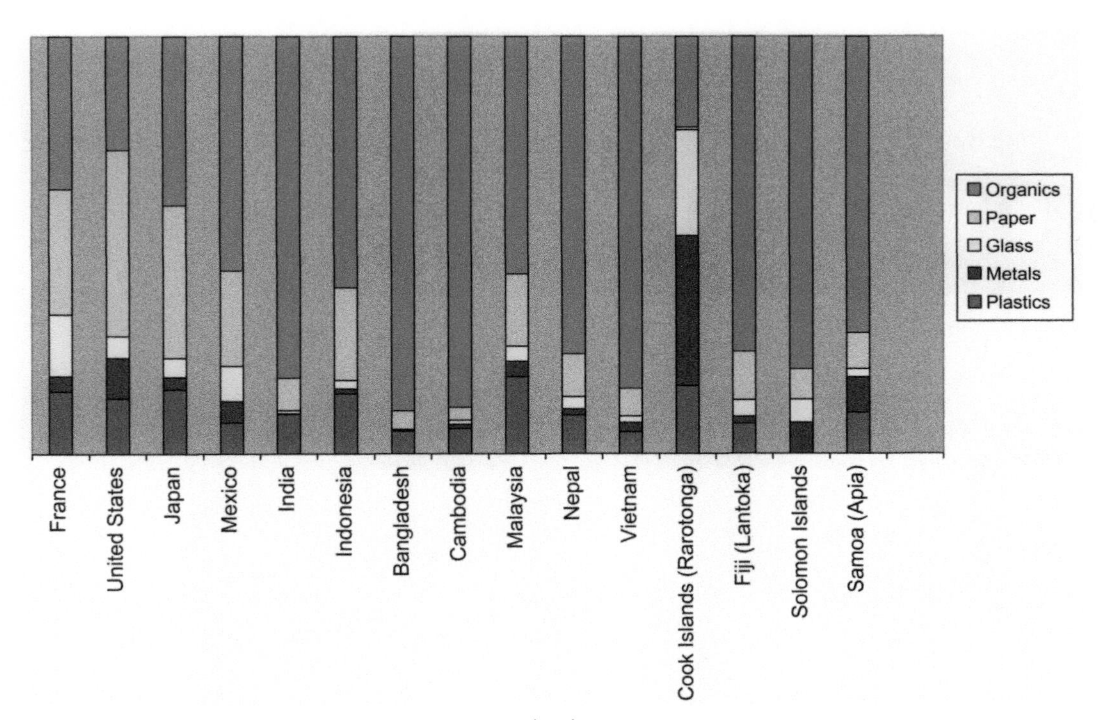

FIGURE 8.3 Top: Composition of the MSW (kg capita $^{-1}$ a $^{-1}$) in 12 countries grouped according to their gross national income (GNI). Bottom: Composition of the MSW (kg capita $^{-1}$ a $^{-1}$) in 23 cities. *Source: eawag: Swiss Federal Institute of Aquatic Science and Technology.*

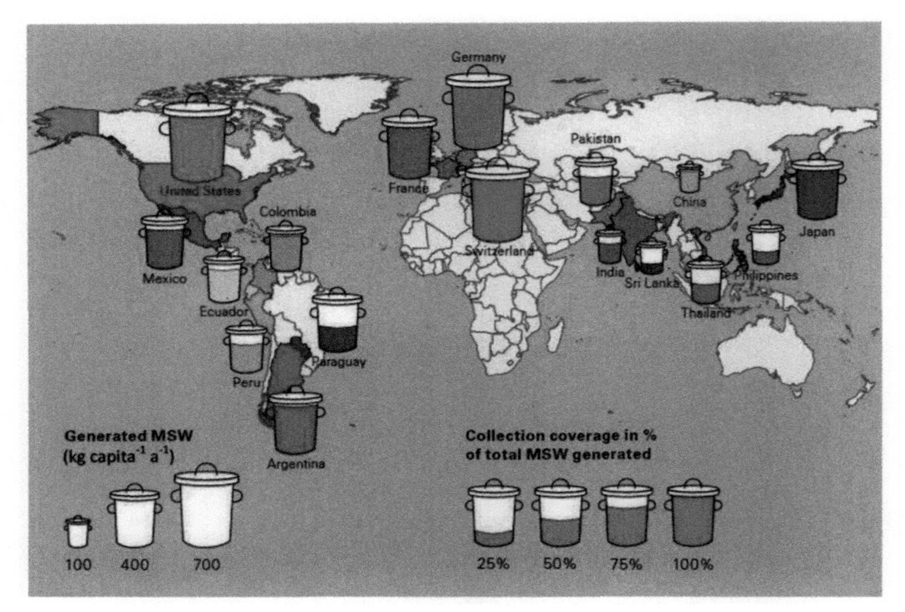

FIGURE 8.4 Total MSW generated (kg capita $^{-1}$ a $^{-1}$) and collection coverage in percentage in 17 countries. *Source: eawag: Swiss Federal Institute of Aquatic Science and Technology.*

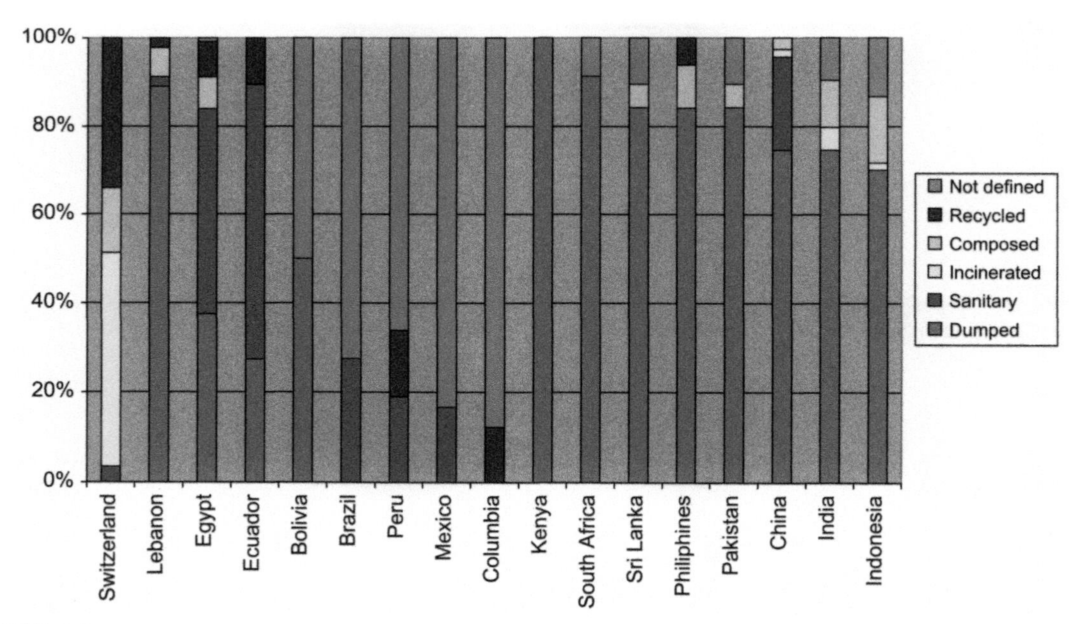

FIGURE 8.5 Percentage of the commonly used MSW treatment and disposal technologies in 21 countries. *Source: eawag: Swiss Federal Institute of Aquatic Science and Technology.*

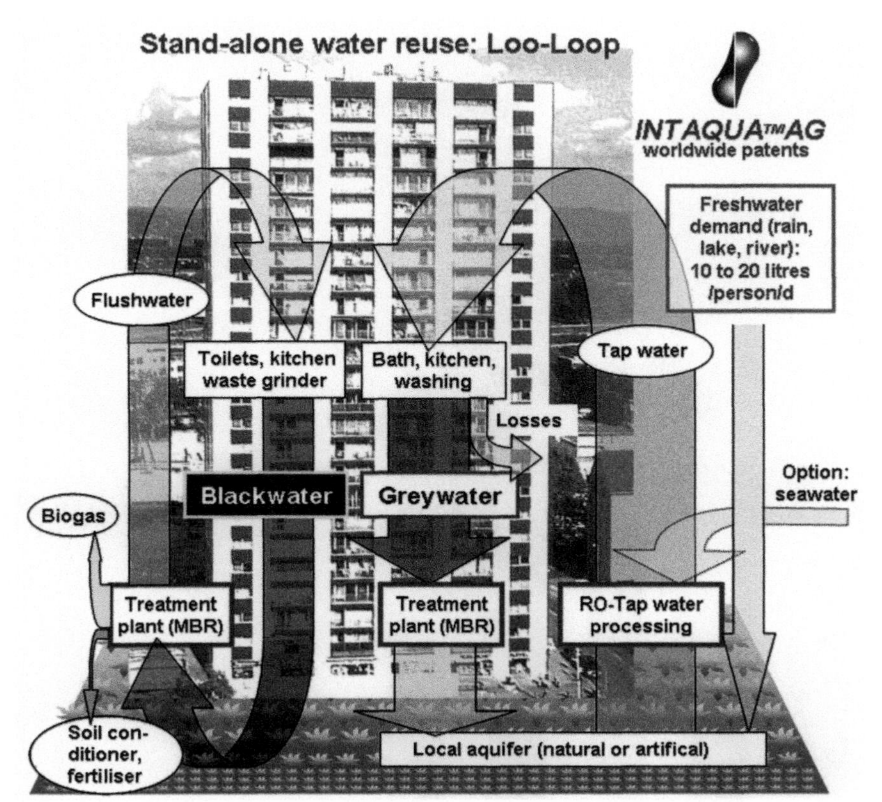

FIGURE 9.4 The blackwater loop combined with greywater reuse for dry areas.

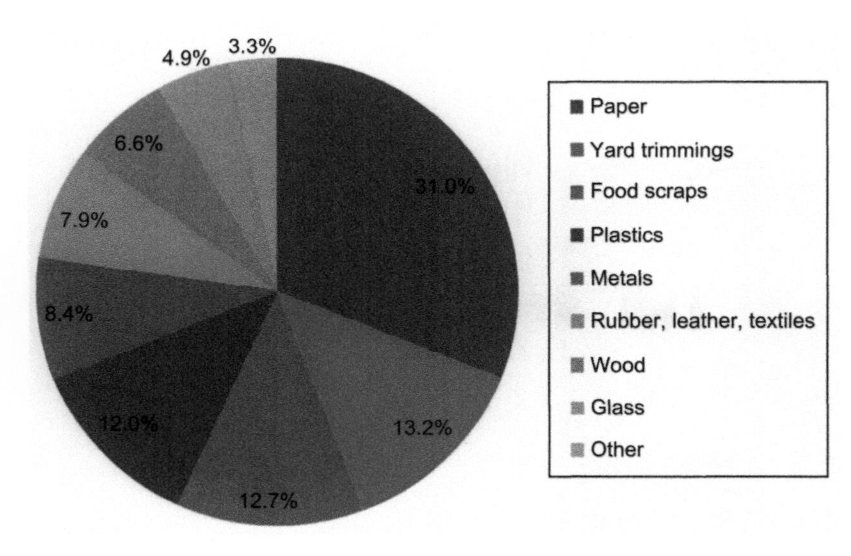

FIGURE 10.2 Composition of municipal solid waste in the United States before recycling in 2008 [15].

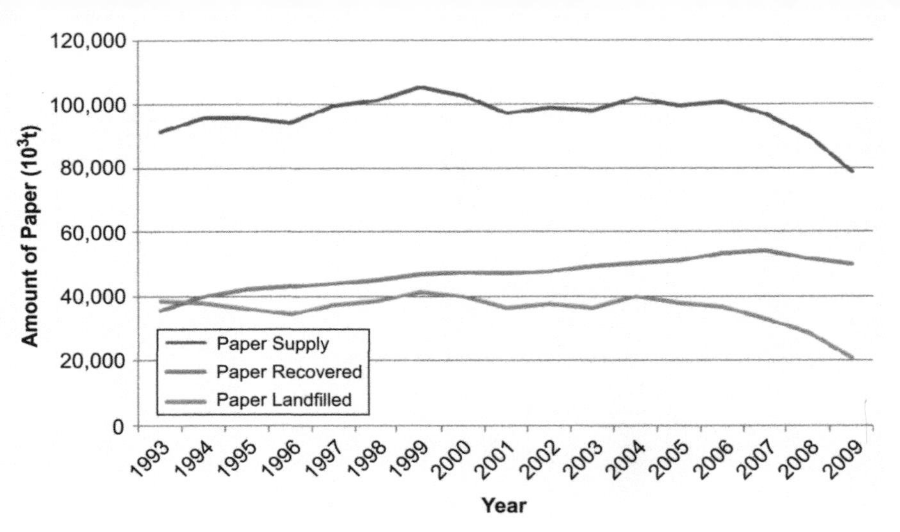

FIGURE 10.4 The paper supply compared with the amount of paper recovered and the amount of paper landfilled in the United States [7].

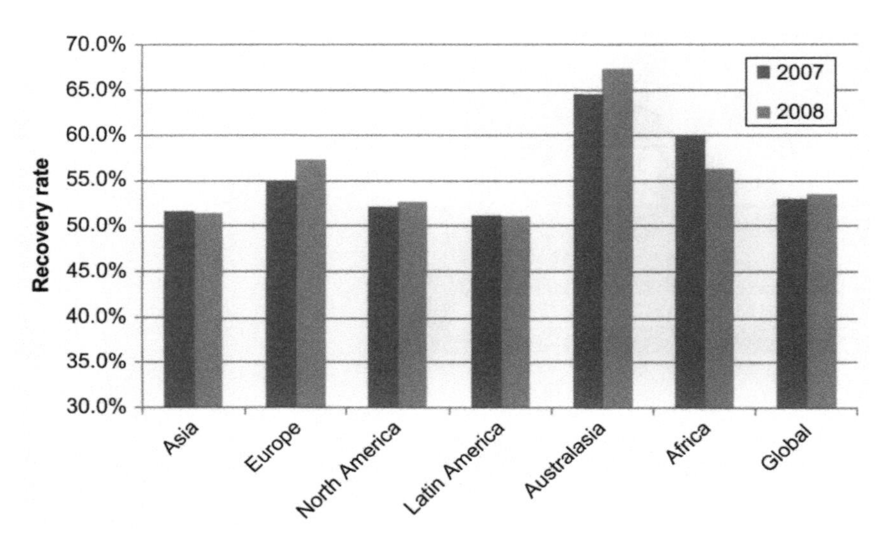

FIGURE 10.5 Paper recovery rates (as a fraction of production) regionally around the world [8].

FIGURE 13.1 Typical household products with potentially hazardous properties.

BOX 13.1B

EUROPEAN CHEMICALS BUREAU (ECB) HAZARD SYMBOLS ACCORDING TO EUROPEAN UNION'S DANGEROUS SUBSTANCES DIRECTIVE 67/548/EEC AND USED ON PACKAGING FOR CONSUMER PRODUCTS WITH POTENTIALLY HAZARDOUS PROPERTIES

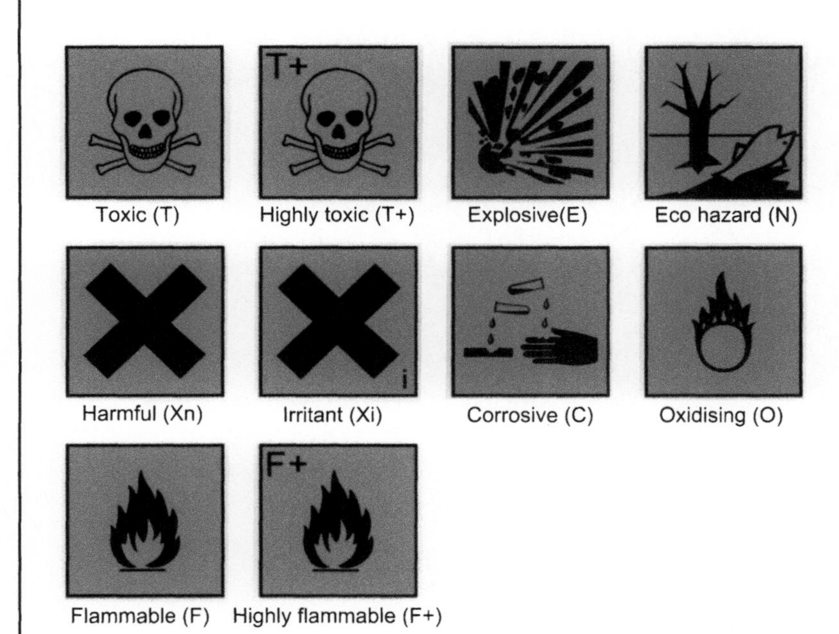

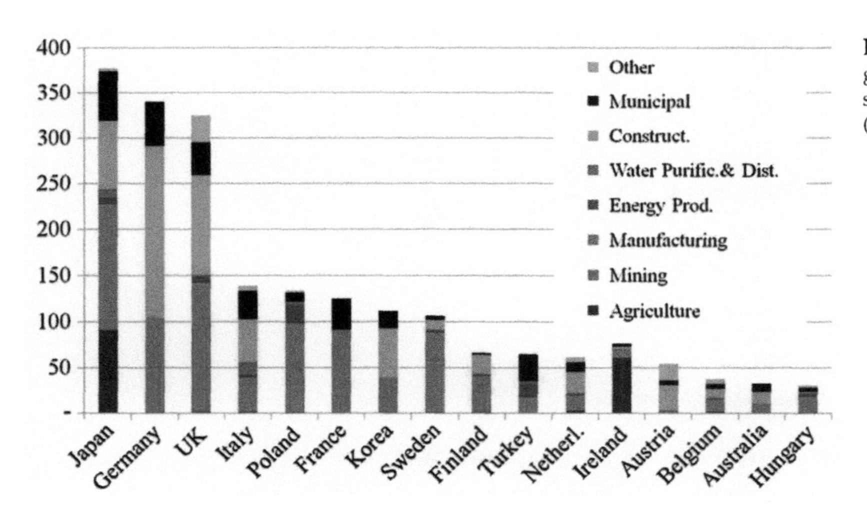

FIGURE 14.1 Total waste generation and its composition in selected countries [2] (OECD 2008) (unit: million metric tonnes).

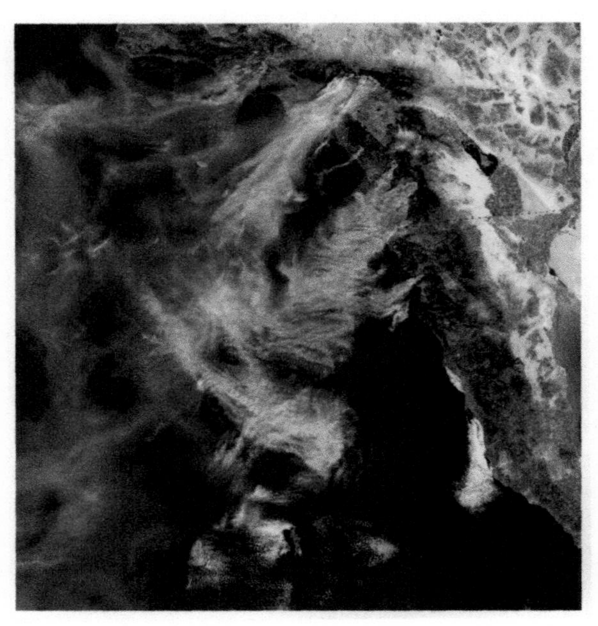

FIGURE 19.5 Satellite image from summer brush fires in California-Baja California region. Large quantities of PAHs are released into the atmosphere and later they reach coastal zones.

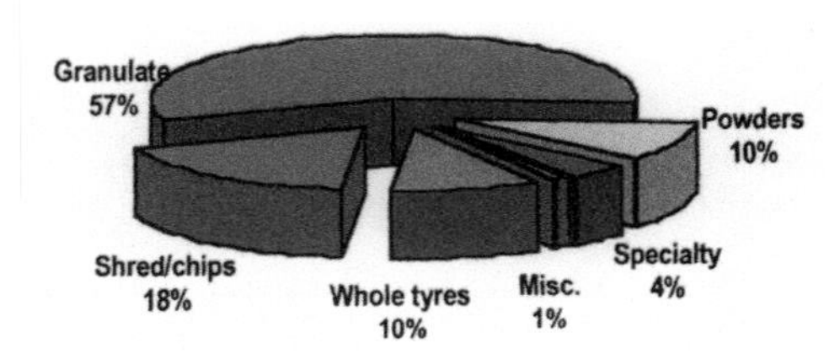

FIGURE 21.5 EU material production by category.

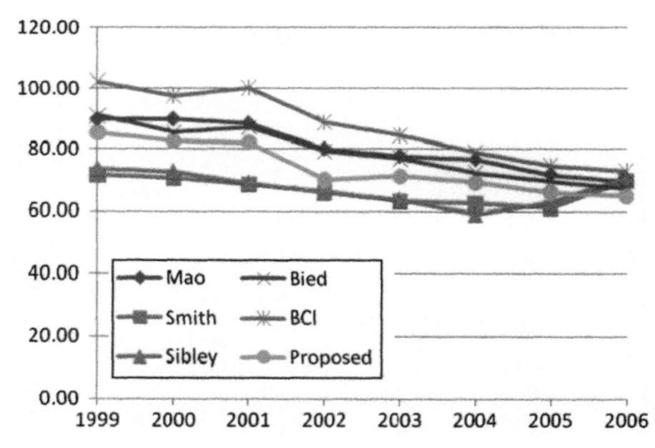

FIGURE 22.4 Revised rates for integrated model in comparison with other models. The y-axis refers to percentage.

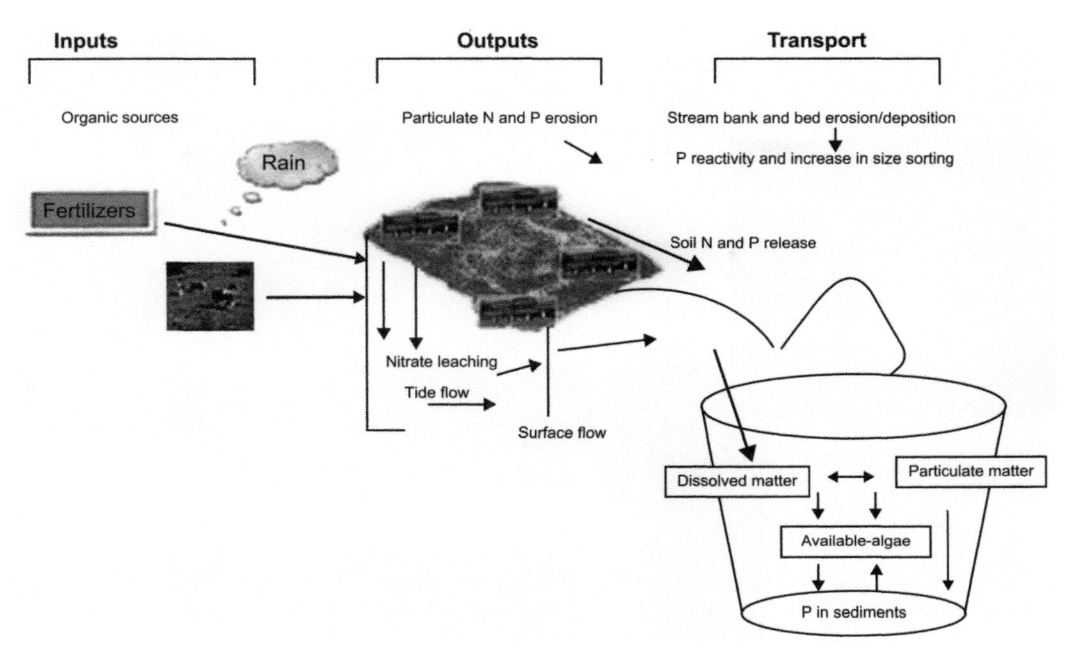

FIGURE 24.4 Inputs, outputs and transport of P and N from agricultural land modified after Carpenter et al. [20].

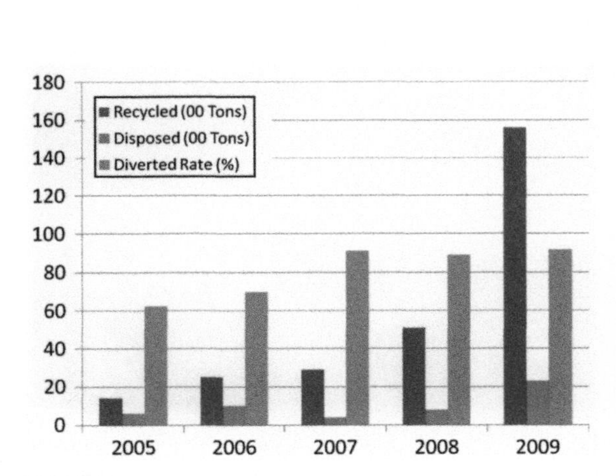

FIGURE 25.1 Solid waste disposal and recycling at the Picatinny Arsenal from 2005 to 2009. The units are U.S. tons (1 ton = 0.907 t).

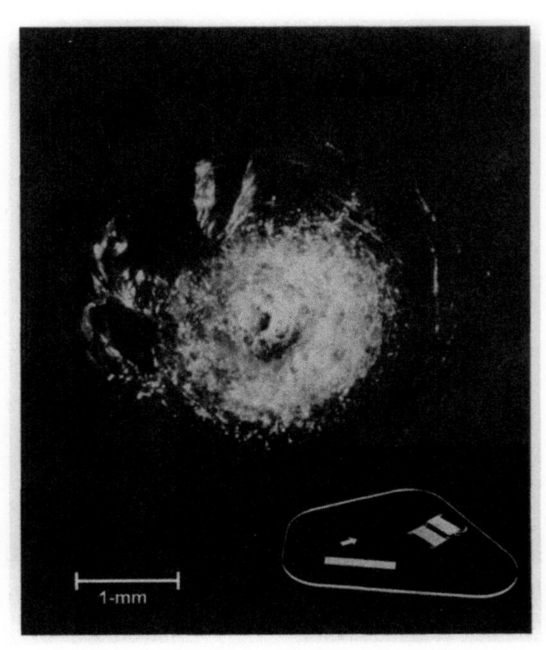

FIGURE 26.1 STS-7 window strike. Damage caused by a paint fleck only 0.2 mm in size.

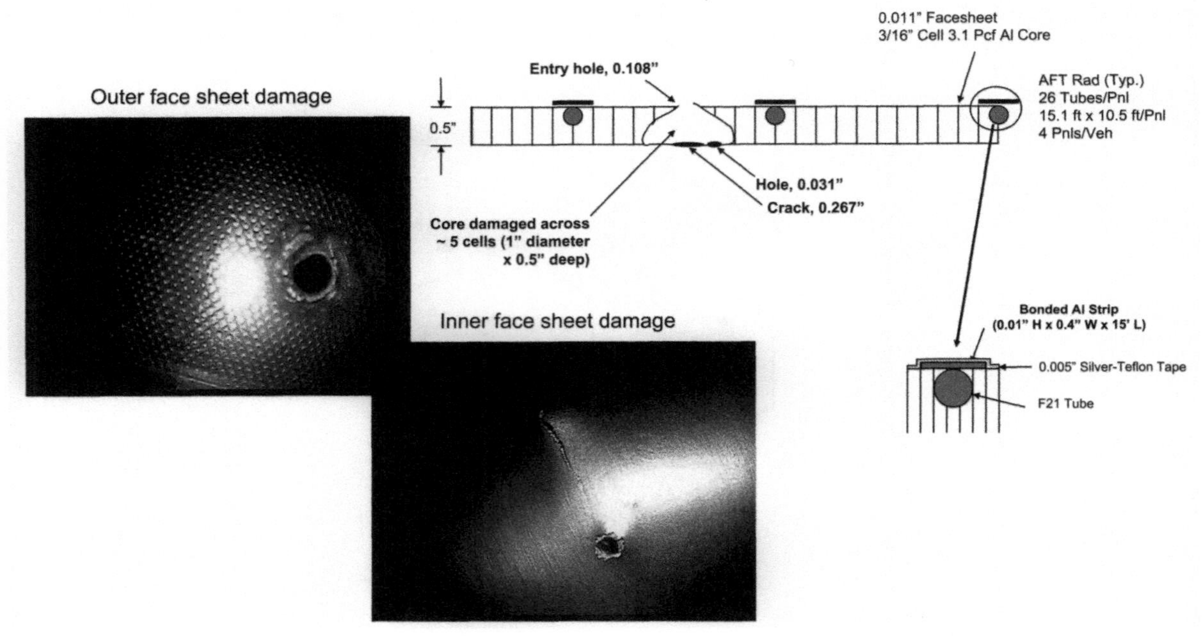

FIGURE 26.2 Radiator damage on STS-115. This damage was caused by a small piece of circuit board material.

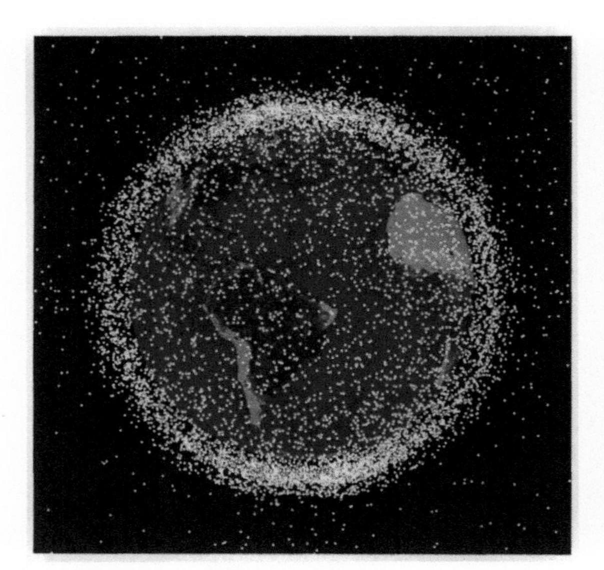

FIGURE 26.6 Representation of the location and distribution of the LEO debris population at a snapshot in time (midnight, January 1, 2010).

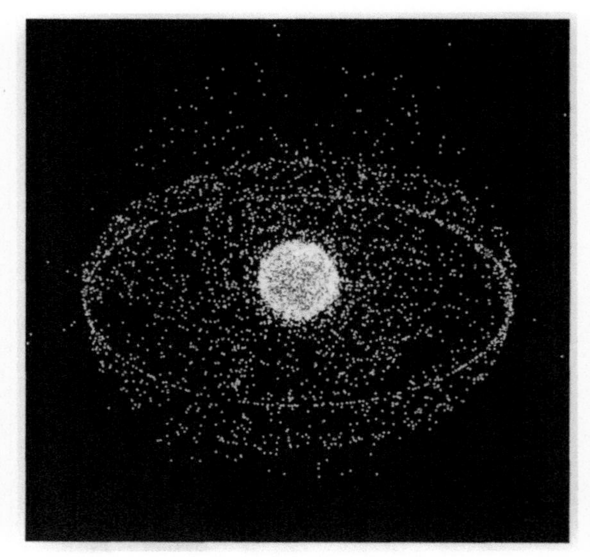

FIGURE 26.7 Representation of the location and distribution of the debris population, including GEO altitudes, at a snapshot in time (midnight, January 1, 2010).

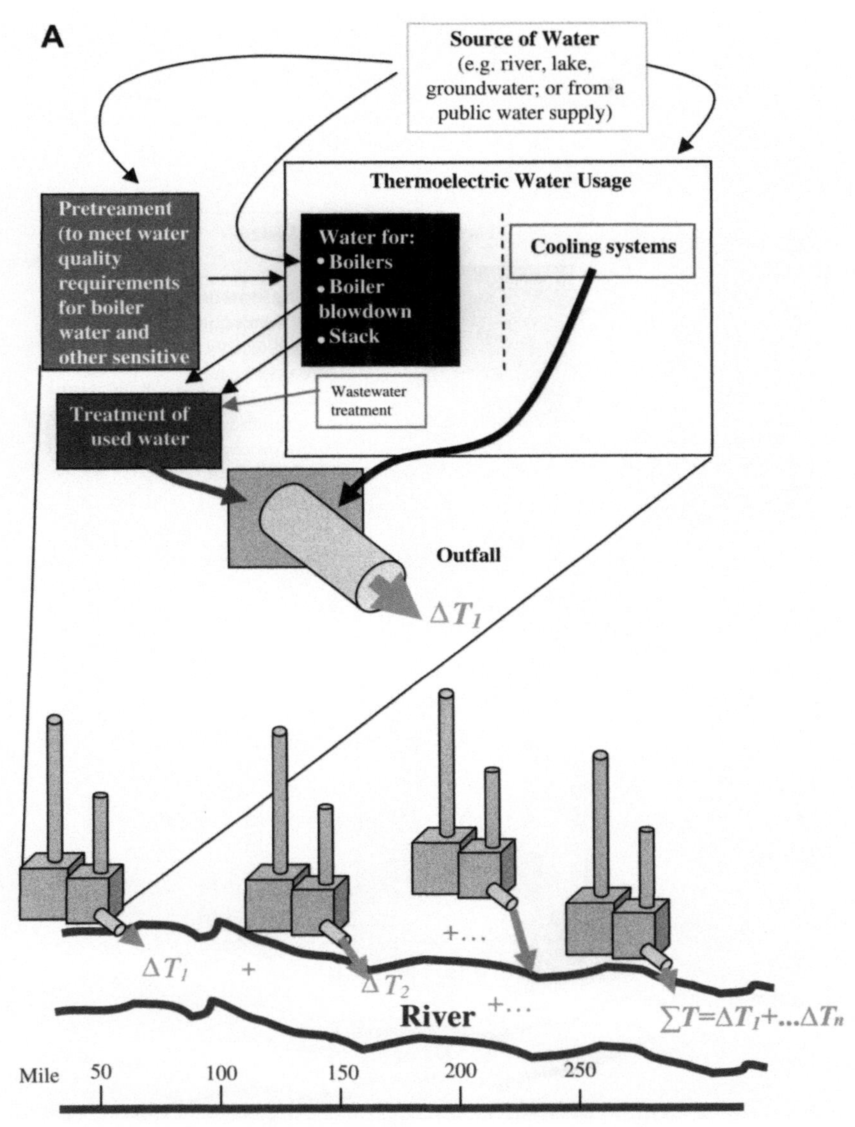

FIGURE 28.6 Difference in cumulative heat contribution to a river from electric generating plants using a once-through cooling system (A) versus the same plants using a cooling water return system (B). The cumulatively added heat is greatly reduced with the closed water return systems compared to the once through cooling systems (hypothetical scenario). *Source: D.A. Vallero (2010). Environmental Biotechnology: A Biosystems Approach. Elsevier Academic Press. Amsterdam, NV.*

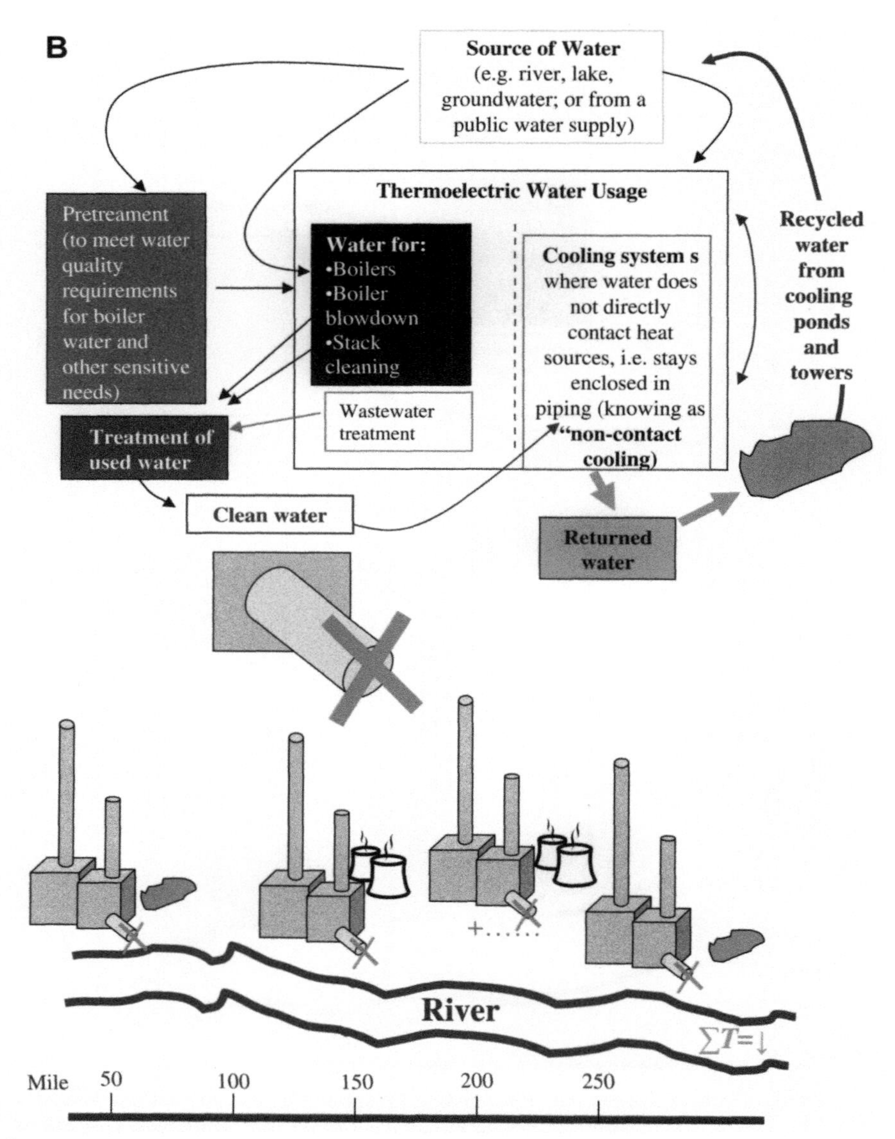

FIGURE 28.6 — (*Continued*).

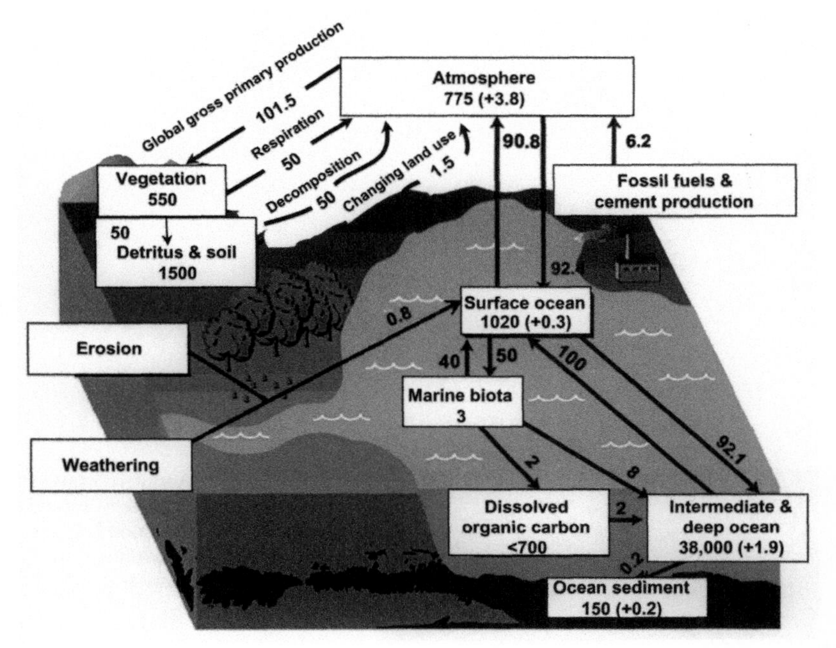

FIGURE 28.7 Global carbon cycle from 1992 to 1997. Carbon pool are boxes, expressed in gigatonnes (Gt) of carbon (Note: Gt = 10^{15} g). Annual increments are expressed in Gt per annum of carbon (shown in parentheses). All fluxes indicated by the arrows are expressed in Gt a^{-1} of carbon. The inferred net terrestrial uptake of 0.7 Gt a^{-1} of carbon considers gross primary production (~101.5), plant respiration (~50), decomposition (~50), and additional removal from the atmosphere directly or indirectly, through vegetation and soil and eventual flow to the ocean through the terrestrial processes of weathering, erosion, and runoff (~0.8). Net ocean uptake (~1.6) considers air/sea exchange (~92.4 gross uptake, −90.8 gross release). As the rate of fossil fuel burning increases the CO_2 released to the atmosphere, it is expected that the fraction of this C remaining in the atmosphere will increase resulting in a doubling or tripling of the atmospheric amount in the coming century. *Source: M. Post. Oak Ridge National Laboratory; http://cdiac.ornl.gov/pns/graphics/c_cycle.htm; accessed on January 29, 2009.*

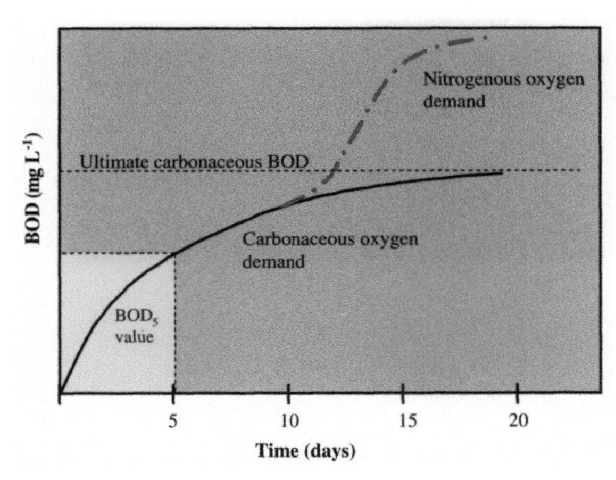

FIGURE 29.1 Biochemical oxygen demand (BOD) curve, showing ultimate carbonaceous BOD and nitrogenous BOD. *Adapted from Ref. [10].*

FIGURE 31.3 Aerial reconnaissance flight taken by N&P Limited showing a small section of the release. Note: home partially buried in ash in southeast corner.

CPI Antony Rowe
Chippenham, UK
2017-08-14 11:43